Collins
World Atlas

Title	Scale	Page
Map Symbols and Time Zones		2
National Statistics		3–5

World		
World Countries		6–7
World Landscapes		8–9

Europe		
Northern Europe	1:10 000 000	10–11
Western Russian Federation	1:7.5 000 000	12–13
Scandinavia and the Baltic States	1:5 000 000	14–15
Inset: Iceland	1:6 000 000	
Inset: Faroe Islands	1:5 000 000	
Northwest Europe	1:5 000 000	16–17
England and Wales	1:2 000 000	18–19
Scotland	1:2 000 000	20
Inset: Shetland Islands	1:2 000 000	
Ireland	1:2 000 000	21
Southern Europe and the Mediterranean	1:10 000 000	22–23
France	1:5 000 000	24
Spain and Portugal	1:5 000 000	25
Italy and the Balkans	1:5 000 000	26–27

Asia		
Northern Asia	1:20 000 000	28–29
Central and Southern Asia	1:20 000 000	30–31
Southwest Asia	1:13 000 000	32–33
Eastern Mediterranean, the Caucasus and Iraq	1:7 000 000	34–35
Northern India, Nepal, Bhutan and Bangladesh	1:7 000 000	36–37
Southern India and Sri Lanka	1:7 000 000	38
Middle East	1:3 000 000	39
Eastern and Southeast Asia	1:20 000 000	40–41
Eastern Asia	1:15 000 000	42–43
Japan, North Korea and South Korea	1:7 000 000	44–45

Africa		
Northern Africa	1:16 000 000	46–47
Inset: Cape Verde	1:16 000 000	
Central and Southern Africa	1:16 000 000	48–49
Republic of South Africa	1:5 000 000	50–51

Oceania		
Australia, New Zealand and Southwest Pacific	1:20 000 000	52–53
Western Australia	1:8 000 000	54–55
Eastern Australia	1:8 000 000	56–57
Inset: Tasmania	1:8 000 000	
Southeast Australia	1:5 000 000	58
New Zealand	1:5 250 000	59

North America		
Canada	1:16 000 000	60–61
United States of America	1:12 000 000	62–63
Inset: Hawaii	1:12 000 000	
Northeast United States	1:3 500 000	64
Southwest United States	1:3 500 000	65
Central America and the Caribbean	1:14 000 000	66–67

South America		
Northern South America	1:14 000 000	68–69
Inset: Galapagos Islands	1:14 000 000	
Southern South America	1:14 000 000	70
Southeast Brazil	1:7 000 000	71

Oceans and Poles		
Atlantic Ocean, Indian Ocean	1:50 000 000	72–73
Pacific Ocean	1:50 000 000	74–75
Antarctica	1:26 000 000	76
The Arctic	1:26 000 000	77

| Index | | 78–127 |

Settlements

Population	National capital	Administrative capital	Other city or town
over 10 million	**BEIJING** ✶	**Karachi** ◉	**New York** ◉
5 million to 10 million	**JAKARTA** ✶	**Tianjin** ◉	**Nova Iguaçu** ◉
1 million to 5 million	**KĀBUL** ✶	**Sydney** ◉	**Kaohsiung** ◉
500 000 to 1 million	**BANGUI** ✶	**Trujillo** ◉	**Jeddah** ◉
100 000 to 500 000	WELLINGTON ✶	Mansa ◉	Apucarana ◉
50 000 to 100 000	PORT OF SPAIN ✶	Potenza ○	Arecibo ◌
10 000 to 50 000	MALABO ✶	Chinhoyi ○	Ceres ◌
under 10 000	VALLETTA ✶	Ati ○	Venta ◌

Built-up area

Boundaries

_____ International boundary

─·─·─ Disputed international boundary or alignment unconfirmed

────── Administrative boundary

········ Ceasefire line

Miscellaneous

----------- National park

············· Reserve or Regional park

✶ Site of specific interest

⌗⌗⌗⌗⌗ Wall

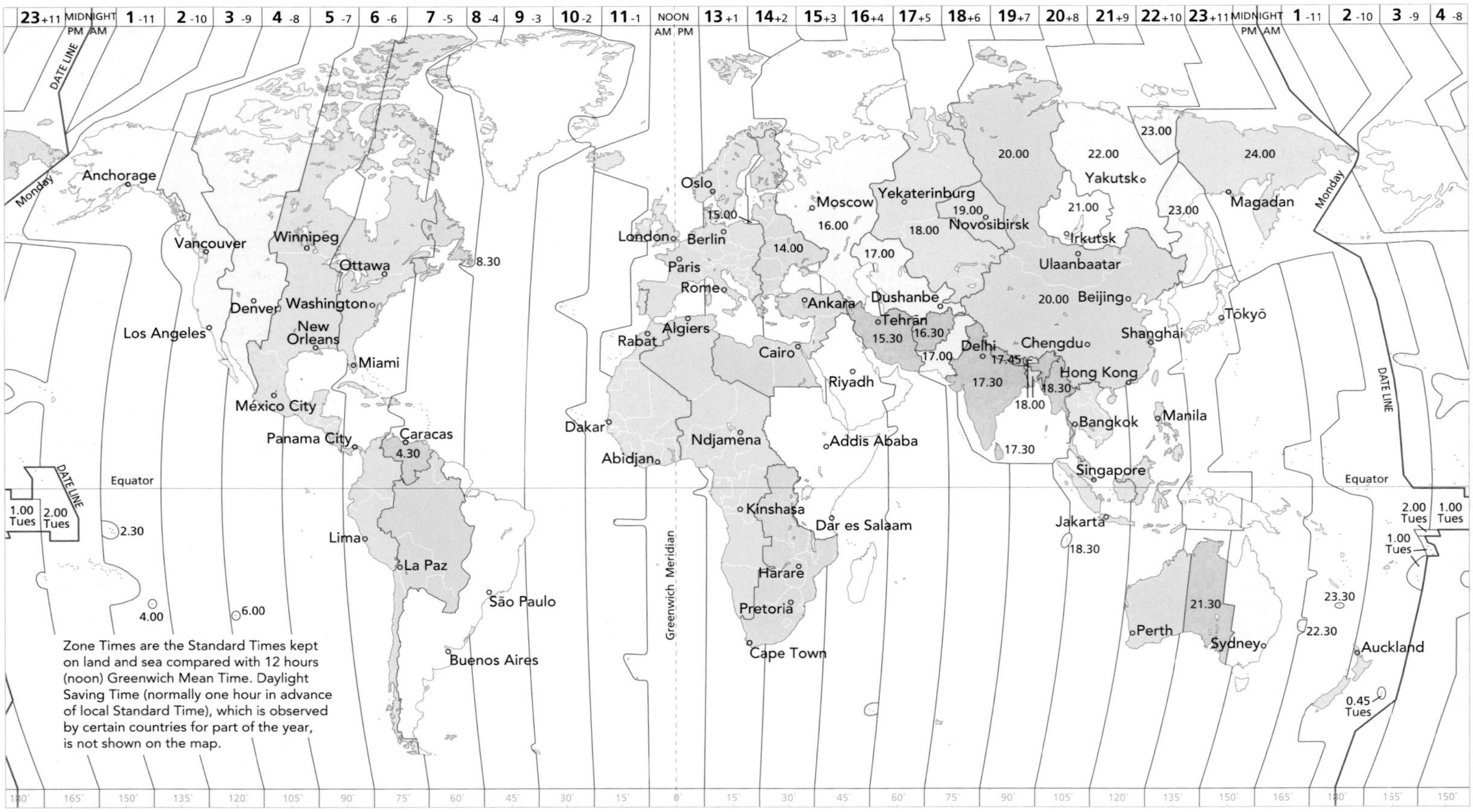

Land and sea features

Desert

Oasis

Lava field

1234 Volcano
height in metres

Marsh

Ice cap or Glacier

Escarpment

Coral reef

1234 Pass
height in metres

Lakes and rivers

Lake

Impermanent lake

Salt lake or lagoon

Impermanent salt lake

Dry salt lake or salt pan

123 Lake height
surface height above
sea level, in metres

River

Impermanent river or watercourse

Waterfall

Dam

Barrage

Relief

Contour intervals and layer colours

Height

metres	feet
5000	16404
3000	9843
2000	6562
1000	3281
500	1640
200	656
0	0

below sea level

0	0
200	656
2000	6562
4000	13124
6000	19686

Depth

1234 ▲ Summit
height in metres

-123 Spot height
height in metres

123 Ocean deep
depth in metres

Transport

→ ⊐ ····· Motorway (tunnel; under construction)

→ ⊐ ---- Main road (tunnel; under construction)

→ ⊐ ---- Secondary road (tunnel; under construction)

········ Track

→ ⊐ ---- Main railway (tunnel; under construction)

→ ⊐ ---- Secondary railway (tunnel; under construction)

→ ⊐ ---- Other railway (tunnel; under construction)

────── Canal

✈ Main airport

✈ Regional airport

Zone Times are the Standard Times kept on land and sea compared with 12 hours (noon) Greenwich Mean Time. Daylight Saving Time (normally one hour in advance of local Standard Time), which is observed by certain countries for part of the year, is not shown on the map.

Map Symbols and Time Zones

Europe

Europe		Area sq km	Area sq miles	Population	Capital	Languages	Religions	Currency	Internet Link
ALBANIA		28 748	11 100	3 155 000	Tirana	Albanian, Greek	Sunni Muslim, Albanian Orthodox, Roman Catholic	Lek	www.km.gov.al
ANDORRA		465	180	86 000	Andorra la Vella	Spanish, Catalan, French	Roman Catholic	Euro	www.govern.ad
AUSTRIA		83 855	32 377	8 364 000	Vienna	German, Croatian, Turkish	Roman Catholic, Protestant	Euro	www.bundeskanzleramt.at
BELARUS		207 600	80 155	9 634 000	Minsk	Belorussian, Russian	Belorussian Orthodox, Roman Catholic	Belarus rouble	www.belarus.by
BELGIUM		30 520	11 784	10 647 000	Brussels	Dutch (Flemish), French (Walloon), German	Roman Catholic, Protestant	Euro	www.belgium.be
BOSNIA-HERZEGOVINA		51 130	19 741	3 767 000	Sarajevo	Bosnian, Serbian, Croatian	Sunni Muslim, Serbian Orthodox, Roman Catholic, Protestant	Marka	www.fbihvlada.gov.ba
BULGARIA		110 994	42 855	7 545 000	Sofia	Bulgarian, Turkish, Romany, Macedonian	Bulgarian Orthodox, Sunni Muslim	Lev	www.government.bg
CROATIA		56 538	21 829	4 416 000	Zagreb	Croatian, Serbian	Roman Catholic, Serbian Orthodox, Sunni Muslim	Kuna	www.vlada.hr
CZECH REPUBLIC		78 864	30 450	10 369 000	Prague	Czech, Moravian, Slovak	Roman Catholic, Protestant	Koruna	www.czech.cz
DENMARK		43 075	16 631	5 470 000	Copenhagen	Danish	Protestant	Danish krone	www.denmark.dk
ESTONIA		45 200	17 452	1 340 000	Tallinn	Estonian, Russian	Protestant, Estonian and Russian Orthodox	Euro	www.valitsus.ee
FINLAND		338 145	130 559	5 326 000	Helsinki	Finnish, Swedish	Protestant, Greek Orthodox	Euro	www.valtioneuvosto.fi
FRANCE		543 965	210 026	62 343 000	Paris	French, Arabic	Roman Catholic, Protestant, Sunni Muslim	Euro	www.premier-ministre.gouv.fr
GERMANY		357 022	137 849	82 167 000	Berlin	German, Turkish	Protestant, Roman Catholic	Euro	www.deutschland.de
GREECE		131 957	50 949	11 161 000	Athens	Greek	Greek Orthodox, Sunni Muslim	Euro	www.primeminister.gr
HUNGARY		93 030	35 919	9 993 000	Budapest	Hungarian	Roman Catholic, Protestant	Forint	www.magyarorszag.hu
ICELAND		102 820	39 699	323 000	Reykjavík	Icelandic	Protestant	Icelandic króna	www.iceland.is
IRELAND		70 282	27 136	4 515 000	Dublin	English, Irish	Roman Catholic, Protestant	Euro	www.gov.ie
ITALY		301 245	116 311	59 870 000	Rome	Italian	Roman Catholic	Euro	www.governo.it
KOSOVO		10 908	4 212	2 153 139	Prishtinë	Albanian, Serbian	Sunni Muslim, Serbian Orthodox	Euro	www.rks-gov.net/en-US
LATVIA		64 589	24 938	2 249 000	Rīga	Latvian, Russian	Protestant, Roman Catholic, Russian Orthodox	Lats	www.saeima.lv
LIECHTENSTEIN		160	62	36 000	Vaduz	German	Roman Catholic, Protestant	Swiss franc	www.liechtenstein.li
LITHUANIA		65 200	25 174	3 287 000	Vilnius	Lithuanian, Russian, Polish	Roman Catholic, Protestant, Russian Orthodox	Litas	www.lrv.lt
LUXEMBOURG		2 586	998	486 000	Luxembourg	Letzeburgish, German, French	Roman Catholic	Euro	www.gouvernement.lu
MACEDONIA (F.Y.R.O.M.)		25 713	9 928	2 042 000	Skopje	Macedonian, Albanian, Turkish	Macedonian Orthodox, Sunni Muslim	Macedonian denar	www.vlada.mk
MALTA		316	122	409 000	Valletta	Maltese, English	Roman Catholic	Euro	www.gov.mt
MOLDOVA		33 700	13 012	3 604 000	Chişinău	Romanian, Ukrainian, Gagauz, Russian	Romanian Orthodox, Russian Orthodox	Moldovan leu	www.moldova.md
MONACO		2	1	33 000	Monaco-Ville	French, Monegasque, Italian	Roman Catholic	Euro	www.monaco.gouv.mc
MONTENEGRO		13 812	5 333	624 000	Podgorica	Serbian (Montenegrin), Albanian	Montenegrin Orthodox, Sunni Muslim	Euro	www.gov.me
NETHERLANDS		41 526	16 033	16 592 000	Amsterdam/The Hague	Dutch, Frisian	Roman Catholic, Protestant, Sunni Muslim	Euro	www.overheid.nl
NORWAY		323 878	125 050	4 812 000	Oslo	Norwegian	Protestant, Roman Catholic	Norwegian krone	www.norway.no
POLAND		312 683	120 728	38 074 000	Warsaw	Polish, German	Roman Catholic, Polish Orthodox	Złoty	www.poland.gov.pl
PORTUGAL		88 940	34 340	10 707 000	Lisbon	Portuguese	Roman Catholic, Protestant	Euro	www.portugal.gov.pt
ROMANIA		237 500	91 699	21 275 000	Bucharest	Romanian, Hungarian	Romanian Orthodox, Protestant, Roman Catholic	Romanian leu	www.guv.ro
RUSSIAN FEDERATION		17 075 400	6 592 849	140 874 000	Moscow	Russian, Tatar, Ukrainian, local languages	Russian Orthodox, Sunni Muslim, Protestant	Russian rouble	www.gov.ru
SAN MARINO		61	24	31 000	San Marino	Italian	Roman Catholic	Euro	www.consigliograndeegenerale.sm
SERBIA		77 453	29 904	7 334 935	Belgrade	Serbian, Hungarian	Serbian Orthodox, Roman Catholic, Sunni Muslim	Serbian dinar,	www.srbija.gov.rs
SLOVAKIA		49 035	18 933	5 406 000	Bratislava	Slovak, Hungarian, Czech	Roman Catholic, Protestant, Orthodox	Euro	www.government.gov.sk
SLOVENIA		20 251	7 819	2 020 000	Ljubljana	Slovene, Croatian, Serbian	Roman Catholic, Protestant	Euro	www.gov.si
SPAIN		504 782	194 897	44 904 000	Madrid	Castilian, Catalan, Galician, Basque	Roman Catholic	Euro	www.la-moncloa.es
SWEDEN		449 964	173 732	9 249 000	Stockholm	Swedish	Protestant, Roman Catholic	Swedish krona	www.sweden.se
SWITZERLAND		41 293	15 943	7 568 000	Bern	German, French, Italian, Romansch	Roman Catholic, Protestant	Swiss franc	www.swissworld.org
UKRAINE		603 700	233 090	45 708 000	Kiev	Ukrainian, Russian	Ukrainian Orthodox, Ukrainian Catholic, Roman Catholic	Hryvnia	www.kmu.gov.ua
UNITED KINGDOM		243 609	94 058	61 565 000	London	English, Welsh, Gaelic	Protestant, Roman Catholic, Muslim	Pound sterling	www.direct.gov.uk
VATICAN CITY		0.5	0.2	557	Vatican City	Italian	Roman Catholic	Euro	www.vaticanstate.va

Asia

Asia		Area sq km	Area sq miles	Population	Capital	Languages	Religions	Currency	Internet Link
AFGHANISTAN		652 225	251 825	28 150 000	Kābul	Dari, Pushtu, Uzbek, Turkmen	Sunni Muslim, Shi'a Muslim	Afghani	www.president.gov.af
ARMENIA		29 800	11 506	3 083 000	Yerevan	Armenian, Azeri	Armenian Orthodox	Dram	www.gov.am
AZERBAIJAN		86 600	33 436	8 832 000	Baku	Azeri, Armenian, Russian, Lezgian	Shi'a Muslim, Sunni Muslim, Russian and Armenian Orthodox	Azerbaijani manat	www.president.az
BAHRAIN		691	267	791 000	Manama	Arabic, English	Shi'a Muslim, Sunni Muslim, Christian	Bahrain dinar	www.bahrain.bh
BANGLADESH		143 998	55 598	162 221 000	Dhaka	Bengali, English	Sunni Muslim, Hindu	Taka	www.bangladesh.gov.bd
BHUTAN		46 620	18 000	697 000	Thimphu	Dzongkha, Nepali, Assamese	Buddhist, Hindu	Ngultrum, Indian rupee	www.bhutan.gov.bt
BRUNEI		5 765	2 226	400 000	Bandar Seri Begawan	Malay, English, Chinese	Sunni Muslim, Buddhist, Christian	Brunei dollar	www.jpm.gov.bn
CAMBODIA		181 035	69 884	14 805 000	Phnom Penh	Khmer, Vietnamese	Buddhist, Roman Catholic, Sunni Muslim	Riel	www.cambodia.gov.kh
CHINA		9 584 492	3 700 593	1 330 265 000	Beijing	Mandarin, Wu, Cantonese, Hsiang, regional languages	Confucian, Taoist, Buddhist, Christian, Sunni Muslim	Yuan, HK dollar*, Macau pataca	www.gov.cn
CYPRUS		9 251	3 572	871 000	Nicosia	Greek, Turkish, English	Greek Orthodox, Sunni Muslim	Euro	www.cyprus.gov.cy
EAST TIMOR		14 874	5 743	1 134 000	Dili	Portuguese, Tetun, English	Roman Catholic	United States dollar	www.timor-leste.gov.tl
GEORGIA		69 700	26 911	4 260 000	T'bilisi	Georgian, Russian, Armenian, Azeri, Ossetian, Abkhaz	Georgian Orthodox, Russian Orthodox, Sunni Muslim	Lari	www.parliament.ge
INDIA		3 064 898	1 183 364	1 198 003 000	New Delhi	Hindi, English, many regional languages	Hindu, Sunni Muslim, Shi'a Muslim, Sikh, Christian	Indian rupee	www.india.gov.in
INDONESIA		1 919 445	741 102	229 965 000	Jakarta	Indonesian, local languages	Sunni Muslim, Protestant, Roman Catholic, Hindu, Buddhist	Rupiah	www.indonesia.go.id
IRAN		1 648 000	636 296	74 196 000	Tehrān	Farsi, Azeri, Kurdish, regional languages	Shi'a Muslim, Sunni Muslim	Iranian rial	www.president.ir
IRAQ		438 317	169 235	30 747 000	Baghdād	Arabic, Kurdish, Turkmen	Shi'a Muslim, Sunni Muslim, Christian	Iraqi dinar	www.cabinet.iq
ISRAEL		20 770	8 019	7 170 000	Jerusalem (Yerushalayim) (El Quds)**	Hebrew, Arabic	Jewish, Sunni Muslim, Christian, Druze	Shekel	www.gov.il
JAPAN		377 727	145 841	127 156 000	Tōkyō	Japanese	Shintoist, Buddhist, Christian	Yen	www.kantei.go.jp
JORDAN		89 206	34 443	6 316 000	'Ammān	Arabic	Sunni Muslim, Christian	Jordanian dinar	www.jordan.gov.jo
KAZAKHSTAN		2 717 300	1 049 155	15 637 000	Astana	Kazakh, Russian, Ukrainian, German, Uzbek, Tatar	Sunni Muslim, Russian Orthodox, Protestant	Tenge	www.government.kz
KUWAIT		17 818	6 880	2 985 000	Kuwait	Arabic	Sunni Muslim, Shi'a Muslim, Christian, Hindu	Kuwaiti dinar	www.e.gov.kw
KYRGYZSTAN		198 500	76 641	5 482 000	Bishkek	Kyrgyz, Russian, Uzbek	Sunni Muslim, Russian Orthodox	Kyrgyz som	www.gov.kg
LAOS		236 800	91 429	6 320 000	Vientiane	Lao, local languages	Buddhist, traditional beliefs	Kip	www.na.gov.la
LEBANON		10 452	4 036	4 224 000	Beirut	Arabic, Armenian, French	Shi'a Muslim, Sunni Muslim, Christian	Lebanese pound	www.presidency.gov.lb
MALAYSIA		332 965	128 559	27 468 000	Kuala Lumpur/Putrajaya	Malay, English, Chinese, Tamil, local languages	Sunni Muslim, Buddhist, Hindu, Christian, traditional beliefs	Ringgit	www.malaysia.gov.my

**De facto capital. Disputed   *Hong Kong dollar

National Statistics   3

Asia continued

		Area sq km	Area sq miles	Population	Capital	Languages	Religions	Currency	Internet Link
MALDIVES		298	115	309 000	Male	Divehi (Maldivian)	Sunni Muslim	Rufiyaa	www.presidencymaldives.gov.mv
MONGOLIA		1 565 000	604 250	2 671 000	Ulan Bator	Khalka (Mongolian), Kazakh, local languages	Buddhist, Sunni Muslim	Tugrik (tögrög)	www.pmis.gov.mn
MYANMAR (BURMA)		676 577	261 228	50 020 000	Nay Pyi Taw/Rangoon	Burmese, Shan, Karen, local languages	Buddhist, Christian, Sunni Muslim	Kyat	www.mofa.gov.mm
NEPAL		147 181	56 827	29 331 000	Kathmandu	Nepali, Maithili, Bhojpuri, English, local languages	Hindu, Buddhist, Sunni Muslim	Nepalese rupee	www.nepalgov.gov.np
NORTH KOREA		120 538	46 540	23 906 000	P'yŏngyang	Korean	Traditional beliefs, Chondoist, Buddhist	North Korean won	www.korea-dpr.com
OMAN		309 500	119 499	2 845 000	Muscat	Arabic, Baluchi, Indian languages	Ibadhi Muslim, Sunni Muslim	Omani riyal	www.omanet.om
PAKISTAN		803 940	310 403	180 808 000	Islamabad	Urdu, Punjabi, Sindhi, Pushtu, English	Sunni Muslim, Shi'a Muslim, Christian, Hindu	Pakistani rupee	www.pakistan.gov.pk
PALAU		497	192	20 000	Melekeok	Palauan, English	Roman Catholic, Protestant, traditional beliefs	United States dollar	www.palaugov.net
PHILIPPINES		300 000	115 831	91 983 000	Manila	English, Filipino, Tagalog, Cebuano, local languages	Roman Catholic, Protestant, Sunni Muslim, Aglipayan	Philippine peso	www.gov.ph
QATAR		11 437	4 416	1 409 000	Doha	Arabic	Sunni Muslim	Qatari riyal	www.mofa.gov.qa
RUSSIAN FEDERATION		17 075 400	6 592 849	140 874 000	Moscow	Russian, Tatar, Ukrainian, local languages	Russian Orthodox, Sunni Muslim, Protestant	Russian rouble	www.gov.ru
SAUDI ARABIA		2 200 000	849 425	25 721 000	Riyadh	Arabic	Sunni Muslim, Shi'a Muslim	Saudi Arabian riyal	www.saudiportal.net
SINGAPORE		639	247	4 737 000	Singapore	Chinese, English, Malay, Tamil	Buddhist, Taoist, Sunni Muslim, Christian, Hindu	Singapore dollar	www.gov.sg
SOUTH KOREA		99 274	38 330	48 333 000	Seoul	Korean	Buddhist, Protestant, Roman Catholic	South Korean won	www.korea.net
SRI LANKA		65 610	25 332	20 238 000	Sri Jayewardenepura Kotte	Sinhalese, Tamil, English	Buddhist, Hindu, Sunni Muslim, Roman Catholic	Sri Lankan rupee	www.priu.gov.lk
SYRIA		185 180	71 498	21 906 000	Damascus	Arabic, Kurdish, Armenian	Sunni Muslim, Shi'a Muslim, Christian	Syrian pound	www.parliament.gov.sy
TAIWAN		36 179	13 969	23 046 000	T'aipei	Mandarin, Min, Hakka, local languages	Buddhist, Taoist, Confucian, Christian	Taiwan dollar	www.gov.tw
TAJIKISTAN		143 100	55 251	6 952 000	Dushanbe	Tajik, Uzbek, Russian	Sunni Muslim	Somoni	www.prezident.tj
THAILAND		513 115	198 115	67 764 000	Bangkok	Thai, Lao, Chinese, Malay, Mon-Khmer languages	Buddhist, Sunni Muslim	Baht	www.mfa.go.th
TURKEY		779 452	300 948	74 816 000	Ankara	Turkish, Kurdish	Sunni Muslim, Shi'a Muslim	Lira	www.tccb.gov.tr
TURKMENISTAN		488 100	188 456	5 110 000	Aşgabat	Turkmen, Uzbek, Russian	Sunni Muslim, Russian Orthodox	Turkmen manat	www.turkmenistan.gov.tm
UNITED ARAB EMIRATES		77 700	30 000	4 599 000	Abu Dhabi	Arabic, English	Sunni Muslim, Shi'a Muslim	United Arab Emirates dirham	www.uae.gov.ae
UZBEKISTAN		447 400	172 742	27 488 000	Toshkent (Tashkent)	Uzbek, Russian, Tajik, Kazakh	Sunni Muslim, Russian Orthodox	Uzbek som	www.gov.uz
VIETNAM		329 565	127 246	88 069 000	Ha Nôi	Vietnamese, Thai, Khmer, Chinese, local languages	Buddhist, Taoist, Roman Catholic, Cao Dai, Hoa Hao	Dong	www.na.gov.vn
YEMEN		527 968	203 850	23 580 000	Şan'ā'	Arabic	Sunni Muslim, Shi'a Muslim	Yemeni rial	www.yemen-nic.info

Africa

		Area sq km	Area sq miles	Population	Capital	Languages	Religions	Currency	Internet Link
ALGERIA		2 381 741	919 595	34 895 000	Algiers	Arabic, French, Berber	Sunni Muslim	Algerian dinar	www.el-mouradia.dz
ANGOLA		1 246 700	481 354	18 498 000	Luanda	Portuguese, Bantu, local languages	Roman Catholic, Protestant, traditional beliefs	Kwanza	www.governo.gov.ao
BENIN		112 620	43 483	8 935 000	Porto-Novo	French, Fon, Yoruba, Adja, local languages	Traditional beliefs, Roman Catholic, Sunni Muslim	CFA franc*	www.gouv.bj
BOTSWANA		581 370	224 468	1 950 000	Gaborone	English, Setswana, Shona, local languages	Traditional beliefs, Protestant, Roman Catholic	Pula	www.gov.bw
BURKINA FASO		274 200	105 869	15 757 000	Ouagadougou	French, Moore (Mossi), Fulani, local languages	Sunni Muslim, traditional beliefs, Roman Catholic	CFA franc*	www.primature.gov.bf
BURUNDI		27 835	10 747	8 303 000	Bujumbura	Kirundi (Hutu, Tutsi), French	Roman Catholic, traditional beliefs, Protestant	Burundian franc	www.burundi.gov.bi
CAMEROON		475 442	183 569	19 522 000	Yaoundé	French, English, Fang, Bamileke, local languages	Roman Catholic, traditional beliefs, Sunni Muslim, Protestant	CFA franc*	www.spm.gov.cm
CAPE VERDE		4 033	1 557	506 000	Praia	Portuguese, creole	Roman Catholic, Protestant	Cape Verde escudo	www.governo.cv
CENTRAL AFRICAN REPUBLIC		622 436	240 324	4 422 000	Bangui	French, Sango, Banda, Baya, local languages	Protestant, Roman Catholic, traditional beliefs, Sunni Muslim	CFA franc*	www.centrafricaine.info
CHAD		1 284 000	495 755	11 206 000	Ndjamena	Arabic, French, Sara, local languages	Sunni Muslim, Roman Catholic, Protestant, traditional beliefs	CFA franc*	www.primature-tchad.org
COMOROS		1 862	719	676 000	Moroni	Comorian, French, Arabic	Sunni Muslim, Roman Catholic	Comoros franc	www.beit-salam.km
CONGO		342 000	132 047	3 683 000	Brazzaville	French, Kongo, Monokutuba, local languages	Roman Catholic, Protestant, traditional beliefs, Sunni Muslim	CFA franc*	www.congo-site.com
CONGO, DEM. REP. OF THE		2 345 410	905 568	66 020 000	Kinshasa	French, Lingala, Swahili, Kongo, local languages	Christian, Sunni Muslim	Congolese franc	www.un.int/drcongo
CÔTE D'IVOIRE (IVORY COAST)		322 463	124 504	21 075 000	Yamoussoukro	French, creole, Akan, local languages	Sunni Muslim, Roman Catholic, traditional beliefs, Protestant	CFA franc*	www.cotedivoirepr.ci
DJIBOUTI		23 200	8 958	864 000	Djibouti	Somali, Afar, French, Arabic	Sunni Muslim, Christian	Djibouti franc	www.presidence.dj
EGYPT		1 000 250	386 199	82 999 000	Cairo	Arabic	Sunni Muslim, Coptic Christian	Egyptian pound	www.egypt.gov.eg
EQUATORIAL GUINEA		28 051	10 831	676 000	Malabo	Spanish, French, Fang	Roman Catholic, traditional beliefs	CFA franc*	www.ceiba-equatorial-guinea.org
ERITREA		117 400	45 328	5 073 000	Asmara	Tigrinya, Tigre	Sunni Muslim, Coptic Christian	Nakfa	www.shabait.com
ETHIOPIA		1 133 880	437 794	82 825 000	Addis Ababa	Oromo, Amharic, Tigrinya, local languages	Ethiopian Orthodox, Sunni Muslim, traditional beliefs	Birr	www.ethiopar.net
GABON		267 667	103 347	1 475 000	Libreville	French, Fang, local languages	Roman Catholic, Protestant, traditional beliefs	CFA franc*	www.legabon.org
THE GAMBIA		11 295	4 361	1 705 000	Banjul	English, Malinke, Fulani, Wolof	Sunni Muslim, Protestant	Dalasi	www.statehouse.gm
GHANA		238 537	92 100	23 837 000	Accra	English, Hausa, Akan, local languages	Christian, Sunni Muslim, traditional beliefs	Cedi	www.ghana.gov.gh
GUINEA		245 857	94 926	10 069 000	Conakry	French, Fulani, Malinke, local languages	Sunni Muslim, traditional beliefs, Christian	Guinea franc	www.guinee.gov.gn
GUINEA-BISSAU		36 125	13 948	1 611 000	Bissau	Portuguese, crioulo, local languages	Traditional beliefs, Sunni Muslim, Christian	CFA franc*	www.gov.gw
KENYA		582 646	224 961	39 802 000	Nairobi	Swahili, English, local languages	Christian, traditional beliefs	Kenyan shilling	www.kenya.go.ke
LESOTHO		30 355	11 720	2 067 000	Maseru	Sesotho, English, Zulu	Christian, traditional beliefs	Loti, S. African rand	www.lesotho.gov.ls
LIBERIA		111 369	43 000	3 955 000	Monrovia	English, creole, local languages	Traditional beliefs, Christian, Sunni Muslim	Liberian dollar	www.emansion.gov.lr
LIBYA		1 759 540	679 362	6 420 000	Tripoli	Arabic, Berber	Sunni Muslim	Libyan dinar	www.libyanmission-un.org
MADAGASCAR		587 041	226 658	19 625 000	Antananarivo	Malagasy, French	Traditional beliefs, Christian, Sunni Muslim	Malagasy Ariary, Malagasy franc	www.madagascar.gov.mg
MALAWI		118 484	45 747	15 263 000	Lilongwe	Chichewa, English, local languages	Christian, traditional beliefs, Sunni Muslim	Malawian kwacha	www.malawi.gov.mw
MALI		1 240 140	478 821	13 010 000	Bamako	French, Bambara, local languages	Sunni Muslim, traditional beliefs, Christian	CFA franc*	www.primature.gov.ml
MAURITANIA		1 030 700	397 955	3 291 000	Nouakchott	Arabic, French, local languages	Sunni Muslim	Ouguiya	www.mauritania.mr
MAURITIUS		2 040	788	1 288 000	Port Louis	English, creole, Hindi, Bhojpurī, French	Hindu, Roman Catholic, Sunni Muslim	Mauritius rupee	www.gov.mu
MOROCCO		446 550	172 414	31 993 000	Rabat	Arabic, Berber, French	Sunni Muslim	Moroccan dirham	www.maroc.ma
MOZAMBIQUE		799 380	308 642	22 894 000	Maputo	Portuguese, Makua, Tsonga, local languages	Traditional beliefs, Roman Catholic, Sunni Muslim	Metical	www.mozambique.mz
NAMIBIA		824 292	318 261	2 171 000	Windhoek	English, Afrikaans, German, Ovambo, local languages	Protestant, Roman Catholic	Namibian dollar	www.grnnet.gov.na
NIGER		1 267 000	489 191	15 290 000	Niamey	French, Hausa, Fulani, local languages	Sunni Muslim, traditional beliefs	CFA franc*	www.presidence.ne
NIGERIA		923 768	356 669	154 729 000	Abuja	English, Hausa, Yoruba, Ibo, Fulani, local languages	Sunni Muslim, Christian, traditional beliefs	Naira	www.nigeria.gov.ng
RWANDA		26 338	10 169	9 998 000	Kigali	Kinyarwanda, French, English	Roman Catholic, traditional beliefs, Protestant	Rwandan franc	www.gov.rw
SÃO TOMÉ AND PRÍNCIPE		964	372	163 000	São Tomé	Portuguese, creole	Roman Catholic, Protestant	Dobra	www.gov.st
SENEGAL		196 720	75 954	12 534 000	Dakar	French, Wolof, Fulani, local languages	Sunni Muslim, Roman Catholic, traditional beliefs	CFA franc*	www.gouv.sn

*Communauté Financière Africaine franc

Africa continued

		Area sq km	Area sq miles	Population	Capital	Languages	Religions	Currency	Internet Link
SEYCHELLES		455	176	84 000	Victoria	English, French, creole	Roman Catholic, Protestant	Seychelles rupee	www.virtualseychelles.sc
SIERRA LEONE		71 740	27 699	5 696 000	Freetown	English, creole, Mende, Temne, local languages	Sunni Muslim, traditional beliefs	Leone	www.statehouse-sl.org
SOMALIA		637 657	246 201	9 133 000	Mogadishu	Somali, Arabic	Sunni Muslim	Somali shilling	www.tfgsomalia.net
SOUTH AFRICA, REPUBLIC OF		1 219 090	470 693	50 110 000	Pretoria/Cape Town	Afrikaans, English, nine official local languages	Protestant, Roman Catholic, Sunni Muslim, Hindu	Rand	www.gov.za
SOUTH SUDAN		644 329	248 775	8 260 490	Juba	Arabic, Dinka, Nubian, Beja, English, local languages	Christian, Sunni Muslim, traditional beliefs	South Sudanese pound	www.goss.org
SUDAN		1 861 484	718 725	36 371 510	Khartoum	Arabic, Dinka, Nubian, Beja, Nuer, local languages	Sunni Muslim, traditional beliefs, Christian	Sudanese pound (Sudani)	www.sudan.gov.sd
SWAZILAND		17 364	6 704	1 185 000	Mbabane	Swazi, English	Christian, traditional beliefs	Emalangeni, South African rand	www.gov.sz
TANZANIA		945 087	364 900	43 739 000	Dodoma	Swahili, English, Nyamwezi, local languages	Shi'a Muslim, Sunni Muslim, traditional beliefs, Christian	Tanzanian shilling	www.tanzania.go.tz
TOGO		56 785	21 925	6 619 000	Lomé	French, Ewe, Kabre, local languages	Traditional beliefs, Christian, Sunni Muslim	CFA franc*	www.republicoftogo.com
TUNISIA		164 150	63 379	10 272 000	Tunis	Arabic, French	Sunni Muslim	Tunisian dinar	www.ministeres.tn
UGANDA		241 038	93 065	32 710 000	Kampala	English, Swahili, Luganda, local languages	Roman Catholic, Protestant, Sunni Muslim, traditional beliefs	Ugandan shilling	www.statehouse.go.ug
ZAMBIA		752 614	290 586	12 935 000	Lusaka	English, Bemba, Nyanja, Tonga, local languages	Christian, traditional beliefs	Zambian kwacha	www.statehouse.gov.zm
ZIMBABWE		390 759	150 873	12 523 000	Harare	English, Shona, Ndebele	Christian, traditional beliefs	Zimbabwean dollar (suspended)	www.parlzim.gov.zw

*Communauté Financière Africaine franc

Oceania

		Area sq km	Area sq miles	Population	Capital	Languages	Religions	Currency	Internet Link
AUSTRALIA		7 692 024	2 969 907	21 293 000	Canberra	English, Italian, Greek	Protestant, Roman Catholic, Orthodox	Australian dollar	www.australia.gov.au
FIJI		18 330	7 077	849 000	Suva	English, Fijian, Hindi	Christian, Hindu, Sunni Muslim	Fiji dollar	www.fiji.gov.fj
KIRIBATI		717	277	98 000	Bairiki	Gilbertese, English	Roman Catholic, Protestant	Australian dollar	www.parliament.gov.ki
MARSHALL ISLANDS		181	70	62 000	Delap-Uliga-Djarrit	English, Marshallese	Protestant, Roman Catholic	United States dollar	www.rmigovernment.org
MICRONESIA, FEDERATED STATES OF		701	271	111 000	Palikir	English, Chuukese, Pohnpeian, local languages	Roman Catholic, Protestant	United States dollar	www.fsmgov.org
NAURU		21	8	10 000	Yaren	Nauruan, English	Protestant, Roman Catholic	Australian dollar	www.naurugov.nr
NEW ZEALAND		270 534	104 454	4 266 000	Wellington	English, Maori	Protestant, Roman Catholic	New Zealand dollar	http://newzealand.govt.nz
PAPUA NEW GUINEA		462 840	178 704	6 732 000	Port Moresby	English, Tok Pisin (creole), local languages	Protestant, Roman Catholic, traditional beliefs	Kina	www.pm.gov.pg
SAMOA		2 831	1 093	179 000	Apia	Samoan, English	Protestant, Roman Catholic	Tala	www.govt.ws
SOLOMON ISLANDS		28 370	10 954	523 000	Honiara	English, creole, local languages	Protestant, Roman Catholic	Solomon Islands dollar	www.pmc.gov.sb
TONGA		748	289	104 000	Nuku'alofa	Tongan, English	Protestant, Roman Catholic	Pa'anga	www.pmo.gov.to
TUVALU		25	10	10 000	Vaiaku	Tuvaluan, English	Protestant	Australian dollar	
VANUATU		12 190	4 707	240 000	Port Vila	English, Bislama (creole), French	Protestant, Roman Catholic, traditional beliefs	Vatu	www.vanuatugovernment.gov.vu

North America

		Area sq km	Area sq miles	Population	Capital	Languages	Religions	Currency	Internet Link
ANTIGUA AND BARBUDA		442	171	88 000	St John's	English, creole	Protestant, Roman Catholic	East Caribbean dollar	www.ab.gov.ag
THE BAHAMAS		13 939	5 382	342 000	Nassau	English, creole	Protestant, Roman Catholic	Bahamian dollar	www.bahamas.gov.bs
BARBADOS		430	166	256 000	Bridgetown	English, creole	Protestant, Roman Catholic	Barbados dollar	www.barbados.gov.bb
BELIZE		22 965	8 867	307 000	Belmopan	English, Spanish, Mayan, creole	Roman Catholic, Protestant	Belize dollar	www.belize.gov.bz
CANADA		9 984 670	3 855 103	33 573 000	Ottawa	English, French, local languages	Roman Catholic, Protestant, Eastern Orthodox, Jewish	Canadian dollar	www.canada.gc.ca
COSTA RICA		51 100	19 730	4 579 000	San José	Spanish	Roman Catholic, Protestant	Costa Rican colón	www.casapres.go.cr
CUBA		110 860	42 803	11 204 000	Havana	Spanish	Roman Catholic, Protestant	Cuban peso	www.cubagob.gov.cu
DOMINICA		750	290	67 000	Roseau	English, creole	Roman Catholic, Protestant	East Caribbean dollar	www.dominica.gov.dm
DOMINICAN REPUBLIC		48 442	18 704	10 090 000	Santo Domingo	Spanish, creole	Roman Catholic, Protestant	Dominican peso	www.cig.gov.do
EL SALVADOR		21 041	8 124	6 163 000	San Salvador	Spanish	Roman Catholic, Protestant	El Salvador colón, United States dollar	www.presidencia.gob.sv
GRENADA		378	146	104 000	St George's	English, creole	Roman Catholic, Protestant	East Caribbean dollar	www.gov.gd
GUATEMALA		108 890	42 043	14 027 000	Guatemala City	Spanish, Mayan languages	Roman Catholic, Protestant	Quetzal, United States dollar	www.congreso.gob.gt
HAITI		27 750	10 714	10 033 000	Port-au-Prince	French, creole	Roman Catholic, Protestant, Voodoo	Gourde	www.haiti.org
HONDURAS		112 088	43 277	7 466 000	Tegucigalpa	Spanish, Amerindian languages	Roman Catholic, Protestant	Lempira	www.congreso.gob.hn
JAMAICA		10 991	4 244	2 719 000	Kingston	English, creole	Protestant, Roman Catholic	Jamaican dollar	www.jis.gov.jm
MEXICO		1 972 545	761 604	109 610 000	Mexico City	Spanish, Amerindian languages	Roman Catholic, Protestant	Mexican peso	www.gob.mx
NICARAGUA		130 000	50 193	5 743 000	Managua	Spanish, Amerindian languages	Roman Catholic, Protestant	Córdoba	www.asamblea.gob.ni
PANAMA		77 082	29 762	3 454 000	Panama City	Spanish, English, Amerindian languages	Roman Catholic, Protestant, Sunni Muslim	Balboa	www.pa
ST KITTS AND NEVIS		261	101	52 000	Basseterre	English, creole	Protestant, Roman Catholic	East Caribbean dollar	www.gov.kn
ST LUCIA		616	238	172 000	Castries	English, creole	Roman Catholic, Protestant	East Caribbean dollar	www.stlucia.gov.lc
ST VINCENT AND THE GRENADINES		389	150	109 000	Kingstown	English, creole	Protestant, Roman Catholic	East Caribbean dollar	www.gov.vc
TRINIDAD AND TOBAGO		5 130	1 981	1 339 000	Port of Spain	English, creole, Hindi	Roman Catholic, Hindu, Protestant, Sunni Muslim	Trinidad and Tobago dollar	www.gov.tt
UNITED STATES OF AMERICA		9 826 635	3 794 085	314 659 000	Washington D.C.	English, Spanish	Protestant, Roman Catholic, Sunni Muslim, Jewish	United States dollar	www.firstgov.gov

South America

		Area sq km	Area sq miles	Population	Capital	Languages	Religions	Currency	Internet Link
ARGENTINA		2 766 889	1 068 302	40 276 000	Buenos Aires	Spanish, Italian, Amerindian languages	Roman Catholic, Protestant	Argentinian peso	www.argentina.gov.ar
BOLIVIA		1 098 581	424 164	9 863 000	La Paz/Sucre	Spanish, Quechua, Aymara	Roman Catholic, Protestant, Baha'i	Boliviano	www.bolivia.gov.bo
BRAZIL		8 514 879	3 287 613	193 734 000	Brasília	Portuguese	Roman Catholic, Protestant	Real	www.brazil.gov.br
CHILE		756 945	292 258	16 970 000	Santiago	Spanish, Amerindian languages	Roman Catholic, Protestant	Chilean peso	www.gobiernodechile.cl
COLOMBIA		1 141 748	440 831	45 660 000	Bogotá	Spanish, Amerindian languages	Roman Catholic, Protestant	Colombian peso	www.gobiernoenlinea.gov.co
ECUADOR		272 045	105 037	13 625 000	Quito	Spanish, Quechua, other Amerindian languages	Roman Catholic	US dollar	www.presidencia.gov.ec
GUYANA		214 969	83 000	762 000	Georgetown	English, creole, Amerindian languages	Protestant, Hindu, Roman Catholic, Sunni Muslim	Guyana dollar	www.gina.gov.gy
PARAGUAY		406 752	157 048	6 349 000	Asunción	Spanish, Guaraní	Roman Catholic, Protestant	Guaraní	www.presidencia.gov.py
PERU		1 285 216	496 225	29 165 000	Lima	Spanish, Quechua, Aymara	Roman Catholic, Protestant	Nuevo sol	www.peru.gob.pe
SURINAME		163 820	63 251	520 000	Paramaribo	Dutch, Surinamese, English, Hindi	Hindu, Roman Catholic, Protestant, Sunni Muslim	Suriname guilder	www.kabinet.sr.org
URUGUAY		176 215	68 037	3 361 000	Montevideo	Spanish	Roman Catholic, Protestant, Jewish	Uruguayan peso	www.presidencia.gub.uy
VENEZUELA		912 050	352 144	28 583 000	Caracas	Spanish, Amerindian languages	Roman Catholic, Protestant	Bolívar fuerte	www.gobiernoenlinea.ve

World
Countries

The current pattern of the world's countries and territories is a result of a long history of exploration, colonialism, conflict and politics. The fact that there are currently 196 independent countries in the world – the most recent, South Sudan, only being created in July 2011 – illustrates the significant political changes which have occurred since 1950 when there were only eighty-two. There has been a steady progression away from colonial influences over the last fifty years, although many dependent overseas territories remain.

The shapes of countries and the pattern of international boundaries reflect both physical and political processes. Some borders follow natural features – rivers, mountain ranges, etc – others are defined according to political agreement or as a result of war. Some are still subject to dispute between two or more countries, and many remain undefined on the ground.

Facts

- The longest single continuous land border stretches for 6 416 kilometres between Canada and the USA
- Both China and the Russian Federation have land borders with 14 different countries
- Vatican City, the smallest independent country, was created in 1929 as an enclave within Rome, the capital of Italy
- All countries of the world are members of the United Nations except Kosovo, Taiwan and Vatican City

Internet Links

United Nations	www.un.org
Foreign and Commonwealth Office	www.fco.gov.uk
International Boundaries Research Unit	www.dur.ac.uk/ibru
Permanent Committee on Geographical Names	www.pcgn.org.uk
U.S. Board on Geographic Names	geonames.usgs.gov

High-resolution satellite image of **Vatican City**, the world's smallest country by both population and area.

World extremes

Countries			
Largest country (area)	**Russian Federation**	17 075 400 sq km	6 592 849 sq miles
Smallest country (area)	**Vatican City**	0.5 sq km	0.2 sq miles
Largest country (population)	**China**	1 330 265 000	
Smallest country (population)	**Vatican City**	557	
Most densely populated country	**Monaco**	17 500 per sq km	35 000 per sq mile
Least densely populated country	**Mongolia**	1.7 per sq km	4.4 per sq mile
Capitals			
Largest national capital (population)	**Tōkyō, Japan**	36 094 000	
Smallest national capital (population)	**Melekeok, Palau**	391	
Most northerly national capital	**Reykjavík, Iceland**	64° 08'N	
Most southerly national capital	**Wellington, New Zealand**	41° 18'S	
Highest national capital	**La Paz, Bolivia**	3 636 m	11 910 ft

World
Landscapes

The earth's physical features, both on land and on the sea bed, closely reflect its geological structure. The current shapes of the continents and oceans have evolved over millions of years. Movements of the tectonic plates which make up the earth's crust have created some of the best-known and most spectacular features. The processes which have shaped the earth continue today with earthquakes, volcanoes, erosion, climatic variations and man's activities all affecting the earth's landscapes.

The total topographic range of the earth's surface is nearly 20 000 metres, from the highest point Mount Everest, to the lowest point in the Mariana Trench. Major mountain ranges include the Himalaya, the Andes and the Rocky Mountains, each of which give rise to some of the world's greatest rivers. In contrast, the deserts of the Sahara, Australia, the Arabian Peninsula and the Gobi cover vast areas and each provide unique landscapes.

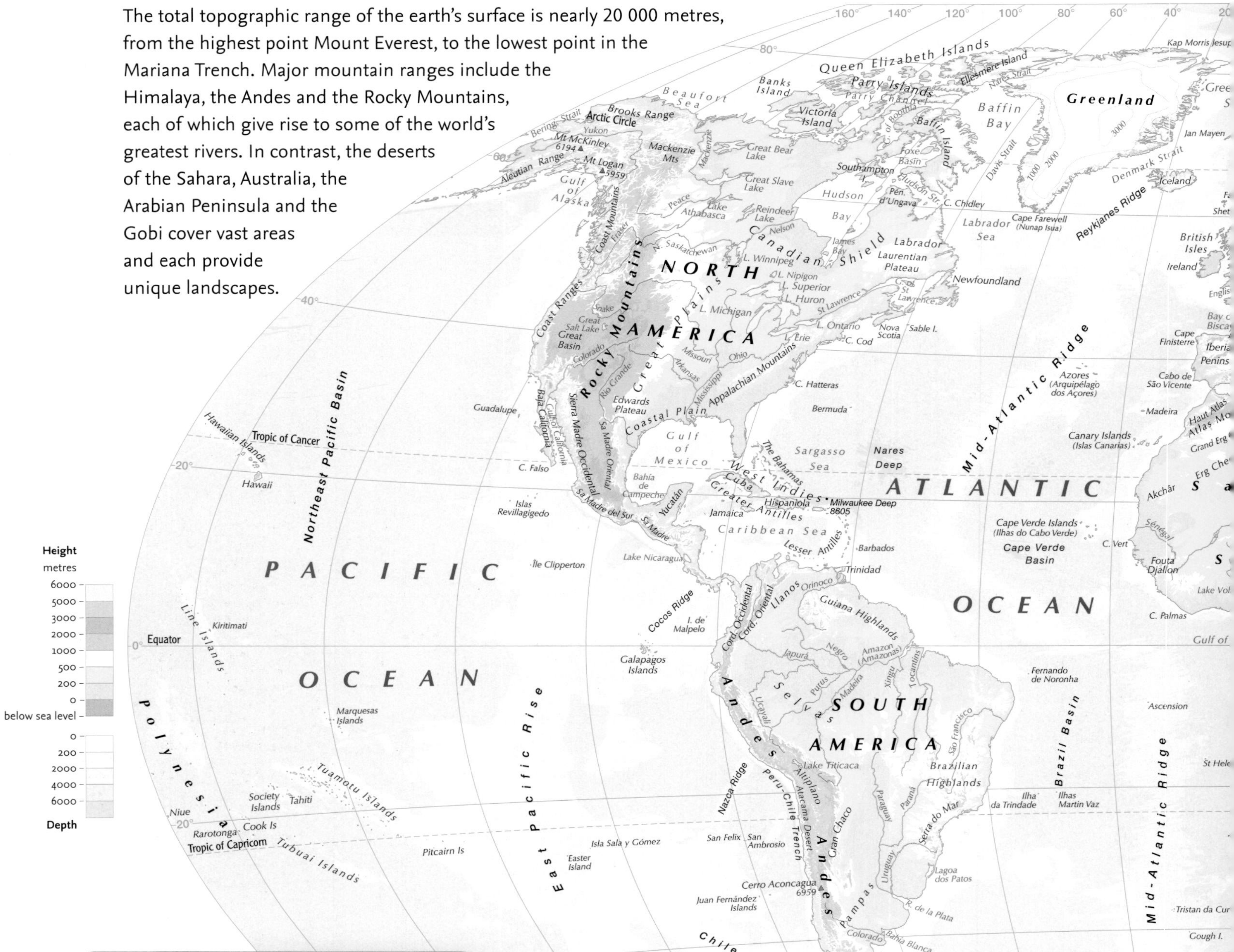

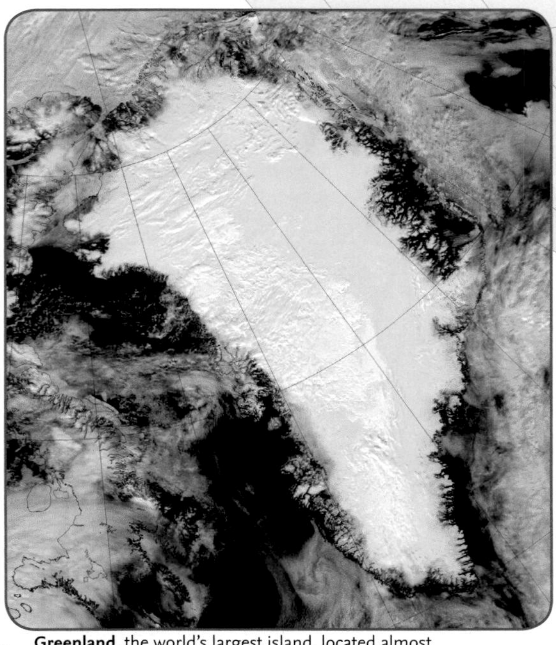

Height
metres
6000
5000
3000
2000
1000
500
200
0
below sea level

Depth
0
200
2000
4000
6000

Greenland, the world's largest island, located almost entirely within the Arctic Circle.

Internet Links

United Nations Environment Programme	**www.unep.org**
IUCN The International Union for Conservation of Nature	**www.iucn.org**
NASA Visible Earth	**visibleearth.nasa.gov**
NASA Earth Observatory	**earthobservatory.nasa.gov**
Earth Resources Observation and Science	**edc.usgs.gov**

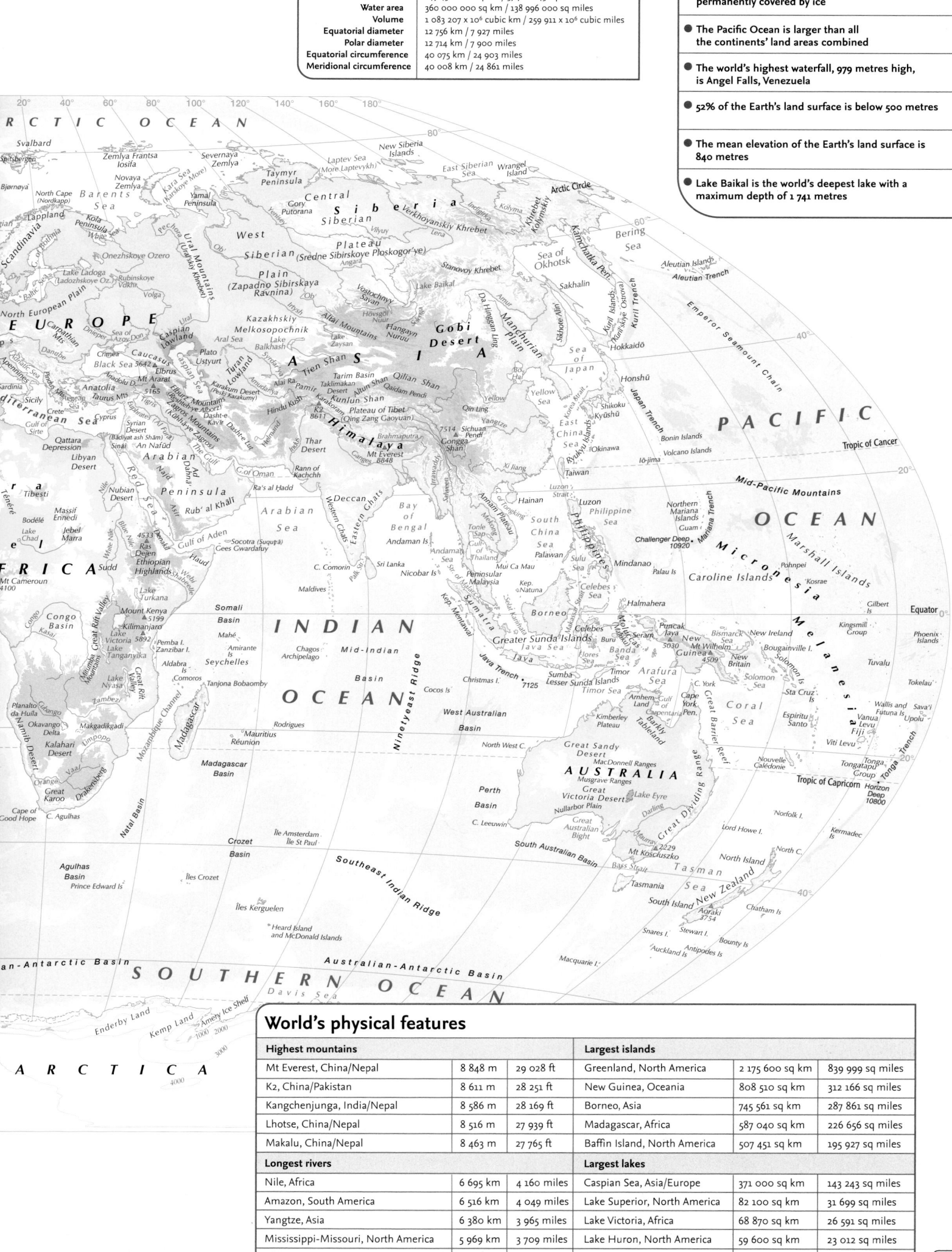

Earth's dimensions

Mass	5.974 x 10²¹ tonnes
Total area	509 450 000 sq km / 196 698 645 sq miles
Land area	149 450 000 sq km / 57 702 645 sq miles
Water area	360 000 000 sq km / 138 996 000 sq miles
Volume	1 083 207 x 10⁶ cubic km / 259 911 x 10⁶ cubic miles
Equatorial diameter	12 756 km / 7 927 miles
Polar diameter	12 714 km / 7 900 miles
Equatorial circumference	40 075 km / 24 903 miles
Meridional circumference	40 008 km / 24 861 miles

Facts

- Approximately 10% of the Earth's land surface is permanently covered by ice
- The Pacific Ocean is larger than all the continents' land areas combined
- The world's highest waterfall, 979 metres high, is Angel Falls, Venezuela
- 52% of the Earth's land surface is below 500 metres
- The mean elevation of the Earth's land surface is 840 metres
- Lake Baikal is the world's deepest lake with a maximum depth of 1 741 metres

World's physical features

Highest mountains			Largest islands		
Mt Everest, China/Nepal	8 848 m	29 028 ft	Greenland, North America	2 175 600 sq km	839 999 sq miles
K2, China/Pakistan	8 611 m	28 251 ft	New Guinea, Oceania	808 510 sq km	312 166 sq miles
Kangchenjunga, India/Nepal	8 586 m	28 169 ft	Borneo, Asia	745 561 sq km	287 861 sq miles
Lhotse, China/Nepal	8 516 m	27 939 ft	Madagascar, Africa	587 040 sq km	226 656 sq miles
Makalu, China/Nepal	8 463 m	27 765 ft	Baffin Island, North America	507 451 sq km	195 927 sq miles

Longest rivers			Largest lakes		
Nile, Africa	6 695 km	4 160 miles	Caspian Sea, Asia/Europe	371 000 sq km	143 243 sq miles
Amazon, South America	6 516 km	4 049 miles	Lake Superior, North America	82 100 sq km	31 699 sq miles
Yangtze, Asia	6 380 km	3 965 miles	Lake Victoria, Africa	68 870 sq km	26 591 sq miles
Mississippi-Missouri, North America	5 969 km	3 709 miles	Lake Huron, North America	59 600 sq km	23 012 sq miles
Ob'-Irtysh, Asia	5 568 km	3 460 miles	Lake Michigan, North America	57 800 sq km	22 317 sq miles

Denmark Strait
Arctic Circle
ICELAND
REYKJAVIK
Vatnajökull
Akureyri
Norwegian Sea
ATLANTIC OCEAN
Faroe Islands (Denmark)
TÓRSHAVN
Rockall (U.K.)
Shetland Islands
Orkney Islands
Outer Hebrides
British Isles
North Sea
Skagerrak
Kattegat
NORWAY
SWEDEN
OSLO
Bergen
Stavanger
Trondheim
Gothenburg
DENMARK
COPENHAGEN
UNITED KINGDOM
Glasgow
Edinburgh
Newcastle upon Tyne
Manchester
Liverpool
Leeds
Birmingham
LONDON
IRELAND
DUBLIN
Cork
Isle of Man (U.K.)
Irish Sea
St George's Channel
English Channel (La Manche)
Channel Islands (U.K.)
NETHERLANDS
AMSTERDAM
Rotterdam
BELGIUM
BRUSSELS
LUX.
LUXEMBOURG
GERMANY
BERLIN
Hamburg
Munich
Frankfurt am Main
FRANCE
PARIS
Lyon
Bordeaux
Nantes
Bay of Biscay
Gulf of Gascony
Massif Central
SWITZERLAND
BERN
Zürich
LIECHTENSTEIN
AUSTRIA
VIENNA
CZECH REPUBLIC
PRAGUE
POLAND
SLOVENIA
ITALY
SPAIN
HUNGARY
BRATISLAVA
Baltic Sea
Öland
Gotland (Sweden)
Bornholm (Denmark)
Alps
Conic Equidistant Projection
1:10 000 000
0 100 200 300 400 miles
0 100 200 300 400 500 600 km

Barents Sea
White Sea
Gulf of Finland
Gulf of Riga
Sea of Azov
Caspian Sea
Kola Peninsula (Kol'skiy Poluostrov)
Lake Ladoga (Ladozhskoye Ozero)
Onezhskoye Ozero
FINLAND
LAPPLAND
ESTONIA
LATVIA
LITHUANIA
BELARUS
UKRAINE
MOLDOVA
ROMANIA
KAZAKHSTAN
RUSSIAN FEDERATION
Ural Mountains (Ural'skiy Khrebet)
Central Russian Upland
Caspian Lowland
MOSCOW
St Petersburg
HELSINKI
TALLINN
RĪGA
VILNIUS
MINSK
KIEV
CHIŞINĂU
Nizhniy Novgorod
Kazan'
Samara
Volgograd
Saratov
Voronezh
Kharkiv
Odessa (Odesa)
Archangel
Murmansk
Perm'
Ufa
Astrakhan'
Rostov-na-Donu
↑ 28
↓ 23

Europe

Northern Europe

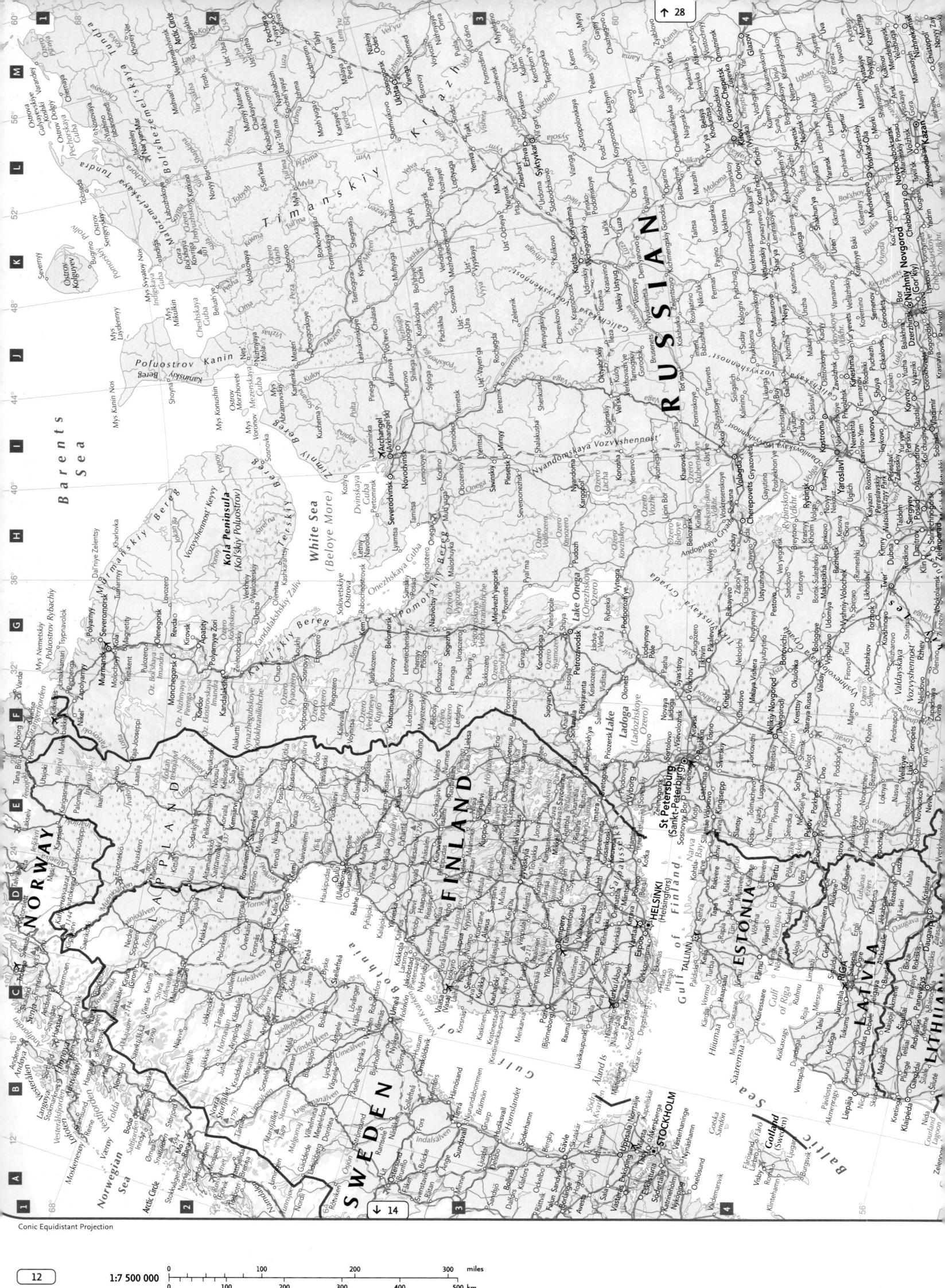

Conic Equidistant Projection

1:7 500 000

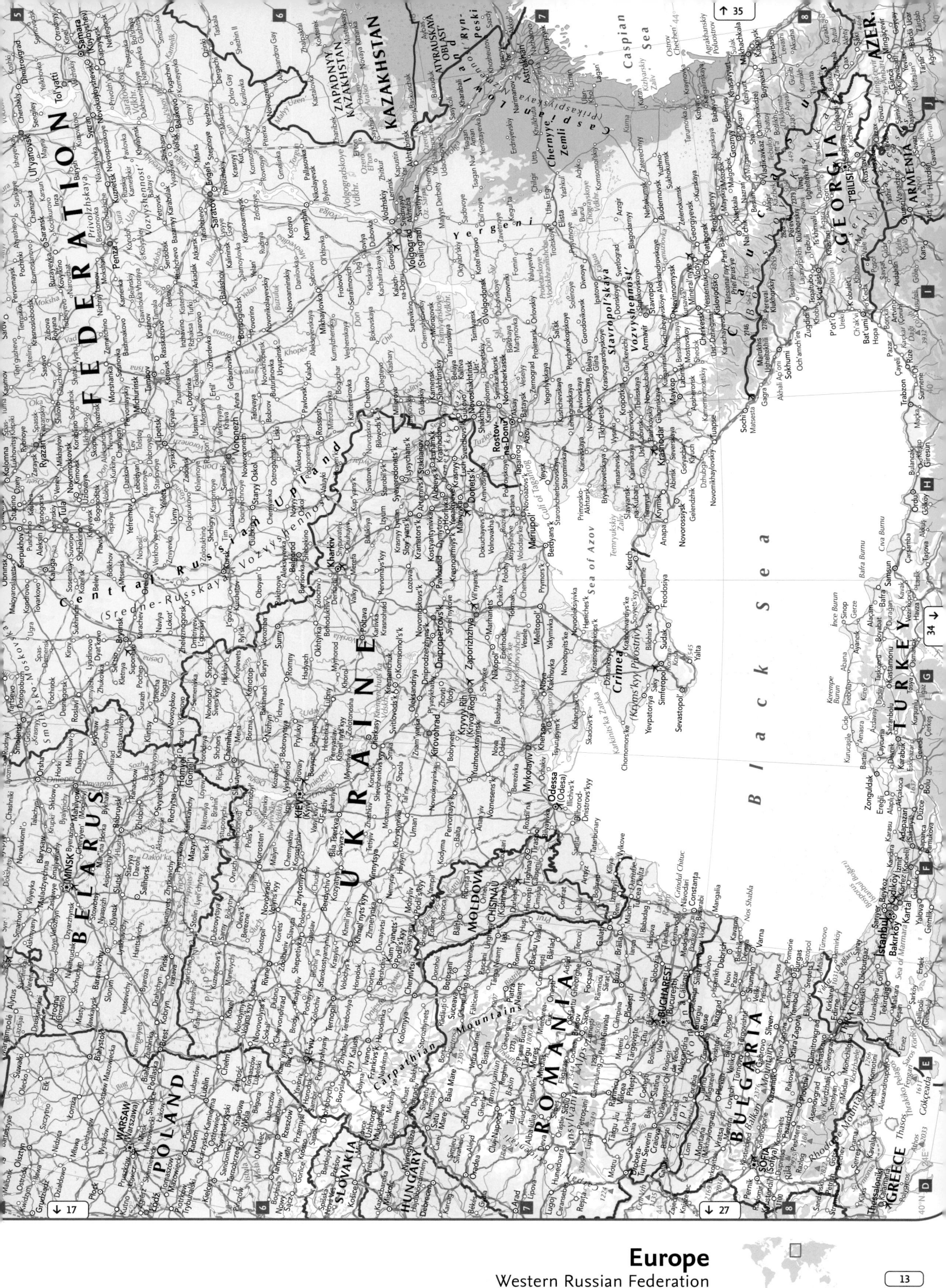
RUSSIAN FEDERATION
UKRAINE
BELARUS
POLAND
SLOVAKIA
HUNGARY
ROMANIA
MOLDOVA
BULGARIA
GREECE
TURKEY
GEORGIA
ARMENIA
AZER.
KAZAKHSTAN
Black Sea
Sea of Azov
Caspian Sea
Crimea
Caucasus
Carpathian Mountains
Central Russian Upland
Stavropol'skaya Vozvyshennost'
Caspian Lowland
WARSAW
MINSK
KIEV
CHIŞINĂU
BUCHAREST
SOFIA
T'BILISI
Moscow
Volgograd
Rostov-na-Donu
Odesa
Kharkiv
Samara
Saratov
Astrakhan'
İstanbul
↑ 35
↓ 17
↓ 27
34 →

Europe

Western Russian Federation

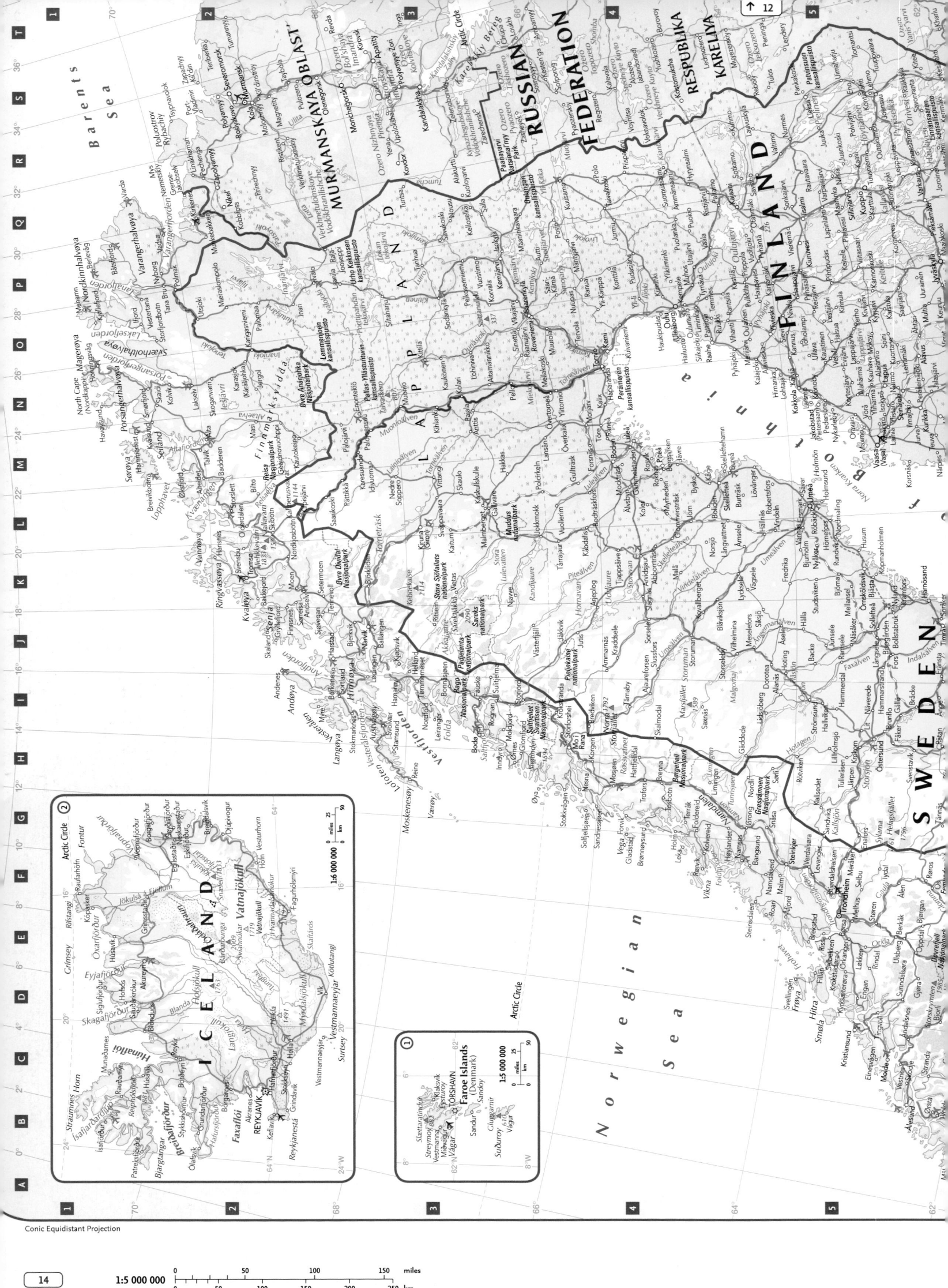

Conic Equidistant Projection

1:5 000 000

miles
0   50   100   150

0   50 100 150 200 250 km

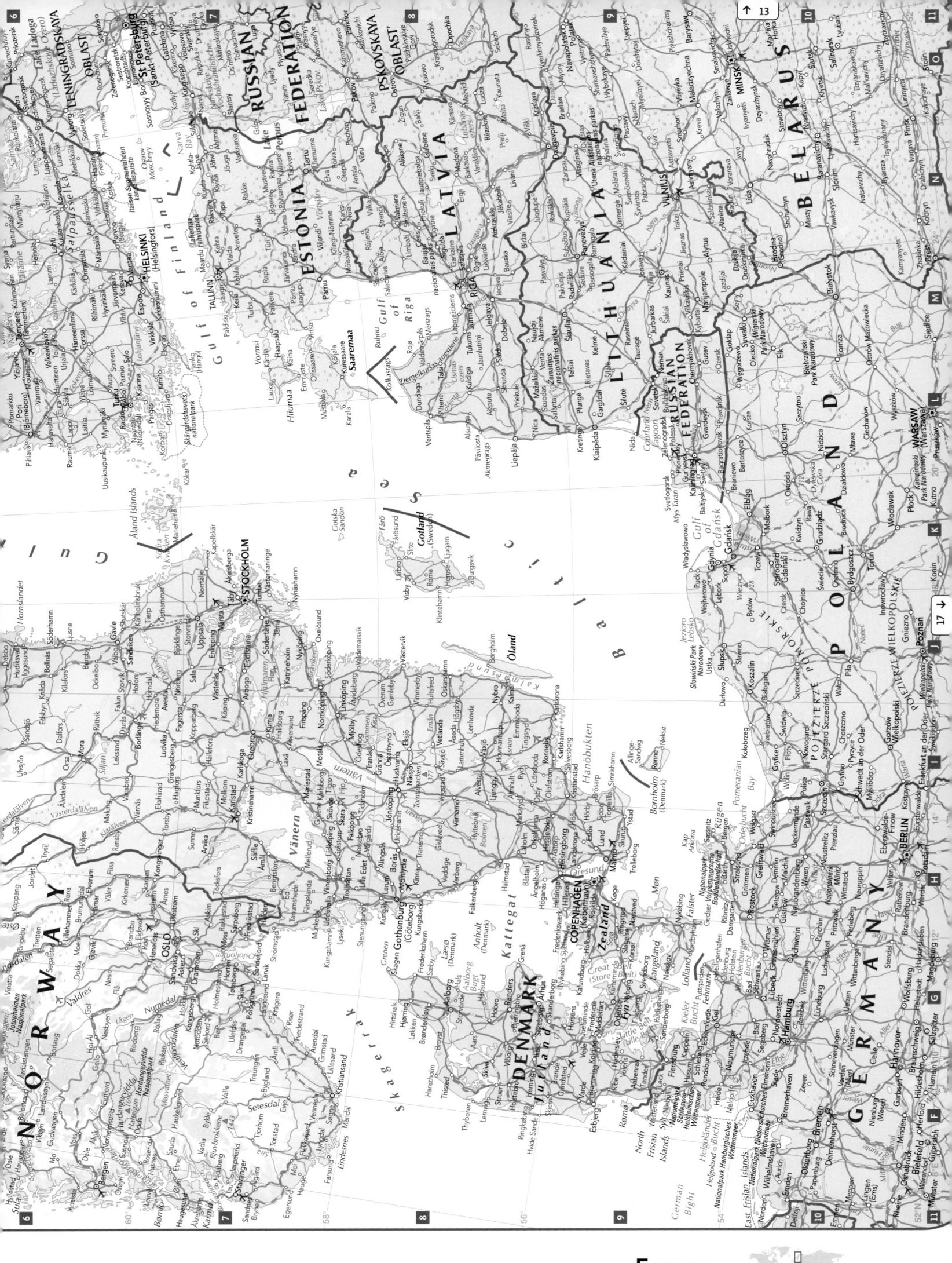

Europe

Scandinavia and the Baltic States

NORWAY
OSLO
Skagerrak
DENMARK
Jutland
Kattegat
COPENHAGEN
(København)
Zealand
Vänern
STOCKHOLM
Åland Islands
Gulf of Finland
HELSINKI
(Helsingfors)
TALLINN
ESTONIA
Saaremaa
Gulf of Riga
RIGA
LATVIA
LITHUANIA
VILNIUS
Baltic Sea
Gotland
(Sweden)
Öland
Bornholm
(Denmark)
RUSSIAN FEDERATION
St Petersburg
(Sankt-Peterburg)
LENINGRADSKAYA OBLAST'
PSKOVSKAYA OBLAST'
BELARUS
MINSK
POLAND
WARSAW
(Warszawa)
Gdańsk
GERMANY
BERLIN
Hamburg
13
17

ATLANTIC OCEAN
North Sea
British Isles
UNITED KINGDOM
IRELAND
Great Britain
FRANCE
BELGIUM
Celtic Sea
Irish Sea
English Channel (La Manche)
Bristol Channel
St George's Channel
North Channel
Strait of Dover (Pas de Calais)
Baie de Seine
Golfe de St-Malo
Cardigan Bay
Galway Bay
Donegal Bay
Moray Firth
Firth of Forth
Solway Firth
The Minch
Little Minch
Outer Hebrides
Orkney Islands
Isle of Lewis
South Harris
North Uist
Benbecula
South Uist
Barra
St Kilda
Skye
Rum
Coll
Mull
Colonsay
Jura
Islay
Arran
Isle of Man (U.K.)
Anglesey
Isle of Wight
Isles of Scilly
Guernsey (U.K.)
Jersey (U.K.)
Channel Islands (Îles Normandes)
Grampian Mountains
Southern Uplands
Pennines
Cambrian Mountains
ULSTER
CONNAUGHT
LEINSTER
MUNSTER
NORMANDY
BRITTANY
PICARDY
ARTOIS
ANJOU
POITOU
BRIE
LONDON
DUBLIN (Baile Átha Cliath)
Edinburgh
Glasgow
Aberdeen
Dundee
Inverness
Belfast
Londonderry
Cork (Corcaigh)
Limerick (Luimneach)
Galway (Gaillimh)
Waterford (Port Láirge)
Newcastle upon Tyne
Leeds
Manchester
Liverpool
Sheffield
Birmingham
Nottingham
Leicester
Bristol
Cardiff (Caerdydd)
Swansea (Abertawe)
Plymouth
Southampton
Portsmouth
Brighton
Norwich
Kingston upon Hull
DOUGLAS
ST PETER PORT
ST HELIER
PARIS
AMSTERDAM
THE HAGUE ('s-Gravenhage) (Den Haag)
Rotterdam
Antwerp
BRUSSELS (Bruxelles)
Le Havre
Rouen
Caen
Cherbourg-Octeville
Rennes
Brest
Nantes
Le Mans
Orléans
Tours
Angers
Lille
Calais
Amiens
Reims
Conic Equidistant Projection
↓ 24
1:5 000 000
0 50 100 150 miles
0 50 100 150 200 250 km

15
NORWAY
SWEDEN
ESTONIA
LATVIA
LITHUANIA
RUSSIAN FEDERATION
DENMARK
POLAND
GERMANY
CZECH REPUBLIC
SLOVAKIA
AUSTRIA
HUNGARY
SWITZERLAND
LIECHTENSTEIN
SLOVENIA
CROATIA
ITALY
SERBIA
Skagerrak
Kattegat
Baltic Sea
Gulf of Riga
Gulf of Gdańsk
German Bight
Pomeranian Bay
Gotland (Sweden)
Öland
Saaremaa
Bornholm (Denmark)
Jutland
Zealand
Fyn
Lolland
Falster
Møn
North Frisian Islands
East Frisian Islands
Helgoland
COPENHAGEN (København)
BERLIN
PRAGUE (Praha)
WARSAW (Warszawa)
VIENNA (Wien)
BRATISLAVA
BUDAPEST
BERN
VADUZ
Gothenburg (Göteborg)
Hamburg
Munich (München)
Nuremberg (Nürnberg)
Cologne (Köln)
Frankfurt am Main
Stuttgart
Leipzig
Dresden
Hannover
Bremen
Kraków
Wrocław
Poznań
Gdańsk
Szczecin
Łódź
Zürich
Salzburg
Graz
Linz
Brno
Ostrava
Kaliningrad
Riga
BAVARIA
Harz
Black Forest
Erzgebirge
Sudety
Böhmer Wald
Schwäbische Alb
Vosges
ALSACE
Jura
13
26

Europe

Northwest Europe

UNITED KINGDOM

North Sea

Irish Sea

Scotland

Grampian Mountains

Cairngorms National Park

Ben Macdui 1309

Ben Nevis 1344

Forest of Atholl

Southern Uplands

Firth of Forth

Firth of Clyde

Glasgow
Edinburgh
Dundee
Aberdeen

Loch Lomond & the Trossachs National Park

NORTHERN IRELAND

Belfast

Antrim Hills

Lough Neagh

Giant's Causeway

IRELAND

DUBLIN (Baile Átha Cliath)

Dún Laoghaire

Drogheda

Isle of Man (U.K.)

DOUGLAS

Snaefell 621

North Channel

Arran

Jura

Islay

Mull

Rum (Rhum)

Tiree

Coll

Anglesey (Ynys Môn)

Holyhead

Pennines

Lake District National Park

Scafell Pike 978

Yorkshire Dales National Park

North York Moors National Park

Newcastle upon Tyne

Sunderland

Middlesbrough

Leeds

Manchester

Liverpool

Sheffield

York

Carlisle

Lancaster

Blackpool

Preston

Bradford

Solway Firth

Northumberland National Park

Kielder Water (Reservoir)

The Cheviot 815

Berwick-upon-Tweed

Holy Island (Lindisfarne)

Farne Islands

Flamborough Head

Kingston upon Hull

Grimsby

Scarborough

Whitby

Hartlepool

Yorkshire Wolds

Lincolnshire Wolds

Mouth of the Humber

Conic Equidistant Projection

1:2 000 000

0   25   50   75   miles
0   25   50   75   100   125   km

18

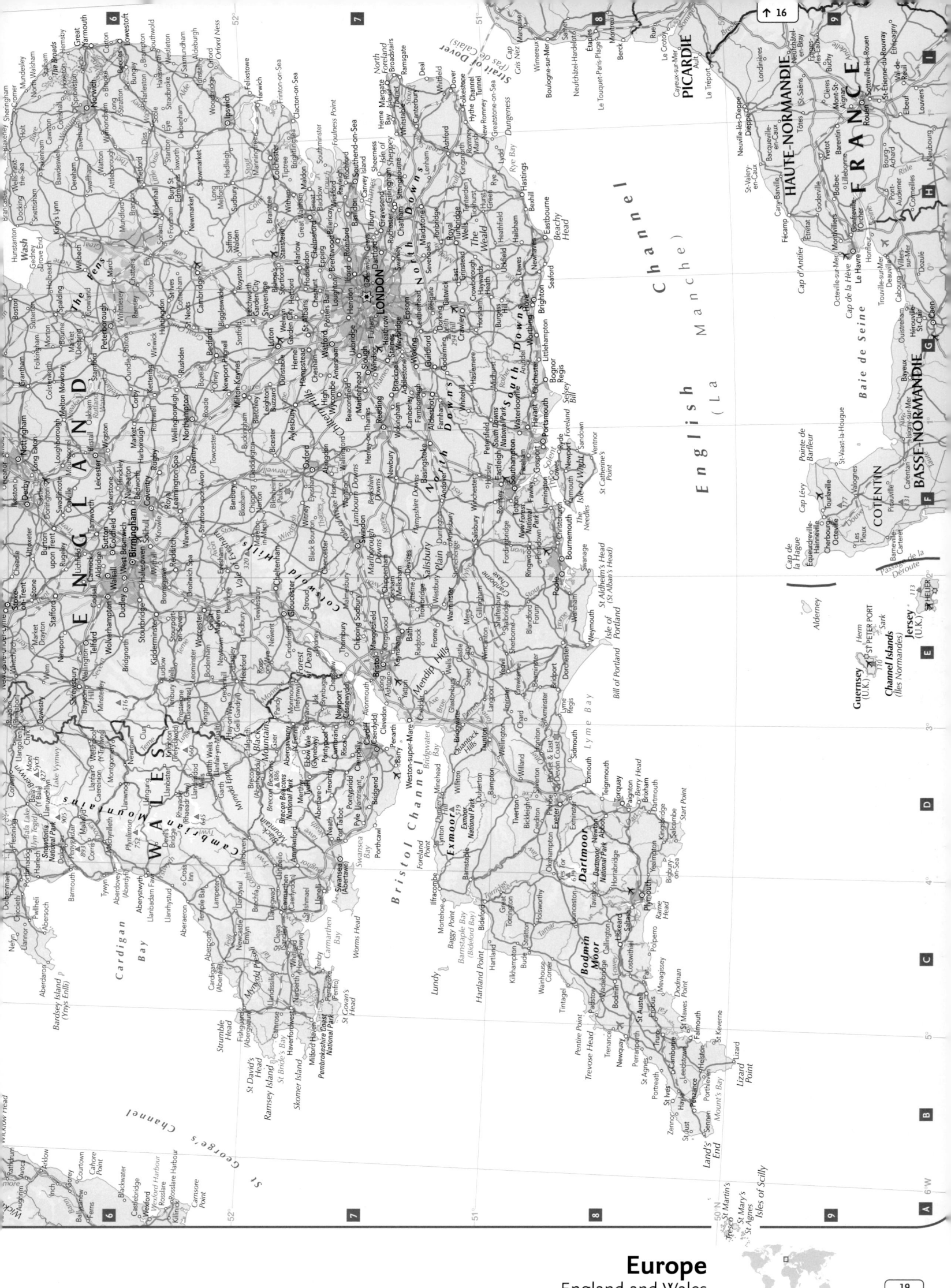

ENGLAND
WALES
FRANCE
LONDON
English Channel
(La Manche)
Bristol Channel
Cardigan Bay
St George's Channel
Strait of Dover
(Pas de Calais)
Isles of Scilly
Channel Islands
(Îles Normandes)
Guernsey (U.K.)
Jersey (U.K.)
Alderney
PICARDIE
HAUTE-NORMANDIE
BASSE-NORMANDIE
COTENTIN
Baie de Seine
Cambrian Mountains
Cotswold Hills
Chiltern Hills
North Downs
South Downs
The Fens
The Wash
Dartmoor
Exmoor
Bodmin Moor
Land's End
Lizard Point
Isle of Wight
Birmingham
Bristol
Cardiff (Caerdydd)
Swansea (Abertawe)
Southampton
Portsmouth
Brighton
Dover
Norwich
Cambridge
Oxford
Plymouth
Exeter
Le Havre
Rouen
Caen
Cherbourg

Europe
England and Wales

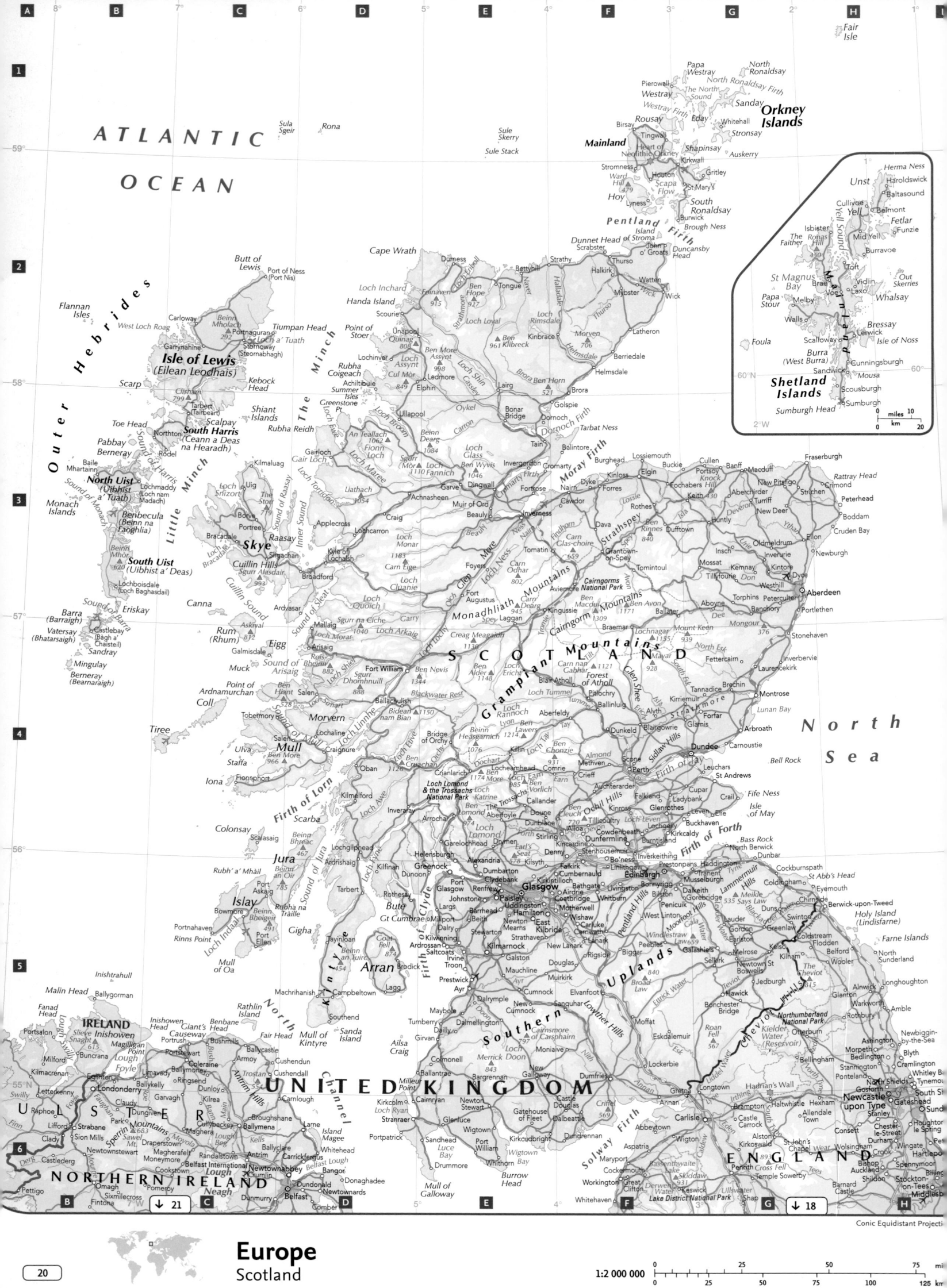

ATLANTIC OCEAN

Orkney Islands

Shetland Islands

SCOTLAND

UNITED KINGDOM

NORTHERN IRELAND

IRELAND

ENGLAND

North Sea

Conic Equidistant Projection

Europe
Scotland

1:2 000 000

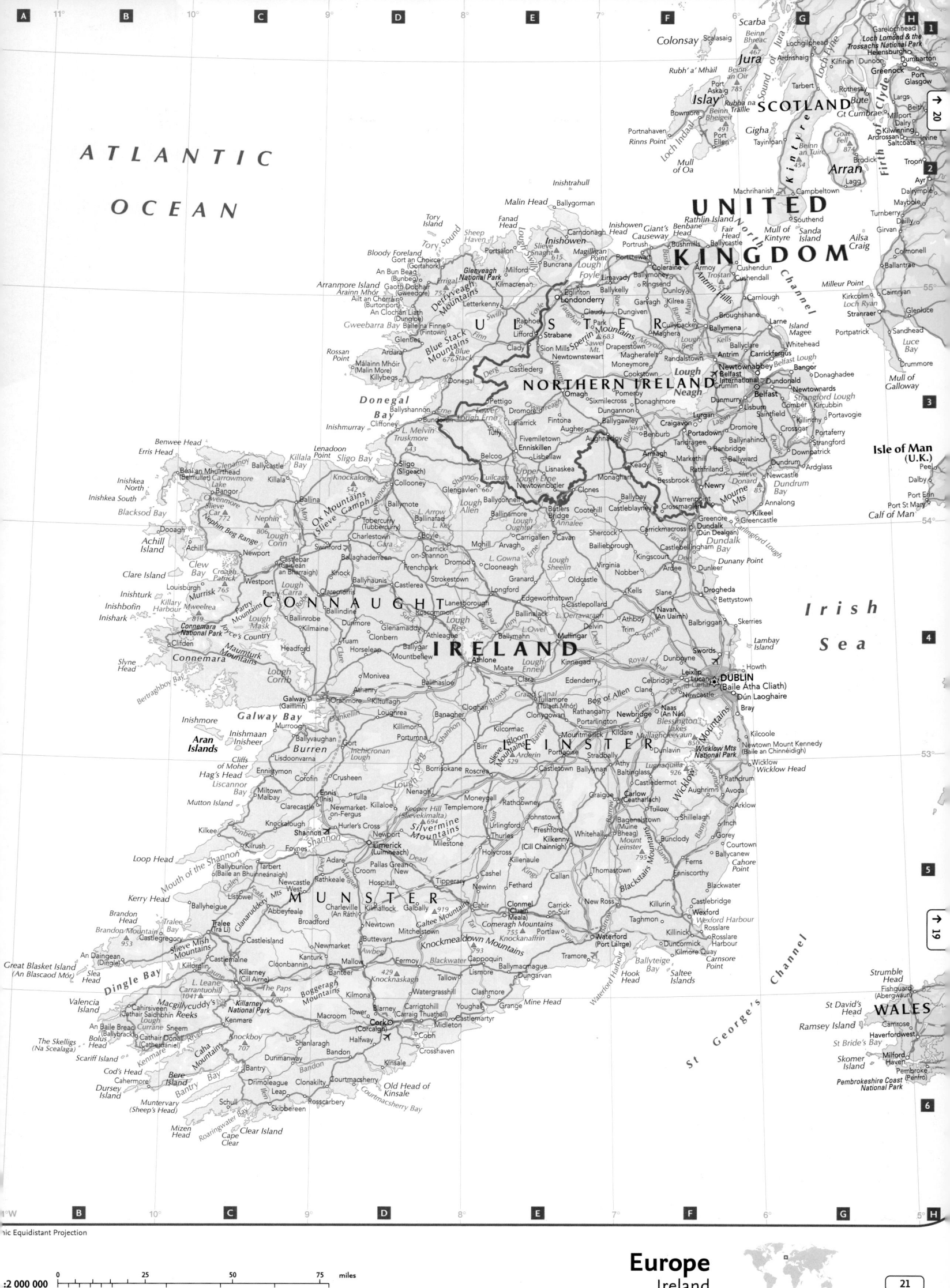

ATLANTIC
OCEAN
UNITED
KINGDOM
SCOTLAND
NORTHERN IRELAND
IRELAND
IRELAND
ULSTER
CONNAUGHT
LEINSTER
MUNSTER
Irish
Sea
North Channel
St George's Channel
WALES
Isle of Man
(U.K.)
Calf of Man
Colonsay
Scalasaig
Scarba
Jura
Islay
Port Askaig
Bowmore
Port Ellen
Rinns Point
Mull of Oa
Gigha
Kintyre
Arran
Lagg
Campbeltown
Machrihanish
Mull of Kintyre
Sanda Island
Ailsa Craig
Lochgilphead
Ardrishaig
Tarbert
Garelochhead
Loch Lomond & the Trossachs National Park
Helensburgh
Dumbarton
Greenock
Port Glasgow
Rothesay
Bute
Gt Cumbrae
Millport
Largs
Dalry
Kilwinning
Ardrossan
Saltcoats
Irvine
Troon
Ayr
Dalrymple
Maybole
Turnberry
Dailly
Girvan
Colmonell
Ballantrae
Milleur Point
Kirkcolm
Cairnryan
Loch Ryan
Stranraer
Glenluce
Portpatrick
Sandhead
Luce Bay
Drummore
Mull of Galloway
Peel
Dalby
Port Erin
Port St Mary
Inishtrahull
Malin Head
Ballygorman
Fanad Head
Tory Island
Tory Sound
Sheep Haven
Inishowen
Carndonagh
Lough Swilly
Buncrana
Lough Foyle
Magilligan Point
Bloody Foreland
Gort an Choirce
(Gortahork)
An Bun Beag
(Bunbeg)
Gaoth Dobhair
(Gweedore)
Arranmore Island
Árainn Mhór
Ailt an Chorráin
(Burtonport)
An Clochán Liath
(Dungloe)
Gweebarra Bay
Glenveagh National Park
Derryveagh Mountains
Errigal
Milford
Kilmacrenan
Letterkenny
Portsalon
Rossan Point
Málainn Mhóir
(Malin More)
Killybegs
Glenties
Blue Stack Mountains
Donegal
Donegal Bay
Ballyshannon
Bundoran
Inishmurray
Cliffoney
Raphoe
Lifford
Strabane
Clady
Castlederg
Sion Mills
Newtownstewart
Sperrin Mountains
Londonderry
Eglinton
Limavady
Coleraine
Portrush
Portstewart
Bushmills
Giant's Causeway
Benbane Head
Ballycastle
Rathlin Island
Fair Head
Cushendun
Cushendall
Ballymoney
Ballymena
Antrim Hills
Carnlough
Larne
Island Magee
Whitehead
Carrickfergus
Antrim
Randalstown
Lough Neagh
Newtownabbey
Belfast
Belfast International
Bangor
Donaghadee
Newtownards
Dundonald
Holywood
Lisburn
Comber
Strangford Lough
Kircubbin
Portavogie
Portaferry
Strangford
Downpatrick
Ardglass
Dundrum Bay
Newcastle
Mourne Mts
Annalong
Kilkeel
Greencastle
Newry
Warrenpoint
Crossmaglen
Armagh
Portadown
Lurgan
Craigavon
Banbridge
Dromore
Ballynahinch
Dungannon
Cookstown
Magherafelt
Maghera
Draperstown
Omagh
Sixmilecross
Pomeroy
Donaghmore
Fintona
Augher
Aughnacloy
Ballygawley
Benburb
Enniskillen
Lisbellaw
Lisnaskea
Lower Lough Erne
Upper Lough Erne
Irvinestown
Pettigo
Belleek
Belcoo
Newtownbutler
Clones
Monaghan
Castleblayney
Carrickmacross
Ballybay
Cootehill
Dundalk
(Dún Dealgan)
Dundalk Bay
Greenore
Carlingford Lough
Dunany Point
Ardee
Drogheda
Bettystown
Navan
(An Uaimh)
Kells
Slane
Balbriggan
Skerries
Lambay Island
Swords
DUBLIN
(Baile Átha Cliath)
Dún Laoghaire
Bray
Leixlip
Lucan
Celbridge
Maynooth
Naas
(An Nás)
Newbridge
Kildare
Blessington Lakes
Wicklow Mountains
Wicklow Mts National Park
Lugnaquilla
Kilcoole
Newtown Mount Kennedy
(Baile an Chinnéidigh)
Wicklow
Wicklow Head
Rathdrum
Avoca
Arklow
Aughrim
Tinahely
Shillelagh
Gorey
Courtown
Ballycanew
Cahore Point
Ferns
Enniscorthy
Blackwater
Castlebridge
Wexford
Wexford Harbour
Rosslare
Rosslare Harbour
Kilmore Quay
Carnsore Point
Saltee Islands
Ballyteige Bay
Hook Head
Waterford Harbour
Waterford
(Port Láirge)
Tramore
New Ross
Thomastown
Kilkenny
(Cill Chainnigh)
Callan
Carlow
(Ceatharlach)
Tullow
Bagenalstown
(Muine Bheag)
Mount Leinster
Blackstairs Mountains
Castledermot
Baltinglass
Athy
Stradbally
Portlaoise
Mountmellick
Portarlington
Tullamore
(Tulach Mhór)
Edenderry
Bog of Allen
Rathangan
Clonygowan
Kilcormac
Slieve Bloom Mountains
Arderin
Birr
Banagher
Clara
Moate
Athlone
Mullingar
Kinnegad
Delvin
Athboy
Trim
Longford
Edgeworthstown
Granard
Castlepollard
Lough Derravaragh
Lough Ennell
Lough Owel
Ballymahon
Lough Ree
Roscommon
Lanesborough
Strokestown
Carrick-on-Shannon
Boyle
Mohill
Arvagh
Cavan
Bailieborough
Kingscourt
Virginia
Nobber
Oldcastle
Lough Sheelin
Lough Gowna
Ballyconnell
Belturbet
Butlers Bridge
Ballinamore
Lough Allen
Glengavlen
Cuilcagh
Dromod
Sligo
(Sligeach)
Sligo Bay
Collooney
Ballymote
Tobercurry
(Tubbercurry)
Ox Mountains
(Slieve Gamph)
Ballina
Killala
Killala Bay
Lenadoon Point
Benwee Head
Erris Head
Belmullet
Carrowmore Lake
Bangor
Inishkea North
Inishkea South
Blacksod Bay
Nephin
Nephin Beg Range
Achill Island
Achill
Newport
Castlebar
Westport
Clew Bay
Croagh Patrick
Clare Island
Louisburgh
Inishturk
Killary Harbour
Mweelrea
Inishbofin
Inishark
Connemara National Park
Clifden
Connemara
Maumturk Mountains
Joyce's Country
Lough Mask
Lough Carra
Lough Corrib
Slyne Head
Bertraghboy Bay
Partry Mountains
Ballinrobe
Claremorris
Knock
Ballyhaunis
Kiltimagh
Swinford
Charlestown
Ballaghaderreen
Frenchpark
Castlerea
Ballindine
Dunmore
Tuam
Glenamaddy
Clonbern
Headford
Athleague
Ballygar
Mountbellew
Horseleap
Monivea
Athenry
Galway
(Gaillimh)
Oranmore
Kiltullagh
Ballinasloe
Loughrea
Portumna
Gort
Galway Bay
Inishmore
Inishmaan
Inisheer
Aran Islands
Ballyvaughan
Burren
Lisdoonvarna
Cliffs of Moher
Hag's Head
Ennistymon
Liscannor Bay
Milltown Malbay
Corofin
Crusheen
Ennis
(Inis)
Tulla
Killaloe
Clarecastle
Newmarket-on-Fergus
Mutton Island
Kilkee
Kilrush
Loop Head
Mouth of the Shannon
Shannon
Limerick
(Luimneach)
Foynes
Askeaton
Adare
Rathkeale
Newcastle West
Abbeyfeale
Listowel
Ballybunion
Tarbert
Kerry Head
Ballyheigue
Tralee
(Trá Lí)
Tralee Bay
Brandon Head
Brandon Mountain
Castlegregory
Slieve Mish Mountains
Dingle Bay
An Daingean
(Dingle)
Great Blasket Island
(An Blascaod Mór)
Slea Head
Castlemaine
Killorglin
Killarney
(Cill Airne)
Killarney National Park
Macgillycuddy's Reeks
Carrantuohill
Valencia Island
Cahirsiveen
(Cathair Saidhbhín)
An Baile Breac
(Ballybrack)
Bolus Head
The Skelligs
(Na Scealaga)
Scariff Island
Cod's Head
Cahermore
Dursey Island
Kenmare
Sneem
Caha Mountains
Bantry
Bantry Bay
Bere Island
Muntervary
(Sheep's Head)
Mizen Head
Schull
Skibbereen
Cape Clear
Clear Island
Roaringwater Bay
Rosscarbery
Clonakilty
Courtmacsherry
Courtmacsherry Bay
Old Head of Kinsale
Kinsale
Bandon
Dunmanway
Macroom
Blarney
Cork
(Corcaigh)
Cobh
Crosshaven
Carrigtohill
(Carraig Thuathail)
Midleton
Youghal
Castlemartyr
Mine Head
Dungarvan
Ballymacmague
Lismore
Cappoquin
Tallow
Knockmealdown Mountains
Knockanaffrin
Comeragh Mountains
Portlaw
Carrick-on-Suir
Clonmel
(Cluain Meala)
Cahir
Galtee Mountains
Mitchelstown
Fermoy
Mallow
Kanturk
Newmarket
Buttevant
Charleville
(An Ráth)
Kilmallock
Galbally
Tipperary
Cashel
Fethard
Newinn
Holycross
Thurles
Templemore
Roscrea
Borrisokane
Nenagh
Newport
Silvermine Mountains
Keeper Hill
(Slievekimalta)
Lough Derg
Hospital
Croom
Pallas Grean
Knocknaskagh
Boggeragh Mountains
Clashmore
Grange
Mullaghcleevaun
Dunlavin
Rathdowney
Johnstown
Freshford
Urlingford
Killenaule
Castletown
Ballylynan
Graigue
Bunclody
Killurin
Taghmon
Killinick
Duncormick
Milford
St David's Head
Ramsey Island
Fishguard
(Abergwaun)
Strumble Head
Camrose
Haverfordwest
St Bride's Bay
Skomer Island
Milford Haven
Pembroke
(Penfro)
Pembrokeshire Coast National Park
→ 20
→ 19
A B C D E F G H
1 2 3 4 5 6
11° 10° 9° 8° 7° 6°
55° 54° 53°
...nic Equidistant Projection
:2 000 000
0 25 50 75 miles
0 25 50 75 100 125 km
Europe
Ireland

ATLANTIC OCEAN

Conic Equidistant Projection

1:10 000 000

0   100   200   300   400 miles
0   100   200   300   400   500   600 km

→ 32

Europe
Southern Europe and the Mediterranean

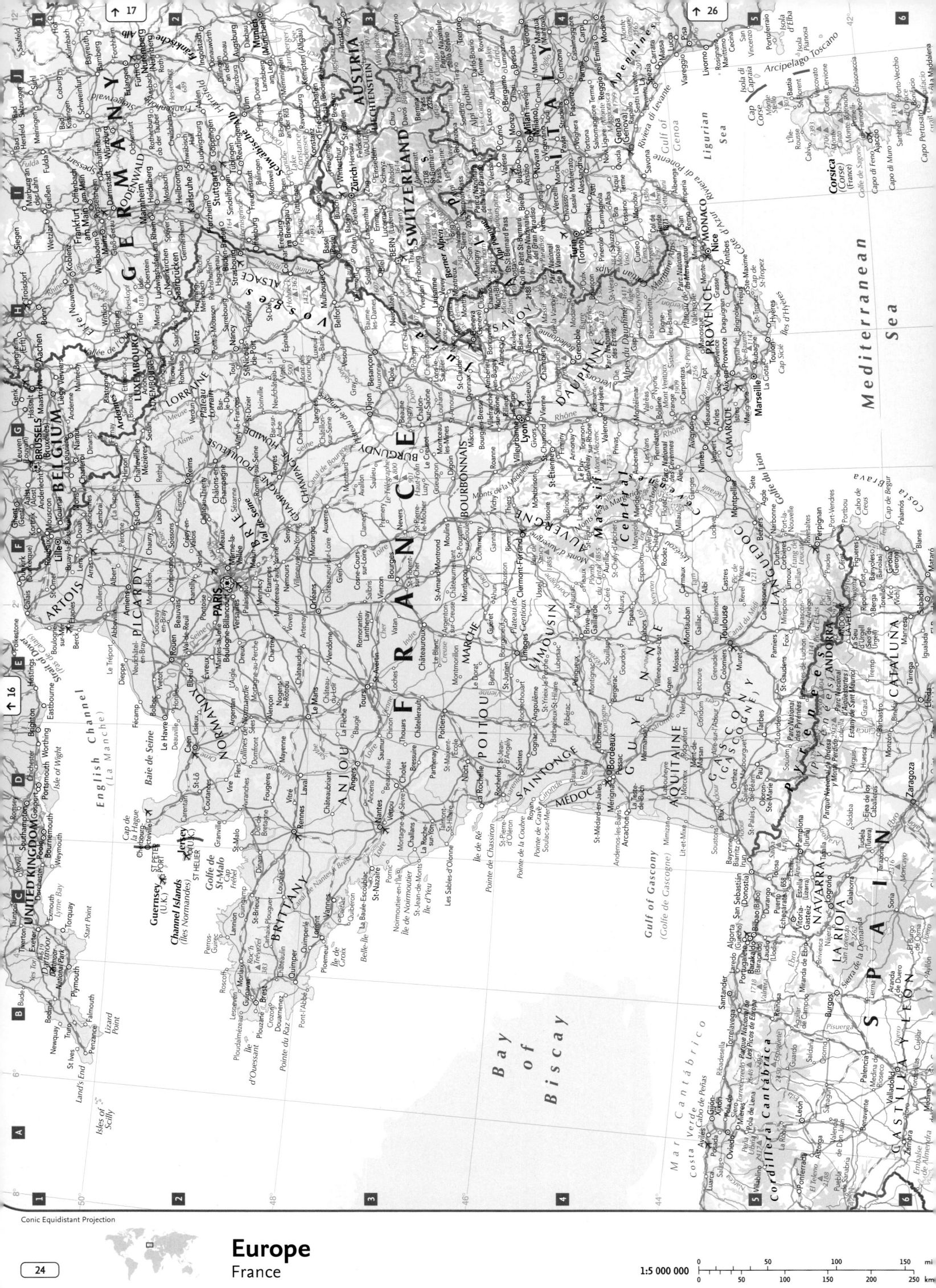

Europe

France

Conic Equidistant Projection

1:5 000 000

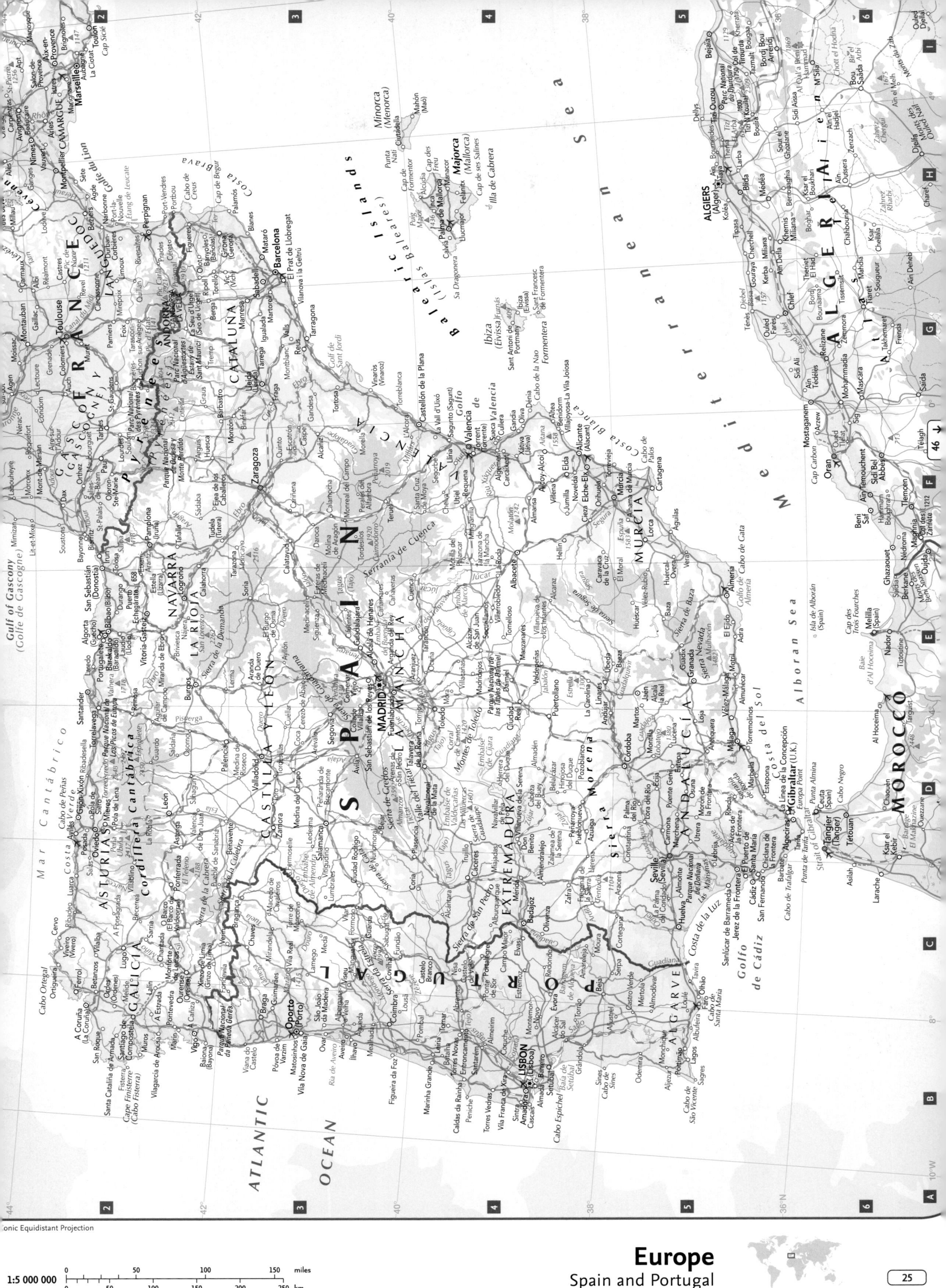

Europe
Spain and Portugal

1:5 000 000

25

Europe
Italy and the Balkans

↑ 13
→ 34
ROMANIA
BULGARIA
SERBIA
KOSOVO
MACEDONIA (F.Y.R.O.M.)
ALBANIA
GREECE
TURKEY
MOLDOVA
UKRAINE
VOJVODINA
Black Sea
Aegean Sea
Sea of Marmara (Marmara Denizi)
Mirtoo Pelagos
Kritiko Pelagos
Transylvanian Alps
Carpathian Mountains
Balkan Mountains
Rhodope Mountains
Pindus Mountains
Transylvanian Basin
DOBRUJA
THRACE
BITHYNIA
MYSIA
LYDIA
PAMPHYLIA
LYCIA
Danube
BUCHAREST (București)
BELGRADE (Beograd)
SOFIA (Sofiya)
SKOPJE
PRISHTINË (Priština)
TIRANA (Tiranë)
ATHENS (Athína)
CHIȘINĂU (Kishinev)
İstanbul
Thessaloniki
Odessa (Odesa)
Constanța
Varna
Burgas
Izmir (Smyrna)
Bursa
Crimea (Kryms'kyy Pivostriv)
Peloponnese
Cyclades (Kyklades)
Dodecanese (Dodekanisa)
Crete (Kriti)
Rhodes (Rodos)
Lesbos (Lesvos)
Chios
Samos
Evvoia
Gökçeada
Limnos
Thasos
Corfu (Kerkyra)
Cephalonia (Kefallonia)
Zakynthos (Zante)
Ionian Islands

ARCTIC
Greenland Sea
Norwegian Sea
North Sea
Barents Sea
Kara Sea (Karskoye More)
White Sea (Beloye More)
Baltic Sea
Black Sea
Sea of Azov
Caspian Sea
Aral Sea
Svalbard (Norway)
Spitsbergen
Zemlya Frantsa-Iosifa
Novaya Zemlya
Faroe Islands (Denmark)
UNITED KINGDOM
NORWAY
SWEDEN
FINLAND
DENMARK
ESTONIA
LATVIA
LITHUANIA
POLAND
BELARUS
UKRAINE
GEORGIA
ARMENIA
AZERBAIJAN
TURKEY
IRAQ
IRAN
TURKMENISTAN
UZBEKISTAN
KAZAKHSTAN
KYRGYZSTAN
TAJIKISTAN
AFGHANISTAN
RUSSIAN FEDERATION
XINJIANG
OSLO
STOCKHOLM
COPENHAGEN
HELSINKI
TALLINN
RIGA
VILNIUS
MINSK
WARSAW
BERLIN
KIEV
CHIŞINĂU
MOSCOW (Moskva)
St Petersburg
TBILISI
YEREVAN
BAKU
TEHRAN
AŞGABAT (Ashkhabad)
TOSHKENT (Tashkent)
BISHKEK (Frunze)
DUSHANBE
ASTANA (Akmola)
Ural Mountains
Caspian Lowland (Prikaspiyskaya Nizmennost')
West Siberian Plain (Zapadno-Sibirskaya Ravnina)
Kazakh Steppe
Karakum Desert (Garagum)
Kyzylkum Desert
Turan Lowland
Tien Shan
Altai Mountains
Tarim Basin (Tarim Pendi)
Conic Equidistant Projection
1:20 000 000
miles
km

Asia

Northern Asia

OCEAN
East Siberian Sea
(Vostochno-Sibirskoye More)
New Siberia Islands
(Novosibirskiye Ostrova)
Ostrov Kotel'nyy
Ostrov Bennetta
Ostrova De-Longa
Laptev Sea
(More Laptevykh)
Severnaya Zemlya
Ostrov Oktyabr'skoy Revolyutsii
Ostrov Bol'shevik
Wrangel Island
(Ostrov Vrangelya)
Chukchi Sea
Bering Strait
Seward Peninsula
Arctic Circle
Norton Sound
U.S.A.
Nome
Kotzebue
Point Hope
St Lawrence Island
St Matthew Island (U.S.A.)
Nunivak Island
Pribilof Islands
Aleutian Islands
Kiska Island (U.S.A.)
Attu Island (U.S.A.)
Bering Sea
Chukotskiy Poluostrov
Anadyr'
Koryakskoye Nagor'ye
Khrebet Kolymskiy
Kolymskaya Nizmennost'
Yukagirskoye Ploskogor'ye
Khrebet Cherskogo
Verkhoyanskiy Khrebet
Khrebet Orulgan
Komandorskiye Ostrova
Kamchatka Peninsula
(Poluostrov Kamchatka)
Sredinnyy Khrebet
Petropavlovsk-Kamchatskiy
Sea of Okhotsk
(Okhotskoye More)
Magadan
Okhotsk
Ayan
Khrebet Dzhugdzhur
Sakhalin
Okha
Yuzhno-Sakhalinsk
Tatarskiy Proliv
La Pérouse Strait
Kuril Islands
(Kuril'skiye Ostrova)
Ostrov Iturup
Ostrov Kunashir
ADMINISTERED BY RUSSIAN FEDERATION, CLAIMED BY JAPAN
PACIFIC OCEAN
Hokkaidō
Sapporo
Asahikawa
Kushiro
Hakodate
Aomori
Morioka
Sendai
Akita
Niigata
Fukushima
JAPAN
TŌKYŌ
Yokohama
Nagoya
Kyoto
Ōsaka
Kobe
Hiroshima
Kita-Kyūshū
Fukuoka
Nagasaki
Kumamoto
Miyazaki
Shikoku
Kyūshū
Sea of Japan
(East Sea)
NORTH KOREA
P'YONGYANG
SOUTH KOREA
SEOUL
Inch'ŏn
Taejŏn
Taegu
Pusan
Kwangju
Yellow Sea
(Huang Hai)
Bo Hai
Korea Bay
Vladivostok
Nakhodka
Khabarovsk
Komsomol'sk-na-Amure
Sikhote-Alin'
Birobidzhan
Blagoveshchensk
Amur
Harbin
Qiqihar
Daqing
Jilin
Changchun
Shenyang
Anshan
Fushun
Dalian
MANCHURIA
Da Hinggan Ling
Xiao Hinggan Ling
Stanovoy Khrebet
Stanovoye Nagor'ye
Yablonovyy Khrebet
Severo-Baykal'skoye Nagor'ye
Baykal'skiy Khrebet
Lake Baikal
Irkutsk
Ulan-Ude
Chita
Bratsk
Tynda
Neryungri
Aldan
Yakutsk
Lena
Vilyuy
Mirnyy
Central Siberian Plateau
(Sredne-Sibirskoye Ploskogor'ye)
SIBERIA
(SIBIR')
RUSSIAN FEDERATION
MONGOLIA
ULAN BATOR
(Ulaanbaatar)
Gobi Desert
INNER MONGOLIA
CHINA
BEIJING
(Peking)
Tianjin
Baotou
Hohhot
Datong
Baoding
Tangshan
Yantai
Dongying

Albers Conic Equal Area Projection

1:20 000 000

RUSSIAN FEDERATION
MONGOLIA
KYRGYZSTAN
CHINA
INNER MONGOLIA
XINJIANG
TIBET
NEPAL
BHUTAN
BANGLADESH
INDIA
MYANMAR (BURMA)
LAOS
THAILAND
CAMBODIA
VIETNAM
MALAYSIA
INDONESIA
SRI LANKA
SINGAPORE
MALDIVES
Gobi Desert
Tarim Basin (Tarim Pendi)
Taklimakan Desert (Taklimakan Shamo)
Plateau of Tibet (Qingzang Gaoyuan)
Kunlun Shan
Altun Shan
Tien Shan
Altai Mountains
Himalaya
Qaidam Pendi
Qilian Shan
Kazakh Steppe
Saryarka
Deccan
Western Ghats
Eastern Ghats
Coromandel Coast
Malabar Coast
Bay of Bengal
Andaman Sea
South China Sea
Gulf of Thailand
Gulf of Tongking
Strait of Malacca
Bo Hai
Andaman Islands (India)
Nicobar Islands (India)
Ten Degree Channel
Nine Degree Channel
Eight Degree Channel
Mouths of the Ganges
Mouths of the Irrawaddy
Mouths of the Mekong
Tropic of Cancer
Equator
ULAN BATOR (Ulaanbaatar)
BEIJING (Peking)
NEW DELHI
KATHMANDU
THIMPHU
DHAKA (Dacca)
NAY PYI TAW
VIENTIANE (Viangchan)
HA NOI
BANGKOK (Krung Thep)
PHNOM PENH
KUALA LUMPUR
PUTRAJAYA
SRI JAYEWARDENEPURA KOTTE
ISLAMABAD
BISHKEK
Almaty (Alma-Ata)
Novosibirsk
Omsk
Ürümqi
Lhasa
Chongqing
Chengdu
Xi'an
Wuhan
Tianjin
Kolkata (Calcutta)
Mumbai (Bombay)
Chennai (Madras)
Bangalore (Bengaluru)
Hyderabad
Delhi
Karachi
Lahore
Chittagong
Mandalay
Yangon (Rangoon)
Ho Chi Minh City (Saigon)
Hong Kong
Guangzhou
Kunming
Colombo
↑ 29
→ 40
→ 41
↓ 73

Albers Conic Equal Area Projection

1:13 000 000

Asia
Southwest Asia

↑ 30
KAZAKHSTAN
TOSHKENT (Tashkent)
KYRGYZSTAN
UZBEKISTAN
TURKMENISTAN
Turan Lowland
Karakum Desert (Peski Karakumy)
AŞGABAT (Ashkhabad)
TAJIKISTAN
DUSHANBE
Pamir
Alai Range
XINJIANG UYGUR ZIZHIQU (SINKIANG)
Taklimakan Desert (Taklimakan Shamo)
Tarim Basin (Tarim Pendi)
CHINA
Kunlun Shan
AKSAI CHIN
UNDER CHINESE ADMINISTRATION CLAIMED BY INDIA
XIZANG ZIZHIQU (TIBET)
Gangdisê Shan
Nganglong Kangri
Karakoram Range
Hindu Kush
LINE OF CONTROL
JAMMU AND KASHMIR
Himalaya
Siwalik Range
NEPAL
↑ 42
KABUL
AFGHANISTAN
ISLAMABAD
Peshawar
Rawalpindi
Lahore
Faisalabad
Multan
Quetta
PAKISTAN
Karachi
Hyderabad
Thar Desert
Mashhad
Herāt
Kandahār
Dasht-e Kavīr
Dasht-e Lūt
Zāhedān
Kermān
Makran Coast Range
MAKRAN
Gulf of Oman
Strait of Hormuz
OMAN
MUSCAT (Masqat)
Sharjah (Ash Shāriqah)
Dubai (Dubayy)
ED ARAB
RATES
Arabian Sea
Socotra (Suquţrā) (Yemen)
Juzur al Ḩalānīyāt (Kuria Muria Islands)
Rann of Kachchh
Gulf of Kachchh
Gulf of Khambhat
INDIA
Delhi
NEW DELHI
Jaipur
Agra
Kanpur
Lucknow
Ahmadabad (Ahmedabad)
Surat
Mumbai (Bombay)
Thane
Navi Mumbai (New Bombay)
Pune (Poona)
Nagpur
Bhopal
Indore
Hyderabad
Bangalore (Bengaluru)
Mysore (Mysuru)
Kozhikode (Calicut)
Kochi (Cochin)
Thiruvananthapuram (Trivandrum)
Coimbatore (Koyamutthoor)
Madurai
Vindhya Range
Satpura Range
Deccan
Western Ghats
Malabar Coast
↑ 31
Laccadive Islands (India)
Amindivi Islands
Nine Degree Channel
Minicoy
Cape Comorin
55° 60° 65° 70° 75° 80° 85°
I J K L M N O
1 2 3 4 5 6 7 8

↑ 13
Administrative divisions in Russian Federation numbered on the map:
1. RESPUBLIKA KALMYKIYA - KHALM'G-TANGCH (G1)
2. RESPUBLIKA DAGESTAN (G2)
3. CHECHENSKAYA RESPUBLIKA (G2)
4. RESPUBLIKA INGUSHETIYA (G2)
5. RESPUBLIKA SEVERNAYA OSETIYA - ALANIYA (G2)
6. KABARDINO-BALKARSKAYA RESPUBLIKA (F2)
7. KARACHAYEVO-CHERKESSKAYA RESPUBLIKA (F2)
8. RESPUBLIKA ADYGEYA (F1)
ROMANIA
BUCHAREST (București)
SERBIA
KOSOVO
PRISHTINE (Priština)
BULGARIA
SOFIA (Sofiya)
Balkan Mountains
MACEDONIA (F.Y.R.O.M.)
SKOPJE
Rhodope Mountains
THRACE
UKRAINE
Crimea
Simferopol'
Sevastopol'
Black Sea
GREECE
Thessaloniki
ATHENS (Athina)
Aegean Sea
Cyclades (Kyklades)
Dodecanese (Dodekanisa)
Kritiko Pelagos
Crete (Kriti)
Rhodes (Rodos)
Lesbos (Lesvos)
Evvoia
Istanbul
Sea of Marmara (Marmara Denizi)
Bosporus (İstanbul Boğazı)
Dardanelles
İzmir (Smyrna)
Bursa
ANKARA
TURKEY
ANADOLU (ANATOLIA)
Konya
Antalya
Adana (Seyhan)
Taurus Mountains (Toros Dağları)
Antalya Körfezi
Aleppo (Halab)
Mediterranean Sea
CYPRUS
NICOSIA (Lefkosia)
Limassol (Lemesos)
LEBANON
BEIRUT (Beyrouth)
Homs
DAMASCUS (Dimashq)
Haifa (Ḥefa)
Tel Aviv-Yafo
WEST BANK
JERUSALEM
GAZA
ISRAEL
JORDAN
'AMMAN
LIBYA
EGYPT
Alexandria (Al Iskandarīyah)
CAIRO (Al Qāhirah)
Port Said (Būr Sa'īd)
Suez (As Suways)
Qattara Depression (Munkhafaḍ al Qaṭṭārah)
Great Sand Sea
Western Desert (Aṣ Ṣaḥrā' al Gharbīyah)
Eastern Desert (Aṣ Ṣaḥrā' ash Sharqīyah)
Sinai (Shibh Jazīrat Sīnā')
Gulf of Suez
Gulf of Aqaba
Red Sea
↑ 27
↑ 47
↓ 47
Conic Equidistant Projection
1:7 000 000
0 100 200 miles
0 100 200 300 400 km

Asia
Eastern Mediterranean, the Caucasus and Iraq

AFGHANISTAN
PAKISTAN
TAJIKISTAN
XINJIANG
AKSAI CHIN
UNDER CHINESE ADMINISTRATION CLAIMED BY INDIA
KHYBER PAKHTUNKHWA (NORTH WEST FRONTIER)
GILGIT-BALTISTAN (NORTHERN AREAS)
TRIBAL AREAS
JAMMU AND KASHMIR
LINE OF CONTROL
HIMACHAL PRADESH
PUNJAB
HARYANA
UTTARAKHAND
UTTAR PRADESH
RAJASTHAN
GUJARAT
MADHYA PRADESH
MAHARASHTRA
BALOCHISTAN
SINDH
INDIA
Thar Desert
Karakoram Range
Hindu Kush
Himalaya
Arabian Sea
Gulf of Kachchh
Gulf of Khambhat
Rann of Kachchh
Tropic of Cancer
Kābul
Islamabad
Karachi
New Delhi
Delhi
Administrative divisions in India numbered on the map:
1. DADRA AND NAGAR HAVELI (C5)
2. DAMAN AND DIU (B5, C5)
Conic Equidistant Projection
1:7 000 000
0 100 200 miles
0 100 200 300 400 km

Asia

Northern India, Nepal, Bhutan and Bangladesh

Asia
Southern India and Sri Lanka

Conic Equidistant Projection

1:7 000 000

Administrative divisions in India
numbered on the map:

1. DADRA AND NAGAR HAVELI (B1)
2. DAMAN AND DIU (A1, B1)
3. PUDUCHERRY (C4)

Asia

Middle East

Asia
Eastern and Southeast Asia

PACIFIC OCEAN
INDIAN OCEAN
PHILIPPINES
MALAYSIA
INDONESIA
AUSTRALIA
THAILAND
LAOS
CAMBODIA
VIETNAM
BRUNEI
SINGAPORE
EAST TIMOR
PAPUA NEW GUINEA
PALAU
FEDERATED STATES OF MICRONESIA
Northern Mariana Islands (U.S.A.)
Guam (U.S.A.)
South China Sea
Philippine Sea
Sulu Sea
Celebes Sea
Laut Jawa (Java Sea)
Laut Banda (Banda Sea)
Laut Flores (Flores Sea)
Arafura Sea
Timor Sea
Andaman Sea
Gulf of Thailand
Gulf of Carpentaria
Strait of Malacca
Luzon Strait
Borneo
Sumatra (Sumatera)
Java (Jawa)
Celebes (Sulawesi)
New Guinea
Halmahera
Seram
Timor
Flores
Sumba
Sumbawa
Bali
Lombok
Palawan
Mindanao
Luzon
Samar
Panay
Negros
Hainan Dao
Moluccas (Maluku)
Greater Sunda Islands
Lesser Sunda Islands
Caroline Islands
Spratly Islands
Paracel Islands (Xisha Qundao)
Christmas Island (Australia)
Cocos Islands (Australia)
Ashmore and Cartier Islands (Australia)
Nicobar Islands (India)
Andaman Islands (India)
MANILA
Quezon City
BANGKOK (Krung Thep)
VIENTIANE (Viangchan)
PHNOM PENH
KUALA LUMPUR
PUTRAJAYA
JAKARTA
BANDAR SERI BEGAWAN
DILI
MELEKEOK
RANGOON (Yangon)
Ho Chi Minh City (Saigon)
Davao
Cebu
Zamboanga
Makassar (Ujung Pandang)
Surabaya
Bandung
Semarang
Medan
Palembang
Kuching
Kota Kinabalu
Darwin
NORTHERN TERRITORY
QUEENSLAND
WESTERN AUSTRALIA
Great Sandy Desert
Tanami Desert
Kimberley Plateau
Barkly Tableland
Arnhem Land
Cape York Peninsula
Tropic of Capricorn
Equator
↑ 52
↓ 52

Albers Conic Equal Area Projection

1:15 000 000

miles
0   200   400

0   200   400   600   800 km

Asia
Eastern Asia

↑ 29
RATION
Sea of Okhotsk
(Okhotskoye More)
Sakhalin
ADMINISTERED BY RUSSIAN FEDERATION, CLAIMED BY JAPAN
Kuril Islands
(Kuril'skiye Ostrova)
Khabarovsk
Komsomol'sk-na-Amure
Blagoveshchensk
Birobidzhan
Vladivostok
Ussuriysk
Nakhodka
MANCHURIA
HEILONGJIANG
Harbin
Qiqihar
Daqing
Mudanjiang
Jiamusi
JILIN
Changchun
Jilin
LIAONING
Shenyang
Anshan
Dalian
NEI MONGOL ZIZHIQU
(INNER MONGOLIA)
Hohhot
Baotou
Chifeng
HEBEI
BEIJING
(Peking)
Tianjin
TIANJIN
Shijiazhuang
Tangshan
Great Wall
Bo Hai
SHANXI
Taiyuan
SHANDONG
Jinan
Qingdao
(Tsingtao)
Yellow Sea
(Huang Hai)
HENAN
Zhengzhou
Luoyang
Xi'an
JIANGSU
Nanjing
Xuzhou
ANHUI
Hefei
Shanghai
SHANGHAI
HUBEI
Wuhan
ZHEJIANG
Hangzhou
Ningbo
Wenzhou
HUNAN
Changsha
JIANGXI
Nanchang
FUJIAN
Fuzhou
Xiamen
(Amoy)
GUANGDONG
Guangzhou
Shenzhen (Bao'an)
Hong Kong
Macao
(Aomen)
Shantou
Guilin
Liuzhou
Zhanjiang
Haikou
HAINAN
Hainan Dao
Sanya
South China Sea
East China Sea
(Dong Hai)
TAIWAN
T'AIPEI
Kaohsiung
The People's Republic of China claims Taiwan as its 23rd province
Taiwan Strait
Bashi Channel
Luzon Strait
Babuyan Islands
PHILIPPINES
Laoag City
NORTH KOREA
P'YŎNGYANG
Ch'ŏngjin
Hamhŭng
Kimch'aek (Sŏngjin)
SOUTH KOREA
SEOUL
(Soul)
Inch'ŏn
Taejŏn
Taegu
Kwangju
Pusan
Ulsan
Cheju
Cheju-do
Korea Strait
Ullŭng-do
Sea of Japan
(East Sea)
JAPAN
Hokkaidō
Sapporo
Asahikawa
Hakodate
Aomori
Sendai
TŌKYŌ
Yokohama
Nagoya
Kyōto
Ōsaka
Kōbe
Hiroshima
Kita-Kyūshū
Fukuoka
Nagasaki
Kumamoto
Kagoshima
Shikoku
Kyūshū
Niigata
Ryukyu Islands
(Nansei-shotō)
(Japan)
Okinawa
Naha
Sakishima-shotō
Amami-Ō-shima
Tokara-rettō
Ōsumi-shotō
Bonin Islands
(Ogasawara-shotō)
(Japan)
Volcano Islands
(Kazan-rettō)
(Japan)
Okino-Tori-shima
(Japan)
Tropic of Cancer
PACIFIC OCEAN
↓ 41

Conic Equidistant Projection

1:7 000 000

Asia
Japan, North Korea and South Korea

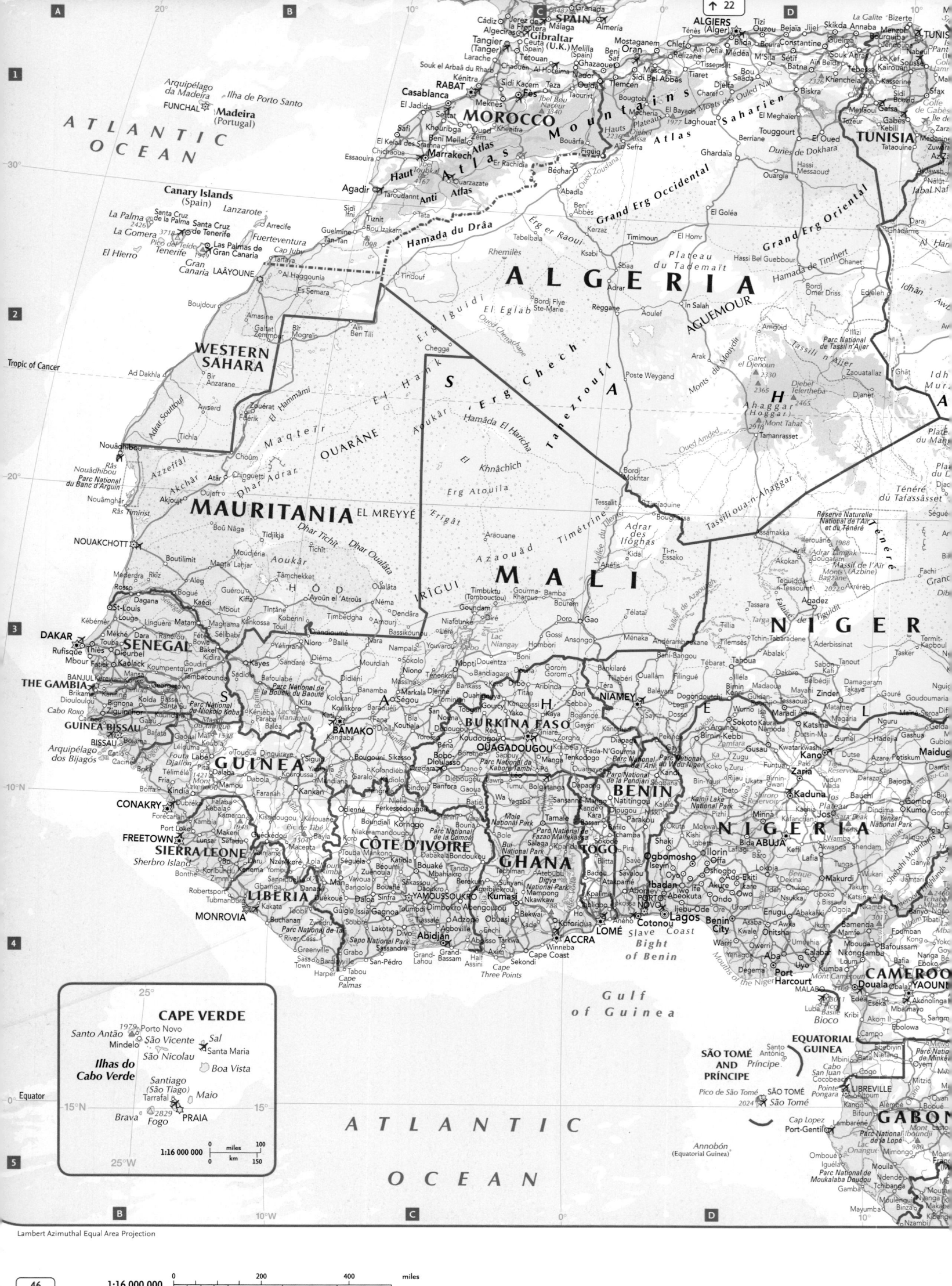

Lambert Azimuthal Equal Area Projection

1:16 000 000

Africa

Northern Africa

ITALY
GREECE
TURKEY
CYPRUS
SYRIA
LEBANON
IRAQ
IRAN
ISRAEL
JORDAN
KUWAIT
Mediterranean Sea
LIBYA
EGYPT
SAUDI ARABIA
YEMEN
CHAD
SUDAN
SOUTH SUDAN
ERITREA
DJIBOUTI
ETHIOPIA
SOMALIA
CENTRAL AFRICAN REPUBLIC
DEMOCRATIC REPUBLIC OF THE CONGO
CONGO
UGANDA
KENYA
RWANDA
BURUNDI
TANZANIA
INDIAN OCEAN
Gulf of Aden
Gulf of Sirte
Red Sea
Libyan Desert
Western Desert
Nubian Desert
Great Sand Sea
Tibesti
Darfur
Ennedi
Ouaddaï
Congo Basin
Great Rift Valley
Lake Victoria
Ogaden
Tropic of Cancer
Equator
ATHENS
NICOSIA
BEIRUT
DAMASCUS
JERUSALEM
'AMMAN
BAGHDAD
KUWAIT
RIYADH
CAIRO
Alexandria
KHARTOUM
ASMARA
ṢAN'Ā'
DJIBOUTI
ADDIS ABABA
MOGADISHU
NAIROBI
KAMPALA
KIGALI
BUJUMBURA
JUBA
BANGUI
NDJAMENA
Mecca
Medina
Jeddah
↑ 23
↑ 32
↓ 49

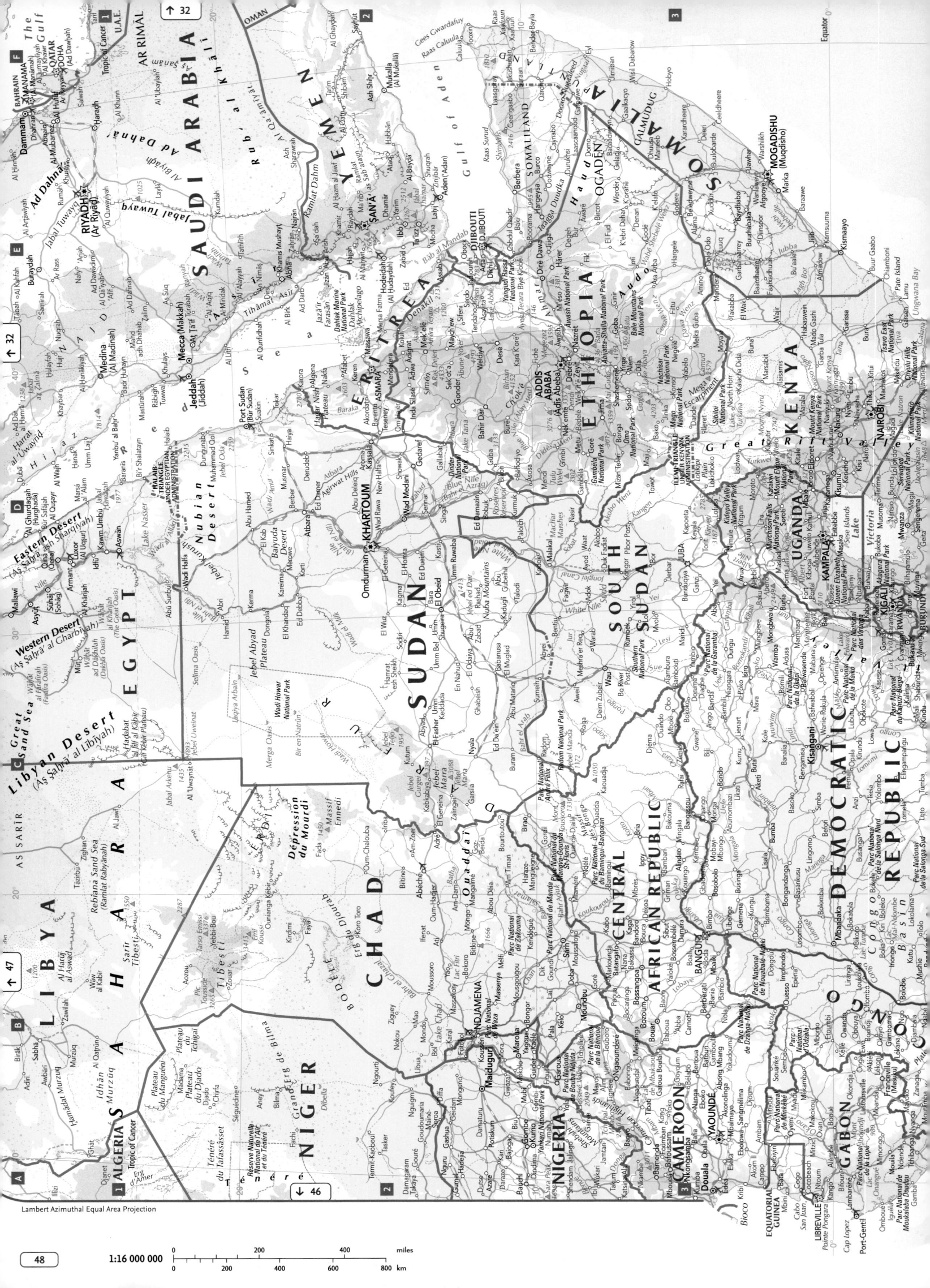

Lambert Azimuthal Equal Area Projection

48

1:16 000 000

Africa

Central and Southern Africa

TANZANIA
CONGO
ZAMBIA
ANGOLA
MALAWI
MOZAMBIQUE
ZIMBABWE
BOTSWANA
NAMIBIA
REPUBLIC OF SOUTH AFRICA
LESOTHO
SWAZILAND
MADAGASCAR
COMOROS
INDIAN OCEAN
ATLANTIC OCEAN
Mozambique Channel
Kalahari Desert
Namib Desert
Tropic of Capricorn
LUANDA
LUSAKA
HARARE
LILONGWE
WINDHOEK
GABORONE
PRETORIA
MAPUTO
MASERU
MBABANE
ANTANANARIVO
MORONI
DODOMA
CAPE TOWN
Dar es Salaam
Johannesburg
Durban
Port Elizabeth

ATLANTIC

OCEAN

NAMIBIA

BOTSWANA

REPUBLIC OF

SOUTH AF

Lambert Azimuthal Equal Area Projection

1:5 000 000

0   50   100   150   miles
0   50 100 150 200 250 km

Africa
Republic of South Africa

ZIMBABWE
MATABELELAND SOUTH
MASVINGO
CENTRAL
LIMPOPO
GAZA
INHAMBANE
MOZAMBIQUE
KGATLENG
GABORONE
MPUMALANGA
PRETORIA (Tshwane)
GAUTENG
Johannesburg
Soweto
MAPUTO
SWAZILAND
MBABANE
Drakensberg
FREE STATE
KWAZULU-NATAL
Bloemfontein
Mangaung
MASERU
LESOTHO
Pietermaritzburg
Durban
GRIQUALAND EAST
EASTERN CAPE
Mthatha (Umtata)
East London
Port Elizabeth
Cape Recife
INDIAN OCEAN
Kruger National Park
Great Limpopo Transfrontier Park
Parque Nacional de Banhine
Tropic of Capricorn
Polokwane (Pietersburg)
Mbombela (Nelspruit)
Vereeniging
Welkom
Newcastle
Richards Bay
Port Shepstone
Port Edward
Lake St Lucia
iSimangaliso Wetland Park
Xai-Xai
Inhambane
Cabo de Santa Maria
Cape Hermes
Great Fish Point

B 120° **C** Celebes Sea Morotai 130° **D** 140° **E** 150°

1

Borneo
Sambaliung
Tanjungselor
Tanjungredeb
Sangkulirang
Samarinda
Balikpapan
Equator
Kotabaru
Laut
Majene
Celebes
(Sulawesi)
Makassar
(Ujung Pandang)
Bontosunggu
Benteng
Salayar
Bulukumba

Tolitoli
Semenanjung Minahasa
Moutong
Gorontalo
Palu
Poso
Kolonedale
Kendari
Wowoni
Raha
Watampone
Sinjai
Baubau
Buton

Manado
Laut Maluku
(Molucca Sea)
Luwuk
Banggai
Kepulauan
Banggai
Kepulauan
Sula
Buru

Tobelo
Ternate Halmahera
Waigeo
Bacan
Obi
Mangole
Wahai
Seram 3019
Ambon
Ambon
Kepulauan
Watubela

Selat Dampir Kwoka
Salawati Sorong
Misool
Fakfak
Kaimana
Semenanjung Bomberai
Adi
Kepulauan
Banda
Kepulauan
Kai
Besar
Kai Kecil
Dobo
Benjina
Kepulauan Aru
Kobroor

Manokwari
Biak
Biak
Numfoor
Teluk
Cenderawasih
Yapen
Nabire
Teluk
Kamrau

Tanjung d'Urville
Sarmi
Pegunungan
Van Rees
Taritatu
Pegunungan Maoke
Tembagapura Puncak
Jaya 5030
Puncak
Trikora
Mandala 4700
Digul

Jayapura
Vanimo
Wewak
Sepik
Aitape
New
Guinea
Mendi
Mount
Hagen
Goroka
Kikori

Wuvulu
Island
Hermit
Islands
Manam
Island
Karkar
Island
Madang
4509
Lae
Morobe

St Matthias
Group
Admiralty Islands
Manus
Island
Rambutyo
Island
Umboi
Island
Long
Island

Mussau
Island
New
Hanover
New
Ireland
Lihir Group
Tabar Islands
Namatanai
Rabaul
New Britain
Kimbe
Gazelle
Channel

Lyra
Reef
Bismarck Archipelago
Bismarck
Sea
Tanga
Islands
Feni
Islands

2

Kepulauan
Kangean
Kepulauan
Tengah
Bali Lombok
Denpasar Mataram
Sumbawa
Raba
Waikabubak Waingapu
Sumba
Dompu
Ruteng
Flores
Ende

Kepulauan Bonerate
Bonerate
Alor
Kalabahi
Larantuka
Laut Flores
(Flores Sea)
Kefamenanu
Kupang
Rote

Wetar
Alor
DILI
Foho
Tatamailau
Atambua
Timor
EAST TIMOR

Damar
Pulau
Romang
Babar
Kepulauan
Leti

Yamdena
Kepulauan
Tanimbar
Selaru

Larat
Sia
Trangan
Pulau
Dolok

Tanjung Deyong
Tanjung
Vals
Merauke

Balimo
Morehead
Daru
Kwikila

Kerema
Gulf
of
Papua
PORT MORESBY
Abau

Mount
Victoria
Kokoda
Owen Stanley Range

Lusancay Islands
and Reefs
Goodenough
Island
Fergusson I.
Normanby
Island
D'Entrecasteaux Islan
Louisiade Archip
Tagula Island
Ross I.
Conflict Group

PAPUA
NEW GUINEA

Bougainville
Solomon
Sohar

INDONESIA

Arafura
Sea
Prince of Wales
Island
Torres Strait

East Timor
Laut Sawu
(Savu Sea)

INDIAN
OCEAN

Ashmore and
Cartier Islands
(Australia)

Timor
Sea

Melville
Island
Bathurst
Island
Van
Diemen
Gulf
Beagle Gulf
Rum Jungle
Darwin
Adelaide
River
Pine Creek
Jabiru

Cobourg
Peninsula
Croker
Island
Goulburn
Islands
Cape Arnhem
Arnhem
Land
Alyangula
Groote
Eylandt

Cape
Wessel
Wessel
Islands
Gulf
of
Carpentaria
Sir Edward
Pellew Group
Mornington
Island

Cape York
Albatross
Bay
Weipa
Cape
York
Peninsula
Coen

Cape
Grenville
Princess
Charlotte
Bay
Cape
Melville
Laura
Cooktown

Great
Barrier

Osprey
Reef

Coral
Islan
Terri
(Austr

3

Cape Leveque
King
Sound
Derby
Broome
Roebuck Bay
Eighty Mile Beach

Bonaparte
Archipelago
Cape
Londonderry
Joseph
Bonaparte
Gulf
Wyndham
Kununurra
Lake
Argyle
Kimberley
Plateau
Mount
Ord 936
King Leopold Ra.
Halls
Creek
Fitzroy
Crossing
Liveringa

Collier
Bay

Drysdale
Durack

Daly
River Downs
Victoria
River Downs
Katherine
Mataranka
Larrimah
Daly
Waters
Timber
Creek

Borroloola

Wellesley
Islands
Normanton
Burketown

Camooweal

Gilbert
Leichhardt

Mossman
Cairns
Bartle Frere
1612
Innisfail
Tully

Mitchell
Forsayth
Gregory
Range
Townsville
Ayr

Flinders
Richmond
Charters
Towers

Whitsunday
Group

Great Dividing Range

Bowen

Port
Hedland
Shay Gap
(abandoned)
Marble
Bar

Great Sandy Desert

Lake
Mackay
Lake
Wills
Lake
White

NORTHERN

Tanami
Desert
Tennant
Creek

Barkly Tableland

Mount Isa
Cloncurry
Dajarra

Selwyn Range
Kajabbi

QUEENSLAND

Mackay
1277 Sarina

Swain
Reefs

Capricorn Channel

4

Barrow
Island
Onslow
Exmouth
Gulf
North
West
Cape
Coral Bay
Tropic of Capricorn
Lake
MacLeod
Shark
Bay
Dirk Hartog
Island
Denham

Karratha
Roebourne
Chichester Range
Pannawonica
Hamersley Range
Paraburdoo
Mount
Meharry
Ashburton
Mount
Augustus
1106
Robinson
Ranges
Meekatharra

Nullagine
Newman

Lake
Disappointment

WESTERN
AUSTRALIA

TERRITORY

Yuendumu
Mount
Liebig
1524
Mount
Zeil
1531
Macdonald
Ranges
Lake
Neale

Barrow Creek

Alice
Springs
Macdonnell Ranges

Simpson
Desert

Boulia
Georgina
Bedourie

Winton
Longreach

Bilpa Morea
Claypan

Windorah

Yaraka
Blackall

Jericho

Clermont
Emerald
Rockhampton
Yeppoon
Curtis I.
Gladstone

Buckland
Tableland
Monto
Bundaberg
Maryborough
Gympie

Robinson
Ranges
Lake
Wells
Wiluna

Gibson
Desert
Lake
Hopkins
Uluru
(Ayers Rock)
867
Mount
Woodroffe 1440
Musgrave Ranges
Everard Range
Mount
Everard

Great
Victoria Desert

SOUTH

Erldunda

Oodnadatta
Macumba
Lake
Eyre
(North)

Birdsville

Strzelecki
Desert

Sturt
Stony
Desert
Cooper Creek

Tibooburra

Lake Yamma
Yamma

Charleville
Quilpie

Mitchell
Roma

Brisbane
Warwick
Casino

5

Kalbarri
Northampton
Mullewa
Houtman
Abrolhos
Geraldton
Dongara

Mount
Magnet
Leonora
Menzies
Lake
Carey

Laverton

Lake
Barlee
Lake
Moore
Lake
Ballard
Southern
Cross
Merredin
Mukinbudin

AUSTRALIA

Lake
Maurice
Maralinga

Nullarbor Plain

Coolgardie
Kalgoorlie
Kambalda
Norseman
Balladonia
Lake
Cowan

Eucla
Mundrabilla

Fowlers Bay
Penong
Ceduna
Streaky Bay
Anxious Bay

Lake
Gairdner
Kyancutta
Eyre
Peninsula

Coober Pedy

Lake
Torrens
Woomera
Island
Lagoon

Flinders Ranges
Lake
Frome

Broken Hill

Bourke
Wilcannia
Cobar

Brewarrina
Barwon
Walgett

Moree
Narrabri

Barraba
Inverell
Glen Innes

Armidale

Coffs
Harbour
Grafton

NEW SOUTH WALES

Perth
Fremantle
Rockingham
Mandurah
Pinjarra
Bunbury
Busselton
Margaret River
Cape Leeuwin
Point D'Entrecasteaux

York
Beverley
Brookton
Narrogin
Katanning
Hyden
Wagin
Kojonup
Manjimup
Albany
Denmark

Esperance

Archipelago of
the Recherche

Great
Australian Bight

Port Lincoln
Whyalla
Port Pirie
Jamestown
Burra
Kadina
Yorke
Pen.
Spencer Gulf
Port Augusta

Adelaide
Murray Bridge

Renmark
Mildura

Swan Hill
Lake
Tyrrell

Balranald
Hay
Deniliquin

Griffith
Narrandera
Wagga
Wagga

Parkes
Forbes
Orange
Bathurst

Dubbo

Tamworth
Gunnedah
Muswellbrook

Newcastle
Sydney
Wollongong

Geographe Bay

Darling Range

VICTORIA

Horsham
Ballarat
Bendigo
Shepparton
Wangaratta
Albury
Wodonga
Mount
Kosciuszko
2229

CANBERRA
A.C.T.
JERVIS BAY
TERR.

Melbourne
Geelong
Moe

Warrnambool
Mount Gambier
Cape
Otway
Portland
Discovery Bay

Sale
Bairnsdale

Wilson's
Promontory
Bass Strait
Furneaux Group
Flinders Island

Eden
Cape Howe
Narooma

Ta

6

King Island
Currie

Hunter Islands

Burnie
Queenstown
Lake Gordon
Mount
Ossa 1617

Devonport
Launceston
Great Lake

TASMANIA
Hobart
South East Cape

Eddystone Point

Port Arthur

40°S

110°E **A** 120° **B** 130° **C** 140° **D** **E** 150°

Lambert Azimuthal Equal Area Projection

1:20 000 000

0   200   400   600   miles
0   200   400   600   800   1000 km

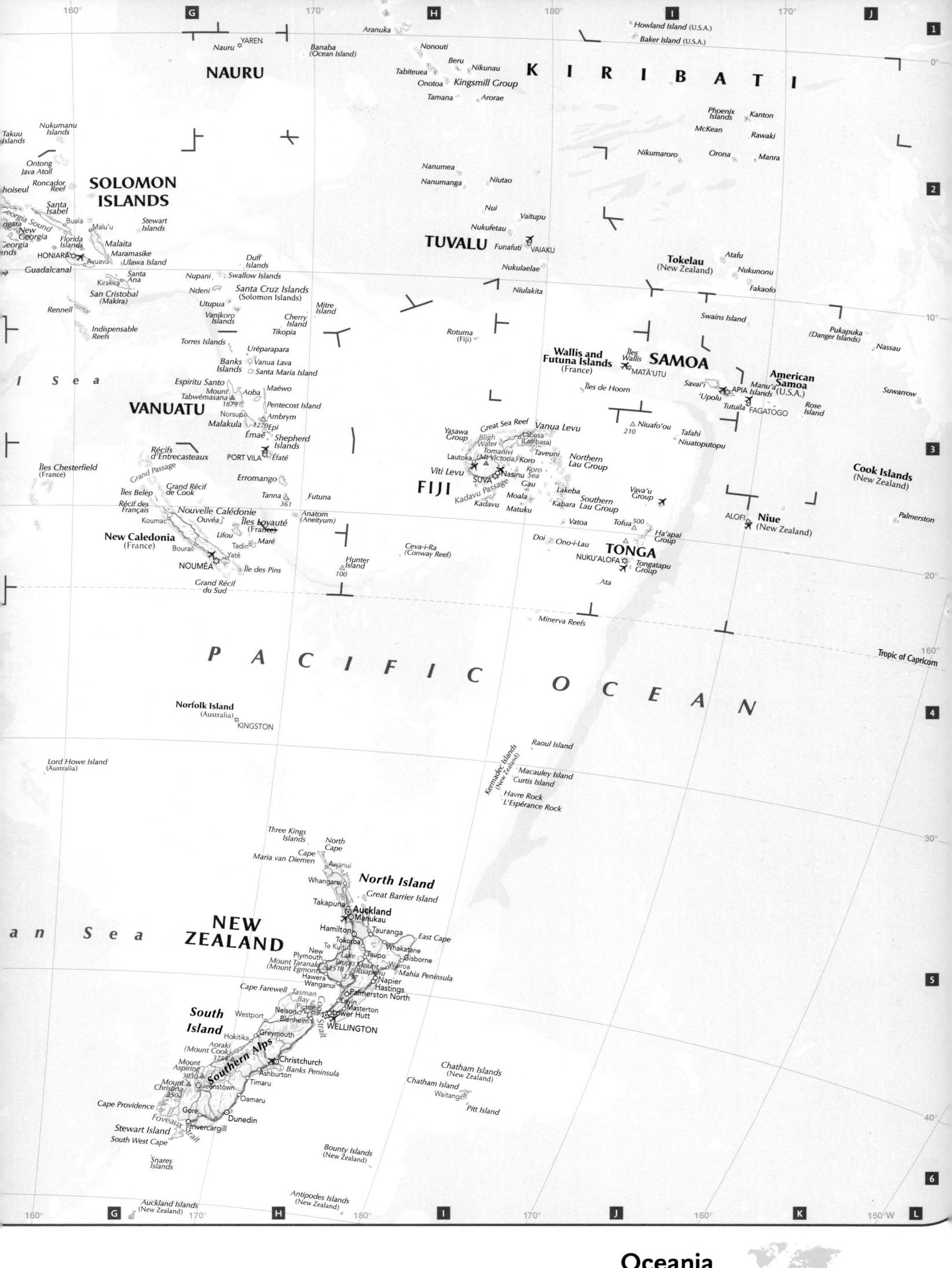

NAURU

KIRIBATI

SOLOMON ISLANDS

TUVALU

Tokelau (New Zealand)

Wallis and Futuna Islands (France)

SAMOA

American Samoa (U.S.A.)

VANUATU

FIJI

TONGA

Niue (New Zealand)

Cook Islands (New Zealand)

New Caledonia (France)

PACIFIC OCEAN

Tropic of Capricorn

Norfolk Island (Australia)

Lord Howe Island (Australia)

NEW ZEALAND

North Island

South Island

Southern Alps

Chatham Islands (New Zealand)

Stewart Island

Lambert Azimuthal Equal Area Projection

1:8 000 000

| 0 | | 100 | | 200 | | 300 | miles |
| 0 | 100 | 200 | 300 | 400 | 500 | km |

Oceania
Western Australia

SOUTH AUSTRALIA
Great Victoria Desert
Gibson Desert
Little Sandy Desert
Simpson Desert
Nullarbor Plain
Great Australian Bight
Darling Range
Macdonnell Ranges
Musgrave Ranges
Robinson Ranges
Gawler Ranges
Eyre Peninsula
Lake Gairdner
Spencer Gulf
Kangaroo Island
Perth
Fremantle
Kalgoorlie
Esperance
Albany
Geraldton
Eucla
Woomera Prohibited Area
Nullarbor National Park
Tropic of Capricorn
↑ 57

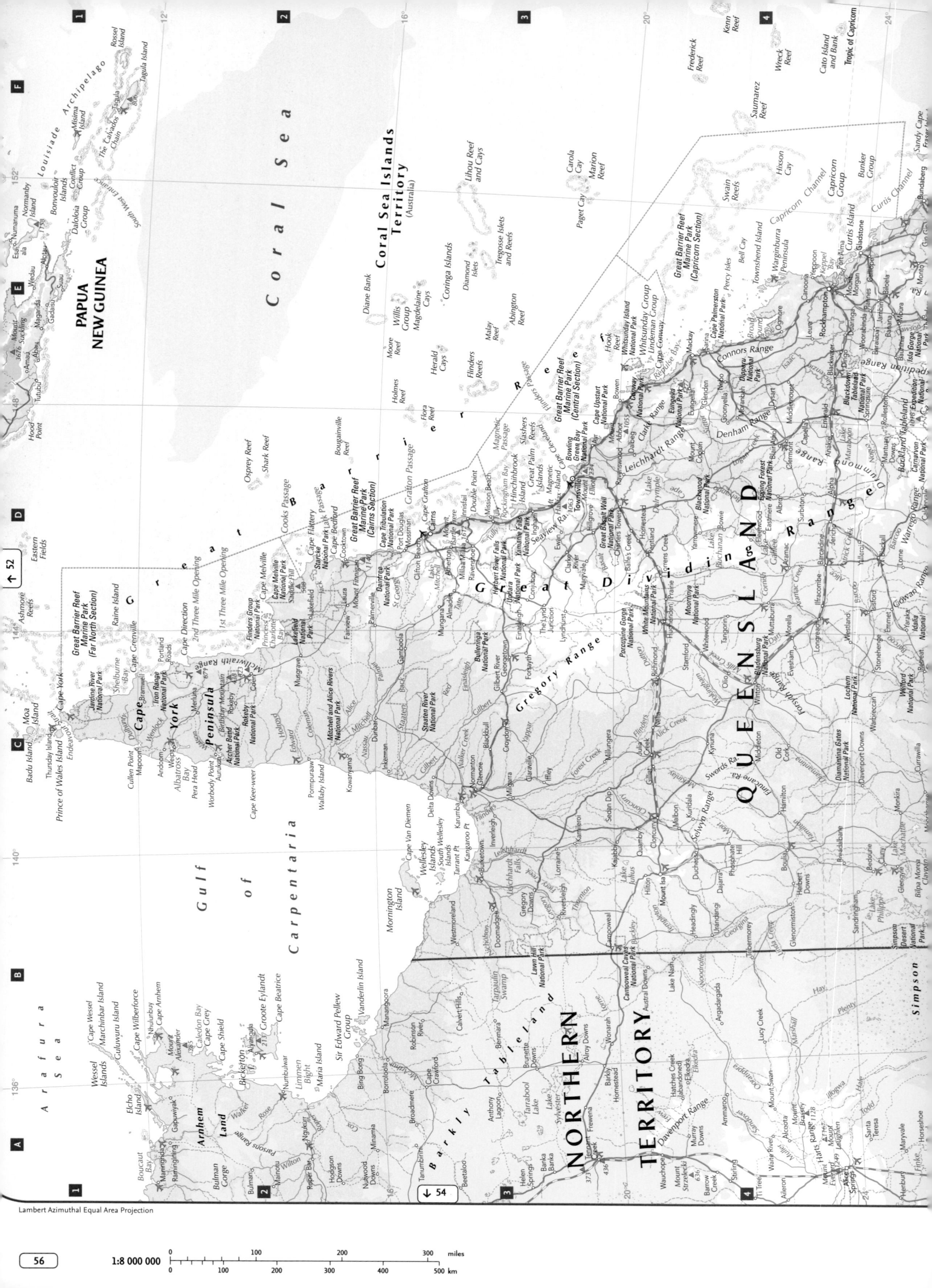

Lambert Azimuthal Equal Area Projection

1:8 000 000

0   100   200   300   miles
0   100   200   300   400   500 km

Oceania
Eastern Australia

SOUTH AUSTRALIA
NEW SOUTH WALES
VICTORIA
TASMANIA
AUSTRALIAN CAPITAL TERRITORY
JERVIS BAY TERRITORY
Tasman Sea
Bass Strait
CANBERRA
Sydney
Melbourne
Adelaide
Brisbane
Hobart
Launceston
Newcastle
Wollongong
Geelong
Gold Coast
Broken Hill
Lake Eyre (North)
Lake Torrens
Spencer Gulf
Gulf St Vincent
Eyre Peninsula
Kangaroo Island
Darling Downs
Simpson Desert Regional Reserve
Sturt Stony Desert
Strzelecki Desert
Flinders Ranges
Grey Range
Snowy Mts
Riverina
Wilson's Promontory
King Island
Flinders Island
Furneaux Group
← 55

Oceania
Southeast Australia

Lambert Azimuthal Equal Area Projection

1:5 000 000

NEW ZEALAND

Tasman Sea

North Island

South Island

PACIFIC OCEAN

...onic Equidistant Projection

1:5 250 000

0   50   100   150   miles
0   50   100   150   200   250 km

Oceania
New Zealand

59

PACIFIC OCEAN

ARCTIC OCEAN

Beaufort Sea

Bering Sea

Gulf of Alaska

U.S.A. ALASKA

YUKON

NORTHWEST TERRITORIES

BRITISH COLUMBIA

ALBERTA

SASKATCHEWAN

CANADA

UNITED STATES OF

WASHINGTON

OREGON

CALIFORNIA

NEVADA

IDAHO

MONTANA

WYOMING

UTAH

NEBRASKA

NORTH DAKOTA

SOUTH DAKOTA

Vancouver Island

Lambert Conformal Conic Projection

1:16 000 000

| 0 | 200 | 400 | miles |
| 0 | 200 | 400 | 600 | 800 km |

↓ 62

Greenland
(Kalaallit Nunaat)
(Denmark)
Nationalparken i Nord- og Østgrønland
ICELAND
REYKJAVÍK
Denmark Strait
Baffin Bay
Davis Strait
Labrador Sea
Hudson Bay
Hudson Strait
James Bay
Foxe Basin
Foxe Channel
Baffin Island
Ellesmere Island
Axel Heiberg Island
Queen Elizabeth Islands
Parry Islands
Devon Island
Somerset Island
Prince of Wales Island
Boothia Peninsula
Melville Peninsula
Southampton Island
Coats Island
Mansel Island
Belcher Islands
Lancaster Sound
Nares Strait
Gulf of Boothia
North Geomagnetic Pole (2010)
NUNAVUT
MANITOBA
ONTARIO
QUÉBEC
NEWFOUNDLAND AND LABRADOR
NEW BRUNSWICK
NOVA SCOTIA
PRINCE EDWARD I.
Péninsule d'Ungava
Labrador
Newfoundland
St Pierre and Miquelon (France)
Île d'Anticosti
Gulf of St Lawrence
Cabot Strait
Bay of Fundy
Gulf of Maine
ATLANTIC OCEAN
CANADA
AMERICA
MINNESOTA
WISCONSIN
MICHIGAN
IOWA
PENNSYLVANIA
NEW YORK
MAINE
Lake Superior
Lake Huron
Lake Michigan
Lake Erie
Lake Ontario
Lake Winnipeg
OTTAWA
Montréal
Québec
Toronto
Winnipeg
Chicago
Detroit
Buffalo
Boston
New York
Halifax
St John's
NUUK (Godthåb)
Iqaluit (Frobisher Bay)
↓ 63
States in the U.S.A. numbered on the map:
1. CONNECTICUT (K5)
2. MASSACHUSETTS (K5)
3. NEW HAMPSHIRE (K5)
4. RHODE ISLAND (K5)
5. VERMONT (K5)

CANADA
UNITED STATES OF AMERICA
ONTARIO
QUÉBEC
NEW BRUNSWICK
NOVA SCOTIA
PRINCE EDWARD ISLAND
MINNESOTA
WISCONSIN
MICHIGAN
IOWA
ILLINOIS
INDIANA
OHIO
PENNSYLVANIA
NEW YORK
VERMONT
NEW HAMPSHIRE
MAINE
MASSACHUSETTS
CONNECTICUT
RHODE ISLAND
NEW JERSEY
DELAWARE
MARYLAND
WASHINGTON D.C.
WEST VIRGINIA
VIRGINIA
KENTUCKY
TENNESSEE
NORTH CAROLINA
SOUTH CAROLINA
GEORGIA
ALABAMA
MISSISSIPPI
LOUISIANA
ARKANSAS
MISSOURI
FLORIDA
THE BAHAMAS
CUBA
James Bay
Lake Superior
Lake Michigan
Lake Huron
Lake Erie
Lake Ontario
Gulf of St Lawrence
Gulf of Maine
Bay of Fundy
ATLANTIC OCEAN
Gulf of Mexico
Straits of Florida
Great Bahama Bank
Appalachian Mountains
Allegheny Mountains
Cumberland Plateau
Ozark Plateau
Coastal Plain
Adirondack Mts
Monts Notre-Dame
Tropic of Cancer
OTTAWA
Montréal
Québec
Toronto
Detroit
Chicago
Milwaukee
Minneapolis
St Louis
Indianapolis
Cincinnati
Columbus
Cleveland
Pittsburgh
Buffalo
Boston
Providence
Hartford
New York
Philadelphia
Baltimore
Richmond
Norfolk
Virginia Beach
Charlotte
Raleigh
Louisville
Nashville
Memphis
Atlanta
Birmingham
Jacksonville
Tampa
Orlando
Miami
New Orleans
Baton Rouge
Houston
Dallas
NASSAU
Cape Cod
Cape Hatteras
Cape Fear
Cape Canaveral
Long Island
Nantucket Island
Mississippi Delta

↑ 61
↓ 67

North America
Northeast United States

Lambert Conformal Conic Project

1:3 500 000

ATLANTIC OCEAN

North America
Southwest United States

Lambert Conformal Conic Projection

500 000

Mexico

UNITED STATES OF AMERICA

CALIFORNIA · ARIZONA · NEW MEXICO · TEXAS · OKLAHOMA · ARKANSAS · KANSAS · MISSOURI · LOUISIANA · MISSISSIPPI · COLORADO · ILLINOIS · TENNESSEE

MEXICO

GUATEMALA · **GUATEMALA CITY** · **BELIZE** · **BELMOPAN** · **EL SALVADOR** · **SAN SALVADOR**

Gulf of Mexico

Bahía de Campeche

Gulf of California

PACIFIC OCEAN

Tropic of Cancer

Sierra Madre Occidental

Sierra Madre Oriental

Sierra Madre del Sur

Baja California

Yucatán

Île Clipperton (France)

Islas Revillagigedo (Mexico)

Guadalupe (Mexico)

Lambert Conformal Conic Projection

1:14 000 000

0   200   400   miles
0   200   400   600   800 km

ATLANTIC

OCEAN

HAMILTON ✈ Bermuda (U.K.)

Tropic of Cancer

THE BAHAMAS

NASSAU

CUBA

HAVANA (La Habana)

Turks and Caicos Islands (U.K.)

GRAND TURK (Cockburn Town)
Turks Islands

Hispaniola

HAITI
PORT-AU-PRINCE

SANTO DOMINGO

DOMINICAN REPUBLIC

SAN JUAN

Puerto Rico (U.S.A.)

Virgin Islands (U.K.)

Virgin Islands (U.S.A.)

CHARLOTTE AMALIE

ROAD TOWN

Anguilla (U.K.)
THE VALLEY

St-Martin (France)
St-Barthélemy (France)

St Maarten (Neth.)

ANTIGUA AND BARBUDA
ST JOHN'S
Antigua
Barbuda

BASSETERRE
ST KITTS AND NEVIS
Montserrat (U.K.)
Plymouth (abandoned)

Guadeloupe (France)
BASSE-TERRE
Pointe-à-Pitre
Marie-Galante

DOMINICA
ROSEAU

FORT-DE-FRANCE
Martinique (France)

CASTRIES
ST LUCIA

BARBADOS
BRIDGETOWN

ST VINCENT AND THE GRENADINES
KINGSTOWN
The Grenadines

GRENADA
ST GEORGE'S

Tobago

PORT OF SPAIN
TRINIDAD AND TOBAGO

JAMAICA
KINGSTON
Montego Bay

Cayman Islands (U.K.)
GEORGE TOWN

Grand Cayman

West Indies

Greater Antilles

Lesser Antilles

Caribbean Sea

Leeward Islands

Windward Islands

Aruba (Neth.)
ORANJESTAD

Curaçao (Neth.)
WILLEMSTAD

Bonaire (Neth.)
Kralendijk

COSTA RICA
SAN JOSÉ

PANAMA
PANAMA CITY

Gulf of Panama

Isthmus of Panama

NICARAGUA
MANAGUA

Isla de San Andrés (Colombia)

Islas del Maíz (Corn Islands) (Nicaragua)

CARACAS

Maracaibo

Lake Maracaibo

Barranquilla
Cartagena

VENEZUELA

Ciudad Guayana
Ciudad Bolívar

GUYANA

COLOMBIA
BOGOTÁ

Medellín

Cali

BRAZIL

North America
Central America and the Caribbean

PACIFIC

OCEAN

NICARAGUA
COSTA RICA
PANAMA
PANAMA CITY
COLOMBIA
ECUADOR
PERU
BOLIVIA
CHILE
VENEZUELA
GRENADA
ARGENTINA

Galapagos Islands
San Salvador (Islas Galápagos)
(Ecuador)

Parque Nacional
Galápagos

Isla Fernandina
Isla Isabela
Isla Santa Cruz
Isla San Cristóbal
Isla Santa María
Isla San Salvador

1:14 000 000

Lambert Azimuthal Equal Area Projection

1:14 000 000

A T L A N T I C

O C E A N

SURINAME

French Guiana

GEORGETOWN

PARAMARIBO

CAYENNE

GUYANA

B R A Z I L

BRASÍLIA

Goiânia

Belo Horizonte

São Paulo

Rio de Janeiro

Salvador (Bahia)

Recife (Pernambuco)

Fortaleza (Ceará)

São Luís

Belém

Natal

João Pessoa

Maceió

Aracaju

Teresina

Vitória

Campinas

PARAGUAY

Equator

Fernando de Noronha (Brazil)

Ilha da Trindade (Brazil)

Brazilian Highlands

Pantanal

↓ 70

South America
Southern South America

Lambert Azimuthal Equal Area Projection

1:14 000 000

Lambert Azimuthal Equal Area Projection

:1 000 000

0   100   200   miles
0   100   200   300   400 km

CANADA
UNITED STATES OF AMERICA
Rocky Mountains
OTTAWA
Montréal
Toronto
Chicago
Milwaukee
Detroit
Cleveland
Pittsburgh
Buffalo
Boston
Providence
New York
Philadelphia
Baltimore
WASHINGTON D.C.
Richmond
Charlotte
Atlanta
Memphis
Nashville
Indianapolis
Columbus
St Louis
Dallas
Austin
San Antonio
Houston
New Orleans
Jacksonville
Orlando
Tampa
Miami
Appalachian Mountains
Newfoundland
St John's
Halifax
Grand Banks of Newfoundland
Labrador
Gulf of St Lawrence
Baja California
Ciudad Juárez
El Paso
Tucson
Albuquerque
Oklahoma City
Hermosillo
Monterrey
Tropic of Cancer
MEXICO
MEXICO CITY
Guadalajara
San Luis Potosí
León
Puebla
Acapulco
Mérida
Tampico
Gulf of Mexico
Florida Keys
Straits of Florida
THE BAHAMAS
NASSAU
HAVANA
CUBA
Cayman Islands
Turks and Caicos Islands
PORT-AU-PRINCE
HAITI
DOMINICAN REP.
SANTO DOMINGO
JAMAICA
KINGSTON
Puerto Rico
BELIZE
BELMOPAN
GUATEMALA
GUATEMALA CITY
SAN SALVADOR
EL SALVADOR
HONDURAS
TEGUCIGALPA
NICARAGUA
MANAGUA
COSTA RICA
SAN JOSÉ
PANAMA
PANAMA CITY
Caribbean Sea
ST KITTS AND NEVIS
ANTIGUA AND BARBUDA
Guadeloupe (France)
DOMINICA
Martinique (France)
ST LUCIA
ST VINCENT AND THE GRENADINES
BARBADOS
GRENADA
PORT OF SPAIN
TRINIDAD AND TOBAGO
Aruba
Curaçao
Bermuda (U.K.)
Bermuda Rise
Sargasso Sea
Nares Deep
Greenland (Denmark)
ICELAND
REYKJAVÍK
Faroe Islands (Denmark)
TÓRSHAVN
Reykjanes Ridge
Iceland Basin
Irminger Basin
Eirik Ridge
UNITED KINGDOM
IRELAND
DUBLIN
LONDON
Edinburgh
PARIS
BRUSSELS
FRANCE
MADRID
SPAIN
LISBON
PORTUGAL
Oporto
Barcelona
Marseille
Azores (Portugal)
PONTA DELGADA
Madeira (Portugal)
FUNCHAL
RABAT
MOROCCO
Casablanca
ALGIERS
TUNIS
ALGERIA
Sahara
WESTERN SAHARA
LAÂYOUNE
Las Palmas de Gran Canaria
MAURITANIA
NOUAKCHOTT
MALI
NIGER
SENEGAL
DAKAR
THE GAMBIA
BANJUL
GUINEA-BISSAU
BISSAU
GUINEA
CONAKRY
SIERRA LEONE
FREETOWN
LIBERIA
MONROVIA
CÔTE D'IVOIRE
YAMOUSSOUKRO
Abidjan
GHANA
ACCRA
BAMAKO
BURKINA FASO
OUAGADOUGOU
NIAMEY
NIGERIA
ABUJA
Lagos
PORTO-NOVO
LOMÉ
CAPE VERDE
PRAIA
Gulf of Guinea
SÃO TOMÉ AND PRÍNCIPE
Mid-Atlantic Ridge
Cape Verde Basin
Guiana Basin
Sierra Leone Basin
Guinea Basin
ATLANTIC OCEAN
PACIFIC OCEAN
Equator
VENEZUELA
CARACAS
Maracaibo
COLOMBIA
BOGOTÁ
Medellín
Cali
GUYANA
GEORGETOWN
SURINAME
PARAMARIBO
French Guiana
CAYENNE
ECUADOR
QUITO
Guayaquil
Galapagos Islands (Ecuador)
PERU
LIMA
BRAZIL
BRASÍLIA
Manaus
Belém
São Luís
Fortaleza
Natal
Recife
Salvador
Belo Horizonte
Rio de Janeiro
São Paulo
Curitiba
Porto Alegre
Amazon
Selvas
BOLIVIA
LA PAZ
SUCRE
PARAGUAY
ASUNCIÓN
URUGUAY
MONTEVIDEO
BUENOS AIRES
Córdoba
Rosario
ARGENTINA
SANTIAGO
Valparaíso
Andes
Falkland Islands
STANLEY
South Georgia (U.K.)
South Sandwich Islands (U.K.)
Scotia Sea
Drake Passage
Cape Horn
Brazil Basin
Angola Basin
Argentine Basin
Rio Grande Rise
St Helena, Ascension and Tristan da Cunha (U.K.)
JAMESTOWN
Tristan da Cunha (U.K.)
Gough Island (U.K.)
Walvis Ridge
Cape Basin
Atlantic-Indian Ridge
American-Antarctic Ridge
Weddell Abyssal Plain
Antarctic Peninsula
Galapagos Rise
Peru Basin
Peru-Chile Trench
Nazca Ridge
Chile Basin
Chile Rise
Cocos Ridge
Tropic of Capricorn
Easter Island (Isla de Pascua) (Chile)
Lambert Azimuthal Equal Area Projection
1:50 000 000
miles
km

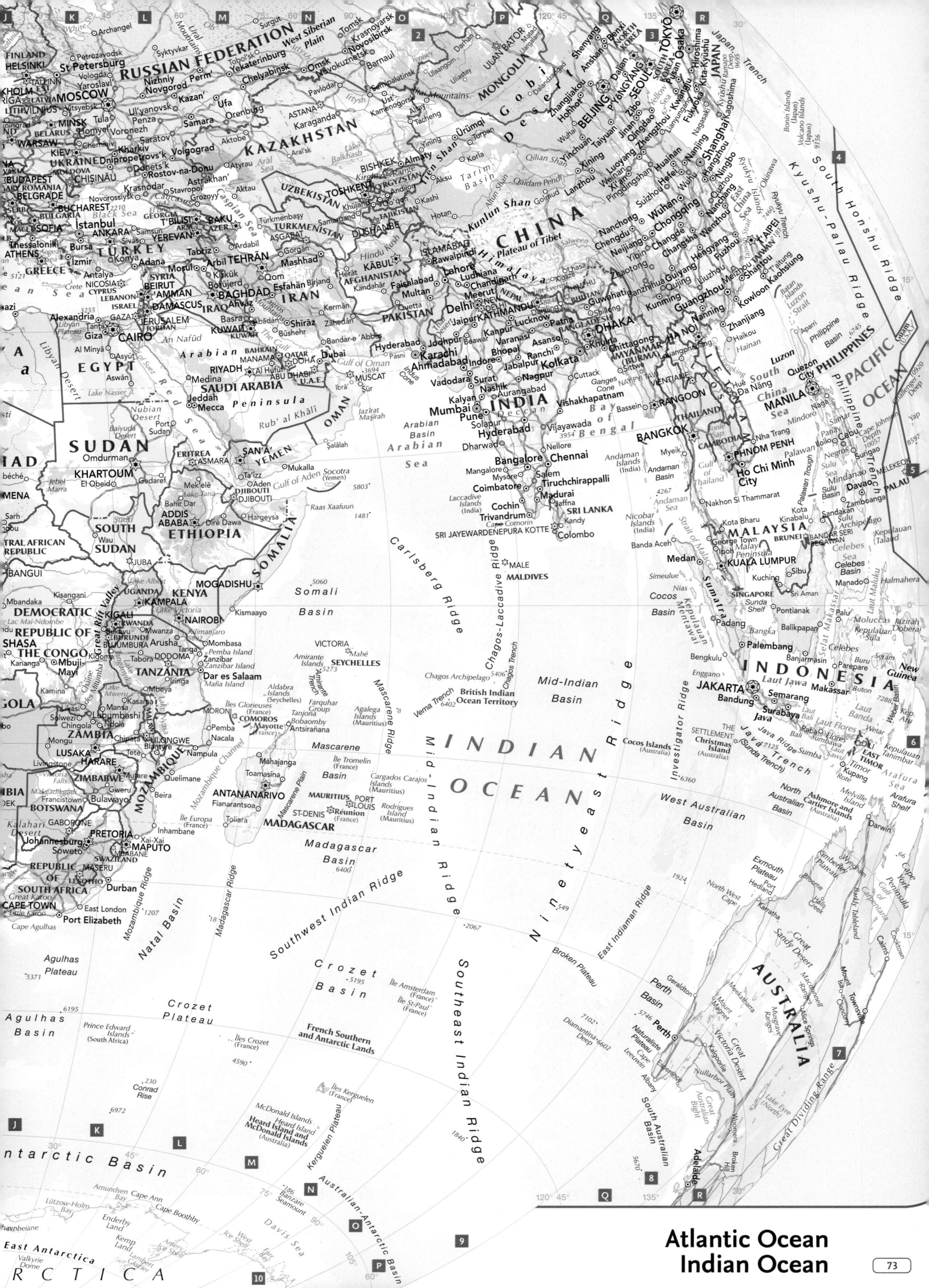
RUSSIAN FEDERATION
West Siberian Plain
MONGOLIA
Gobi Desert
KAZAKHSTAN
UZBEKISTAN
TURKMENISTAN
CHINA
Plateau of Tibet
Kunlun Shan
Himalaya
Tarim Basin
Qaidam Pendi
SOUTH KOREA
NORTH KOREA
JAPAN
Japan Trench
Kyushu-Palau Ridge
South Honshu Ridge
Bonin Islands (Japan)
Volcano Islands (Japan)
BEIJING
SEOUL
TŌKYŌ
Shanghai
TAIPEI
PYONGYANG
ULAN BATOR
ASTANA
BISHKEK
Almaty
TOSHKENT
DUSHANBE
KABUL
ISLAMABAD
AFGHANISTAN
PAKISTAN
IRAN
IRAQ
TURKEY
SYRIA
EGYPT
SAUDI ARABIA
Arabian Peninsula
YEMEN
OMAN
U.A.E.
QATAR
KUWAIT
SUDAN
SOUTH SUDAN
ETHIOPIA
SOMALIA
KENYA
UGANDA
TANZANIA
DEMOCRATIC REPUBLIC OF THE CONGO
ZAMBIA
ZIMBABWE
MOZAMBIQUE
BOTSWANA
SOUTH AFRICA
MADAGASCAR
INDIA
NEPAL
BHUTAN
BANGLADESH
MYANMAR (BURMA)
THAILAND
LAOS
VIETNAM
CAMBODIA
MALAYSIA
BRUNEI
SINGAPORE
INDONESIA
PHILIPPINES
EAST TIMOR
SRI LANKA
MALDIVES
SEYCHELLES
COMOROS
MAURITIUS
AUSTRALIA
Great Sandy Desert
Great Victoria Desert
Nullarbor Plain
Great Dividing Range
Cape York Peninsula
Arabian Sea
Bay of Bengal
Red Sea
Caspian Sea
Black Sea
Gulf of Aden
Gulf of Oman
South China Sea
Arafura Sea
Timor Sea
Java Sea
PACIFIC OCEAN
INDIAN OCEAN
Arabian Basin
Somali Basin
Carlsberg Ridge
Chagos-Laccadive Ridge
Chagos Trench
Mid-Indian Basin
Mid-Indian Ridge
Ninetyeast Ridge
Mascarene Basin
Mascarene Ridge
Madagascar Basin
Madagascar Ridge
Mozambique Channel
Mozambique Ridge
Natal Basin
Agulhas Plateau
Agulhas Basin
Southwest Indian Ridge
Southeast Indian Ridge
Crozet Basin
Crozet Plateau
Broken Plateau
East Indiaman Ridge
Investigator Ridge
Cocos Basin
West Australian Basin
North Australian Basin
Perth Basin
South Australian Basin
Diamantina Deep
Java Trench (Sunda Trench)
Wharton Basin
Kerguelen Plateau
Îles Kerguelen (France)
Heard Island and McDonald Islands (Australia)
French Southern and Antarctic Lands
Îles Crozet (France)
Prince Edward Islands (South Africa)
Île Amsterdam (France)
Île St-Paul (France)
Conrad Rise
British Indian Ocean Territory
Chagos Archipelago
Cocos Islands (Australia)
Christmas Island (Australia)
Ashmore and Cartier Islands (Australia)
Australian-Antarctic Basin
Antarctic Basin
Davis Sea
East Antarctica
ANTARCTICA
Amundsen Bay
Cape Ann
Cape Boothby
Enderby Land
Kemp Land
Amery Ice Shelf
West Ice Shelf
Banzare Seamount
Lützow-Holm Bay

Atlantic Ocean
Indian Ocean

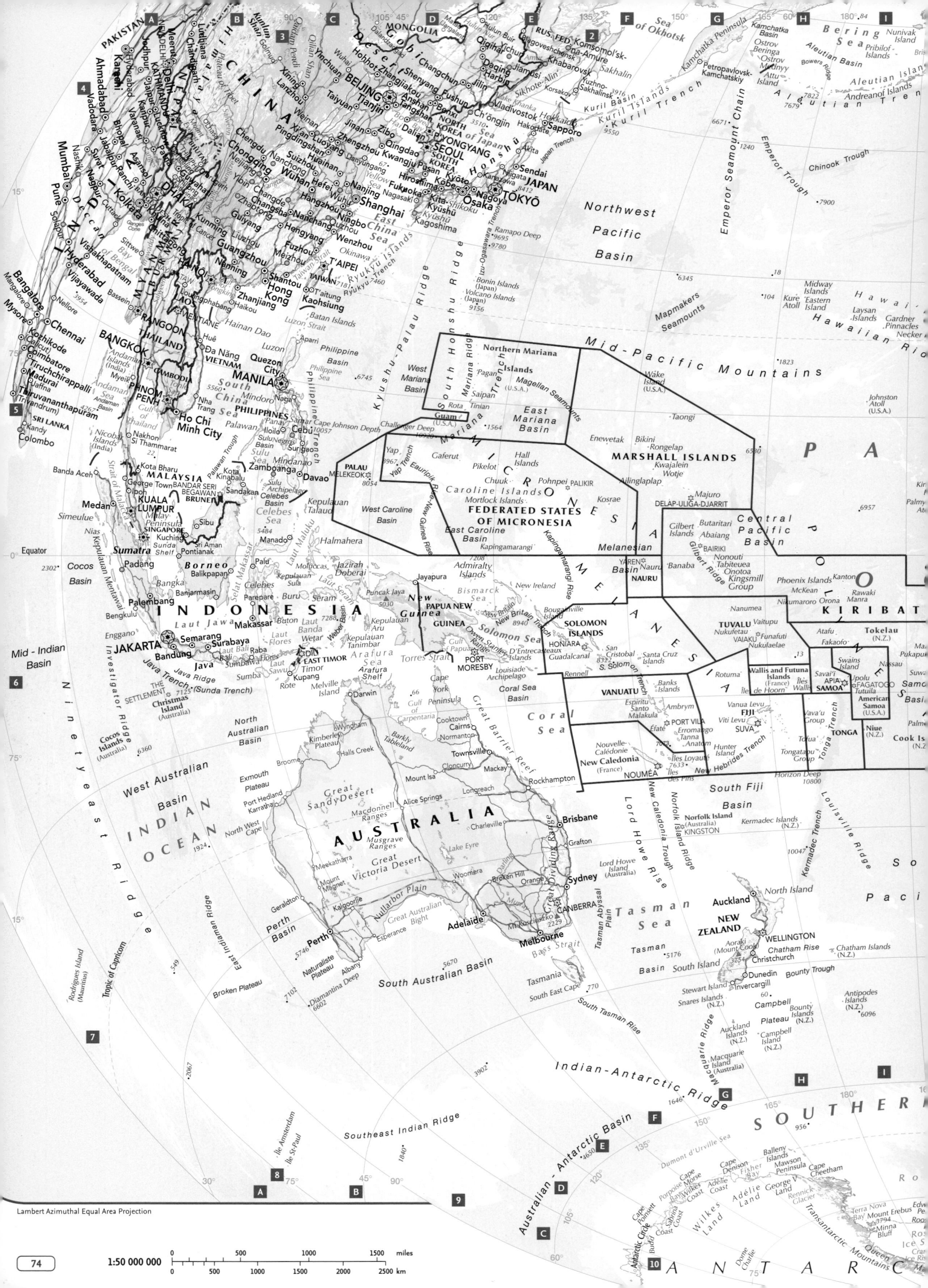

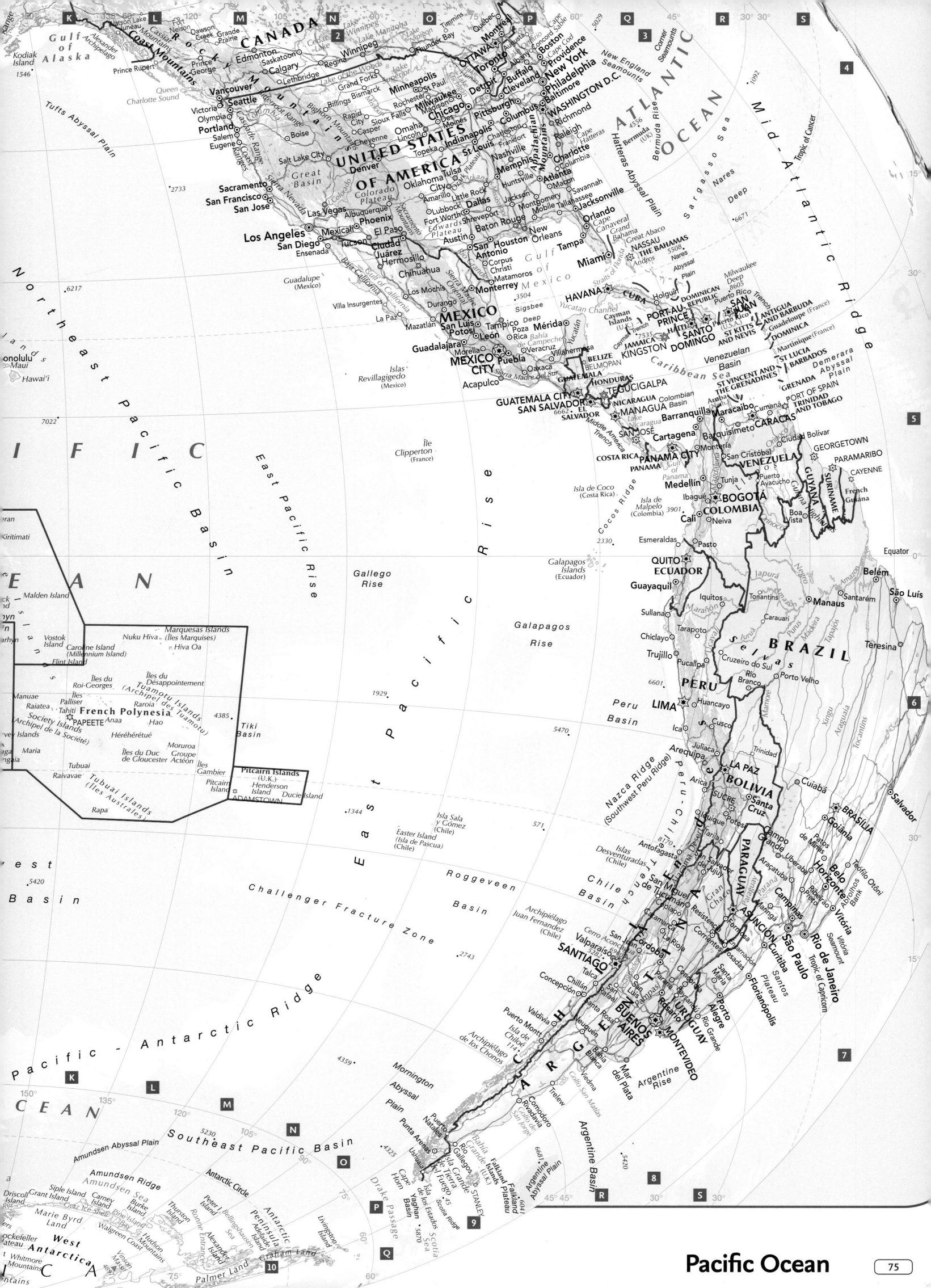

CANADA
UNITED STATES OF AMERICA
MEXICO
ATLANTIC OCEAN
Gulf of Alaska
Kodiak Island
Coast Mountains
Rocky Mountains
Edmonton
Calgary
Saskatoon
Regina
Winnipeg
Prince Rupert
Prince George
Vancouver
Victoria
Seattle
Olympia
Portland
Salem
Eugene
Boise
Salt Lake City
Great Basin
Sacramento
San Francisco
San Jose
Las Vegas
Los Angeles
San Diego
Ensenada
Mexicali
Tucson
Phoenix
Albuquerque
Denver
Cheyenne
Casper
Rapid City
Billings
Bismarck
Grand Forks
Minneapolis
Sioux Falls
Omaha
Lincoln
Topeka
Kansas City
Oklahoma City
Tulsa
Amarillo
Lubbock
Dallas
Fort Worth
El Paso
Ciudad Juárez
Hermosillo
Chihuahua
Austin
San Antonio
Houston
Corpus Christi
Baton Rouge
New Orleans
Shreveport
Jackson
Little Rock
Memphis
Nashville
St Louis
Chicago
Milwaukee
Madison
Des Moines
Indianapolis
Detroit
Cleveland
Columbus
Pittsburgh
Buffalo
Toronto
OTTAWA
Montréal
Québec
Boston
Providence
New York
Philadelphia
Baltimore
WASHINGTON D.C.
Richmond
Raleigh
Charlotte
Atlanta
Columbia
Savannah
Jacksonville
Tallahassee
Montgomery
Mobile
Huntsville
Orlando
Tampa
Miami
Gulf of Mexico
Monterrey
Matamoros
Los Mochis
Durango
Mazatlán
La Paz
Villa Insurgentes
Guadalajara
San Luis Potosí
León
Tampico
Mérida
Veracruz
Puebla
MEXICO CITY
Morelia
Oaxaca
Acapulco
Villahermosa
Islas Revillagigedo (Mexico)
Guadalupe (Mexico)
HAVANA
CUBA
THE BAHAMAS
NASSAU
Bermuda (UK)
PORT-AU-PRINCE
HAITI
SANTO DOMINGO
DOMINICAN REPUBLIC
SAN JUAN
Puerto Rico (U.S.A.)
KINGSTON
JAMAICA
ST KITTS AND NEVIS
ANTIGUA AND BARBUDA
Guadeloupe (France)
DOMINICA
Martinique (France)
ST LUCIA
BARBADOS
ST VINCENT AND THE GRENADINES
GRENADA
PORT OF SPAIN
TRINIDAD AND TOBAGO
Caribbean Sea
BELIZE
BELMOPAN
GUATEMALA
GUATEMALA CITY
SAN SALVADOR
EL SALVADOR
HONDURAS
TEGUCIGALPA
NICARAGUA
MANAGUA
SAN JOSÉ
COSTA RICA
PANAMA
PANAMA CITY
Barranquilla
Cartagena
Maracaibo
CARACAS
Barquisimeto
VENEZUELA
Ciudad Bolívar
GEORGETOWN
GUYANA
PARAMARIBO
SURINAME
CAYENNE
French Guiana
Medellín
BOGOTÁ
COLOMBIA
Cali
Neiva
Pasto
QUITO
ECUADOR
Guayaquil
Galapagos Islands (Ecuador)
Isla de Coco (Costa Rica)
Isla de Malpelo (Colombia)
Île Clipperton (France)
Equator
Belém
São Luís
Manaus
Santarém
Iquitos
Sullana
Chiclayo
Trujillo
Tarapoto
Pucallpa
PERU
LIMA
Huancayo
Cusco
Ica
Arequipa
Juliaca
Arica
Iquique
LA PAZ
BOLIVIA
SUCRE
Santa Cruz
Potosí
Trinidad
Cuiabá
Campo Grande
BRAZIL
BRASÍLIA
Goiânia
Salvador
Teresina
Porto Velho
Rio Branco
Cruzeiro do Sul
Belo Horizonte
Vitória
Campinas
São Paulo
Rio de Janeiro
Curitiba
Florianópolis
Porto Alegre
PARAGUAY
ASUNCIÓN
Antofagasta
San Miguel de Tucumán
Salta
Córdoba
Mendoza
Rosario
Santa Fe
Resistencia
Corrientes
Posadas
URUGUAY
MONTEVIDEO
BUENOS AIRES
Mar del Plata
Bahía Blanca
Santa Rosa
SANTIAGO
Valparaíso
Talca
Chillán
Concepción
Valdivia
Puerto Montt
Isla de Chiloé
Archipiélago de los Chonos
Archipiélago Juan Fernandez (Chile)
Islas Desventuradas (Chile)
CHILE
ARGENTINA
Neuquén
Viedma
Trelew
Comodoro Rivadavia
Puerto Natales
Punta Arenas
Río Gallegos
Ushuaia
Isla Grande de Tierra del Fuego
Isla de los Estados
Cape Horn
Falkland Islands (U.K.)
STANLEY
Drake Passage
Scotia Sea
Argentine Basin
Argentine Abyssal Plain
Argentine Rise
Falkland Plateau
Easter Island (Isla de Pascua) (Chile)
Isla Sala y Gómez (Chile)
Pitcairn Islands (U.K.)
Henderson Island
Ducie Island
ADAMSTOWN
French Polynesia
PAPEETE
Tahiti
Marquesas Islands (Îles Marquises)
Nuku Hiva
Hiva Oa
Tuamotu Islands (Archipel des Tuamotu)
Society Islands (Archipel de la Société)
Tubuai Islands (Îles Australes)
Îles Gambier
Caroline Island (Millennium Island)
Malden Island
Vostok Island
Flint Island
Kiritimati
Honolulu
Maui
Hawai'i
Northeast Pacific Basin
East Pacific Rise
Galapagos Rise
Gallego Rise
Peru Basin
Chile Basin
Roggeveen Basin
Challenger Fracture Zone
Nazca Ridge (Southwest Peru Ridge)
Peru-Chile Trench
Middle America Trench
Cocos Ridge
Tufts Abyssal Plain
Tiki Basin
Pacific - Antarctic Ridge
Southeast Pacific Basin
Mornington Abyssal Plain
Amundsen Abyssal Plain
Amundsen Ridge
Amundsen Sea
Antarctic Circle
Antarctic Peninsula
Marie Byrd Land
West Antarctica
Thurston Island
Peter I Island
Alexander Island
Bellingshausen Sea
Graham Land
Palmer Land
Livingston Island
Mid-Atlantic Ridge
Sargasso Sea
Nares Deep
Hatteras Abyssal Plain
Bermuda Rise
New England Seamounts
Corner Seamounts
Venezuelan Basin
Colombian Basin
Demerara Abyssal Plain
Tropic of Cancer
Tropic of Capricorn

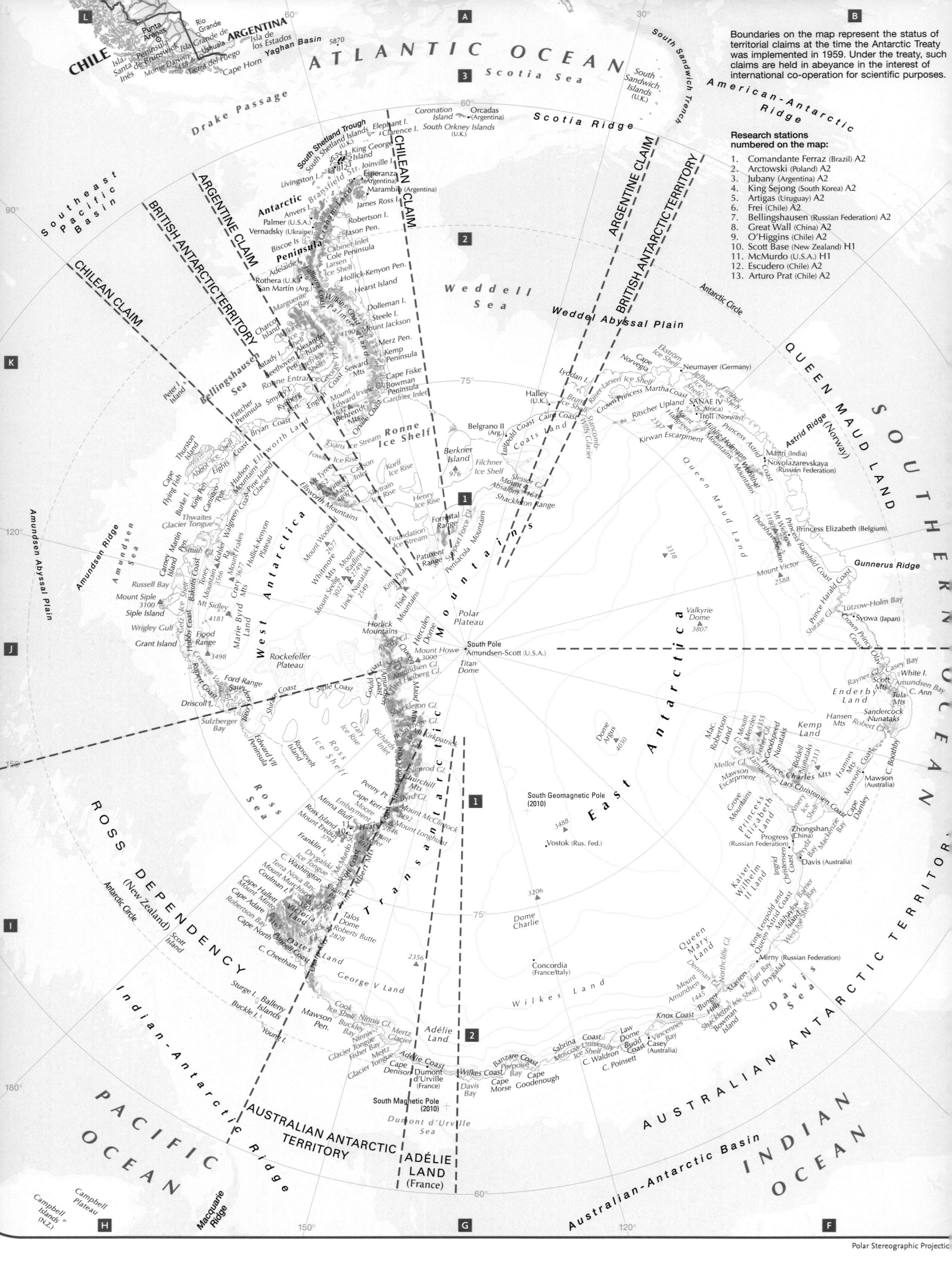

Research stations numbered on the map:

1. Comandante Ferraz (Brazil) A2
2. Arctowski (Poland) A2
3. Jubany (Argentina) A2
4. King Sejong (South Korea) A2
5. Artigas (Uruguay) A2
6. Frei (Chile) A2
7. Bellingshausen (Russian Federation) A2
8. Great Wall (China) A2
9. O'Higgins (Chile) A2
10. Scott Base (New Zealand) H1
11. McMurdo (U.S.A.) H1
12. Escudero (Chile) A2
13. Arturo Prat (Chile) A2

Polar Stereographic Projection

Antarctica

1:26 000 000

| 0 | 200 | 400 | 600 | 800 | 1000 mile |

| 0 | 200 | 400 | 600 | 800 | 1000 | 1200 | 1400 | 1600 km |

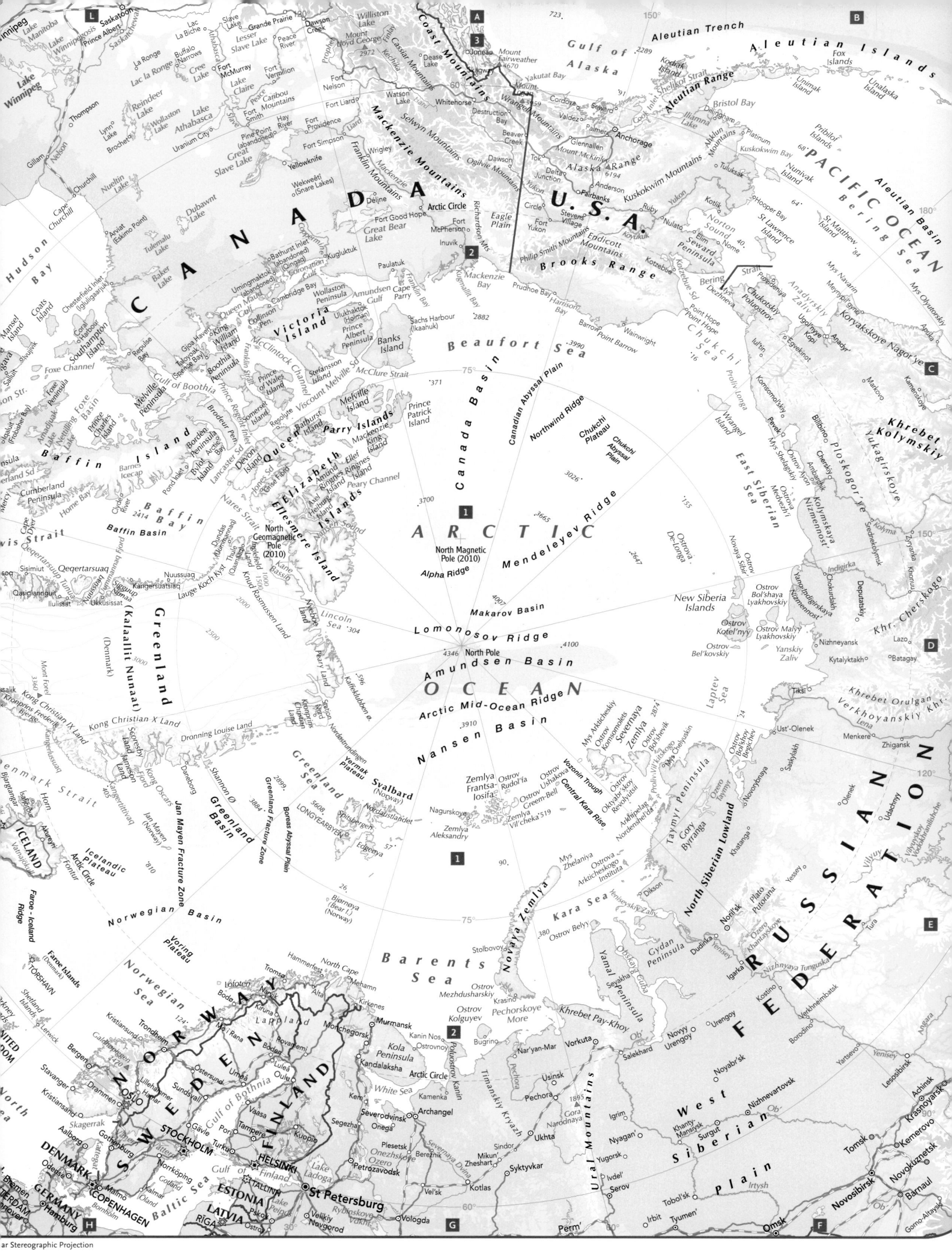

The Arctic

ar Stereographic Projection

6 000 000

| 0 | 200 | 400 | 600 | 800 | 1000 miles |
| 0 | 200 | 400 | 600 | 800 | 1000 | 1200 | 1400 | 1600 km |

Index

Introduction to the index

The index includes all names shown on the reference maps in the atlas. Each entry includes the country or geographical area in which the feature is located, a page number and an alphanumeric reference. Additional entry details and aspects of the index are explained below.

Name forms

The names policy in this atlas is generally to use local name forms which are officially recognized by the governments of the countries concerned. Rules established by the Permanent Committee on Geographical Names for British Official Use (PCGN) are applied to the conversion of non-roman alphabet names, for example in the Russian Federation, into the roman alphabet used in English.

However, English conventional name forms are used for the most well-known places for which such a form is in common use. In these cases, the local form is included in brackets on the map and appears as a cross-reference in the index. Other alternative names, such as well-known historical names or those in other languages, may also be included in brackets on the map and as cross-references in the index. All country names and those for international physical features appear in their English forms. Names appear in full in the index, although they may appear in abbreviated form on the maps.

Referencing

Names are referenced by page number and by grid reference. The grid reference relates to the alphanumeric values which appear on the edges of each map. These reflect the graticule on the map – the letter relates to longitude divisions, the number to latitude divisions.

Names are generally referenced to the largest scale map page on which they appear. For large geographical features, including countries, the reference is to the largest scale map on which the feature appears in its entirety, or on which the majority of it appears.

Rivers are referenced to their lowest downstream point – either their mouth or their confluence with another river. The river name will generally be positioned as close to this point as possible.

Alternative names

Alternative names appear as cross-references and refer the user to the index entry for the form of the name used on the map.

For rivers with multiple names - for example those which flow through several countries - all alternative name forms are included within the main index entries, with details of the countries in which each form applies.

Administrative qualifiers

Administrative divisions are included in entries to differentiate duplicate names - entries of exactly the same name and feature type within the one country - where these division names are shown on the maps. In such cases, duplicate names are alphabetized in the order of the administrative division names.

Additional qualifiers are included for names within selected geographical areas, to indicate more clearly their location.

Descriptors

Entries, other than those for towns and cities, include a descriptor indicating the type of geographical feature. Descriptors are not included where the type of feature is implicit in the name itself, unless there is a town or city of exactly the same name.

Insets

Where relevant, the index clearly indicates [inset] if a feature appears on an inset map.

Alphabetical order

The Icelandic characters Þ and þ are transliterated and alphabetized as 'Th' and 'th'. The German character ß is alphabetized as 'ss'. Names beginning with Mac or Mc are alphabetized exactly as they appear. The terms Saint, Sainte, etc, are abbreviated to St, Ste, etc, but alphabetized as if in the full form.

Numerical entries

Entries beginning with numerals appear at the beginning of the index, in numerical order. Elsewhere, numerals are alphabetized before 'a'.

Permuted terms

Names beginning with generic geographical terms are permuted - the descriptive term is placed after, and the index alphabetized by, the main part of the name. For example, Mount Everest is indexed as Everest, Mount; Lake Superior as Superior, Lake. This policy is applied to all languages. Permuting has not been applied to names of towns, cities or administrative divisions beginning with such geographical terms. These remain in their full form, for example, Lake Isabella, USA.

Abbreviations

admin. dist.	administrative district	IL	Illinois	Phil.	Philippines
admin. div.	administrative division	imp. l.	impermanent lake	plat.	plateau
admin. reg.	administrative region	IN	Indiana	P.N.G.	Papua New Guinea
Afgh.	Afghanistan	Indon.	Indonesia	Port.	Portugal
AK	Alaska	Kazakh.	Kazakhstan	pref.	prefecture
AL	Alabama	KS	Kansas	prov.	province
Alg.	Algeria	KY	Kentucky	pt	point
AR	Arkansas	Kyrg.	Kyrgyzstan	Qld	Queensland
Arg.	Argentina	l.	lake	Que.	Québec
aut. comm.	autonomous community	LA	Louisiana	r.	river
aut. reg.	autonomous region	lag.	lagoon	reg.	region
aut. rep.	autonomous republic	Lith.	Lithuania	res.	reserve
AZ	Arizona	Lux.	Luxembourg	resr	reservoir
Azer.	Azerbaijan	MA	Massachusetts	RI	Rhode Island
b.	bay	Madag.	Madagascar	Rus. Fed.	Russian Federation
Bangl.	Bangladesh	Man.	Manitoba	S.	South, Southern
B.C.	British Columbia	MD	Maryland	S.A.	South Australia
Bol.	Bolivia	ME	Maine	salt l.	salt lake
Bos.-Herz.	Bosnia-Herzegovina	Mex.	Mexico	Sask.	Saskatchewan
Bulg.	Bulgaria	MI	Michigan	SC	South Carolina
c.	cape	MN	Minnesota	SD	South Dakota
CA	California	MO	Missouri	sea chan.	sea channel
Cent. Afr. Rep.	Central African Republic	Mont.	Montenegro	Sing.	Singapore
CO	Colorado	Moz.	Mozambique	Switz.	Switzerland
Col.	Colombia	MS	Mississippi	Tajik.	Tajikistan
CT	Connecticut	MT	Montana	Tanz.	Tanzania
Czech Rep.	Czech Republic	mt.	mountain	Tas.	Tasmania
DC	District of Columbia	mts	mountains	terr.	territory
DE	Delaware	N.	North, Northern	Thai.	Thailand
Dem. Rep. Congo	Democratic Republic of the Congo	nat. park	national park	TN	Tennessee
depr.	depression	N.B.	New Brunswick	Trin. and Tob.	Trinidad and Tobago
des.	desert	NC	North Carolina	Turkm.	Turkmenistan
Dom. Rep.	Dominican Republic	ND	North Dakota	TX	Texas
E.	East, Eastern	NE	Nebraska	U.A.E.	United Arab Emirates
Equat. Guinea	Equatorial Guinea	Neth.	Netherlands	U.K.	United Kingdom
esc.	escarpment	NH	New Hampshire	Ukr.	Ukraine
est.	estuary	NJ	New Jersey	U.S.A.	United States of America
Eth.	Ethiopia	NM	New Mexico	UT	Utah
Fin.	Finland	N.S.	Nova Scotia	Uzbek.	Uzbekistan
FL	Florida	N.S.W.	New South Wales	VA	Virginia
for.	forest	N.T.	Northern Territory	Venez.	Venezuela
Fr. Guiana	French Guiana	NV	Nevada	Vic.	Victoria
F.Y.R.O.M.	Former Yugoslav Republic of Macedonia	N.W.T.	Northwest Territories	vol.	volcano
g.	gulf	NY	New York	vol. crater	volcanic crater
GA	Georgia	N.Z.	New Zealand	VT	Vermont
Guat.	Guatemala	OH	Ohio	W.	West, Western
HI	Hawaii	OK	Oklahoma	WA	Washington
H.K.	Hong Kong	OR	Oregon	W.A.	Western Australia
Hond.	Honduras	PA	Pennsylvania	WI	Wisconsin
i.	island	Para.	Paraguay	WV	West Virginia
IA	Iowa	P.E.I.	Prince Edward Island	WY	Wyoming
ID	Idaho	pen.	peninsula	Y.T.	Yukon

1

t Three Mile Opening sea chan.
Australia 56 D2
d Three Mile Opening sea chan.
Australia 56 C2
y Severnyy Rus. Fed. 11 S3
le Outubro Angola see Xá-Muteba
le Julio Arg. 70 D5
de Mayo Buenos Aires Arg. 70 D5
de Mayo La Pampa Arg. 70 C5
0 Mile House Canada 62 C1

A

benraa Denmark 15 F9
chen Germany 17 K5
lborg Denmark 15 F8
lborg Bugt b. Denmark 15 G8
len Germany 17 M6
lesund Norway see Ålesund
ley Lebanon see Aley
lst Belgium 16 J5
naar Fin. see Inari
rhus Denmark see Århus
len Belgium see Arlon
rs Denmark 15 F8
siaat Greenland 61 M3
a China 42 I6
a Dem. Rep. Congo 48 D3
a Nigeria 46 D4
acaxis r. Brazil 69 G4
ādān Iran 35 H5
adan Turkm. 33 I2
ādeh Iran 35 I5
ādeh Ţashk Iran 35 I5
adla Alg. 22 D5
aeté Brazil 71 B2
aetetuba Brazil 69 I4
agnar Qi China see Xilinhot
aiang atoll Kiribati 74 H5
akaliki Nigeria 46 D4
akan Rus. Fed. 42 G2
akanskiy Khrebet mts Rus. Fed. 42 F2
alak Niger 46 D3
ana Turkey 34 D2
ancay Peru 68 D6
ariringa atoll Kiribati see Kanton
arkūh Iran 35 I5
arkūh, Kavīr-e des. Iran 35 I5
arshahr Iran see Neyshābūr
ashiri Japan 44 G3
ashiri-wan b. Japan 44 G3
au P.N.G. 56 E1
aya, Lake Eth. 48 D3
aya Hāyk' l. Eth. see Abaya, Lake
ay Wenz r. Eth./Sudan see Blue Nile
ay Wenz r. Eth./Sudan 48 D2 see
Blue Nile
aza Rus. Fed. 42 G2
ba Cent. Afr. Rep. 48 B3
bāsābād Iran 35 I4
basanta Sardegna Italy 26 C4
batis Villa France see Abbeville
be, Lake Djibouti/Eth. 32 F7
beville France 24 E1
beville LA U.S.A. 63 I6
beyfeale Ireland 21 C5
beytown U.K. 18 D4
oorträsk Sweden 14 K4
bot, Mount Australia 56 D4
bot Ice Shelf Antarctica 76 K2
bott VA U.S.A. 64 A4
bottabad Pak. 33 L3
d al 'Azīz, Jabal h. Syria 35 F3
d al Kūrī i. Yemen 32 H7
d Allah, Khawr sea chan. Iraq/Kuwait
35 H5
d al Ma'asīr well Saudi Arabia 39 D4
dānān Iran 35 G4
dollāhābād Iran 35 I4
dulino Rus. Fed. 11 Q5
éché Chad 47 F3
ellinum Italy see Avellino
el Tasman National Park N.Z. 59 D5
engourou Côte d'Ivoire 46 C4
enrå Denmark see Aabenraa
eokuta Nigeria 46 D4
eraeron U.K. 19 C6
erchirder U.K. 20 G3
ercorn Zambia see Mbala
ercrombie r. Australia 58 D4
erdare U.K. 19 D7
erdaron U.K. 19 C6
erdaugleddau U.K. see Milford Haven
erdeen Australia 58 E4
erdeen S. Africa 50 G7
erdeen U.K. 20 G3
erdeen U.S.A. 62 H2
erdeen Lake Canada 61 I3
erdovey U.K. 19 C6
erfeldy U.K. 20 F4
erford U.K. 18 F5
erfoyle U.K. 20 E4
ergavenny U.K. 19 D7
ergwaun U.K. see Fishguard
erhonddu U.K. see Brecon
ermaw U.K. see Barmouth
erporth U.K. 19 C6
ersoch U.K. 19 C6
ertawe U.K. see Swansea
erteifi U.K. see Cardigan
erystwyth U.K. 19 C6
eshr Chad see Abéché
ez' Rus. Fed. 11 S2
hā Saudi Arabia 32 F6
har Iran 35 H3
iad, Bahr el r. Sudan/Uganda 32 D6 see
White Nile
djan Côte d'Ivoire 46 C4
ijatta-Shalla National Park Eth. 48 D3
ilene TX U.S.A. 62 H5
ingdon U.K. 19 F7
ington Reef Australia 56 E3
insk Rus. Fed. 34 E1
itibi, Lake Canada 63 K2
minga Australia 55 F6
nūb Egypt 34 C6
o Fin. see Turku
ohar India 36 C3
oisso Côte d'Ivoire 46 C4
omey Benin 46 D4
ong Mbang Cameroon 46 E4
ou Déia Chad 47 E3
ovyan Armenia 35 G2
oyne U.K. 20 G3
qaiq Saudi Arabia 48 E1
rantes Port. 25 B4
ra Pampa Arg. 70 C2
ri Sudan 32 D5
rolhos Bank sea feature S. Atlatic Ocean
72 F7
ruzzo, Parco Nazionale d' nat. park Italy
26 E4
salom, Mount Antarctica 76 B1
saroka Range mts U.S.A. 62 E3
tar, Jabal al hills Syria 39 C2

Abū aḍ Ḍuhūr Syria 39 C2
Abū al Ḥusayn, Qā' imp. l. Jordan 39 D3
Abū 'Alī i. Saudi Arabia 35 H6
Abū 'Āmūd, Wādī watercourse Jordan
39 C4
Abū 'Arīsh Saudi Arabia 32 F6
Abu 'Aweigîla well Egypt see Abū
'Uwayqilah
Abu Deleiq Sudan 32 D6
Abu Dhabi U.A.E. 33 H5
Abū Du'ān Syria 39 D1
Abu Gubeiha Sudan 32 D7
Abū Ḥafnah, Wādī watercourse Jordan
39 D3
Abu Haggag Egypt see Ra's al Ḥikmah
Abū Ḥallūfah, Jabal h. Jordan 39 C4
Abu Hamed Sudan 32 D6
Abuja Nigeria 46 D4
Abū Jurdhān Jordan 39 B4
Abū Kamāl Syria 35 F4
Abu Matariq Sudan 47 F3
Abumombazi Dem. Rep. Congo 48 C3
Abunã r. Bol. 68 E5
Abunã Brazil 68 E5
Ābune Yosēf mt. Eth. 32 E7
Abū Nujaym Libya 47 E1
Abū Qa'ṭūr Syria 39 C2
Abū Rawthah, Jabal mt. Egypt 39 B5
Aburo mt. Dem. Rep. Congo 48 D3
Abu Road India 31 G4
Abū Rujmayn, Jabal mts Syria 39 D2
Abū Rûtha, Gebel mt. Egypt see Abū
Rawthah, Jabal
Abu Simbil Egypt see Abū Sunbul
Abū Sunbul Egypt 32 D5
Abū Ṭarfā', Wādī watercourse Egypt 39 A5
Abut Head hd N.Z. 59 C6
Abū 'Uwayqilah well Egypt 39 B4
Abu Zabad Sudan 32 C7
Abū Ẓabī U.A.E. see Abu Dhabi
Abū Zanīmah Egypt 34 D5
Abu Zenîma Egypt see Abū Zanīmah
Abyad Sudan 32 C7
Abyaḍ, Jabal al mts Syria 39 C2
Abyār al Hakīm well Libya 34 A5
Abydos Australia 54 B5
Abyei Sudan 32 C8
Abyssinia country Africa see Ethiopia
Academician Vernadskiy research stn
Antarctica see Vernadsky
Academy Bay Rus. Fed. see Akademii, Zaliv
Acadia prov. Canada see Nova Scotia
Açailândia Brazil 69 I5
Acamarachi mt. Chile see Pili, Cerro
Acampamento de Caça do Mucusso
Angola 49 C5
Acandí Col. 68 C2
A Cañiza Spain 25 B2
Acaponeta Mex. 66 C4
Acapulco Mex. 66 E5
Acapulco de Juárez Mex. see Acapulco
Acará Brazil 69 I4
Acarai Mountains hills Brazil/Guyana
69 G3
Acaraú Brazil 69 J4
Acaray, Represa de resr Para. 70 E3
Acarigua Venez. 68 E2
Acatlán Mex. 66 E5
Accho Israel see 'Akko
Accomac VA U.S.A. 64 D4
Accra Ghana 46 C4
Accrington U.K. 18 E5
Achacachi Bol. 68 E7
Achaguas Venez. 68 E2
Achalpur India 36 D5
Achampet India 38 C2
Achan Rus. Fed. 44 E2
Achayvayam Rus. Fed. 29 S3
Acheng China 44 B3
Achhota India 38 D1
Achill Ireland 21 C4
Achillbeg Island Ireland 21 C4
Achill Island Ireland 21 B4
Achiltibuie U.K. 20 D2
Achinsk Rus. Fed. 28 K4
Achit Rus. Fed. 11 R4
Achit Nuur l. Mongolia 42 G3
Achkhoy-Martan Rus. Fed. 35 G2
Achna Cyprus 39 A2
Achnasheen U.K. 20 D3
Acıgöl l. Turkey 27 M6
Acıpayam Turkey 27 M6
Acireale Sicilia Italy 26 F6
Acklins Island Bahamas 67 J4
Acle U.K. 19 I6
Aconcagua, Cerro mt. Arg. 70 B4
Acopiara Brazil 69 K5
Açores terr. N. Atlantic Ocean see Azores
Açores, Arquipélago dos terr. N.
Atlantic Ocean see Azores
A Coruña Spain 25 B2
Acqui Terme Italy 26 C2
Acra NY U.S.A. 64 D1
Acragas Sicilia Italy see Agrigento
Acraman, Lake imp. l. Australia 57 A7
Acre r. Brazil 68 E6
Acre Israel see 'Akko
Acre, Bay of Israel see Haifa, Bay of
Acri Italy 26 G5
Ács Hungary 17 Q7
Actaeon Group is Fr. Polynesia see
Actéon, Groupe
Actéon, Groupe is Fr. Polynesia 75 K7
Acton Ont. Canada 64 A1
Acton CA U.S.A. 65 C3
Acungui Brazil 71 A4
Acunum Acusio France see Montélimar
Ada OK U.S.A. 63 H5
Adaja r. Spain 25 D3
Adalia Turkey see Antalya
Adam Oman 33 I5
Adam, Mount h. Falkland Is 70 E8
Adamantina Brazil 71 A3
Adams MA U.S.A. 64 E1
Adam's Peak Sri Lanka 38 D5
Adamstown Pitcairn Is 75 L7
'Adan Yemen see Aden
Adana Turkey 39 B1
Adana prov. Turkey 39 B1
Adana Yemen see Aden
Adapazarı Turkey 27 N4
Adare Ireland 21 D5
Adare, Cape Antarctica 76 H2
Adavale Australia 57 D5
Ad Dabbah Sudan see Ed Debba
Ad Dafīnah Saudi Arabia 32 F5
Ad Dahnā' des. Saudi Arabia 32 G5
Ad Dakhla W. Sahara 46 B2
Ad Damir Sudan see Ed Damer
Ad Dammām Saudi Arabia see Dammam
Addanki India 38 C3
Ad Dār al Ḥamrā' Saudi Arabia 32 E4
Ad Darb Saudi Arabia 32 F6
Ad Dawādimī Saudi Arabia 32 F5
Ad Dawḥah Qatar see Doha
Ad Dawr Iraq 35 F4
Ad Daww plain Syria 39 C2
Ad Dayr Iraq 35 G5
Ad Dibdibah plain Saudi Arabia 35 G6
Aḍ Ḍiffah plat. Egypt see Libyan Plateau
Addis Ababa Eth. 48 D3

Addison NY U.S.A. 64 C1
Ad Dīwānīyah Iraq 35 G5
Addlestone U.K. 19 G7
Addo Elephant National Park S. Africa
51 G7
Ad Duwayd well Saudi Arabia 35 F5
Ad Duwaym Sudan see Ed Dueim
Adegaon India 36 D5
Adelaide Australia 57 B7
Adelaide r. Australia 54 E3
Adelaide Island Antarctica 76 L2
Adelaide River Australia 54 E3
Adele Island Australia 54 C3
Adélie Coast Antarctica 76 G2
Adélie Land Antarctica 76 G2
Adelong Australia 58 D5
Aden Yemen 32 F7
Aden, Gulf of Somalia/Yemen 32 G7
Adena OH U.S.A. 64 A2
Aderbissinat Niger 46 D3
Aderno Sicilia Italy see Adrano
Adesar India 36 B5
Adh Dhāyūf well Saudi Arabia 35 G6
'Adhfā' well Saudi Arabia 35 F5
'Adhiriyāt, Jibāl al mts Jordan 39 C4
Adi i. Indon. 41 F8
Ādī Ārk'ay Eth. 32 E7
Adige r. Italy 26 E2
Ādīgrat Eth. 48 D2
Adilabad India 38 C2
Adilcevaz Turkey 35 F3
Adīrī Libya 47 E2
Adirondack Mountains NY U.S.A. 64 D1
Ādīs Ābeba Eth. see Addis Ababa
Adi Ugri Eritrea see Mendefera
Adıyaman Turkey 34 E3
Adjud Romania 27 L1
Admiralty Island Canada 61 H3
Admiralty Island National Monument-
Kootznoowoo Wilderness
nat. park U.S.A. 60 E4
Admiralty Islands P.N.G. 52 E2
Ado-Ekiti Nigeria 46 D4
Adok South Sudan 32 D8
Adonara i. Indon. 54 C2
Adoni India 38 C3
Ado-Tymovo Rus. Fed. 44 F2
Adour r. France 24 D5
Adra Spain 25 E5
Adramyttium Turkey see Edremit
Adramyttium, Gulf of Turkey see
Edremit Körfezi
Adrano Sicilia Italy 26 F6
Adrar Alg. 46 C2
Adrar hills Mali see Ifôghas, Adrar des
Adré Chad 47 F3
Adrian TX U.S.A. 62 G4
Adrianople Turkey see Edirne
Adrianopolis Turkey see Edirne
Adriatic Sea Europe 26 E2
Adua Eth. see Ādwa
Adunara i. Indon. see Adonara
Adusa Dem. Rep. Congo 48 C3
Aduwa Eth. see Ādwa
Adverse Well Australia 54 C5
Ādwa Eth. 48 D2
Adycha r. Rus. Fed. 29 O3
Adyk Rus. Fed. 13 J7
Adzopé Côte d'Ivoire 46 C4
Aegean Sea Greece/Turkey 27 K5
Aegina i. Greece see Aigina
Aegyptus country Africa see Egypt
Aela Jordan see Al 'Aqabah
Aelana Jordan see Al 'Aqabah
Aelia Capitolina Israel/West Bank see
Jerusalem
Aelōnlaplap atoll Marshall Is see
Ailinglaplap
Aenus Turkey see Enez
Aesernia Italy see Isernia
A Estrada Spain 25 B2
Afabet Eritrea 32 E6
Afanas'yevo Rus. Fed. 12 L4
Affreville Alg. see Khemis Miliana
Afghānestān country Asia see Afghanistan
Afghanistan country Asia 33 K3
Afgooye Somalia 48 E3
'Afīf Saudi Arabia 32 F5
Afiun Karahissar Turkey see Afyon
Åfjord Norway 14 G5
Aflou Alg. 22 E5
Afmadow Somalia 48 E3
Afogados da Ingazoira Brazil 69 K5
A Fonsagrada Spain 25 C2
Afonso Cláudio Brazil 71 C3
Āfrēra Terara vol. Eth. 32 F7
Africa Nova country Africa see Tunisia
'Afrīn Syria 39 C1
'Afrīn, Nahr r. Syria/Turkey 39 C1
Afşin Turkey 34 E3
Afuá Brazil 69 H4
'Afula Israel 39 B3
Afyon Turkey 27 N5
Afyonkarahisar Turkey see Afyon
Agadès Niger see Agadez
Agadez Niger 46 D3
Agadir Morocco 46 C1
Agadyr' Kazakh. 42 C3
Agalega Islands Mauritius 73 L6
Agara Georgia 35 F2
Agartala India 37 G5
Agashi India 38 B2
Agathe France see Agde
Agathonisi i. Greece 27 L6
Agatti i. India 38 B4
Agboville Côte d'Ivoire 46 C4
Ağcabādi Azer. 35 G2
Ağdam (abandoned) Azer. 35 G3
Ağdaş Azer. 35 G2
Agdash Azer. see Ağdaş
Agde France 24 F5
Agdzhabedi Azer. see Ağcabādi
Agedabia Libya see Ajdābiyā
Agen France 24 E4
Aggeneys S. Africa 50 D5
Aggteleki nat. park Hungary 17 R6
Aghil Dawan China 36 D1
Ağın Turkey 34 E3
Aginskoye Rus. Fed. 42 G1
Aginum France see Agen
Agios Dimitrios Greece 27 J6
Agios Efstratios i. Greece 27 K5
Agios Georgios i. Greece 27 J6
Agios Nikolaos Greece 27 K7
Agios Theodoros Cyprus 39 B2
Agiou Orous, Kolpos b. Greece 27 J4
Agirwat Hills Sudan 32 E6
Agisanang S. Africa 51 G4
Agnes, Mount h. Australia 55 E6
Agnew Australia 55 C6
Agnibilékrou Côte d'Ivoire 46 C4
Agnita Romania 27 K2
Agniye-Afanas'yevsk Rus. Fed. 44 E2
Agra India 36 D4
Agrakhanskiy Poluostrov pen. Rus. Fed.
35 G2
Agram Croatia see Zagreb
Ağrı Turkey 35 F3
Agria Gramvousa i. Greece 27 J7
Agrigento Sicilia Italy 26 E6
Agrigentum Sicilia Italy see Agrigento

Agrinio Greece 27 I5
Agropoli Italy 26 F4
Agryz Rus. Fed. 11 Q4
Ağsu Azer. 35 H2
Agua, Volcán de vol. Guat. 66 F6
Agua Clara Brazil 70 F2
Aguadilla Puerto Rico 67 K5
Agua Escondida Arg. 70 C5
Aguanga CA U.S.A. 65 D4
Aguapeí r. Brazil 71 A3
Agua Prieta Mex. 66 C2
Aguaro-Guariquito, Parque Nacional
nat. park Venez. 68 E2
Aguascalientes Mex. 66 D4
Agudos Brazil 71 A3
Águeda Port. 25 B3
Aguemour reg. Alg. 46 D2
Aguié Niger 46 D3
Aguilar de Campoo Spain 25 D2
Águilas Spain 25 F5
Agulhas, Cape S. Africa 50 E8
Agulhas Basin sea feature Southern Ocean
73 J9
Agulhas Negras mt. Brazil 71 B3
Agulhas Plateau sea feature Soutern Ocean
73 J8
Agulhas Ridge sea feature S. Atlatic Ocean
72 I8
Ağva Turkey 27 M4
Agvali Rus. Fed. 35 G2
Ahaggar plat. Alg. 46 D2
Ahar Iran 35 G3
Ahaura N.Z. 59 C6
Ahipara Bay N.Z. 59 D2
Ahiri India 38 D2
Ahklun Mountains U.S.A. 60 B4
Ahmadabad India 36 C5
Aḥmad al Bāqir, Jabal mt. Jordan 39 B5
Ahmadnagar India 38 B2
Ahmadpur East Pak. 33 L4
Ahmar mts Eth. 48 E3
Ahmar Mountains Eth. see Ahmar
Ahmedabad India see Ahmadabad
Ahmednagar India see Ahmadnagar
Ahram Iran 35 H5
Ähtäri Fin. 14 N5
Ahtme Estonia 15 O7
Āhū Iran 35 H5
Ahun France 24 F3
Ahvāz Iran 35 H5
Ahwa India 38 B1
Ahwāz Iran see Ahvāz
Ai-Ais Namibia 50 C4
Ai-Ais Hot Springs and Fish River Canyon
Park nature res. Namibia 50 C4
Aichwara India 36 D4
Aigialousa Cyprus 39 B2
Aigina i. Greece 27 J6
Aigio Greece 27 J5
Aigle de Chambeyron mt. France 24 H4
Aigüestortes i Estany de Sant Maurici,
Parc Nacional d' nat. park Spain 25 G2
Ai He r. China 44 B4
Aihui China see Heihe
Aijal India see Aizawl
Aikawa Japan 45 E5
Aiken U.S.A. 63 K5
Aileron Australia 54 F5
Ailinglabelab atoll Marshall Is see
Ailinglaplap
Ailinglaplap atoll Marshall Is 74 H5
Ailsa Craig Ont. Canada 64 A1
Ailsa Craig i. U.K. 20 D5
Ailt an Chorráin Ireland 21 D3
Aimorés, Serra dos hills Brazil 71 C2
Aïn Beïda Alg. 26 B7
'Aîn Ben Tili Mauritania 46 C2
'Ain Dâlla spring Egypt see 'Ayn Dāllah
Aïn Defla Alg. 25 H5
Aïn Deheb Alg. 25 G6
Aïn el Hadjel Alg. 25 H6
'Ain el Maqfi spring Egypt see 'Ayn al Maqfī
Aïn el Melh Alg. 25 I6
Aïn Mdila well Alg. 26 B7
Aïn-M'Lila Alg. 22 F4
Aïn Oussera Alg. 25 H6
Aïn Salah Alg. see In Salah
Aïn Sefra Alg. 22 D5
Ainsworth U.S.A. 62 H3
Aintab Turkey see Gaziantep
Aïn Taya Alg. 25 H5
Aïn Tédélès Alg. 25 G6
Aïn Temouchent Alg. 25 F6
'Ain Tibaghbagh spring Egypt see
'Ayn Tabaghbugh
'Ain Timeira spring Egypt see 'Ayn
Tumayrah
'Ain Zeitûn Egypt see 'Ayn Zaytūn
Aiquile Bol. 68 E7
Airdrie Canada 62 E1
Airdrie U.K. 20 F5
Aire-sur-l'Adour France 24 D5
Air Force Island Canada 61 K3
Airpanas Indon. 54 D1
Aisatung Mountain Myanmar 37 H5
Aisne r. France 24 F2
Aïssa, Djebel mt. Alg. 22 D5
Aitamännikkö Fin. 14 N3
Aitana mt. Spain 25 F4
Aït Benhaddou tourist site Morocco 22 C5
Aiud Romania 27 J1
Aix France see Aix-en-Provence
Aix-en-Provence France 24 G5
Aix-la-Chapelle Germany see Aachen
Aix-les-Bains France 24 G4
Aiyina i. Greece see Aigina
Aíyion Greece see Aigio
Aizawl India 37 H5
Aizkraukle Latvia 15 N8
Aizpute Latvia 15 L8
Aizu-Wakamatsu Japan 45 E5
Ajaccio Corse France 24 I6
Ajanta India 38 B1
Ajanta Range hills India see
Sahyadriparvat Range
Ajaureforsen Sweden 14 I4
Ajayameru India see Ajmer
Aj Bogd Uul mt. Mongolia 42 H4
Ajdābiyā Libya 47 F1
a-Jiddét des. Oman see Ḥarāsīs, Jiddat al
'Ajlūn Jordan 39 B3
Ajmer India 36 C4
Ajmer-Merwara India see Ajmer
Ajnala India 36 C3
Ajo AZ U.S.A. 65 F4
Ajrestān Afgh. 36 A2
Ajyyap Turkm. 35 I3
Akademii, Zaliv b. Rus. Fed. 44 E1
Akagera National Park Rwanda 48 D4
Akalkot India 38 C2
Akama, Akra c. Cyprus see Arnauti, Cape
Akamagaseki Japan see Shimonoseki
Akan Kokuritsu-kōen Japan 44 G4
Akaroa N.Z. 59 D6
Akāshat Iraq 35 E4
Akbarābād Iran 35 I5
Akbarpur Uttar Prad. India 36 E4
Akbarpur Uttar Prad. India 37 E4
Akbez Turkey 39 C1
Akçadağ Turkey 34 E3
Akçakale Turkey 39 D1

Akçakoca Turkey 27 N4
Akçakoca Dağları mts Turkey 27 N4
Akçakoyunlu Turkey 39 C1
Akçalı Dağları mts Turkey 39 A1
Akchâr reg. Mauritania 46 B3
Akdağlar mts Turkey 27 M6
Akdağmadeni Turkey 34 D3
Akdere Turkey 39 A1
Åkersberga Sweden 15 K7
Aketi Dem. Rep. Congo 48 C3
Akgyr Erezi hills Turkm. 35 I2
Akhali-Afoni Georgia see Akhali Ap'oni
Akhali Ap'oni Georgia 35 F2
Akhḍar, Al Jabal al mts Libya 47 F1
Akhḍar, Jabal mts Oman 33 I5
Akhisar Turkey 27 L5
Akhnoor India 36 C2
Akhsu Azer. see Ağsu
Akhta Armenia see Hrazdan
Akhtarīn Syria 39 C1
Akhtubinsk Rus. Fed. 13 J6
Akhty Rus. Fed. 35 G2
Akhtyrka Ukr. see Okhtyrka
Aki Japan 45 D6
Akiéni Gabon 48 B4
Akimiski Island Canada 63 K1
Akishma r. Rus. Fed. 44 D1
Akita Japan 45 F5
Akjoujt Mauritania 46 B3
Akkajaure l. Sweden 14 J3
Akkerman Ukr. see Bilhorod-Dnistrovs'kyy
Akkeshi Japan 44 G4
'Akko Israel 39 B3
Akkol' Akmolinskaya Oblast' Kazakh. 28 I4
Akkol' Atyrauskaya Oblast' Kazakh. 13 K7
Akku Kazakh. 42 D2
Akkul' Kazakh. see Akkol'
Akkuş Turkey 34 E2
Akkyr, Gory hills Turkm. see Akgyr Erezi
Aklavik Canada 60 E3
Aklera India 36 D4
Ak-Mechet Kazakh. see Kyzylorda
Akmeņrags pt Latvia 15 L8
Akmeqit China 36 D1
Akmola Kazakh. see Astana
Akmolinsk Kazakh. see Astana
Akobo South Sudan 47 G4
Akobo Wenz r. Eth./South Sudan 48 D3
Akokan Niger 46 D3
Akola India 38 C1
Akom II Cameroon 46 E4
Akonolinga Cameroon 46 E4
Akordat Eritrea 32 E6
Akōren Turkey 34 D3
Akot India 36 D5
Akpatok Island Canada 61 L3
Akqi China 42 D4
Akra, Jabal mt. Syria/Turkey see
Aqra', Jabal al
Akranes Iceland 14 [inset]
Åkrehamn Norway 15 D7
Akréréb Niger 46 D3
Akron CO U.S.A. 62 G3
Akron OH U.S.A. 64 A2
Akrotiri Bay Cyprus 39 A2
Akrotirion Bay Cyprus see Akrotiri Bay
Akrotiriou, Kolpos b. Cyprus see
Akrotiri Bay
Akrotiri Sovereign Base Area military base
Cyprus 39 A2
Aksai Chin terr. Asia 36 D2
Aksaray Turkey 34 D3
Aksay China 42 G5
Aksay Kazakh. 11 Q5
Ak-Say r. Kyrg. 33 M1
Aksay Rus. Fed. 13 H7
Akşehir Turkey 27 N5
Akşehir Gölü l. Turkey 27 N5
Akseki Turkey 34 C3
Aksha Rus. Fed. 43 K2
Akshiganak Kazakh. 42 A3
Akshukur Kazakh. 35 H2
Aksu China 42 E4
Aksu Kazakh. 42 D2
Aksu r. Turkey 27 N6
Aksuat Kazakh. 42 E3
Aksu-Ayuly Kazakh. 42 C3
Aksubayevo Rus. Fed. 13 K5
Āksum Eth. 32 E7
Aktag mt. China 37 F1
Aktaş Dağı mt. Turkey 35 G3
Aktau Kazakh. 30 E2
Aktobe Kazakh. 30 E1
Aktogay Karagandinskaya Oblast' Kazakh.
42 D3
Aktogay Vostochnyy Kazakhstan Kazakh.
42 D3
Aktsyabrski Belarus 13 F5
Aktyubinsk Kazakh. see Aktobe
Akulivik Canada 61 K3
Akune Japan 45 C6
Akure Nigeria 46 D4
Akuressa Sri Lanka 38 D5
Akureyri Iceland 14 [inset]
Akusha Rus. Fed. 13 J8
Akwanga Nigeria 46 D4
Akxokesay China 37 G1
Akyab Myanmar see Sittwe
Akyatan Gölü salt l. Turkey 39 B1
Akyazı Turkey 27 N4
Akzhaykyn, Ozero salt l. Kazakh. 42 B4
Ål Norway 15 F6
'Alā, Jabal al hills Syria 39 C2
Alabama r. U.S.A. 63 J5
Alabama state U.S.A. 63 J5
Al 'Abṭīyah well Iraq 35 G5
Alaca Turkey 34 D2
Alacahan Turkey 34 E3
Alaçam Turkey 34 D2
Alaçam Dağları mts Turkey 27 M5
Alacant Valencia Spain see Alicante
Alaçatı Turkey 27 L5
Aladağ Turkey 34 D3
Ala Dağlar mts Turkey 35 F3
Ala Dağları mts Turkey 34 D3
Al 'Adam Libya 34 A5
Ala'er China 42 E4
Alag Hu l. China 37 I2
Alagir Rus. Fed. 35 G2
Alagoinhas Brazil 71 D1
Alahärmä Fin. 14 M5
Al Aḥmadī Kuwait 35 H5
Alai Range mts Asia 33 L2
Ālājah Syria 39 B2
Alajärvi Fin. 14 M5
Al 'Ajrūd well Egypt 39 B4
Alakanuk U.S.A. 60 B3
Al Akhḍar Saudi Arabia 34 E5
Alakol', Ozero salt l. Kazakh. 28 J5
Ala Kul salt l. Kazakh. see Alakol', Ozero
Alakurtti Rus. Fed. 12 F2
Al 'Alamayn Egypt 34 C5
Al 'Alayyah Saudi Arabia 32 F6
Alama Somalia 48 E3
Al 'Amādīyah Iraq 35 F3
Al 'Amārah Iraq 35 G5
'Alam ar Rūm, Ra's pt Egypt 34 B5
'Alāmarvdasht watercourse Iran 35 I5
Alameda CA U.S.A. 65 A2
'Alam el Rûm, Râs pt Egypt see
'Alam ar Rūm, Ra's

Al Amghar waterhole Iraq 35 G5
Al 'Āmirīyah Egypt 34 C5
Alamo NV U.S.A. 65 E2
Alamo Dam AZ U.S.A. 65 F3
Alamogordo U.S.A. 62 F5
Álamos Sonora Mex. 66 B3
Álamos Sonora Mex. 66 C3
Alamos r. Mex. 62 G6
Alamosa U.S.A. 62 F4
Alampur India 38 C3
Alan Myanmar see Aunglan
Alanäs Sweden 14 I4
Åland is Fin. see Åland Islands
Aland India 38 C2
Al Andarīn Syria 39 C2
Åland Islands Fin. 15 K6
Alando China 37 H3
Alandur India 38 D3
Alanya Turkey 34 D3
Alapli Turkey 27 N4
Alappuzha India 38 C4
Alapuzha India see Alappuzha
Al 'Aqabah Jordan 39 B5
Al 'Aqīq Saudi Arabia 32 F5
Al 'Arabīyah as Sa'ūdīyah country Asia see
Saudi Arabia
Alarcón, Embalse de resr Spain 25 E4
Al 'Arīsh Egypt 39 A4
Al Arṭāwīyah Saudi Arabia 32 G4
Alas, Selat sea chan. Indon. 54 B2
Alaşehir Turkey 27 M5
Alashiya country Asia see Cyprus
Al Ashmūnayn Egypt 34 C6
Alaska state U.S.A. 60 D3
Alaska, Gulf of U.S.A. 60 D4
Alaska Peninsula U.S.A. 60 B4
Alaska Range mts U.S.A. 60 D3
Älät Azer. 35 H3
Al Atwā' well Saudi Arabia 35 F5
Alatyr' Rus. Fed. 13 J5
Alatyr' r. Rus. Fed. 13 J5
Alausí Ecuador 68 C4
Alaverdi Armenia 35 G2
Alavieska Fin. 14 N4
'Alavījeh Iran 35 H4
Alavus Fin. 14 M5
Alawoona Australia 57 C7
Alay Kyrka Toosu mts Asia see Alai Range
Alayskiy Khrebet mts Asia see Alai Range
Al 'Azīzīyah Iraq 35 G4
Al 'Azīzīyah Libya 23 G5
Al Azraq al Janūbī Jordan 39 C4
Alba Italy 26 C2
Al Bāb Syria 39 C1
Albacete Spain 25 F4
Al Bādiyah al Janūbīyah des. Iraq 35 G5
Al Baḥrayn country Asia see Bahrain
Alba Iulia Romania 27 J1
Al Bājī Iran 35 H5
Albania country Europe 27 H4
Albany Australia 55 B8
Albany r. Canada 63 K1
Albany GA U.S.A. 63 K5
Albany NY U.S.A. 64 E1
Albany OR U.S.A. 62 C3
Albany Downs Australia 58 D1
Albardão do João Maria coastal area Brazil
70 F4
Al Bardī Libya 34 B5
Al Bāridah hills Saudi Arabia 39 D5
Al Başrah Iraq see Basra
Al Baṭḥa' marsh Iraq 35 G5
Albatross Bay Australia 56 C2
Albatross Island Australia 57 [inset]
Al Bawīṭī Egypt 34 C5
Al Bayḍā' Libya 32 B3
Al Bayḍā' Yemen 32 G7
Albemarle Island Galápagos Ecuador see
Isabela, Isla
Albenga Italy 26 C2
Alberche r. Spain 25 D4
Alberga Australia 57 A5
Alberga watercourse Australia 57 A5
Albergaria-a-Velha Port. 25 B3
Albert Australia 58 C4
Albert France 24 F2
Albert, Lake Dem. Rep. Congo/Uganda
48 D3
Albert, Parc National nat. park
Dem. Rep. Congo see Virunga, Parc
National des
Alberta prov. Canada 62 E1
Alberta VA U.S.A. 64 C4
Albert Lea U.S.A. 63 I3
Albert Nile r. Sudan/Uganda 47 G4
Alberto de Agostini, Parque Nacional
nat. park Chile 70 B8
Alberton S. Africa 51 I4
Albertville Dem. Rep. Congo see Kalemie
Albertville France 24 H4
Albi France 24 F5
Albina Suriname 69 H2
Albino Italy 26 C2
Albion NY U.S.A. 64 B1
Albion PA U.S.A. 64 A2
Al Biqā' val. Lebanon see El Béqaa
Al Bi'r Saudi Arabia 34 E5
Al Birk Saudi Arabia 32 F6
Al Biyāḍh reg. Saudi Arabia 32 G5
Alborán, Isla de i. Spain 25 E6
Alboran Sea Europe 25
Ålborg Denmark see Aalborg
Ålborg Bugt b. Denmark see Aalborg Bugt
Albro Australia 56 D4
Albufeira Port. 25 B5
Al Buḥayrāt al Murrah lakes Egypt see
Bitter Lakes
Albuquerque U.S.A. 62 F4
Al Burayj Syria 39 C2
Al Buraymī Oman 33 I5
Al Burj Jordan 39 B5
Alburquerque Spain 25 C4
Albury Australia 58 C6
Al Buşayrah Syria 35 F4
Al Busayṭā' plain Saudi Arabia 39 D4
Al Bushūk well Saudi Arabia 35 G5
Alcácer do Sal Port. 25 B4
Alcalá de Henares Spain 25 E3
Alcalá la Real Spain 25 E5
Alcamo Sicilia Italy 26 E6
Alcañiz Spain 25 F3
Alcántara Spain 25 C4
Alcaraz Spain 25 E4
Alcázar de San Juan Spain 25 E4
Alcazarquivir Morocco see Ksar el Kebir
Alchevs'k Ukr. 13 H6
Alcobaça Brazil 71 D2
Alcoi Spain see Alcoy-Alcoi
Alcoota Australia 54 F5
Alcoy Spain see Alcoy-Alcoi
Alcoy-Alcoi Spain 25 F4
Alcúdia Spain 25 H4
Aldabra Islands Seychelles 49 E4
Aldan Rus. Fed. 29 N4
Aldan r. Rus. Fed. 29 N3
Alde r. U.K. 19 I6
Aldeburgh U.K. 19 I6
Alder Creek NY U.S.A. 64 D1
Alderney i. Channel Is 19 E9
Alder Peak CA U.S.A. 65 B3
Aldershot U.K. 19 G7

Aldingham U.K. 18 D4
Aldridge U.K. 19 F6
Aleg Mauritania 46 B3
Alegre *Espírito Santo* Brazil 71 C3
Alegre *Minas Gerais* Brazil 71 B2
Alegrete Brazil 70 E3
Aleksandra, Mys *hd* Rus. Fed. 44 E1
Aleksandriya Ukr. *see* Oleksandriya
Aleksandro-Nevskiy Rus. Fed. 13 I5
Aleksandropol Armenia *see* Gyumri
Aleksandrov Rus. Fed. 12 H4
Aleksandrov Gay Rus. Fed. 13 K6
Aleksandrovsk Rus. Fed. 11 R4
Aleksandrovsk Ukr. *see* Zaporizhzhya
Aleksandrovskiy Rus. Fed. *see* Aleksandrovsk
Aleksandrovskoye Rus. Fed. 35 F1
Aleksandrovsk-Sakhalinskiy Rus. Fed. 44 F2
Aleksandry, Zemlya *i.* Rus. Fed. 28 F1
Alekseyevka *Akmolinskaya Oblast'* Kazakh. *see* Akkol'
Alekseyevka *Vostochnyy Kazakhstan* Kazakh. *see* Terekty
Alekseyevka *Amurskaya Oblast'* Rus. Fed. 44 B1
Alekseyevka *Belgorodskaya Oblast'* Rus. Fed. 13 H6
Alekseyevka *Belgorodskaya Oblast'* Rus. Fed. 13 H6
Alekseyevskaya Rus. Fed. 13 I6
Alekseyevskoye Rus. Fed. 12 K5
Aleksin Rus. Fed. 13 H5
Aleksinac Serbia 27 I3
Alèmbé Gabon 48 B4
Ålen Norway 14 G5
Alençon France 24 E2
Alenquer Brazil 69 H4
Alep Syria *see* Aleppo
Aleppo Syria 39 C1
Alert Canada 61 L1
Alerta Peru 68 D6
Alès France 24 G4
Aleşd Romania 27 J1
Aleshki Ukr. *see* Tsyurupyns'k
Aleşkirt Turkey *see* Eleşkirt
Alessandria Italy 26 C2
Alessio Albania *see* Lezhë
Ålesund Norway 14 E5
Aleutian Basin *sea feature* Bering Sea 74 H2
Aleutian Islands U.S.A. 60 A4
Aleutian Range *mts* U.S.A. 60 C4
Aleutian Trench *sea feature* N. Pacific Ocean 74 I2
Alevina, Mys *c.* Rus. Fed. 29 Q4
Alevişik Turkey *see* Samandağı
Alexander, Kap *c.* Greenland *see* Ullersuaq
Alexander, Mount *h.* Australia 56 B2
Alexander Archipelago *is* U.S.A. 60 E4
Alexander Bay *b.* Namibia/S. Africa 50 C5
Alexander Bay S. Africa 50 C5
Alexander Island Antarctica 76 L2
Alexandra Australia 58 B6
Alexandra N.Z. 59 B7
Alexandra, Cape S. Georgia 70 I8
Alexandra Land *i.* Rus. Fed. *see* Aleksandry, Zemlya
Alexandreia Greece 27 J4
Alexandretta Turkey *see* İskenderun
Alexandria Afgh. *see* Ghaznī
Alexandria Egypt 34 C5
Alexandria Romania 27 K3
Alexandria S. Africa 51 H7
Alexandria Turkm. *see* Mary
Alexandria U.K. 20 E5
Alexandria *LA* U.S.A. 63 I5
Alexandria *VA* U.S.A. 64 C3
Alexandria Arachoton Afgh. *see* Kandahār
Alexandria Areion Afgh. *see* Herāt
Alexandria Prophthasia Afgh. *see* Farāh
Alexandrina, Lake Australia 57 B7
Alexandroupoli Greece 27 K4
Alexis Creek Canada 62 C1
Aley Lebanon 39 B3
Aleysk Rus. Fed. 42 E2
Al Farwānīyah Kuwait 35 G5
Al Fas Morocco *see* Fès
Al Fatḥah Iraq 35 F4
Al Fāw Iraq 35 H5
Al Fayyūm Egypt 34 C5
Alfenas Brazil 71 B3
Alford U.K. 18 H5
Alfred *ME* U.S.A. 64 F1
Alfred *NY* U.S.A. 64 C1
Alfred and Marie Range *hills* Australia 55 D6
Al Fujayrah U.A.E. *see* Fujairah
Al Fuqahā' Libya 47 E2
Al Furāt *r.* Asia *see* Euphrates
Al Furāt *r.* Asia 39 D2 *see* Euphrates
Ålgård Norway 15 D7
Algarrobo del Aguilla Arg. 70 C5
Algarve *reg.* Port. 25 B5
Algeciras Spain 25 D5
Algemesí Spain 25 F4
Algena Eritrea 32 E6
Alger Alg. *see* Algiers
Algeria *country* Africa 46 C2
Algérie *country* Africa *see* Algeria
Al Ghammās Iraq 35 G5
Al Ghardaqah Egypt *see* Al Ghurdaqah
Al Ghawr *plain* Jordan/West Bank 39 B4
Al Ghayḍah Yemen 32 H6
Alghero *Sardegna* Italy 26 C4
Al Ghurdaqah Egypt 32 D4
Algiers Alg. 25 H5
Algoa Bay S. Africa 51 G7
Algona U.S.A. 63 I3
Algorta Spain 25 E2
Algueirao Moz. *see* Hacufera
Al Habakah *well* Saudi Arabia 35 F5
Al Ḥabbānīyah Iraq 35 F4
Al Ḥadaqah *well* Saudi Arabia 35 G5
Al Hadhālīl *plat.* Saudi Arabia 35 F5
Al Ḥadīdīyah Syria 39 C2
Al Ḥadīthah Iraq 35 F4
Al Ḥadīthah Saudi Arabia 39 C4
Al Ḥaḍr Iraq *see* Hatra
Al Ḥafār *well* Saudi Arabia 35 F5
Al Ḥaffah Syria 39 C2
Al Haggounia W. Sahara 46 B2
Al Ḥajar al Gharbī *mts* Oman 33 I5
Al Ḥamād *plain* Asia 34 E5
Al Ḥamādah al Ḥamrā' *plat.* Libya 46 E2
Alhama de Murcia Spain 25 F5
Al Ḥamīdīyah Syria 39 B2
Al Ḥammām Egypt 34 C5
Al Ḥanākīyah Saudi Arabia 32 F5
Al Haniyah *esc.* Iraq 35 G5
Al Ḥarrah Egypt 34 C5
Al Harūj al Aswad *hills* Libya 47 E2
Al Ḥasakah Syria 35 F3
Al Hawi *salt pan* Saudi Arabia 39 D5
Al Hawjā' Saudi Arabia 34 E5
Al Ḥayy Iraq 35 G4
Al Ḥayz Egypt 34 C5
Al Ḥazīm Jordan 39 C4
Al Ḥazm Saudi Arabia 34 E5
Al Ḥazm al Jawf Yemen 32 F6
Al Ḥibāk *des.* Saudi Arabia 33 H6
Al Hījānah Syria 39 C3
Al Ḥillah Iraq *see* Hillah
Al Ḥillah Saudi Arabia 32 G5
Al Ḥinnāh Saudi Arabia 48 E1
Al Ḥinw *mt.* Saudi Arabia 39 D4
Al Ḥishah Syria 39 D1
Al Ḥismā *plain* Saudi Arabia 34 D5
Al Ḥiṣn Jordan 39 B3
Al Hoceima Morocco 25 E6
Al Ḥudaydah Yemen *see* Hodeidah
Al Ḥufrah *reg.* Saudi Arabia 34 E5
Al Ḥūj *hills* Saudi Arabia 34 E5
Ali China 36 D2
'Alīābād *Golestān* Iran 35 I3
'Alīābād *Hormozgān* Iran 33 I4
'Alīābād *Kordestān* Iran 35 G3
'Alīābād, Kūh-e *mt.* Iran 35 H4
Aliağa Turkey 27 L5
Aliakmonas *r.* Greece 27 J4
Alibag India 38 B2
Alicante Spain 25 F4
Alice *r.* Australia 56 C2
Alice *watercourse* Australia 56 D5
Alice U.S.A. 62 H6
Alice, Punta *pt* Italy 26 G5
Alice Springs Australia 55 F5
Alichur Tajik. 33 L2
Alick Creek *r.* Australia 56 C4
Aliganj India 36 D4
Aligarh *Rajasthan* India 36 D4
Aligarh *Uttar Prad.* India 36 D4
Alīgūdarz Iran 35 H4
Alihe China 44 A2
'Alī Kheyl Afgh. 36 B2
Al Imārāt al 'Arabīyah at Muttaḥidah *country* Asia *see* United Arab Emirates
Alimia *i.* Greece 27 L6
Alindao Cent. Afr. Rep. 48 C3
Alingsås Sweden 15 H8
Aliova *r.* Turkey 27 M5
Alipura India 36 D4
Alipur Duar India 37 G4
Aliquippa *PA* U.S.A. 64 A2
Alirajpur India 36 C5
Al 'Irāq *country* Asia *see* Iraq
Al 'Īsāwīyah Saudi Arabia 39 C4
Al Iskandarīyah Egypt *see* Alexandria
Al Iskandarīyah Iraq 35 G4
Al Ismā'īlīyah Egypt 34 D5
Al Ismā'īlīyah *governorate* Egypt 39 A4
Aliveri Greece 27 K5
Aliwal North S. Africa 51 H6
Al Jafr Jordan 39 C4
Al Jaghbūb Libya 34 B5
Al Jahrah Kuwait 35 G5
Al Jarāwī *well* Saudi Arabia 39 D4
Al Jauf Saudi Arabia *see* Dumat al Jandal
Al Jawf Libya 47 F2
Al Jawsh Libya 46 E1
Al Jaza'ir *country* Africa *see* Algeria
Al Jaza'ir Alg. *see* Algiers
Aljezur Port. 25 B5
Al Jil *well* Iraq 35 F5
Al Jithāmīyah Saudi Arabia 35 F6
Al Jīzah Egypt *see* Giza
Al Jīzah Jordan 39 B4
Al Jufrah Libya 47 E2
Al Julayqah *well* Saudi Arabia 35 H6
Al Jumaylīyah Qatar 32 H4
Aljustrel Port. 25 B5
Al Juwayf *depr.* Syria 39 C3
Al Kahfah *Al Qaṣīm* Saudi Arabia 32 F4
Al Kahfah *Ash Sharqīyah* Saudi Arabia 35 H6
Al Karak Jordan 39 B4
Al Khalīl West Bank *see* Hebron
Al Khāliṣ Iraq 35 G4
Al Khārijah Egypt 32 D4
Al Kharrūbah Egypt 39 A4
Al Khaṣab Oman 33 I4
Al Khawkhah Yemen 32 F7
Al Khawr Qatar 32 H4
Al Khums Libya 47 E1
Al Khunfah *des.* Saudi Arabia 34 E5
Al Khunn Saudi Arabia 48 E1
Al Kifl Iraq 35 G4
Al Kiswah Syria 39 C3
Alkmaar Neth. 16 J4
Al Kūbrī Egypt 39 A4
Al Kūfah Iraq 35 G4
Al Kumayt Iraq 35 G4
Al Kuntillah Egypt 39 B5
Al Kusūr *hills* Saudi Arabia 39 D4
Al Kūt Iraq 35 G4
Al Kuwayt *country* Asia *see* Kuwait
Al Kuwayt Kuwait *see* Kuwait
Al Labbah *plain* Saudi Arabia 35 F5
Al Lādhiqīyah Syria *see* Latakia
Allagadda India 38 C3
Allahabad India 37 E4
Al Lajā *lava field* Syria 39 C3
Allakaket U.S.A. 60 C3
Allakh-Yun' Rus. Fed. 29 O3
Allanmyo Myanmar *see* Aunglan
Allanridge S. Africa 51 H4
Allapalli India 38 D2
'Allāqī, Wādī al *watercourse* Egypt 32 D5
'Allāqi, Wādi el *watercourse* Egypt *see* 'Allāqī, Wādī al
Alldays S. Africa 51 I2
Allegheny *r.* *PA* U.S.A. 64 B2
Allegheny Mountains U.S.A. 64 A4
Allegheny Reservoir *PA* U.S.A. 64 B2
Allen, Lough *l.* Ireland 21 D3
Allendale Town U.K. 18 E4
Allende *Coahuila* Mex. 62 G6
Allenstein Poland *see* Olsztyn
Allentown *PA* U.S.A. 64 D2
Alleppey India *see* Alappuzha
Aller *r.* Germany 17 L4
Alliance *NE* U.S.A. 62 G3
Alliance *OH* U.S.A. 64 A2
Al Lībīyah *country* Africa *see* Libya
Allier *r.* France 24 F3
Al Lihābah *well* Saudi Arabia 35 G6
Allinge-Sandvig Denmark 15 I9
Al Liṣāfah *well* Saudi Arabia 35 G6
Al Lisān *pen.* Jordan 39 B4
Al Līth Saudi Arabia 32 F5
Alloa U.K. 20 F4
Allora Australia 58 F2
Allur India 38 D3
Alluru Kottapatnam India 38 D3
Al Lussuf *well* Iraq 35 F5
Alma Canada 63 M2
Al Ma'ānīyah Iraq 35 F5
Alma-Ata Kazakh. *see* Almaty
Almada Port. 25 B4
Al Madāfi' *plat.* Saudi Arabia 34 E5
Al Ma'danīyāt *well* Iraq 35 G5
Almaden Australia 56 D3
Almadén Spain 25 D4
Al Madīnah Saudi Arabia *see* Medina
Al Mafraq Jordan 39 C3
Al Maghrib *country* Africa *see* Morocco
Al Mahdum Syria 39 C1
Al Maḥiā *depr.* Saudi Arabia 34 E6
Al Maḥwīt Yemen 32 F6
Al Manāmah Bahrain *see* Manama
Almansa Spain 25 F4
Al Manṣūrah Egypt 34 C5
Almanzor *mt.* Spain 25 D3
Al Mariyyah U.A.E. 33 H5
Al Marj Libya 47 F1
Almas, Rio das *r.* Brazil 71 A1
Al Maṭarīyah Egypt 34 D5
Almaty Kazakh. 42 D4
Al Mawṣil Iraq *see* Mosul
Al Mayādīn Syria 35 F4
Al Mazār Egypt 39 A4
Almaznyy Rus. Fed. 29 M3
Almeirim Brazil 69 H4
Almeirim Port. 25 B4
Almelo Neth. 17 K4
Almenara Brazil 71 C2
Almendra, Embalse de *resr* Spain 25 C3
Almendralejo Spain 25 C4
Almería Spain 25 E5
Almería, Golfo de *b.* Spain 25 E5
Al'metievsk Rus. Fed. *see* Al'met'yevsk
Al'met'yevsk Rus. Fed. 11 Q5
Älmhult Sweden 15 I8
Almina, Punta *pt* Spain 25 D6
Al Mindak Saudi Arabia 32 F5
Al Minyā Egypt 34 C5
Almirós Greece *see* Almyros
Al Mish'āb Saudi Arabia 35 H5
Almodôvar Port. 25 B5
Almond *r.* U.K. 20 F4
Almonte Spain 25 C5
Almora India 36 D3
Al Mu'ayzīlah *h.* Saudi Arabia 39 D5
Al Mubarrez Saudi Arabia 32 G4
Al Muḍaibī Oman 33 I5
Al Mukallā Yemen *see* Mukalla
Al Mukhā Yemen *see* Mocha
Al Mukhaylī Libya 32 B3
Almuñécar Spain 25 E5
Al Muqdādīyah Iraq 35 G4
Al Mūrītānīyah *country* Africa *see* Mauritania
Al Murūt *well* Saudi Arabia 35 E5
Almus Turkey 34 E2
Al Musannāh *ridge* Saudi Arabia 35 G5
Al Muwaqqar Jordan 39 C4
Almyros Greece 27 J5
Almyrou, Ormos *b.* Greece 27 K7
Alnwick U.K. 18 F3
Alofi Niue 53 J3
Aloja Latvia 15 N8
Along India 37 H3
Alongshan China 44 A2
Alonnisos *i.* Greece 27 J5
Alor *i.* Indon. 41 E8
Alor, Kepulauan *is* Indon. 41 E8
Alor Setar Malaysia 41 C7
Alor Star Malaysia *see* Alor Setar
Alost Belgium *see* Aalst
Aloysius, Mount Australia 55 E6
Alozero (abandoned) Rus. Fed. 14 Q4
Alpena U.S.A. 63 K2
Alpercatas, Serra das *hills* Brazil 69 J5
Alpha Australia 56 D4
Alpha Ridge *sea feature* Arctic Ocean 77 A1
Alpine *NY* U.S.A. 64 C1
Alpine *TX* U.S.A. 62 G5
Alpine National Park Australia 58 C6
Alps *mts* Europe 24 H4
Al Qa'āmīyāt *reg.* Saudi Arabia 32 G6
Al Qaddāḥīyah Libya 47 E1
Al Qadmūs Syria 39 C2
Al Qāhirah Egypt *see* Cairo
Al Qā'īyah Saudi Arabia 32 F5
Al Qal'a Beni Hammad *tourist site* Alg. 25 I6
Al Qalībah Saudi Arabia 34 E5
Al Qāmishlī Syria 35 F3
Al Qar'ah Libya 34 B5
Al Qar'ah *lava field* Syria 39 C3
Al Qardāḥah Syria 39 C2
Al Qarqar Saudi Arabia 39 C4
Al Qaryatayn Syria 39 C2
Al Qaṭn Yemen 32 G6
Al Qaṭrānah Jordan 39 C4
Al Qaṭrūn Libya 47 E2
Al Qayṣūmah *well* Saudi Arabia 35 F5
Al Qumur *country* Africa *see* Comoros
Al Qunayṭirah (abandoned) Israel 39 B3
Al Qunfidhah Saudi Arabia 32 F6
Al Qurayyāt Saudi Arabia 39 C4
Al Qurnah Iraq 35 G5
Al Quṣaymah Egypt 39 B4
Al Quṣayr Egypt 32 D4
Al Quṣayr Syria 39 C2
Al Qūṣīyah Egypt 34 C6
Al Quṭayfah Syria 39 C3
Al Quwayīyah Saudi Arabia 32 G5
Al Quwayrah Jordan 39 B5
Alroy Downs Australia 56 B3
Alsace *reg.* France 24 H2
Alsager U.K. 19 E5
Al Samīt *well* Iraq 35 F5
Alsatia *reg.* France *see* Alsace
Alston U.K. 18 E4
Alstonville Australia 58 F2
Alsunga Latvia 15 L8
Alta Norway 14 M2
Alta, Mount N.Z. 59 B7
Altaelva *r.* Norway 14 M2
Altafjorden *sea chan.* Norway 14 M1
Alta Floresta Brazil 69 G5
Altai Mountains Asia 42 F3
Altamaha *r.* U.S.A. 63 K5
Altamira Brazil 69 H4
Altamura Italy 26 G4
Altan Shiret China 43 J5
Altan Xiret China *see* Altan Shiret
Altavista *VA* U.S.A. 64 B4
Altay China 42 F3
Altay Mongolia 42 G4
Altay Mongolia 42 H3
Altayskiy Khrebet *mts* Asia *see* Altai Mountains
Altdorf Switz. 24 I3
Altea Spain 25 F4
Alteidet Norway 14 M1
Altenqoke China 37 H1
Altin Köprü Iraq 35 G4
Altınoluk Turkey 27 L5
Altınözü Turkey 39 C1
Altıntaş Turkey 27 N5
Altiplano *plain* Bol. 68 E7
Altmühl *r.* Germany 17 M6
Alto, Monte *h.* Italy 26 D2
Alto Chicapa Angola 49 B5
Alto de Pencoso *hills* Arg. 70 C4
Alto Garças Brazil 69 H7
Alto Madidi, Parque Nacional *nat. park* Bol. 68 E6
Alton *NH* U.S.A. 64 F1
Altoona *PA* U.S.A. 64 B2
Alto Paraíso de Goiás Brazil 71 B1
Alto Parnaíba Brazil 69 I5
Alto Taquarí *Mato Grosso* Brazil 69 H7
Altötting Germany 17 N6
Altrincham U.K. 18 E5
Altūn Kūbrī Iraq *see* Altin Köprü
Altun Shan *mts* China 42 F5
Alturas U.S.A. 62 C3
Altus U.S.A. 62 H5
Al 'Ubaylah Saudi Arabia 48 F1
Alucra Turkey 34 E2
Alūksne Latvia 15 O8
Alūm Iran 35 H4
Alum Bridge *WV* U.S.A. 64 A3
Al 'Uqaylah Libya 47 E1
Al 'Uqaylah Saudi Arabia *see* An Nabk
Al Uqṣur Egypt *see* Luxor
Alur India 38 C3
Al 'Urayq *des.* Saudi Arabia 34 E5
Al 'Urdun *country* Asia *see* Jordan
Alur Setar Malaysia *see* Alor Setar
'Ālūt Iran 35 G4
Aluva India *see* Alwaye
Al 'Uwaynāt Libya 32 B5
Al 'Uwayqīlah Saudi Arabia 35 F5
Al 'Uzayr Iraq 35 G5
Alva U.S.A. 62 H4
Alvand, Kūh-e *mt.* Iran 35 H4
Alvarães Brazil 68 F4
Alvdal Norway 14 G5
Älvdalen Sweden 15 I6
Alvesta Sweden 15 I8
Ålvik Norway 15 E6
Alvik Sweden 14 J5
Alvorada do Norte Brazil 71 B1
Älvsbyn Sweden 14 L4
Al Wafrah Kuwait 35 G5
Al Wajh Saudi Arabia 32 E4
Al Waqbá *well* Saudi Arabia 35 G5
Alwar India 36 D4
Al Warī'ah Saudi Arabia 32 G4
Al Wāṭīyah *well* Egypt 34 B5
Alwaye India 38 C4
Al Widyān *plat.* Iraq/Saudi Arabia 35 F4
Al Wusayṭ *well* Saudi Arabia 35 G5
Alxa Youqi China *see* Ehen Hudag
Alxa Zuoqi China *see* Bayan Hot
Al Yaman *country* Asia *see* Yemen
Alyangula Australia 56 B2
Alyth U.K. 20 F4
Alytus Lith. 15 N9
Amacayacu, Parque Nacional *nat. park* Col. 68 D4
Amadeus, Lake *imp. l.* Australia 55 E6
Amadjuak Lake Canada 61 K3
Amadora Port. 25 B4
Amakusa-nada *b.* Japan 45 C6
Åmål Sweden 15 H7
Amalia S. Africa 51 G4
Amaliada Greece 27 I6
Amalner India 36 C5
Amamapare Indon. 41 F8
Amambaí Brazil 70 E2
Amambaí, Serra de *hills* Brazil/Para. 70 E2
Amami-Ō-shima *i.* Japan 45 C7
Amami-shotō *is* Japan 45 C8
Amamula Dem. Rep. Congo 48 C4
Amantea Italy 26 G5
Amanzimtoti S. Africa 51 J6
Amapá Brazil 69 H3
Amarante Brazil 69 J5
Amarapura Myanmar 37 I5
Amareleja Port. 25 C4
Amargosa Brazil 71 D1
Amargosa *watercourse CA* U.S.A. 65 D2
Amargosa Desert *NV* U.S.A. 65 D2
Amargosa Range *mts CA* U.S.A. 65 D2
Amargosa Valley *NV* U.S.A. 65 D2
Amarillo U.S.A. 62 G4
Amarillo, Cerro *mt.* Arg. 70 C4
Amarkantak India 37 E5
Amarpur *Madh. Prad.* India 36 E5
Amasia Turkey *see* Amasya
Amasine W. Sahara 46 B2
Amasra Turkey 34 D2
Amasya Turkey 34 D2
Amata Australia 55 E6
Amatulla India 37 H4
Amau P.N.G. 56 E1
Amazar Rus. Fed. 44 A1
Amazar *r.* Rus. Fed. 44 A1
Amazon *r.* S. America 68 F4
Amazon, Mouths of the Brazil 69 I3
Amazonas *r.* S. America 68 F4 *see* Amazon
Amazon Cone *sea feature* S. Atlantic Ocean 72 E5
Amazônia, Parque Nacional *nat. park* Brazil 69 G4
Ambajogai India 38 C2
Ambala India 36 D3
Ambalangoda Sri Lanka 38 D5
Ambalavao Madag. 49 E6
Ambam Cameroon 48 B3
Ambarchik Rus. Fed. 29 R3
Ambarnyy Rus. Fed. 14 R4
Ambasa India *see* Ambassa
Ambasamudram India 38 C4
Ambassa India 37 G5
Ambathala Australia 57 D5
Ambato Ecuador 68 C4
Ambato Boeny Madag. 49 E5
Ambato Finandrahana Madag. 49 E6
Ambatolampy Madag. 49 E5
Ambatomainty Madag. 49 E5
Ambatondrazaka Madag. 49 E5
Ambejogai India *see* Ambajogai
Ambeno *enclave* East Timor *see* Ocussi
Amberg Germany 17 M6
Ambergris Caye *i.* Belize 66 G5
Ambérieu-en-Bugey France 24 G4
Ambgaon India 38 D1
Ambianum France *see* Amiens
Ambikapur India 37 E5
Ambilobe Madag. 49 E5
Amble U.K. 18 F3
Ambler U.S.A. 60 C3
Ambleside U.K. 18 E4
Ambo India 37 F5
Amboasary Madag. 49 E6
Ambodifotatra Madag. 49 E5
Ambohimahasoa Madag. 49 E6
Ambohitra *mt.* Madag. 49 E5
Amboina Indon. *see* Ambon
Ambon Indon. 41 E8
Ambon *i.* Indon. 41 E8
Amboró, Parque Nacional *nat. park* Bol. 68 F7
Ambositra Madag. 49 E6
Ambovombe Madag. 49 E6
Amboy *CA* U.S.A. 65 E3
Ambre, Cap d' *c.* Madag. *see* Bobaomby, Tanjona
Ambrim *i.* Vanuatu *see* Ambrym
Ambriz Angola 49 B4
Ambrizete Angola *see* N'zeto
Ambrym *i.* Vanuatu 53 G3
Ambur India 38 C3
Am-Dam Chad 47 F3
Amded, Oued *watercourse* Alg. 46 D2
Amdo China *see* Lharigarbo
Amelia Court House *VA* U.S.A. 64 C4
Amenia *NY* U.S.A. 64 E2
Amer, Erg d' *des.* Alg. 48 A1
Amereli India *see* Amreli
American, North Fork *r. CA* U.S.A. 65 B1
Americana Brazil 71 B3
American-Antarctic Ridge *sea feature* S. Atlantic Ocean 72 G9
American Falls U.S.A. 62 E3
American Falls Reservoir U.S.A. 62 E3
American Fork U.S.A. 62 E3
American Samoa *terr.* S. Pacific Ocean 53 J3
Americus U.S.A. 63 K5
Amersfoort Neth. 16 J4
Amersfoort S. Africa 51 I4
Amersham U.K. 19 G7
Amery Ice Shelf Antarctica 76 E2
Ames U.S.A. 63 I3
Amesbury U.K. 19 F7
Amesbury *MA* U.S.A. 64 F1
Amet India 36 C4
Amethi India 37 E4
Amfissa Greece 27 J5
Amga Rus. Fed. 29 O3
Amgalang China 43 L3
Amgu Rus. Fed. 44 E3
Amguid Alg. 46 D2
Amgun' *r.* Rus. Fed. 44 E1
Amherst *MA* U.S.A. 64 E1
Amherst *VA* U.S.A. 64 B4
Amiata, Monte *mt.* Italy 26 D3
Amida Turkey *see* Diyarbakır
Amiens France 24 F2
'Amīj, Wādī *watercourse* Iraq 35 F4
Amik Ovası *marsh* Turkey 39 C1
Amīnābād Iran 35 I5
Amindivi *atoll* India *see* Amini
Amindivi Islands India 38 B4
Amini *atoll* India 38 B4
Amino Eth. 48 E3
Aminuis Namibia 50 D2
Amīrābād Iran 35 G4
Amirante Islands Seychelles 73 L6
Amirante Trench *sea feature* Indian Ocean 73 L6
Amisk Lake Canada 62 G1
Amistad, Represa de *resr* Mex./U.S.A. *see* Amistad Reservoir
Amistad Reservoir Mex./U.S.A. 62 G6
Amisus Turkey *see* Samsun
Amity Point Australia 58 F1
Amla India 36 D5
Amlapura Indon. *see* Karangasem
Amlash Iran 35 H3
Amlekhganj Nepal 37 F4
Åmli Norway 15 F7
Amlia Island U.S.A. 60 A4
Amlwch U.K. 18 C5
'Ammān Jordan 39 B4
Ammanazar Turkm. 35 I3
Ammanford U.K. 19 D7
Ämmänsaari Fin. 14 P4
'Ammār, Tall *h.* Syria 39 C3
Ammarnäs Sweden 14 J4
Ammaroo Australia 56 A4
Ammassalik Greenland 77 J2
Ammochostos Cyprus *see* Famagusta
Ammochostos Bay Cyprus 39 B2
Am Nābiyah Yemen 32 F7
Amne Machin Range *mts* China *see* A'nyêmaqên Shan
Amnok-kang *r.* China/N. Korea *see* Yalu Jiang
Āmol Iran 35 I3
Amorgos *i.* Greece 27 K6
Amory U.S.A. 63 J5
Amos Canada 63 L2
Amourj Mauritania 46 C3
Amoy China *see* Xiamen
Ampani India 38 D2
Ampanihy Madag. 49 E6
Ampara Sri Lanka 38 D5
Amparo Brazil 71 B3
Ampasimanolotra Madag. 49 E5
Amphitheatre Australia 58 A6
Amraoti India *see* Amravati
Amravati India 38 C1
Amrawad India 36 D5
Amreli India 36 B5
Amri Pak. 36 B4
Amring India 37 H4
'Amrīt Syria 39 B2
Amritsar India 36 C3
Amroha India 36 D3
Åmsele Sweden 14 K4
Amstelveen Neth. 16 J4
Amsterdam Neth. 16 J4
Amsterdam S. Africa 51 J4
Amsterdam *NY* U.S.A. 64 D1
Amsterdam, Île *i.* Indian Ocean 73 N8
Amstetten Austria 17 O6
Am Timan Chad 47 F3
Amudar'ya *r.* Asia 33 I1
Amudaryo *r.* Asia *see* Amudar'ya
Amund Ringnes Island Canada 61 I2
Amundsen, Mount Antarctica 76 F2
Amundsen Abyssal Plain *sea feature* Southern Ocean 76 J2
Amundsen Basin *sea feature* Arctic Ocean 77 H1
Amundsen Bay Antarctica 76 D2
Amundsen Coast Antarctica 76 J1
Amundsen Glacier Antarctica 76 I1
Amundsen Gulf Canada 60 F2
Amundsen Ridges *sea feature* Southern Ocean 76 J2
Amundsen-Scott *research stn* Antarctica 76 C1
Amundsen Sea Antarctica 76 K2
Amuntai Indon. 41 D8
Amur *r.* China/Rus. Fed. *see* Heilong Jiang
Amur *r.* Rus. Fed. *see* Heilong Jiang
'Amur, Wadi *watercourse* Sudan 32 D6
Amuro-Baltiysk Rus. Fed. 44 B1
Amur Oblast *admin. div.* Rus. Fed. *see* Amurskaya Oblast'
Amursk Rus. Fed. 44 E2
Amurskaya Oblast' *admin. div.* Rus. Fed. 44 C1
Amurskiy Liman *str.* Rus. Fed. 44 F1
Amurzet Rus. Fed. 44 C3
Amvrosiyivka Ukr. 13 H7
Amyderya *r.* Asia *see* Amudar'ya
Am-Zoer Chad 47 F3
Anaa *atoll* Fr. Polynesia 75 K7
Anabanua Indon. 41 E8
Anabar *r.* Rus. Fed. 29 M2
Anacapa Islands *CA* U.S.A. 65 C3
Anaconda U.S.A. 62 E2
Anadolu *reg.* Turkey 34 D3
Anadolu Dağları *mts* Turkey 34 E2
Anadyr' Rus. Fed. 29 S3
Anadyr, Gulf of Rus. Fed. *see* Anadyrskiy Zaliv
Anadyrskiy Zaliv *b.* Rus. Fed. 29 T3
Anafi *i.* Greece 27 K6
Anagé Brazil 71 C1
'Ānah Iraq 35 F4
Anaheim *CA* U.S.A. 65 D4
Anaimalai Hills India 38 C4
Anaiteum *i.* Vanuatu *see* Anatom
Anajás Brazil 69 I4
Anakie Australia 56 D4
Analalava Madag. 49 E5
Anamã Brazil 68 F4
Anambas, Kepulauan *is* Indon. 41 C7
Anamur Turkey 39 A1
Anan Japan 45 D6
Anand India 36 C5
Anandapur India 37 F5
Anantapur India 38 C3
Ananthapur India *see* Anantapur
Anantnag India 36 C2
Anant Peth India 36 D4
Anantpur India *see* Anantapur
Ananyev Ukr. *see* Anan'yiv
Anan'yiv Ukr. 13 F7
Anapa Rus. Fed. 34 E1
Anápolis Brazil 71 A2
Anār Fin. *see* Inari
Anār Iran 35 I5
Anatom *i.* Vanuatu 53 G4
Añatuya Arg. 70 D3
Anayapazari Turkey *see* Gülnar
An Baile Breac Ireland 21 B6
An Blascaod Mór Ireland *see* Great Blasket Island
An Bun Beag Ireland 21 D2
Anbyon N. Korea 45 B5
Ancenis France 24 D3
Anchorage U.S.A. 60 D3
Anchorage Island *atoll* Cook Is *see* Suwarrow
Anchuthengu India 38 C4
Anci China *see* Langfang
An Clochán Liath Ireland 21 D3
An Cóbh Ireland *see* Cobh
Ancona Italy 26 E3
Ancud, Golfo de *g.* Chile 70 B6
Ancyra Turkey *see* Ankara
Anda *Heilong.* China *see* Daqing
Anda *Heilong.* China 44 B3
Andacollo Chile 70 B4
Andado Australia 56 A5
Andahuaylas Peru 68 D6
An Daingean Ireland 21 B5
Andal India 37 F5
Åndalsnes Norway 14 E5
Andalucía *aut. comm.* Spain 25 D5
Andalusia *aut. comm.* Spain *see* Andaluc
Andaman Basin *sea feature* Indian Ocea
73 O5
Andaman Islands India 31 I5
Andaman Sea Indian Ocean 41 B6
Andamooka Australia 57 B6
Andapa Madag. 49 E5
Andegavum France *see* Angers
Andelle *r.* France 19 I9
Andenes Norway 14 J2
Andenne Belgium 16 J5
Andéramboukane Mali 46 D3
Anderlecht Belgium 16 J5
Andermatt Switz. 24 I3
Andernos-les-Bains France 24 D4
Anderson *r.* Canada 60 F3
Anderson *AK* U.S.A. 60 D3
Anderson *IN* U.S.A. 63 J3
Anderson *SC* U.S.A. 63 K5
Anderson Bay Australia 57 [inset]
Andes *mts* S. America 70 C4
Andfjorden *sea chan.* Norway 14 J2
Andhíparos *i.* Greece *see* Antiparos
Andhra Lake India 38 B2
Andhra Pradesh *state* India 38 C2
Andijon Uzbek. 33 L1
Andikithira *i.* Greece *see* Antikythira
Andilamena Madag. 49 E5
Andilanatoby Madag. 49 E5
Andimeshk Iran 35 H4
Andímilos *i.* Greece *see* Antimilos
Andípsara *i.* Greece *see* Antipsara
Andırın Turkey 34 E3
Andirlangar China 37 E1
Andkhvoy Afgh. 36 A1
Andoany Madag. 49 E5
Andoas Peru 68 C4
Andogskaya Gryada *hills* Rus. Fed. 12 H
Andol India 38 C2
Andong China *see* Dandong
Andong S. Korea 45 C5
Andoom Australia 56 C2
Andorra *country* Europe 25 G2
Andorra la Vella Andorra 25 G2
Andorra la Vieja Andorra *see* Andorra la Vella
Andover U.K. 19 F7
Andover *NY* U.S.A. 64 C1
Andover *OH* U.S.A. 64 A2
Andøya *i.* Norway 14 I2
Andrade *CA* U.S.A. 65 E4
Andradina Brazil 71 A3
Andranomavo Madag. 49 E5
Andranopasy Madag. 49 E6
Andreanof Islands U.S.A. 74 I2
Andreapol' Rus. Fed. 12 G4
Andreas Isle of Man 18 C4
André Félix, Parc National *nat. park* Cer
Afr. Rep. 48 C3
Andrelândia Brazil 71 B3
Andrews *TX* U.S.A. 62 G5
Andria Italy 26 G4
Androka Madag. 49 E6
Andropov Rus. Fed. *see* Rybinsk
Andros *i.* Bahamas 67 I4
Andros *i.* Greece 27 K6
Andros Town Bahamas 67 I4
Androtti *i.* India 38 B4
Andselv Norway 14 K2
Andújar Spain 25 D4
Andulo Angola 49 B5
Anec, Lake *imp. l.* Australia 55 E5
Äneen-Kio *terr.* N. Pacific Ocean *see* Wa
Island
Anéfis Mali 46 D3
Anegada, Bahía *b.* Arg. 70 D6
Anegada Passage Virgin Is (U.K.) 67 L5
Aného Togo 46 D4
Aneityum *i.* Vanuatu *see* Anatom
'Aneiza, Jabal *h.* Iraq *see* 'Unayzah, Jabal
Anemourion *tourist site* Turkey 39 A1
Anetchom, Île *i.* Vanuatu *see* Anatom
Aneto *mt.* Spain 25 G2
Änewetak *atoll* Marshall Is *see* Enewetak
Aney Niger 46 E3
Aneytioum, Île *i.* Vanuatu *see* Anatom
Angalarri *r.* Australia 54 E3
Angamos, Punta *pt* Chile 70 B2
Ang'angxi China 44 A3
Angara *r.* Rus. Fed. 42 G1
Angarsk Rus. Fed. 42 I2
Angas Downs Australia 55 F6
Angatuba Brazil 71 A3
Ånge Sweden 14 I5
Angel, Salto *waterfall* Venez. *see* Angel Falls
Ángel de la Guarda, Isla *i.* Mex. 66 B3
Angel Falls *waterfall* Venez. 68 F2
Ängelholm Sweden 15 H8
Angellala Creek *r.* Australia 58 C1
Angels Camp *CA* U.S.A. 65 B1
Ångermanälven *r.* Sweden 14 J5
Angers France 24 D3
Angikuni Lake Canada 61 I3
Angiola *CA* U.S.A. 65 C3
Anglesea Australia 58 B7
Anglesey *i.* U.K. 18 C5
Anglo-Egyptian Sudan *country* Africa *see* Sudan
Angmagssalik Greenland *see* Ammassa
Ango Dem. Rep. Congo 48 C3
Angoche Moz. 49 D5
Angol Chile 70 B5

gola country Africa 49 B5
gola NY U.S.A. 64 B1
gola Basin sea feature S. Atlantic Ocean 2 H7
gora Turkey see Ankara
goulême France 24 E4
gra dos Reis Brazil 71 B3
gren Uzbek. 33 L1
guang China 44 A3
guilla terr. West Indies 67 L5
gul India 38 E1
gutia Char i. Bangl. 37 G5
holt i. Denmark 15 G8
hui prov. China 43 L6
humas Brazil 69 H7
hwei prov. China see Anhui
ak U.S.A. 60 C3
akchak National Monument and Preserve nat. park U.S.A. 60 C4
tápolis Brazil 71 A4
tlı Turkey 39 A1
iva Rus. Fed. 44 F3
va, Mys c. Rus. Fed. 44 F3
va, Zaliv b. Rus. Fed. 44 F3
adip i. India 38 B3
alankoski Fin. 15 O6
engo India see Anchuthengu
ir Avand Iran 35 I4
ou reg. France 24 D3
ouan i. Comoros see Nzwani
ozorobe Madag. 49 E5
uthengu India see Anchuthengu
kang China 43 J6
kara Turkey 34 D3
karatra mt. Madag. 49 E5
kazoabo Madag. 49 E6
kleshwar India 36 C5
klesvar India see Ankleshwar
kola India 38 B3
u China 43 K6
moore WV U.S.A. 64 A3
Muileann gCearr Ireland see Mullingar
myŏn-do i. S. Korea 45 B5
n, Cape Antarctica 76 D2
n, Cape MA U.S.A. 64 F1
na Rus. Fed. 13 I6
na, Lake VA U.S.A. 64 C3
naba Alg. 26 B6
Nabk Saudi Arabia 39 C4
Nabk Syria 39 C2
Nafūd des. Saudi Arabia 35 F5
Najaf Iraq 35 G5
nalee r. Ireland 21 E3
nalong U.K. 21 G3
nan U.K. 20 F6
nan r. U.K. 20 F6
nān, Wādī al watercourse Syria 39 D2
nandale VA U.S.A. 64 C3
na Plains Australia 54 C4
napolis MD U.S.A. 64 C3
napurna Conservation Area nature res. Nepal 37 F3
napurna I mt. Nepal 37 E3
n Arbor U.S.A. 63 K3
na Regina Guyana 69 G2
Nās Ireland see Naas
Nāṣirīyah Iraq 35 G5
Naṣrānī, Jabal mts Syria 39 C3
nean, Lake imp. l. Australia 55 B6
necy France 24 H4
Nimārah Syria 39 C3
Nimāṣ Saudi Arabia 32 F6
niston U.S.A. 63 J5
nobón i. Equat. Guinea 46 D5
nonay France 24 G4
Nu'mānīyah Iraq 35 G4
Nuṣayrīyah, Jabal mts Syria 39 C2
ón de Sardinas, Bahía de b. Col. 68 C3
orontany, Tanjona hd Madag. 49 E5
qing China 43 L6
sbach Germany 17 M6
ser Group is Australia 58 C7
shan China 44 A4
shun China 42 J7
shun China 42 I7
Sirhān, Wādī watercourse Saudi Arabia 4 E5
son Bay Australia 54 E3
songo Mali 46 D3
sted WV U.S.A. 64 A3
abamba Peru 68 D6
akya Turkey 39 C1
alaha Madag. 49 F5
alya Turkey 27 N6
alya prov. Turkey 39 A1
alya Körfezi g. Turkey 27 N6
ananarivo Madag. 49 E5
tAonach Ireland see Nenagh
arctica 76
arctic Peninsula Antarctica 76 L2
as r. Brazil 71 A5
Teallach mt. U.K. 20 D3
elope Range mts NV U.S.A. 65 D1
equera Spain 25 D5
hony Lagoon Australia 56 A3
i Atlas mts Morocco 22 C6
ibes France 24 H5
icosti, Île d' i. Canada 61 L5
icosti Island Canada see Anticosti, Île d'
ifer, Cap d' c. France 19 H9
igua i. Antigua and Barbuda 67 L5
igua country West Indies see Antigua and Barbuda
igua and Barbuda country West Indies 7 L5
ikythira i. Greece 27 J7
ikythiro, Steno sea chan. Greece 27 J7
i Lebanon mts Lebanon/Syria see Sharqī, Jabal ash
imilos i. Greece 27 K6
Inbhear Mór Ireland see Arklow
ioch Turkey see Antakya
ioch CA U.S.A. 65 B1
iocheia ad Cragum tourist site Turkey 9 A1
iochia Turkey see Antakya
iparos i. Greece 27 K6
ipodes Islands N.Z. 53 H6
ipsara i. Greece 27 K5
ium Italy see Anzio
ofagasta Chile 70 B2
ofagasta de la Sierra Arg. 70 C3
ofalla, Volcán vol. Arg. 70 C3
ónio Enes Moz. see Angoche
ri India 36 D4
rim U.K. 21 F3
rim Hills U.K. 21 F2
rim Plateau Australia 54 E4
ropovo Rus. Fed. 12 I4
salova Madag. 49 E5
seranana Madag. see Antsirañana
sirabe Madag. 49 E5
sirañana Madag. 49 E5
sla Estonia 15 O8
sohihy Madag. 49 E5
tis Sweden 14 M3
tola Fin. 15 O6
werp Belgium 16 J5
werpen Belgium see Antwerp
Uaimh Ireland see Navan
Anuchino Rus. Fed. 44 D4
Anugul India see Angul
Anupgarh India 36 C3
Anuradhapura Sri Lanka 38 D4
Anveh Iran 35 I6
Anvers Island Antarctica 76 L2
Anvik U.S.A. 60 B3
Anxi Gansu China 42 H4
Anxious Bay Australia 55 F8
Anyang Henan China 43 K5
Anyang S. Korea 45 B5
A'nyêmaqên Shan mts China 42 H6
Anyuy r. Rus. Fed. 44 E2
Anyuysk Rus. Fed. 29 R3
Anzhero-Sudzhensk Rus. Fed. 28 J4
Anzi Dem. Rep. Congo 48 C4
Anzio Italy 26 E4
Aoba i. Vanuatu 53 G3
Aoga-shima i. Japan 45 E6
Aomen China see Macao
Aomori Japan 44 F4
Aoraki mt. N.Z. 59 C6
Aoraki/Mount Cook National Park N.Z. 59 C6
Aorangi mt. N.Z. see Aoraki
Aosta Italy 26 B2
Aotearoa country Oceania see New Zealand
Aouk, Bahr r. Cent. Afr. Rep./Chad 47 E4
Aoukâr reg. Mali/Mauritania 46 C2
Aoulef Alg. 46 D2
Aozou Chad 47 E2
Apa r. Brazil 70 E2
Apaiang atoll Kiribati see Abaiang
Apalachee Bay U.S.A. 63 K6
Apalachin NY U.S.A. 64 C1
Apamea Turkey see Dinar
Apaporis r. Col. 68 E4
Aparecida do Tabuado Brazil 71 A3
Aparima N.Z. see Riverton
Aparri Phil. 74 E4
Apatity Rus. Fed. 14 R3
Apatzingán Mex. 66 D5
Ape Latvia 15 O8
Apeldoorn Neth. 17 J4
Apennines mts Italy 26 C2
Api mt. Nepal 36 E3
Api i. Vanuatu see Epi
Apia atoll Kiribati see Abaiang
Apia Samoa 53 I3
Apiacas, Serra dos hills Brazil 69 G6
Apiaí Brazil 71 A4
Apiti N.Z. 59 E4
Aplao Peru 68 D7
Apoera Suriname 69 G2
Apollo Bay Australia 58 A7
Apollonia Bulg. see Sozopol
Apolo Bol. 68 E6
Aporé Brazil 71 A2
Aporé r. Brazil 71 A2
Apostolens Tommelfinger mt. Greenland 61 N3
Apostolos Andreas, Cape Cyprus 39 B2
Apoteri Guyana 69 G3
Apozai Pak. 36 B3
Appalachian Mountains U.S.A. 63 K4
Appalla i. Fiji see Kabara
Appennino mts Italy see Apennines
Appennino Abruzzese mts Italy 26 E3
Appennino Tosco-Emiliano mts Italy 26 D3
Appennino Umbro-Marchigiano mts Italy 26 E3
Applecross U.K. 20 D3
Appleton WI U.S.A. 63 J3
Apple Valley CA U.S.A. 65 D3
Appomattox VA U.S.A. 64 B4
Aprilia Italy 26 E4
Apsheronsk Rus. Fed. 13 H7
Apsheronskaya Rus. Fed. see Apsheronsk
Apt France 24 G5
Apucarana Brazil 71 A3
Apucarana, Serra da hills Brazil 71 A3
Apuka Rus. Fed. 29 R3
Apulum Romania see Alba Iulia
Aq"a Georgia see Sokhumi
'Aqaba Jordan see Al 'Aqabah
Aqaba, Gulf of Asia 34 D5
'Aqaba, Wâdi el watercourse Egypt see 'Aqabah, Wādī al
'Aqabah, Wādī al watercourse Egypt 39 A4
Aqadyr Kazakh. see Agadyr'
Aqdoghmish r. Iran 35 G3
Aqköl Akmolinskaya Oblast' Kazakh. see Akkol'
Aqköl Atyrauskaya Oblast' Kazakh. see Akkol'
Aqmola Kazakh. see Astana
Aqqan China 37 F1
Aqqikkol Hu salt l. China 37 G1
Aqra', Jabal al mt. Syria/Turkey 39 B2
'Aqran h. Saudi Arabia 39 D4
Aqsay Kazakh. see Aksay
Aqsayqin Hit terr. Asia see Aksai Chin
Aqshuqyr Kazakh. see Akshukur
Aqsū Kazakh. see Aksu
Aqsüat Kazakh. see Aksuat
Aqsū-Ayuly Kazakh. see Aksu-Ayuly
Aqtaū Kazakh. see Aktau
Aqtöbe Kazakh. see Aktobe
Aqtoghay Kazakh. see Aktogay
Aquae Grani Germany see Aachen
Aquae Gratianae France see Aix-les-Bains
Aquae Sextiae France see Aix-en-Provence
Aquae Statiellae Italy see Acqui Terme
Aquarius Mountains AZ U.S.A. 65 F3
Aquaviva delle Fonti Italy 26 G4
Aquidauana Brazil 70 E2
Aquincum Hungary see Budapest
Aquiry r. Brazil see Acre
Aquisgranum Germany see Aachen
Aquitaine reg. France 24 D5
Aquitania reg. France see Aquitaine
Aqzhayqyn Köli salt l. Kazakh. see Akzhaykyn, Ozero
Ara India 37 F4
Āra Ārba Eth. 48 E3
Arab, Bahr el watercourse Sudan 47 F4
'Arab, Khalîg el b. Egypt see 'Arab, Khalīj al
'Arab, Khalīj al b. Egypt 34 C5
'Arabah, Wādī al watercourse Israel/Jordan 39 B5
Arabian Basin sea feature Indian Ocean 73 M5
Arabian Gulf Asia see The Gulf
Arabian Peninsula Asia 32 G5
Arabian Sea Indian Ocean 33 K6
Araç Turkey 34 D2
Araça r. Brazil 68 F4
Aracaju Brazil 69 K6
Aracati Brazil 69 K4
Aracatu Brazil 71 C1
Araçatuba Brazil 71 A3
Aracena Spain 25 C5
Aracruz Brazil 71 C2
Araçuaí Brazil 71 C2
Araçuaí r. Brazil 71 C2
'Arad Israel 39 B4
Arad Romania 27 I1
Arafura Sea Australia/Indon. 52 D2
Arafura Shelf sea feature Australia/Indon. 74 E6
Aragarças Brazil 69 H7
Aragón r. Spain 25 F2
Araguaçu Brazil 71 A1
Araguaia r. Brazil 71 A1
Araguaia, Parque Nacional do nat. park Brazil 69 H6
Araguaiana Brazil 71 A1
Araguaína Brazil 69 I5
Araguari Brazil 71 A2
Araguari r. Brazil 69 H3
Araguatins Brazil 69 I5
Arai Brazil 71 B1
'Arâif el Naga, Gebel h. Egypt see 'Urayf an Nāqah, Jabal
Araioses Brazil 69 J4
Arak Alg. 46 D2
Arāk Iran 35 H4
Arak Syria 39 D2
Arakan Yoma mts Myanmar 37 H5
Arakkonam India 38 C3
Araks r. Azer. see Araz
Araku India 38 D2
Aral Kazakh. see Aral'sk
Aral Tajik. see Vose
Aral Sea salt l. Kazakh./Uzbek. 30 F2
Aral'sk Kazakh. 28 H5
Aral'skoye More salt l. Kazakh./Uzbek. see Aral Sea
Aralsor, Ozero l. Kazakh. 13 K6
Aral Tengizi salt l. Kazakh./Uzbek. see Aral Sea
Aramac Australia 56 D4
Aramac Creek watercourse Australia 56 D4
Aran r. India 38 C2
Aranda de Duero Spain 25 E3
Aranđelovac Serbia 27 I2
Arandis Namibia 50 B2
Arang India 37 E5
Arani India 38 C3
Aran Islands Ireland 21 C4
Aranjuez Spain 25 E3
Aranos Namibia 50 D3
Aransas Pass U.S.A. 63 H6
Arantangi India 38 C4
Aranuka atoll Kiribati 53 H1
Arao Japan 45 C6
Araouane Mali 46 C3
Arapgir Turkey 34 E3
Arapiraca Brazil 69 K5
Arapis, Akrotirio pt Greece 27 K4
Arapkir Turkey see Arapgir
Arapongas Brazil 71 A3
Araquari Brazil 71 A4
'Ar'ar Saudi Arabia 35 F5
Araracuara Col. 68 D4
Araranguá Brazil 71 A5
Araraquara Brazil 71 A3
Araras Brazil 69 H5
Ararat Armenia 35 G3
Ararat Australia 58 A6
Ararat, Mount Turkey 35 G3
Araria India 37 F4
Araripina Brazil 69 J5
Aras r. Azer. see Araz
Aras Turkey 35 F3
Arataca Brazil 71 D1
Arauca Col. 68 D2
Arauca r. Venez. 68 E2
Aravalli Range mts India 36 C4
Aravete Estonia 15 N7
Arawa P.N.G. 52 F2
Araxá Brazil 71 B2
Araxes r. Azer. see Araz
Arayıt Dağı mt. Turkey 27 N5
Araz r. Azer. 35 H2
Arbailu Iraq see Arbīl
Arbat Iraq 35 G4
Arbela Iraq see Arbīl
Arberth U.K. see Narberth
Arbīl Iraq 35 G3
Arboga Sweden 15 I7
Arbroath U.K. 20 G4
Arbuckle CA U.S.A. 65 A1
Arcachon France 24 D4
Arcade NY U.S.A. 64 B1
Arcadia FL U.S.A. 63 K6
Arc Dome mt. NV U.S.A. 65 D1
Arcelia Mex. 66 D5
Archangel Rus. Fed. 12 I2
Archer r. Australia 41 G9
Archer Bend National Park Australia 56 C2
Archipiélago Los Roques, Parque Nacional nat. park Venez. 68 E1
Ärçivan Azer. 35 H3
Arckaringa watercourse Australia 57 A6
Arco U.S.A. 62 E3
Arcos Brazil 71 B3
Arcos de la Frontera Spain 25 D5
Arctic Bay Canada 61 J2
Arctic Institute Islands Rus. Fed. see Arkticheskogo Instituta, Ostrova
Arctic Mid-Ocean Ridge sea feature Arctic Ocean 77 H1
Arctic Ocean 77 B1
Arctic Red r. Canada 60 E3
Arctowski research stn Antarctica 76 A2
Arda r. Bulg. 27 K4
Ardabīl Iran 35 H3
Ardahan Turkey 35 F2
Ardakān Iran 35 I4
Årdalstangen Norway 15 E6
Ardara Ireland 21 D3
Ardas r. Bulg. see Arda
Arḍ aş Şawwān plain Jordan 39 C4
Ardatov Nizhegorodskaya Oblast' Rus. Fed. 13 I5
Ardatov Respublika Mordoviya Rus. Fed. 13 J5
Ardee Ireland 21 F4
Ardennes plat. Belgium 16 J6
Arden Town CA U.S.A. 65 B1
Arderin h. Ireland 21 E4
Ardestān Iran 35 I4
Ardglass U.K. 21 G3
Ardila r. Port. 25 C4
Ardlethan Australia 58 C5
Ardmore U.S.A. 63 H5
Ardnamurchan, Point of U.K. 20 C4
Ardon Rus. Fed. 35 G2
Ardrishaig U.K. 20 D4
Ardrossan U.K. 20 E5
Ardvasar U.K. 20 D3
Areia Branca Brazil 69 K4
Arel Belgium see Arlon
Arelas France see Arles
Arelate France see Arles
Arena, Point U.S.A. 62 C4
Arenas de San Pedro Spain 25 D3
Arendal Norway 15 F7
Areopoli Greece 27 J6
Arequipa Peru 68 D7
Arere Brazil 69 H4
Arévalo Spain 25 D3
Arezzo Italy 26 D3
'Arfajah well Saudi Arabia 39 D4
Argadargada Australia 56 B4
Arganda del Rey Spain 25 E3
Argel Alg. see Algiers
Argentan France 24 D2
Argentario, Monte h. Italy 26 D3
Argentera, Cima dell' mt. Italy 26 B2
Argentina country S. America 70 C5
Argentine Abyssal Plain sea feature S. Atlantic Ocean 72 E9
Argentine Basin sea feature S. Atlantic Ocean 72 F8
Argentine Republic country S. America see Argentina
Argentine Rise sea feature S. Atlantic Ocean 72 E8
Argentino, Lago l. Arg. 70 B8
Argenton-sur-Creuse France 24 E3
Argentoratum France see Strasbourg
Argeș r. Romania 27 L2
Arg-e Zārī Afgh. 36 A2
Argi r. Rus. Fed. 44 C1
Argolikos Kolpos b. Greece 27 J6
Argos Greece 27 J6
Argostoli Greece 27 I5
Arguis Spain 25 F2
Argun' r. China/Rus. Fed. 43 M2
Argun Rus. Fed. 35 G2
Argungu Nigeria 46 D3
Argus Range mts CA U.S.A. 65 D3
Argyle, Lake Australia 54 E4
Argyrokastron Albania see Gjirokastër
Ar Horqin Qi China see Tianshan
Århus Denmark 15 G8
Ariah Park Australia 58 C5
Ariamsvlei Namibia 50 D5
Ariana Tunisia see L'Ariana
Ariano Irpino Italy 26 F4
Aribinda Burkina Faso 46 C3
Arica Chile 68 D7
Arid, Cape Australia 55 C8
Arīḥā Syria 39 C2
Arīḥā West Bank see Jericho
Arima Trin. and Tob. 67 L6
Ariminum Italy see Rimini
Arinos Brazil 71 B1
Aripuanã Brazil 69 G6
Aripuanã r. Brazil 68 F5
Ariquemes Brazil 68 F5
Aris Namibia 50 C2
Arisaig U.K. 20 D4
Arisaig, Sound of sea chan. U.K. 20 D4
'Arîsh, Wâdi el watercourse Egypt 39 A4
Arixang China see Wenquan
Ariyalur India 38 C4
Arizaro, Salar de salt flat Arg. 70 C2
Arizona Arg. 70 C5
Arizona state U.S.A. 62 E5
Arizpe Mex. 62 E5
'Arjah Saudi Arabia 32 F5
Arjeplog Sweden 14 J3
Arjuni Chhattisgarh India 38 D1
Arjuni India 36 E5
Arkadak Rus. Fed. 13 I6
Arkadelphia U.S.A. 63 I5
Arkaig, Loch l. U.K. 20 D4
Arkalyk Kazakh. 42 B2
Arkansas r. U.S.A. 63 I5
Arkansas state U.S.A. 63 I4
Arkansas City KS U.S.A. 63 H4
Arkatag Shan mts China 37 G1
Arkenu, Jabal mt. Libya 32 B5
Arkhangel'sk Rus. Fed. see Archangel
Arkhara Rus. Fed. 44 C2
Arkhipovka Rus. Fed. 44 D4
Árki i. Greece see Arkoi
Arklow Ireland 21 F5
Arkoi i. Greece 27 L6
Arkona Ont. Canada 64 A1
Arkona, Kap c. Germany 17 N3
Arkonam India see Arakkonam
Arkport NY U.S.A. 64 C1
Arkticheskogo Instituta, Ostrova is Rus. Fed. 28 J2
Arkul' Rus. Fed. 12 K4
Arlandag mt. Turkm. 35 I3
Arles France 24 G5
Arlington S. Africa 51 H5
Arlington NY U.S.A. 64 E2
Arlington VA U.S.A. 64 C3
Arlit Niger 46 D3
Arlon Belgium 17 J6
Armadale Australia 55 A8
Armagh U.K. 21 F3
Armant Egypt 32 D4
Armavir Armenia 35 G2
Armavir Rus. Fed. 13 I7
Armenia country Asia 35 G2
Armenia Col. 68 C3
Armenopolis Romania see Gherla
Armeria Mex. 66 D5
Armidale Australia 58 E3
Armori India 38 D1
Armoy U.K. 21 F2
Armstrong r. Australia 54 E4
Armstrong Island Cook Is see Rarotonga
Armu r. Rus. Fed. 44 E3
Armur India 38 C2
Armutçuk Dağı mts Turkey 27 L5
Armyanskaya S.S.R. country Asia see Armenia
Arnaoutis, Cape Cyprus see Arnauti, Cape
Arnauti, Cape Cyprus 39 A2
Årnes Norway 15 G6
Arnhem Neth. 17 J5
Arnhem, Cape Australia 56 B2
Arnhem Land reg. Australia 54 F3
Arno r. Italy 26 D3
Arno Bay Australia 57 B7
Arnold U.K. 19 F5
Arnon r. Jordan see Mawjib, Wādī al
Arnprior Canada 63 L2
Arnsberg Germany 17 L5
Aroab Namibia 50 D4
Aroma Sudan 32 E6
Arona Italy 26 C2
Arorae i. Kiribati 53 H2
Arore i. Kiribati see Arorae
Aros r. Mex. 62 F6
Arossi i. Solomon Is see San Cristobal
Arqalyq Kazakh. see Arkalyk
Arquipélago da Madeira aut. reg. Port. 46 B1
Arrabury Australia 57 C5
Arrah India see Ara
Arraias Brazil 71 B1
Arraias, Serra de hills Brazil 71 B1
Ar Ramādī Iraq 35 F4
Ar Ramlah Jordan 39 B5
Ar Ramthā Jordan 39 C3
Arran i. U.K. 20 D5
Arranmore Island Ireland 21 D3
Ar Raqqah Syria 39 D2
Arras France 24 F1
Ar Rass Saudi Arabia 32 F4
Ar Rastān Syria 39 C2
Ar Rayyān Qatar 32 H4
Arrecife Canary Is 46 B2
Arretium Italy see Arezzo
Arriagá Mex. 66 F5
Ar Rifā'ī Iraq 35 G5
Ar Riḥāb salt flat Iraq 35 G5
Ar Rimāl reg. Saudi Arabia 48 F1
Arrington VA U.S.A. 64 B4
Ar Riyāḍ Saudi Arabia see Riyadh
Arrochar U.K. 20 E4
Arrojado r. Brazil 71 B1
Arrow, Lough l. Ireland 21 D3
Arrowsmith, Mount N.Z. 59 C6
Arroyo Grande CA U.S.A. 65 B3
Ar Rummān Jordan 39 B3
Ar Ruq'ī well Saudi Arabia 35 G5
Ar Ruṣāfah Syria 39 D2
Ar Ruṣayfah Jordan 39 C3
Ar Rustāq Oman 33 I5
Ar Ruṭbah Iraq 35 F4
Ar Ruwayḍah Syria 39 C2
Års Denmark see Aars
Ars Iran 35 G3
Arsen'yev Rus. Fed. 44 D3
Arsk Rus. Fed. 12 K4
Arta Greece 27 I5
Artem Rus. Fed. 44 D4
Artemivs'k Ukr. 13 H6
Artemovsk Ukr. see Artemivs'k
Artenay France 24 E2
Artesia NM U.S.A. 62 G5
Arthur, Lake PA U.S.A. 64 A2
Arthur's Pass National Park N.Z. 59 C6
Arti Rus. Fed. 11 R4
Artigas Uruguay 70 E4
Artigas (Uruguay) research stn Antarctica 76 A2
Art'ik Armenia 35 F2
Artillery Lake Canada 60 H3
Artisia Botswana 51 H3
Artois reg. France 24 E1
Artos Dağı mt. Turkey 35 F3
Artova Turkey 34 E2
Artsakh aut. reg. Azer. see Dağlıq Qarabağ
Artsiz Ukr. see Artsyz
Artsyz Ukr. 27 M2
Artur de Paiva Angola see Kuvango
Arturo Prat (Chile) research stn Antarctica 76 A2
Artux China 42 D5
Artvin Turkey 35 F2
Aru, Kepulauan is Indon. 41 F8
Arua Uganda 48 D3
Aruanã Brazil 71 A1
Aruba terr. West Indies 67 K6
Arumã Brazil 68 F4
Arunachal Pradesh state India 37 H4
Arundel U.K. 19 G8
Arun Gol r. China 44 B3
Arun He r. China see Arun Gol
Arun Qi China see Naji
Aruppukkottai India 38 C4
Arusha Tanz. 48 D4
Aruwimi r. Dem. Rep. Congo 48 C3
Arvagh Ireland 21 E4
Arvayheer Mongolia 42 I3
Arviat Canada 61 I3
Arvidsjaur Sweden 14 K4
Arvika Sweden 15 H7
Arvonia VA U.S.A. 64 B4
Arwād i. Syria 39 B2
Arwala Indon. 54 D1
Arxan China 43 L3
Aryanah Tunisia see L'Ariana
Arys Kazakh. 42 B4
Arzamas Rus. Fed. 13 I5
Arzew Alg. 25 F6
Arzgir Rus. Fed. 35 G1
Arzila Morocco see Asilah
Asaba Nigeria 46 D4
Asad, Buḥayrat al resr Syria 39 D1
Asadābād Afgh. 36 B2
Asadābād Iran 35 H4
Asahi-dake vol. Japan 44 F4
Asahikawa Japan 44 F4
'Asal Egypt 39 A5
Āsalē l. Eth. 48 E2
Asālem Iran 35 H3
'Asalūyeh Iran 35 I6
Asan-man b. S. Korea 45 B5
Asansol India 37 F5
Āsayita Eth. 48 E2
Asbestos Mountains S. Africa 50 F5
Asbury Park NJ U.S.A. 64 D2
Ascalon Israel see Ashqelon
Ascea Italy 26 F4
Ascensión Bol. 68 F7
Ascension atoll Micronesia see Pohnpei
Ascension i. S. Atlantic Ocean 72 H6
Aschaffenburg Germany 17 L6
Ascoli Piceno Italy 26 E3
Asculum Italy see Ascoli Piceno
Asculum Picenum Italy see Ascoli Piceno
Ascutney VT U.S.A. 64 E1
Āseb Eritrea see Assab
Åseda Sweden 15 I8
Åsele Sweden 14 J4
Asenovgrad Bulg. 27 K3
Aṣfar, Jabal al mt. Jordan 39 C3
Aṣfar, Tall al h. Syria 39 C3
Aşgabat Turkm. see Aşgabat
Aşgabat Turkm. 30 E3
Asha Rus. Fed. 11 R5
Ashburton watercourse Australia 54 A5
Ashburton N.Z. 59 C6
Ashburton Range hills Australia 54 F4
Ashdod Israel 39 B4
Asheville U.S.A. 63 K4
Ashford Australia 58 E2
Ashford U.K. 19 H7
Ashibetsu Japan 44 F4
Ashikaga Japan 45 E5
Ashington U.K. 18 F3
Ashizuri-misaki pt Japan 45 D6
Ashkelon Israel see Ashqelon
Ashkhabad Turkm. see Aşgabat
Ashland OR U.S.A. 62 C3
Ashland VA U.S.A. 64 C4
Ashland WI U.S.A. 63 I2
Ashley Australia 58 D2
Ashmore and Cartier Islands terr. Australia 54 C3
Ashmore Reef Australia 54 C3
Ashmore Reefs Australia 56 D1
Ashmyany Belarus 15 N9
Ashqelon Israel 39 B4
Ash Shabakah Iraq 35 F5
Ash Shaddādah Syria 35 F3
Ash Shallūfah Egypt 39 A4
Ash Sham Syria see Damascus
Ash Shanāfīyah Iraq 35 G5
Ash Shaqiq well Saudi Arabia 35 F5
Ash Sharāh reg. Jordan 39 B4
Ash Sharawrah Saudi Arabia 32 G6
Ash Shāriqah U.A.E. see Sharjah
Ash Sharqāṭ Iraq 35 F4
Ash Shaṭrah Iraq 35 G5
Ash Shaṭṭ Egypt 39 A5
Ash Shawbak Jordan 39 B4
Ash Shaybānī well Saudi Arabia 35 F5
Ash Shaykh Ibrāhīm Syria 39 D2
Ash Shiblīyāt h. Saudi Arabia 39 C5
Ash Shiḥr Yemen 32 G7
Ash Shu'aybah Saudi Arabia 35 F6
Ash Shu'bah Saudi Arabia 32 F4
Ash Shuruf Saudi Arabia see Khaybar
Ashta India 36 D5
Ashtabula OH U.S.A. 64 A2
Ashtarak Armenia 35 G2
Ashti Mahar. India 36 D5
Ashti Mahar. India 38 B2
Ashti Mahar. India 38 C2
Ashtiān Iran 35 H4
Ashton S. Africa 50 E7
Ashton-under-Lyne U.K. 18 E5
Ashuanipi Lake Canada 61 L4
Ashur Iraq see Ash Sharqāṭ
Asi r. Asia 34 E3 see 'Āṣī, Nahr al
'Āṣī r. Lebanon/Syria see Orontes
'Āṣī, Nahr al r. Asia 34 E3
Āsīā Bak Iran 35 H4
Asifabad India 38 C2
Asika India 38 E2
Asilah Morocco 25 C6
Asinara, Golfo dell' b. Sardegna Italy 26 C4
Asino Rus. Fed. 28 J4
Asipovichy Belarus 13 F5
Asīr Iran 35 I6
'Asīr reg. Saudi Arabia 32 F5
Asisium Italy see Assisi
Askale Pak. 36 C2
Aşkale Turkey 35 F3
Asker Norway 15 G7
Askersund Sweden 15 I7
Askim Norway 15 G7
Askī Mawşil Iraq 35 F3
Askino Rus. Fed. 11 R4
Askival h. U.K. 20 C4
Asl Egypt see 'Asal
Aslanköy r. Turkey 39 B1
Asmara Eritrea 32 E6
Āsmera Eritrea see Asmara
Åsnen l. Sweden 15 I8
Aso-Kuju Kokuritsu-kōen Japan 45 C6
Asop India 36 C4
Āsosa Eth. 48 D2
Aspang-Markt Austria 17 P7
Aspatria U.K. 18 D4
Aspen U.S.A. 62 F4
Aspiring, Mount N.Z. 59 B7
Aspro, Cape Cyprus 39 A2
Aspromonte, Parco Nazionale dell' nat. park Italy 26 F5
Aspron, Cape Cyprus see Aspro, Cape
As Sa'an Syria 39 C2
Assab Eritrea 32 F7
Assad, Lake resr Syria see Asad, Buḥayrat al
Aş Şafā lava field Syria 39 C3
Aş Şafāqis Tunisia see Sfax
Aş Şaff Egypt 34 C5
As Safīrah Syria 39 C1
Aş Şaḥrā' al Gharbīyah des. Egypt see Western Desert
Aş Şaḥrā' ash Sharqīyah des. Egypt see Eastern Desert
Assake-Audan, Vpadina depr. Kazakh./Uzbek. 35 J2
'Assal, Lac l. Djibouti see Assal, Lake
Assal, Lake Djibouti 32 F7
Aş Şāliḥīyah Syria 35 F4
As Sallūm Egypt 34 B5
As Salmān Iraq 35 G5
As Salṭ Jordan 39 B3
Assam state India 37 G4
Assamakka Niger 46 D3
As Samāwah Iraq 35 G5
As Samrā' Jordan 39 C3
Aş Şanām reg. Saudi Arabia 32 H5
As Sarīr reg. Libya 47 F2
Assateague Island MD U.S.A. 64 D3
Assayeta Eth. see Āsayita
Assen Neth. 17 K4
As Sidrah Libya 47 E1
Assiniboia Canada 62 F2
Assiniboine r. Canada 62 H2
Assiniboine, Mount Canada 60 G4
Assis Brazil 71 A3
Assisi Italy 26 E3
Aş Şubayḥīyah Kuwait 35 G5
Aş Şufayrī well Saudi Arabia 35 G5
As Sukhnah Syria 39 D2
As Sulaymānīyah Iraq 35 G4
As Sulaymī Saudi Arabia 32 F4
As Sūq Saudi Arabia 32 F5
As Sūrīyah country Asia see Syria
Aş Şuwar Syria 35 F4
As Suwaydā' Syria 39 C3
As Suways Egypt see Suez
As Suways governorate Egypt 39 A4
Assynt, Loch l. U.K. 20 D2
Astacus Kocaeli Turkey see İzmit
Astakida i. Greece 27 L7
Astakos Greece 27 I5
Astana Kazakh. 42 C2
Astaneh Iran 35 H3
Astara Azer. 35 H3
Āstārā Iran 32 G2
Asterabad Iran see Gorgān
Asti Italy 26 C2
Astillero Peru 68 E6
Astin Tag mts China see Altun Shan
Astipálaia i. Greece see Astypalaia
Astor r. Pak. 36 C2
Astorga Spain 25 C2
Astoria U.S.A. 62 C2
Åstorp Sweden 15 H8
Astrabad Iran see Gorgān
Astrakhan' Rus. Fed. 13 K7
Astrakhan' Bazar Azer. see Cälilabad
Astravyets Belarus 15 N9
Astrida Rwanda see Butare
Asturias aut. comm. Spain 25 C2
Asturias, Principado de aut. comm. Spain see Asturias
Asturica Augusta Spain see Astorga
Astypalaia i. Greece 27 L6
Asunción Para. 70 E3
Aswān Egypt see Aswân
Aswân Egypt 32 D5
Asyūṭ Egypt 34 C6
Asyût Egypt see Asyūṭ
Ata i. Tonga 53 I4
Atacama, Desierto de des. Chile see Atacama Desert
Atacama, Salar de salt flat Chile 70 C2
Atacama Desert Chile 70 C3
Atafu atoll Tokelau 53 I2
Atafu i. Tokelau 74 I6
'Aṭā'iṭah, Jabal al mt. Jordan 39 B4
Atakent Turkey 39 B1
Atakpamé Togo 46 D4
Atalándi Greece see Atalanti
Atalanti Greece 27 J5
Atalaya Peru 68 D6
Ataléia Brazil 71 C2
Atambua Indon. 54 D2
Atamyrat Turkm. 30 F3
Ataniya Turkey see Adana
Atanur India 38 C2
'Ataq Yemen 32 G7
Atâr Mauritania 46 B2
Atari Pak. 36 C3
Atascadero CA U.S.A. 65 B3
Ātashān Iran 35 I4
Atasu Kazakh. 42 C3
Ataúro, Ilha de i. East Timor 54 D2
Atáviros mt. Greece see Attavyros
Atayurt Turkey 39 A1
Atbara Sudan 32 D6
Atbara r. Sudan 32 D6
Atbasar Kazakh. 28 H4

Atchison U.S.A. 63 H4
Atebubu Ghana 46 C4
Ateransk Kazakh. see Atyrau
Atessa Italy 26 F3
Athabasca r. Canada 60 G4
Athabasca, Lake Canada 60 H4
Atharan Hazari Pak. 36 C3
Athboy Ireland 21 F4
Athenae Greece see Athens
Athenry Ireland 21 D4
Athens Greece 27 J6
Athens GA U.S.A. 63 K5
Athens OH U.S.A. 63 K4
Athens PA U.S.A. 64 C2
Athens TN U.S.A. 63 K4
Atherstone U.K. 19 F6
Atherton Australia 56 D3
Athina Greece see Athens
Athínai Greece see Athens
Athleague Ireland 21 D4
Athlone Ireland 21 E4
Athnā', Wādī al watercourse Jordan 39 D3
Athni India 38 B2
Athol N.Z. 59 B7
Athol MA U.S.A. 64 E1
Atholl, Forest of reg. U.K. 20 E4
Athos mt. Greece 27 K4
Ath Thamad Egypt 39 B5
Ath Thāyat mt. Saudi Arabia 39 C5
Ath Thumāmī well Saudi Arabia 35 G6
Athy Ireland 21 F5
Ati Chad 47 E3
Atico Peru 68 D7
Atikokan Canada 61 I5
Atka Rus. Fed. 29 Q3
Atka Island U.S.A. 60 A4
Atkarsk Rus. Fed. 13 J6
Atlanta GA U.S.A. 63 K5
Atlantic IA U.S.A. 63 H3
Atlantic City NJ U.S.A. 64 D3
Atlantic-Indian-Antarctic Basin sea feature S. Atlantic Ocean 72 H10
Atlantic-Indian Ridge sea feature Southern Ocean 72 H9
Atlantic Ocean 72
Atlantis S. Africa 50 D7
Atlas Méditerranéen mts Alg. see Atlas Tellien
Atlas Mountains Africa 22 C5
Atlas Saharien mts Alg. 22 E5
Atlas Tellien mts Alg. 25 H6
Atmakur India 38 C3
Atmore U.S.A. 63 J5
Atocha Bol. 68 E8
Atouila, Erg des. Mali 46 C2
Atqan China see Aqqan
Atrak r. Iran/Turkm. see Atrek
Atrato r. Col. 68 C2
Atrek r. Iran/Turkm. 33 H2
Atrek r. Iran/Turkm. 33 I2
Atropatene country Asia see Azerbaijan
Atsonupuri vol. Rus. Fed. 44 G3
Aţ Ţafīlah Jordan 39 B4
Aţ Ţā'if Saudi Arabia 32 F5
Attalea Turkey see Antalya
Attalia Turkey see Antalya
At Tamīmī Libya 34 A4
Attavyros mt. Greece 27 L6
Attawapiskat Canada 63 K1
Attawapiskat r. Canada 63 K1
Attawapiskat Lake Canada 63 J1
Aţ Ţawīl mts Saudi Arabia 35 E5
At Taysīyah plat. Saudi Arabia 35 F5
Attersee l. Austria 17 N7
Attila Line Cyprus 39 A2
Attleborough U.K. 19 I6
Attu Greenland 61 M3
Aţ Ţubayq reg. Saudi Arabia 39 C5
Attu Island U.S.A. 29 S4
At Tūnisīyah country Africa see Tunisia
Aţ Ţūr Egypt 34 D5
Attur India 38 C4
Aţ Ţuwayyah well Saudi Arabia 35 F6
Atuk Mountain h. U.S.A. 60 A3
Åtvidaberg Sweden 15 I7
Atwater CA U.S.A. 65 B2
Atyashevo Rus. Fed. 13 J5
Atyrau Kazakh. 30 E2
Atyraū admin. div. Kazakh. see Atyrauskaya Oblast'
Atyrau Oblast admin. div. Kazakh. see Atyrauskaya Oblast'
Atyrauskaya Oblast' admin. div. Kazakh. 11 Q6
Aubagne France 24 G5
Aubenas France 24 G4
Aubrey Cliffs mts AZ U.S.A. 65 F3
Aubry Lake Canada 60 F3
Auburn r. Australia 57 E5
Auburn CA U.S.A. 65 B1
Auburn NE U.S.A. 63 H3
Auburn NY U.S.A. 64 C1
Auburn Range hills Australia 56 E5
Aubusson France 24 F4
Auch France 24 E5
Auchterarder U.K. 20 F4
Auckland N.Z. 59 E3
Auckland Islands N.Z. 53 G7
Audo mts Eth. 48 E3
Audo Range mts Eth. see Audo
Augathella Australia 57 D5
Augher U.K. 21 E3
Aughnacloy U.K. 21 F3
Aughrim Ireland 21 F5
Augrabies S. Africa 50 E5
Augrabies Falls S. Africa 50 E5
Augrabies Falls National Park S. Africa 50 E5
Augsburg Germany 17 M6
Augusta Australia 55 A8
Augusta Sicilia Italy 26 F6
Augusta GA U.S.A. 63 K5
Augusta ME U.S.A. 63 N3
Augusta Auscorum France see Auch
Augusta Taurinorum Italy see Turin
Augusta Treverorum Germany see Trier
Augusta Vindelicorum Germany see Augsburg
Augusto de Lima Brazil 71 B2
Augustus, Mount Australia 55 B6
Aukštaitijos nacionalinis parkas nat. park Lith. 15 O9
Aulavik National Park Canada 60 G2
Auld, Lake imp. l. Australia 54 C5
Auliye Ata Kazakh. see Taraz
Aulon Albania see Vlorë
Ault France 19 I8
Aumale Alg. see Sour el Ghozlane
Aundh India 38 B2
Aundhi India 38 D1
Aunglan Myanmar 37 H6
Auob watercourse Namibia/S. Africa 50 E4
Aura Fin. 15 M6
Auraiya India 36 D4
Aurangabad Bihar India 37 F4
Aurangabad Mahar. India 38 B2
Aure r. France 19 F9
Aurich Germany 17 K4
Aurigny i. Channel Is see Alderney
Aurilândia Brazil 71 A2
Aurillac France 24 F4
Aurora CO U.S.A. 62 G4
Aurora IL U.S.A. 63 J3
Aurora Island Vanuatu see Maéwo
Aurukun Australia 56 C2
Aus Namibia 50 C4
Auskerry i. U.K. 20 G1
Austin MN U.S.A. 63 I3
Austin NV U.S.A. 62 D4
Austin TX U.S.A. 62 H5
Austin, Lake imp. l. Australia 55 B6
Austintown OH U.S.A. 64 A2
Austral Downs Australia 56 B4
Australes, Îles is Fr. Polynesia see Tubuai Islands
Australia country Oceania 52 C4
Australian-Antarctic Basin sea feature S. Atlantic Ocean 74 C9
Australian Antarctic Territory Antarctica 76 G2
Australian Capital Territory admin. div. Australia 58 D5
Austria country Europe 17 N7
Austvågøy i. Norway 14 I2
Autazes Brazil 69 G4
Autesiodorum France see Auxerre
Autti Fin. 14 O3
Auvergne reg. France 24 F4
Auvergne, Monts d' mts France 24 F4
Auxerre France 24 F3
Auxonne France 24 G3
Auyuittuq National Park Canada 61 L3
Auzangate, Nevado mt. Peru 68 D6
Ava NY U.S.A. 64 D1
Avallon France 24 F3
Avalon CA U.S.A. 65 C4
Avalon Peninsula Canada 61 M5
Avān Iran 35 G3
Avarau atoll Cook Is see Palmerston
Avaré Brazil 71 A3
Avaricum France see Bourges
Avarua Cook Is 75 J7
Aveiro Port. 25 B3
Aveiro, Ria de est. Port. 25 B3
Āvej Iran 35 H4
Avellino Italy 26 F4
Avenal CA U.S.A. 65 B2
Avenio France see Avignon
Aversa Italy 26 F4
Avesta Sweden 15 J6
Aveyron r. France 24 E4
Avezzano Italy 26 E3
Aviemore U.K. 20 F3
Avignon France 24 G5
Ávila Spain 25 D3
Avilés Spain 25 D2
Avis PA U.S.A. 64 C2
Avlama Dağı mt. Turkey 39 A1
Avlama Dağı mts Turkey 39 A1
Avlona Albania see Vlorë
Avnyugskiy Rus. Fed. 12 J3
Avoca Australia 58 A6
Avoca r. Australia 58 A5
Avoca Ireland 21 F5
Avoca NY U.S.A. 64 C1
Avola Sicilia Italy 26 F6
Avon r. England U.K. 19 E6
Avon r. England U.K. 19 E7
Avon r. England U.K. 19 F8
Avon r. Scotland U.K. 20 F3
Avon NY U.S.A. 64 C1
Avonmore r. Ireland 21 F5
Avonmore PA U.S.A. 64 B2
Avonmouth U.K. 19 E7
Avranches France 24 D2
Avsuyu Turkey 39 C1
Avuavu Solomon Is 53 G2
Avveel Fin. see Ivalo
Avvil Fin. see Ivalo
A'waj r. Syria 39 B3
Awakino N.Z. 59 E4
Awanui N.Z. 59 D2
Āwarē Eth. 48 E3
'Awāriḍ, Wādī al watercourse Syria 39 D2
Awarua Point N.Z. 59 B7
Āwash Eth. 48 E3
Āwash r. Eth. 48 E2
Awa-shima i. Japan 45 E5
Āwash National Park Eth. 48 D3
Awasib Mountains Namibia 50 B3
Awat China 42 E4
Awatere r. N.Z. 59 E5
Awbārī Libya 46 E2
Awbeg r. Ireland 21 D5
'Awdah, Hawr al imp. l. Iraq 35 G5
Aw Dheegle Somalia 47 H4
Awe, Loch l. U.K. 20 D4
Aweil South Sudan 47 F4
Awka Nigeria 46 D4
Awserd W. Sahara 46 B2
Axe r. England U.K. 19 D8
Axe r. England U.K. 19 E7
Axedale Australia 58 B6
Axel Heiberg Glacier Antarctica 76 I1
Axel Heiberg Island Canada 61 I2
Axim Ghana 46 C4
Axminster U.K. 19 E8
Axum Eth. see Āksum
Ayachi, Jbel mt. Morocco 22 D5
Ayacucho Arg. 70 E5
Ayacucho Peru 68 D6
Ayadaw Myanmar 37 H5
Ayagoz Kazakh. 42 E3
Ayaguz Kazakh. see Ayagoz
Ayakkum Hu salt l. China 37 G1
Ayaköz Kazakh. see Ayagoz
Ayan Rus. Fed. 29 O4
Ayancık Turkey 34 D2
Ayang N. Korea 45 B5
Ayaş Turkey 34 D2
Aybak Afgh. 36 B1
Aybas Kazakh. 13 K7
Aydar r. Ukr. 13 H6
Aydarko'l ko'li l. Uzbek. 33 K1
Aydere Turkm. 35 J3
Aydın Turkey 27 L6
Aydıncık Turkey 39 A1
Aydın Dağları mts Turkey 27 L5
Aýdyň Turkm. 35 I3
Āyelu Terara vol. Eth. 32 F7
Ayer MA U.S.A. 64 F1
Ayers Rock h. Australia see Uluru
Ayeyarwady r. Myanmar see Irrawaddy
Ayila Ri'gyü mts China 36 D2
Áyios Dhimítrios Greece see Agios Dimitrios
Áyios Evstrátios i. Greece see Agios Efstratios
Áyios Nikólaos Greece see Agios Nikolaos
Áyios Yeóryios i. Greece see Agios Georgios
Aylesbury N.Z. 59 D6
Aylesbury U.K. 19 G7
Aylett VA U.S.A. 64 C4
Ayllón Spain 25 E3
Aylmer Ont. Canada 64 A1
Aylmer Lake Canada 60 H3
Aymangala India 38 C3
'Ayn al 'Abd well Saudi Arabia 35 H5
'Ayn al Baidā' Saudi Arabia 39 C4
'Ayn al Bayḍā' well Syria 39 C2
'Ayn al Ghazalah well Libya 34 A4
'Ayn al Maqfī spring Egypt 34 C6
'Ayn Dāllah spring Egypt 34 B6
'Ayn 'Īsá Syria 39 D1
'Ayn Tabaghbugh spring Egypt 34 B5
'Ayn Tumayrah spring Egypt 34 B5
'Ayn Zaytūn Egypt 34 B5
Ayod South Sudan 32 D8
Ayon, Ostrov i. Rus. Fed. 29 R3
'Ayoûn el 'Atroûs Mauritania 46 C3
Ayr Australia 56 D3
Ayr U.K. 20 E5
Ayr r. U.K. 20 E5
Ayr, Point of U.K. 18 D5
Ayrancı Turkey 34 D3
Ayre, Point of Isle of Man 18 C4
Aytos Bulg. 27 L3
Ayuthia Thai. see Ayutthaya
Ayutthaya Thai. 31 J5
Ayvacık Turkey 27 L5
Ayvalı Turkey 34 E3
Ayvalık Turkey 27 L5
Azak Rus. Fed. see Azov
Azamgarh India 37 E4
Azaouâd reg. Mali 46 C3
Azaouagh, Vallée de watercourse Mali/Niger 46 D3
Azaran Iran see Hashtrud
Azārbaycan country Asia see Azerbaijan
Azărbayjan country Asia see Azerbaijan
Azare Nigeria 46 E3
A'zāz Syria 39 C1
Azbine mts Niger see L'Aïr, Massif de
Azdavay Turkey 34 D2
Azerbaijan country Asia 35 G2
Azerbaydzhanskaya S.S.R. country Asia see Azerbaijan
Azhikode India 38 B4
Aziziye Turkey see Pınarbaşı
Azogues Ecuador 68 C4
Azores terr. N. Atlantic Ocean 72 G3
Azores-Biscay Rise sea feature N. Atlantic Ocean 72 G3
Azotus Israel see Ashdod
Azov Rus. Fed. 13 H7
Azov, Sea of Rus. Fed./Ukr. 13 H7
Azovs'ke More sea Rus. Fed./Ukr. see Azov, Sea of
Azovskoye More sea Rus. Fed./Ukr. see Azov, Sea of
Azraq, Bahr el r. Eth./Sudan 32 D6 see Blue Nile
Azraq ash Shīshān Jordan 39 C4
Azrou Morocco 22 C5
Azuaga Spain 25 D4
Azuero, Península de pen. Panama 67 H7
Azul, Cordillera mts Peru 68 C5
Azuma-san vol. Japan 45 F5
'Azza Gaza see Gaza
Azzaba Alg. 26 B6
Aẓ Ẓahrān Saudi Arabia see Dhahran
Az Zaqāzīq Egypt 34 C5
Az Zarbah Syria 39 C1
Az Zarqā' Jordan 39 C3
Az Zawr, Ra's pt Saudi Arabia 35 H6
Azzeffâl hills Mauritania/W. Sahara 46 B2
Az Zubayr Iraq 35 G5
Az Zuqur i. Yemen 32 F7

B

Baa Indon. 54 C2
Baabda Lebanon 39 B3
Ba'albek Lebanon 39 C2
Ba'al Hazor mt. West Bank 39 B4
Baan Baa Australia 58 D3
Baardheere Somalia 48 E3
Bab India 36 D4
Bābā, Kūh-e mts Afgh. 36 B2
Baba Burnu pt Turkey 27 L5
Babadağ mt. Azer. 35 H2
Babadag Romania 27 M2
Babaeski Turkey 27 L4
Babahoyo Ecuador 68 C4
Babai India 36 D5
Babai r. Nepal 37 E3
Bābā Kalān Iran 35 H5
Bāb al Mandab str. Africa/Asia 32 F7
Babanusa Sudan 32 C7
Babar i. Indon. 41 E8
Babar, Kepulauan is Indon. 54 E1
Babati Tanz. 49 D4
Babayevo Rus. Fed. 12 G4
Babayurt Rus. Fed. 13 J8
B'abdā Lebanon see Baabda
Bab el Mandeb, Straits of Africa/Asia see Bāb al Mandab
Babine Lake Canada 60 F4
Bābol Iran 35 I3
Bābol Sar Iran 35 I3
Babongo Cameroon 47 E4
Baboon Point S. Africa 50 D7
Baboua Cent. Afr. Rep. 48 B3
Babruysk Belarus 13 F5
Babstovo Rus. Fed. 44 D2
Babu China see Hezhou
Babuhri India 36 B4
Babusar Pass Pak. 36 C2
Babuyan i. Phil. 43 M9
Babuyan Channel Phil. 43 M9
Babuyan Islands Phil. 41 E6
Bacaadweyn Somalia 48 E3
Bacabal Brazil 69 J4
Bacan i. Indon. 41 E8
Bacău Romania 27 L1
Bacha China 44 D2
Bach Ice Shelf Antarctica 76 L2
Bachu China 42 D5
Back r. Australia 56 C3
Back r. Canada 61 I3
Bačka Palanka Serbia 27 H2
Backbone Mountain MD U.S.A. 64 B3
Backe Sweden 14 J5
Backstairs Passage Australia 57 B7
Bac Liêu Vietnam 31 J6
Bacolod Phil. 41 E6
Bacqueville-en-Caux France 19 H9
Bada mt. Eth. 48 D3
Badain Jaran Shamo des. China 42 I4
Badajoz Spain 25 C4
Badami India 38 B3
Badampaharh India 37 F5
Badanah Saudi Arabia 35 F5
Badanjilin Shamo des. China see Badain Jaran Shamo
Badaojiang China see Baishan
Badarpur India 37 H4
Badaun India see Budaun
Badderen Norway 14 M2
Bademli Turkey see Aladağ
Bademli Geçidi pass Turkey 34 C3
Baden Austria 17 P6
Baden Switz. 24 I3
Baden-Baden Germany 17 L6
Bad Hersfeld Germany 17 L5
Bad Hofgastein Austria 17 N7
Badia Polesine Italy 26 D2
Badin Pak. 33 K5
Bad Ischl Austria 17 N7
Bādiyat ash Shām des. Asia see Syrian Desert
Bad Kissingen Germany 17 M5
Bad Königsdorff Poland see Jastrzębie-Zdrój
Badnawar India 36 C5
Badnera India 38 C1
Badnor India 36 C4
Badou Togo 46 D4
Badrah Iraq 35 G4
Bad Reichenhall Germany 17 N7
Badr Ḥunayn Saudi Arabia 32 E5
Bād Rūd Iran 35 I4
Bad Salzungen Germany 17 M5
Bad Schwartau Germany 17 M4
Bad Segeberg Germany 17 M4
Badu Island Australia 56 C1
Badulla Sri Lanka 38 D5
Badzhal Rus. Fed. 44 D2
Badzhal'skiy Khrebet mts Rus. Fed. 44 D2
Bae Colwyn U.K. see Colwyn Bay
Baeza Spain 25 E5
Bafatá Guinea-Bissau 46 B3
Baffa Pak. 36 C2
Baffin Bay sea Canada/Greenland 61 L2
Baffin Island Canada 61 L3
Bafia Cameroon 46 E4
Bafilo Togo 46 D4
Bafing r. Africa 46 B3
Bafoulabé Mali 46 B3
Bafoussam Cameroon 46 E4
Bāfq Iran 35 I5
Bafra Turkey 34 D2
Bafra Burnu pt Turkey 34 D2
Bāft Iran 33 I4
Bafwaboli Dem. Rep. Congo 48 C3
Bafwasende Dem. Rep. Congo 48 C3
Bagaha India 37 F4
Bagalkot India 38 B2
Bagalkote India see Bagalkot
Bagamoyo Tanz. 49 D4
Bagan China 37 I2
Bagata Dem. Rep. Congo 48 B4
Bagdad AZ U.S.A. 65 F3
Bagdarin Rus. Fed. 43 K2
Bagé Brazil 70 F4
Bagenalstown Ireland 21 F5
Bagerhat Bangl. 37 G5
Bageshwar India 36 D3
Baggy Point U.K. 19 C7
Bagh India 36 C5
Bàgh a' Chaisteil U.K. see Castlebay
Baghak Pak. 36 A3
Baghdād Iraq 35 G4
Bāgh-e Malek Iran 35 H5
Bagherhat Bangl. see Bagerhat
Baghlān Afgh. 36 B1
Baghrān Afgh. 36 A2
Bağırsak r. Turkey 39 C1
Bağırsak Deresi r. Syria/Turkey see Sājūr, Nahr
Baglung Nepal 37 E3
Bagnères-de-Luchon France 24 E5
Bago Myanmar see Pegu
Bagrationovsk Rus. Fed. 15 L9
Bagrax China see Bohu
Bagrax Hu l. China see Bosten Hu
Bagur, Cabo c. Spain see Begur, Cap de
Bagzane, Monts mts Niger 46 D3
Bahalda India 37 F5
Bahāmabād Iran see Rafsanjān
Bahara Pak. 36 A4
Baharampur India 37 G4
Bahardipur Pak. 36 B4
Bahariya Oasis oasis Egypt see Baḥrīyah, Wāḥāt al
Bahawalnagar Pak. 36 C3
Bahawalpur Pak. 36 B3
Bahçe Adana Turkey 39 B1
Bahçe Osmaniye Turkey 34 E3
Baher Dar Eth. see Bahir Dar
Baheri India 36 D3
Bahia Brazil see Salvador
Bahia state Brazil 71 C1
Bahía, Islas de la is Hond. 67 G5
Bahía Blanca Arg. 70 D5
Bahía Laura Arg. 70 C7
Bahía Negra Para. 70 E2
Bahía Tortugas Mex. 66 B3
Bahir Dar Eth. 48 D2
Bahl India 36 C3
Bahlā Oman 33 I5
Bahraich India 37 E4
Bahrain country Asia 32 H4
Bahrām Beyg Iran 35 H3
Baḥrīyah, Wāḥāt al oasis Egypt 34 C6
Bahuaja-Sonene, Parque Nacional nat. park Peru 68 E6
Baia Mare Romania 27 J1
Baiazeh Iran 35 I4
Baicang China 37 G3
Baicheng Jilin China 44 A3
Baicheng Xinjiang China 42 E4
Baidoa Somalia see Baydhabo
Baidoi Co l. China 37 F2
Baie-aux-Feuilles Canada see Tasiujaq
Baie-Comeau Canada 63 N2
Baie-du-Poste Canada see Mistissini
Baie-St-Paul Canada 63 M2
Baihar India 36 E5
Baihe Jilin China 44 C4
Baiji Iraq see Bayjī
Baikal, Lake Rus. Fed. 42 J2
Baikalu Shan mt. China 44 A1
Baikunthpur India 37 E5
Baile Átha Cliath Ireland see Dublin
Baile Átha Luain Ireland see Athlone
Baile Mhartainn U.K. 20 B3
Baile na Finne Ireland 21 D3
Băileşti Romania 27 J2
Bailey Range hills Australia 55 C7
Bailieborough Ireland 21 F4
Baima Qinghai China 42 I6
Baima Xizang China see Baxoi
Bain r. U.K. 18 G5
Bainang China see Norkyung
Bainbridge GA U.S.A. 63 K5
Bainbridge NY U.S.A. 64 D1
Bainduru India 38 B3
Baingoin China see Porong
Baiona Spain 25 B2
Baiquan China 44 B3
Bā'ir Jordan 39 C4
Bā'ir, Wādī watercourse Jordan/Saudi Arabia 39 C4
Bairab Co l. China 37 E2
Bairat India 36 D4
Baird Mountains U.S.A. 60 C3
Bairiki Kiribati 74 H5
Bairin Youqi China see Daban
Bairnsdale Australia 58 C6
Baise China 42 J8
Baishan Jilin China see Baishanzhen
Baishan Jilin China 44 B4
Baishanzhen China 44 B4
Baisogala Lith. 15 M9
Baitadi Nepal 36 E3
Baitang China 37 I2
Baitou Shan mt. China/N. Korea 44 C4
Baiyin China 42 I5
Baiyuda Desert Sudan 32 D6
Baja Hungary 26 H1
Baja California pen. Mex. 66 A2
Bajawa Indon. 54 C2
Baj Baj India 37 G5
Bājil Yemen 32 F7
Bajo Caracoles Arg. 70 B7
Bajoga Nigeria 46 E3
Bajrakot India 37 F5
Bakala Cent. Afr. Rep. 47 F4
Bakanas Kazakh. 42 D4
Bakar Pak. 36 B4
Bakel Senegal 46 B3
Baker CA U.S.A. 65 D3
Baker MT U.S.A. 62 G2
Baker NV U.S.A. 65 E1
Baker OR U.S.A. 62 D3
Baker WV U.S.A. 64 B3
Baker, Mount vol. U.S.A. 62 C2
Baker Island terr. N. Pacific Ocean 53 I1
Baker Lake imp. l. Australia 55 D6
Baker Lake Canada 61 I3
Baker Lake Canada 61 I3
Bakersfield CA U.S.A. 65 C3
Bakhardok Turkm. see Bokurdak
Bakhasar India 36 B4
Bakhirevo Rus. Fed. 44 C2
Bakhmach Ukr. 13 G6
Bakhma Dam Iraq see Bēkma, Sadd
Bakhmut Ukr. see Artemivs'k
Bākhtarān Iran see Kermānshāh
Bakhtegan, Daryācheh-ye l. Iran 35 I5
Bakı Azer. see Baku
Baki Awdal 48 E2
Bakırköy Turkey 27 M4
Bakkejord Norway 14 K2
Bakloh India 36 C2
Bako Eth. 48 D3
Bakouma Cent. Afr. Rep. 48 C3
Baksan Rus. Fed. 35 F2
Baku Azer. 35 H2
Baku Dem. Rep. Congo 48 D3
Bakutis Coast Antarctica 76 J2
Baky Azer. see Baku
Balâ Turkey 34 D3
Bala U.K. 19 D6
Bala, Cerros de mts Bol. 68 E6
Balabac Strait Malaysia/Phil. 41 D7
Baladeh Māzandarān Iran 35 H3
Baladeh Māzandarān Iran 35 H3
Baladek Rus. Fed. 44 D1
Balaghat India 36 E5
Balaghat Range hills India 38 B2
Balaka Malawi 49 D5
Balakän Azer. 35 G2
Balakhna Rus. Fed. 12 I4
Balakhta Rus. Fed. 42 G1
Balaklava Australia 57 B7
Balaklava Ukr. 34 D1
Balakleya Ukr. see Balakliya
Balakliya Ukr. 13 H6
Balakovo Rus. Fed. 13 J5
Bala Lake l. U.K. 19 D6
Balaman India 36 E4
Balan India 36 B4
Balanda Rus. Fed. see Kalininsk
Balanda r. Rus. Fed. 13 J6
Balan Dağı h. Turkey 27 M6
Balanga Phil. 41 E6
Balangir India see Bolangir
Balaözen r. Kazakh./Rus. Fed. see Malyy Uzen'
Balarampur India see Balrampur
Balashov Rus. Fed. 13 I6
Balasore India see Baleshwar
Balaton, Lake Hungary 26 G1
Balatonboglár Hungary 26 G1
Balatonfüred Hungary 26 G1
Balbina Brazil 69 G4
Balbina, Represa de resr Brazil 69 G4
Balbriggan Ireland 21 F4
Balchik Bulg. 27 M3
Balclutha N.Z. 59 B8
Bald Mountain NV U.S.A. 65 E2
Baldwin PA U.S.A. 64 B2
Baldy Mountain h. Canada 62 G1
Baldy Peak U.S.A. 62 F5
Bâle Switz. see Basel
Baléa Mali 46 B3
Baleares is Spain see Balearic Islands
Baleares, Islas is Spain see Balearic Islands
Baleares Insulae is Spain see Balearic Islands
Balearic Islands is Spain 25 G4
Balears is Spain see Balearic Islands
Balears, Illes is Spain see Balearic Islands
Baleia, Ponta da pt Brazil 71 D2
Bale Mountains National Park Eth. 48 D3
Baleshwar India 37 F5
Balestrand Norway 15 E6
Baléyara Niger 46 D3
Balezino Rus. Fed. 11 Q4
Balfe's Creek Australia 56 D4
Balfour Downs Australia 54 C5
Balgo Australia 54 D5
Balguntay China 42 F4
Bali India 36 C4
Bali i. Indon. 41 D8
Bali, Laut sea Indon. 41 D8
Balia India see Ballia
Baliapal India 37 F5
Balige Indon. 41 B7
Baliguda India 38 D1
Balıkesir Turkey 27 L5
Balīkh r. Syria/Turkey 39 D2
Balikpapan Indon. 41 D8
Balimila Reservoir India 38 D2
Balimo P.N.G. 52 E2
Balin China 44 A2
Balingen Germany 17 L6
Balintore U.K. 20 F3
Balkanabat Turkm. 35 I3
Balkan Mountains Bulg./Serbia 27 J3
Balkassar Pak. 36 C2
Balkhash Kazakh. 42 C3
Balkhash, Lake Kazakh. 42 C3
Balkhash, Ozero l. Kazakh. see Balkhash, Lake
Balkuduk Kazakh. 13 J7
Ballachulish U.K. 20 D4
Balladonia Australia 55 C8
Balladoran Australia 58 D3
Ballaghaderreen Ireland 21 D4
Ballan Australia 58 B6
Ballangen Norway 14 J2
Ballantrae U.K. 20 E5
Ballarat Australia 58 A6
Ballard, Lake imp. l. Australia 55 C7
Ballarpur India 38 C2
Ballater U.K. 20 F3
Ballé Mali 46 C3
Ballena, Punta pt Chile 70 B3
Balleny Islands Antarctica 76 H2
Ballia India 37 F4
Ballina Australia 58 F2
Ballina Ireland 21 C3
Ballinafad Ireland 21 D3
Ballinalack Ireland 21 E4
Ballinamore Ireland 21 E3
Ballinasloe Ireland 21 D4
Ballindine Ireland 21 D4
Ballinger U.S.A. 62 H5
Ballinluig U.K. 20 F4
Ballinrobe Ireland 21 C4
Ballston Spa NY U.S.A. 64 E1
Ballybay Ireland 21 F3
Ballybunion Ireland 21 C5
Ballycanew Ireland 21 F5
Ballycastle Ireland 21 C3
Ballycastle U.K. 21 F2
Ballyclare U.K. 21 G3
Ballyconnell Ireland 21 E3
Ballygar Ireland 21 D4
Ballygawley U.K. 21 E3
Ballygorman Ireland 21 E2
Ballyhaunis Ireland 21 D4
Ballyheigue Ireland 21 C5
Ballykelly U.K. 21 E2
Ballylynan Ireland 21 E5
Ballymacmague Ireland 21 E5
Ballymahon Ireland 21 E4
Ballymena U.K. 21 F3
Ballymoney U.K. 21 F2
Ballymote Ireland 21 D3
Ballynahinch U.K. 21 G3
Ballyshannon Ireland 21 D3
Ballyteige Bay Ireland 21 F5
Ballyvaughan Ireland 21 C4
Ballyward U.K. 21 F3
Balmartin U.K. see Baile Mhartainn
Balmer India see Barmer
Balochistan prov. Pak. 36 A3
Balombo Angola 49 B5
Balonne r. Australia 58 D2
Balotra India 36 C4
Balqash Kazakh. see Balkhash
Balqash Köli l. Kazakh. see Balkhash, La
Balrampur India 37 E4
Balranald Australia 58 A5
Balş Romania 27 K2
Balsas Brazil 69 I5
Balta Ukr. 13 F7
Baltasound U.K. 20 [inset]
Baltay Rus. Fed. 13 J5
Bălţi Moldova 13 E7
Baltic Sea g. Europe 15 J9
Balṭīm Egypt 34 C5
Baltîm Egypt see Balṭīm
Baltimore S. Africa 51 I2
Baltimore MD U.S.A. 64 C3
Baltinglass Ireland 21 F5
Baltistan reg. Pak. 36 C2
Baltiysk Rus. Fed. 12 C5
Balurghat India 37 G4
Balvi Latvia 15 O8
Balya Turkey 27 L5
Balykchy Kyrg. 42 D4
Balykshi Kazakh. 30 E2
Balyqshy Kazakh. see Balykshi
Bam Iran 33 I4
Bamako Mali 46 C3
Bamba Mali 46 C3
Bambari Cent. Afr. Rep. 48 C3
Bamberg Germany 17 M6
Bambili Dem. Rep. Congo 48 C3
Bambio Cent. Afr. Rep. 48 B3
Bamboesberg mts S. Africa 51 H6
Bamboo Creek Australia 54 C5
Bambouti Cent. Afr. Rep. 48 C3
Bambuí Brazil 71 B3
Bamda China 37 I3
Bamenda Cameroon 46 E4
Bāmīān Afgh. 36 A2
Bamiantong China see Muling
Bamingui Cent. Afr. Rep. 48 C3
Bamingui-Bangoran, Parc National du nat. park Cent. Afr. Rep. 48 B3
Bamor India 36 D4
Bamori India 38 C1
Bampton U.K. 19 D8
Bamrūd Iran 33 J3
Bam Tso l. China 37 G3
Bamyili Australia 54 F3
Banaba i. Kiribati 53 G2
Banabuiu, Açude resr Brazil 69 K5
Bañados del Izozog swamp Bol. 68 F7
Banagher Ireland 21 E4
Banalia Dem. Rep. Congo 48 C3
Banamana, Lagoa l. Moz. 51 K2
Banamba Mali 46 C3
Banana Australia 56 E5
Bananal, Ilha do i. Brazil 69 H6
Banapur India 38 E2
Banas r. India 36 D4
Banaz Turkey 27 M5
Ban Ban Laos 42 I9
Banbar China see Domartang
Banbridge U.K. 21 F3
Banbury U.K. 19 F6
Ban Cang Vietnam 42 I8
Banc d'Arguin, Parc National du nat. p Mauritania 46 B2
Banchory U.K. 20 G3
Bancroft Zambia see Chililabombwe
Banda Dem. Rep. Congo 48 C3
Banda India 36 E4
Banda, Kepulauan is Indon. 41 E8
Banda, Laut sea Indon. 41 F8
Banda Aceh Indon. 41 B7
Banda Banda, Mount Australia 58 F3
Banda Daud Shah Pak. 36 B2
Bandama r. Côte d'Ivoire 46 C4
Bandar India see Machilipatnam
Bandar Moz. 49 D5
Bandar Abbas Iran see Bandar-e 'Abbās
Bandarban Bangl. 37 H5
Bandar-e 'Abbās Iran 33 I4
Bandar-e Anzalī Iran 35 H3
Bandar-e Deylam Iran 35 H5
Bandar-e Emām Khomeynī Iran 35 H5
Bandar-e Lengeh Iran 33 H4
Bandar-e Ma'shur Iran 35 H5
Bandar-e Nakhīlū Iran 35 I6
Bandar-e Pahlavī Iran see Bandar-e Anz
Bandar-e Shāh Iran see Bandar-e Torkeman
Bandar-e Shāhpūr Iran see Bandar-e Emām Khomeynī
Bandar-e Shīū' Iran 35 I6
Bandar-e Torkeman Iran 35 I3
Bandar Lampung Indon. 41 C8
Bandarpunch mt. India 36 D3
Bandar Seri Begawan Brunei 41 D7
Banda Sea sea Indon. see Banda, Laut
Band-e Amīr lakes Afgh. 36 A2
Band-e Amīr, Daryā-ye r. Afgh. 36 A1
Bandeira Brazil 71 C1
Bandeirante Brazil 71 A1
Bandeiras, Pico de mt. Brazil 71 C3
Bandelierkop S. Africa 51 I2
Banderas, Bahía de b. Mex. 66 C4
Band-e Sar Qom Iran 35 I4
Band-e Torkestān mts Afgh. 36 A2
Bandhi Pak. 36 B4
Bandhogarh India 36 E5
Bandi r. India 36 C4
Bandiagara Mali 46 C3
Bandikui India 36 D4
Bandipur National Park India 38 C4
Bandırma Turkey 27 L4
Bandjarmasin Indon. see Banjarmasin
Bandon Ireland 21 D6
Bandon r. Ireland 21 D6

an Don Thai. see Surat Thani
and Qīr Iran 35 H5
andra India 38 B2
andundu Dem. Rep. Congo 48 B4
andung Indon. 41 C8
andya Australia 55 C6
ineh Iran 35 G4
anera India 36 C4
anes Cuba 67 I4
anff Canada 62 D1
anff U.K. 20 G3
anfora Burkina Faso 46 C3
anga Dem. Rep. Congo 49 C4
angalore India 38 C3
angalow Australia 58 F2
angaon India 37 G5
angassou Cent. Afr. Rep. 48 C3
angdag Co salt l. China 37 E2
anggai Indon. 52 C2
anggai, Kepulauan is Indon. 41 E8
anggi i. Malaysia 41 D7
anghāzī Libya see Benghazi
angka i. Indon. 41 C8
angko Indon. 41 C8
angkok Thai. 31 J5
angkor China 37 F3
angla state India see West Bengal
angladesh country Asia 37 G4
angolo Côte d'Ivoire 46 C4
angong Co salt l. China/India 36 D2
angor Ireland 21 C3
angor Northern Ireland U.K. 21 G3
angor Wales U.K. 18 C5
angor ME U.S.A. 63 N3
angor PA U.S.A. 64 D2
angs, Mount AZ U.S.A. 65 F2
angsund Norway 14 G4
angued Phil. 43 M9
angui Cent. Afr. Rep. 48 B3
angweulu, Lake Zambia 49 C5
anhā Egypt 34 C5
anhine, Parque Nacional de nat. park Moz. 51 K2
an Houei Sai Laos see Huayxay
ani Cent. Afr. Rep. 48 B3
ani-Bangou Niger 46 D3
anifing r. Mali 46 C3
anihal Pass and Tunnel India 36 C2
anister r. VA U.S.A. 64 B4
anī Suwayf Egypt 34 C5
anī Walīd Libya 47 E1
aniyās Al Qunayṭirah Israel 39 B3
aniyās Ṭarṭūs Syria 39 B2
anja Luka Bos.-Herz. 26 G2
anjarmasin Indon. 41 D8
anjul Gambia 46 B3
anka India 37 F4
anka Banka Australia 54 F4
ankapur India 38 B3
ankass Mali 46 C3
ankilaré Niger 46 D3
anks Island N.W.T. Canada 60 F2
anks Islands Vanuatu 53 G3
anks Peninsula N.Z. 59 D6
anks Strait Australia 57 [inset]
ankura India 37 F5
anmaw Myanmar see Bhamo
anmo Myanmar see Bhamo
ann r. Ireland 21 F5
ann r. U.K. 21 F2
anning CA U.S.A. 65 D4
anningville Dem. Rep. Congo see Bandundu
annu Pak. 36 B2
ano India 37 F5
añolas Spain see Banyoles
an Phôn-Hông Laos 42 I9
ansi Bihar India 37 F4
ansi Rajasthan India 36 C4
ansi Uttar Prad. India 36 D4
ansi Uttar Prad. India 37 E4
ansihari India 37 G4
anská Bystrica Slovakia 17 Q6
anspani India 37 F5
ansur India 36 D4
answara India 36 C5
anteer Ireland 21 D5
antry Ireland 21 C6
antry Bay Ireland 21 C6
antval India 38 B3
anyo Cameroon 46 E4
anyoles Spain 25 H2
anyuwangi Indon. 54 A2
anzare Coast Antarctica 76 G2
anzare Seamount sea feature Indian Ocean 73 N9
anzart Tunisia see Bizerte
anzyville Dem. Rep. Congo see Mobayi-Mbongo
ao'an China see Shenzhen
aochang China 43 L4
aoding China 43 L5
aoji Shaanxi China 42 J6
aokang Nei Mongol China 44 A3
aolin China 44 C3
aoqing China 44 D3
aoro Cent. Afr. Rep. 48 B3
aoshan China 42 H7
aotou China 43 K4
aoulé r. Mali 46 C3
ap India 36 C4
apatla India 38 D3
aq'ā' Saudi Arabia 35 F6
aqbaq Egypt see Buqbuq
aqên Xizang China 37 H2
aqên Xizang China 37 H3
a'qūbah Iraq 35 G4
ar Montenegro 27 H3
ara Sudan 32 D7
araawe Somalia 48 E3
arabanki India 36 E4
ara Banki India see Barabanki
aracaju r. Brazil 71 A1
aracaldo Spain see Barakaldo
aracoa Cuba 67 J4
aradá, Nahr r. Syria 39 C3
aradine Australia 58 D3
aradine r. Australia 58 D3
aragarh India see Bargarh
arahona Dom. Rep. 67 J5
arail Range mts India 37 H4
araka watercourse Eritrea/Sudan 47 G3
arakaldo Spain 25 E2
araki Barak Afgh. 36 B2
aralaba Australia 56 E5
ara Lacha Pass India 36 D2
aram India 37 F5
aramati India 38 B2
aramula India see Baramulla
aramulla India 36 C2
aran India 36 D4
aran r. Pak. 36 B4
arana Pak. 36 C3
aranavichy Belarus 15 O10
aranikha Rus. Fed. 29 R3
aranīs Egypt see Baranīs
aranīs Egypt 32 E5
arannda India 36 E4
aranof Island U.S.A. 60 E4
aranovichi Belarus see Baranavichy
aranowicze Belarus see Baranavichy
Baraouéli Mali 46 C3
Barasat India 37 G5
Barat Daya, Kepulauan is Indon. 41 E8
Baraut India 36 D3
Barbacena Brazil 71 C3
Barbados country West Indies 67 M6
Barbar, Gebel el mt. Egypt see Barbar, Jabal
Barbar, Jabal mt. Egypt 39 A5
Barbastro Spain 25 G2
Barbate Spain 25 D5
Barberton S. Africa 51 J3
Barbezieux-St-Hilaire France 24 D4
Barbuda i. Antigua and Barbuda 67 L5
Barcaldine Australia 56 D4
Barce Libya see Al Marj
Barcelona Spain 25 H3
Barcelona Venez. 68 F1
Barcelonnette France 24 H4
Barcelos Brazil 68 F4
Barcino Spain see Barcelona
Barclay de Tolly atoll Fr. Polynesia see Raroia
Barclayville Liberia 46 C4
Barcoo watercourse Australia 56 C5
Barcoo Creek watercourse Australia see Cooper Creek
Barcoo National Park Australia see Welford National Park
Barcs Hungary 26 G2
Bärdä Azer. 35 G2
Bárðarbunga mt. Iceland 14 [inset]
Bardawil, Khabrat al salt pan Saudi Arabia 39 D4
Bardawīl, Sabkhat al lag. Egypt 39 A4
Barddhaman India 37 F5
Bardejov Slovakia 13 D6
Bardera Somalia see Baardheere
Bardhaman India see Barddhaman
Bardsey Island U.K. 19 C6
Bardsīr Iran 33 I4
Barðsneshorn pt Iceland 10 D2
Barduli Italy see Barletta
Bareilly India 36 D3
Barellan Australia 58 C5
Barentin France 19 H9
Barentsburg Svalbard 28 C2
Barents Sea Arctic Ocean 12 I1
Barentu Eritrea 32 E6
Barfleur, Pointe de pt France 19 F9
Bargarh India 37 E5
Bargrennan U.K. 20 E5
Barguna Bangl. 37 G5
Barhaj India 37 E4
Barham Australia 58 B5
Bari Italy 26 G4
Bari Doab lowland Pak. 33 L3
Barika Alg. 22 F4
Barinas Venez. 68 D2
Baripada India 37 F5
Bariri Brazil 71 A3
Bari Sadri India 36 C4
Barisal Bangl. 37 G5
Barisan, Pegunungan mts Indon. 41 C8
Barito r. Indon. 41 D8
Barium Italy see Bari
Barkal Bangl. 37 H5
Barkam China 42 I6
Barkan, Ra's-e pt Iran 35 H5
Barkava Latvia 15 O8
Barkly East S. Africa 51 H6
Barkly Homestead Australia 56 A3
Barkly-Oos S. Africa see Barkly East
Barkly Tableland reg. Australia 56 A3
Barkly-Wes S. Africa see Barkly West
Barkly West S. Africa 50 G5
Barkol China 42 G4
Barla Turkey 27 N5
Bârlad Romania 27 L1
Bar-le-Duc France 24 G2
Barlee, Lake imp. l. Australia 55 B7
Barlee Range hills Australia 55 A5
Barletta Italy 26 G4
Barlow Canada 60 E3
Barmah Forest Australia 58 B5
Barmedman Australia 58 C5
Barmen-Elberfeld Germany see Wuppertal
Barmer India 36 B4
Barm Fīrūz, Kūh-e mt. Iran 35 H5
Barmouth U.K. 19 C6
Barnala India 36 C3
Barnard Castle U.K. 18 F4
Barnato Australia 58 B3
Barnaul Rus. Fed. 42 E2
Barnegat Bay NJ U.S.A. 64 D3
Barnes Icecap Canada 61 K2
Barneville-Carteret France 19 F9
Barneys Lake imp. l. Australia 58 B4
Barnsley U.K. 18 F5
Barnstable MA U.S.A. 64 F2
Barnstaple U.K. 19 C7
Barnstaple Bay U.K. 19 C7
Baro Nigeria 46 D4
Baroda Gujarat India see Vadodara
Baroda Madh. Prad. India 36 D4
Barons Range hills Australia 55 D6
Barowghil, Kowtal-e Afgh. 36 C1
Barpeta India 37 G4
Bar Pla Soi Thai. see Chon Buri
Barquisimeto Venez. 68 E1
Barra Brazil 69 J6
Barra i. U.K. 20 B4
Barra, Ponta da pt Moz. 51 L2
Barra, Sound of sea chan. U.K. 20 B3
Barraba Australia 58 E3
Barra Bonita Brazil 71 A3
Barracão do Barreto Brazil 69 G5
Barra do Bugres Brazil 69 G7
Barra do Corda Brazil 69 I5
Barra do Cuieté Brazil 71 C2
Barra do Garças Brazil 69 H7
Barra do Piraí Brazil 71 C3
Barra do São Manuel Brazil 69 G5
Barra do Turvo Brazil 71 A4
Barra Falsa, Ponta da pt Moz. 51 L2
Barraigh i. U.K. see Barra
Barra Mansa Brazil 71 B3
Barranca Peru 68 C4
Barranqueras Arg. 70 E3
Barranquilla Col. 68 D1
Barre MA U.S.A. 64 E1
Barre des Écrins mt. France 24 H4
Barreiras Brazil 69 J6
Barreirinha Brazil 69 G4
Barreirinhas Brazil 69 J4
Barreiro Port. 25 B4
Barreiros Brazil 69 K5
Barren Island Kiribati see Starbuck Island
Barretos Brazil 71 A3
Barrett, Mount h. Australia 54 D4
Barrhead U.K. 20 E5
Barrier Bay Antarctica 76 E2
Barrière Canada 62 C1
Barrier Range hills Australia 57 C6
Barrington, Mount Australia 58 E4
Barrington Tops National Park Australia 58 E4
Barringun Australia 58 B2
Barro Alto Brazil 71 A1
Barrocão Brazil 71 C2
Barron U.S.A. 63 I2
Barrow r. Ireland 21 F5
Barrow U.S.A. 60 C2
Barrow, Point U.S.A. 60 C2
Barrow Creek Australia 54 F5
Barrow-in-Furness U.K. 18 D4
Barrow Island Australia 54 A5
Barrow Range hills Australia 55 D6
Barrow Strait Canada 61 I2
Barr Smith Range hills Australia 55 C6
Barry U.K. 19 D7
Barrydale S. Africa 50 E7
Barry Mountains Australia 58 C6
Barryville NY U.S.A. 64 D2
Barsalpur India 36 C3
Barshatas Kazakh. 42 D3
Barshi India see Barsi
Barsi India 38 B2
Barstow CA U.S.A. 65 D3
Barsur India 38 D2
Bar-sur-Aube France 24 G2
Barth Germany 17 N3
Bartica Guyana 69 G2
Bartın Turkey 34 D2
Bartle Frere, Mount Australia 56 D3
Barton-upon-Humber U.K. 18 G5
Bartoszyce Poland 17 R3
Barú, Volcán vol. Panama 67 H7
Barunga Australia see Bamyili
Barun-Torey, Ozero l. Rus. Fed. 43 L2
Baruunturuun Mongolia 42 G3
Baruun-Urt Mongolia 43 K3
Baruva India 38 E2
Barwani India 36 C5
Barwéli Mali see Baraouéli
Barwon r. Australia 58 C3
Barygaza India see Bharuch
Barysaw Belarus 15 P9
Barysh Rus. Fed. 13 J5
Basalt r. Australia 56 D3
Basankusu Dem. Rep. Congo 48 B3
Basar India 38 C2
Basarabi Romania 27 M2
Basargechar Armenia see Vardenis
Bascuñán, Cabo c. Chile 70 B3
Basel Switz. 24 H3
Bashanta Rus. Fed. see Gorodovikovsk
Bashee r. S. Africa 51 I7
Bāshī Iran 35 H5
Bashi Channel Phil./Taiwan 43 M8
Bashmakovo Rus. Fed. 13 I5
Bāsht Iran 35 H5
Bashtanka Ukr. 13 G7
Basi Punjab India 36 D3
Basi Rajasthan India 36 D4
Basia India 37 F5
Basilan i. Phil. 41 E7
Basildon U.K. 19 H7
Basile, Pico vol. Equat. Guinea 46 D4
Basingstoke U.K. 19 F7
Basirhat India 37 G5
Basīt, Ra's al pt Syria 39 B2
Başkale Turkey 35 G3
Baskatong, Réservoir resr Canada 63 L2
Baskerville, Cape Australia 54 C4
Başkomutan Tarıhı Milli Parkı nat. park Turkey 27 N5
Başköy Turkey 39 A1
Baskunchak, Ozero l. Rus. Fed. 13 J6
Basle Switz. see Basel
Basmat India 38 C2
Basoko Dem. Rep. Congo 48 C3
Basra Iraq 35 G5
Bassano Canada 62 E1
Bassano del Grappa Italy 26 D2
Bassar Togo 46 D4
Bassas da India rf Indian Ocean 49 D6
Bassas de Pedro Padua Bank sea feature India 38 B3
Bassein Myanmar 42 G9
Basse-Normandie admin. reg. France 19 F9
Bassenthwaite Lake U.K. 18 D4
Basse Santa Su Gambia 46 B3
Basse-Terre Guadeloupe 67 L5
Basseterre St Kitts and Nevis 67 L5
Bassikounou Mauritania 46 C3
Bass Rock i. U.K. 20 G4
Bass Strait Australia 57 D8
Båstad Sweden 15 H8
Bāstānābād Iran 35 G3
Basti India 37 E4
Bastia Corse France 24 I5
Bastioes r. Brazil 69 K5
Bastogne Belgium 17 J5
Bastrop LA U.S.A. 63 I5
Basuo China see Dongfang
Basutoland country Africa see Lesotho
Başyayla Turkey 39 A1
Bata Equat. Guinea 46 D4
Batabanó, Golfo de b. Cuba 67 H4
Batagay Rus. Fed. 29 O3
Batala India 36 C3
Batalha Port. 25 B4
Batamay Rus. Fed. 29 N3
Batan i. Phil. 43 M8
Batangafo Cent. Afr. Rep. 48 B3
Batangas Phil. 41 E6
Batan Islands Phil. 41 E5
Batavia Indon. see Jakarta
Batavia NY U.S.A. 64 B1
Bataysk Rus. Fed. 13 H7
Batchawana Mountain h. Canada 63 K2
Bătdâmbâng Cambodia 31 J5
Batéké, Plateaux Congo 48 B4
Batemans Bay Australia 58 E5
Bates Range hills Australia 55 C6
Batesville AR U.S.A. 63 I4
Batetskiy Rus. Fed. 12 F4
Bath U.K. 19 E7
Bath NY U.S.A. 64 C1
Batha watercourse Chad 47 E3
Bathgate U.K. 20 F5
Bathinda India 36 C3
Bathurst Australia 58 D4
Bathurst Canada 63 N2
Bathurst Gambia see Banjul
Bathurst S. Africa 51 H7
Bathurst, Cape Canada 60 F2
Bathurst, Lake Australia 58 D5
Bathurst Inlet inlet Canada 60 H3
Bathurst Inlet (abandoned) Canada 60 H3
Bathurst Island Australia 54 E2
Bathurst Island Canada 61 I2
Batié Burkina Faso 46 C4
Batı Menteşe Dağları mts Turkey 27 L6
Batı Toroslar mts Turkey 27 N6
Batken Kyrg. 42 C5
Batkes Indon. 54 E1
Bāṭlāq-e Gavkhūnī marsh Iran 35 I4
Batley U.K. 18 F5
Batlow Australia 58 D5
Batman Turkey 35 F3
Batna Alg. 22 F4
Baton Rouge U.S.A. 63 I5
Batouri Cameroon 47 E4
Batrā' tourist site Jordan see Petra
Batrā', Jabal al mt. Jordan 39 B5
Batroûn Lebanon 39 B2
Båtsfjord Norway 14 P1
Battambang Cambodia see Bătdâmbâng
Batticaloa Sri Lanka 38 D5
Battipaglia Italy 26 F4
Battle r. Canada 62 F1
Battle Creek U.S.A. 63 J3
Battle Mountain U.S.A. 62 D3
Battura Glacier Pak. 36 C1
Batu mt. Eth. 48 D3
Batu, Pulau-pulau is Indon. 41 B8
Batum Georgia see Bat'umi
Bat'umi Georgia 35 F2
Baturité Brazil 69 K4
Batyrevo Rus. Fed. 13 J5
Batys Qazaqstan admin. div. Kazakh. see Zapadnyy Kazakhstan
Baubau Indon. 41 E8
Baucau East Timor 54 D2
Bauchi Nigeria 46 D3
Bauda India see Boudh
Baudh India see Boudh
Baugé France 24 D3
Bauhinia Australia 56 E5
Baukau East Timor see Baucau
Baume-les-Dames France 24 H3
Baundal India 36 C2
Baura Bangl. 37 G4
Bauru Brazil 71 A3
Bauska Latvia 15 N8
Bautino Kazakh. 35 H1
Bautzen Germany 17 O5
Bavānāt Iran 35 I5
Bavaria reg. Germany 17 M6
Bavda India 38 B2
Baviaanskloofberge mts S. Africa 50 F7
Bavla India 36 C5
Bavly Rus. Fed. 11 Q5
Baw Myanmar 37 H5
Bawal India 36 D3
Baw Baw National Park Australia 58 C6
Bawdeswell U.K. 19 I6
Baxoi China 42 H6
Bay China see Baicheng
Bayamo Cuba 67 I4
Bayan Heilong. China 44 B3
Bayan Qinghai China 37 I2
Bayana India 36 D4
Bayan-Adraga Mongolia 43 K3
Bayanaul Kazakh. 42 D2
Bayanbulak China 42 E4
Bayanday Rus. Fed. 42 J2
Bayan Gol China see Dengkou
Bayan Har Shan mts China 42 G5
Bayan Har Shankou pass China 37 I2
Bayanhongor Mongolia 42 I3
Bayan Hot China 42 J5
Bayan Kuang China 43 J4
Bayan Mod China 42 I4
Bayantsagaan Mongolia 42 H3
Bayan Ul Hot China 43 L4
Bayan-Uul Mongolia 43 K3
Bayat Turkey 27 N5
Bayburt Turkey 35 F2
Bay City MI U.S.A. 63 K3
Bay City TX U.S.A. 63 H6
Baydaratskaya Guba Rus. Fed. 28 H3
Baydhabo Somalia 48 E3
Bayerischer Wald mts Germany 17 N6
Bayerischer Wald, Nationalpark nat. park Germany 17 N6
Bayeux France 19 G9
Bayfield Ont. Canada 64 A1
Bayındır Turkey 27 L5
Bay Islands is Hond. see Bahía, Islas de la
Bayizhen China see Nyingchi
Bayjī Iraq 35 F4
Baykal, Ozero l. Rus. Fed. see Baikal, Lake
Baykal-Amur Magistral Rus. Fed. 44 C1
Baykal Range mts Rus. Fed. see Baykal'skiy Khrebet
Baykal'skiy Khrebet mts Rus. Fed. 43 J2
Baykan Turkey 35 F3
Bay-Khaak Rus. Fed. 42 G2
Baykibashevo Rus. Fed. 11 R4
Baykonur Kazakh. see Baykonyr
Baykonyr Kazakh. 42 A3
Baymak Rus. Fed. 28 G4
Bayombong Phil. 41 E6
Bayona Spain see Baiona
Bayonne France 24 D5
Bayonne NJ U.S.A. 64 D2
Bayqongyr Kazakh. see Baykonyr
Bayram-Ali Turkm. see Bayramaly
Bayramaly Turkm. 33 J2
Bayramiç Turkey 27 L5
Bayreuth Germany 17 M6
Bayrūt Lebanon see Beirut
Bay Shore NY U.S.A. 64 E2
Bayston Hill U.K. 19 E6
Bayt Lahm West Bank see Bethlehem
Bay View N.Z. 59 F4
Bayy al Kabīr, Wādī watercourse Libya 47 E1
Baza Spain 25 E5
Baza, Sierra de mts Spain 25 E5
Bazardüzü Dağı mt. Azer./Rus. Fed. see Bazardyuzyu, Gora
Bazardyuzyu, Gora mt. Azer./Rus. Fed. 35 G2
Bāzār-e Māsāl Iran 35 H3
Bazarnyy Karabulak Rus. Fed. 13 J5
Bazaruto, Ilha do i. Moz. 49 D6
Bazdar Pak. 33 K4
Bazhong China 42 J6
Bazhou China see Bazhong
Bazmān Iran 33 J4
Bazmān, Kūh-e mt. Iran 33 J4
Bcharré Lebanon 39 C2
Beachy Head hd U.K. 19 H8
Beacon NY U.S.A. 64 E2
Beacon Bay S. Africa 51 H7
Beaconsfield U.K. 19 G7
Beagle, Canal sea chan. Arg. 70 C8
Beagle Bank rf Australia 54 C3
Beagle Bay Australia 54 C4
Beagle Gulf Australia 54 E3
Bealanana Madag. 49 E5
Béal an Átha Ireland see Ballina
Béal an Mhuirthead Ireland 21 C3
Béal Átha na Sluaighe Ireland see Ballinasloe
Beale, Lake India 38 B2
Beaminster U.K. 19 E8
Bearalváhki Norway see Berlevåg
Beardmore Glacier Antarctica 76 H1
Bear Island Arctic Ocean see Bjørnøya
Bearma r. India 36 D4
Bearnaraigh i. U.K. see Berneray
Bear Paw Mountain U.S.A. 62 F2
Bearpaw Mountains U.S.A. 62 F2
Beas Dam India 36 C3
Beata, Cabo c. Dom. Rep. 67 J5
Beatrice U.S.A. 63 H3
Beatrice, Cape Australia 56 B2
Beatty NV U.S.A. 65 D2
Beaucaire France 24 G5
Beauchene Island Falkland Is 70 E8
Beaufort Australia 58 A6
Beaufort Sea Canada/U.S.A. 60 D2
Beaufort West S. Africa 50 F7
Beauly U.K. 20 E3
Beauly r. U.K. 20 E3
Beaumaris U.K. 18 C5
Beaumont N.Z. 59 B7
Beaumont TX U.S.A. 63 I5
Beaune France 24 G3
Beaupréau France 24 D3
Beauséjour Canada 63 H1
Beauvais France 24 F2
Beaver r. Alberta/Saskatchewan Canada 60 H4
Beaver PA U.S.A. 64 A2
Beaver UT U.S.A. 65 F1
Beaver r. UT U.S.A. 65 F1
Beaver Creek Canada 77 A2
Beaver Falls PA U.S.A. 64 A2
Beaver Hill Lake Canada 63 H1
Beaver Island U.S.A. 63 J2
Beawar India 36 C4
Beazley Arg. 70 C4
Bebedouro Brazil 71 A3
Bebington U.K. 18 D5
Bêca China 37 I3
Beccles U.K. 19 I6
Bečej Serbia 27 I2
Becerreá Spain 25 C2
Béchar Alg. 22 D5
Bechuanaland country Africa see Botswana
Beckley WV U.S.A. 64 A4
Bedale U.K. 18 F4
Bedelē Eth. 48 D3
Bedford E. Cape S. Africa 51 H7
Bedford KwaZulu-Natal S. Africa 51 J5
Bedford U.K. 19 G6
Bedford IN U.S.A. 63 J4
Bedford PA U.S.A. 64 B2
Bedford VA U.S.A. 64 B4
Bedford, Cape Australia 56 D2
Bedford Downs Australia 54 D4
Bedgerebong Australia 58 C4
Bedi India 36 B5
Bedla India 36 C4
Bedlington U.K. 18 F3
Bedourie Australia 56 B5
Bedworth U.K. 19 F6
Beechworth Australia 58 C6
Beecroft Peninsula Australia 58 E5
Beed India see Bid
Beenleigh Australia 58 F1
Beersheba Israel 39 B4
Be'ér Sheva' Israel see Beersheba
Be'ér Sheva' watercourse Israel 39 B4
Beervlei Dam S. Africa 50 F7
Beerwah Australia 58 F1
Beetaloo Australia 54 F4
Beethoven Peninsula Antarctica 76 L2
Beeville U.S.A. 62 H6
Befori Dem. Rep. Congo 48 C3
Beg, Lough l. U.K. 21 F3
Bega Australia 58 D6
Begari r. Pak. 36 B3
Begicheva, Ostrov i. Rus. Fed. see Bol'shoy Begichev, Ostrov
Begur, Cap de c. Spain 25 H3
Begusarai India 37 F4
Béhague, Pointe pt Fr. Guiana 69 H3
Behbehān Iran 35 H5
Behchokǫ̀ Canada 60 G3
Behrendt Mountains Antarctica 76 L2
Behrūsī Iran 35 I5
Behshahr Iran 35 I3
Behsūd Afgh. 36 A2
Bei'an China 44 B2
Beida Libya see Al Bayḍā'
Beiguan China see Anyang
Beihai China 43 J8
Bei Hulsan Hu salt l. China 37 H1
Beijing China 43 L5
Beijing mun. China 43 L4
Beik Myanmar see Myeik
Beinn an Oir h. U.K. 20 D5
Beinn an Tuirc h. U.K. 20 D5
Beinn Bheigeir h. U.K. 20 C5
Beinn Bhreac h. U.K. 20 D4
Beinn Dearg mt. U.K. 20 E3
Beinn Heasgarnich mt. U.K. 20 E4
Beinn Mholach h. U.K. 20 C2
Beinn Mhòr h. U.K. 20 B3
Beinn na Faoghlia i. U.K. see Benbecula
Beipiao China 43 M4
Beira Moz. 49 D5
Beirut Lebanon 39 B3
Bei Shan mts China 42 H4
Beitbridge Zimbabwe 49 C6
Beith U.K. 20 E5
Beit Jālā West Bank 39 B4
Beja Port. 25 C4
Béja Tunisia 26 C6
Bejaïa Alg. 25 I5
Béjar Spain 25 D3
Bekaa val. Lebanon see El Béqaa
Békés Hungary 27 I1
Békéscsaba Hungary 27 I1
Bekily Madag. 49 E6
Bekkai Japan 44 G4
Bēkma, Sadd dam Iraq 35 G3
Bekovo Rus. Fed. 13 I5
Bekwai Ghana 46 C4
Bela India 37 E4
Bela Pak. 33 K4
Belab r. Pak. 36 B3
Bela-Bela S. Africa 51 I3
Bélabo Cameroon 46 E4
Bela Crkva Serbia 27 I2
Bel Air MD U.S.A. 64 C3
Belalcázar Spain 25 D4
Belapur India 38 B2
Belarus country Europe 13 E5
Belau country N. Pacific Ocean see Palau
Bela Vista Brazil 70 E2
Bela Vista Moz. 51 K4
Bela Vista de Goiás Brazil 71 A2
Belaya r. Rus. Fed. 29 S3
Belaya Glina Rus. Fed. 13 I7
Belaya Kalitva Rus. Fed. 13 I6
Belaya Kholunitsa Rus. Fed. 12 K4
Belaya Tserkva Ukr. see Bila Tserkva
Belbédji Niger 46 D3
Bełchatów Poland 17 Q5
Belcher Islands Canada 61 K4
Belcoo U.K. 21 E3
Beleapani rf India see Cherbaniani Reef
Belebey Rus. Fed. 11 Q5
Beledweyne Somalia 48 E3
Belém Brazil 69 I4
Belém Novo Brazil 71 A5
Belén Arg. 70 C3
Belen Antalya Turkey 39 A1
Belen Hatay Turkey 39 C1
Belen U.S.A. 62 F5
Belep, Îles is New Caledonia 53 G3
Belev Rus. Fed. 13 H5
Belfast U.K. 21 G3
Belfast U.S.A. 63 N3
Belfast Lough inlet U.K. 21 G3
Bēlfodiyo Eth. 48 D2
Belford U.K. 18 F3
Belfort France 24 H3
Belgaum India 38 B3
Belgian Congo country Africa see Congo, Democratic Republic of the
België country Europe see Belgium
Belgique country Europe see Belgium
Belgium country Europe 16 J5
Belgorod Rus. Fed. 13 H6
Belgorod-Dnestrovskyy Ukr. see Bilhorod-Dnistrovs'kyy
Belgrade Serbia 27 I2
Belgrano II research stn Antarctica 76 A1
Belice r. Sicilia Italy 26 E6
Belinskiy Rus. Fed. 13 I5
Belinyu Indon. 41 C8
Belitung i. Indon. 41 C8
Belize Angola 49 B4
Belize Belize 66 G5
Belize country Central America 66 G5
Beljak Austria see Villach
Belkina, Mys pt Rus. Fed. 44 E3
Bel'kovskiy, Ostrov i. Rus. Fed. 29 O2
Bell Australia 58 E1
Bell r. Australia 58 D4
Bellac France 24 E3
Bellary India 38 C3
Bellata Australia 58 D2
Bella Unión Uruguay 70 E4
Bellbrook Australia 58 F3
Bell Cay rf Australia 56 E4
Belledonne mts France 24 G4
Bellefonte PA U.S.A. 64 C2
Belle Fourche U.S.A. 62 G3
Belle Fourche r. U.S.A. 62 G3
Belle Glade U.S.A. 63 K6
Belle-Île i. France 24 C3
Belle Isle i. Canada 61 M4
Belle Isle, Strait of Canada 61 M4
Belleville IL U.S.A. 63 J4
Bellevue WA U.S.A. 62 C2
Bellin Canada see Kangirsuk
Bellingham U.K. 18 E3
Bellingham U.S.A. 62 C2
Bellingshausen (Russian Federation) research stn Antarctica 76 A2
Bellingshausen Sea Antarctica 76 L2
Bellinzona Switz. 24 I3
Bellows Falls VT U.S.A. 64 E1
Bellpat Pak. 36 B3
Belluno Italy 26 E1
Bell Ville Arg. 70 D4
Bellville S. Africa 50 D7
Belmont Australia 58 E4
Belmont U.K. 20 [inset]
Belmont NY U.S.A. 64 B1
Belmonte Brazil 71 D1
Belmopan Belize 66 G5
Belmore, Mount h. Australia 58 F2
Belo Madag. 49 E6
Belo Campo Brazil 71 C1
Belogorsk Rus. Fed. 44 C2
Belogorsk Ukr. see Bilohirs'k
Beloha Madag. 49 E6
Belo Horizonte Brazil 71 C2
Beloit WI U.S.A. 63 J3
Belokurikha Rus. Fed. 42 E2
Belo Monte Brazil 69 H4
Belomorsk Rus. Fed. 12 G2
Belonia India 37 G5
Belorechensk Rus. Fed. 35 E1
Belorechenskaya Rus. Fed. see Belorechensk
Belören Turkey 34 D3
Beloretsk Rus. Fed. 28 G4
Belorussia country Europe see Belarus
Belorusskaya S.S.R. country Europe see Belarus
Belostok Poland see Białystok
Belot, Lac l. Canada 60 F3
Belo Tsiribihina Madag. 49 E5
Belovo Rus. Fed. 42 F2
Beloyarskiy Rus. Fed. 11 T3
Beloye, Ozero l. Rus. Fed. 12 H3
Beloye More sea Rus. Fed. see White Sea
Belozersk Rus. Fed. 12 H3
Belpre OH U.S.A. 64 A3
Beltana Australia 57 B6
Belted Range mts NV U.S.A. 65 D2
Bel'ts' Moldova see Bălți
Bel'tsy Moldova see Bălți
Belukha, Gora mt. Kazakh./Rus. Fed. 42 F3
Belush'ye Rus. Fed. 12 J2
Belvidere NJ U.S.A. 64 D2
Belyando r. Australia 56 D4
Belyayevka Ukr. see Bilyayivka
Belyy Rus. Fed. 12 G5
Belyy, Ostrov i. Rus. Fed. 28 I2
Bemaraha, Plateau du Madag. 49 E5
Bembe Angola 49 B4
Bemidji U.S.A. 63 I2
Béna Burkina Faso 46 C3
Bena Dibele Dem. Rep. Congo 48 C4
Ben Alder mt. U.K. 20 E4
Benalla Australia 58 B6
Benares India see Varanasi
Ben Arous Tunisia 26 D6
Benavente Spain 25 D2
Ben Avon mt. U.K. 20 F3
Benbane Head hd U.K. 21 F2
Benbecula i. U.K. 20 B3
Ben Boyd National Park Australia 58 E6
Benburb U.K. 21 F3
Ben Chonzie h. U.K. 20 F4
Ben Cleuch h. U.K. 20 F4
Ben Cruachan mt. U.K. 20 D4
Bend U.S.A. 62 C3
Bendearg mt. S. Africa 51 H6
Bender Moldova see Tighina
Bender-Bayla Somalia 48 F3
Bendery Moldova see Tighina
Bendigo Australia 58 B6
Bendoc Australia 58 D6
Bene Moz. 49 D5
Benenitra Madag. 49 E6
Benešov Czech Rep. 17 O6
Benevento Italy 26 F4
Beneventum Italy see Benevento
Benezette PA U.S.A. 64 B2
Bengal, Bay of sea Indian Ocean 31 H5
Bengaluru India see Bangalore
Bengamisa Dem. Rep. Congo 48 C3
Bengbu China 43 L6
Benghazi Libya 47 F1
Bengkulu Indon. 41 C8
Bengtsfors Sweden 15 H7
Benguela Angola 49 B5
Benha Egypt see Banhā
Ben Hiant h. U.K. 20 C4
Ben Hope h. U.K. 20 E2
Ben Horn h. U.K. 20 E2
Beni r. Bol. 68 E6
Beni Dem. Rep. Congo 48 C3
Beni Nepal 37 E3
Beni Abbès Alg. 22 D5
Benidorm Spain 25 F4
Beni Mellal Morocco 22 C5
Benin country Africa 46 D4
Benin, Bight of g. Africa 46 D4
Benin City Nigeria 46 D4
Beni Saf Alg. 25 F6
Beni Snassen, Monts des mts Morocco 25 E6
Beni Suef Egypt see Banī Suwayf
Benito Juárez Arg. 70 E5
Benito Juárez Baja California Mex. 65 E4
Benjamim Constant Brazil 68 E4
Benjamín Hill Mex. 66 B2
Benjina Indon. 52 D2
Ben Klibreck h. U.K. 20 E2

Ben Lavin Nature Reserve S. Africa 51 I2
Ben Lawers mt. U.K. 20 E4
Ben Lomond mt. Australia 58 E3
Ben Lomond h. U.K. 20 E4
Ben Lomond National Park Australia 57 [inset]
Ben Macdui mt. U.K. 20 F3
Benmara Australia 56 B3
Ben More h. U.K. 20 C4
Ben More mt. U.K. 20 E4
Benmore, Lake N.Z. 59 C7
Ben More Assynt h. U.K. 20 E2
Bennetta, Ostrov i. Rus. Fed. 29 P2
Bennett Island Rus. Fed. see Bennetta, Ostrov
Ben Nevis mt. U.K. 20 D4
Bennington NH U.S.A. 64 F1
Bennington VT U.S.A. 64 E1
Benoni S. Africa 51 I4
Ben Rinnes h. U.K. 20 F3
Benson AZ U.S.A. 62 E5
Benteng Indon. 41 E8
Bentiu South Sudan 32 C8
Bent Jbaïl Lebanon 39 B3
Bentley U.K. 18 F5
Bento Gonçalves Brazil 71 A5
Benton CA U.S.A. 65 C2
Benton Harbor U.S.A. 63 J3
Bentonville U.S.A. 63 I4
Benue r. Nigeria 46 D4
Ben Vorlich h. U.K. 20 E4
Benwee Head hd Ireland 21 C3
Benwood WV U.S.A. 64 A2
Ben Wyvis mt. U.K. 20 E3
Benxi Liaoning China 44 A4
Benxi Liaoning China 44 B4
Beograd Serbia see Belgrade
Béoumi Côte d'Ivoire 46 C4
Beppu Japan 45 C6
Béqaa val. Lebanon see El Béqaa
Berach r. India 36 C4
Beraketa Madag. 49 E6
Berasia India 36 D5
Berat Albania 27 H4
Beravina Madag. 49 E5
Berber Sudan 32 D6
Berbera Somalia 48 E2
Berbérati Cent. Afr. Rep. 48 B3
Berchtesgaden, Nationalpark nat. park Germany 17 N7
Berck France 24 E1
Berdichev Ukr. see Berdychiv
Berdigestyakh Rus. Fed. 29 N3
Berdyans'k Ukr. 13 H7
Berdychiv Ukr. 13 F6
Beregovo Ukr. see Berehove
Beregovoy Rus. Fed. 44 B1
Berehove Ukr. 13 D6
Bereina P.N.G. 52 E2
Bere Island Ireland 21 C6
Bereket Turkm. 35 I3
Berekum Ghana 46 C4
Berenice Egypt see Baranīs
Berenice Libya see Benghazi
Berens River Canada 63 H1
Bereza Belarus see Byaroza
Berezino Belarus see Byerazino
Berezivka Ukr. 13 F7
Berezne Ukr. 13 E6
Bereznik Rus. Fed. 12 I3
Berezniki Rus. Fed. 11 R4
Berezov Rus. Fed. see Berezovo
Berezovka Rus. Fed. 44 B2
Berezovka Ukr. see Berezivka
Berezovo Rus. Fed. 11 T3
Berezovyy Rus. Fed. 44 D2
Berga Spain 25 G2
Bergama Turkey 27 L5
Bergamo Italy 26 C2
Bergby Sweden 15 J6
Bergen Mecklenburg-Vorpommern Germany 17 N3
Bergen Norway 15 D6
Bergen NY U.S.A. 64 C1
Bergerac France 24 E4
Bergheim (Erft) Germany 17 K5
Bergland Namibia 50 C2
Bergomum Italy see Bergamo
Bergoo WV U.S.A. 64 A3
Bergsjö Sweden 15 J6
Bergsviken Sweden 14 L4
Bergville S. Africa 51 I5
Berhampur India see Baharampur
Beringa, Ostrov i. Rus. Fed. 29 R4
Beringovskiy Rus. Fed. 29 S3
Bering Sea N. Pacific Ocean 29 S4
Bering Strait Rus. Fed./U.S.A. 29 U3
Berislav Ukr. see Beryslav
Berkåk Norway 14 G5
Berkane Morocco 25 E6
Berkeley CA U.S.A. 65 A2
Berkeley Springs WV U.S.A. 64 B3
Berkner Island Antarctica 76 A1
Berkovitsa Bulg. 27 J3
Berkshire Downs hills U.K. 19 F7
Berkshire Hills MA U.S.A. 64 E1
Berlevåg Norway 14 P1
Berlin Germany 17 N4
Berlin MD U.S.A. 64 D3
Berlin PA U.S.A. 64 B3
Berlin Lake OH U.S.A. 64 A2
Bermagui Australia 58 E5
Bermejo r. Arg./Bol. 70 E3
Bermejo Bol. 68 F8
Bermen, Lac l. Canada 61 L4
Bermuda terr. N. Atlantic Ocean 67 L2
Bermuda Rise sea feature N. Atlantic Ocean 72 D4
Bern Switz. 24 H3
Bernardino de Campos Brazil 71 A3
Bernardo O'Higgins, Parque Nacional nat. park Chile 70 B7
Bernasconi Arg. 70 D5
Berne Switz. see Bern
Berner Alpen mts Switz. 24 H3
Berneray i. Scotland U.K. 20 B3
Berneray i. Scotland U.K. 20 B4
Bernier Island Australia 55 A6
Bernina Pass Switz. 24 J3
Beroea Greece see Veroia
Beroea Syria see Aleppo
Beroroha Madag. 49 E6
Beroun Czech Rep. 17 O6
Berounka r. Czech Rep. 17 O6
Berovina Madag. see Beravina
Berri Australia 57 C7
Berriane Alg. 22 E5
Berridale Australia 58 D6
Berriedale U.K. 20 F2
Berrigan Australia 58 B5
Berrima Australia 58 E5
Berrouaghia Alg. 25 H5
Berry Australia 58 E5
Berryessa, Lake CA U.S.A. 65 A1
Berry Head hd U.K. 19 D8
Berry Islands Bahamas 67 I3
Berryville VA U.S.A. 64 C3
Berseba Namibia 50 C4
Berté, Lac l. Canada 63 N1
Berthoud Pass U.S.A. 62 F4
Bertolínia Brazil 69 J5
Bertoua Cameroon 46 E4
Bertraghboy Bay Ireland 21 C4
Beru atoll Kiribati 53 H2
Beruri Brazil 68 F4
Beruwala Sri Lanka 38 C5
Berwick Australia 58 B7
Berwick-upon-Tweed U.K. 18 E3
Berwyn hills U.K. 19 D6
Beryslav Ukr. 27 O1
Berytus Lebanon see Beirut
Besalampy Madag. 49 E5
Besançon France 24 H3
Beshneh Iran 35 I5
Besikama Indon. 54 D2
Beskra Alg. see Biskra
Beslan Rus. Fed. 35 G2
Besnard Lake Canada 60 H4
Besni Turkey 34 E3
Besor watercourse Israel 39 B4
Beşparmak Dağları mts Cyprus see Pentadaktylos Range
Bessbrook U.K. 21 F3
Bessemer U.S.A. 63 J5
Besshoky, Gora h. Kazakh. 35 I1
Besskorbnaya Rus. Fed. 13 I7
Bessonovka Rus. Fed. 13 J5
Betanzos Spain 25 B2
Bethal S. Africa 51 I4
Bethanie Namibia 50 C4
Bethel Park PA U.S.A. 64 A2
Bethesda U.K. 18 C5
Bethesda MD U.S.A. 64 C3
Bethesda OH U.S.A. 64 A2
Bethlehem S. Africa 51 I5
Bethlehem PA U.S.A. 64 D2
Bethlehem West Bank 39 B4
Bethulie S. Africa 51 G6
Beti Pak. 36 B3
Betim Brazil 71 B2
Bet Lehem West Bank see Bethlehem
Betma India 36 C5
Betoota Australia 56 C5
Betpakdala plain Kazakh. 42 C3
Betroka Madag. 49 E6
Bet She'an Israel 39 B3
Betsiamites Canada 63 N2
Bettiah India 37 F4
Bettyhill U.K. 20 E2
Bettystown Ireland 21 F4
Betul India 36 D5
Betwa r. India 36 D4
Betws-y-coed U.K. 19 D5
Beulah Australia 57 C7
Beult r. U.K. 19 H7
Beuthen Poland see Bytom
Beverley U.K. 18 G5
Beverly MA U.S.A. 64 F1
Beverly Hills CA U.S.A. 65 C3
Bexhill U.K. 19 H8
Bexley, Cape Canada 60 G3
Beyānlū Iran 35 G4
Beyce Turkey see Orhaneli
Bey Dağları mts Turkey 27 N6
Beykoz Turkey 27 M4
Beyla Guinea 46 C4
Beylagan Azer. see Beyläqan
Beyläqan Azer. 35 G3
Beyneu Kazakh. 30 E2
Beypazarı Turkey 27 N4
Beypınarı Turkey 34 E3
Beypore India 38 B4
Beyrouth Lebanon see Beirut
Beyşehir Turkey 34 C3
Beyşehir Gölü l. Turkey 34 C3
Beytonovo Rus. Fed. 44 B1
Beytüşşebap Turkey 35 F3
Bezbozhnik Rus. Fed. 12 K4
Bezhanitsy Rus. Fed. 12 F4
Bezhetsk Rus. Fed. 12 H4
Béziers France 24 F5
Bezmein Turkm. see Abadan
Bezwada India see Vijayawada
Bhabha India see Bhabhua
Bhabhar India 36 B4
Bhabhua India 37 E4
Bhabua India see Bhabhua
Bhachau India 36 B5
Bhachbhar India 36 B4
Bhadgaon Nepal see Bhaktapur
Bhadohi India 37 E4
Bhadra India 36 C3
Bhadrachalam Road Station India see Kottagudem
Bhadrak India 37 F5
Bhadrakh India see Bhadrak
Bhadravati India 38 B3
Bhag Pak. 36 A3
Bhagalpur India 37 F4
Bhainsa India 38 C2
Bhainsdehi India 36 D5
Bhairab Bazar Bangl. 37 G4
Bhairi Hol mt. Pak. see ...
Bhaktapur Nepal 37 F4
Bhalki India 38 C2
Bhamo Myanmar 42 H8
Bhamragarh India 38 D2
Bhandara India 36 D5
Bhanjanagar India 38 E2
Bhanrer Range hills India 36 D5
Bhaptiahi India 37 F4
Bharat country Asia see India
Bharatpur India 36 D4
Bhareli r. India 37 H4
Bharuch India 36 C5
Bhatapara India 37 E5
Bhatarsaigh i. U.K. see Vatersay
Bhatghar Lake India 38 B2
Bhatinda India see Bathinda
Bhatnair India see Hanumangarh
Bhatpara India 37 G5
Bhaunagar India see Bhavnagar
Bhavani r. India 38 C4
Bhavani Sagar l. India 38 C4
Bhavnagar India 36 C5
Bhawana Pak. 36 C3
Bhawanipatna India 38 D2
Bhearnaraigh, Eilean i. U.K. see Berneray
Bheemavaram India see Bhimavaram
Bhekuzulu S. Africa 51 J4
Bhera Pak. 36 C2
Bhigvan India 38 B2
Bhikhna Thori Nepal 37 F4
Bhilai India 36 E5
Bhildi India 36 C4
Bhilwara India 36 C4
Bhima r. India 38 C2
Bhimar India 36 B4
Bhimavaram India 38 D2
Bhimlath India 36 E5
Bhind India 36 D4
Bhinga India 37 E4
Bhisho S. Africa 51 H7
Bhiwandi India 38 B2
Bhiwani India 36 D3
Bhogaipur India 36 D4
Bhojpur Nepal 37 F4
Bhola Bangl. 37 G5
Bhongweni S. Africa 51 I6
Bhopal India 36 D5
Bhopalpatnam India 38 D2
Bhrigukaccha India see Bharuch
Bhuban India 38 E1
Bhubaneshwar India 38 E1
Bhubaneswar India see Bhubaneshwar
Bhuj India 36 B5
Bhusawal India 36 C5
Bhutan country Asia 37 G4
Bhuttewala India 36 B4
Bia r. Ghana 46 C4
Bia, Phou mt. Laos 42 I9
Biafo Glacier Pak. 36 C2
Biafra, Bight of g. Africa see Benin, Bight of
Biak Indon. 41 F8
Biak i. Indon. 41 F8
Biała Podlaska Poland 13 D5
Białogard Poland 17 O4
Białystok Poland 13 D5
Bianco, Monte mt. France/Italy see Mont Blanc
Bianzhao China 44 A3
Biaora India 36 D5
Biarritz France 24 D5
Bibai Japan 44 F4
Bibbenluke Australia 58 D6
Bibbiena Italy 26 D3
Biberach an der Riß Germany 17 L6
Bibile Sri Lanka 38 D5
Biblos Lebanon see Jbail
Bicas Brazil 71 C3
Biçer Turkey 27 N5
Bicester U.K. 19 F7
Bichabhera India 36 C4
Bichevaya Rus. Fed. 44 D3
Bichi r. Rus. Fed. 44 E1
Bickerton Island Australia 56 B2
Bickleigh U.K. 19 D8
Bicuari, Parque Nacional do nat. park Angola 49 B5
Bid India 38 B2
Bida Nigeria 46 D4
Bidar India 38 C2
Biddeford ME U.S.A. 64 F1
Bidean nam Bian mt. U.K. 20 D4
Bideford U.K. 19 C7
Bideford Bay U.K. see Barnstaple Bay
Bidzhan Rus. Fed. 44 C3
Bié Angola see Kuito
Bié, Planalto do Angola 49 B5
Biebrzański Park Narodowy nat. park Poland 15 M10
Biel Switz. 24 H3
Bielawa Poland 17 P5
Bielefeld Germany 17 L4
Bielitz Poland see Bielsko-Biała
Biella Italy 26 C2
Bielsko-Biała Poland 17 Q6
Biên Hoa Vietnam 31 J5
Bienne Switz. see Biel
Bienville, Lac l. Canada 61 K4
Bierbank Australia 58 B1
Biesiesvlei S. Africa 51 G4
Bifoun Gabon 48 B4
Biga Turkey 27 L4
Bigadiç Turkey 27 M5
Biga Yarımadası pen. Turkey 27 L5
Big Bear Lake CA U.S.A. 65 D3
Big Bend Swaziland 51 J4
Bigbury-on-Sea U.K. 19 D8
Biger Nuur salt l. Mongolia 42 H3
Biggar Canada 62 F1
Biggar U.K. 20 F5
Bigge Island Australia 54 D3
Biggenden Australia 57 F5
Biggleswade U.K. 19 G6
Big Hole r. U.S.A. 62 E2
Bighorn r. U.S.A. 62 F2
Bighorn Mountains U.S.A. 62 F3
Big Island Nunavut Canada 61 K3
Big Lake U.S.A. 62 G5
Bignona Senegal 46 B3
Big Pine CA U.S.A. 65 C2
Big Pine Peak CA U.S.A. 65 C3
Big Rapids U.S.A. 63 J3
Big River Canada 62 F1
Big Smokey Valley val. NV U.S.A. 65 D1
Big Spring U.S.A. 62 G5
Bigstone Lake Canada 63 H1
Big Timber U.S.A. 62 F2
Big Trout Lake Canada 63 J1
Big Trout Lake Canada 63 I1
Bihać Bos.-Herz. 26 F2
Bihar state India 37 F4
Bihariganj India 37 F4
Bihar Sharif India 37 F4
Bihor, Vârful mt. Romania 27 J1
Bihoro Japan 44 G4
Bijagós, Arquipélago dos is Guinea-Bissau 46 B3
Bijaipur India 36 D4
Bijapur India 38 B2
Bījār Iran 35 G4
Bijbehara India 36 C2
Bijeljina Bos.-Herz. 27 H2
Bijelo Polje Montenegro 27 H3
Bijeraghogarh India 36 E5
Bijie China 42 J7
Bijji India 38 D2
Bijnor India 36 D3
Bijnore India see Bijnor
Bikampur India 36 C4
Bikaner India 36 C3
Bikin Rus. Fed. 44 D3
Bikin r. Rus. Fed. 44 D3
Bikini atoll Marshall Is 74 H5
Bikori Sudan 32 D7
Bikoro Dem. Rep. Congo 48 B4
Bikou China 42 J6
Bikramganj India 37 F4
Bilād Banī Bū 'Alī Oman 33 I5
Bilaigarh India 38 D1
Bilara India 36 C4
Bilaspur Chhattisgarh India 37 E5
Bilaspur Hima. Prad. India 36 D3
Biläsuvar Azer. 35 H3
Bila Tserkva Ukr. 13 F6
Bilbao Spain 25 E2
Bilbays Egypt 34 C5
Bilbeis Egypt see Bilbays
Bilbo Spain see Bilbao
Bilecik Turkey 27 M4
Biłgoraj Poland 13 D6
Bilharamulo Tanz. 48 D4
Bilhaur India 36 E4
Bilhorod-Dnistrovs'kyy Ukr. 27 N1
Bili Dem. Rep. Congo 48 C3
Bilibino Rus. Fed. 29 R3
Billabalong Australia 55 A6
Billabong Creek r. Australia see Moulamein Creek
Billericay U.K. 19 H7
Billiluna Australia 54 D4
Billingham U.K. 18 F4
Billings U.S.A. 62 F2
Billiton i. Indon. see Belitung
Bill of Portland hd U.K. 19 E8
Bill Williams r. AZ U.S.A. 65 E3
Bilma Niger 46 E3
Bilo r. Rus. Fed. see Belaya
Biloela Australia 56 E5
Bilohirs'k Ukr. 34 D1
Bilohir"ya Ukr. 13 E6
Biloku Guyana 69 G3
Biloli India 38 C2
Bilovods'k Ukr. 13 H6
Biloxi U.S.A. 63 J5
Bilpa Morea Claypan salt flat Australia 56 B5
Bilston U.K. 20 F5
Biltine Chad 47 F3
Bilto Norway 14 L2
Bilyayivka Ukr. 27 N1
Bima Indon. 54 B2
Bimberi, Mount Australia 58 D5
Bimbo Ombella-Mpoko 47 E4
Bimini Islands Bahamas 67 I3
Bimlipatam India 38 D2
Bināb Iran 35 H3
Bina-Etawa India 36 D4
Binaija, Gunung mt. Indon. 41 E8
Binboğa Dağı mt. Turkey 34 E3
Bincheng China see Binzhou
Bindebango Australia 58 C1
Bindle Australia 58 D1
Bindu Dem. Rep. Congo 49 B4
Bindura Zimbabwe 49 D5
Binéfar Spain 25 G3
Binga Zimbabwe 49 C5
Binga, Monte mt. Moz. 49 D5
Bingara Australia 58 B2
Bingaram i. India 38 B4
Bing Bong Australia 56 B2
Binghamton NY U.S.A. 64 D1
Bingöl Turkey 35 F3
Bingol Dağı mt. Turkey 35 F3
Binika India 37 E5
Binnaway Australia 58 D3
Binpur India 37 F5
Bint Jbeil Lebanon see Bent Jbaïl
Bintulu Sarawak Malaysia 41 D7
Binxian Heilong. China 44 B3
Binxian Shaanxi China 43 J6
Binya Australia 58 C5
Bin-Yauri Nigeria 46 D3
Binzhou Heilong. China see Binxian
Binzhou Shandong China 43 L5
Bioco i. Equat. Guinea 46 D4
Biograd na Moru Croatia 26 F3
Bioko i. Equat. Guinea see Bioco
Biokovo mts Croatia 26 G3
Biquinhas Brazil 71 B2
Bir India see Bid
Bira Rus. Fed. 44 D2
Bi'r Abū Jady oasis Syria 39 D1
Birāk Libya 47 E2
Birakan Rus. Fed. 44 C2
Bi'r al 'Abd Egypt 39 A4
Bi'r al Ḥalbā well Syria 39 D2
Bi'r al Jifjāfah well Egypt 39 A4
Bi'r al Khamsah well Egypt 34 B5
Bi'r al Māliḩah well Egypt 39 A5
Bi'r al Mulūsi Iraq 35 F4
Bi'r al Munbaṭiḩ well Syria 39 D2
Bi'r al Qaṭrānī well Egypt 34 B5
Bi'r al Ubbayiḑ well Egypt 34 B6
Bi'r an Nuṣf well Egypt see Bi'r an Nuṣṣ
Bi'r an Nuṣṣ well Egypt 34 B5
Bir Anzarane W. Sahara 46 B2
Birao Cent. Afr. Rep. 48 C2
Bi'r ar Rābiyah well Egypt 34 B5
Birata Turkm. 33 J1
Biratnagar Nepal 37 F4
Bi'r aṭ Ṭarfāwī well Libya 34 B5
Bi'r Başīrī well Syria 39 C2
Bi'r Bayḑā' well Egypt 39 B4
Bi'r Baylī well Egypt 34 B5
Bîr Beida well Egypt see Bi'r Bayḑā'
Bîr Bellī well Egypt see ...
Bi'r Buṭaymān Syria 35 E3
Birch Mountains Canada 60 G4
Birch River WV U.S.A. 64 A3
Bircot Eth. 48 E3
Bîr Dignâsh well Egypt see Bi'r Diqnāsh
Bi'r Diqnāsh well Egypt 34 B5
Birdsville Australia 57 B5
Birecik Turkey 34 E3
Bîr el 'Abd Egypt see Bi'r al 'Abd
Bir el Arbi well Alg. 25 I6
Bîr el Istabl well Egypt see Bi'r Iṣṭabl
Bîr el Khamsa well Egypt see Bi'r al Khamsah
Bîr el Nuṣṣ well Egypt see Bi'r an Nuṣṣ
Bîr el Obeiyiḍ well Egypt see Bi'r al Ubbayiḑ
Bîr el Qatrâni well Egypt see Bi'r al Qaṭrānī
Bîr el Râbia well Egypt see Bi'r ar Rābiyah
Birendranagar Nepal see Surkhet
Bir en Natrûn well Sudan 32 C6
Bireun Indon. 41 B7
Bi'r Fajr well Saudi Arabia 34 E5
Bi'r Fu'ād well Egypt 34 B5
Bîr Gifgâfa well Egypt see Bi'r al Jifjāfah
Bi'r Ḥajal well Syria 39 D2
Birhan mt. Eth. 48 D2
Bi'r Ḥasanah well Egypt 39 A4
Bi'r Ḥayzān well Saudi Arabia 34 E6
Bi'r Ibn Hirmās Saudi Arabia see Al Bi'r
Birigüi Brazil 71 A3
Bīrīn Syria 39 C2
Bi'r Iṣṭabl well Egypt 34 B5
Bīrjand Iran 33 I3
Bi'r Jubnī well Libya 34 B5
Birkát Hamad well Iraq 35 G5
Birkenhead U.K. 18 D5
Birkirkara Malta 26 F7
Birksgate Range hills Australia 55 E6
Bîrlad Romania see Bârlad
Bi'r Lahfān well Egypt 39 A4
Birlik Kazakh. 42 C4
Birmingham U.K. 19 F6
Birmingham U.S.A. 63 J5
Bîr Mogreïn Mauritania 46 B2
Bi'r Muḩaymid al Wazwaz well Syria 39 D2
Bi'r Nāḩid oasis Egypt 34 C5
Birnin-Gwari Nigeria 46 D3
Birnin-Kebbi Nigeria 46 D3
Birnin Konni Niger 46 D3
Birobidzhan Rus. Fed. 44 D2
Bi'r Qaşīr as Sirr well Egypt 34 B5
Birr Ireland 21 E4
Bi'r Rawḑ Sālim well Egypt 39 A4
Birrie r. Australia 58 C2
Birrindudu Australia 54 E4
Bîr Rôḍ Sâlim well Egypt see Bi'r Rawḑ Sālim
Birsay U.K. 20 F1
Bi'r Shalatayn Egypt 32 E5
Bîr Shalatein Egypt see Bi'r Shalatayn
Birsk Rus. Fed. 11 R4
Birstall U.K. 19 F6
Birthday Mountain h. Australia 56 C2
Biru China 37 H3
Birur India 38 B3
Biruxiong China see Biru
Biržai Lith. 15 N8
Bisalpur India 36 D3
Bisau India 36 C3
Bisbee U.S.A. 62 F5
Biscay, Bay of sea France/Spain 24 B4
Biscay Abyssal Plain sea feature N. Atlantic Ocean 72 H3
Biscoe Islands Antarctica 76 L2
Biscotasi Lake Canada 63 K2
Bishbek Kyrg. see Bishkek
Bishenpur India see Bishnupur
Bishkek Kyrg. 42 C4
Bishnupur Manipur India 37 H4
Bishnupur W. Bengal India 37 F5
Bishop CA U.S.A. 65 C2
Bishop Auckland U.K. 18 F4
Bishop's Stortford U.K. 19 H7
Bishrī, Jabal hills Syria 39 D2
Bishui Heilong. China 44 A1
Biskra Alg. 22 F5
Bismarck U.S.A. 62 G2
Bismarck Archipelago is P.N.G. 52 E2
Bismarck Range mts P.N.G. 52 E2
Bismarck Sea P.N.G. 52 E2
Bismil Turkey 35 F3
Bismo Norway 14 F6
Bispgården Sweden 14 J5
Bissa, Djebel mt. Alg. 25 G5
Bissamcuttak India 38 D2
Bissau Guinea-Bissau 46 B3
Bissaula Nigeria 46 E4
Bissett Canada 63 H1
Bistcho Lake Canada 60 G4
Bistriţa Romania 27 K1
Bistriţa r. Romania 27 L1
Bitburg Germany 17 K6
Bitche France 24 H2
Bithur India 36 E4
Bithynia reg. Turkey 27 M4
Bitkine Chad 47 E3
Bitlis Turkey 35 F3
Bitola Macedonia 27 I4
Bitolj Macedonia see Bitola
Bitonto Italy 26 G4
Bitra Par rf India 38 B4
Bitterfontein S. Africa 50 D6
Bitter Lakes Egypt 34 D5
Bitterroot r. U.S.A. 62 E2
Bitterroot Range mts U.S.A. 62 D2
Bitterwater CA U.S.A. 65 B2
Biu Nigeria 46 E3
Biwa-ko l. Japan 45 D6
Biwmaris U.K. see Beaumaris
Bïye K'obē Eth. 48 E2
Biysk Rus. Fed. 42 F2
Bizana S. Africa 51 I6
Bizerta Tunisia see Bizerte
Bizerte Tunisia 26 C6
Bjargtangar hd Iceland 14 [inset]
Bjästa Sweden 14 K5
Bjelovar Croatia 26 G2
Bjerkvik Norway 14 J2
Bjerringbro Denmark 15 F8
Bjørgan Norway 14 G5
Björkliden Sweden 14 K2
Björklinge Sweden 15 J6
Bjorli Norway 14 F5
Björna Sweden 14 K5
Björneborg Fin. see Pori
Bjørnøya i. Arctic Ocean 28 C2
Bjurholm Sweden 14 K5
Bla Mali 46 C3
Black r. Vietnam 42 J8
Blackadder Water r. U.K. 20 G5
Blackall Australia 56 D5
Black Bourton U.K. 19 F7
Blackbull Australia 56 C3
Blackburn U.K. 18 E5
Blackbutt Australia 58 F1
Black Canyon gorge AZ U.S.A. 65 E3
Black Combe h. U.K. 18 D4
Blackdown Tableland National Park Australia 56 E4
Black Forest mts Germany 17 L7
Black Hill h. U.K. 18 F5
Black Hills SD U.S.A. 62 G3
Black Lake Canada 60 H4
Black Lake Canada 60 H4
Black Mountain Pak. 36 C2
Black Mountain hills U.K. 19 D7
Black Mountain AK U.S.A. 60 B3
Black Mountain CA U.S.A. 65 D3
Black Mountains hills U.K. 19 D7
Black Mountains AZ U.S.A. 65 E3
Black Nossob watercourse Namibia 50 D2
Black Pagoda India see Konarka
Blackpool U.K. 18 D5
Black Rock h. Jordan see 'Unāb, Jabal al
Blacksburg VA U.S.A. 64 A4
Black Sea Asia/Europe 13 H8
Blacksod Bay Ireland 21 B3
Blackstairs Mountains hills Ireland 21 F5
Blackstone VA U.S.A. 64 B4
Black Sugarloaf mt. Australia 58 E3
Blackville Australia 58 E3
Blackwater Australia 56 E4
Blackwater Ireland 21 F5
Blackwater r. Ireland 21 E5
Blackwater r. Ireland/U.K. 21 F3
Blackwater Reservoir U.K. 20 E4
Blackwood r. Australia 55 A8
Blackwood National Park Australia 56 D4
Bladensburg National Park Australia 56 C4
Blaenavon U.K. 19 D7
Blagodarnyy Rus. Fed. 13 I7
Blagoevgrad Bulg. 27 J3
Blagoveshchensk Amurskaya Oblast' Rus. Fed. 44 B2
Blagoveshchensk Respublika Bashkortostan Rus. Fed. 11 R4
Blaine Lake Canada 62 F1
Blair Athol Australia 56 D4
Blair Atholl U.K. 20 F4
Blairgowrie U.K. 20 F4
Blakeney U.K. 19 I6
Blanca, Bahía b. Arg. 70 D5
Blanche, Lake imp. l. S.A. Australia 57 B6
Blanche, Lake imp. l. W.A. Australia 54 C5
Blanco r. Bol. 68 F6
Blanco, Cape U.S.A. 62 C3
Blanc-Sablon Canada 61 M4
Bland r. Australia 58 C4
Bland VA U.S.A. 64 A4
Blanda r. Iceland 14 [inset]
Blandford Forum U.K. 19 E8
Blanes Spain 25 H3
Blanquilla, Isla i. Venez. 68 F1
Blansko Czech Rep. 17 P6
Blantyre Malawi 49 D5
Blarney Ireland 21 D6
Blåviksjön Sweden 14 K4
Blaye France 24 D4
Blayney Australia 58 D4
Blaze, Point Australia 54 E3
Blenheim N.Z. 59 D5
Blenheim Palace tourist site U.K. 19 F7
Blessington Lakes Ireland 21 F4
Bletchley U.K. 19 G6
Blida Alg. 25 H5
Bligh Water b. Fiji 53 H3
Blitta Togo 46 D4
Block Island RI U.S.A. 64 F2
Block Island Sound sea chan. RI U.S.A. 64 F2
Bloemfontein S. Africa 51 H5
Bloemhof S. Africa 51 G4
Bloemhof Dam S. Africa 51 G4
Bloemhof Dam Nature Reserve S. Africa 51 G4
Blönduós Iceland 14 [inset]
Blongas Indon. 54 B2
Bloods Range mts Australia 55 E6
Bloodsworth Island MD U.S.A. 64 C3
Bloody Foreland pt Ireland 21 D2
Bloomington IL U.S.A. 63 J3
Bloomington IN U.S.A. 63 J4
Bloomsburg PA U.S.A. 64 C2
Blossburg PA U.S.A. 64 C2
Blosseville Kyst coastal area Greenland 61 P3
Blouberg S. Africa 51 I2
Blouberg Nature Reserve S. Africa 51 I2
Bloxham U.K. 19 F6
Blue Diamond NV U.S.A. 65 E2
Bluefield WV U.S.A. 64 A4
Bluefields Nicaragua 67 H6
Blue Knob h. PA U.S.A. 64 B2
Blue Mountain India 37 H5
Blue Mountain Pass Lesotho 51 H5
Blue Mountains Australia 58 D4
Blue Mountains U.S.A. 62 D2
Blue Mountains National Park Australia 58 E4
Blue Nile r. Eth./Sudan 32 D6
Blue Nile r. Eth./Sudan 47 G3
Bluenose Lake Canada 60 G3
Blue Ridge VA U.S.A. 64 B4
Blue Ridge mts VA U.S.A. 64 A4
Blue Stack h. Ireland 21 D3
Blue Stack Mountains hills Ireland 21 D3
Bluestone Lake WV U.S.A. 64 A4
Bluff N.Z. 59 B8
Bluff Knoll mt. Australia 55 B8
Blumenau Brazil 71 A4
Blyde River Canyon Nature Reserve S. Africa 51 J3
Blyth England U.K. 18 F3
Blyth England U.K. 18 F5
Blythe CA U.S.A. 65 E4
Blytheville U.S.A. 63 J4
Bø Norway 15 F7
Bo Sierra Leone 46 B4
Boa Esperança Brazil 71 B3
Boali Cent. Afr. Rep. 48 B3
Boane Moz. 51 K4
Boa Nova Brazil 71 C1
Boatlaname Botswana 51 G2
Boa Viagem Brazil 69 K5
Boa Vista Brazil 68 F3
Boa Vista i. Cape Verde 46 [inset]
Bobadah Australia 58 C4
Bobai China 43 K8
Bobaomby, Tanjona c. Madag. 49 E5
Bobbili India 38 D2
Bobo-Dioulasso Burkina Faso 46 C3
Bobotov Kuk mt. Montenegro see Durmitor
Bobriki Rus. Fed. see Novomoskovsk
Bobrinets Ukr. see Bobrynets'
Bobrov Rus. Fed. 13 I6
Bobrovitsa Ukr. see Bobrovytsya
Bobrovytsya Ukr. 13 F6
Bobruysk Belarus see Babruysk
Bobrynets' Ukr. 13 G6
Bobuk Sudan 32 D7
Bobures Venez. 68 D2
Boby mt. Madag. 49 E6
Boca de Macareo Venez. 68 F2
Boca do Acre Brazil 68 E5
Boca do Jari Brazil 69 H4
Bocaiúva Brazil 71 C2
Bocaranga Cent. Afr. Rep. 48 B3
Bocas del Toro Panama 67 H7
Bochnia Poland 17 R6
Bochum Germany 17 K5
Bocoio Angola 49 B5
Boda Cent. Afr. Rep. 48 B3
Bodalla Australia 58 E6
Bodallin Australia 55 B7
Bodaybo Rus. Fed. 29 M4
Boddam U.K. 20 H3
Bodega Head CA U.S.A. 65 A1
Bodélé reg. Chad 47 E3
Boden Sweden 14 L4
Bodenham U.K. 19 E6
Bodensee l. Germany/Switz. see Constance, Lake
Bodie (abandoned) CA U.S.A. 65 C1
Bodinayakkanur India 38 C4
Bodmin U.K. 19 C8
Bodmin Moor moorland U.K. 19 C8
Bodø Norway 14 I3
Bodoquena Brazil 69 G7
Bodoquena, Serra da hills Brazil 70 E2
Bodrum Turkey 27 L6
Bodträskfors Sweden 14 L3
Boende Dem. Rep. Congo 47 F5
Boffa Guinea 46 B3
Bogalusa U.S.A. 63 J5
Bogan r. Australia 58 C2
Bogandé Burkina Faso 46 C3
Bogan Gate Australia 58 C4
Boğazlıyan Turkey 34 D3
Bogcang Zangbo r. China 37 F3
Bogd Övörhangay Mongolia 42 I4
Bogda Shan mts China 42 F4
Boggabilla Australia 58 E2
Boggabri Australia 58 E3
Boggeragh Mountains hills Ireland 21 C5
Boghar Alg. 25 H6
Boghari Alg. see Ksar el Boukhari
Bognor Regis U.K. 19 G8
Bogodukhov Ukr. see Bohodukhiv
Bog of Allen reg. Ireland 21 E4
Bogong, Mount Australia 58 C6
Bogopol' Rus. Fed. 44 D3
Bogoroditsk Rus. Fed. 13 H5
Bogorodsk Rus. Fed. 12 I4
Bogorodskoye Khabarovskiy Kray Rus. Fed. 44 F1
Bogorodskoye Kirovskaya Oblast' Rus. Fed. 12 K4
Bogotá Col. 68 D3
Bogotol Rus. Fed. 28 J4
Bogoyavlenskoye Rus. Fed. see Pervomayskiy
Bogra Bangl. 37 G4
Boguchany Rus. Fed. 29 K4
Boguchar Rus. Fed. 13 I6
Bogué Mauritania 46 B3
Bo Hai g. China 43 L5
Bohai Wan b. China 40 D4
Bohemian Forest mts Germany see Böhmer Wald
Bohlokong S. Africa 51 I5
Böhmer Wald mts Germany 17 N6
Bohodukhiv Ukr. 13 G6
Bohol Sea Phil. 41 E7
Bohu China 42 F4
Boiaçu Brazil 68 F4
Boichoko S. Africa 50 F5
Boikhutso S. Africa 51 H4
Boileau, Cape Australia 54 C4
Boim Brazil 69 G4
Boipeba, Ilha i. Brazil 71 D1
Bois r. Brazil 71 A2
Boise U.S.A. 62 D3
Boise City U.S.A. 62 G4
Boitumelong S. Africa 51 G4
Bojnūrd Iran 33 I2
Bokaak atoll Marshall Is see Taongi
Bokajan India 37 H4

›karo India 37 F5
›karo Reservoir India 37 F5
›katola Dem. Rep. Congo 48 B4
›ké Guinea 46 B3
›kele Dem. Rep. Congo 48 C4
›khara r. Australia 58 C2
›knafjorden sea chan. Norway 15 D7
›koko Dem. Rep. Congo 48 C3
›koro Chad 47 E3
›kovskaya Rus. Fed. 13 I6
›kspits S. Africa 50 E4
›ktor Rus. Fed. 44 E2
›kurdak Turkm. 33 I2
›l Chad 47 E3
›laiti Dem. Rep. Congo 47 F5
›lama Guinea-Bissau 46 B3
›langir India 38 D1
›lan Pass Pak. 36 A3
›lbec France 24 E2
›le China 42 E4
›le Ghana 46 C4
›leko Dem. Rep. Congo 48 B4
›len Rus. Fed. 44 D2
›lgar Rus. Fed. 13 K5
›lgatanga Ghana 46 C3
›lgrad Ukr. see Bolhrad
›lhrad Ukr. 27 M2
›li China 44 C3
›lia Dem. Rep. Congo 48 B4
›liden Sweden 14 L4
›lintin-Vale Romania 27 K2
›lívar Peru 68 C5
›livar NY U.S.A. 64 B1
›lívar, Pico mt. Venez. 68 D2
›livia country S. America 68 E7
›lkhov Rus. Fed. 13 H5
›llène France 24 G4
›llnäs Sweden 15 J6
›llon Australia 58 C2
›llstabruk Sweden 14 J5
›lmen l. Sweden 15 H8
›lobo Dem. Rep. Congo 48 B4
›logna Italy 26 D2
›lognesi Peru 68 D5
›logoye Rus. Fed. 12 G4
›lokanang S. Africa 51 G5
›lomba Dem. Rep. Congo 48 B3
›lon' Rus. Fed. see Achan
›lpur India 37 F5
›lsena, Lago di l. Italy 26 D3
›l'shakovo Rus. Fed. 15 L9
›l'shaya Chernigovka Rus. Fed. 11 Q5
›l'shaya Glushitsa Rus. Fed. 13 K5
›l'shaya Imandra, Ozero l. Rus. Fed. 14 R3
›l'shaya Martynovka Rus. Fed. 13 I7
›l'shaya Tsarevshchina Rus. Fed. see Volzhskiy
›l'shevik, Ostrov i. Rus. Fed. 29 L2
›l'shezemel'skaya Tundra lowland Rus. Fed. 12 L2
›l'shiye Chirki Rus. Fed. 12 J3
›l'shoy Begichev, Ostrov i. Rus. Fed. 77 E2
›l'shoye Murashkino Rus. Fed. 12 J5
›l'shoy Irgiz r. Rus. Fed. 13 J6
›l'shoy Kamen' Rus. Fed. 44 D4
›l'shoy Kavkaz mts Asia/Europe see Caucasus
›l'shoy Kundysh r. Rus. Fed. 12 J4
›l'shoy Lyakhovskiy, Ostrov i. Rus. Fed. 29 P2
›l'shoy Tokmak Kyrg. see Tokmok
›l'shoy Tokmak Ukr. see Tokmak
›lsón de Mapimí des. Mex. 66 D3
›lton U.K. 18 E5
›lu Turkey 27 N4
›luntay China 37 H1
›lus Head hd Ireland 21 B6
›lvadin Turkey 27 N5
›lzano Italy 26 D1
›ma Dem. Rep. Congo 49 B4
›maderry Australia 58 E5
›mbala Australia 58 D6
›mbay India see Mumbai
›mbay Beach CA U.S.A. 65 E4
›mberai, Semenanjung pen. Indon. 41 F8
›mboma Dem. Rep. Congo 48 B3
›m Comércio Brazil 68 E5
›mdila India 37 H4
›mi China 42 H7
›mili Dem. Rep. Congo 48 C3
›m Jardim Brazil 71 D1
›m Jardim de Goiás Brazil 71 A2
›m Jesus Brazil 71 A5
›m Jesus da Gurgueia, Serra do hills Brazil 69 J5
›m Jesus da Lapa Brazil 71 C1
›m Jesus do Norte Brazil 71 C3
›mlo i. Norway 15 D7
›mokandi r. Dem. Rep. Congo 48 C3
›m Retiro Brazil 71 A4
›m Sucesso Brazil 71 B3
›n, Cap c. Tunisia 26 D6
›na Alg. see Annaba
›nāb Iran 35 G3
›n Air VA U.S.A. 64 C4
›naire i. West Indies 67 K6
›naparte Archipelago is Australia 54 D3
›nar Bridge U.K. 20 E3
›navista Bay Canada 61 M5
›nchester Bridge U.K. 20 G5
›ndo Dem. Rep. Congo 48 C3
›ndoukou Côte d'Ivoire 46 C4
›ndyuzhskiy Rus. Fed. see Mendeleyevsk
›ne Alg. see Annaba
›ne, Teluk b. Indon. 41 E8
›nerate, Kepulauan is Indon. 41 E8
›'ness U.K. 20 F4
›nete, Cerro mt. Arg. 70 C3
›nga Eth. 48 D3
›ngaigaon India 37 G4
›ngandanga Dem. Rep. Congo 48 C3
›ngani S. Africa 50 F5
›ngba China 36 E2
›ng Co l. China 37 G3
›ngo, Massif des mts Cent. Afr. Rep. 48 C3
›ngo, Serra do mts Angola 49 B4
›ngolava mts Madag. 49 E5
›ngor Chad 47 E3
›ni Mali 46 C3
›nifacio Corse France 24 I6
›nifacio, Bocche di str. France/Italy see Bonifacio, Strait of
›nifacio, Bouches de str. France/Italy see Bonifacio, Strait of
›nifacio, Strait of France/Italy 24 I6
›nin Islands Japan 45 F8
›nn Germany 17 K5
›nna Germany see Bonn
›nnåsjøen Norway 14 I3
›nners Ferry U.S.A. 62 D2
›nneville France 24 H3
›nnie Rock Australia 55 B7
›nnyrigg U.K. 20 F5
›nnyville Canada 62 E1
›norva Italy see Bologna
›norva Sardegna Italy 26 C4
›nshaw Australia 58 E2
Bontebok National Park S. Africa 50 E8
Bonthe Sierra Leone 46 B4
Bontoc Phil. 41 E6
Bontosunggu Indon. 52 B2
Bontrug S. Africa 51 G7
Bonvouloir Islands P.N.G. 56 E1
Bonwapitse Botswana 51 H2
Boolba Australia 58 D2
Booligal Australia 58 B4
Boomer WV U.S.A. 64 A3
Boomi Australia 58 D2
Boonah Australia 58 F1
Boones Mill VA U.S.A. 64 B4
Booneville MS U.S.A. 63 J5
Böön Tsagaan Nuur salt l. Mongolia 42 H3
Boonville CA U.S.A. 65 A1
Boonville NY U.S.A. 64 D1
Boorabin National Park Australia 55 C7
Boorama Somalia 48 E3
Booroorban Australia 58 B5
Boorowa Australia 58 D5
Boort Australia 58 A6
Boothby, Cape Antarctica 76 D2
Boothia, Gulf of Canada 61 J3
Boothia Peninsula Canada 61 I2
Bootle U.K. 18 E5
Booué Gabon 48 B4
Boqê China 37 G3
Boqueirão, Serra do hills Brazil 69 J6
Bor Rus. Fed. 12 J4
Bor Serbia 27 J2
Bor South Sudan 47 G4
Bor Turkey 34 D3
Boraha, Nosy i. Madag. 49 F5
Borai India 38 D1
Borakalalo Nature Reserve S. Africa 51 H3
Boran Kazakh. see Buran
Borås Sweden 15 H8
Borasambar India 38 D1
Borāzjān Iran 35 H5
Borba Brazil 69 G4
Borborema, Planalto da plat. Brazil 69 K5
Borçka Turkey 35 F2
Bor Dağı mt. Turkey 27 M6
Bordeaux France 24 D4
Borden Island Canada 61 G2
Borden Peninsula Canada 61 J2
Border Ranges National Park Australia 58 F2
Borðeyri Iceland 14 [inset]
Bordj Bou Arréridj Alg. 25 I5
Bordj Bounaama Alg. 25 G6
Bordj Flye Ste-Marie Alg. 46 C2
Bordj Messaouda Alg. 22 F5
Bordj Mokhtar Alg. 46 D2
Bordj Omar Driss Alg. see Bordj Omer Driss
Bordj Omer Driss Alg. 46 D2
Boreas Abyssal Plain sea feature Arctic Ocean 77 H1
Borgå Fin. see Porvoo
Borgarfjörður Iceland 14 [inset]
Borgarnes Iceland 14 [inset]
Børgefjell Nasjonalpark nat. park Norway 14 H4
Borgholm Sweden 15 J8
Borgo San Lorenzo Italy 26 D3
Bori India 38 C1
Bori r. India 36 C5
Borislav Ukr. see Boryslav
Borisoglebsk Rus. Fed. 13 I6
Borisov Belarus see Barysaw
Borisovka Rus. Fed. 13 H6
Borispol' Ukr. see Boryspil'
Bo River Post South Sudan 47 F4
Borja Peru 68 C4
Borkenes Norway 14 J2
Borkovskaya Rus. Fed. 12 K2
Borlänge Sweden 15 I6
Borlaug Norway 15 E6
Borlu Turkey 27 M5
Borneo i. Asia 41 D7
Bornholm i. Denmark 15 I9
Bornholm Denmark 15 I9
Bornova Turkey 27 L5
Borodino Rus. Fed. 28 J3
Borodinskoye Rus. Fed. 15 P6
Borogontsy Rus. Fed. 29 O3
Borohoro Shan mts China 42 E4
Borok-Sulezhskiy Rus. Fed. 12 H4
Boromo Burkina Faso 46 C3
Boron CA U.S.A. 65 D3
Borondi India 38 D2
Boroughbridge U.K. 18 F4
Borovichi Rus. Fed. 12 G4
Borovoy Kirovskaya Oblast' Rus. Fed. 12 K4
Borovoy Respublika Kareliya Rus. Fed. 14 R4
Borovoy Respublika Komi Rus. Fed. 12 L3
Borpeta India see Barpeta
Borrisokane Ireland 21 D5
Borroloola Australia 56 B3
Børsa Norway 14 G5
Borşa Romania 13 E7
Borsakelmas sho'rxogi salt marsh Uzbek. 35 J2
Borshchiv Ukr. 13 E6
Borshchovochnyy Khrebet mts Rus. Fed. 43 J3
Bortala China see Bole
Borūjen Iran 35 H5
Borūjerd Iran 35 H4
Borve U.K. 20 C3
Boryslav Ukr. 13 D6
Boryspil' Ukr. 13 F6
Borzna Ukr. 13 G6
Borzya Rus. Fed. 43 L2
Bosanska Dubica Bos.-Herz. 26 G2
Bosanska Gradiška Bos.-Herz. 26 G2
Bosanska Krupa Bos.-Herz. 26 G2
Bosanski Novi Bos.-Herz. 26 G2
Bosansko Grahovo Bos.-Herz. 26 G2
Boscawen Island Tonga see Niuatoputapu
Bose China see Baise
Boshof S. Africa 51 G5
Bosna i Hercegovina country Europe see Bosnia-Herzegovina
Bosna Saray Bos.-Herz. see Sarajevo
Bosnia-Herzegovina country Europe 26 G2
Bosobogolo Pan salt pan Botswana 50 F3
Bosobolo Dem. Rep. Congo 48 B3
Bōsō-hantō pen. Japan 45 F6
Bosporus str. Turkey 27 M4
Bossangoa Cent. Afr. Rep. 48 B3
Bossembélé Cent. Afr. Rep. 48 B3
Bossiesvlei Namibia 50 C3
Bossut, Cape Australia 54 C4
Bostan China 37 F1
Bostān Iran 35 G5
Bostan Pak. 36 A3
Bosten Hu l. China 42 F4
Boston U.K. 19 G6
Boston MA U.S.A. 64 F1
Boston Mountains U.S.A. 63 I4
Boston Spa U.K. 18 F5
Botad India 36 B5
Botany Bay Australia 58 E4
Botev mt. Bulg. 27 K3
Botevgrad Bulg. 27 J3
Bothaville S. Africa 51 H4
Bothnia, Gulf of Fin./Sweden 15 K6
Botlikh Rus. Fed. 35 G2
Botoşani Romania 13 E7
Botou China 43 L5
Botshabelo S. Africa 51 H5
Botswana country Africa 49 C6
Botte Donato, Monte mt. Italy 26 G5
Bottesford U.K. 18 G5
Bottrop Germany 17 K5
Botucatu Brazil 71 A3
Botuporã Brazil 71 C1
Bouaflé Côte d'Ivoire 46 C4
Bouaké Côte d'Ivoire 46 C4
Bouar Cent. Afr. Rep. 48 B3
Bou Arfa Morocco 22 D5
Bouba Ndjida, Parc National de nat. park Cameroon 47 E4
Bouca Cent. Afr. Rep. 48 B3
Boucaut Bay Australia 54 F3
Boudh India 38 E1
Bougaa Alg. 25 I5
Bougainville, Cape Australia 54 D3
Bougainville Island P.N.G. 52 F2
Bougainville Reef Australia 56 D2
Boughessa Mali 46 D3
Bougie Alg. see Bejaïa
Bougouni Mali 46 C3
Bougtob Alg. 22 E5
Bouillon Belgium 16 J6
Bouira Alg. 25 H5
Bou Izakarn Morocco 46 C2
Boujdour W. Sahara 46 B2
Boulder Australia 55 C7
Boulder CO U.S.A. 62 F3
Boulder Canyon gorge NV U.S.A. 65 E2
Boulder City NV U.S.A. 65 E3
Boulevard CA U.S.A. 65 D4
Boulia Australia 56 B4
Boulogne France see Boulogne-sur-Mer
Boulogne-Billancourt France 24 F2
Boulogne-sur-Mer France 24 E1
Boumerdes Alg. 25 H5
Bouna Côte d'Ivoire 46 C4
Bou Naceur, Jbel mt. Morocco 22 D5
Boû Nâga Mauritania 46 B3
Boundary Peak NV U.S.A. 65 C2
Boundiali Côte d'Ivoire 46 C4
Boundji Congo 48 B4
Bounty Islands N.Z. 53 H6
Bounty Trough sea feature S. Pacific Ocean 74 H9
Bourail New Caledonia 53 G4
Bourbon reg. France see Bourbonnais
Bourbon terr. Indian Ocean see Réunion
Bourbonnais reg. France 24 F3
Bourem Mali 46 C3
Bouressa Mali see Boughessa
Bourg-Achard France 19 H9
Bourganeuf France 24 E4
Bourg-en-Bresse France 24 G3
Bourges France 24 F3
Bourgogne reg. France see Burgundy
Bourgogne, Canal de France 24 G3
Bourke Australia 58 B3
Bourne U.K. 19 G6
Bournemouth U.K. 19 F8
Bourtoutou Chad 47 F3
Bou Saâda Alg. 25 I6
Bou Salem Tunisia 26 C6
Bouse AZ U.S.A. 65 E4
Bouse Wash watercourse AZ U.S.A. 65 E3
Boutilimit Mauritania 46 B3
Bouvet Island terr. S. Atlantic Ocean see Bouvetøya
Bouvetøya terr. S. Atlantic Ocean 72 I9
Bova Marina Italy 26 F6
Bow r. Alta Canada 62 E2
Bowa China see Muli
Bowden WV U.S.A. 64 B3
Bowditch atoll Tokelau see Fakaofo
Bowen Australia 56 E4
Bowen, Mount Australia 58 D6
Bowenville Australia 58 E1
Bowers Ridge sea feature Bering Sea 74 H2
Bowie Australia 56 D4
Bow Island Canada 62 E2
Bowling Green KY U.S.A. 63 J4
Bowling Green OH U.S.A. 63 K3
Bowling Green VA U.S.A. 64 C3
Bowling Green Bay National Park Australia 56 D3
Bowman U.S.A. 62 G2
Bowman Island Antarctica 76 F2
Bowman Peninsula Antarctica 76 L2
Bowmore U.K. 20 C5
Bowo China see Bomi
Bowral Australia 58 E5
Boyabat Turkey 34 D2
Boyana tourist site Bulg. 27 J3
Boyd r. Australia 58 F2
Boyd Lagoon imp. l. Australia 55 D6
Boyers PA U.S.A. 64 B2
Boyle Ireland 21 D4
Boyne r. Ireland 21 F4
Boysun Uzbek. 33 K2
Boyuibe Bol. 68 F8
Böyük Qafqaz mts Asia/Europe see Caucasus
Bozcaada i. Turkey 27 L5
Bozdağ mt. Turkey 27 L5
Bozdağ mt. Turkey 39 C1
Boz Dağları mts Turkey 27 L5
Bozdoğan Turkey 27 M6
Bozeat U.K. 19 G6
Bozeman U.S.A. 62 E2
Bozen Italy see Bolzano
Bozoum Cent. Afr. Rep. 48 B3
Bozova Turkey 34 E3
Bozqūsh, Kūh-e mts Iran 35 G3
Bozüyük Turkey 27 N5
Bozyazı Turkey 39 A1
Bra Italy 26 B2
Brač i. Croatia 26 G3
Bracadale U.K. 20 C3
Bracadale, Loch b. U.K. 20 C3
Bracara Port. see Braga
Bracciano, Lago di l. Italy 26 E3
Bracebridge Canada 63 L2
Bräcke Sweden 14 I5
Bracknell U.K. 19 G7
Bradano r. Italy 26 G4
Bradenton U.S.A. 63 K6
Brades Germany 67 L5
Bradford U.K. 18 F5
Bradford PA U.S.A. 64 B2
Brady U.S.A. 62 H5
Brae U.K. 20 [inset]
Braemar U.K. 20 F3
Braga Port. 25 B3
Bragado Arg. 70 D5
Bragança Brazil 69 I4
Bragança Port. 25 C3
Bragança Paulista Brazil 71 B3
Brahin Belarus 13 F6
Brahmanbaria Bangl. 37 G5
Brahmapur India 38 E2
Brahmaputra r. China 42 F7
Brahmaputra r. China/India 40 B5
Brahmaputra r. India 37 H4
Brahmaur India 36 D2
Brăila Romania 27 L2
Brainerd U.S.A. 63 I2
Braintree U.K. 19 H7
Braithwaite Point Australia 54 F2
Brak r. S. Africa 51 I2
Brakwater Namibia 50 C2
Bramfield Australia 55 F8
Bramming Denmark 15 F9
Brämön i. Sweden 14 J5
Brampton England U.K. 18 E4
Brampton England U.K. 19 I6
Bramwell Australia 56 C2
Brancaster U.K. 19 H6
Branco r. Brazil 68 F4
Brandberg mt. Namibia 49 B6
Brandbu Norway 15 G6
Brande Denmark 15 F9
Brandenburg Germany 17 N4
Brandfort S. Africa 51 H5
Brandon Canada 62 H2
Brandon U.K. 19 H6
Brandon Head hd Ireland 21 B5
Brandon Mountain h. Ireland 21 B5
Brandvlei S. Africa 50 E6
Braniewo Poland 17 Q3
Bransfield Strait Antarctica 76 L2
Brantford Ont. Canada 64 A1
Branxton Australia 58 E4
Brasil country S. America see Brazil
Brasileia Brazil 68 E6
Brasília Brazil 71 B1
Brasília de Minas Brazil 71 B2
Braslav Belarus see Braslaw
Braslaw Belarus 15 O9
Braşov Romania 27 K2
Brassey, Mount Australia 55 F5
Brassey Range hills Australia 55 C6
Bratislava Slovakia 17 P6
Bratsk Rus. Fed. 42 I1
Bratskoye Vodokhranilishche resr Rus. Fed. 42 I1
Brattleboro VT U.S.A. 64 E1
Braunau am Inn Austria 17 N6
Braunschweig Germany 17 M4
Brava i. Cape Verde 46 [inset]
Brave PA U.S.A. 64 A3
Bråviken inlet Sweden 15 J7
Bravo, Cerro mt. Bol. 68 F7
Bravo del Norte, Río r. Mex. 62 H6
Bravo del Norte, Río r. Mex./U.S.A. see Rio Grande
Brawley CA U.S.A. 65 E4
Bray Ireland 21 F4
Bray Island Canada 61 K3
Brazil country S. America 69 G5
Brazil Basin sea feature S. Atlantic Ocean 72 G7
Brazos r. U.S.A. 63 H6
Brazzaville Congo 49 B4
Brčko Bos.-Herz. 26 H2
Bré Ireland see Bray
Breadalbane Australia 56 B4
Breaksea Sound inlet N.Z. 59 A7
Bream Bay N.Z. 59 E2
Brechfa U.K. 19 C7
Brechin U.K. 20 G4
Břeclav Czech Rep. 17 P6
Brecon U.K. 19 D7
Brecon Beacons reg. U.K. 19 D7
Brecon Beacons National Park U.K. 19 D7
Breda Neth. 16 J5
Bredasdorp S. Africa 50 E8
Bredbo Australia 58 D5
Bredviken Sweden 14 I3
Bregenz Austria 17 L7
Breiðafjörður b. Iceland 14 [inset]
Breiðdalsvík Iceland 14 [inset]
Breivikbotn Norway 14 M1
Breizh reg. France see Brittany
Brejo Velho Brazil 71 C1
Brekstad Norway 14 F5
Bremen Germany 17 L4
Bremer Bay Australia 55 B8
Bremerhaven Germany 17 L4
Bremer Range hills Australia 55 C8
Bremersdorp Swaziland see Manzini
Brenham U.S.A. 63 H5
Brenna Norway 14 H4
Brennero, Passo di pass Austria/Italy see Brenner Pass
Brennerpaß pass Austria/Italy see Brenner Pass
Brenner Pass Austria/Italy 26 D1
Brentwood U.K. 19 H7
Brescia Italy 26 D2
Breslau Poland see Wrocław
Bresle r. France 19 I8
Bressanone Italy 26 D1
Bressay i. U.K. 20 [inset]
Bressuire France 24 D3
Brest Belarus 15 M10
Brest France 24 B2
Brest-Litovsk Belarus see Brest
Bretagne reg. France see Brittany
Breton Sound b. U.S.A. 63 J6
Brett, Cape N.Z. 59 E2
Bretton U.K. 18 E5
Breves Brazil 69 H4
Brewarrina Australia 58 C2
Brewster OH U.S.A. 64 A2
Brewster, Kap c. Greenland see Kangikajik
Brewster, Lake imp. l. Australia 58 B4
Breyten S. Africa 51 I4
Breytovo Rus. Fed. 12 H4
Brezhnev Rus. Fed. see Naberezhnyye Chelny
Brezno Slovakia 17 Q6
Brezovo Bulg. 27 K3
Brezovo Polje plain Croatia 26 G2
Bria Cent. Afr. Rep. 48 C3
Briançon France 24 H4
Brian Head mt. UT U.S.A. 65 F2
Bribbaree Australia 58 C5
Bribie Island Australia 58 F1
Briceni Moldova 13 E6
Brichany Moldova see Briceni
Brichen' Moldova see Briceni
Bridgend U.K. 19 D7
Bridge of Orchy U.K. 20 E4
Bridgeport CA U.S.A. 65 C1
Bridgeport CT U.S.A. 64 E2
Bridgeport NE U.S.A. 62 G3
Bridgeton NJ U.S.A. 64 D3
Bridgetown Australia 55 B8
Bridgetown Barbados 67 M6
Bridgeville DE U.S.A. 64 D3
Bridgewater Canada 63 O3
Bridgewater NY U.S.A. 64 D1
Bridgnorth U.K. 19 E6
Bridgwater U.K. 19 D7
Bridgwater Bay U.K. 19 D7
Bridlington U.K. 18 G4
Bridlington Bay U.K. 18 G4
Bridport Australia 57 [inset]
Bridport U.K. 19 E8
Brie reg. France 24 F2
Brieg Poland see Brzeg
Briery Knob mt. WV U.S.A. 64 A3
Brig Switz. 24 H3
Brigg U.K. 18 G5
Brigham City U.S.A. 62 E3
Brightlingsea U.K. 19 I7
Brighton U.K. 19 G8
Brighton NY U.S.A. 64 C1
Brignoles France 24 H5
Brikama Gambia 46 B3
Brindisi Italy 26 G4
Brioude France 24 F4
Brisbane Australia 58 F1
Brisbane Ranges National Park Australia 58 B6
Bristol U.K. 19 E7
Bristol CT U.S.A. 64 E2
Bristol NH U.S.A. 64 F1
Bristol RI U.S.A. 64 F2
Bristol TN U.S.A. 63 K4
Bristol Bay U.S.A. 60 B4
Bristol Channel est. U.K. 19 C7
Bristol Lake CA U.S.A. 65 E3
Britannia Island New Caledonia see Maré
British Antarctic Territory Antarctica 76 L2
British Columbia prov. Canada 62 C1
British Empire Range mts Canada 61 J1
British Guiana country S. America see Guyana
British Honduras country Central America see Belize
British Indian Ocean Territory terr. Indian Ocean 73 M6
British Solomon Islands country S. Pacific Ocean see Solomon Islands
Brito Godins Angola see Kiwaba N'zogi
Brits S. Africa 51 H3
Britstown S. Africa 50 F6
Brittany reg. France 24 C2
Brive-la-Gaillarde France 24 E4
Briviesca Spain 25 E2
Brixham U.K. 19 D8
Brixia Italy see Brescia
Brlik Kazakh. see Birlik
Brno Czech Rep. 17 P6
Broach India see Bharuch
Broad r. U.S.A. 63 K5
Broadalbin NY U.S.A. 64 D1
Broad Arrow Australia 55 C7
Broadback r. Canada 63 L1
Broad Bay U.K. see Tuath, Loch a'
Broadford Australia 58 B6
Broadford Ireland 21 D5
Broadford U.K. 20 D3
Broad Law h. U.K. 20 F5
Broadmere Australia 56 A3
Broad Sound sea chan. Australia 56 E4
Broadstairs U.K. 19 I7
Broadus U.S.A. 62 F2
Broadway VA U.S.A. 64 B3
Broadwood N.Z. 59 D2
Brochet Canada 77 L3
Brochet, Lac l. Canada 61 H4
Brochet, Lac au l. Canada 63 N2
Brockman, Mount Australia 54 B5
Brockton MA U.S.A. 64 F1
Brockway PA U.S.A. 64 B2
Brodeur Peninsula Canada 61 J2
Brodick U.K. 20 D5
Brodnica Poland 17 Q4
Brody Ukr. 13 E6
Broken Arrow U.S.A. 63 H4
Broken Bay Australia 58 E4
Broken Hill Australia 57 C6
Broken Hill Zambia see Kabwe
Broken Plateau sea feature Indian Ocean 73 O8
Brokopondo Suriname 69 G2
Brokopondo Stuwmeer resr Suriname see Professor van Blommestein Meer
Bromberg Poland see Bydgoszcz
Bromsgrove U.K. 19 E6
Brønderslev Denmark 15 F8
Brønnøysund Norway 14 H4
Brooke U.K. 19 I6
Brookhaven U.S.A. 63 I5
Brookings OR U.S.A. 62 C3
Brookings SD U.S.A. 63 H3
Brookline MA U.S.A. 64 F1
Brookneal VA U.S.A. 64 B4
Brooks Canada 62 E1
Brooks Range mts U.S.A. 60 D3
Brookton Australia 55 B8
Brookville PA U.S.A. 64 B2
Broom, Loch inlet U.K. 20 D3
Broome Australia 54 C4
Brora U.K. 20 F2
Brora r. U.K. 20 F2
Brösarp Sweden 15 I9
Brosna r. Ireland 21 E4
Brough U.K. 18 E4
Brough Ness pt U.K. 20 G2
Broughshane U.K. 21 F3
Broughton Island Canada see Qikiqtarjuaq
Broughton Islands Australia 58 F4
Brovary Ukr. 13 F6
Brovinia Australia 57 E5
Brovst Denmark 15 F8
Browne Range hills Australia 55 D6
Brownfield U.S.A. 62 G5
Brown Mountain CA U.S.A. 65 D3
Brownsville PA U.S.A. 64 B2
Brownsville TN U.S.A. 63 J4
Brownsville TX U.S.A. 63 H6
Brownwood U.S.A. 62 H5
Browse Island Australia 54 C3
Bruay-la-Bussière France 24 F1
Bruce Rock Australia 55 B7
Bruck an der Mur Austria 17 O7
Brue r. U.K. 19 E7
Bruges Belgium see Brugge
Brugge Belgium 16 I5
Bruin PA U.S.A. 64 B2
Bruint India 37 I3
Brûk, Wâdi el watercourse Egypt see Burūk, Wādī al
Brukkaros Namibia 50 D3
Brûlé Canada 62 D1
Brumado Brazil 71 C1
Brumunddal Norway 15 G6
Brundisium Italy see Brindisi
Brunei country Asia 41 D7
Brunei Brunei see Bandar Seri Begawan
Brunette Downs Australia 56 A3
Brunflo Sweden 14 I5
Brunico Italy 26 D1
Brünn Czech Rep. see Brno
Brunner, Lake N.Z. 59 C6
Brunswick Germany see Braunschweig
Brunswick GA U.S.A. 63 K5
Brunswick MD U.S.A. 64 C3
Brunswick ME U.S.A. 63 N3
Brunswick, Península de pen. Chile 70 B8
Brunswick Bay Australia 54 D3
Bruntál Czech Rep. 17 P6
Brunt Ice Shelf Antarctica 76 B2
Bruntville S. Africa 51 J5
Bruny Island Australia 57 [inset]
Brusa Turkey see Bursa
Brusenets Rus. Fed. 12 I3
Brusque Brazil 71 A4
Brussel Belgium see Brussels
Brussels Belgium 16 J5
Bruthen Australia 58 C6
Bruxelles Belgium see Brussels
Bruzual Venez. 68 E2
Bryan TX U.S.A. 63 H5
Bryan, Mount h. Australia 57 B7
Bryan Coast Antarctica 76 L2
Bryansk Rus. Fed. 13 G5
Bryansk Rus. Fed. 35 G1
Brynbuga U.K. see Usk
Bryne Norway 15 D7
Bryukhovetskaya Rus. Fed. 13 H7
Brzeg Poland 17 P5
Brześć nad Bugiem Belarus see Brest
Bua r. Malawi 49 D5
Bu'aale Somalia 48 E3
Buala Solomon Is 53 F2
Būbīyān, Jazīrat Kuwait 35 H5
Bucak Turkey 27 N6
Bucaramanga Col. 68 D2
Buccaneer Archipelago is Australia 54 C4
Buchanan Liberia 46 B4
Buchanan VA U.S.A. 64 B4
Buchanan, Lake imp. l. Australia 56 D4
Buchan Gulf Canada 61 K2
Bucharest Romania 27 L2
Buchon, Point CA U.S.A. 65 B3
Buchy France 19 I9
Bucin, Pasul pass Romania 27 K1
Buckambool Mountain h. Australia 58 B3
Buckeye U.S.A. 62 E5
Buckhannon WV U.S.A. 64 A3
Buckhaven U.K. 20 F4
Buckie U.K. 20 G3
Buckingham U.K. 19 G6
Buckingham VA U.S.A. 64 B4
Buckingham Bay Australia 41 F9
Buckland Tableland reg. Australia 56 E5
Buckleboo Australia 55 G8
Buckle Island Antarctica 76 H2
Buckley watercourse Australia 56 B4
Buckley Bay Antarctica 76 G2
Buckskin Mountains AZ U.S.A. 65 F3
Bucureşti Romania see Bucharest
Buda-Kashalyova Belarus 13 F5
Budalin Myanmar 37 H5
Budapest Hungary 27 H1
Budaun India 36 D3
Budawang National Park Australia 58 E5
Budda Australia 58 B3
Budd Coast Antarctica 76 F2
Buddusò Sardegna Italy 26 C4
Bude U.K. 19 C8
Budennovsk Rus. Fed. 13 J7
Buderim Australia 58 F1
Budīyah, Jabal mts Egypt 39 A5
Budongquan China 37 H2
Budoni Sardegna Italy 26 C4
Budweis Czech Rep. see České Budějovice
Buenaventura Col. 68 C3
Buena Vista i. N. Mariana Is see Tinian
Buena Vista VA U.S.A. 64 B4
Buenos Aires Arg. 70 E4
Buenos Aires, Lago l. Arg./Chile 70 B7
Buerarema Brazil 71 D1
Buffalo NY U.S.A. 64 B1
Buffalo SD U.S.A. 62 G2
Buffalo WY U.S.A. 62 F3
Buffalo Narrows Canada 77 L3
Buffels watercourse S. Africa 50 C5
Buffels Drift S. Africa 51 H2
Buftea Romania 27 K2
Bug r. Poland 17 S5
Buga Col. 68 C3
Bugaldie Australia 58 D3
Bugdaýly Turkm. 35 I3
Bugojno Bos.-Herz. 26 G2
Bugrino Rus. Fed. 12 K1
Bugt China 44 A2
Bugul'ma Rus. Fed. 11 Q5
Büğür China see Luntai
Buguruslan Rus. Fed. 11 Q5
Buhera Zimbabwe 49 D5
Buhuşi Romania 27 L1
Builth Wells U.K. 19 D6
Bui National Park Ghana 46 C4
Buinsk Rus. Fed. 13 K5
Bu'in Zahrā Iran 35 H4
Buir Nur l. Mongolia 43 L3
Buitepos Namibia 50 D2
Bujanovac Serbia 27 I3
Bujumbura Burundi 48 C4
Bukachacha Rus. Fed. 43 L2
Bukadaban Feng mt. China 37 G1
Buka Island P.N.G. 52 F2
Būkān Iran 35 G3
Būkand Iran 35 I5
Bukavu Dem. Rep. Congo 48 C4
Bukhoro Uzbek. see Buxoro
Bukittinggi Indon. 41 C8
Bukkapatnam India 38 C3
Bukoba Tanz. 48 D4
Bükreş Romania see Bucharest
Būl, Kūh-e mt. Iran 35 I5
Bülach Switz. 24 I3
Bulancak Turkey 34 E2
Bulandshahr India 36 D3
Bulanık Turkey 35 F3
Bulava Rus. Fed. 44 F2
Bulawayo Zimbabwe 49 C6
Buldan Turkey 27 M5
Buldana India see Buldhana
Buldhana India 38 C1
Bulembu Swaziland 51 J3
Bulgan Bulgan Mongolia 42 I3
Bulgan Mongolia 42 G3
Bulgar Rus. Fed. see Bolgar
Bŭlgariya country Europe 27 K3
Bŭlgariya country Europe see Bulgaria
Bullawarra, Lake imp. l. Australia 58 A1
Buller r. N.Z. 59 C5
Buller, Mount Australia 58 C6
Bulleringa National Park Australia 56 C3
Bullfinch Australia 55 B7
Bullhead City AZ U.S.A. 65 E3
Bulli Australia 58 E5
Bullion Mountains CA U.S.A. 65 D3
Bullo r. Australia 54 E3
Bulloo Downs Australia 57 C6
Bulloo Lake imp. l. Australia 57 C6
Büllsport Namibia 50 C3
Bulman Australia 54 F3
Bulman Gorge Australia 54 F3
Buloke, Lake dry lake Australia 58 A6
Bulsar India see Valsad
Bultfontein S. Africa 51 H5
Bulukumba Indon. 41 E8
Bulun Rus. Fed. 29 N2
Bulungu Dem. Rep. Congo 49 C4
Bulung'ur Uzbek. 42 B5
Bumba Dem. Rep. Congo 48 C3
Būmbah Libya 34 A4
Bumbah, Khalīj b. Libya 34 A4
Bumpha Bum mt. Myanmar 37 I4
Buna Dem. Rep. Congo 48 B4
Buna Kenya 48 D3
Bunazi Tanz. 48 D4
Bunbury Australia 55 A8
Bunclody Ireland 21 F5
Buncrana Ireland 21 E2
Bunda Tanz. 48 D4
Bundaberg Australia 56 F5

Bundaleer Australia 58 C2
Bundarra Australia 58 E3
Bundi India 36 C4
Bundjalung National Park Australia 58 F2
Bundoran Ireland 21 D3
Bunduqiya South Sudan 47 G4
Bungay U.K. 19 I6
Bungendore Australia 58 D5
Bunger Hills Antarctica 76 F2
Bungle Bungle National Park Australia see Purnululu National Park
Bungo-suidō sea chan. Japan 45 D6
Bunguran, Kepulauan is Indon. see Natuna, Kepulauan
Bunguran, Pulau i. Indon. see Natuna Besar
Bunia Dem. Rep. Congo 48 D3
Bunianga Dem. Rep. Congo 48 C4
Buningonia well Australia 55 C7
Bunji Pak. 36 C2
Bunker Group atolls Australia 56 F4
Bunkeya Dem. Rep. Congo 49 C5
Bünsum China 37 E3
Bunya Mountains National Park Australia 58 E1
Bünyan Turkey 34 D3
Buôn Ma Thuôt Vietnam 31 J5
Bup r. China 37 F3
Buqayq Saudi Arabia see Abqaiq
Buqbuq Egypt 34 B5
Bura Kenya 48 D4
Buraan Somalia 48 E2
Buram Sudan 47 F3
Buran Kazakh. 42 F3
Buranhaém Brazil 71 C2
Buranhaém r. Brazil 71 D2
Burāq Syria 39 C3
Buray r. India 36 C5
Buraydah Saudi Arabia 32 F4
Burbank CA U.S.A. 65 C3
Burcher Australia 58 C4
Burco Somalia 48 E3
Burdigala France see Bordeaux
Burdur Turkey 27 N6
Burdur Gölü l. Turkey 27 N6
Burdwan India see Barddhaman
Burē Eth. 48 D2
Bure r. U.K. 19 I6
Bureå Sweden 14 L4
Bureinskiy Khrebet mts Rus. Fed. 44 D2
Bureinskiy Zapovednik nature res. Rus. Fed. 44 D2
Burewala Pak. 33 L3
Bureya r. Rus. Fed. 44 C2
Bureya Range mts Rus. Fed. see Bureinskiy Khrebet
Burford Ont. Canada 64 A1
Burgas Bulg. 27 L3
Burgeo Canada 61 M5
Burgersdorp S. Africa 51 H6
Burgersfort S. Africa 51 J3
Burges, Mount h. Australia 55 C7
Burgess Hill U.K. 19 G8
Burghausen Germany 17 N6
Burghead U.K. 20 F3
Burgio, Serra di h. Sicilia Italy 26 F6
Burgos Mex. 62 H7
Burgos Spain 25 E2
Burgsvik Sweden 15 K8
Burgundy reg. France 24 G3
Burhan Budai Shan mts China 42 G5
Burhaniye Turkey 27 L5
Burhanpur India 36 D5
Burhar-Dhanpuri India 37 E5
Buri Brazil 71 A3
Buritama Brazil 71 A3
Buriti Alegre Brazil 71 A2
Buriti Bravo Brazil 69 J5
Buritirama Brazil 69 J6
Buritis Brazil 71 B1
Burj Aziz Khan Pak. 36 A3
Burke Island Antarctica 76 K2
Burke Pass N.Z. see Burkes Pass
Burkes Pass N.Z. 59 C7
Burketown Australia 56 B3
Burkeville VA U.S.A. 64 B4
Burkina country Africa see Burkina Faso
Burkina Faso country Africa 46 C3
Burley U.S.A. 62 E3
Burlington Ont. Canada 64 B1
Burlington CO U.S.A. 62 G4
Burlington IA U.S.A. 63 I3
Burlington VT U.S.A. 63 M3
Burma country Asia see Myanmar
Burmantovo Rus. Fed. 11 S3
Burney, Monte vol. Chile 70 B8
Burnie Australia 57 [inset]
Burniston U.K. 18 G4
Burnley U.K. 18 E5
Burns U.S.A. 62 D3
Burnside r. Canada 60 H3
Burnside, Lake imp. l. Australia 55 C6
Burns Lake Canada 60 F4
Burntisland U.K. 20 F4
Burntwood r. Canada 61 I4
Burog Co l. China 37 F2
Burqin China 42 F3
Burqu' Jordan 39 D3
Burra Australia 57 B7
Burra i. U.K. 20 [inset]
Burravoe U.K. 20 [inset]
Burrel Albania 27 I4
Burrel CA U.S.A. 65 C2
Burren reg. Ireland 21 C4
Burrendong, Lake Australia 58 D4
Burren Junction Australia 58 D3
Burrewarra Point Australia 58 E5
Burrinjuck Australia 58 D5
Burrinjuck Reservoir Australia 58 D5
Burro, Serranías del mts Mex. 66 D3
Burro Creek watercourse AZ U.S.A. 65 F3
Burrowa Pine Mountain National Park Australia 58 C6
Burrow Head hd U.K. 20 E6
Burrundie Australia 54 E3
Bursa Turkey 27 M4
Būr Safāga Egypt see Būr Safājah
Būr Safājah Egypt 32 D4
Būr Sa'īd Egypt see Port Said
Būr Sa'īd Egypt see Port Said
Būr Sa'īd governorate Egypt see Būr Sa'īd
Būr Sa'īd governorate Egypt 39 A4
Bursinskoye Vodokhranilishche resr Rus. Fed. 44 C2
Būr Sudan Sudan see Port Sudan
Burton upon Trent U.K. 19 F6
Burträsk Sweden 14 L4
Burt Well Australia 55 F5
Buru i. Indon. 41 E8
Burūk, Wādī al watercourse Egypt 39 A4
Burullus, Baḥra el lag. Egypt see Burullus, Lake
Burullus, Buḥayrat al lag. Egypt see Burullus, Lake
Burullus, Lake lag. Egypt 34 C5
Burultokay China see Fuhai
Burūn, Ra's pt Egypt 39 A4
Burundi country Africa 48 C4
Burunniy Rus. Fed. see Tsagan Aman
Bururi Burundi 48 C4
Burwash Landing Canada 60 E3
Burwick U.K. 20 G2
Buryn' Ukr. 13 G6
Bury St Edmunds U.K. 19 H6
Burzil Pass Pak. 36 C2
Busan S. Korea see Pusan
Busanga Dem. Rep. Congo 48 C4
Buseire Syria see Al Buşayrah
Bush r. U.K. 21 F2
Būshehr Iran 35 H5
Bushêngcaka China 37 E2
Bushenyi Uganda 48 D4
Bushire Iran see Būshehr
Bushmills U.K. 21 F2
Businga Dem. Rep. Congo 48 C3
Buşrá ash Shām Syria 39 C3
Busse Rus. Fed. 44 B2
Busselton Australia 55 A8
Busto Arsizio Italy 26 C2
Buta Dem. Rep. Congo 48 C3
Butare Rwanda 48 C4
Butaritari atoll Kiribati 74 H5
Bute Australia 57 B7
Bute i. U.K. 20 D5
Butha Buthe Lesotho 51 I5
Butha Qi China see Zalantun
Buthidaung Myanmar 37 H5
Butler PA U.S.A. 64 B2
Butlers Bridge Ireland 21 E3
Buton i. Indon. 41 E8
Butte MT U.S.A. 62 E2
Butterworth S. Africa 51 I7
Buttevant Ireland 21 D5
Butt of Lewis hd U.K. 20 C2
Button Bay Canada 61 I4
Butuan Phil. 41 E7
Buturlinovka Rus. Fed. 13 I6
Butwal Nepal 37 E4
Buulobarde Somalia 48 E3
Buur Gaabo Somalia 48 E4
Buurhabaka Somalia 48 E3
Buutsagaan Mongolia 42 H3
Buxar India 37 F4
Buxoro Uzbek. 33 J2
Buxton U.K. 18 F5
Buy Rus. Fed. 12 I4
Buynaksk Rus. Fed. 13 J8
Büyükçekmece Turkey 34 C2
Büyük Egri Dağ mt. Turkey 39 A1
Büyükmenderes r. Turkey 27 L6
Buzău Romania 27 L2
Buzdyak Rus. Fed. 11 Q5
Búzi Moz. 49 D5
Büzmeÿin Turkm. see Abadan
Buzuluk Rus. Fed. 11 Q5
Buzuluk r. Rus. Fed. 13 I6
Buzzards Bay MA U.S.A. 64 F2
Byakar Bhutan see Jakar
Byala Bulg. 27 K3
Byala Slatina Bulg. 27 J3
Byalynichy Belarus 13 F5
Byarezina r. Belarus 13 F5
Byaroza Belarus 15 N10
Byblos tourist site Lebanon 39 B2
Bydgoszcz Poland 17 Q4
Byelorussia country Europe see Belarus
Byerazino Belarus 13 F5
Byeshankovichy Belarus 13 F5
Byesville OH U.S.A. 64 A3
Bygland Norway 15 E7
Bykhaw Belarus 13 F5
Bykhov Belarus see Bykhaw
Bykle Norway 15 E7
Bykovo Rus. Fed. 13 J6
Bylot Island Canada 61 K2
Byramgore Reef India 38 A4
Byrd Glacier Antarctica 76 H1
Byrkjelo Norway 15 E6
Byrock Australia 58 C3
Byron, Cape Australia 58 F2
Byron Bay Australia 58 F2
Byron Island Kiribati see Nikunau
Byrranga, Gory mts Rus. Fed. 29 K2
Byske Sweden 14 L4
Byssa Rus. Fed. 44 C1
Byssa r. Rus. Fed. 44 C1
Bytom Poland 17 Q5
Bytów Poland 17 P3
Byurgyutli Turkm. 35 I3
Byzantium Turkey see İstanbul

C

Caacupé Para. 70 E3
Caatinga Brazil 71 B2
Caazapá Para. 70 E3
Caballas Peru 68 C6
Caballococha Peru 68 D4
Caballos Mesteños, Llano de los plain Mex. 66 D3
Cabanaconde Peru 68 D7
Cabanatuan Phil. 41 E6
Cabdul Qaadir Somalia 48 E2
Cabeceira Rio Manso Brazil 69 G7
Cabeceiras Brazil 71 B1
Cabeza del Buey Spain 25 D4
Cabezas Bol. 68 F7
Cabimas Venez. 68 D1
Cabinda Angola 49 B4
Cabinda prov. Angola 49 B5
Cabinet Inlet Antarctica 76 L2
Cabistra Turkey see Ereğli
Cabo Frio Brazil 71 C3
Cabo Frio, Ilha do i. Brazil 71 C3
Cabonga, Réservoir resr Canada 63 L2
Caboolture Australia 58 F1
Cabo Orange, Parque Nacional de nat. park Brazil 69 H3
Cabo Pantoja Peru 68 C4
Cabora Bassa, Lake resr Moz. 49 D5
Cabo Raso Arg. 70 C6
Caborca Mex. 66 B2
Cabo San Lucas Baja California Sur Mex. 66 C4
Cabot Strait Canada 61 L5
Cabourg France 19 G9
Cabo Verde country N. Atlantic Ocean see Cape Verde
Cabo Verde, Ilhas do is N. Atlantic Ocean 46 [inset]
Cabo Yubi Morocco see Tarfaya
Cabral, Serra do mts Brazil 71 B2
Cäbrayıl Azer. 35 G3
Cabrera, Illa de i. Spain 25 H4
Caçador Brazil 71 A4
Čačak Serbia 27 I3
Caccia, Capo c. Sardegna Italy 26 C4
Cacequi Brazil 70 F3
Cáceres Brazil 69 G7
Cáceres Spain 25 C4
Cache Creek Canada 62 C1
Cacheu Guinea-Bissau 46 B3
Cachi, Nevados de mts Arg. 70 C2
Cachimbo, Serra do hills Brazil 69 H5
Cachoeira Brazil 71 D1
Cachoeira Alta Brazil 71 A2
Cachoeira de Goiás Brazil 71 A2
Cachoeira do Arari Brazil 69 I4
Cachoeiro de Itapemirim Brazil 71 C3
Cacine Guinea-Bissau 46 B3
Caciporé, Cabo c. Brazil 69 H3
Cacolo Angola 49 B5
Cacongo Angola 49 B4
Caçu Brazil 71 A2
Caculé Brazil 71 C1
Čadca Slovakia 17 Q6
Cadereyta Mex. 62 G6
Cadibarrawirracanna, Lake imp. l. Australia 57 A6
Cadillac U.S.A. 63 J3
Cádiz Spain 25 C5
Cadiz OH U.S.A. 64 A2
Cádiz, Golfo de g. Spain 25 C5
Cadiz Lake CA U.S.A. 65 E3
Cadotte Lake Canada 60 G4
Caen France 24 D2
Caerdydd U.K. see Cardiff
Caerffili U.K. see Caerphilly
Caerfyrddin U.K. see Carmarthen
Caergybi U.K. see Holyhead
Caernarfon U.K. 19 C5
Caernarfon Bay U.K. 19 C5
Caernarvon U.K. see Caernarfon
Caerphilly U.K. 19 D7
Caesaraugusta Spain see Zaragoza
Caesarea Alg. see Cherchell
Caesarea Cappadociae Turkey see Kayseri
Caesarea Philippi Syria see Bāniyās
Caesarodunum France see Tours
Caesaromagus U.K. see Chelmsford
Caetité Brazil 71 C1
Cafayate Arg. 70 C3
Cafelândia Brazil 71 A3
Caffa Ukr. see Feodosiya
Cagayan de Oro Phil. 41 E7
Cagli Italy 26 E3
Cagliari Sardegna Italy 26 C5
Cagliari, Golfo di b. Sardegna Italy 26 C5
Çagyl Turkm. 35 I2
Cahama Angola 49 B5
Caha Mountains hills Ireland 21 C6
Cahermore Ireland 21 B6
Cahersiveen Ireland see Cahirsiveen
Cahir Ireland 21 E5
Cahirsiveen Ireland 21 B6
Cahora Bassa, Lago de resr Moz. see Cabora Bassa, Lake
Cahore Point Ireland 21 F5
Cahors France 24 E4
Cahuapanas Peru 68 C5
Cahul Moldova 27 M2
Caia Moz. 49 D5
Caiabis, Serra dos hills Brazil 69 G6
Caianda Angola 49 C5
Caiapó r. Brazil 71 A1
Caiapó, Serra do mts Brazil 71 A2
Caiapônia Brazil 71 A2
Caicara Venez. 68 E2
Caicos Islands Turks and Caicos Is 67 J4
Caicos Passage Bahamas/Turks and Caicos Is 67 J4
Caiguna Australia 55 D8
Caimodorro mt. Spain 25 F3
Caipe Arg. 70 C2
Caird Coast Antarctica 76 B1
Cairngorm Mountains U.K. 20 F3
Cairnryan U.K. 20 D6
Cairns Australia 56 D3
Cairnsmore of Carsphairn h. U.K. 20 E5
Cairo Egypt 34 C5
Caisleán an Bharraigh Ireland see Castlebar
Caiundo Angola 49 B5
Caiwarro (abandoned) Australia 58 B2
Cajamarca Peru 68 C5
Cajati Brazil 71 A4
Cajuru Brazil 71 B3
Čakovec Croatia 26 G1
Çal Denizli Turkey 27 M5
Cala S. Africa 51 H6
Calabar Nigeria 46 D4
Calabria, Parco Nazionale della nat. park Italy 26 G5
Calafat Romania 27 J3
Calagurris Spain see Calahorra
Calahorra Spain 25 F2
Calai Angola 49 B5
Calais France 24 E1
Calais U.S.A. 63 N2
Calalasteo, Sierra de mts Arg. 70 C3
Calama Brazil 68 F5
Calama Chile 70 C2
Calamar Col. 68 D1
Calamian Group is Phil. 41 D6
Calamocha Spain 25 F3
Calandula Angola 49 B4
Calapan Phil. 41 E6
Călărași Romania 27 L2
Calatayud Spain 25 F3
Calayan i. Phil. 43 M9
Calbayog Phil. 41 E6
Calçoene Brazil 69 H3
Calcutta India see Kolkata
Caldas da Rainha Port. 25 B4
Caldas Novas Brazil 69 I7
Caldera Chile 70 B3
Caldervale Australia 56 D5
Caldew r. U.K. 18 E4
Caldwell ID U.S.A. 62 D3
Caldwell OH U.S.A. 64 A3
Caledon r. Lesotho/S. Africa 51 H6
Caledon S. Africa 50 D8
Caledon Bay Australia 56 B2
Caledonia Ont. Canada 64 B1
Caledonia admin. div. U.K. see Scotland
Caleta el Cobre Chile 70 B2
Calexico CA U.S.A. 65 E4
Calf of Man i. Isle of Man 18 C4
Calgary Canada 62 E1
Cali Col. 68 C3
Calicut India see Kozhikode
Caliente NV U.S.A. 65 E2
California PA U.S.A. 64 B2
California state U.S.A. 62 C3
California, Golfo de g. Mex. see California, Gulf of
California, Gulf of Mex. 66 B2
California Aqueduct canal CA U.S.A. 65 B2
Cälilabad Azer. 35 H3
Calingasta Arg. 70 C4
Calipatria CA U.S.A. 65 E4
Calistoga CA U.S.A. 65 A1
Calkiní Mex. 66 F4
Callabonna, Lake imp. l. Australia 57 C6
Callan Ireland 21 E5
Callan r. U.K. 21 F3
Callander U.K. 20 E4
Callao Peru 68 C6
Callicoon NY U.S.A. 64 D2
Callington U.K. 19 C8
Calliope Australia 56 E5
Callipolis Turkey see Gallipoli
Caloundra Australia 58 F1
Caltagirone Sicilia Italy 26 F6
Caltanissetta Sicilia Italy 26 F6
Calucinga Angola 49 B5
Calulo Angola 49 B4
Calunga Angola 49 B5
Caluquembe Angola 49 B5
Caluula Somalia 48 F2
Caluula, Raas pt Somalia 48 F2
Calvert Hills Australia 56 B3
Calvi Corse France 24 I5
Calvià Spain 25 H4
Calvinia S. Africa 50 D6
Calvo, Monte mt. Italy 26 F4
Cam r. U.K. 19 H6
Camaçari Brazil 71 D1
Camache Reservoir CA U.S.A. 65 B1
Camacho Mex. 66 D4
Camacuio Angola 49 B5
Camacupa Angola 49 B5
Camagüey Cuba 67 I4
Camagüey, Archipiélago de is Cuba 67 I4
Camamu Brazil 71 D1
Camana Peru 68 D7
Camanongue Angola 49 C5
Camapuã Brazil 69 H7
Camaquã Brazil 70 F4
Çamardı Turkey 34 D3
Camargo Bol. 68 E8
Camargue reg. France 24 G5
Camarillo CA U.S.A. 65 C3
Camarones Arg. 70 C6
Camarones, Bahía b. Arg. 70 C6
Ca Mau Vietnam 31 J6
Cambay India see Khambhat
Cambay, Gulf of India see Khambhat, Gulf of
Camberley U.K. 19 G7
Cambodia country Asia 31 J5
Camboriú Brazil 71 A4
Camborne U.K. 19 B8
Cambrai France 24 F1
Cambria admin. div. U.K. see Wales
Cambrian Mountains hills U.K. 19 D6
Cambridge Ont. Canada 64 A1
Cambridge N.Z. 59 E3
Cambridge U.K. 19 H6
Cambridge MA U.S.A. 64 F1
Cambridge MD U.S.A. 64 C3
Cambridge MN U.S.A. 63 I2
Cambridge NY U.S.A. 64 E1
Cambridge OH U.S.A. 63 K3
Cambridge Bay Canada 61 H3
Cambrien, Lac l. Canada 61 L4
Cambulo Angola 49 C4
Cambundi-Catembo Angola 49 B5
Cambuquira Brazil 71 B3
Cam Co l. China 37 E2
Camden AR U.S.A. 63 I5
Camden NJ U.S.A. 64 D3
Camden NY U.S.A. 64 D1
Cameia Angola 49 C5
Cameia, Parque Nacional da nat. park Angola 49 C5
Cameron Island Canada 61 H2
Cameron Park CA U.S.A. 65 B1
Cameroon country Africa 46 E4
Cameroon, Mount vol. Cameroon see Cameroun, Mont
Cameroon Highlands slope Cameroon/Nigeria 46 E4
Cameroun country Africa see Cameroon
Cameroun, Mont vol. Cameroon 46 D4
Cametá Brazil 69 I4
Camiña Chile 68 E7
Camiri Bol. 68 F8
Camisea Peru 68 D6
Camocim Brazil 69 J4
Camooweal Australia 56 B3
Camooweal Caves National Park Australia 56 B4
Campana, Isla i. Chile 70 A7
Campbell S. Africa 50 F5
Campbell, Cape N.Z. 59 E5
Campbell, Mount h. Australia 54 E5
Campbell Island N.Z. 74 H9
Campbell Plateau sea feature S. Pacific Ocean 74 H9
Campbell Range hills Australia 54 D3
Campbell River Canada 62 B1
Campbellton Canada 63 N2
Campbelltown Australia 58 E5
Campbeltown U.K. 20 D5
Campeche Mex. 66 F5
Campeche, Bahía de g. Mex. 66 F5
Camperdown Australia 58 A7
Câmpina Romania 27 K2
Campina Grande Brazil 69 K5
Campinas Brazil 71 B3
Campina Verde Brazil 71 A2
Campo Cameroon 46 D4
Campobasso Italy 26 F4
Campo Belo Brazil 71 B3
Campo Belo do Sul Brazil 71 A4
Campo de Diauarum Brazil 69 H6
Campo Florido Brazil 71 A2
Campo Gallo Arg. 70 D3
Campo Grande Brazil 70 F2
Campo Largo Brazil 71 A4
Campo Maior Brazil 69 J4
Campo Maior Port. 25 C4
Campo Mourão Brazil 70 F2
Campos Brazil 71 C3
Campos Altos Brazil 71 B2
Campos Novos Brazil 71 A4
Campos Sales Brazil 69 J5
Câmpulung Romania 27 K2
Câmpulung Moldovenesc Romania 27 K1
Camrose Canada 62 E1
Camrose U.K. 19 B7
Camsell Portage Canada 60 H4
Camulodunum U.K. see Colchester
Çan Turkey 27 L4
Canaan CT U.S.A. 64 E1
Canabrava Brazil 71 B2
Canacona India 38 B3
Canada country N. America 60 H4
Canada Basin sea feature Arctic Ocean 77 A1
Canadian U.S.A. 62 G4
Canadian r. U.S.A. 62 H4
Canadian Abyssal Plain sea feature Antarctica 77 A1
Cañadon Grande, Sierra mts Arg. 70 C7
Canaima, Parque Nacional nat. park Venez. 68 F2
Çanakkale Turkey 27 L4
Çanakkale Boğazı str. Turkey see Dardanelles
Canalejas Arg. 70 C5
Cañamares Spain 25 E3
Canandaigua NY U.S.A. 64 C1
Cananea Mex. 66 B2
Cananéia Brazil 71 B4
Canápolis Brazil 71 A2
Cañar Ecuador 68 C4
Canarias terr. N. Atlantic Ocean see Canary Islands
Canárias, Ilha das i. Brazil 69 J4
Canarias, Islas terr. N. Atlantic Ocean see Canary Islands
Canary Islands terr. N. Atlantic Ocean 46 B2
Canastota NY U.S.A. 64 D1
Canastra, Serra da mts Brazil 71 A1
Canastra, Serra da mts Brazil 71 B2
Canatiba Brazil 71 C1
Canatlán Mex. 66 D4
Canaveral, Cape U.S.A. 63 K6
Cañaveras Spain 25 E3
Canavieiras Brazil 71 D1
Canbelego Australia 58 C3
Canberra Australia 58 D5
Cancún Mex. 67 G4
Çandar Turkey see Kastamonu
Çandarlı Turkey 27 L5
Candia Greece see Iraklion
Cândido de Abreu Brazil 71 A4
Çandır Turkey 34 D2
Candle Lake Canada 62 F1
Candlewood, Lake CT U.S.A. 64 E2
Cane r. Australia 54 A5
Canea Greece see Chania
Canela Brazil 71 A5
Canelones Uruguay 70 E4
Cangallo Peru 68 D6
Cangamba Angola 49 B5
Cangandala, Parque Nacional de nat. park Angola 49 B4
Cangbu r. China see Brahmaputra
Cango Caves S. Africa 50 F7
Cangola Angola 49 B4
Canguaretama Brazil 69 K5
Canguçu Brazil 70 F4
Canguçu, Serra do hills Brazil 70 F4
Cangzhou China 43 L5
Caniapiscau Canada 61 L4
Caniapiscau r. Canada 61 L4
Caniapiscau, Réservoir de resr Canada 61 K4
Caniçado Moz. see Guija
Canicattì Sicilia Italy 26 E6
Canindé Brazil 69 K4
Canisteo r. NY U.S.A. 64 C1
Canisteo Peninsula Antarctica 76 K2
Çankırı Turkey 34 D2
Canna Australia 55 A7
Canna i. U.K. 20 C3
Cannanore India see Kannur
Cannanore Islands India 38 B4
Cannes France 24 H5
Cannock U.K. 19 E6
Cann River Australia 58 D6
Canoas Brazil 71 A5
Canōas, Rio das r. Brazil 71 A4
Canoeiros Brazil 71 B2
Canoinhas Brazil 71 A4
Canoona Australia 56 E4
Canora Canada 62 G1
Canowindra Australia 58 D4
Cantabrian Mountains Spain see Cantábrica, Cordillera
Cantábrica, Cordillera mts Spain 25 D2
Cantábrico, Mar sea Spain 25 C2
Canterbury U.K. 19 I7
Canterbury Bight b. N.Z. 59 C7
Canterbury Plains N.Z. 59 C6
Cân Thơ Vietnam 31 J5
Cantil CA U.S.A. 65 D3
Canton MS U.S.A. 63 I5
Canton OH U.S.A. 64 A2
Canton PA U.S.A. 64 C2
Canton Island atoll Kiribati see Kanton
Cantuaria U.K. see Canterbury
Canunda National Park Australia 57 C8
Canutama Brazil 68 F5
Canvey Island U.K. 19 H7
Cany-Barville France 19 H9
Canyon U.S.A. 62 G4
Canyon Ferry Lake U.S.A. 62 E2
Cao Băng Vietnam 31 J4
Caohu China 42 E4
Caoshi China 44 B4
Caozhou China see Heze
Çapakçur Turkey see Bingöl
Capanaparo r. Venez. 68 E2
Capanema Brazil 69 I4
Capão Bonito Brazil 71 A4
Caparaó, Serra do mts Brazil 71 C3
Cape r. Australia 56 D4
Cape Arid National Park Australia 55 C8
Cape Barren Island Australia 57 [inset]
Cape Basin sea feature S. Atlantic Ocean 72 I8
Cape Breton Island Canada 61 L5
Cape Charles VA U.S.A. 64 C4
Cape Coast Ghana 46 C4
Cape Coast Castle Ghana see Cape Coast
Cape Cod Bay MA U.S.A. 64 F2
Cape Cod National Seashore nature res. MA U.S.A. 64 G2
Cape Crawford Australia 56 A3
Cape Dorset Canada 61 K3
Cape Girardeau U.S.A. 63 J4
Cape Johnson Depth sea feature N. Pacific Ocean 74 E5
Cape Juby Morocco see Tarfaya
Cape Krusenstern National Monument nat. park U.S.A. 60 B3
Capel Australia 55 A8
Cape Le Grand National Park Australia 55 C8
Capelinha Brazil 71 C2
Capella Australia 56 E4
Capelongo Angola see Kuvango
Cape May NJ U.S.A. 64 D3
Cape May Court House NJ U.S.A. 64 D3
Cape May Point NJ U.S.A. 64 D3
Cape Melville National Park Australia 56 D2
Capenda-Camulemba Angola 49 B4
Cape Palmerston National Park Australia 56 E4
Cape Range National Park Australia 54 A5
Cape Town S. Africa 50 D7
Cape Tribulation National Park Australia 56 D2
Cape Upstart National Park Australia 56 D3
Cape Verde country N. Atlantic Ocean 46 [inset]
Cape Verde Basin sea feature N. Atlantic Ocean 72 F5
Cape Verde Plateau sea feature N. Atlantic Ocean 72 F4
Cape York Peninsula Australia 56 C2
Cap-Haïtien Haiti 67 J5
Capim r. Brazil 69 I4
Capivara, Represa resr Brazil 71 A3
Čapljina Bos.-Herz. 26 G3
Cappoquin Ireland 21 E5
Capraia, Isola di i. Italy 26 C3
Caprara, Punta pt Sardegna Italy 26 C4
Capri, Isola di i. Italy 26 F4
Capricorn Channel Australia 56 E4
Capricorn Group atolls Australia 56 F4
Caprivi Strip reg. Namibia 49 C5
Capsa Tunisia see Gafsa
Captina r. OH U.S.A. 64 A3
Capuava Brazil 71 B4
Caquetá r. Col. 68 E4
Caracal Romania 27 K2
Caracas Venez. 68 E1
Caraguatatuba Brazil 71 B3
Caraí Brazil 71 C2
Carajás Brazil 69 H5
Carajás, Serra dos hills Brazil 69 H5
Carales Sardegna Italy see Cagliari
Caralis Sardegna Italy see Cagliari
Carandaí Brazil 71 C3
Caransebeş Romania 27 J2
Caraquet Canada 63 N2
Caratasca, Laguna de lag. Hond. 67 H5
Caratinga Brazil 71 C2
Carauari Brazil 68 E4
Caravaca de la Cruz Spain 25 F4
Caravelas Brazil 71 D2
Carbó Mex. 66 B3
Carbon, Cap c. Alg. 25 F6
Carbonara, Capo c. Sardegna Italy 26 C5
Carbondale PA U.S.A. 64 D2
Carbonia Sardegna Italy 26 C5
Carbonita Brazil 71 C2
Carcaixent Spain 25 F4
Carcassonne France 24 F5
Cardamom Hills India 38 C4
Cárdenas Cuba 67 H4
Cárdenas Mex. 66 E4
Cardenyabba watercourse Australia 58 A2
Çardi Turkey see Harmancık
Cardiel, Lago l. Arg. 70 B7
Cardiff U.K. 19 D7
Cardiff MD U.S.A. 64 C3
Cardigan U.K. 19 C6
Cardigan Bay U.K. 19 C6
Cardoso Brazil 71 A3
Cardoso, Ilha do i. Brazil 71 B4
Carei Romania 27 J1
Carentan France 24 D2
Carey, Lake imp. l. Australia 55 C7
Cargados Carajos Islands Mauritius 73 L7
Carhaix-Plouguer France 24 C2
Cariacica Brazil 71 C3
Cariamanga Ecuador 68 C4
Caribbean Sea N. Atlantic Ocean 67 H5
Caribou U.S.A. 63 N2
Caribou Lake Canada 61 J4
Caribou Mountains Canada 60 G4
Carinda Australia 58 C3
Cariñena Spain 25 F3
Carinhanha r. Brazil 71 C1
Carlabhagh U.K. see Carloway
Carletonville S. Africa 51 H4
Carlingford Lough inlet Ireland/U.K. 21
Carlisle U.K. 18 E4
Carlisle NY U.S.A. 64 D1
Carlisle PA U.S.A. 64 C2
Carlisle Lakes imp. l. Australia 55 D7
Carlit, Pic mt. France 24 E5
Carlos Chagas Brazil 71 C2
Carlow Ireland 21 F5
Carloway U.K. 20 C2
Carlsbad Czech Rep. see Karlovy Vary
Carlsbad CA U.S.A. 65 D4
Carlsbad NM U.S.A. 62 G5
Carlsberg Ridge sea feature Indian Ocea 73 L5
Carlson Inlet Antarctica 76 L1
Carlton Hill Australia 54 E3
Carluke U.K. 20 F5
Carlyle Canada 62 G2
Carmacks Canada 60 E3
Carmagnola Italy 26 B2
Carman Canada 62 H2
Carmana Iran see Kermān
Carmarthen U.K. 19 C7
Carmarthen Bay U.K. 19 C7
Carmaux France 24 F4
Carmel NY U.S.A. 64 E2
Carmel, Mount h. Israel 39 B3
Carmel Head hd U.K. 18 C5
Carmel Valley CA U.S.A. 65 B2
Carmen, Isla i. Mex. 66 B3
Carmen de Patagones Arg. 70 D6
Carmichael CA U.S.A. 65 B1
Carmo da Cachoeira Brazil 71 B3
Carmo do Paranaíba Brazil 71 B2
Carmona Angola see Uíge
Carmona Spain 25 D5
Carnac France 24 C3
Carnamah Australia 55 A7
Carnarvon Australia 55 A6
Carnarvon S. Africa 50 F6
Carnarvon National Park Australia 56 D5
Carnarvon Range hills Australia 55 C6
Carnarvon Range mts Australia 56 E5
Carn Dearg h. U.K. 20 E3
Carndonagh Ireland 21 E2
Carnegie Australia 55 C6
Carnegie, Lake imp. l. Australia 55 C6
Carn Eige mt. U.K. 20 D3
Carnes Australia 55 F7
Carney Island Antarctica 76 J2
Carnforth U.K. 18 E4
Carn Glas-choire h. U.K. 20 F3
Carnlough U.K. 21 G3
Carn nan Gabhar mt. U.K. 20 F4
Carn Odhar h. U.K. 20 E3
Carnot Cent. Afr. Rep. 48 B3
Carnoustie U.K. 20 G4
Carnsore Point Ireland 21 F5
Carnwath U.K. 20 F5
Carola Cay rf Australia 56 F3
Carolina Brazil 69 I5
Carolina S. Africa 51 J4
Caroline Island atoll Kiribati 75 J6
Caroline Islands N. Pacific Ocean 41 G7
Caroline Peak N.Z. 59 A7
Caroline Range hills Australia 54 D4
Caroní r. Venez. 68 F2
Carpathian Mountains Europe 13 C6
Carpații mts Europe see Carpathian Mountains
Carpații Meridionali mts Romania see Transylvanian Alps
Carpații Occidentali mts Romania 27 J2
Carpentaria, Gulf of Australia 56 B2
Carpentras France 24 G4
Carpi Italy 26 D2
Carpinteria CA U.S.A. 65 C3
Carra, Lough l. Ireland 21 C4
Carraig na Siuire Ireland see Carrick-on-Suir
Carrantuohill mt. Ireland 21 C6
Carrara Italy 26 D2
Carrasco, Parque Nacional nat. park Bol 68 F7
Carrathool Australia 58 B5
Carrhae Turkey see Harran
Carrickfergus U.K. 21 G3
Carrickmacross Ireland 21 F4
Carrick-on-Shannon Ireland 21 D4
Carrick-on-Suir Ireland 21 E5
Carrigallen Ireland 21 E4
Carrigtohill Ireland 21 D6
Carrington U.S.A. 62 H2
Carrizal Bajo Chile 70 B3
Carrizo Springs U.S.A. 62 H6
Carrizozo U.S.A. 62 F5
Carroll U.S.A. 63 I3
Carrollton GA U.S.A. 63 J5
Carrollton OH U.S.A. 64 A2
Carrolltown PA U.S.A. 64 B2
Carron r. U.K. 20 E3
Carrowmore Lake Ireland 21 C3
Çarşamba Turkey 34 E2
Carson City NV U.S.A. 65 C1
Carson Escarpment Australia 54 D3
Carson Lake NV U.S.A. 65 C1
Carstensz Pyramid mt. Indon. see Jaya, Puncak

rstensz-top mt. Indon. see Jaya, Puncak
Cartagena Col. 68 C1
-tagena Spain 25 F5
-teret Group is P.N.G. see
Kilinailau Islands
-teret Island Solomon Is see Malaita
rthage tourist site Tunisia 26 D6
rthage MO U.S.A. 63 I4
rthago tourist site Tunisia see Carthage
rthago Nova Spain see Cartagena
rtier Island Australia 54 C3
rtmel U.K. 18 E4
rtwright Nfld. and Lab. Canada 61 M4
ruaru Brazil 69 K5
rúpano Venez. 68 F1
ryapundy Swamp Australia 57 C6
sablanca Morocco 22 C5
sa Branca Brazil 71 B3
sa de Piedra, Embalse resr Arg. 70 C5
sa Grande U.S.A. 62 E5
sale Monferrato Italy 26 C2
salmaggiore Italy 26 D2
sca Brazil 71 A5
scade Australia 55 C8
scade r. N.Z. 59 B7
scade Point N.Z. 59 B7
scade Range mts Canada/U.S.A. 60 F5
scais Port. 25 B4
scavel Brazil 70 F2
serta Italy 26 F4
sey (Australia) research stn Antarctica
76 F2
sey Bay Antarctica 76 D2
seyr, Raas c. Somalia see
Gwardafuy, Gees
shel Ireland 21 E5
shmere Australia 58 D1
sino Australia 58 F2
siquiare, Canal r. Venez. 68 E3
snewydd U.K. see Newport
spe Spain 25 F3
sper U.S.A. 62 F3
spian Lowland Kazakh./Rus. Fed. 30 D2
spian Sea l. Asia/Europe 35 H1
ss WV U.S.A. 64 B3
ssacatiza Moz. 49 D5
ssadaga NY U.S.A. 64 B1
ssaigne Alg. see Sidi Ali
ssamba Angola 49 C5
ssia Brazil 71 B3
ssiar Mountains Canada 60 E3
ssilândia Brazil 71 A2
ssilis Australia 58 D4
ssino Italy 26 E4
ssley r. U.K. 20 E3
ssongue Angola 49 B5
stanhal Brazil 69 I4
stanho Brazil 68 F5
staños Mex. 66 D3
stelfranco Veneto Italy 26 D2
stell-nedd U.K. see Neath
stell Newydd Emlyn U.K. see
Newcastle Emlyn
stellón Spain see Castellón de la Plana
stellón de la Plana Spain 25 F4
stelo Branco Port. 25 C4
stelo de Vide Port. 25 C4
steltermini Sicilia Italy 26 E6
stelvetrano Sicilia Italy 26 E6
stiglione della Pescaia Italy 26 D3
stilla y León reg. Spain 24 B6
stlebar Ireland 21 C4
stlebay U.K. 20 B4
stlebellingham Ireland 21 F4
stleblayney Ireland 21 F3
stlebridge Ireland 21 F5
stle Carrock U.K. 18 E4
stle Cary U.K. 19 E7
stlederg U.K. 21 E3
stledermot Ireland 21 F5
stle Dome Mountains AZ U.S.A. 65 E4
stle Donington U.K. 19 F6
stle Douglas U.K. 20 F6
stleford U.K. 18 F5
stlegar Canada 62 D2
stlegregory Ireland 21 B5
stleisland Ireland 21 C5
stlemaine Australia 58 B6
stlemaine Ireland 21 C5
stlemartyr Ireland 21 D6
stle Mountain CA U.S.A. 65 B3
stlepoint N.Z. 59 F5
stlepollard Ireland 21 E4
stlerea Ireland 21 D4
stlereagh r. Australia 58 C3
stle Rock U.S.A. 62 G4
stletown Ireland 21 E5
stletown Isle of Man 18 C4
stra Regina Germany see Regensburg
stres France 24 F5
stries St Lucia 67 L6
stro Brazil 71 A4
stro Chile 70 B6
stro Alves Brazil 71 D1
stro Verde Port. 25 B5
stroville CA U.S.A. 65 B2
t Turkey 35 F3
tacaos Peru 68 B5
taguases Brazil 71 C3
tak Turkey 35 F3
talão Brazil 71 B2
talca Yarımadası pen. Turkey 27 M4
talonia aut. comm. Spain see Cataluña
taluña aut. comm. Spain 25 G3
talunya aut. comm. Spain see Cataluña
tamarca Arg. 70 C3
tana Sicilia Italy see Catania
tanduanes i. Phil. 41 E6
tanduva Brazil 71 A3
tania Sicilia Italy 26 F6
tanzaro Italy 26 G5
tarman Phil. 41 E6
tastrophe, Cape Australia 57 A7
taxa Moz. 49 D5
tbalogan Phil. 41 E6
tembe Moz. 51 K4
tengue Angola 49 B5
tete Angola 49 B4
thair Dónall Ireland 21 B6
thcart Australia 58 D6
thcart S. Africa 51 H7
thedral Peak S. Africa 51 I5
thedral Rock National Park Australia
58 F3
theys Valley CA U.S.A. 65 B2
tió Guinea-Bissau 46 B3
tisimiña Venez. 68 F3
t Island Bahamas 67 I4
t Lake Canada 63 I1
toche, Cabo c. Mex. 67 G4
to Island and Bank rf Australia 56 F4
triló Arg. 70 D5
tskill NY U.S.A. 64 E1
tskill Mountains NY U.S.A. 64 D1
tuane Moz. 51 K4
ubvick, Mount Canada 61 L4
uca r. Col. 67 J7
ucaia Brazil 69 K4
ucasia Col. 68 D2
ucasus mts Asia/Europe 35 F2
u Giat Vietnam 42 J9
Caulonia Italy 26 G5
Caungula Angola 49 B4
Cauquenes Chile 70 B5
Cavaglià Italy 26 C2
Cavalcante, Serra do hills Brazil 71 B1
Cavan Ireland 21 E4
Çavdır Turkey 27 M6
Caveira r. Brazil 71 C1
Caviana, Ilha i. Brazil 69 H3
Cawdor U.K. 20 F3
Cawnpore India see Kanpur
Cawston U.K. 19 I6
Caxias Brazil 69 J4
Caxias do Sul Brazil 71 A5
Caxito Angola 49 B4
Çay Turkey 27 N5
Cayambe, Volcán vol. Ecuador 68 C3
Çaybaşı Turkey see Çayeli
Çaycuma Turkey 27 O4
Çayeli Turkey 35 F2
Cayenne Fr. Guiana 69 H3
Cayeux-sur-Mer France 19 I8
Çayırhan Turkey 27 N4
Cayman Brac i. Cayman Is 67 I5
Cayman Islands terr. West Indies 67 H5
Cayman Trench sea feature Caribbean Sea
72 C4
Caynabo Somalia 48 E3
Cayucos CA U.S.A. 65 B3
Cayuga Ont. Canada 64 B1
Cayuga Lake NY U.S.A. 64 C1
Cazê China 37 F3
Cazenovia NY U.S.A. 64 D1
Cazombo Angola 49 C5
Ceadâr-Lunga Moldova see Ciadîr-Lunga
Ceanannus Mór Ireland see Kells
Ceann a Deas na Hearadh pen. U.K. see
South Harris
Ceará Brazil see Fortaleza
Ceara Abyssal Plain sea feature
S. Atlantic Ocean 72 F6
Ceatharlach Ireland see Carlow
Ceballos Mex. 66 D3
Cebu Phil. 41 E6
Čechy reg. Czech Rep. 17 N6
Cecil Plains Australia 58 E1
Cecil Rhodes, Mount h. Australia 55 C6
Cecina Italy 26 D3
Cedar City UT U.S.A. 65 F2
Cedar Island VA U.S.A. 64 D4
Cedar Lake Canada 62 G1
Cedar Rapids U.S.A. 63 I3
Cedar Run NJ U.S.A. 64 D3
Cedarville S. Africa 51 I6
Cedros, Isla i. Mex. 66 A3
Ceduna Australia 55 F8
Ceeldheere Somalia 48 E3
Ceerigaabo Somalia 48 E2
Cefalù Sicilia Italy 26 F5
Cegléd Hungary 27 H1
Cêgnê China 37 H2
Çekerek Turkey 34 D2
Çekiçler Turkm. 35 I3
Celaya Mex. 66 D4
Celbridge Ireland 21 F4
Celebes i. Indon. 41 E8
Celebes Basin sea feature Pacific Ocean
74 E5
Celebes Sea Indon./Phil. 41 E7
Celestún Mex. 66 F4
Celje Slovenia 26 F1
Celle Germany 17 M4
Celovec Austria see Klagenfurt
Celtic Sea Ireland/U.K. 16 D5
Celtic Shelf sea feature N. Atlantic Ocean
72 H2
Cenderawasih, Teluk b. Indon. 41 F8
Çendir r. Turkm. 35 I3
Centane S. Africa see Kentani
Centenary Zimbabwe 49 D5
Centereach NY U.S.A. 64 E2
Centerville WV U.S.A. 64 A3
Centrafricaine, République country Africa
see Central African Republic
Central admin. dist. Botswana 51 H2
Central, Cordillera mts Col. 68 C3
Central, Cordillera mts Peru 68 C6
Central African Empire country Africa see
Central African Republic
Central African Republic country Africa
48 B3
Central Brahui Range mts Pak. 33 K4
Central City U.S.A. 62 H3
Central Kalahari Game Reserve nature res.
Botswana 50 F2
Central Kara Rise sea feature Arctic Ocean
77 F1
Central Makran Range mts Pak. 33 J4
Central Mount Stuart h. Australia 54 F5
Central Pacific Basin sea feature
Pacific Ocean 74 H5
Central Provinces state India see
Madhya Pradesh
Central Range mts P.N.G. 52 E2
Central Russian Upland hills Rus. Fed.
13 H5
Central Siberian Plateau Rus. Fed. 29 M3
Central Square NY U.S.A. 64 C1
Centreville MD U.S.A. 64 C3
Ceos i. Greece see Tzia
Cephaloedium Sicilia Italy see Cefalù
Cephalonia i. Greece 27 I5
Ceram i. Indon. see Seram
Cerbat Mountains AZ U.S.A. 65 E3
Ceres Arg. 70 D3
Ceres Brazil 71 A1
Ceres S. Africa 50 D7
Ceres CA U.S.A. 65 B2
Céret France 24 F5
Cerezo de Abajo Spain 25 E3
Cerignola Italy 26 F4
Cerigo i. Greece see Kythira
Cêringgolêb China see Dongco
Çerkeş Turkey 34 D2
Çerkeşli Turkey 27 M4
Çermik Turkey 35 E3
Cernăuţi Ukr. see Chernivtsi
Cernavodă Romania 27 M2
Cerralvo, Isla i. Mex. 66 C4
Cërrik Albania 27 H4
Cerritos Mex. 66 D4
Cerro Azul Brazil 71 A4
Cerro de Pasco Peru 68 C6
Cerros Colorados, Embalse resr Arg.
70 C5
Cervantes, Cerro mt. Arg. 70 B8
Cervati, Monte mt. Italy 26 F4
Cervione Corse France 24 I5
Cervo Spain 25 C2
Cesena Italy 26 E2
Cēsis Latvia 15 N8
Česká Republika country Europe see
Czech Republic
České Budějovice Czech Rep. 17 O6
Českomoravská vysočina hills Czech Rep.
17 O6
Český Krumlov Czech Rep. 17 O6
Český les mts Czech Rep./Germany
17 N6
Çeşme Turkey 27 L5
Cessnock Australia 58 E4
Cetatea Albă Ukr. see
Bilhorod-Dnistrovs'kyy
Cetinje Montenegro 26 H3
Cetraro Italy 26 F5
Ceuta N. Africa 25 D6
Ceva-i-Ra rf Fiji 53 H4
Cévennes mts France 24 F5
Cévennes, Parc National des nat. park
France 24 F4
Cevizli Turkey 39 C1
Cevizlik Turkey see Maçka
Ceyhan Turkey 34 D3
Ceyhan r. Turkey 39 B1
Ceyhan Boğazı r. mouth Turkey 39 B1
Ceylanpınar Turkey 35 F3
Ceylon country Asia see Sri Lanka
Chābahār Iran 33 J4
Chabrol i. New Caledonia see Lifou
Chabug China 37 E2
Chabyêr Caka salt l. China 37 F3
Chachapoyas Peru 68 C5
Chacharan Pak. 36 B3
Chaco Boreal reg. Para. 70 E2
Chad country Africa 47 E3
Chad, Lake Africa 47 E3
Chadaasan Mongolia 42 I3
Chadan Rus. Fed. 42 G2
Chadibe Botswana 51 H2
Chadron U.S.A. 62 G3
Chadyr-Lunga Moldova see Ciadîr-Lunga
Chaeryŏng N. Korea 45 B5
Chagai Pak. 33 J4
Chagdo Kangri mt. China 37 F2
Chaghcharān Afgh. 36 A2
Chagny France 24 G3
Chagoda Rus. Fed. 12 G4
Chagos Archipelago is B.I.O.T. 73 M6
Chagos-Laccadive Ridge sea feature
Indian Ocean 73 M6
Chagos Trench sea feature Indian Ocean
73 M6
Chagoyan Rus. Fed. 44 C1
Chahbounia Alg. 25 H6
Chāh-e Bāgh well Iran 35 I5
Chāh-e Bāzargānī Iran 35 I5
Chāh-e Gonbad well Iran 35 I4
Chāh-e Khorāsān well Iran 35 I4
Chāh-e Malek well Iran 35 I5
Chāh-e Mūjān well Iran 35 I4
Chāh-e Qobād well Iran 35 I4
Chāh-e Shūr well Iran 35 I4
Chāh Kūh Iran 35 I5
Chāh Pās well Iran 35 I4
Chaibasa India 37 F5
Chainjoin Co l. China 37 F2
Chaitén Chile 70 B6
Chajarí Arg. 70 E4
Chakai India 37 F4
Chak Amru Pak. 36 C2
Chakar r. Pak. 36 B3
Chakaria Bangl. 37 H5
Chakdarra Pak. 36 C2
Chakku Pak. 36 A4
Chakonipau, Lac l. Canada 61 L4
Chakoria Bangl. see Chakaria
Ch'ak'vi Georgia 35 F2
Chala Peru 68 D7
Chalatenango El Salvador 66 G6
Chaláua Moz. 49 D5
Chalaxung China 37 I2
Chalcedon Turkey see Kadıköy
Chalengkou China 42 G5
Chaleur Bay inlet Canada 63 N2
Chaleurs, Baie des inlet Canada see
Chaleur Bay
Chalisgaon India 38 B1
Chalki i. Greece 27 L6
Chalkida Greece 27 J5
Challakere India 38 C3
Challans France 24 D3
Challapata Bol. 68 E7
Challenger Deep sea feature
N. Pacific Ocean 74 F5
Challenger Fracture Zone sea feature
S. Pacific Ocean 74 M8
Challis U.S.A. 62 E3
Châlons-en-Champagne France 24 G2
Châlons-sur-Marne France see
Châlons-en-Champagne
Chalon-sur-Saône France 24 G3
Chaltan Pass Azer. 35 H2
Chālūs Iran 35 H3
Cham, Kūh-e h. Iran 35 H4
Chamaico Arg. 70 D5
Chamais Bay Namibia 50 B4
Chaman Pak. 30 F3
Chaman Bīd Iran 35 J3
Chamba India 36 D2
Chamba Tanz. 49 D5
Chambal r. India 36 D4
Chamberlain r. Australia 54 D4
Chamberlain U.S.A. 62 H3
Chambersburg PA U.S.A. 64 C3
Chambéry France 24 G4
Chambeshi r. Zambia 49 C5
Chambi, Jebel mt. Tunisia 26 C7
Chamdo China see Qamdo
Chamechaude mt. France 24 G4
Chamoli India see Gopeshwar
Chamonix-Mont-Blanc France 24 H4
Champa India 37 E5
Champagne Castle mt. S. Africa 51 I5
Champagne Humide reg. France 24 G2
Champagne Pouilleuse reg. France 24 F2
Champagnole France 24 G3
Champagny Islands Australia 54 D3
Champaign U.S.A. 63 J3
Champhai India 37 H5
Champlain VA U.S.A. 64 C3
Champlain, Lake Canada/U.S.A. 61 K5
Champotón Mex. 66 F5
Chamrajnagar India 38 C4
Chamzinka Rus. Fed. 13 J5
Chanak Turkey see Çanakkale
Chañaral Chile 70 B3
Chanda India see Chandrapur
Chandalar r. U.S.A. 60 D3
Chandausi India 36 D3
Chandbali India 37 F5
Chanderi India 36 D4
Chandigarh India 36 D3
Chandil India 37 F5
Chandler AZ U.S.A. 62 E5
Chandod India 36 C5
Chandpur Bangl. 37 G5
Chandpur India 36 D3
Chandragiri India 38 C3
Chandrapur India 38 C2
Chandvad India 38 B1
Changane r. Moz. 51 K3
Changbai China 44 C4
Changbai Shan mts China/N. Korea 44 B4
Chang Cheng research stn Antarctica see
Great Wall
Changchow Fujian China see Zhangzhou
Changchow Jiangsu China see Changzhou
Changchun China 44 B4
Changchunling China 44 B3
Changde China 43 K7
Changgi-ap pt S. Korea 45 C5
Changgo China 37 F3
Changhua Taiwan 43 M8
Changhŭng S. Korea 45 B6
Changhwa Taiwan see Changhua
Changji China 42 F4
Changjin-ho resr N. Korea 45 B4
Changkiang China see Zhanjiang
Changlang India 37 H4
Changling China 44 A3
Changlung India 33 M3
Changma China 42 H5
Changnyŏn N. Korea 45 B5
Ch'ang-pai Shan mts China/N. Korea see
Changbai Shan
Changp'yŏng S. Korea 45 C5
Changsan-got pt N. Korea 45 B5
Changsha China 43 K7
Changteh China see Changde
Changting Heilong. China 44 C3
Ch'angwŏn S. Korea 45 C6
Changyŏn N. Korea 45 B5
Changzhi China 43 K5
Changzhou China 43 L6
Chañi, Nevado de mt. Arg. 70 C2
Chania Greece 27 K7
Chanion, Kolpos b. Greece 27 J7
Channapatna India 38 C3
Channel Islands English Chan. 19 E9
Channel Islands CA U.S.A. 65 C4
Channel Islands National Park CA U.S.A.
65 C3
Channel-Port-aux-Basques Canada 61 M5
Channel Tunnel France/U.K. 19 I7
Chantada Spain 25 C2
Chanthaburi Thai. 31 J5
Chantilly France 24 F2
Chanuwala Pak. 36 C2
Chany, Ozero salt l. Rus. Fed. 28 I4
Chaor He r. China see Qulin Gol
Chaowula Shan mt. China 37 I2
Chaoyang Heilong. China see Jiayin
Chaoyang Liaoning China 43 M4
Chaoyangcun China 44 B2
Chaoyang Hu l. China 37 F2
Chaozhong China 44 A2
Chaozhou China 43 L8
Chapada Diamantina, Parque Nacional
nat. park Brazil 71 C1
Chapada dos Veadeiros, Parque Nacional
da nat. park Brazil 71 B1
Chapak Gozār Afgh. 36 A1
Chapala, Laguna de l. Mex. 66 D4
Chāpārī, Kowtal-e Afgh. 36 A2
Chapayevo Kazakh. 30 E1
Chapayevsk Rus. Fed. 13 K5
Chapecó Brazil 70 F3
Chapecó r. Brazil 70 F3
Chapel-en-le-Frith U.K. 18 F5
Chapeltown U.K. 18 F5
Chapleau Canada 63 K2
Chaplygin Rus. Fed. 13 H5
Chappell Islands Australia 57 [inset]
Chapra Bihar India see Chhapra
Chapra Jharkhand India see Chatra
Charagua Bol. 68 F7
Charcas Mex. 66 D4
Charcot Island Antarctica 76 L2
Chard U.K. 19 E8
Chardara Kazakh. see Shardara
Chardara, Step' plain Kazakh. 42 B4
Chardon OH U.S.A. 64 A2
Chardzhev Turkm. see Türkmenabat
Chardzhou Turkm. see Türkmenabat
Charef Alg. 25 H6
Charef, Oued watercourse Morocco 22 D5
Charente r. France 24 D4
Chari r. Cameroon/Chad 47 E3
Chārīkār Afgh. 36 B2
Chārjew Turkm. see Türkmenabat
Charkayuvom Rus. Fed. 12 L2
Chār Kent Afgh. 36 A1
Charkhlik China see Ruoqiang
Charleroi Belgium 16 J5
Charles, Cape VA U.S.A. 64 D4
Charles City IA U.S.A. 63 I3
Charles City VA U.S.A. 64 C4
Charles Hill Botswana 50 E2
Charles Island Galápagos Ecuador see
Santa María, Isla
Charles Point Australia 54 E3
Charleston N.Z. 59 C5
Charleston SC U.S.A. 63 L5
Charleston WV U.S.A. 63 K4
Charleston Peak NV U.S.A. 65 E2
Charlestown Ireland 21 D4
Charlestown NH U.S.A. 64 E1
Charlestown RI U.S.A. 64 F2
Charles Town WV U.S.A. 64 C3
Charleville Australia 57 D5
Charleville Ireland 21 D5
Charleville-Mézières France 24 G2
Charlotte NC U.S.A. 63 K4
Charlotte Amalie Virgin Is (U.S.A.) 67 L5
Charlotte Harbor b. U.S.A. 63 K6
Charlottesville VA U.S.A. 64 B3
Charlottetown Canada 61 L5
Charlton Australia 58 A6
Charlton Island Canada 63 L1
Charsadda Pak. 36 B2
Charshangngy Turkm. see Köýtendag
Charters Towers Australia 56 D4
Chartres France 24 E2
Chas India 37 F5
Chase Canada 62 D1
Chashmeh-ye Palasi Iran 35 I4
Chashmeh-ye Shotoran well Iran 35 I4
Chashniki Belarus 13 F5
Chaslands Mistake c. N.Z. 59 B8
Chasŏng N. Korea 44 B4
Chasseral mt. Switz. 17 K7
Chassiron, Pointe de pt France 24 D3
Chastab, Kūh-e mts Iran 35 I4
Chāt Iran 35 I3
Chatanika U.S.A. 60 D3
Châteaubriant France 24 D3
Château-du-Loir France 24 E3
Châteaudun France 24 E2
Châteaulin France 24 B2
Châteaumeillant France 24 F3
Châteauneuf-sur-Loire France 24 F3
Châteauroux France 24 E3
Château-Thierry France 24 F2
Châtellerault France 24 E3
Chatham U.K. 19 H7
Chatham MA U.S.A. 64 G2
Chatham NY U.S.A. 64 E1
Chatham PA U.S.A. 64 D3
Chatham VA U.S.A. 64 B4
Chatham, Isla i. Chile 70 B8
Chatham Island Galápagos Ecuador see
San Cristóbal, Isla
Chatham Island N.Z. 53 I6
Chatham Island Samoa see Savai'i
Chatham Islands N.Z. 53 I6
Chatham Rise sea feature S. Pacific Ocean
74 I8
Châtillon-sur-Seine France 24 G3
Chatkal Range mts Kyrg./Uzbek. 33 L1
Chatra India 37 F4
Chatra Nepal 37 F4
Chatsworth NJ U.S.A. 64 D3
Chattagam Bangl. see Chittagong
Chattanooga U.S.A. 63 J4
Chattarpur India see Chhatarpur
Chatteris U.K. 19 H6
Chattisgarh state India see Chhattisgarh
Chatyr-Tash Kyrg. 33 M1
Chauhtan India 36 B4
Chauk Myanmar 37 H5
Chaumont France 24 G2
Chaunskaya Guba b. Rus. Fed. 29 R3
Chauny France 24 F2
Chausy Belarus see Chavusy
Chautauqua, Lake NY U.S.A. 64 B1
Chauter Pak. 36 A3
Chavakachcheri Sri Lanka 38 D4
Chavakkad India 38 B4
Chaves Port. 25 C3
Chavusy Belarus 13 F5
Chayatyn, Khrebet ridge Rus. Fed. 44 E1
Chayevo Rus. Fed. 12 H4
Chaykovskiy Rus. Fed. 11 Q4
Chazhegovo Rus. Fed. 12 L3
Cheadle U.K. 19 F6
Cheat r. WV U.S.A. 64 B3
Cheb Czech Rep. 17 N5
Chebba Tunisia 26 D7
Cheboksarskoye Vodokhranilishche resr
Rus. Fed. 12 J5
Cheboksary Rus. Fed. 12 J4
Cheboygan U.S.A. 63 K2
Chechen', Ostrov i. Rus. Fed. 13 J8
Chech'ŏn S. Korea 45 C5
Cheddar U.K. 19 E7
Cheektowaga NY U.S.A. 64 B1
Cheepie Australia 58 B1
Cheetham, Cape Antarctica 76 H2
Chefchaouene Morocco 25 D6
Chefoo China see Yantai
Chefornak U.S.A. 60 B3
Chefu Moz. 51 K2
Chegdomyn Rus. Fed. 44 D2
Chegga Mauritania 46 C2
Chegutu Zimbabwe 49 D5
Chehalis U.S.A. 62 C2
Chehel Chashmeh, Kūh-e mt. Iran 35 G4
Cheju S. Korea 45 B6
Cheju-do i. S. Korea 45 B6
Cheju-haehyŏp sea chan. S. Korea 45 B6
Chekhov Moskovskaya Oblast' Rus. Fed.
13 H5
Chekhov Sakhalinskaya Oblast' Rus. Fed.
44 F3
Chekiang prov. China see Zhejiang
Chekichler Turkm. see Çekiçler
Chekunda Rus. Fed. 44 D2
Chela, Serra da mts Angola 49 B5
Chelan, Lake U.S.A. 62 C2
Cheleken Turkm. see Hazar
Cheline Moz. 51 L2
Chelkar Kazakh. see Shalkar
Chełm Poland 13 D6
Chelmer r. U.K. 19 H7
Chełmno Poland 17 Q4
Chelmsford U.K. 19 H7
Cheltenham U.K. 19 E7
Chelva Spain 25 F4
Chelyabinsk Rus. Fed. 28 H4
Chemba Moz. 49 D5
Chêm Co l. China 36 D2
Chemnitz Germany 17 N5
Chemulpo S. Korea see Inch'ŏn
Chenab r. India/Pak. 36 B3
Chenachane, Oued watercourse Alg. 46 C2
Chendir r. Turkm. see Çendir
Chengalpattu India 38 D3
Chengchow China see Zhengzhou
Chengde China 43 L4
Chengdu China 42 I6
Chengjiang China see Taihe
Chengtu China see Chengdu
Chengxian China 42 J6
Chenkaladi Sri Lanka 38 D5
Chennai India 38 D3
Chenqing China 44 B2
Chenqingqiao China see Chenqing
Chenstokhov Poland see Częstochowa
Chentejn Nuruu mts Mongolia 43 J3
Chenzhou China 43 K7
Chepén Peru 68 C5
Chepes Arg. 70 C4
Chepo Panama 67 I7
Chepstow U.K. 19 E7
Cheptsa r. Rus. Fed. 12 K4
Cher r. France 24 E3
Chera state India see Kerala
Cherbaniani Reef India 38 A3
Cherbourg-Octeville France 24 D2
Cherchell Alg. 25 H5
Cherchen China see Qiemo
Cherdakly Rus. Fed. 13 K5
Cherdyn' Rus. Fed. 11 R3
Chereapani rf India see Byramgore Reef
Cheremkhovo Rus. Fed. 42 I2
Cheremshany Rus. Fed. 44 D3
Cheremukhovka Rus. Fed. 12 K4
Cherepanovo Rus. Fed. 42 E2
Cherepovets Rus. Fed. 12 H4
Cherevkovo Rus. Fed. 12 J3
Chergui, Chott ech imp. l. Alg. 22 D5
Chéria Alg. 26 B7
Cheriton VA U.S.A. 64 D4
Cheriyam atoll India 38 B4
Cherkassy Ukr. see Cherkasy
Cherkasy Ukr. 13 G6
Cherkessk Rus. Fed. 13 I7
Cherla India 38 D2
Chernaya Rus. Fed. 12 M1
Chernaya r. Rus. Fed. 12 M1
Chernigov Ukr. see Chernihiv
Chernigovka Rus. Fed. 44 D3
Chernihiv Ukr. 13 F6
Cherninivka Ukr. 13 H7
Chernivtsi Ukr. 13 E6
Chernobyl' Ukr. see Chornobyl'
Chernogorsk Rus. Fed. 42 G2
Chernovtsy Ukr. see Chernivtsi
Chernoye More sea Asia/Europe see
Black Sea
Chernushka Rus. Fed. 11 R4
Chernyakhiv Ukr. 13 F6
Chernyakhovsk Rus. Fed. 13 D5
Chernyanka Rus. Fed. 13 H6
Chernyayevo Rus. Fed. 44 B1
Chernyshevsk Rus. Fed. 43 L2
Chernyshevskiy Rus. Fed. 29 M3
Chernyshkovskiy Rus. Fed. 13 I6
Chernyye Zemli reg. Rus. Fed. 13 J7
Chernyy Irtysh r. China/Kazakh. see
Ertix He
Chernyy Porog Rus. Fed. 12 G3
Chernyy Yar Rus. Fed. 13 J6
Cherrapunji India 37 G4
Cherry Hill NJ U.S.A. 64 D3
Cherry Island Solomon Is 53 G3
Cherry Lake CA U.S.A. 65 C1
Cherskiy Rus. Fed. 77 C2
Cherskiy Range mts Rus. Fed. see
Cherskogo, Khrebet
Cherskogo, Khrebet mts Rus. Fed. 29 P3
Cherskogo, Khrebet mts Rus. Fed. 43 K2
Chertkov Ukr. see Chortkiv
Chertkovo Rus. Fed. 13 I6
Cherven Bryag Bulg. 27 K3
Chervonoarmeyskoye Ukr. see Vil'nyans'k
Chervonoarmiys'k Donets'ka Oblast' Ukr.
see Krasnoarmiys'k
Chervonoarmiys'k Rivnens'ka Oblast' Ukr.
see Radyvyliv
Chervonograd Ukr. see Chervonohrad
Chervonohrad Ukr. 13 E6
Chervyen' Belarus 13 F5
Cherwell r. U.K. 19 F7
Cherykaw Belarus 13 F5
Chesapeake VA U.S.A. 63 L4
Chesapeake Bay Maryland/Virginia U.S.A.
64 C3
Chesham U.K. 19 G7
Cheshire Plain U.K. 18 E5
Cheshskaya Guba b. Rus. Fed. 12 J2
Cheshunt U.K. 19 G7
Chesnokovka Rus. Fed. see Novoaltaysk
Chester U.K. 18 E5
Chester SC U.S.A. 63 K5
Chester r. MD U.S.A. 64 C3
Chesterfield U.K. 18 F5
Chesterfield VA U.S.A. 64 C4
Chesterfield, Îles is New Caledonia 53 F3
Chesterfield Inlet Canada 61 I3
Chesterfield Inlet inlet Canada 61 I3
Chester-le-Street U.K. 18 F4
Chestertown MD U.S.A. 64 C3
Chestertown NY U.S.A. 64 E1
Chestnut Ridge PA U.S.A. 64 B2
Chesuncook Lake U.S.A. 63 N2
Chetaïbi Alg. 26 B6
Chetlat i. India 38 B4
Chetumal Mex. 66 G5
Chetwynd Canada 60 F4
Cheviot N.Z. 59 D6
Cheviot Hills U.K. 18 E3
Cheviot Range hills Australia 56 C5
Cheyenne WY U.S.A. 62 G3
Cheyenne r. U.S.A. 62 G3
Cheyenne Wells U.S.A. 62 G4
Cheyne Bay Australia 55 B8
Cheyur India 38 D3
Chhapra India 37 F4
Chhata India 36 D4
Chhatak Bangl. 37 G4
Chhatarpur Jharkhand India 37 F4
Chhatarpur Madh. Prad. India 36 D4
Chhatr Pak. 36 B3
Chhatrapur India 38 E2
Chhattisgarh state India 37 E5
Chhindwara India 36 D5
Chhitkul India 36 D3
Chhukha Bhutan 37 G4
Chiai Taiwan 43 M8
Chiamboni Somalia 48 E4
Chiange Angola 49 B5
Chiang Mai Thai. 31 I5
Chiang Rai Thai. 31 I5
Chiari Italy 26 C2
Chiautla Mex. 66 E5
Chiavenna Italy 26 C1
Chiayi Taiwan see Chiai
Chiba Japan 45 F6
Chibi China 43 K7
Chibia Angola 49 B5
Chibizovka Rus. Fed. see Zherdevka
Chiboma Moz. 49 D6
Chibougamau Canada 63 M2
Chibuto Moz. 51 K3
Chibuzhang Co l. China 37 G2
Chicacole India see Srikakulam
Chicago U.S.A. 63 J3
Chichagof Island U.S.A. 60 E4
Chichaoua Morocco 22 C5
Chichatka Rus. Fed. 44 A1
Chichester U.K. 19 G8
Chichester Range mts Australia 54 B5
Chichgarh India 38 D1
Chichibu Japan 45 E6
Chichibu-Tama Kokuritsu-kōen Japan
45 E6
Chichijima-rettō is Japan 45 F8
Chickasha U.S.A. 62 H4
Chiclana de la Frontera Spain 25 C5
Chiclayo Peru 68 C5
Chico r. Arg. 70 C6
Chico U.S.A. 62 C4
Chicomo Moz. 51 L3
Chicopee MA U.S.A. 64 E1
Chicoutimi Canada 61 K5
Chicualacuala Moz. 51 J2
Chidambaram India 38 C4
Chidenguele Moz. 51 L3
Chidley, Cape Canada 61 L3
Chido China see Sêndo
Chido S. Korea 45 B6
Chiducuane Moz. 51 L3
Chiemsee l. Germany 17 N7
Chiengmai Thai. see Chiang Mai
Chieti Italy 26 F3
Chifeng China 43 L4
Chifre, Serra do mts Brazil 71 C2
Chiganak Kazakh. 42 C3
Chiginagak Volcano, Mount U.S.A. 60 C4
Chigu China 37 G3
Chigubo Moz. 51 K2
Chigu Co l. China 37 G3
Chihli, Gulf of China see Bo Hai
Chihuahua Mex. 66 C3
Chiili Kazakh. 42 B4
Chikalda India 36 D5
Chikhali Kalan Parasia India 36 D5
Chikhli India 38 C1
Chikishlyar Turkm. see Çekiçler
Chikmagalur India 38 B3
Chikodi India 38 B2
Chilanko r. Canada 62 C1
Chilas Pak. 36 C2
Chilaw Sri Lanka 38 C5
Childers Australia 56 F5
Childress U.S.A. 62 G5
Chile country S. America 70 B4
Chile Basin sea feature S. Pacific Ocean
75 O8
Chile Chico Chile 70 B7
Chile Rise sea feature S. Pacific Ocean
75 O8
Chilgir Rus. Fed. 13 J7
Chilia-Nouă Ukr. see Kiliya
Chilika Lake India 38 E2
Chililabombwe Zambia 49 C5
Chilko Lake Canada 62 C1
Chillán Chile 70 B5
Chillicothe MO U.S.A. 63 I4
Chillicothe OH U.S.A. 63 K4
Chilliwack Canada 62 C2
Chilo India 36 C4
Chiloé, Isla de i. Chile 70 B6
Chiloé, Isla Grande de i. Chile see
Chiloé, Isla de
Chilpancingo Mex. 66 E5
Chilpancingo de los Bravos Mex. see
Chilpancingo
Chilpi Pak. 36 C1
Chiltern Hills U.K. 19 G7
Chiluage Angola 49 C4
Chilubi Zambia 49 C5
Chilung Taiwan 43 M7
Chilwa, Lake Malawi 49 D5

Chimala Tanz. 49 D4
Chimaltenango Guat. 66 F6
Chimbas Arg. 70 C4
Chimborazo mt. Ecuador 68 C4
Chimbote Peru 68 C5
Chimboy Uzbek. 33 I1
Chimishliya Moldova see Cimişlia
Chimkent Kazakh. see Shymkent
Chimoio Moz. 49 D5
Chimtargha, Qullai mt. Tajik. 33 K2
Chimtorga, Gora mt. Tajik. see Chimtargha, Qullai
China country Asia 42 H5
China, Republic of country Asia see Taiwan
China Lake CA U.S.A. 65 D3
Chinandega Nicaragua 66 G6
China Point CA U.S.A. 65 C4
Chincha Alta Peru 68 C6
Chinchaga r. Canada 60 G4
Chinchilla Australia 58 E1
Chincholi India 38 C2
Chinchorro, Banco sea feature Mex. 67 G5
Chincoteague Bay Maryland/Virginia U.S.A. 64 D4
Chinde Moz. 49 D5
Chindo S. Korea 45 B6
Chin-do i. S. Korea 45 B6
Chindwin r. Myanmar 37 H5
Chinese Turkestan aut. reg. China see Xinjiang Uygur Zizhiqu
Chinghai prov. China see Qinghai
Chingleput India see Chengalpattu
Chingola Zambia 49 C5
Chinguar Angola 49 B5
Chinguetti Mauritania 46 B2
Chinhae S. Korea 45 C6
Chinhoyi Zimbabwe 49 D5
Chini India see Kalpa
Chining China see Jining
Chiniot Pak. 33 L3
Chinju S. Korea 45 C6
Chinle U.S.A. 62 F4
Chinnamp'o N. Korea see Namp'o
Chinnur India 38 C2
Chino Creek watercourse AZ U.S.A. 65 F3
Chinon France 24 E3
Chinook Trough sea feature N. Pacific Ocean 74 I3
Chino Valley U.S.A. 62 E5
Chintamani India 38 C3
Chioggia Italy 26 E2
Chios Greece 27 L5
Chios i. Greece 27 K5
Chipata Zambia 49 D5
Chipchihua, Sierra de mts Arg. 70 C6
Chipindo Angola 49 B5
Chipinga Zimbabwe see Chipinge
Chipinge Zimbabwe 49 D6
Chippenham U.K. 19 E7
Chipping Norton U.K. 19 F7
Chipping Sodbury U.K. 19 E7
Chipurupalle Andhra Prad. India 38 D2
Chipurupalle Andhra Prad. India 38 D2
Chiquinquira Col. 68 D2
Chir r. Rus. Fed. 13 I6
Chirada India 38 D3
Chirala India 38 D3
Chīras Afgh. 36 A2
Chirchiq Uzbek. 33 K1
Chiredzi Zimbabwe 49 D6
Chirfa Niger 46 E2
Chiricahua Peak U.S.A. 62 F5
Chirikof Island U.S.A. 60 C4
Chiriquí, Golfo de b. Panama 67 H7
Chiriquí, Volcán de vol. Panama see Barú, Volcán
Chiri-san mt. S. Korea 45 B6
Chirk U.K. 19 D6
Chirnside U.K. 20 G5
Chirripó mt. Costa Rica 67 H7
Chisamba Zambia 49 C5
Chisasibi Canada 61 K4
Chishima-retto is Rus. Fed. see Kuril Islands
Chishtian Pak. 33 L4
Chishui China 42 J7
Chisimaio Somalia see Kismaayo
Chişinău Moldova 27 M1
Chistopol' Rus. Fed. 12 K5
Chita Rus. Fed. 43 K2
Chitado Angola 49 B5
Chitaldrug India see Chitradurga
Chitalwana India 36 B4
Chitambo Zambia 49 D5
Chita Oblast Rus. Fed. see Chitinskaya Oblast'
Chitato Angola 49 C4
Chitembo Angola 49 B5
Chitina U.S.A. 60 D3
Chitinskaya Oblast' Rus. Fed. 44 A1
Chitipa Malawi 49 D4
Chitkul India see Chhitkul
Chitobe Moz. 49 D6
Chitoor India see Chittoor
Chitor India see Chittaurgarh
Chitose Japan 44 F4
Chitradurga India 38 C3
Chitrakoot India 36 E4
Chitrakut India see Chitrakoot
Chitral Pak. 33 L2
Chitral r. Pak. 36 B2
Chitravati r. India 38 C3
Chitré Panama 67 H7
Chitrod India 36 B5
Chittagong Bangl. 37 G5
Chittaurgarh India 36 C4
Chittoor India 38 C3
Chittor India see Chittoor
Chittorgarh India see Chittaurgarh
Chittur India 38 C4
Chitungwiza Zimbabwe 49 D5
Chiu Lung H.K. China see Kowloon
Chiume Angola 49 C5
Chivasso Italy 26 B2
Chivhu Zimbabwe 49 D5
Chizarira National Park Zimbabwe 49 C5
Chizu Japan 45 D6
Chkalov Rus. Fed. see Orenburg
Chkalovsk Rus. Fed. 12 I4
Chkalovskoye Rus. Fed. 44 D3
Chlef Alg. 25 G5
Chlef, Oued r. Alg. 25 G5
Chloride AZ U.S.A. 65 E3
Chlya, Ozero l. Rus. Fed. 44 F1
Chobe National Park Botswana 49 C5
Choele Choel Arg. 70 C5
Chogar r. Rus. Fed. 44 D1
Chogori Feng mt. China/Pakistan see K2
Chograyskoye Vodokhranilishche resr Rus. Fed. 13 J7
Choiseul i. Solomon Is 53 F2
Choix Mex. 66 C3
Chojnice Poland 17 P4
Chōkai-san vol. Japan 45 F5
Ch'ok'ē mts Eth. 48 D2
Ch'ok'ē Mountains Eth. see Ch'ok'ē
Chokola mt. China 36 E3
Choksum China 37 F3
Chokue Moz. see Chókwé
Chokurdakh Rus. Fed. 29 P2
Chókwé Moz. 51 K3
Cholame CA U.S.A. 65 B3
Cholet France 24 D3
Choluteca Hond. 67 G6
Choma Zambia 49 C5
Chomo Ganggar mt. China 37 G3
Chomo Lhari mt. China/Bhutan 37 G4
Chomutov Czech Rep. 17 N5
Ch'ŏnan S. Korea 45 B5
Chon Buri Thai. 31 J5
Ch'ŏnch'ŏn N. Korea 44 B4
Chone Ecuador 68 B4
Ch'ŏngch'ŏn-gang r. N. Korea 45 B5
Ch'ongdo S. Korea 45 C6
Chonggye China see Qonggyai
Ch'ŏngjin N. Korea 44 C4
Ch'ŏngju S. Korea 45 B5
Chongkü China 37 I3
Chongming Dao i. China 43 M6
Chongoroi Angola 49 B5
Chŏngp'yŏng N. Korea 45 B5
Chongqing China 42 J7
Chongqing mun. China 42 J6
Chonguene Moz. 51 K3
Chŏngŭp S. Korea 45 B6
Chŏnju S. Korea 45 B6
Cho Oyu mt. China/Nepal 37 F3
Chopda India 36 C5
Chor Pak. 36 B4
Chora Sfakion Greece 27 K7
Chorley U.K. 18 E5
Chornobyl' Ukr. 13 F6
Chornomors'ke Ukr. 27 O2
Chortkiv Ukr. 13 E6
Ch'osan N. Korea 44 B4
Chōshi Japan 45 F6
Chosŏn country Asia see South Korea
Chosŏn-minjujuŭi-inmin-konghwaguk country Asia see North Korea
Choszczno Poland 17 O4
Chota Peru 68 C5
Chota Sinchula mt. India 37 G4
Choti Pak. 36 B3
Choûm Mauritania 46 B2
Chowchilla CA U.S.A. 65 B2
Choybalsan Mongolia 43 K3
Choyr Mongolia 43 J3
Chřiby hills Czech Rep. 17 P6
Chrissiesmeer S. Africa 51 J4
Christchurch N.Z. 59 D6
Christchurch U.K. 19 F8
Christian, Cape Canada 61 L2
Christiana S. Africa 51 G4
Christiania Norway see Oslo
Christiansburg VA U.S.A. 64 A4
Christianshåb Greenland see Qasigiannguit
Christina, Mount N.Z. 59 B7
Christmas Island terr. Indian Ocean 41 C9
Christopher, Lake imp. l. Australia 55 D6
Chrudim Czech Rep. 17 O6
Chrysochou Bay Cyprus 39 A2
Chrysochous, Kolpos b. Cyprus see Chrysochou Bay
Chu Kazakh. see Shu
Chu r. Kazakh./Kyrg. 42 B4
Chuadanga Bangl. 37 G5
Chuali, Lago l. Moz. 51 K3
Chuanhui China see Zhoukou
Chubarovka Ukr. see Polohy
Chubartau Kazakh. see Barshatas
Chūbu-Sangaku Kokuritsu-kōen Japan 45 E5
Chuchkovo Rus. Fed. 13 I5
Chuckwalla Mountains CA U.S.A. 65 E4
Chudniv Ukr. 13 F6
Chudovo Rus. Fed. 12 F4
Chudskoye, Ozero l. Estonia/Rus. Fed. see Peipus, Lake
Chugach Mountains U.S.A. 60 D3
Chūgoku-sanchi mts Japan 45 D6
Chugqênsumdo China see Jigzhi
Chuguchak China see Tacheng
Chuguyev Ukr. see Chuhuyiv
Chuguyevka Rus. Fed. 44 D3
Chuhuyiv Ukr. 13 H6
Chujiang China see Shimen
Chukchagirskoye, Ozero l. Rus. Fed. 44 E1
Chukchi Abyssal Plain sea feature Arctic Ocean 77 B1
Chukchi Peninsula Rus. Fed. see Chukotskiy Poluostrov
Chukchi Plateau sea feature Arctic Ocean 77 B1
Chukchi Sea Rus. Fed./U.S.A. 29 T3
Chukhloma Rus. Fed. 12 I4
Chukotskiy, Mys c. Rus. Fed. 60 A3
Chukotskiy Poluostrov pen. Rus. Fed. 29 T3
Chulakkurgan Kazakh. see Sholakkorgan
Chulaktau Kazakh. see Karatau
Chulasa Rus. Fed. 12 J2
Chula Vista CA U.S.A. 65 D4
Chulucanas Peru 68 B5
Chulung Pass Pak. 36 D2
Chulym Rus. Fed. 28 J4
Chumar India 36 D2
Chumbicha Arg. 70 C3
Chumda China 37 I2
Chumikan Rus. Fed. 29 O4
Chumphon Thai. 31 I5
Chunar India 37 E4
Ch'unch'ŏn S. Korea 45 B5
Chundzha Kazakh. 42 D4
Chunga Zambia 49 C5
Chung-hua Jen-min Kung-ho-kuo country Asia see China
Chung-hua Min-kuo country Asia see Taiwan
Ch'ungju S. Korea 45 B5
Chungking China see Chongqing
Ch'ungmu S. Korea see T'ongyŏng
Chŭngsan N. Korea 45 B5
Chunian Pak. 36 C3
Chunskiy Rus. Fed. 42 H1
Chunya r. Rus. Fed. 29 K3
Chupa Rus. Fed. 14 R3
Chūplū Iran 35 G3
Chuquicamata Chile 70 C2
Chur Switz. 24 I3
Churachandpur India 37 H4
Churapcha Rus. Fed. 29 O3
Churchill Canada 61 I4
Churchill r. Man. Canada 61 I4
Churchill r. Nfld. and Lab. Canada 61 L4
Churchill, Cape Canada 61 I4
Churchill Mountains Antarctica 76 H1
Churchville VA U.S.A. 64 B3
Churia Ghati Hills Nepal 37 F4
Churu India 36 C3
Churún-Merú waterfall Venez. see Angel Falls
Chushul India 36 D2
Chusovaya r. Rus. Fed. 11 R4
Chusovoy Rus. Fed. 11 R4
Chust Ukr. see Khust
Chute-des-Passes Canada 63 M2
Chutia Assam India 37 H4
Chutia Jharkhand India 37 F5
Chuuk is Micronesia 74 G5
Chuxiong China 42 I7
Chüy r. Kazakh./Kyrg. see Chu
Chymyshliya Moldova see Cimişlia
Chyulu Hills National Park Kenya 48 D4
Ciadâr-Lunga Moldova see Ciadîr-Lunga
Ciadîr-Lunga Moldova 27 M1
Cianorte Brazil 70 F2
Čičarija mts Croatia 26 E2
Cide Turkey 34 D2
Ciechanów Poland 17 R4
Ciego de Ávila Cuba 67 I4
Ciénaga Col. 68 D1
Cienfuegos Cuba 67 H4
Cieza Spain 25 F4
Çiftlik Turkey see Kelkit
Cifuentes Spain 25 E3
Cigüela r. Spain 25 E4
Cihanbeyli Turkey 34 D3
Cilacap Indon. 41 C8
Çıldır Turkey 35 F2
Çıldır Gölü l. Turkey 35 F2
Çıldıroba Turkey 39 C1
Cilento e del Vallo di Diano, Parco Nazionale del nat. park Italy 26 F4
Cilician Gates pass Turkey see Gülek Boğazı
Cill Airne Ireland see Killarney
Cill Chainnigh Ireland see Kilkenny
Cill Mhantáin Ireland see Wicklow
Çilmämmetgum des. Turkm. 35 I2
Cilo Dağı mt. Turkey 35 G3
Çiloy Adası i. Azer. 35 H2
Cimarron r. U.S.A. 62 H4
Cimişlia Moldova 27 M1
Cimone, Monte mt. Italy 26 D2
Cîmpina Romania see Câmpina
Cîmpulung Romania see Câmpulung
Cîmpulung Moldovenesc Romania see Câmpulung Moldovenesc
Çınar Turkey 35 F3
Cinca r. Spain 25 G3
Cincinnati U.S.A. 63 K4
Cinco de Outubro Angola see Xá-Muteba
Cinderford U.K. 19 E7
Çine Turkey 27 M6
Cintalapa Mex. 66 F5
Cinto, Monte mt. France 24 I5
Ciping China 43 K7
Circeo, Parco Nazionale del nat. park Italy 26 E4
Circle AK U.S.A. 60 D3
Cirebon Indon. 41 C8
Cirencester U.K. 19 F7
Cirò Marina Italy 26 G5
Cirta Alg. see Constantine
Cisne, Islas del is Caribbean Sea 67 H5
Citlaltépetl vol. Mex. see Orizaba, Pico de
Čitluk Bos.-Herz. 26 G3
Citrus Heights CA U.S.A. 65 B1
Città di Castello Italy 26 E3
Ciucaş, Vârful mt. Romania 27 K2
Ciudad Acuña Mex. 62 G6
Ciudad Altamirano Mex. 66 D5
Ciudad Bolívar Venez. 68 F2
Ciudad Camargo Mex. 66 C3
Ciudad Constitución Mex. 66 B3
Ciudad del Carmen Mex. 66 F5
Ciudad Delicias Mex. 66 C3
Ciudad de Panamá Panama see Panama City
Ciudad de Valles Mex. 66 E4
Ciudad Flores Guat. see Flores
Ciudad Guayana Venez. 68 F2
Ciudad Guzmán Mex. 66 D5
Ciudad Juárez Mex. 66 C2
Ciudad Mante Mex. 66 E4
Ciudad Obregón Mex. 66 C3
Ciudad Real Spain 25 E4
Ciudad Río Bravo Mex. 62 H6
Ciudad Rodrigo Spain 25 C3
Ciudad Trujillo Dom. Rep. see Santo Domingo
Ciudad Victoria Mex. 66 E4
Ciutadella Spain 25 H3
Civa Burnu pt Turkey 34 E2
Cividale del Friuli Italy 26 E1
Civitanova Marche Italy 26 E3
Civitavecchia Italy 26 D3
Çivril Turkey 27 M5
Cizre Turkey 35 F3
Clacton-on-Sea U.K. 19 I7
Clady U.K. 21 E3
Claire, Lake Canada 60 G4
Clairfontaine Alg. see El Aouinet
Clamecy France 24 F3
Clane Ireland 21 F4
Clanwilliam Dam S. Africa 50 D7
Clara Ireland 21 E4
Claraville Australia 56 C3
Clare N.S.W. Australia 58 A4
Clare S.A. Australia 57 B7
Clare r. Ireland 21 C4
Clarecastle Ireland 21 D5
Clare Island Ireland 21 B4
Claremont NH U.S.A. 64 E1
Claremorris Ireland 21 D4
Clarence r. Australia 58 F2
Clarence N.Z. 59 D6
Clarence Island Antarctica 76 A2
Clarence Town Bahamas 63 M7
Clarendon PA U.S.A. 64 B2
Clarenville Canada 61 M5
Claresholm Canada 62 E1
Clarie Coast Antarctica see Wilkes Coast
Clarington OH U.S.A. 64 A3
Clarion PA U.S.A. 64 B2
Clarion r. PA U.S.A. 64 B2
Clarión, Isla i. Mex. 66 B5
Clarkebury S. Africa 51 I6
Clarke Range mts Australia 56 D4
Clarke River Australia 56 D3
Clark Mountain CA U.S.A. 65 E3
Clarksburg WV U.S.A. 64 A3
Clarksdale U.S.A. 63 I5
Clarksville AR U.S.A. 63 I4
Clarksville TN U.S.A. 63 J4
Claro r. Goiás Brazil 71 A2
Claro r. Mato Grosso Brazil 71 A1
Clashmore Ireland 21 E5
Claudy U.K. 21 E3
Clay WV U.S.A. 64 A3
Clayhole Wash watercourse AZ U.S.A. 65 F2
Clayton DE U.S.A. 64 D3
Clayton NM U.S.A. 62 G4
Claytor Lake VA U.S.A. 64 A4
Clear, Cape Ireland 21 C6
Clearco WV U.S.A. 64 A3
Clear Creek Ont. Canada 64 A1
Cleare, Cape U.S.A. 60 D4
Clearfield PA U.S.A. 64 B2
Clear Island Ireland 21 C6
Clear Lake IA U.S.A. 63 I3
Clear Lake CA U.S.A. 65 A1
Clear Lake UT U.S.A. 65 F1
Clearwater r. Alberta/Saskatchewan Canada 60 G4
Clearwater U.S.A. 63 K6
Cleburne U.S.A. 63 H5
Cleethorpes U.K. 18 G5
Clendenin WV U.S.A. 64 A3
Clendening Lake OH U.S.A. 64 A2
Clères France 19 I9
Clerke Reef Australia 54 B4
Clermont Australia 56 D4
Clermont-Ferrand France 24 F4
Cles Italy 26 D1
Clevedon U.K. 19 E7
Cleveland MS U.S.A. 63 I5
Cleveland OH U.S.A. 64 A2
Cleveland TN U.S.A. 63 K4
Cleveland, Cape Australia 56 D3
Cleveland, Mount U.S.A. 62 E2
Cleveland Heights OH U.S.A. 64 A2
Cleveland Hills U.K. 18 F4
Cleveleys U.K. 18 D5
Clew Bay Ireland 21 C4
Clifden Ireland 21 B4
Cliffoney Ireland 21 D3
Clifton Australia 58 E1
Clifton U.S.A. 62 F5
Clifton Beach Australia 56 D3
Clifton Forge VA U.S.A. 64 B4
Clifton Park NY U.S.A. 64 E1
Clinton Ont. Canada 64 A1
Clinton IA U.S.A. 63 I3
Clinton OK U.S.A. 62 H4
Clipperton, Île terr. N. Pacific Ocean 75 M5
Clisham h. U.K. 20 C3
Clitheroe U.K. 18 E5
Cliza Bol. 68 E7
Clocolan S. Africa 51 H5
Cloghan Ireland 21 E4
Clonakilty Ireland 21 D6
Clonbern Ireland 21 D4
Cloncurry Australia 56 C4
Cloncurry r. Australia 56 C3
Clones Ireland 21 E3
Clonmel Ireland 21 E5
Clonygowan Ireland 21 E4
Cloonbannin Ireland 21 C5
Clooneagh Ireland 21 E4
Cloud Peak WY U.S.A. 62 F3
Cloverdale CA U.S.A. 65 A1
Clovis CA U.S.A. 65 C2
Clovis NM U.S.A. 62 G5
Cluain Meala Ireland see Clonmel
Cluanie, Loch l. U.K. 20 D3
Cluff Lake Mine Canada 60 H4
Cluj-Napoca Romania 27 J1
Clun U.K. 19 D6
Clunes Australia 58 A6
Cluny Australia 56 B5
Cluses France 24 H3
Clutterbuck Hills h. Australia 55 D6
Clwydian Range hills U.K. 18 D5
Clyde r. U.K. 20 E5
Clyde NY U.S.A. 64 C1
Clyde, Firth of est. U.K. 20 E5
Clydebank U.K. 20 E5
Clyde River Canada 61 L2
Côa r. Port. 25 C3
Coachella CA U.S.A. 65 D4
Coaldale NV U.S.A. 65 D1
Coalinga CA U.S.A. 65 B2
Coalport PA U.S.A. 64 B2
Coal River Canada 60 F4
Coal Valley val. NV U.S.A. 65 E2
Coalville U.K. 19 F6
Coari Brazil 68 F4
Coari r. Brazil 68 F4
Coarsegold CA U.S.A. 65 C2
Coastal Plain U.S.A. 63 I5
Coast Mountains Canada 60 F4
Coast Range hills Australia 57 E5
Coast Ranges mts CA U.S.A. 65 B2
Coatbridge U.K. 20 E5
Coatesville PA U.S.A. 64 D3
Coats Island Canada 61 J3
Coats Land reg. Antarctica 76 A1
Coatzacoalcos Mex. 66 F5
Cobar Australia 58 B3
Cobargo Australia 58 D6
Cobden Australia 58 A7
Cobh Ireland 21 D6
Cobija Bol. 68 E6
Coblenz Germany see Koblenz
Cobleskill NY U.S.A. 64 D1
Cobourg Peninsula Australia 54 F2
Cobra Australia 55 B6
Cobram Australia 58 B5
Coburg Germany 17 M5
Coburg Island Canada 61 K2
Coca Ecuador 68 C4
Coca Spain 25 D3
Cocalinho Brazil 71 A1
Cocanada India see Kakinada
Cochabamba Bol. 68 E7
Cochin India see Kochi
Cochrane Alta Canada 62 E1
Cochrane Ont. Canada 63 K2
Cockburn Australia 57 C7
Cockburnspath U.K. 20 G5
Cockburn Town Turks and Caicos Is see Grand Turk
Cockermouth U.K. 18 D5
Cocklebiddy Australia 55 D8
Cockscomb mt. S. Africa 50 G7
Coco r. Hond./Nicaragua 67 H6
Coco, Isla de i. N. Pacific Ocean 67 G7
Cocobeach Gabon 48 A3
Coconino Plateau AZ U.S.A. 65 F3
Cocopara National Park Australia 58 C5
Cocos Brazil 71 B1
Cocos Basin sea feature Indian Ocean 73 O5
Cocos Islands terr. Indian Ocean 41 B9
Cocos Ridge sea feature N. Pacific Ocean 75 O5
Cocuy, Sierra Nevada del mt. Col. 68 D2
Cod, Cape MA U.S.A. 64 F2
Codajás Brazil 68 F4
Codfish Island N.Z. 59 A8
Codigoro Italy 26 E2
Cod Island Canada 61 L4
Codlea Romania 27 K2
Codó Brazil 69 J4
Codsall U.K. 19 E6
Cod's Head hd Ireland 21 B6
Cody U.S.A. 62 F3
Coen Australia 56 C2
Coeur d'Alene U.S.A. 62 D2
Coffee Bay S. Africa 51 I6
Coffeyville U.S.A. 63 H4
Coffin Bay Australia 57 A7
Coffin Bay National Park Australia 57 A7
Coffs Harbour Australia 58 F3
Cofimvaba S. Africa 51 H7
Cognac France 24 D4
Cogo Equat. Guinea 46 D4
Coguno Moz. 51 L3
Cohoes NY U.S.A. 64 E1
Cohuna Australia 58 B5
Coiba, Isla de i. Panama 67 H7
Coigeach, Rubha pt U.K. 20 D2
Coihaique Chile 70 B7
Coimbatore India 38 C4
Coimbra Port. 25 B3
Coipasa, Salar de salt flat Bol. 68 E7
Coire Switz. see Chur
Colac Australia 58 A7
Colair Lake India see Kolleru Lake
Colatina Brazil 71 C2
Colby U.S.A. 62 G4
Colchester U.K. 19 H7
Colchester CT U.S.A. 64 E2
Cold Bay U.S.A. 60 B4
Coldingham U.K. 20 G5
Coldstream U.K. 20 G5
Coleambally Australia 58 B5
Coleman r. Australia 56 C2
Coleman U.S.A. 62 H5
Çölemerik Turkey see Hakkâri
Colenso S. Africa 51 I5
Cole Peninsula Antarctica 76 L2
Coleraine Australia 57 C8
Coleraine U.K. 21 F2
Coles, Punta de pt Peru 68 D7
Coles Bay Australia 57 [inset]
Colesberg S. Africa 51 G6
Colfax CA U.S.A. 65 B1
Colhué Huapí, Lago l. Arg. 70 C7
Coligny S. Africa 51 H4
Colima Mex. 66 D5
Colima, Nevado de vol. Mex. 66 D5
Coll i. U.K. 20 C4
Collado Villalba Spain 25 E3
Collarenebri Australia 58 D2
Collerina Australia 58 C2
Collie N.S.W. Australia 58 D3
Collie W.A. Australia 55 B8
Collier Bay Australia 54 D4
Collier Range National Park Australia 55 B6
Collingwood N.Z. 59 D5
Collins Glacier Antarctica 76 E2
Collinson Peninsula Canada 61 H2
Collipulli Chile 70 B5
Collooney Ireland 21 D3
Colmar France 24 H2
Colmenar Viejo Spain 25 E3
Colmonell U.K. 20 E5
Colne r. U.K. 19 H7
Cologne Germany 17 K5
Colomb-Béchar Alg. see Béchar
Colômbia Brazil 71 A3
Colombia country S. America 68 D3
Colombian Basin sea feature S. Atlantic Ocean 72 C5
Colombo Sri Lanka 38 C5
Colomiers France 24 E5
Colón Buenos Aires Arg. 70 D4
Colón Entre Ríos Arg. 70 E4
Colón Panama 67 I7
Colón, Archipiélago de is Ecuador see Galapagos Islands
Colona Australia 55 F7
Colonelganj India 37 E4
Colônia r. Brazil 71 D1
Colonia Agrippina Germany see Cologne
Colonia Julia Fenestris Italy see Fano
Colonia Las Heras Arg. 70 C7
Colonial Heights VA U.S.A. 64 C4
Colonna, Capo c. Italy 26 G5
Colonsay i. U.K. 20 C4
Colorado r. Arg. 70 D5
Colorado r. Mex./U.S.A. 65 F2
Colorado r. U.S.A. 62 H6
Colorado state U.S.A. 62 F4
Colorado City AZ U.S.A. 65 F2
Colorado Desert CA U.S.A. 65 D4
Colorado Plateau U.S.A. 62 F4
Colorado River Aqueduct canal CA U.S.A. 65 E3
Colorado Springs U.S.A. 62 G4
Colossae Turkey see Honaz
Colotlán Mex. 66 D4
Colquiri Bol. 68 E7
Colsterworth U.K. 19 G6
Colstrip U.S.A. 62 F2
Coltishall U.K. 19 I6
Colton CA U.S.A. 65 D3
Columbia MD U.S.A. 64 C3
Columbia MO U.S.A. 63 I4
Columbia PA U.S.A. 64 C2
Columbia SC U.S.A. 63 K5
Columbia TN U.S.A. 63 J4
Columbia r. U.S.A. 62 C2
Columbia, District of admin. dist. U.S.A. 64 C3
Columbia, Mount Canada 62 D1
Columbia Mountains Canada 62 C1
Columbia Plateau U.S.A. 62 D2
Columbine, Cape S. Africa 50 C7
Columbus GA U.S.A. 63 K5
Columbus IN U.S.A. 63 J4
Columbus MS U.S.A. 63 J5
Columbus NE U.S.A. 63 H3
Columbus NM U.S.A. 62 F5
Columbus OH U.S.A. 63 K4
Columbus Salt Marsh NV U.S.A. 65 C1
Colusa CA U.S.A. 65 A1
Colville N.Z. 59 E3
Colville r. U.S.A. 60 C2
Colville Channel N.Z. 59 E3
Colville Lake Canada 60 F3
Colwyn Bay U.K. 18 D5
Comacchio Italy 26 E2
Comacchio, Valli di lag. Italy 26 E2
Comai China 37 G3
Comalcalco Mex. 66 F5
Comandante Ferraz (Brazil) research stn Antarctica 76 A2
Comandante Salas Arg. 70 C4
Comănești Romania 27 L1
Combarbalá Chile 70 B4
Comber U.K. 21 G3
Combermere Bay Myanmar 37 H6
Combomune Moz. 51 K2
Comboyne Australia 58 F3
Comencho, Lac l. Canada 63 L1
Comendador Dom. Rep. see Elías Piña
Comendador Gomes Brazil 71 A2
Comeragh Mountains hills Ireland 21 E5
Comercinho Brazil 71 C2
Cometela Moz. 51 L1
Comilla Bangl. 37 G5
Comino, Capo c. Sardegna Italy 26 C4
Comitán de Domínguez Mex. 66 F5
Commack NY U.S.A. 64 E2
Commentry France 24 F3
Committee Bay Canada 61 J3
Commonwealth Territory admin. div. Australia see Jervis Bay Territory
Como Italy 26 C2
Como, Lago di Italy see Como, Lake
Como, Lake Italy 26 C2
Como Chamling l. China 37 G3
Comodoro Rivadavia Arg. 70 C7
Comores country Africa see Comoros
Comorin, Cape India 38 C4
Comoro Islands country Africa see Comoros
Comoros country Africa 49 E5
Compiègne France 24 F2
Comprida, Ilha i. Brazil 71 B4
Comrat Moldova 27 M1
Comrie U.K. 20 F4
Cona China 37 G4
Conakry Guinea 46 B4
Cona Niyeo Arg. 70 C6
Conceição r. Brazil 71 B2
Conceição da Barra Brazil 71 D2
Conceição do Araguaia Brazil 69 I5
Conceição do Mato Dentro Brazil 71 C2
Concepción Chile 70 B5
Concepción Mex. 66 D4
Concepción Para. 70 E2
Concepción de la Vega Dom. Rep. see La Vega
Conception, Point CA U.S.A. 65 B3
Conchos r. Mex. 66 D3
Conchos r. Nuevo León/Tamaulipas Mex. 66 E4
Concord CA U.S.A. 65 A2
Concord NH U.S.A. 64 F1
Concordia Arg. 70 E4
Concordia Peru 68 D4
Concordia S. Africa 50 C5
Concordia KS U.S.A. 62 H4
Concordia research stn 76 G2
Concord Peak Afgh. 36 C1
Condamine Australia 58 E1
Condamine r. Australia 58 D1
Condeúba Brazil 71 C1
Condobolin Australia 58 C4
Condom France 24 E5
Condor, Cordillera del mts Ecuador/Per 68 C4
Conegliano Italy 26 E2
Conemaugh r. PA U.S.A. 64 B2
Conesus Lake NY U.S.A. 64 C1
Conflict Group is P.N.G. 56 E1
Confoederatio Helvetica country Europe see Switzerland
Confusion Range mts UT U.S.A. 65 F1
Congdü China 37 F3
Congleton U.K. 18 E5
Congo country Africa 48 B4
Congo r. Congo/Dem. Rep. Congo 48 B
Congo (Brazzaville) country Africa see Congo
Congo (Kinshasa) country Africa see Congo, Democratic Republic of the
Congo, Democratic Republic of the country Africa 48 C4
Congo, Republic of country Africa see Congo
Congo Basin Dem. Rep. Congo 48 C4
Congo Cone sea feature S. Atlantic Ocea 72 I6
Congo Free State country Africa see Congo, Democratic Republic of the
Congonhas Brazil 71 C3
Congress AZ U.S.A. 65 F3
Conimbla National Park Australia 58 D4
Coningsby U.K. 19 G5
Coniston U.K. 18 D4
Conjuboy Australia 56 D3
Conn, Lough l. Ireland 21 C3
Connacht reg. Ireland see Connaught
Connaught reg. Ireland 21 C4
Conneaut OH U.S.A. 64 A2
Connecticut state U.S.A. 64 E2
Connemara reg. Ireland 21 C4
Connemara National Park Ireland 21 C4
Connors Range hills Australia 56 E4
Conoble Australia 58 B4
Conquista Brazil 71 B2
Conrad U.S.A. 62 E2
Conrad Rise sea feature Southern Ocean 73 K9
Conroe U.S.A. 63 H5
Conselheiro Lafaiete Brazil 71 C3
Consett U.K. 18 F4
Constance Germany see Konstanz
Constance, Lake Germany/Switz. 17 L7
Constância dos Baetas Brazil 68 F5
Constanța Romania 27 M2
Constantia tourist site Cyprus see Salami
Constantia Germany see Konstanz
Constantina Spain 25 D5
Constantine Alg. 22 F4
Constantine, Cape U.S.A. 60 C4
Constantinople Turkey see İstanbul
Contagalo Brazil 71 C3
Contamana Peru 68 C5
Contas r. Brazil 71 D1
Contria Brazil 71 B2
Contwoyto Lake Canada 60 G3
Convención Col. 68 D2
Conway U.K. see Conwy
Conway AR U.S.A. 63 I4
Conway, Cape Australia 56 E4
Conway, Lake imp. l. Australia 57 A6
Conway National Park Australia 56 E4
Conway Reef Fiji see Ceva-i-Ra
Conwy U.K. 18 D5
Conwy r. U.K. 19 D5
Coober Pedy Australia 55 F7
Coochbehar India see Koch Bihar
Cooch Behar India see Koch Bihar
Cook Australia 55 E7
Cook, Grand Récif de rf New Caledonia 53 G3
Cook, Mount N.Z. see Aoraki
Cookhouse S. Africa 51 G7
Cook Ice Shelf Antarctica 76 H2
Cook Inlet sea chan. U.S.A. 60 C3
Cook Islands terr. S. Pacific Ocean 74 J7
Cooksburg NY U.S.A. 64 D1
Cooks Passage Australia 56 D2
Cookstown U.K. 21 F3
Cook Strait N.Z. 59 E5
Cooktown Australia 56 D2
Coolabah Australia 58 C3
Cooladdi Australia 58 B1
Coolah Australia 58 D3
Coolamon Australia 58 C5
Coolgardie Australia 55 C7
Coolibah Australia 54 E3
Cooloola National Park Australia 57 F5
Coolum Beach Australia 57 F5
Cooma Australia 58 D6
Coombah Australia 57 C7
Coonabarabran Australia 58 D3
Coonamble Australia 58 D3
Coondambo Australia 57 A6
Coondapoor India see Kundapura
Coongoola Australia 58 B1
Cooper Creek watercourse Australia 57 B6
Coopernook Australia 58 F3
Cooperstown NY U.S.A. 64 D1
Coopracambra National Park Australia 58 D6
Coorabie Australia 55 F7
Coorong National Park Australia 57 B8
Coorow Australia 55 B7
Coos Bay U.S.A. 62 C3
Cootamundra Australia 58 D5
Cootehill Ireland 21 E3
Cooyar Australia 58 E1
Copala Mex. 66 E5
Copenhagen Denmark 15 H9
Copertino Italy 26 H4
Copeton Reservoir Australia 58 E2
Copiapó Chile 70 B3
Copley Australia 57 B6
Copparo Italy 26 D2
Coppermine Canada see Kugluktuk
Coppermine r. Canada 77 L2
Copperton S. Africa 50 F5
Coqên Xizang China 37 F3
Coqên China 37 F3

oquilhatville Dem. Rep. Congo see Mbandaka
oquille i. Micronesia see Pikelot
oquimbo Chile 70 B3
orabia Romania 27 K3
oração de Jesus Brazil 71 B2
oracesium Turkey see Alanya
oraki Australia 58 F2
oral Bay Australia 55 A5
oral Harbour Canada 61 J3
oral Sea S. Pacific Ocean 52 F3
oral Sea Basin S. Pacific Ocean 74 G6
oral Sea Islands Territory terr. Australia 52 F3
orangamite, Lake Australia 58 A7
orat Azer. 35 H2
orbett National Park India 36 D3
orby U.K. 19 G6
orcaigh Ireland see Cork
orcoran CA U.S.A. 65 C2
orcovado, Golfo de sea chan. Chile 70 B6
orcyra i. Greece see Corfu
ordele U.S.A. 63 K5
ordelia CA U.S.A. 65 A1
ordilheiras, Serra das hills Brazil 69 I5
ordillera Azul, Parque Nacional nat. park Peru 68 C5
ordillera de los Picachos, Parque Nacional nat. park Col. 68 D3
ordillo Downs Australia 57 C5
ordisburgo Brazil 71 B2
órdoba Arg. 70 D4
órdoba Veracruz Mex. 66 E5
órdoba Spain 25 D5
órdoba, Sierras de mts Arg. 70 D4
ordova Spain see Córdoba
ordova U.S.A. 60 D3
orduba Spain see Córdoba
orfu i. Greece 27 H5
oria Spain 25 C4
oribe Brazil 71 B1
oricudgy mt. Australia 58 E4
origliano Calabro Italy 26 G5
oringa Islands Australia 56 E3
orinium U.K. see Cirencester
orinth Greece 27 J6
orinth MS U.S.A. 63 J5
orinth NY U.S.A. 64 E1
orinth, Gulf of sea chan. Greece 27 J5
orinthus Greece see Corinth
orinto Brazil 71 B2
ork Ireland 21 D6
orleone Sicilia Italy 26 E6
orlu Turkey 27 L4
ormeilles France 19 H9
ornelia S. Africa 51 I4
ornélio Procópio Brazil 71 A3
ornélios Brazil 71 A5
orner Brook Canada 61 M5
orner Inlet b. Australia 58 C7
orner Seamounts sea feature N. Atlantic Ocean 72 E3
orneto Italy see Tarquinia
orning NY U.S.A. 64 C1
ornish watercourse Australia 56 D4
orn Islands is Nicaragua see Maíz, Islas del
orno di Campo mt. Italy/Switz. 24 J3
orno Grande mt. Italy 26 E3
ornwall Canada 63 M2
ornwallis Island Canada 61 I2
ornwall Island Canada 61 I2
oro Venez. 68 E1
oroaci Brazil 71 C2
oroatá Brazil 69 J4
orofin Ireland 21 C5
oromandel Brazil 71 B2
oromandel Coast India 38 D4
oromandel Peninsula N.Z. 59 E3
oromandel Range hills N.Z. 59 E3
orona CA U.S.A. 65 D4
oronado CA U.S.A. 65 D4
oronado, Bahía de b. Costa Rica 67 H7
oronation Gulf Canada 60 G3
oronation Island S. Atlantic Ocean 76 A2
oronda Arg. 70 D4
oronel Fabriciano Brazil 71 C2
oronel Oviedo Para. 70 E3
oronel Pringles Arg. 70 D5
oronel Suárez Arg. 70 D5
orovodë Albania 27 I4
orowa Australia 58 C5
orpus Christi U.S.A. 63 H6
orque Bol. 68 E7
orral de Cantos mt. Spain 25 D4
orrandibby Range hills Australia 55 A6
orrente Brazil 69 I6
orrente r. Bahia Brazil 71 C1
orrente r. Minas Gerais Brazil 71 A2
orrentes Brazil 69 H7
orrentina Brazil 71 B1
orrentina r. Brazil see Éguas
orrib, Lough l. Ireland 21 C4
orrientes Arg. 70 E3
orrientes, Cabo c. Col. 68 C2
orrientes, Cabo c. Mex. 66 C4
orrigin Australia 55 B8
orris U.K. 19 D6
orry PA U.S.A. 64 B2
orse i. France see Corsica
orse, Cap c. Corse France 24 I5
orsham U.K. 19 E7
orsica i. France 24 I5
orsicana U.S.A. 63 H5
orte Corse France 24 I5
ortegana Spain 25 C5
ortes, Sea of g. Mex. see California, Gulf of
ortez U.S.A. 62 F4
ortina d'Ampezzo Italy 26 E1
ortland NY U.S.A. 64 C1
orton U.K. 19 I6
ortona Italy 26 D3
oruche Port. 25 B4
oruh Turkey see Artvin
oruh r. Turkey 35 F2
orum Turkey 34 D2
orumbá Brazil 69 G7
orumbá r. Brazil 71 A2
orumbá de Goiás Brazil 71 A1
orumbaíba Brazil 71 A2
orumbaú, Ponta pt Brazil 71 D2
orunna Spain see A Coruña
orvallis U.S.A. 62 C3
orwen U.K. 19 D6
oryville PA U.S.A. 64 B2
os i. Greece see Kos
osentia Italy see Cosenza
osenza Italy 26 G5
osne-Cours-sur-Loire France 24 F3
osta Blanca coastal area Spain 25 F4
osta Brava coastal area Spain 25 H3
osta de la Luz coastal area Spain 25 C5
osta del Sol coastal area Spain 25 D5
osta de Miskitos coastal area Nicaragua see Costa de Mosquitos
osta de Mosquitos coastal area Nicaragua 67 H6
osta Marques Brazil 68 F6
osta Rica Brazil 69 H7
osta Rica country Central America 67 H6
Costa Rica Mex. 66 C4
Costa Verde coastal area Spain 25 C2
Costermansville Dem. Rep. Congo see Bukavu
Costești Romania 27 K2
Cotabato Phil. 41 E7
Cotagaita Bol. 68 E8
Cotahuasi Peru 68 D7
Côte d'Azur coastal area France 24 H5
Côte d'Ivoire country Africa 46 C4
Côte Française de Somalis country Africa see Djibouti
Cotentin pen. France 19 F9
Cothi r. U.K. 19 C7
Cotiaeum Turkey see Kütahya
Cotiella mt. Spain 25 G2
Cotonou Benin 46 D4
Cotopaxi, Volcán vol. Ecuador 68 C4
Cotswold Hills U.K. 19 E7
Cottage Grove U.S.A. 62 C3
Cottbus Germany 17 O5
Cottenham U.K. 19 H6
Cottian Alps mts France/Italy 24 H4
Cottica Suriname 69 H3
Cottiennes, Alpes mts France/Italy see Cottian Alps
Coudersport PA U.S.A. 64 B2
Couedic, Cape du Australia 57 B8
Coulman Island Antarctica 76 H2
Coulterville CA U.S.A. 65 B2
Council Bluffs U.S.A. 63 H3
Councillor Island Australia 57 [inset]
Courland Lagoon b. Lith./Rus. Fed. 15 L9
Courtenay Canada 62 C2
Courtmacsherry Ireland 21 D6
Courtmacsherry Bay Ireland 21 D6
Courtown Ireland 21 F5
Courtrai Belgium see Kortrijk
Coutances France 24 D2
Cove Fort UT U.S.A. 65 F1
Cove Mountains hills PA U.S.A. 64 B3
Coventry U.K. 19 F6
Covesville VA U.S.A. 64 B4
Covilhã Port. 25 C3
Covington VA U.S.A. 64 A4
Cowal, Lake dry lake Australia 58 C4
Cowan, Lake imp. l. Australia 55 C7
Cowargarzê China 37 I2
Cowcowing Lakes imp. l. Australia 55 B7
Cowdenbeath U.K. 20 F4
Cowell Australia 57 B7
Cowes U.K. 19 F8
Cowley Australia 58 B1
Cowper Point Canada 61 G2
Cowra Australia 58 D4
Cox r. Australia 56 A2
Coxá r. Brazil 71 B1
Coxen Hole Hond. see Roatán
Coxilha de Santana hills Brazil/Uruguay 70 E4
Coxilha Grande hills Brazil 70 F3
Coxim Brazil 69 H7
Cox's Bazar Bangl. 37 G5
Coyhaique Chile see Coihaique
Coyote Lake CA U.S.A. 65 D3
Coyote Peak h. AZ U.S.A. 65 E4
Cozhê China 37 F2
Cozie, Alpi mts France/Italy see Cottian Alps
Cozumel Mex. 67 G4
Cozumel, Isla de i. Mex. 67 G4
Craboon Australia 58 D4
Cracovia Poland see Kraków
Cracow Australia 56 E5
Cracow Poland see Kraków
Cradle Mountain Lake St Clair National Park Australia 57 [inset]
Cradock S. Africa 51 G7
Craig U.K. 20 D3
Craig CO U.S.A. 62 F3
Craigavon U.K. 21 F3
Craigieburn Australia 58 B6
Craignure U.K. 20 D4
Craigsville WV U.S.A. 64 A3
Crail U.K. 20 G4
Crailsheim Germany 17 M6
Craiova Romania 27 J2
Cramlington U.K. 18 F3
Cranberry Portage Canada 62 G1
Cranborne Chase for. U.K. 19 E8
Cranbourne Australia 58 B7
Cranbrook Canada 62 D2
Cranston RI U.S.A. 64 F2
Cranz Rus. Fed. see Zelenogradsk
Crary Ice Rise Antarctica 76 I1
Crary Mountains Antarctica 76 J1
Crateús Brazil 69 J5
Crato Brazil 69 K5
Crawley U.K. 19 G7
Creag Meagaidh mt. U.K. 20 E4
Credenhill U.K. 19 E6
Crediton U.K. 19 D8
Creel Mex. 66 C3
Cree Lake Canada 60 H4
Crema Italy 26 C2
Cremona Italy 26 D2
Crépy-en-Valois France 24 F2
Cres i. Croatia 26 F2
Crescent City CA U.S.A. 62 C3
Crescent Head Australia 58 F3
Cressy Australia 58 A7
Creston Canada 62 D2
Creston IA U.S.A. 63 I3
Crestview U.S.A. 63 J5
Creswick Australia 58 A6
Creta i. Greece see Crete
Crete i. Greece 27 K7
Creus, Cabo de c. Spain 25 H2
Creuse r. France 24 E3
Crevasse Valley Glacier Antarctica 76 J1
Crewe U.K. 19 E5
Crewe VA U.S.A. 64 B4
Crewkerne U.K. 19 E8
Crianlarich U.K. 20 E4
Criccieth U.K. 19 C6
Criciúma Brazil 71 A5
Crieff U.K. 20 F4
Criffel h. U.K. 20 F6
Criffell h. U.K. see Criffel
Crikvenica Croatia 26 F2
Crimea pen. Ukr. 34 D1
Crimond U.K. 20 H3
Crisfield MD U.S.A. 64 D4
Cristalândia Brazil 69 I6
Cristalina Brazil 71 B2
Cristalino r. Brazil see Mariembero
Cristóbal Colón, Pico mt. Col. 68 D1
Crixás Brazil 71 A1
Crixás Açu r. Brazil 71 A1
Crixás Mirim r. Brazil 71 A1
Crna Gora country Europe see Montenegro
Crni Vrh mt. Serbia 27 J2
Črnomelj Slovenia 26 F2
Croagh Patrick h. Ireland 21 C4
Croajingolong National Park Australia 58 D6
Croatia country Europe 26 G2
Crocker, Banjaran mts Malaysia 41 D7
Croker Island Australia 54 F2
Cromarty U.K. 20 E3
Cromarty Firth est. U.K. 20 E3
Cromer U.K. 19 I6
Crook U.K. 18 F4
Crooked Island Bahamas 67 J4
Crooked Island Passage Bahamas 67 J4
Crookston U.S.A. 63 H2
Crookwell Australia 58 D5
Croom Ireland 21 D5
Croppa Creek Australia 58 E2
Crosby U.K. 18 D5
Crosby ND U.S.A. 62 G2
Cross City U.S.A. 63 K6
Cross Fell h. U.K. 18 E4
Crossgar U.K. 21 G3
Crosshaven Ireland 21 D6
Cross Inn U.K. 19 C6
Cross Lake Canada 61 I4
Cross Lake NY U.S.A. 64 C1
Crossmaglen U.K. 21 F3
Crossman Peak AZ U.S.A. 65 E3
Croton Italy see Crotone
Crotone Italy 26 G5
Crouch r. U.K. 19 H7
Crowal watercourse Australia 58 C3
Crowborough U.K. 19 H7
Crowdy Bay National Park Australia 58 F3
Crowland U.K. 19 G6
Crowley, Lake CA U.S.A. 65 C2
Crown Prince Olav Coast Antarctica 76 D2
Crown Princess Martha Coast Antarctica 76 B1
Crows Nest Australia 58 F1
Croydon Australia 56 C3
Crozet VA U.S.A. 64 B3
Crozet, Îles is Indian Ocean 73 L9
Crozet Basin sea feature Indian Ocean 73 M8
Crozet Plateau sea feature Indian Ocean 73 K8
Crozon France 24 B2
Cruden Bay U.K. 20 H3
Crumlin U.K. 21 F3
Crusheen Ireland 21 D5
Cruz Alta Brazil 70 F3
Cruz del Eje Arg. 70 D4
Cruzeiro Brazil 71 B3
Cruzeiro do Sul Brazil 68 D5
Crystal City U.S.A. 62 H6
Crystal Falls U.S.A. 63 J2
Csongrád Hungary 27 I1
Cuamba Moz. 49 D5
Cuando r. Angola/Zambia 49 C5
Cuangar Angola 49 B5
Cuango Angola 49 B4
Cuanza r. Angola 49 B4
Cuatro Ciénegas Mex. 66 D3
Cuauhtémoc Mex. 62 F6
Cuba NY U.S.A. 64 B1
Cuba country West Indies 67 H4
Cubal Angola 49 B5
Cubango r. Angola/Namibia 49 C5
Cubatão Brazil 71 B3
Çubuk Turkey 34 D2
Cuchi Angola 49 B5
Cuchilla Grande hills Uruguay 70 E4
Cucuí Brazil 68 E3
Cúcuta Col. 68 D2
Cudal Australia 58 D4
Cuddalore India 38 C4
Cuddapah India 38 C3
Cuddeback Lake CA U.S.A. 65 D3
Cue Australia 55 B6
Cuéllar Spain 25 D3
Cuemba Angola 49 B5
Cuenca Ecuador 68 C4
Cuenca Spain 25 E3
Cuenca, Serranía de mts Spain 25 E3
Cuernavaca Mex. 66 E5
Cuervos Baja California Mex. 65 E4
Cugir Romania 27 J2
Cuiabá Amazonas Brazil 69 G5
Cuiabá Mato Grosso Brazil 69 G7
Cuiabá r. Brazil 69 G7
Cuilcagh h. Ireland/U.K. 21 E3
Cuillin Hills U.K. 20 C3
Cuillin Sound sea chan. U.K. 20 C3
Cuilo Angola 49 B4
Cuiluan China 44 C3
Cuité r. Brazil 71 C2
Cuito r. Angola 49 C5
Cuito Cuanavale Angola 49 B5
Çukurova plat. Turkey 39 B1
Culcairn Australia 58 C5
Culfa Azer. 35 G3
Culgoa r. Australia 58 C2
Culiacán Mex. 66 C4
Culiacán Rosales Mex. see Culiacán
Cullen U.K. 20 G3
Cullen Point Australia 56 C1
Cullera Spain 25 F4
Cullivoe U.K. 20 [inset]
Cullman U.S.A. 63 J5
Cullybackey U.K. 21 F3
Cul Mòr h. U.K. 20 D2
Culpeper VA U.S.A. 64 C3
Culuene r. Brazil 69 H6
Culver, Point Australia 55 D8
Culverden N.Z. 59 D6
Cumaná Venez. 68 F1
Cumari Brazil 71 A2
Cumbal, Nevado de vol. Col. 68 C3
Cumberland MD U.S.A. 64 B3
Cumberland VA U.S.A. 64 B4
Cumberland Lake Canada 62 G1
Cumberland Peninsula Canada 61 L3
Cumberland Plateau U.S.A. 63 J4
Cumberland Sound sea chan. Canada 61 L3
Cumbernauld U.K. 20 F5
Cumbum India 38 C3
Cummins Australia 57 A7
Cummins Range hills Australia 54 D4
Cumnock Australia 58 D4
Cumnock U.K. 20 E5
Çumra Turkey 34 D3
Cumuruxatiba Brazil 71 D2
Cunderdin Australia 55 B7
Cunene r. Angola 49 B5
Cuneo Italy 26 B2
Cunnamulla Australia 58 B2
Cunningsburgh U.K. 20 [inset]
Cupar U.K. 20 F4
Cupica, Golfo de b. Col. 68 C2
Curaçá Brazil 69 K5
Curaçá r. Brazil 68 D4
Curaçao i. West Indies 67 K6
Curaray r. Ecuador 68 D4
Curdlawidny Lagoon imp. l. Australia 57 B6
Curia Switz. see Chur
Curicó Chile 70 B4
Curitiba Brazil 71 A4
Curitibanos Brazil 71 A4
Curlewis Australia 58 E3
Curnamona Australia 57 B6
Currabubula Australia 58 E3
Currais Novos Brazil 69 K5
Currane, Lough l. Ireland 21 B6
Currant NV U.S.A. 65 E1
Curranyalpa Australia 58 B3
Currawilla Australia 56 C5
Currawinya National Park Australia 58 B2
Currie Australia 52 E5
Currockbilly, Mount Australia 58 E5
Curtis Channel Australia 56 F5
Curtis Island Australia 56 E4
Curtis Island N.Z. 53 I5
Curuá r. Brazil 69 H5
Curupira, Serra mts Brazil/Venez. 68 F3
Cururupu Brazil 69 J4
Curvelo Brazil 71 B2
Cusco Peru 68 D6
Cushendall U.K. 21 F2
Cushendun U.K. 21 F2
Cut Bank U.S.A. 62 E2
Cuthbertson Falls Australia 54 F3
Cuttaburra Creek r. Australia 58 B2
Cuttack India 38 E1
Cuvelai Angola 49 B5
Cuxhaven Germany 17 L4
Cuya Chile 68 D7
Cuyahoga Falls OH U.S.A. 64 A2
Cuyama CA U.S.A. 65 C3
Cuyama r. CA U.S.A. 65 B3
Cuyuni r. Guyana 69 G2
Cuzco Peru see Cusco
Cwmbrân U.K. 19 D7
Cyangugu Rwanda 48 C4
Cyclades is Greece 27 K6
Cydonia Greece see Chania
Cymru admin. div. U.K. see Wales
Cypress Hills Canada 62 E2
Cyprus country Asia 39 A2
Cyrenaica reg. Libya 47 F2
Cythera i. Greece see Kythira
Czechia country Europe see Czech Republic
Czech Republic country Europe 17 O6
Czernowitz Ukr. see Chernivtsi
Czersk Poland 17 P4
Częstochowa Poland 17 Q5

D

Đa, Sông r. Vietnam see Black
Da'an China 44 B3
Ḍabāb, Jabal aḍ mt. Jordan 39 B4
Dabakala Côte d'Ivoire 46 C4
Daban China 43 L4
Daba Shan mts China 43 J6
Dabhoi India 36 C5
Ḍab'ī, Wādī aḍ watercourse Jordan 39 C4
Dablana India 36 C4
Dabola Guinea 46 B3
Dabqig China 43 J5
Dąbrowa Górnicza Poland 17 Q5
Dabs Nur l. China 44 A3
Dabusu Pao l. China see Dabs Nur
Dacca Bangl. see Dhaka
Dachau Germany 17 M6
Dachuan China see Dazhou
Daday Turkey 34 D2
Dadong China see Donggang
Dadra India see Achalpur
Dadu Pak. 33 K4
Daegu S. Korea see Taegu
Daejŏn S. Korea see Taejŏn
Daet Phil. 41 E6
Dafla Hills India 37 H4
Dagana Senegal 46 B3
Dagcagoin China see Zoigê
Dağlıq Qarabağ aut. reg. Azer. 35 G3
Daglung China 37 G3
Dagö i. Estonia see Hiiumaa
Dagon Myanmar see Rangoon
Daguokui Shan mt. China 44 C3
Dagupan Phil. 41 E6
Dagzê China 37 G3
Dagzê Co salt l. China 37 F3
Dahalach, Isole is Eritrea see Dahlak Archipelago
Dahana des. Saudi Arabia see Ad Dahnā'
Daheiding Shan mt. China 44 C3
Dahei Shan mts China 44 B4
Dahej India 36 C5
Da Hinggan Ling mts China 44 A2
Dahlak Archipelago is Eritrea 32 F6
Dahlak Marine National Park Eritrea 32 F6
Dahm, Ramlat des. Saudi Arabia/Yemen 32 G6
Dahmani Tunisia 26 C7
Dahod India 36 C5
Dahomey country Africa see Benin
Dahongliutan Aksai Chin 36 D2
Dahra Senegal see Dara
Dahūk Iraq 35 F3
Dai i. Indon. 54 E1
Dailekh Nepal 37 E3
Dailly U.K. 20 E5
Daimiel Spain 25 E4
Dainkognubma China 37 I2
Daintree National Park Australia 56 D3
Dair, Jebel ed mt. Sudan 32 D7
Dairen China see Dalian
Dai-sen vol. Japan 45 D6
Daisetsu-zan Kokuritsu-kōen Japan 44 F4
Daiyun Shan mts China 43 L7
Dajarra Australia 56 B4
Da Juh China 37 H1
Dakar Senegal 46 B3
Dākhilah, Wāḥāt ad oasis Egypt 32 C4
Dakhla W. Sahara see Ad Dakhla
Dakhla Oasis oasis Egypt see Dākhilah, Wāḥāt ad
Dakol'ka r. Belarus 13 F5
Dakor India 36 C5
Dakoro Niger 46 D3
Đakovica Kosovo see Gjakovë
Đakovo Croatia 26 H2
Daktuy Rus. Fed. 44 B1
Dala Angola 49 C5
Dalaba Guinea 46 B3
Dalai China see Da'an
Dalain Hob China 42 I4
Dālakī Iran 35 H5
Dalälven r. Sweden 15 J6
Dalaman Turkey 27 M6
Dalandzadgad Mongolia 42 I4
Dalap-Uliga-Darrit Marshall Is see Delap-Uliga-Djarrit
Đa Lat Vietnam 31 J5
Dalatando Angola see N'dalatando
Dalaud India 36 C5
Dalauda India 36 C5
Dalbandin Pak. 33 J4
Dalbeattie U.K. 20 F6
Dalbeg Australia 56 D4
Dalby Australia 58 E1
Dalby Isle of Man 18 C4
Dale Hordaland Norway 15 D6
Dale Sogn og Fjordane Norway 15 D6
Dale City VA U.S.A. 64 C3
Dale Hollow Lake U.S.A. 63 J4
Dalet Myanmar 37 H6
Daletme Myanmar 37 H5
Dalfors Sweden 15 I6
Dalgety Australia 58 D6
Dalgety r. Australia 55 A6
Dalhart U.S.A. 62 G4
Dalhousie, Cape Canada 60 F2
Dali Yunnan China 42 I7
Dalian China 43 M5
Dalin China 44 A4
Dalizi China 44 B4
Dalkeith U.K. 20 F5
Dallas TX U.S.A. 63 H5
Dalles City U.S.A. see The Dalles
Dall Island U.S.A. 60 E4
Dalmacija reg. Bos.-Herz./Croatia see Dalmatia
Dalmatia reg. Bos.-Herz./Croatia 30 A2
Dalmau India 36 E4
Dalmellington U.K. 20 E5
Dalmi India 37 F5
Dal'negorsk Rus. Fed. 44 D3
Dal'nerechensk Rus. Fed. 44 D3
Dal'niye Zelentsy Rus. Fed. 12 H1
Dalny China see Dalian
Daloa Côte d'Ivoire 46 C4
Dalol Eth. 32 F7
Daloloia Group is P.N.G. 56 E1
Dalry U.K. 20 E5
Dalrymple U.K. 20 E5
Dalrymple, Lake Australia 56 D4
Daltenganj India 37 F4
Dalton S. Africa 51 J5
Dalton GA U.S.A. 63 K5
Dalton MA U.S.A. 64 E1
Dalton PA U.S.A. 64 D2
Daltonganj India see Daltenganj
Dalton-in-Furness U.K. 18 D4
Daly r. Australia 54 E3
Daly City CA U.S.A. 65 A2
Daly River Australia 54 E3
Daly Waters Australia 54 F4
Damagaram Takaya Niger 46 D3
Daman India 38 B1
Daman and Diu union terr. India 38 A1
Damanhûr Egypt see Damanhūr
Damanhūr Egypt 34 C5
Damão India see Daman
Damar i. Indon. 41 E8
Damara Cent. Afr. Rep. 48 B3
Damaraland reg. Namibia 49 B6
Damas Syria see Damascus
Damascus Syria 39 C3
Damaturu Nigeria 46 E3
Damāvand Iran 35 I4
Damāvand, Qolleh-ye mt. Iran 35 I4
Dambulla Sri Lanka 38 D5
Dāmghān Iran 35 I3
Damianópolis Brazil 71 B1
Damietta Egypt see Dumyāṭ
Dāmiyā Jordan 39 B3
Damjong China 37 H2
Damlasu Turkey 39 D1
Dammam Saudi Arabia 32 H4
Damoh India 36 D5
Damour Lebanon 39 B3
Dampier Archipelago is Australia 54 B5
Dampier Island P.N.G. see Karkar Island
Dampier Land pen. Australia 54 C4
Dampier Strait P.N.G. 52 E2
Dampir, Selat sea chan. Indon. 41 F8
Damqoq Zangbo r. China see Maquan He
Dam Qu r. China 37 H2
Damxoi China see Comai
Damxung China 37 G3
Ḍānā Jordan 39 B4
Dana Nepal 37 E3
Danakil reg. Africa see Denakil
Danané Côte d'Ivoire 46 C4
Đa Năng Vietnam 31 J5
Danata Turkm. 35 I3
Danbury CT U.S.A. 64 E2
Danby VT U.S.A. 64 E1
Danby Lake CA U.S.A. 65 E3
Dandaragan Australia 55 A7
Dande Eth. 48 D3
Dandeldhura Nepal 36 E3
Dandeli India 38 B3
Dandong China 45 B4
Dandot Pak. 36 C2
Dane r. U.K. 18 E5
Daneborg Greenland 77 I2
Danese WV U.S.A. 64 A4
Dangara Tajik. see Danghara
Dangbi Rus. Fed. 44 C3
Dangchengwan China see Subei
Danger Islands atoll Cook Is see Pukapuka
Danger Point S. Africa 50 D8
Danghara Tajik. 33 K2
Danghe Nanshan mts China 42 G5
Dang La pass China see Tanggula Shankou
Dangla Shan mts China see Tanggula Shan
Dangqên China 37 G3
Dangriga Belize 66 G5
Daniëlskuil S. Africa 50 F5
Danilov Rus. Fed. 12 I4
Danilovka Rus. Fed. 13 J6
Danilovskaya Vozvyshennost' hills Rus. Fed. 12 H4
Danjiangkou China 43 K6
Danjo-guntō is Japan 45 C6
Dankhar India 36 D2
Dankov Rus. Fed. 13 H5
Danlí Hond. 67 G6
Danmark country Europe see Denmark
Dannebrog Ø i. Greenland see Qillak
Dannevirke N.Z. 59 F5
Dannhauser S. Africa 51 J5
Dano Burkina Faso 46 C3
Dansville NY U.S.A. 64 C1
Danta India 36 C4
Dantan India 37 F5
Dantevada India see Dantewara
Dantewada India see Dantewara
Dantewara India 38 D2
Danube r. Austria/Germany 24 I2
Danube r. Bulg./Croatia/S.M. 13 F7
Danube r. Europe 17 P6
Danube r. Hungary 17 Q8
Danube Delta Romania/Ukr. 27 M2
Danville IL U.S.A. 63 J3
Danville PA U.S.A. 64 C2
Danville VA U.S.A. 63 L4
Danzig Poland see Gdańsk
Danzig, Gulf of Poland/Rus. Fed. see Gdańsk, Gulf of
Dao Tay Sa is S. China Sea see Paracel Islands
Daoud Alg. see Aïn Beïda
Daoukro Côte d'Ivoire 46 C4
Dapaong Togo 46 D3
Daphabum mt. India 37 I4
Dapitan Phil. 41 E7
Daporijo India 37 H4
Da Qaidam China 42 H5
Daqing China 44 B3
Daqq-e Sorkh salt flat Iran 35 I4
Dara Senegal 46 B3
Dar'ā Syria 39 C3
Dâra, Gebel mt. Egypt see Dārah, Jabal
Dārāb Iran 35 I5
Dārah, Jabal mt. Egypt 34 D6
Daraj Libya 46 E1
Dārākūyeh Iran 35 I5
Dārān Iran 35 H4
Darazo Nigeria 46 E3
Darbhanga India 37 F4
Dardanelles str. Turkey 27 L4
Dardania country Europe see Kosovo
Dardo China see Kangding
Dar el Beida Morocco see Casablanca
Darende Turkey 34 E3
Dar es Salaam Tanz. 49 D4
Darfo Boario Terme Italy 26 D2
Dargai Pak. 36 B2
Dargaville N.Z. 59 D2
Dargo Australia 58 C6
Dargo Zangbo r. China 37 F3
Darhan Mongolia 42 J3
Darién, Golfo del g. Col. 68 C2
Darién, Parque Nacional de nat. park Panama 67 I7
Darīganga Mongolia 43 K3
Darjeeling India see Darjiling
Darjiling India 37 G4
Darkhazineh Iran 35 H5
Darlag China 42 H6
Darling r. Australia 58 B3
Darling Downs hills Australia 58 D1
Darling Range hills Australia 55 A8
Darlington U.K. 18 F4
Darlington Point Australia 58 C5
Darlot, Lake imp. l. Australia 55 C6
Darłowo Poland 17 P3
Darma Pass China/India 36 E3
Darmstadt Germany 17 L6
Darnah Libya 34 A4
Darnall S. Africa 51 J5
Darnick Australia 58 A4
Darnley, Cape Antarctica 76 E2
Daroca Spain 25 F3
Darovskoy Rus. Fed. 12 J4
Darr watercourse Australia 56 C4
Darreh-ye Bāhābād Iran 35 I5
Darreh-ye Shahr Iran 35 G4
Darsi India 38 C3
Dart r. U.K. 19 D8
Dartang China see Baqên
Dartford U.K. 19 H7
Dartmoor Australia 57 C8
Dartmoor hills U.K. 19 C8
Dartmoor National Park U.K. 19 D8
Dartmouth Canada 63 O3
Dartmouth U.K. 19 D8
Dartmouth, Lake imp. l. Australia 57 D5
Dartmouth Reservoir Australia 58 C6
Darton U.K. 18 F5
Daru P.N.G. 52 E2
Daru Sierra Leone 46 B4
Darvāzgay Afgh. 36 A3
Darwen U.K. 18 E5
Darwin Australia 54 E3
Darwin, Monte mt. Chile 70 C8
Daryācheh-ye Orūmīyeh salt l. Iran see Urmia, Lake
Dar'yalyktakyr, Ravnina plain Kazakh. 42 A3
Daryā-ye Kholm r. Afgh. 36 A1
Dar"yoi Amu r. Asia see Amudar'ya
Dārzāb Afgh. 36 A2
Dasada India 36 B5
Dashhowuz Turkm. see Daşoguz
Dashkesan Azer. see Daşkäsän
Dashkhovuz Turkm. see Daşoguz
Dashkōpri Turkm. see Daşköpri
Dasht Iran 35 J3
Dasht-e Arbū Shomālī des. Afgh. 33 J4
Dasht-e Nāomīd des. Afgh./Iran 33 J3
Daska Pak. 36 C2
Daşkäsän Azer. 35 G2
Daşköpri Turkm. 33 J2
Daşoguz Turkm. 30 E2
Daşoguz Turkm. see Daşoguz
Daspar mt. Pak. 36 C1
Datça Turkey 27 L6
Date Japan 44 F4
Date Creek watercourse AZ U.S.A. 65 F3
Dateland AZ U.S.A. 65 F4
Datha India 36 C5
Datia India 36 D4
Datong Heilong. China 44 B3
Datong Shanxi China 43 K4
Datong He r. China 42 I5
Dattapur India 38 C1
Daudkandi Bangl. 37 G5
Daugava r. Latvia 15 N8
Daugavpils Latvia 15 O9
Daulatabad India 38 B2
Daulatabad Iran see Malāyer
Daulatpur Bangl. 37 G5
Daungyu r. Myanmar 37 H5
Dauphin Canada 62 G1
Dauphiné reg. France 24 G4
Dauphiné, Alpes du mts France 24 G4
Dauphin Lake Canada 62 H1
Daurie Creek r. Australia 55 A6
Dausa India 36 D4
Dava U.K. 20 F3
Däväçi Azer. 35 H2
Davanagere India see Davangere
Davangere India 38 B3
Davao Phil. 41 E7
Davel S. Africa 51 I4
Davenport IA U.S.A. 63 I3
Davenport Downs Australia 56 C5
Davenport Range hills Australia 54 F5
Daventry U.K. 19 F6
Daveyton S. Africa 51 I4
David Panama 67 H7
Davidson Canada 62 F1
Davidson, Mount h. Australia 54 E5
Davis r. Australia 54 C5
Davis CA U.S.A. 65 B1
Davis WV U.S.A. 64 B3
Davis (Australia) research stn Antarctica 76 E2
Davis, Mount h. PA U.S.A. 64 B3
Davis Bay Antarctica 76 G2
Davis Dam AZ U.S.A. 65 E3
Davis Inlet (abandoned) Canada 61 L4
Davis Sea Antarctica 76 F2
Davis Strait Canada/Greenland 61 M3
Davlekanovo Rus. Fed. 11 Q5
Davos Switz. 24 I3
Dawa Co l. China 37 F3
Dawa Wenz r. Eth. 48 E3
Dawaxung China 37 F3
Dawei Myanmar see Tavoy
Dawera i. Indon. 54 E1
Dawo China see Maqên
Dawqah Oman 33 H6
Dawson r. Australia 56 E4
Dawson Canada 60 E3
Dawson Creek Canada 60 F4
Dawsons Landing Canada 62 B1
Dawu Qinghai China see Maqên
Dawukou China see Shizuishan
Dax France 24 D5
Daxian China see Dazhou
Daxing'an Ling mts China see Da Hinggan Ling
Da Xueshan mts China 31 J3
Dayan China see Lijiang
Dayangshu China 44 B2
Dayao China 42 I7
Dāykundī Afgh. 36 A2
Daylesford Australia 58 B6
Daylight Pass NV U.S.A. 65 D2
Dayong China see Zhangjiajie

Dayr Abū Saʿīd Jordan 39 B3
Dayr az Zawr Syria 35 F4
Dayr Ḥāfir Syria 39 C1
Dayton OH U.S.A. 63 K4
Dayton VA U.S.A. 64 B3
Daytona Beach U.S.A. 63 K6
Dayyer Iran 35 H6
Dazhou China 43 J6
De Aar S. Africa 50 G6
Dead r. Ireland 21 D5
Deadman Lake CA U.S.A. 65 D3
Dead Mountains NV U.S.A. 65 E3
Dead Sea salt l. Asia 39 B4
Deakin Australia 55 D7
Deal U.K. 19 I7
Dealesville S. Africa 51 G5
Dean, Forest of U.K. 19 E7
Deán Funes Arg. 70 C4
Deanuvuotna inlet Norway see
    Tanafjorden
Dearne r. U.K. 18 F5
Dease Lake Canada 60 F4
Dease Strait Canada 60 H3
Death Valley CA U.S.A. 65 D3
Death Valley Junction CA U.S.A. 65 D2
Death Valley National Park CA U.S.A.
    65 D2
Deauville France 24 E2
De Baai S. Africa see Port Elizabeth
Debar Macedonia 27 I4
Debenham U.K. 19 I6
Débo, Lac l. Mali 46 C3
Deborah East, Lake imp. l. Australia 55 B7
Deborah West, Lake imp. l. Australia
    55 B7
Debrecen Hungary 27 I1
Debre Markos Eth. 32 E7
Debre Tabor Eth. 32 E7
Debre Zeyit Eth. 48 D3
Decatur AL U.S.A. 63 J5
Decatur IL U.S.A. 63 J4
Deccan plat. India 38 C2
Deception Bay Australia 58 F1
Děčín Czech Rep. 17 O5
Decorah U.S.A. 63 I3
Deddington U.K. 19 I6
Dede de Deus mt. Brazil 71 B4
Dedovichi Rus. Fed. 12 F4
Dedu China see Wudalianchi
Dee r. Ireland 21 F4
Dee est. U.K. 18 D5
Dee r. England/Wales U.K. 19 D5
Dee r. Scotland U.K. 20 G3
Deel r. Ireland 21 D5
Deel r. Ireland 21 D5
Deep Creek Lake MD U.S.A. 64 D3
Deepwater Australia 58 E2
Deeri Somalia 48 E3
Deering, Mount Australia 55 E6
Deering U.S.A. 60 B3
Deer Island U.S.A. 60 B4
Deer Lodge U.S.A. 62 E2
Deesa India see Disa
Defensores del Chaco, Parque Nacional
    nat. park Para. 70 D2
Degana India 36 C4
Degeh Bur Eth. 48 E3
Degema Nigeria 46 D4
Deggendorf Germany 17 N6
Degh r. Pak. 36 C3
Degtevo Rus. Fed. 13 I6
De Hadī Ghar Afgh. 36 A3
Dehaq Iran 35 H4
Deh Golān Iran 35 G4
Dehi Afgh. 36 A2
Dehküyeh Iran 35 I4
Dehlorān Iran 35 G4
De Hoop Nature Reserve S. Africa 50 E8
Dehradun India see Dehra Dun
Dehra Dun India 36 D3
Dehri India 37 F4
Deim Zubeir South Sudan 47 F4
Deir-ez-Zor Syria see Dayr az Zawr
Dej Romania 27 J1
Deji China see Rinbung
Dejiang China 43 F3
De Kalb IL U.S.A. 63 J3
De-Kastri Rus. Fed. 44 F2
Dekemhare Eritrea 32 E6
Dekina Nigeria 46 D4
Dékoa Cent. Afr. Rep. 48 B3
Delaki Indon. 54 D2
Delamar Lake NV U.S.A. 65 E2
Delano CA U.S.A. 65 D3
Delano Peak UT U.S.A. 65 F1
Delap-Uliga-Djarrit Marshall Is 74 H5
Delārām Afgh. 33 J3
Delareyville S. Africa 51 G4
Delaronde Lake Canada 62 F1
Delaware r. New Jersey/Pennsylvania U.S.A.
    64 D3
Delaware state U.S.A. 64 D3
Delaware, East Branch r. NY U.S.A. 64 D2
Delaware Bay Delaware/New Jersey U.S.A.
    64 D3
Delaware Water Gap National
    Recreational Area park New Jersey/
    Pennsylvania U.S.A. 64 D2
Dêlêg China 37 F3
Delegate Australia 58 D6
Delémont Switz. 24 H3
Delfinópolis Brazil 71 B3
Delft Neth. 16 J4
Delfzijl Neth. 15 E10
Delgado, Cabo c. Moz. 49 E5
Delgerhaan Mongolia 42 I3
Delhi Ont. Canada 64 A1
Delhi India 36 D3
Delhi China 36 D3
Delhi NY U.S.A. 64 D1
Delice Turkey 34 D3
Delice r. Turkey 34 D2
Delījān Iran 35 H4
Déline Canada 60 F3
Delingha China see Delhi
Dellys Alg. 25 H5
Del Mar CA U.S.A. 65 D4
Delmenhorst Germany 17 L4
Delnice Croatia 26 J2
De Long Islands Rus. Fed. see
    De-Longa, Ostrova
De Long Mountains U.S.A. 60 B3
De Long Strait Rus. Fed. see Longa, Proliv
Deloraine Canada 62 F1
Delportshoop S. Africa 50 G5
Del Rio U.S.A. 62 G6
Delsbo Sweden 15 J6
Delta CO U.S.A. 62 F4
Delta UT U.S.A. 62 E4
Delta del Orinoco Venez. 68 F2
Delta Downs Australia 56 C3
Delta Junction U.S.A. 60 D3
Delvin Ireland 21 E4
Delungra Australia 58 E2
Delvinë Albania 27 I5
Delwara India 36 C4

Demavend mt. Iran see
    Damāvand, Qolleh-ye
Demba Dem. Rep. Congo 49 C4
Dembī Dolo Eth. 32 D8
Demerara Guyana see Georgetown
Demerara Abyssal Plain sea feature
    S. Atlantic Ocean 72 F6
Demidov Rus. Fed. 13 F5
Deming U.S.A. 62 F5
Demirci Turkey 27 M5
Demirköy Turkey 27 L4
Demirtaş Turkey 39 A1
Demmin Germany 17 N4
Dempo, Gunung vol. Indon. 41 C8
Dem'yanovo Rus. Fed. 12 J3
Dem'yanskoye Rus. Fed. 12 I4
De Naawte S. Africa 50 E6
Denakil reg. Africa see Danakil
Denali mt. U.S.A. see McKinley, Mount
Denali National Park and Preserve U.S.A.
    60 C3
Denan Eth. 48 E3
Denbigh U.K. 18 D5
Den Bosch Neth. see 's-Hertogenbosch
Dendâra Mauritania 46 C3
Dendi mt. Eth. 48 D3
Dendron S. Africa see Mogwadi
Denezhkin Kamen', Gora mt. Rus. Fed.
    11 R3
Dêngka China see Têwo
Dêngkagoin China see Têwo
Dengkou China 42 J4
Dêngqên China 37 H3
Den Haag Neth. see The Hague
Denham r. Australia 54 E3
Denham Australia 55 A6
Denham Range mts Australia 56 E4
Den Helder Neth. 16 J4
Dénia Spain 25 G4
Denial Bay Australia 55 A7
Deniliquin Australia 58 B5
Denison IA U.S.A. 63 H3
Denison, Cape Antarctica 76 G2
Denison Plains Australia 54 E4
Deniyaya Sri Lanka 38 D5
Denizli Turkey 27 M6
Denman Australia 58 E4
Denman Glacier Antarctica 76 F2
Denmark Australia 52 B5
Denmark country Europe 15 G8
Denmark Strait Greenland/Iceland 10 A2
Dennis, Lake imp. l. Australia 54 E5
Denny U.K. 20 F4
Denpasar Indon. 41 D8
Denton MD U.S.A. 64 D3
Denton TX U.S.A. 62 H5
D'Entrecasteaux, Point Australia 55 A8
D'Entrecasteaux, Récifs rf New Caledonia
    53 G3
D'Entrecasteaux Islands P.N.G. 52 F2
D'Entrecasteaux National Park Australia
    55 A8
Denver CO U.S.A. 62 F4
Denver PA U.S.A. 64 C2
Deo India 37 F4
Deoband India see Deoband
Deogarh Jharkhand India see Deoghar
Deogarh Orissa India 37 F5
Deogarh Rajasthan India 36 C4
Deogarh Uttar Prad. India 36 D4
Deogarh India 37 E5
Deoghar India 37 F4
Deolali India 38 B2
Deoli India 37 F5
Deori Madh. Prad. India 36 D5
Deoria India 37 E4
Deosil India 37 E5
Deothang Bhutan 37 G4
Deposit NY U.S.A. 64 D2
Depsang Point h. Aksai Chin 36 D2
Deputatskiy Rus. Fed. 29 O3
Dêqên Xizang China see Dagzê
Dêqên Xizang China 37 G3
Dêqên Xizang China see Dagzê
De Queen U.S.A. 63 I5
Dera Ghazi Khan Pak. 33 L3
Dera Ismail Khan Pak. 33 L3
Derawar Fort Pak. 36 B3
Derbent Rus. Fed. 35 H2
Derbesiye Turkey see Şenyurt
Derbur China 44 A2
Derby Australia 54 C4
Derby U.K. 19 F6
Derby CT U.S.A. 64 E2
Derby NY U.S.A. 64 B1
Đerdap, Nacionalni Park nat. park Serbia
    27 J2
Dereham U.K. 19 H6
Derg r. Ireland/U.K. 21 E3
Derg, Lough l. Ireland 21 D5
Dergachi Rus. Fed. 13 K6
Dergachi Ukr. see Derhachi
Derhachi Ukr. 13 H6
De Ridder U.S.A. 63 I5
Derik Turkey 35 F3
Derm Namibia 50 D2
Derna Libya see Darnah
Dernberg, Cape Namibia 50 B4
Dêrong China 42 H7
Derravaragh, Lough l. Ireland 21 E4
Derry U.K. see Londonderry
Derry NH U.S.A. 64 F1
Derryveagh Mountains hills Ireland 21 D3
Derub U.K. 19 H6
Derudeb Sudan 32 E6
De Rust S. Africa 50 F7
Derventa Bos.-Herz. 26 G2
Derwent r. England U.K. 18 F6
Derwent r. England U.K. 18 G5
Derwent Water l. U.K. 18 D4
Derzhavinsk Kazakh. 28 H4
Derzhavinskiy Kazakh. see Derzhavinsk
Desaguadero r. Arg. 70 C4
Désappointement, Îles du is Fr. Polynesia
    75 K6
Desē Eth. 48 D2
Deseado Arg. 70 C7
Deseado r. Arg. 70 C7
Desengaño, Punta pt Arg. 70 C7
Desert Center CA U.S.A. 65 E4
Desert Lake NV U.S.A. 65 E3
Des Moines IA U.S.A. 63 I3
Des Moines r. U.S.A. 63 I3
Desna r. Rus. Fed./Ukr. 13 F6
Desnogorsk Rus. Fed. 13 G5
Desolación, Isla i. Chile 70 B8
Dessau Germany 17 N5
Dessye Eth. see Desē
Destruction Bay Canada 77 A2
Desventuradas, Islas is S. Pacific Ocean
    75 O7
Dete Zimbabwe 49 C5
Detmold Germany 17 L5
Detroit U.S.A. 64 C2
Detroit Lakes U.S.A. 63 H2
Dett Zimbabwe see Dete
Deua National Park Australia 58 D5
Deutschland country Europe see Germany

Deutschlandsberg Austria 17 O7
Deva Romania 27 J2
Deva U.K. see Chester
Devana U.K. see Aberdeen
Devangere India see Davangere
Devanhalli India 38 C3
Deve Bair pass Bulg./Macedonia see
    Velbŭzhdki Prokhod
Develi Turkey 34 D3
Deventer Neth. 16 K4
Deveron r. U.K. 20 G3
Devgarh India 38 B2
Devghar India see Deoghar
Devikot India 36 B4
Devil's Bridge U.K. 19 D6
Devil's Gate pass CA U.S.A. 65 C1
Devil's Lake U.S.A. 62 F1
Devil's Paw mt. U.S.A. 60 E4
Devil's Peak U.S.A. 65 C2
Devizes U.K. 19 F7
Devli India 36 C4
Devnya Bulg. 27 L3
Devon r. U.K. 20 F4
Devon Island Canada 61 I2
Devonport Australia 57 [inset]
Devrek Turkey 27 N4
Devrukh India 38 B2
Dewas India 36 C5
Dewetsdorp S. Africa 51 H5
Dewsbury U.K. 18 F5
Dexter MO U.S.A. 63 J4
Deyang China 42 I6
Dey-Dey Lake salt l. Australia 55 E7
Deyong, Tanjung pt Indon. 41 J8
Deyhūk Iran 33 I3
Dez r. Iran 32 G3
Dezfūl Iran 35 H4
Dezhneva, Mys c. Rus. Fed. 29 T3
Dezhou Shandong China 43 L5
Dezh Shāhpūr Iran see Marīvān
Dhabarau India 37 E4
Dhahab, Wādī adh r. Syria 39 B3
Dhāhiriya West Bank 39 B4
Dhahran Saudi Arabia 30 E4
Dhaka Bangl. 37 G5
Dhalbhum reg. India 37 F5
Dhalgaon India 38 B2
Dhamār Yemen 32 F7
Dhamoni India 36 D4
Dhamtari India 38 D1
Dhana Pak. 36 B4
Dhana Sar r. Pak. 36 B3
Dhanbad India 37 F5
Dhanera India 36 C4
Dhang Range mts Nepal 37 E3
Dhankuta Nepal 37 F4
Dhansia India 36 C3
Dhar India 36 C5
Dhar Adrar hills Mauritania 46 B3
Dharampur India 38 B1
Dharan Bazar Nepal 37 F4
Dharashiv India see Osmanabad
Dhari India 36 B5
Dharmapuri India 38 C3
Dharmavaram India 38 C3
Dharmsala Hima. Prad. India see
    Dharmshala
Dharmsala Orissa India 37 F5
Dharmshala India 36 D2
Dharnaoda India 36 D4
Dhar Oualâta hills Mauritania 46 C3
Dhar Tîchît hills Mauritania 46 C3
Dharug National Park Australia 58 E4
Dharur India 38 C2
Dharwad India 38 B3
Dharwar India see Dharwad
Dharwas India 36 D2
Dhasan r. India 36 D4
Dhāt al Ḥājj Saudi Arabia 34 E5
Dhaulagiri mt. Nepal 37 E3
Dhaulpur India see Dholpur
Dhaura India 36 D4
Dhaurahra India 36 E4
Dhawlagiri mt. Nepal see Dhaulagiri
Dhebar Lake India see Jaisamand Lake
Dhekelia Sovereign Base Area
    military base Cyprus 39 A2
Dhemaji India 37 H4
Dhenkanal India 37 F5
Dhībān Jordan 39 B4
Dhidhimótikhon Greece see
    Didymoteicho
Dhing India 37 H4
Dhirwah, Wādī adh watercourse Jordan
    39 C4
Dhodhekánisos is Greece see Dodecanese
Dhola India 36 B5
Dholera India 36 C5
Dholpur India 36 D4
Dhomokós Greece see Domokos
Dhone India 38 C3
Dhoraji India 36 B5
Dhori India 36 B5
Dhrangadhra India 36 B5
Dhubāb Yemen 32 F7
Dhubri India see Dhubri
Dhudial Pak. 36 C2
Dhule India 38 B1
Dhulia India see Dhule
Dhulian India 37 F4
Dhulian Pak. 36 C2
Dhuma India 36 D5
Dhund r. India 36 D4
Dhurwai India 36 D4
Dhuusa Marreeb Somalia 48 E3
Dia i. Greece 27 K7
Diablo, Mount CA U.S.A. 65 B2
Diablo, Picacho del mt. Mex. 66 A2
Diablo Range mts CA U.S.A. 65 B2
Diagbe Dem. Rep. Congo 48 C3
Diamante Arg. 70 D4
Diamantina watercourse Australia 56 B5
Diamantina Brazil 71 C2
Diamantina, Chapada plat. Brazil 71 C1
Diamantina Deep sea feature
    Indian Ocean 73 O8
Diamantina Gates National Park Australia
    56 C4
Diamantino Brazil 69 G6
Diamond Islets Australia 56 E3
Dian Chi l. China 42 I5
Diandioumé Mali 46 C3
Diane Bank sea feature Australia 56 E2
Dianópolis Brazil 69 I6
Diaobingshan China 44 A4
Diaoling China 44 C3
Diapaga Burkina Faso 46 D3
Diarizos r. Cyprus 39 A2
Diaz Point Namibia 50 B4
Dibaya Dem. Rep. Congo 49 C4
Dibba well Niger 46 E3
Dibella well Niger 46 E3
Dibeng S. Africa 50 F4
Dibete Botswana 51 H2
Dibrugarh India 37 H4
Dibsê Syria see Dibsī
Dibsī Syria 39 D2
Dickinson U.S.A. 62 G2
Diclê r. Asia see Tigris
Diclê r. Asia 35 F3 see Tigris

Dídesa Wenz r. Eth. 48 D3
Didiéni Mali 46 C3
Didwana India 36 C4
Didymoteicho Greece 27 L4
Die France 24 G4
Diébougou Burkina Faso 46 C3
Diedenhofen France see Thionville
Diefenbaker, Lake Canada 62 F1
Diego de Almagro, Isla i. Chile 70 A8
Diégo Suarez Madag. see Antsirañana
Diéma Mali 46 C3
Dieppe France 24 E2
Di'er Songhua Jiang r. China 44 B3
Dietikon Switz. 24 I3
Diffa Niger 46 E3
Digby Canada 63 N3
Diggi India 36 C4
Diglur India 38 C2
Digne France see Digne-les-Bains
Digne-les-Bains France 24 H4
Digoin France 24 H4
Digras India 38 C1
Digri Pak. 36 B4
Digul r. Indon. 52 F2
Digya National Park Ghana 46 C4
Dihang r. China see Brahmaputra
Dihôk Iraq see Dahūk
Diinsoor Somalia 48 E3
Dijon France 24 G3
Dik Chad 47 E4
Diken India 36 C4
Dikhil Djibouti 32 F7
Dikili Turkey 27 L5
Diklosmta mt. Rus. Fed. 13 J8
Dikson Rus. Fed. 28 J2
Dīla Eth. 48 D3
Dili East Timor 54 D2
Dillenburg Germany 17 L5
Dillingen an der Donau Germany 17 M6
Dillingham U.S.A. 60 C4
Dillon U.S.A. 62 E3
Dillwyn VA U.S.A. 64 B4
Dilolo Dem. Rep. Congo 49 C5
Dimapur India 37 H4
Dimashq Syria see Damascus
Dimbokro Côte d'Ivoire 46 C4
Dimboola Australia 57 C8
Dimitrov Ukr. see Dymytrov
Dimitrovgrad Bulg. 27 K3
Dimitrovgrad Rus. Fed. 13 K5
Dimitrovo Bulg. see Pernik
Dīmona Israel 39 B4
Dimpho Pan salt pan Botswana 50 E3
Dinajpur Bangl. 37 G4
Dinan France 24 C2
Dinant Belgium 16 J5
Dinapur India 37 F4
Dinar Turkey 27 N5
Dīnār, Kūh-e mt. Iran 35 H5
Dinara Planina mts Bos.-Herz./Croatia see
    Dinaric Alps
Dinaric Alps mts Bos.-Herz./Croatia 26 G2
Dinbych U.K. see Denbigh
Dinbych-y-pysgod U.K. see Tenby
Dinder National Park Sudan 47 G3
Dindi r. India 38 C2
Dindigul India 38 C4
Dindima Nigeria 46 E3
Dindiza Moz. 51 K2
Dindori India 36 E5
Dingan China 43 L8
Dingbian China 43 J4
Dingcheng China see Dingyuan
Dingelstädt Germany 17 M5
Dingin, Bukit mt. Indon. see Daik, Bukit
Dingle Bay Ireland 21 B5
Dingnan China 43 F3
Dingo Australia 56 E4
Dinguiraye Guinea 46 B3
Dingwall U.K. 20 E3
Dingxi China 42 I5
Dinngyê China 37 F3
Dinokwe Botswana 51 H2
Dinuba CA U.S.A. 65 C3
Dinwiddie VA U.S.A. 64 C4
Diŏlla Mali 46 C4
Dionísio Cerqueira Brazil 70 F3
Diorama Brazil 71 A2
Dioscurias Georgia see Sokhumi
Diouloulou Senegal 46 B3
Diourbel Senegal 46 B3
Dipkarpaz Cyprus see Rizokarpason
Diplo Pak. 36 B4
Dipperu National Park Australia 56 E4
Dirang India 37 H4
Diré Mali 46 C3
Direction, Cape Australia 56 C2
Dirê Dawa Eth. 48 E3
Dirico Angola 49 C5
Dirk Hartog Island Australia 55 A6
Dirranbandi Australia 58 D2
Dirs Saudi Arabia 48 E2
Dirschau Poland see Tczew
Disa India 36 C4
Disang r. India 37 H4
Disappointment, Cape S. Georgia 70 I8
Disappointment, Lake imp. l. Australia
    55 C5
Disappointment Islands Fr. Polynesia see
    Désappointement, Îles du
Disaster Bay Australia 58 D6
Discovery Bay Australia 57 C8
Disko i. Greenland see Qeqertarsuaq
Disko Bugt b. Greenland see
    Qeqertarsuup Tunua
Dispur India 37 G4
Disputanta VA U.S.A. 64 C4
Diss U.K. 19 I6
Distrito Federal admin. dist. Brazil 71 B1
Disûq Egypt 34 C5
Ditloung S. Africa 50 F5
Dittaino r. Sicilia Italy 26 F6
Diu India 38 A1
Dīvān Darreh Iran 35 G4
Divehi country Indian Ocean see Maldives
Divi, Point India 38 D3
Divichi Azer. see Dǝvǝçi
Divinópolis Brazil 71 B3
Divnoye Rus. Fed. 13 I7
Divo Côte d'Ivoire 46 C4
Diwana Pak. 36 A4
Dixon CA U.S.A. 65 B1
Dixon IL U.S.A. 63 J3
Dixon Entrance sea chan. Canada/U.S.A.
    60 E4
Diyadin Turkey 35 F3
Diyarbakır Turkey 35 F3
Diz Chah Iran 35 I4
Dize Turkey see Yüksekova
Djado Niger 46 E2
Djado, Plateau du Niger 46 E2
Djaja, Puntjak mt. Indon. see Jaya, Puncak
Djakarta Indon. see Jakarta
Djambala Congo 48 B4
Djanet Alg. 46 D2
Djebel Bou Kornine, Parc National du
    nat. park Tunisia 26 C6
Djelfa Alg. 25 H6
Djenné Mali 46 C3
Djibo Burkina Faso 46 C3

Djibouti country Africa 32 F7
Djibouti Djibouti 32 F7
Djidjelli Alg. see Jijel
Djougou Benin 46 D4
Djoum Cameroon 46 E4
Djourab, Erg du des. Chad 47 E3
Djúpivogur Iceland 14 [inset]
Djúrás Sweden 15 I6
Djurdjura, Parc National du Alg. 25 I5
Dmitriya Lapteva, Proliv sea chan.
    Rus. Fed. 29 P2
Dmitriyev-L'govskiy Rus. Fed. 13 G5
Dmitriyevsk Ukr. see Makiyivka
Dmitrov Rus. Fed. 12 H4
Dmytriyev's Ukr. see Makiyivka
Dnepr r. Europe 13 F5 see Dnieper
Dneprodzerzhinsk Ukr. see
    Dniprodzerzhyns'k
Dnepropetrovsk Ukr. see Dnipropetrovs'k
Dnestr r. Moldova 13 F7
Dnieper r. Belarus 23 K1
Dnieper r. Europe 13 G7
Dnieper r. Ukr. 13 G6
Dniester r. Moldova 13 F7
Dnipro r. Europe 13 G7 see Dnieper
Dnipro r. Ukr. 13 G6
Dniprodzerzhyns'k Ukr. 13 G6
Dnipropetrovs'k Ukr. 13 G6
Dnister r. Ukr. 13 F6 see Dniester
Dno Rus. Fed. 12 F4
Dnyapro r. Belarus see Dnieper
Dnyapro r. Europe 13 F6 see Dnieper
Doāb Afgh. 36 A2
Doaba Pak. 36 A2
Doaktown Canada 63 N2
Doba Chad 47 E4
Dobele Latvia 15 M8
Doberai, Jazirah pen. Indon. 41 F8
Doberai Peninsula Indon. see
    Doberai, Jazirah
Dobo Indon. 41 F8
Doboj Bos.-Herz. 26 H2
Dobrich Bulg. 27 L3
Dobrinka Rus. Fed. 13 I5
Dobroye Rus. Fed. 13 H5
Dobrudja reg. Romania see Dobruja
Dobrudzha reg. Romania 27 L3
Dobrush Belarus 13 F5
Dobryanka Rus. Fed. 11 R4
Dobzha China 37 G3
Doce r. Brazil 71 D2
Dochart r. U.K. 20 E4
Docking U.K. 19 H6
Doctor Hicks Range hills Australia 55 D7
Doctor Pedro P. Peña Para. 70 D2
Doda India 36 C2
Doda Betta mt. India 38 C4
Dod Ballapur India 38 C3
Dodecanese is Greece 27 L7
Dodekanisos is Greece see Dodecanese
Dodge City U.S.A. 62 G4
Dodman Point U.K. 19 C8
Dodoma Tanz. 49 D4
Dogai Coring salt l. China 37 G2
Dogaicoring Qangco salt l. China 37 G2
Doğanşehir Turkey 34 E3
Dogên Co l. Xizang China see Bam Tso
Dogên Co l. Xizang China 37 G3
Dōgo i. Japan 45 D5
Dogondoutchi Niger 46 D3
Doğubeyazıt Turkey 35 G3
Doğu Menteşe Dağları mts Turkey 27 M6
Dogxung Zangbo r. China 37 G3
Do'gyaling China 37 G3
Doha Qatar 32 H4
Dohad India see Dahod
Dohazari Bangl. 37 H5
Dohrighat India 37 E4
Doi i. Fiji 53 I4
Doire U.K. see Londonderry
Dois Irmãos, Serra dos hills Brazil 69 J5
Dokan, Sad Iraq 35 G4
Dok-do i. N. Pacific Ocean see
    Liancourt Rocks
Dokhara, Dunes de des. Alg. 22 F5
Dokka Norway 15 G6
Dokkum Neth. 17 J4
Dokri Pak. 36 B4
Dokshukino Rus. Fed. see Nartkala
Dokshytsy Belarus 15 O9
Dokuchayeva, Mys c. Rus. Fed. 44 G3
Dokuchayevs'k Ukr. 13 H7
Dolbenmaen U.K. 19 C6
Dol-de-Bretagne France 24 D2
Dole France 24 G3
Dolgellau U.K. 19 D6
Dolgiy, Ostrov i. Rus. Fed. 12 L1
Dolgorukovo Rus. Fed. 13 H5
Dolina Ukr. see Dolyna
Dolinsk Rus. Fed. 44 F3
Dolisie Congo see Loubomo
Dolleman Island Antarctica 76 L2
Dolo Odo Eth. 48 E3
Dolok, Pulau i. Indon. 41 J8
Dolomites mts Italy see Dolomites
Dolomiti mts Italy see Dolomites
Dolomiti Bellunesi, Parco Nazionale delle
    nat. park Italy 26 D1
Dolomitiche, Alpi mts Italy see Dolomites
Dolonnur China 43 L4
Dolo Odo Eth. 48 E3
Dolores Arg. 70 E5
Dolores Uruguay 70 E4
Dolphin and Union Strait Canada 60 G3
Dolphin Head hd Namibia 50 B3
Dolyna Ukr. 13 D6
Domaila India 36 D3
Domar China 42 G6
Domartang China 42 G6
Domažlice Czech Rep. 17 N6
Domba China 37 H2
Dom Bākh Iran 35 G4
Dombås Norway 14 F5
Dombóvár Hungary 26 H1
Dombrau Poland see Dąbrowa Górnicza
Dombrovitsa Ukr. see Dubrovytsya
Dombrowa Poland see Dąbrowa Górnicza
Domda China 37 H2
Dome Argus Antarctica 76 E1
Dome Charlie Antarctica 76 F2
Dome Creek Canada 60 F4
Dome Rock Mountains AZ U.S.A. 65 E4
Domeyko Chile 70 B3
Domfront France 24 D2
Dominica country West Indies 67 L5
Dominicana, República country
    West Indies see Dominican Republic
Dominican Republic country West Indies
    67 J5
Dominion, Cape Canada 61 K3
Dominique i. Fr. Polynesia see Hiva Oa
Dom Joaquim Brazil 71 C2
Domo Eth. 48 E3
Domokos Greece 27 J5
Dompu Indon. 41 D8
Domuyo, Volcán vol. Arg. 70 B5
Domville, Mount h. Australia 58 E2
Don r. Rus. Fed. 13 H7
Don r. U.K. 20 G3

Donaghadee U.K. 21 G3
Donaghmore U.K. 21 F3
Donald Australia 58 A6
Donald Australia 58 A6
Doñana, Parque Nacional de nat. park
    Spain 25 C5
Donau r. Austria/Germany see Danube
Donau r. Europe 17 P6 see Danube
Donauwörth Germany 17 M6
Don Benito Spain 25 D4
Doncaster U.K. 18 F5
Dondo Angola 49 B4
Dondo Moz. 49 D5
Dondra Head hd Sri Lanka 38 D5
Donegal Ireland 21 D3
Donegal Bay Ireland 21 D3
Donets'k Ukr. 13 H7
Donetsko-Amrovsiyevka Ukr. see
    Amvrosiyivka
Donets'kyy Kryazh hills Rus. Fed./Ukr.
    13 H7
Donga r. Cameroon/Nigeria 46 D4
Dongane, Lagoa lag. Moz. 51 L3
Dongara Australia 55 A7
Dongbo China see Médog
Dongchuan Yunnan China 42 I7
Dongco China 37 F2
Dong Dabsan Hu salt l. China 37 H1
Dongfang China 43 J9
Dongfanghong China 44 D3
Donggang China 45 B5
Donggang Shandong China 43 L5
Donggou China see Donggang
Dongguan China see Donggang
Donggi Conag l. China 37 I2
Dongguan China 43 F4
Donggou Congo 48 B3
Dongjiao China 43 M6
Dongting China see Toiba
Dong Hai sea N. Pacific Ocean see
    East China Sea
Đông Hơi Vietnam 31 J5
Dongjug China 37 H3
Dongliao He r. China 44 A4
Dongminzhutun China 44 A3
Dongning China 44 C3
Dongo Angola 49 B5
Dongo Dem. Rep. Congo 48 B3
Dongola Sudan 32 C5
Dongou Congo 48 B3
Dongqiao China 37 G3
Dongtai China 43 M6
Dongting China 43 M6
Dongting Hu l. China 43 K7
Dong Ujimqin Qi China see Uliastai
Dongxing Heilong. China 44 B3
Dongxiang China 43 L6
Dongxing China 44 D3
Dongying China 43 L5
Donnellys Crossing N.Z. 59 D2
Donner Pass U.S.A. 65 B1
Donostia Spain see San Sebastián
Donousa i. Greece 27 K6
Donskoye Rus. Fed. 13 I7
Dooagh Ireland 21 B4
Doomadgee Australia 56 B3
Doon r. U.K. 20 E5
Doon, Loch l. U.K. 20 E5
Doonbeg r. Ireland 21 C5
Dooxo Nugaaleed val. Somalia 48 E3
Dor Israel 39 B3
Dora, Lake imp. l. Australia 54 C5
Dorah Pass Pak. 36 B1
Dorbiljin China see Emin
Dorbod China see Taikang
Dorbod Qi China see Ulan Hua
Dorchester U.K. 19 E8
Dordabis Namibia 50 C2
Dordogne r. France 24 D4
Dordrecht Neth. 16 J5
Dordrecht S. Africa 51 H6
Doreenville Namibia 50 D2
Doré Lake Canada 60 H4
Dores do Indaiá Brazil 71 B2
Dorgê Co l. China 37 H2
Dori Burkina Faso 46 C3
Doring r. S. Africa 50 D6
Dorisvale Australia 54 E3
Dorking U.K. 19 G7
Dormidontovka Rus. Fed. 44 D3
Dornoch U.K. 20 E3
Dornoch Firth est. U.K. 20 E3
Doro Mali 46 C3
Dorogobuzh Rus. Fed. 13 G5
Dorogorskoye Rus. Fed. 12 J2
Dorohoi Romania 13 E7
Döröö Nuur salt l. Mongolia 42 G3
Dorostol Bulg. see Silistra
Dorotea Sweden 14 J4
Dorre Island Australia 55 A6
Dorrigo Australia 58 F3
Dorsoidong Co l. China 37 G2
Dortmund Germany 17 K5
Dörtyol Turkey 39 C1
Dorūd Iran 35 H4
Doruma Dem. Rep. Congo 48 C3
Dorylaeum Turkey see Eskişehir
Dos Bahías, Cabo c. Arg. 70 C6
Dos de Mayo Peru 68 C5
Dos Palos CA U.S.A. 65 B2
Dosso Niger 46 D3
Dothan U.S.A. 63 J5
Douai France 24 F1
Douala Cameroon 46 D4
Douarnenez France 24 B2
Double Island Point Australia 57 F5
Double Peak CA U.S.A. 65 E5
Double Point Australia 56 D3
Doubs r. France/Switz. 24 G3
Doubtful Sound inlet N.Z. 59 A7
Doubtless Bay N.Z. 59 D2
Douentza Mali 46 C3
Dougga tourist site Tunisia 26 C6
Douglas Isle of Man 18 C4
Douglas S. Africa 50 F5
Douglas U.K. 20 F5
Douglas AZ U.S.A. 62 F5
Douglas GA U.S.A. 63 K5
Douglas WY U.S.A. 62 F3
Douglas Reef i. Japan see
    Okino-Tori-shima
Doulatpur Bangl. see Daulatpur
Doullens France 24 F1
Douna Mali 46 C3
Doune U.K. 20 E4
Dourada, Serra hills Brazil 71 A2
Dourada, Serra mts Brazil 71 A1
Douro r. Port. 25 B3
Douve r. France 19 F9
Dove r. U.K. 19 F6
Dover DE U.S.A. 64 D3
Dover U.K. 19 I7
Dover NH U.S.A. 64 F1
Dover NJ U.S.A. 64 D2
Dover OH U.S.A. 64 A2
Dover, Strait of France/U.K. 24 C1
Dovey r. U.K. 19 D6
Dovrefjell Nasjonalpark nat. park Norway
    14 F5
Dowah China Qala Afgh. 36 B3
Dowlaiswaram India 38 D2
Dowlatābād Fārs Iran 35 H5
Dowlatābād Fārs Iran 35 H5
Dowl at Yār Afgh. 36 A2
Downpatrick U.K. 21 G3

90

wnsville NY U.S.A. 64 D1
wrey Rūd r. Afgh. 36 A3
ylestown PA U.S.A. 64 D2
zen is Japan 45 D5
zois, Réservoir resr Canada 63 L2
zulé France 19 G9
ia, Hamada du plat. Alg. 22 C6
acena Brazil 71 A3
achten Neth. 17 K4
ăgănești-Olt Romania 27 K2
ăgășani Romania 27 K2
agonera, Isla i. Spain see Sa Dragonera
agsfjärd Fin. 15 M6
aguignan France 24 H5
ahichyn Belarus 15 N10
ake Australia 58 F2
akensberg mts S. Africa 51 I3
ake Passage S. Atlantic Ocean 72 D9
akes Bay CA U.S.A. 65 A2
ama Greece 27 K4
ammen Norway 15 G7
angedal Norway 15 F7
aperstown U.K. 21 F3
apsaca Afgh. see Kunduz
as India 36 C2
asan Pak. 36 C1
au r. Europe 17 O7 see Drava
ava r. Europe 26 H2
iva r. Europe see Drava
ave r. Europe see Drava
ayton Valley Canada 62 E1
azinda Pak. 36 B3
ëan Alg. 26 B6
esden Germany 17 N5
eux France 24 E2
evsjø Norway 15 H6
China 37 I3
ffield U.K. 18 G4
ftwood PA U.S.A. 64 B2
llham Australia 58 E1
moleague Ireland 21 C6
na r. Bosnia-Herzegovina/Serbia 27 H2
scoll Island Antarctica 76 J1
ssa Belarus see Vyerkhnyadzvinsk
niš Croatia 26 G3
obeta-Turnu Severin Romania 27 J2
ogheda Ireland 21 F4
ogichin Belarus see Drahichyn
ogobych Ukr. see Drohobych
ohobych Ukr. 13 D6
oichead Átha Ireland see Drogheda
oichead Nua Ireland see Newbridge
oitwich U.K. see Droitwich Spa
oitwich Spa U.K. 19 E6
omedary, Cape Australia 58 E6
omod Ireland 21 E4
omore Northern Ireland U.K. 21 E3
omore Northern Ireland U.K. 21 F3
onfield U.K. 18 F5
onning Louise Land reg. Greenland
7 I1
onning Maud Land reg. Antarctica see
Queen Maud Land
ık-Yul country Asia see Bhutan
ımheller Canada 62 E1
ımmond atoll Kiribati see Tabiteuea
ımmond Island Kiribati see McKean
ımmond Range hills Australia 56 D5
ımmondville Canada 63 M2
ımmore U.K. 20 E6
ıskieniki Lith. see Druskininkai
ıskininkai Lith. 15 N10
ızhina Rus. Fed. 29 P3
ızhnaya Gorka Rus. Fed. 15 Q7
y r. Australia 54 F3
yanovo Bulg. 27 K3
yden Canada 63 I2
yden NY U.S.A. 64 C1
ygalski Ice Tongue Antarctica 76 H1
ygalski Island Antarctica 76 F2
y Lake NV U.S.A. 65 E2
ymen U.K. 20 E4
ysdale r. Australia 54 D3
ysdale River National Park Australia
4 D3
aringa Australia 56 E4
arte, Pico mt. Dom. Rep. 67 J5
artina Brazil 71 A3
bā Saudi Arabia 32 E4
bai U.A.E. 33 I4
bawnt Lake Canada 61 H3
bayy U.A.E. see Dubai
bbo Australia 58 D4
blin Ireland 21 F4
blin U.S.A. 63 K5
bna Rus. Fed. 12 H4
bno Ukr. 13 E6
Bois PA U.S.A. 64 B2
bovka Rus. Fed. 13 J6
bovskoye Rus. Fed. 13 I7
bréka Guinea 46 B4
bris U.K. see Dover
brovnik Croatia 26 H3
brovytsya Ukr. 13 E6
buque U.S.A. 63 I3
bysa r. Lith. 15 M9
c de Gloucester, Îles du is Fr. Polynesia
5 K7
chess Australia 56 B4
cie Island atoll Pitcairn Is 75 L7
ck Bay Canada 62 G1
ck Creek r. Australia 54 B5
ckwater Peak NV U.S.A. 65 E1
dhi India 37 E4
dhwa India 36 E3
dinka Rus. Fed. 28 J3
dley U.K. 19 E6
dna r. India 38 C2
du India 36 C4
ékoué Côte d'Ivoire 46 C4
ero r. Spain 25 C3
ff Islands Solomon Is 53 G2
ffreboy, Lac l. Canada 61 K4
fftown U.K. 20 F3
fourspitze mt. Italy/Switz. 24 H4
gi Otok i. Croatia 26 F2
gi Rat Croatia 26 G3
ida-Marahuaca, Parque Nacional
nat. park Venez. 68 E3
isburg Germany 17 K5
iwelskloof S. Africa 51 J2
kathole S. Africa 51 H6
ke of Clarence atoll Tokelau see
Nukunonu
ke of Gloucester Islands Fr. Polynesia
see Duc de Gloucester, Îles du
ke of York atoll Tokelau see Atafu
k Fadiat South Sudan 47 G4
ukhovnitskoye Rus. Fed. 13 K5
ki Pak. 36 B3
ki Rus. Fed. 44 D2
ki r. Rus. Fed. 44 D2
kou China see Panzhihua
ıkštas Lith. 15 O9
lan China 42 H5
lce r. Arg. 70 D4
l'durga Rus. Fed. 43 K2
lhunty r. Australia 56 C1
lishi Hu salt l. China 37 E2
llewala Pak. 36 B3
llstroom S. Africa 51 J3
Dulmera India 36 C3
Dulovo Bulg. 27 L3
Duluth U.S.A. 63 I2
Dulverton U.K. 19 D7
Dūmā Syria 39 C3
Dumaguete Phil. 41 E7
Dumai Indon. 41 C7
Dumaresq r. Australia 58 E2
Dumas U.S.A. 62 G4
Dumat al Jandal Saudi Arabia 35 E5
Dumayr Syria 39 C3
Ḑumayr, Jabal mts Syria 39 C3
Dumbakh Iran see Dom Bākh
Dumbarton U.K. 20 E5
Dumbe S. Africa 51 J4
Ďumbier mt. Slovakia 17 Q6
Dumchele India 36 D2
Dum Duma India 37 H4
Dumfries U.K. 20 F5
Dumka India 37 F4
Dumont d'Urville (France) research stn
Antarctica 76 G2
Dumont d'Urville Sea Antarctica 76 G2
Dumyāṭ Egypt see Dumyāṭ
Dumyāṭ Egypt 34 C5
Duna r. Europe 26 H2 see Danube
Duna r. Hungary see Danube
Dünaburg Latvia see Daugavpils
Dunaj r. Europe see Danube
Dunajská Streda Slovakia 17 P7
Dunakeszi Hungary 27 H1
Dunany Point Ireland 21 F4
Dunărea r. Europe see Danube
Dunării, Delta Romania/Ukr. see
Danube Delta
Dunaújváros Hungary 26 H1
Dunav r. Bulg./Croatia/S.M. see Danube
Dunav r. Europe 26 L2 see Danube
Dunay r. Europe see Danube
Dunayivtsi Ukr. 13 E6
Dunbar Australia 56 C3
Dunbar U.K. 20 G4
Dunblane U.K. 20 F4
Dunboyne Ireland 21 F4
Duncan OK U.S.A. 62 H5
Duncansby Head hd U.K. 20 F2
Duncormick Ireland 21 F5
Dundaga Latvia 15 M8
Dundalk Ireland 21 F3
Dundalk MD U.S.A. 64 C3
Dundalk Bay Ireland 21 F4
Dundas Ont. Canada 64 B1
Dundas Greenland 61 L2
Dundas, Lake imp. l. Australia 55 C8
Dundas Strait Australia 54 E2
Dún Dealgan Ireland see Dundalk
Dundee S. Africa 51 J5
Dundee U.K. 20 G4
Dundee NY U.S.A. 64 C1
Dundonald U.K. 21 G3
Dundoo Australia 58 B1
Dundrennan U.K. 20 F6
Dundrum U.K. 21 G3
Dundrum Bay U.K. 21 G3
Dundwa Range mts India/Nepal 37 E4
Dunedin N.Z. 59 C7
Dunfermline U.K. 20 F4
Dungannon U.K. 21 F3
Dún Garbhán Ireland see Dungarvan
Dungarpur India 36 C5
Dungarvan Ireland 21 E5
Dung Co l. China 37 G3
Dungeness hd U.K. 19 H8
Dungeness, Punta pt Arg. 70 C8
Dungiven U.K. 21 F3
Dungog Australia 58 E4
Dungu Dem. Rep. Congo 48 C3
Dungun Malaysia 41 C7
Dungunab Sudan 32 E5
Dunhua China 44 C4
Dunhuang China 42 G4
Dunkeld Australia 58 D1
Dunkeld U.K. 20 F4
Dunkellin r. Ireland 21 D4
Dunkerque France see Dunkirk
Dunkery Hill h. U.K. 19 D7
Dunkirk France 24 F1
Dunkirk NY U.S.A. 64 B1
Dún Laoghaire Ireland 21 F4
Dunlavin Ireland 21 F4
Dunleer Ireland 21 F4
Dunloy U.K. 21 F2
Dunmanway Ireland 21 C6
Dunmarra Australia 54 F4
Dunmore Ireland 21 D4
Dunmore PA U.S.A. 64 D2
Dunmurry U.K. 21 G3
Dunnet Head hd U.K. 20 F2
Dunnigan CA U.S.A. 65 B1
Dunnville Ont. Canada 64 B1
Dunolly Australia 58 A6
Dunoon U.K. 20 E5
Duns U.K. 20 G5
Dunstable U.K. 19 G7
Dunstan Mountains N.Z. 59 B7
Duntroon N.Z. 59 C7
Dunyapur Pak. 36 B3
Duolun Nei Mongol China see Dolonnur
Duomula China 37 E2
Dupang Ling mts China 43 K7
Duperré Alg. see Aïn Defla
Dupnitsa Bulg. 27 J3
Duque de Bragança Angola see Calandula
Dūrā West Bank 39 B4
Durack r. Australia 54 D3
Durack Range hills Australia 54 D4
Dura Europos Syria see Aş Şāliḩīyah
Durağan Turkey 34 D2
Durance r. France 24 G5
Durango Mex. 66 D4
Durango Spain 25 E2
Durango U.S.A. 62 F4
Durant U.S.A. 63 H5
Durazno Uruguay 70 E4
Durazzo Albania see Durrës
Durban S. Africa 51 J5
Durban-Corbières France 24 F5
Durbanville S. Africa 50 D7
Durbin WV U.S.A. 64 B3
Durg India 36 E5
Durgapur Bangl. 37 G4
Durgapur India 37 F5
Durham U.K. 18 F4
Durham U.S.A. 63 L4
Durham Downs Australia 57 C5
Durlas Ireland see Thurles
Durleşti Moldova 27 M1
Durmitor mt. Montenegro 27 H3
Durmitor, Nacionalni Park nat. park
Montenegro 26 H3
Durness U.K. 20 E2
Durocortorum France see Reims
Durong South Australia 57 E5
Durostorum Bulg. see Silistra
Durovernum U.K. see Canterbury
Durrës Albania 27 H4
Durrie Australia 56 C5
Durrington U.K. 19 F7
Dursey Island Ireland 21 B6
Dursunbey Turkey 27 M5
Durukhsi Somalia 48 E3
Durusu Gölü l. Turkey 27 M4
Durūz, Jabal ad mt. Syria 39 C3
D'Urville, Tanjung pt Indon. 41 F8
D'Urville Island N.Z. 59 D5
Duşak Turkm. 33 I2
Dushai Pak. 36 A3
Dushanbe Tajik. 33 K2
Dushet'i Georgia 35 G2
Dushore PA U.S.A. 64 C2
Dusse-Alin', Khrebet mts Rus. Fed. 44 D2
Düsseldorf Germany 17 K5
Dutch East Indies country Asia see
Indonesia
Dutch Guiana country S. America see
Suriname
Dutlwe Botswana 50 F2
Dutse Nigeria 46 D3
Dutsin-Ma Nigeria 46 D3
Dutton r. Australia 56 C4
Dutton Ont. Canada 64 A1
Dutton, Lake imp. l. Australia 57 B6
Duvno Bos.-Herz. see Tomislavgrad
Duwin Iraq 35 G3
Düxanbibazar China 36 E1
Duyun China 43 J7
Düzce Turkey 27 N4
Duzdab Iran see Zāhedān
Dvina r. Europe see Zapadnaya Dvina
Dvina r. Rus. Fed. see Severnaya Dvina
Dvinsk Latvia see Daugavpils
Dvinskaya Guba g. Rus. Fed. 12 H2
Dwarka India 36 B5
Dwarsberg S. Africa 51 H3
Dwyka S. Africa 50 E7
Dyat'kovo Rus. Fed. 13 G5
Dyce U.K. 20 G3
Dyer, Cape Canada 61 L3
Dyersburg U.S.A. 63 J4
Dyffryn U.K. see Valley
Dyfi r. U.K. see Dovey
Dyfrdwy r. U.K. see Dee
Dyje r. Austria/Czech Rep. 17 P6
Dyke U.K. 20 F3
Dykh-Tau, Gora mt. Rus. Fed. 35 F2
Dylewska Góra h. Poland 17 Q4
Dymytrov Ukr. 13 H6
Dynevor Downs Australia 58 B2
Dyoki S. Africa 51 I6
Dyrrhachium Albania see Durrës
Dysart Australia 56 E4
Dysselsdorp S. Africa 50 F7
Dyurtyuli Rus. Fed. 11 Q4
Dzamīn Üüd Mongolia 43 K4
Dzanga-Ndoki, Parc National de nat. park
Cent. Afr. Rep. 48 B3
Dzaoudzi Mayotte 49 E5
Dzaudzhikau Rus. Fed. see Vladikavkaz
Dzavhan Mongolia 42 G3
Dzerzhinsk Belarus see Dzyarzhynsk
Dzerzhinsk Rus. Fed. 12 I4
Dzhagdy, Khrebet mts Rus. Fed. 44 C1
Dzhaki-Unakhta Yakbyyana, Khrebet mts
Rus. Fed. 44 D2
Dzhalalabad Azer. see Cälilabad
Dzhalal-Abad Kyrg. see Jalal-Abad
Dzhalil' Rus. Fed. 11 Q4
Dzhalinda Rus. Fed. 44 A1
Dzhaltyr Kazakh. see Zhaltyr
Dzhambeyty Kazakh. see Zhympity
Dzhambul Kazakh. see Taraz
Dzhangala Kazakh. 11 Q6
Dzhankoy Ukr. 13 G7
Dzhanybek Kazakh. see Zhanibek
Dzharkent Kazakh. see Zharkent
Dzhava Georgia see Java
Dzhetygara Kazakh. see Zhitikara
Dzhezkazgan Kazakh. see Zhezkazgan
Dzhidinskiy Khrebet mts Mongolia/
Rus. Fed. 42 I2
Dzhokhar Ghala Rus. Fed. see Groznyy
Dzhubga Rus. Fed. 34 E1
Dzhugdzhur, Khrebet mts Rus. Fed.
29 O4
Dzhul'fa Azer. see Culfa
Dzhusaly Kazakh. 42 A3
Działdowo Poland 17 R4
Dzūkija nat. park Lith. 15 N9
Dzungarian Basin China see
Junggar Pendi
Dzüünharaa Mongolia 42 J3
Dzuunmod Mongolia 42 J3
Dzyaniskavichy Belarus 15 O10
Dzyarzhynsk Belarus 15 O10
Dzyatlavichy Belarus 15 O10

E

Eagle AK U.S.A. 60 D3
Eagle Cap mt. U.S.A. 62 D2
Eagle Crags mt. CA U.S.A. 65 D3
Eagle Lake Canada 63 I2
Eagle Lake CA U.S.A. 62 C3
Eagle Mountain CA U.S.A. 65 E4
Eagle Pass U.S.A. 62 G6
Eagle Plain Canada 60 E3
Eagle Rock VA U.S.A. 64 B4
Eap i. Micronesia see Yap
Ear Falls Canada 63 I1
Earlimart CA U.S.A. 65 C3
Earl's Seat h. U.K. 20 E4
Earlston U.K. 20 G5
Earn r. U.K. 20 F4
Earn, Loch l. U.K. 20 E4
Earp CA U.S.A. 65 E3
Easington U.K. 18 H5
East Alligator r. Australia 54 F3
East Antarctica reg. Antarctica 76 F1
East Ararat PA U.S.A. 64 D2
East Aurora NY U.S.A. 64 B1
East Bengal country Asia see Bangladesh
Eastbourne U.K. 19 H8
East Branch Clarion River Reservoir PA
U.S.A. 64 B2
East Cape N.Z. 59 G3
East Caroline Basin sea feature
N. Pacific Ocean 74 F5
East China Sea N. Pacific Ocean 43 N6
East Coast Bays N.Z. 59 E3
Eastend Canada 62 F2
Easter Island S. Pacific Ocean 75 M7
Eastern Cape prov. S. Africa 51 H6
Eastern Desert Egypt 32 D4
Eastern Desert Egypt 32 D4
Eastern Fields rf Australia 56 D1
Eastern Ghats mts India 38 C4
Eastern Island U.S.A. 74 I4
Eastern Nara canal Pak. 36 B4
Eastern Samoa terr. S. Pacific Ocean see
American Samoa
Eastern Sayan Mountains Rus. Fed. see
Vostochnyy Sayan
Eastern Taurus mts Turkey see
Güneydoğu Toroslar
Eastern Transvaal prov. S. Africa see
Mpumalanga
Easterville Canada 62 H1
East Falkland i. Falkland Is 70 E8
East Falmouth MA U.S.A. 64 F2
East Frisian Islands Germany 17 K4
East Greenwich RI U.S.A. 64 F2
East Grinstead U.K. 19 G7
Easthampton MA U.S.A. 64 E1
East Hampton NY U.S.A. 64 E2
East Hartford CT U.S.A. 64 E2
East Indiaman Ridge sea feature
Indian Ocean 73 O7
East Kilbride U.K. 20 E5
Eastlake OH U.S.A. 64 A2
Eastleigh U.K. 19 F8
East Liverpool OH U.S.A. 64 A2
East London S. Africa 51 H7
Eastmain Canada 63 L1
Eastmain r. Canada 63 L1
East Mariana Basin sea feature
N. Pacific Ocean 74 G5
Eastmere Australia 56 D4
Easton MD U.S.A. 64 C3
Easton PA U.S.A. 64 D2
East Orange NJ U.S.A. 64 D2
East Pacific Rise sea feature
N. Pacific Ocean 75 M4
East Pakistan country Asia see Bangladesh
East Palestine OH U.S.A. 64 A2
East Providence RI U.S.A. 64 F2
East Retford U.K. see Retford
East Sea N. Pacific Ocean see Japan, Sea of
East Siberian Sea Rus. Fed. 29 P2
East Side Canal canal CA U.S.A. 65 C3
East Stroudsburg PA U.S.A. 64 D2
East Timor country Asia 54 D2
East Toorale Australia 58 B3
Eastville VA U.S.A. 64 D4
East York Ont. Canada 64 B1
Eau Claire U.S.A. 63 I3
Eauripik atoll Micronesia 41 G7
Eauripik Rise-New Guinea Rise sea feature
N. Pacific Ocean 74 F5
Eaurypyg atoll Micronesia see Eauripik
Ebbw Vale U.K. 19 D7
Ebebiyin Equat. Guinea 46 E4
Ebenerde Namibia 50 C3
Ebensburg PA U.S.A. 64 B2
Eber Gölü l. Turkey 27 N5
Eberswalde-Finow Germany 17 N4
Ebetsu Japan 44 F4
Ebi Nor salt l. China see Ebinur Hu
Ebinur Hu salt l. China 42 E4
Eboli Italy 26 F4
Ebolowa Cameroon 46 E4
Ebony Namibia 50 B2
Ebre r. Spain see Ebro
Ebro r. Spain 25 G3
Eburacum U.K. see York
Ebusus i. Spain see Ibiza
Ecbatana Iran see Hamadān
Eceabat Turkey 27 L4
Ech Chélif Alg. see Chlef
Echegárate, Puerto pass Spain 25 E2
Echeng China see Ezhou
Echeverria, Pico mt. Mex. 66 B3
Echmiadzin Armenia see Ejmiatsin
Echuca Australia 58 B6
Écija Spain 25 D5
Eckernförde Germany 17 L3
Eclipse Sound sea chan. Canada 61 J2
Écrins, Parc National des nat. park France
24 H4
Ecuador country S. America 68 C4
Ed Eritrea 32 F7
Ed Sweden 15 G7
Eday i. U.K. 20 G1
Ed Da'ein Sudan 47 F3
Ed Damazin Sudan 32 D7
Ed Damer Sudan 32 D6
Ed Debba Sudan 32 D6
Ed Dueim Sudan 32 D7
Eddystone Point Australia 57 [inset]
Edéa Cameroon 46 E4
Edéia Brazil 71 A2
Eden Australia 58 D6
Eden r. U.K. 18 D4
Edenburg S. Africa 51 G5
Edendale N.Z. 59 B8
Edenderry Ireland 21 E4
Edenville S. Africa 51 H4
Edessa Greece 27 J4
Edessa Turkey see Şanlıurfa
Edfu Egypt see Idfū
Edgar Ranges hills Australia 54 C4
Edgartown MA U.S.A. 64 F2
Edgecumbe Island Solomon Is see Utupua
Edge Island Svalbard see Edgeøya
Edgeøya i. Svalbard 28 D2
Edgeworthstown Ireland 21 E4
Édhessa Greece see Edessa
Edinboro PA U.S.A. 64 A2
Edinburg TX U.S.A. 62 H6
Edinburg VA U.S.A. 64 B3
Edinburgh U.K. 20 F5
Edirne Turkey 27 L4
Edith Ronne Land Antarctica see
Ronne Ice Shelf
Edjeleh Libya 46 D2
Edjudina Australia 55 C7
Edku Egypt see Idkū
Edmonton Canada 62 E1
Edmundston Canada 63 N2
Edo Japan see Tōkyō
Edom reg. Israel/Jordan 39 B4
Édouard, Lac l. Dem. Rep. Congo/Uganda
see Edward, Lake
Edremit Turkey 27 L5
Edremit Körfezi b. Turkey 27 L5
Edrengiyn Nuruu mts Mongolia 42 H4
Edsbyn Sweden 15 I6
Edson Canada 60 G4
Edward r. N.S.W. Australia 58 B5
Edward r. Qld Australia 56 C2
Edward, Lake Dem. Rep. Congo/Uganda
48 C4
Edward, Mount Antarctica 76 L1
Edwardesabad Pak. see Bannu
Edward's Creek Australia 57 A6
Edwards Plateau U.S.A. 62 G5
Edward VII Peninsula Antarctica 76 I1
Eenzamheid Pan salt pan S. Africa 50 E4
Eesti country Europe see Estonia
Éfaté i. Vanuatu 53 G3
Effingham U.S.A. 63 J4
Efsus Turkey see Afşin
Egadi, Isole is Sicilia Italy 26 D5
Egedesminde Greenland see Aasiaat
Eger Hungary 17 R7
Egersund Norway 15 E7
Egerton, Mount h. Australia 55 B6
Egilsstaðir Iceland 14 [inset]
Eğin Turkey see Kemaliye
Eğirdir Turkey 27 N6
Eğirdir Gölü l. Turkey 27 N6
Eglinton U.K. 21 E2
Egmont, Cape N.Z. 59 D4
Egmont, Mount vol. N.Z. see
Taranaki, Mount
Egmont National Park N.Z. 59 E4
eGoli S. Africa see Johannesburg
Eğrigöz Dağı mts Turkey 27 M5
Egton U.K. 18 G4
Éguas r. Brazil 71 B1
Egvekinot Rus. Fed. 29 T3
Egypt country Africa 32 C4
Ehden Lebanon 39 B2
Ehen Hudag China 42 I5
Ehingen (Donau) Germany 17 L6
Ehrenberg AZ U.S.A. 65 E4
Ehrenberg Range hills Australia 55 E5
Eidfjord Norway 15 E6
Eidsvold Australia 56 E5
Eidsvoll Norway 15 G6
Eifel hills Germany 17 K5
Eigg i. U.K. 20 C4
Eight Degree Channel India/Maldives
38 B5
Eights Coast Antarctica 76 K2
Eighty Mile Beach Australia 54 C4
Eilat Israel 39 B5
Eildon Australia 58 B6
Eildon, Lake Australia 58 C6
Einasleigh Australia 56 D3
Einasleigh r. Australia 56 C3
Eindhoven Neth. 16 J5
Einsiedeln Switz. 24 I3
Eirik Ridge sea feature N. Atlantic Ocean
72 F2
Eiriosgaigh i. U.K. see Eriskay
Eirunepé Brazil 68 E5
Eiseb watercourse Namibia 49 C5
Eisenach Germany 17 M5
Eisenhüttenstadt Germany 17 O4
Eisenstadt Austria 17 P7
Eite, Loch inlet U.K. see Etive, Loch
Eivissa Spain see Ibiza
Eivissa i. Spain see Ibiza
Ejea de los Caballeros Spain 25 F2
Ejeda Madag. 49 E6
Ejin Horo Qi China see Altan Shiret
Ejin Qi China see Dalain Hob
Ejmiadzin Armenia see Ejmiatsin
Ejmiatsin Armenia 35 G2
Ekenäs Fin. 15 M7
Ekerem Turkm. 35 I3
Eketahuna N.Z. 59 E5
Ekimchan Rus. Fed. 44 D1
Ekinyazı Turkey 39 D1
Ekonda Rus. Fed. 29 L3
Ekostrovskaya Imandra, Ozero l. Rus. Fed.
12 G2
Ekshärad Sweden 15 H6
Eksjö Sweden 15 I8
Eksteenfontein S. Africa 50 C5
Ekström Ice Shelf Antarctica 76 B2
Ekwan r. Canada 63 K1
El Aaiún W. Sahara see Laâyoune
Elafonisou, Steno sea chan. Greece 27 J6
El 'Agrûd well Egypt see Al 'Ajrūd
Elaia, Cape Cyprus 39 B2
El 'Alamein Egypt see Al 'Alamayn
El 'Âmirîya Egypt see Al 'Āmirīyah
Elands r. S. Africa 51 I3
Elandsdoorn S. Africa 51 I3
El Aouinet Alg. 26 B7
Elar Armenia see Abovyan
El Araïche Morocco see Larache
El Ariana Tunisia see L'Ariana
El Aricha Alg. 22 D5
El 'Arîsh Egypt see Al 'Arīsh
El Arrouch Alg. 26 B6
El Ashmûnein Egypt see Al Ashmūnayn
El Asnam Alg. see Chlef
Elassona Greece 27 J5
Elat Israel see Eilat
Elazığ Turkey 35 E3
Elba, Isola d' i. Italy 26 D3
El'ban Rus. Fed. 44 E2
El Barco de Valdeorras Spain see O Barco
Elbasan Albania 27 I4
El Batroun Lebanon see Batroûn
El Baúl Venez. 68 E2
El Bawîti Egypt see Al Bawīṭī
El Bayadh Alg. 22 E5
Elbe r. Germany 17 L4
El Béqaa val. Lebanon 39 C2
Elbert, Mount U.S.A. 62 F4
Elbeuf France 24 E2
Elbeyli Turkey 39 C1
Elbing Poland see Elbląg
Elbistan Turkey 34 E3
Elbląg Poland 17 Q3
El Boulaïda Alg. see Blida
El'brus mt. Rus. Fed. 35 F2
El Burgo de Osma Spain 25 E3
Elburz Mountains Iran 35 H3
El Cajon CA U.S.A. 65 D4
El Callao Venez. 68 F2
El Campo U.S.A. 63 H6
El Centro CA U.S.A. 65 E4
El Cerro Bol. 68 F7
Elche Spain see Elche-Elx
Elche-Elx Spain 25 F4
Elcho Island Australia 56 A1
El Coca Ecuador see Coca
El Cocuy, Parque Nacional nat. park Col.
68 D2
Elda Spain 25 F4
El'dikan Rus. Fed. 29 O3
El Djezair country Africa see Algeria
El Djezair Alg. see Algiers
Eldorado Arg. 70 F3
Eldorado Brazil 71 A4
El Dorado Col. 68 D3
El Dorado Mex. 62 F7
El Dorado AR U.S.A. 63 I5
El Dorado KS U.S.A. 63 H4
El Dorado Venez. 68 F2
Eldorado Mountains NV U.S.A. 65 E3
Eldoret Kenya 48 D3
Elea, Cape Cyprus see Elaia, Cape
Elefantes r. Moz./S. Africa see Olifants
El Eglab plat. Alg. 46 C2
El Ejido Spain 25 E5
Elemi Triangle terr. Africa see
Ilemi Triangle
El Encanto Col. 68 D4
Elephanta Caves caves India 38 B2
Elephant Island Antarctica 76 A2
Elephant Pass Sri Lanka 38 D4
Elephant Point Bangl. 37 H5
Eleşkirt Turkey 35 F3
El Eulma Alg. 22 F4
Eleuthera i. Bahamas 67 I3
El Fahs Tunisia 26 C6
El Faiyûm Egypt see Al Fayyūm
El Fasher Sudan 47 F3
El Ferrol Spain see Ferrol
El Ferrol del Caudillo Spain see Ferrol
El Fud Eth. 48 E3
El Fuerte Mex. 66 C3
El Gara Egypt see Qārah
El Geneina Sudan 47 F3
El Geteina Sudan 32 D7
El Ghardaqa Egypt see Al Ghurdaqah
El Ghor plain Jordan/West Bank see
Al Ghawr
Elgin U.K. 20 F3
Elgin IL U.S.A. 63 J3
Elgin NV U.S.A. 65 E2
El'ginskiy Rus. Fed. 29 P3
El Gîza Egypt see Giza
El Goléa Alg. 22 E5
El Golfo de Santa Clara Mex. 62 E5
Elgon, Mount Kenya/Uganda 30 C6
El Hadjar Alg. 26 B6
El Ḥammâm Egypt see Al Ḥammām
El Hammâmi reg. Mauritania 46 B2
El Hank esc. Mali/Mauritania 46 C2
El Ḥarra Egypt see Al Ḥarrah
El Hazim Jordan see Al Hazim
El Heiz Egypt see Al Ḥayz
El Hierro i. Canary Is 46 B2
El Homr Alg. 22 E6
El Homra Sudan 32 D7
Elias Ireland. 54 E2
Elías Piña Dom. Rep. 67 J5
Elichpur India see Achalpur
Elie U.K. 20 G4
Elila r. Dem. Rep. Congo 48 C4
Elim U.S.A. 60 B3
Elimberrum France see Auch
Elingampangu Dem. Rep. Congo 48 C4
Élisabethville Dem. Rep. Congo see
Lubumbashi
Eliseu Martins Brazil 69 J5
El Iskandarîya Egypt see Alexandria
Elista Rus. Fed. 13 J7
Elizabeth NJ U.S.A. 64 D2
Elizabeth WV U.S.A. 64 A3
Elizabeth, Mount h. Australia 54 D4
Elizabeth Bay Namibia 50 B4
Elizabeth City U.S.A. 63 L4
Elizabeth Island Pitcairn Is see
Henderson Island
Elizabeth Point Namibia 50 B4
Elizabethtown KY U.S.A. 63 J4
El Jadida Morocco 22 C5
El Jem Tunisia 26 D7
Ełk Poland 17 S4
Elk r. MD U.S.A. 64 D3
El Kaa Lebanon see Qaa
El Kab Sudan 32 D6
El Kala Alg. 26 C6
Elk City U.S.A. 62 H4
Elkedra Australia 56 A4
Elkedra watercourse Australia 56 B4
El Kef Tunisia see Le Kef
El Kelaâ des Srarhna Morocco 22 C5
Elkford Canada 62 E2
Elk Grove CA U.S.A. 65 B1
El Khalil West Bank see Hebron
El Khandaq Sudan 32 D6
El Khârga Egypt see Al Khārijah
El Kharrûba Egypt see Al Kharrūbah
Elkhart IN U.S.A. 63 J3
El Khartûm Sudan see Khartoum
El Khenachich esc. Mali see El Khnâchîch
El Khnâchîch esc. Mali 46 C2
Elkhovo Bulg. 27 L3
Elki Turkey see Beytüşşebap
Elkins WV U.S.A. 64 B3
Elkland PA U.S.A. 64 C2
Elko Canada 62 D2
Elko U.S.A. 62 D3
Elkton MD U.S.A. 64 D3
Elkton VA U.S.A. 64 B3
El Kûbri Egypt see Al Kubrī
El Kuntilla Egypt see Al Kuntillah
Elkview WV U.S.A. 64 A3
Ellas country Europe see Greece
Ellef Ringnes Island Canada 61 H2
Ellendale U.S.A. 62 H2
Ellensburg U.S.A. 62 C2
Ellenville NY U.S.A. 64 D2
Ellesmere, Lake N.Z. 59 D6
Ellesmere Island Canada 61 J2
Ellesmere Island National Park Reserve
Canada see Quttinirpaaq National Park
Ellesmere Port U.K. 18 E5
Ellice r. Canada 61 H3
Ellice Island atoll Tuvalu see Funafuti
Ellice Islands country S. Pacific Ocean see
Tuvalu
Ellicott City MD U.S.A. 64 C3
Elliot S. Africa 51 H6
Elliot, Mount Australia 56 D3
Elliotdale S. Africa 51 I6
Elliot Knob mt. VA U.S.A. 64 B3
Elliott Australia 54 F4
Elliston VA U.S.A. 64 A4
Ellon U.K. 20 G3
Ellora Caves caves India 38 B1
Ellsworth ME U.S.A. 63 N3
Ellsworth Land reg. Antarctica 76 K1
Ellsworth Mountains Antarctica 76 L1
El Maghreb country Africa see Morocco
Elmakuz Dağı mt. Turkey 39 A1
Elmalı Turkey 27 M6
El Manşûra Egypt see Al Manşūrah
El Maṭarîya Egypt see Al Maṭarīyah
El Mazâr Egypt see Al Mazār
El Meghaïer Alg. 22 F5
El Milia Alg. 22 F4
El Minya Egypt see Al Minyā
Elmira Ont. Canada 64 A1
Elmira NY U.S.A. 64 C1
El Moral Spain 25 E5
Elmore Australia 58 B6
El Mreyyé reg. Mauritania 46 C3
Elmshorn Germany 17 L4
El Muglad Sudan 32 C7
Elnesvågen Norway 14 E5
El Nevado, Cerro mt. Col. 68 D3
El Oasis Baja California Mex. 65 E4
El Obeid Sudan 32 D7
El Odaiya Sudan 32 C7
El Oro Mex. 66 D4
Elorza Venez. 68 E2
El Oued Alg. 22 F5
El Paso TX U.S.A. 62 F5
Elphin U.K. 20 D2
El Portal CA U.S.A. 65 C2
El Porvenir Mex. 66 C2
El Porvenir Panama 67 I7
El Prat de Llobregat Spain 25 H3
El Progreso Hond. 66 G5
El Puerto de Santa María Spain 25 C5
El Qâhira Egypt see Cairo
El Qasimiye r. Lebanon 39 B3
El Quds Israel/West Bank see Jerusalem
El Quseima Egypt see Al Quşaymah
El Quseir Egypt see Al Quşayr
El Qûşîya Egypt see Al Qūşīyah
El Reno U.S.A. 62 H4
Elsa Canada 60 E3
El Ṣaff Egypt see Aş Şaff
El Sahuaro Mex. 62 E5
El Salado Mex. 66 D4
El Salto Mex. 66 C4
El Salvador country Central America 66 G6
El Salvador Chile 70 C3
El Salvador Mex. 62 G7
Elsass reg. France see Alsace
El Sellûm Egypt see As Sallūm
Elsey Australia 54 F3
El Shallûfa Egypt see Ash Shallūfah
El Sharana Australia 54 F3
El Shaṭṭ Egypt see Ash Shaṭṭ
Elsinore Denmark see Helsingør
Elsinore CA U.S.A. 65 D4
Elsinore Lake CA U.S.A. 65 D4
El Suweis Egypt see Suez
El Suweis governorate Egypt see As Suways
El Tama, Parque Nacional nat. park Venez.
68 D2
El Tarf Alg. 26 C6

El Teleno *mt.* Spain 25 C2
El Temascal Mex. 66 E4
El Thamad Egypt *see* Ath Thamad
El Tigre Venez. 68 F2
El'ton Rus. Fed. 13 J6
El'ton, Ozero *l.* Rus. Fed. 13 J6
El Tuparro, Parque Nacional *nat. park* Col. 68 E2
El Tûr Egypt *see* Aţ Ţūr
El Turbio Arg. 70 B8
El Uqşur Egypt *see* Luxor
Eluru India 38 D2
Elva Estonia 15 O7
Elvanfoot U.K. 20 F5
Elvas Port. 25 C4
Elverum Norway 15 G6
Elvira Brazil 68 D5
El Wak Kenya 48 E3
El Wâtya *well* Egypt *see* Al Wāţiyah
El Wuz Sudan 32 D7
Elx Spain *see* Elche-Elx
Ely U.K. 19 H6
Ely *MN* U.S.A. 63 I2
Ely *NV* U.S.A. 65 E1
Elyria U.S.A. 63 K3
El Zagâzîg Egypt *see* Az Zaqāzīq
Émaé *i.* Vanuatu 53 G3
eMakhazeni S. Africa 51 J3
Emāmrūd Iran 35 I3
Emām Şaḩēb Afgh. 36 B1
Emån *r.* Sweden 15 J8
Emas, Parque Nacional das *nat. park* Brazil 69 H7
Emba Kazakh. 28 G5
Embalenhle S. Africa 51 I4
Embalse de Buendía *resr* Spain 25 E3
Embalse de Cíjara *resr* Spain 25 D4
Embarcación Arg. 70 D2
Embi Kazakh. *see* Emba
Embira *r.* Brazil *see* Envira
Emborcação, Represa de *resr* Brazil 71 B2
Embu Kenya 48 D4
Emden Germany 17 K4
Emden Deep *sea feature* N. Pacific Ocean *see* Cape Johnson Depth
Emerald Australia 56 E4
Emerita Augusta Spain *see* Mérida
Emerson Canada 63 H2
Emesa Syria *see* Homs
Emet Turkey 27 M5
Emgwenya S. Africa 51 J3
eMgwenya S. Africa 51 J3
Emigrant Valley *val.* *NV* U.S.A. 65 E2
Emi Koussi *mt.* Chad 47 E3
Emiliano Zapata Mex. 66 F5
Emin China 42 E3
Emine, Nos *pt* Bulg. 27 L3
Eminska Planina *hills* Bulg. 27 L3
Emirdağ Turkey 27 N5
Emir Dağı *mt.* Turkey 27 N5
Emir Dağları *mts* Turkey 27 N5
eMjindini S. Africa 51 J3
eMkhondo S. Africa 51 J4
Emmaboda Sweden 15 I8
Emmaste Estonia 15 M7
Emmaus *PA* U.S.A. 64 D2
Emmaville Australia 58 E2
Emmen Neth. 17 K4
Emmen Switz. 24 I3
Emmet Australia 56 D5
Emmiganuru India 38 C3
Emona Slovenia *see* Ljubljana
Emory Peak U.S.A. 62 G6
Empangeni S. Africa 51 J5
Emperor Seamount Chain *sea feature* N. Pacific Ocean 74 H2
Emperor Trough *sea feature* N. Pacific Ocean 74 H2
Empingham Reservoir U.K. *see* Rutland Water
Emplawas Indon. 54 E2
Empoli Italy 26 D3
Emporia *KS* U.S.A. 63 H4
Emporia *VA* U.S.A. 63 L4
Emporium *PA* U.S.A. 64 B2
Empty Quarter *des.* Saudi Arabia *see* Rub' al Khālī
Ems *r.* Germany 17 K4
eMzinoni S. Africa 51 I4
Enafors Sweden 14 H5
Encantadas, Serra das *hills* Brazil 70 F4
Encarnación Para. 70 E3
Enchi Ghana 46 C4
Encinitas *CA* U.S.A. 65 D4
Encruzilhada Brazil 71 C1
Ende Indon. 41 E8
Endeavour Strait Australia 56 C1
Endeh Indon. *see* Ende
Enderby Land *reg.* Antarctica 76 D2
Endicott *NY* U.S.A. 64 C1
Endicott Mountains U.S.A. 60 C3
EnenKio *terr.* N. Pacific Ocean *see* Wake Island
Energodar Ukr. *see* Enerhodar
Enerhodar Ukr. 13 G7
Enewetak *atoll* Marshall Is 74 G5
Enez Turkey 27 L4
Enfe Lebanon 39 B2
Enfião, Ponta do *pt* Angola 49 B5
Enfidaville Tunisia 26 D6
Engan Norway 14 F5
Engaru Japan 44 F3
Engcobo S. Africa 51 H6
En Gedi Israel 39 B4
Engel's Rus. Fed. 13 J6
Enggano *i.* Indon. 41 C8
England *admin. div.* U.K. 19 E6
English Bazar India *see* Ingraj Bazar
English Channel France/U.K. 19 F9
English Coast Antarctica 76 L2
Engozero Rus. Fed. 12 G2
Enhlalakahle S. Africa 51 J5
Enid U.S.A. 62 H4
Eniwa Japan 44 F4
Eniwetok *atoll* Marshall Is *see* Enewetak
Enkeldoorn Zimbabwe *see* Chivhu
Enköping Sweden 15 J7
Enna *Sicilia* Italy 26 F6
Ennadai Lake Canada 61 H3
En Nahud Sudan 32 C7
Ennedi, Massif *mts* Chad 47 F3
Ennell, Lough *l.* Ireland 21 E4
Enngonia Australia 58 B2
Ennis Ireland 21 D5
Ennis *TX* U.S.A. 63 H5
Enniscorthy Ireland 21 F5
Enniskillen U.K. 21 E3
Ennistymon Ireland 21 C5
Enn Nâqoûra Lebanon 39 B3
Enns *r.* Austria 17 O6
Eno Fin. 14 Q5
Enoch *UT* U.S.A. 65 F2
Enontekiö Fin. 14 M2
Ensay Australia 58 C6
Enschede Neth. 17 K4
Ensenada Mex. 66 A2
Enshi China 43 J6
Entebbe Uganda 48 D3
Enterprise Canada 60 G3
Enterprise *UT* U.S.A. 65 F2
Entre Ríos Bol. 68 F8
Entre Rios Brazil 69 H5
Entre Rios de Minas Brazil 71 B3
Entroncamento Port. 25 B4
Enugu Nigeria 46 D4
Enurmino Rus. Fed. 29 T3
Envira Brazil 68 D5
Envira *r.* Brazil 68 D5
'En Yahav Israel 39 B4
Enyamba Dem. Rep. Congo 48 C4
Eochaill Ireland *see* Youghal
Epéna Congo 48 B3
Ephrata *PA* U.S.A. 64 C2
Epi *i.* Vanuatu 53 G3
Epidamnus Albania *see* Durrës
Épinal France 24 H2
Episkopi Bay Cyprus 39 A2
Episkopis, Kolpos *b.* Cyprus *see* Episkopi Bay
ePitoli S. Africa *see* Pretoria
Epomeo, Monte *vol.* Italy 26 E4
Epping U.K. 19 H7
Epping Forest National Park Australia 56 D4
Eppynt, Mynydd *hills* U.K. 19 D6
Epsom U.K. 19 G7
Eqlīd Iran 35 I5
Equatorial Guinea *country* Africa 46 D4
Équeurdreville-Hainneville France 19 F9
Erac Creek *watercourse* Australia 58 B1
Erandol India 38 B1
Erawadi *r.* Myanmar *see* Irrawaddy
Erbaa Turkey 34 E2
Erbeskopf *h.* Germany 17 K6
Ercan *airport* Cyprus 39 A2
Erciş Turkey 35 F3
Erciyes Dağı *mt.* Turkey 34 D3
Érd Hungary 26 H1
Erdaobaihe China *see* Baihe
Erdaogou Bingzhan China 37 H2
Erdao Jiang *r.* China 44 B4
Erdek Turkey 27 L4
Erdemli Turkey 39 B1
Erdenedalay Mongolia 42 I3
Erdenet Mongolia 42 I3
Erdenetsagaan Mongolia 43 L3
Erdi *reg.* Chad 47 F3
Erdniyevskiy Rus. Fed. 13 J7
Erebus, Mount *vol.* Antarctica 76 H1
Erechim Brazil 70 F3
Ereentsav Mongolia 43 L3
Ereğli *Konya* Turkey 34 D3
Ereğli *Zonguldak* Turkey 27 N4
Erego Moz. *see* Errego
Erei, Monti *mts* *Sicilia* Italy 26 F6
Erementaū Kazakh. *see* Yereymentau
Erenhot China 43 K4
Erepucu, Lago de *l.* Brazil 69 G4
Erevan Armenia *see* Yerevan
Erfurt Germany 17 M5
Ergani Turkey 35 E3
'Erg Chech *des.* Alg./Mali 46 C2
Ergene *r.* Turkey 27 L4
Erg er Raoui *des.* Alg. 22 D6
Ērgļi Latvia 15 N8
Ergu China 44 C3
Ergun China 43 M2
Ergun He *r.* China/Rus. Fed. *see* Argun'
Ergun Youqi China *see* Ergun
Ergun Zuoqi China *see* Gegen Gol
Erhulai China 44 B4
Eriboll, Loch *inlet* U.K. 20 E2
Ericht *r.* U.K. 20 F4
Ericht, Loch *l.* U.K. 20 E4
Erie *PA* U.S.A. 64 A1
Erie, Lake Canada/U.S.A. 64 A1
'Erîgât *des.* Mali 46 C3
Erik Eriksenstretet *sea chan.* Svalbard 28 D2
Erimo-misaki *c.* Japan 44 F4
Erinpura Road India 36 C4
Eriskay *i.* U.K. 20 B3
Eritrea *country* Africa 32 E6
Erlangen Germany 17 M6
Erldunda Australia 55 F6
Erlistoun *watercourse* Australia 55 C6
Erlong Shan *mt.* China 44 C4
Erlongshan Shuiku *resr* China 44 B4
Ermak Kazakh. *see* Aksu
Ermelo S. Africa 51 I4
Ermenek Turkey 39 A1
Ermenek *r.* Turkey 39 A1
Ermont Egypt *see* Armant
Ermoupoli Greece 27 K6
Ernakulam India 38 C4
Erne *r.* Ireland/U.K. 21 D3
Ernest Giles Range *hills* Australia 55 C6
Erode India 38 C4
Eromanga Australia 57 C5
Erongo *admin. reg.* Namibia 50 B1
Errabiddy Hills Australia 55 A6
Er Rachidia Morocco 22 D5
Errego Moz. 49 D5
Er Remla Tunisia 26 D7
Er Renk Sudan 32 D7
Errigal *h.* Ireland 21 D2
Errinundra National Park Australia 58 D6
Erris Head *hd* Ireland 21 B3
Erromango *i.* Vanuatu 53 G3
Erronan *i.* Vanuatu *see* Futuna
Erseka Albania *see* Ersekë
Ersekë Albania 27 I4
Ersmark Sweden 14 L5
Ertai China 42 G3
Ertil' Rus. Fed. 13 I6
Ertis *r.* Kazakh./Rus. Fed. *see* Irtysh
Ertix He *r.* China/Kazakh. 42 F3
Ērtra *country* Africa *see* Eritrea
Eruh Turkey 35 F3
Eryuan China 42 H7
Erzerum Turkey *see* Erzurum
Erzgebirge *mts* Czech Rep./Germany 17 N5
Erzhan China 44 B2
Erzin Turkey 39 C1
Erzincan Turkey 35 E3
Erzurum Turkey 35 F3
Esa-ala P.N.G. 56 E1
Esan-misaki *pt* Japan 44 F4
Esashi Japan 44 F3
Esbjerg Denmark 15 F9
Esbo Fin. *see* Espoo
Escalante Desert *UT* U.S.A. 65 F2
Escalón Mex. 66 D3
Escanaba U.S.A. 63 J2
Escárcega Mex. 66 F5
Escatrón Spain 25 F3
Eschscholtz *atoll* Marshall Is *see* Bikini
Eschwege Germany 17 M5
Escondido *CA* U.S.A. 65 D4
Escuinapa Mex. 66 C4
Escuintla Guat. 66 F6
Eséka Cameroon 46 E4
Eşen Turkey 27 M6
Esenguly Turkm. 35 I3
Esenguly Döwlet Gorugy *nature res.* Turkm. 35 I3
Eşfahān Iran 35 H4
Eshkamesh Afgh. 36 B1
Eshkanān Iran 35 I6
Eshowe S. Africa 51 J5
Esikhawini S. Africa 51 K5
Esil Kazakh. *see* Yesil'
Esil *r.* Kazakh./Rus. Fed. *see* Ishim
Esk Australia 58 F1
Esk *r.* Australia 57 [inset]
Esk *r.* U.K. 18 D4
Eskdalemuir U.K. 20 F5
Esker Canada 61 L4
Eskifjörður Iceland 14 [inset]
Eski Gediz Turkey 27 M5
Eskilstuna Sweden 15 J7
Eskimo Lakes Canada 60 E3
Eskimo Point Canada *see* Arviat
Eskipazar Turkey 34 D2
Eskişehir Turkey 27 N5
Esla *r.* Spain 25 C3
Eslāmābād-e Gharb Iran 35 G4
Esler Dağı *mt.* Turkey 27 M6
Eslöv Sweden 15 H9
Eşme Turkey 27 M5
Esmeraldas Ecuador 68 C3
Esmont *VA* U.S.A. 64 B4
Espakeh Iran 33 J4
Espalion France 24 F4
España *country* Europe *see* Spain
Espanola Canada 63 K2
Esperance Australia 55 C8
Esperance Bay Australia 55 C8
Esperanza Arg. 70 B8
Esperanza Mex. 62 F6
Esperanza (Argentina) *research stn* Antarctica 76 A2
Espichel, Cabo *c.* Port. 25 B4
Espigão, Serra do *mts* Brazil 71 A4
Espigüete *mt.* Spain 25 D2
Espinhaço, Serra do *mts* Brazil 71 C2
Espinosa Brazil 71 C1
Espírito Santo Brazil *see* Vila Velha
Espírito Santo *state* Brazil 71 C2
Espiritu Santo *i.* Vanuatu 53 G3
Espíritu Santo, Isla *i.* Mex. 62 E7
Espoo Fin. 15 N6
Espuña *mt.* Spain 25 F5
Esquel Arg. 70 B6
Essaouira Morocco 46 C1
Es Semara W. Sahara 46 B2
Essen Germany 17 K5
Essequibo *r.* Guyana 69 G2
Essex *CA* U.S.A. 65 E3
Essex *MD* U.S.A. 64 C3
Esso Rus. Fed. 29 Q4
Essoyla Rus. Fed. 12 G3
Eşţahbān Iran 35 I5
Estālef Afgh. 36 B2
Estância Brazil 69 K6
Estats, Pic d' *mt.* France/Spain 24 E5
Estcourt S. Africa 51 I5
Estelí Nicaragua 67 G6
Estella Spain 25 E2
Estepa Spain 25 D5
Estepona Spain 25 D5
Esteras de Medinaceli Spain 25 E3
Esterhazy Canada 62 G1
Estero Bay *CA* U.S.A. 65 B3
Esteros Para. 70 D2
Estevan Canada 62 G2
Estherville U.S.A. 63 I3
Eston Canada 62 F1
Estonia *country* Europe 15 N7
Estonskaya S.S.R. *country* Europe *see* Estonia
Estrela Brazil 71 A5
Estrela, Serra da *mts* Port. 25 C3
Estrela do Sul Brazil 71 B2
Estrella *mt.* Spain 25 E4
Estremoz Port. 25 C4
Estrondo, Serra *hills* Brazil 69 I5
Etadunna Australia 57 B6
Etah India 36 D4
Étampes France 24 F2
Étaples France 24 E1
Etawah *Rajasthan* India 36 D4
Etawah *Uttar Prad.* India 36 D4
Ethandakukhanya S. Africa 51 J4
Ethel Creek Australia 55 C5
E'Thembini S. Africa 50 F5
Ethiopia *country* Africa 48 D3
Ethnikos Drymos Prespon *nat. park* Greece 27 I4
Etimesğut Turkey 34 D3
Etive, Loch *inlet* U.K. 20 D4
Etna, Mount *vol.* *Sicilia* Italy 26 F6
Etne Norway 15 D7
Etobicoke *Ont.* Canada 64 B1
Etolin Strait U.S.A. 60 B3
Etorofu-tō *i.* Rus. Fed. *see* Iturup, Ostrov
Etosha National Park Namibia 49 B5
Etosha Pan *salt pan* Namibia 49 B5
Etoumbi Congo 48 B3
Etrek *r.* Iran/Turkm. *see* Atrek
Etrek Turkm. 35 I3
Étrépagny France 19 I9
Étretat France 19 H9
Ettelbruck Lux. 17 K6
Ettrick Water *r.* U.K. 20 F5
Euabalong Australia 58 C4
Euboea *i.* Greece *see* Evvoia
Eucla Australia 55 E7
Euclid *OH* U.S.A. 64 A2
Euclides da Cunha Brazil 69 K6
Eucumbene, Lake Australia 58 D6
Eudunda Australia 57 B7
Eufaula Lake *resr* U.S.A. 63 H4
Eugene U.S.A. 62 C3
Eugenia, Punta *pt* Mex. 66 A3
Eugowra Australia 58 D4
Eulo Australia 58 B2
Eumungerie Australia 58 D3
Eungella Australia 56 E4
Eungella National Park Australia 56 E4
Euphrates *r.* Asia 23 L4
Euphrates *r.* Asia 32 F3
Euphrates *r.* Asia 35 G5
Eura Fin. 15 M6
Eureka *CA* U.S.A. 62 C3
Eureka *MT* U.S.A. 62 D2
Eureka *NV* U.S.A. 62 D4
Eureka Sound *sea chan.* Canada 61 J2
Eureka Valley *val.* *CA* U.S.A. 65 D2
Euriowie Australia 57 C6
Euroa Australia 58 B6
Eurombah Australia 57 E5
Eurombah Creek *r.* Australia 57 E5
Europa, Île *i.* Indian Ocean 49 E6
Europa, Punta de *pt* Gibraltar *see* Europa Point
Europa Point Gibraltar 25 D5
Eva Downs Australia 54 F4
Evans, Lac *l.* Canada 63 L1
Evans City *PA* U.S.A. 64 A2
Evans Head Australia 58 F2
Evans Head *hd* Australia 58 F2
Evans Ice Stream Antarctica 76 L1
Evanston *WY* U.S.A. 62 E3
Evansville *IN* U.S.A. 63 J4
Eva Perón Arg. *see* La Plata
Evaton S. Africa 51 H4
Evaz Iran 35 I6
Evensk Rus. Fed. 29 Q3
Everard, Lake *imp. l.* Australia 57 A6
Everard, Mount Australia 55 F5
Everard Range *hills* Australia 55 F6
Everek Turkey *see* Develi
Everest, Mount China/Nepal 37 F4
Everett *PA* U.S.A. 64 B2
Everett *WA* U.S.A. 62 C2
Everglades *swamp* U.S.A. 63 K6
Evesham Australia 56 C4
Evesham U.K. 19 F6
Evesham, Vale of *val.* U.K. 19 F6
Evijärvi Fin. 14 M5
Evje Norway 15 E7
Évora Port. 25 C4
Evoron, Ozero *l.* Rus. Fed. 44 E2
Évreux France 24 E2
Evros *r.* Bulg. *see* Maritsa
Evros *r.* Turkey *see* Meriç
Evrotas *r.* Greece 27 J6
Evrychou Cyprus 39 A2
Evrykhou Cyprus *see* Evrychou
Evvoia *i.* Greece 27 K5
Ewan Australia 56 D3
Ewaso Ngiro *r.* Kenya 48 E3
Ewe, Loch *b.* U.K. 20 D3
Ewo Congo 48 B4
Exaltación Bol. 68 E6
Excelsior S. Africa 51 H5
Excelsior Mountain *CA* U.S.A. 65 C1
Excelsior Mountains *NV* U.S.A. 65 C1
Exe *r.* U.K. 19 D8
Exeter Australia 58 E5
Exeter *Ont.* Canada 64 A1
Exeter U.K. 19 D8
Exeter *CA* U.S.A. 65 C2
Exeter *NH* U.S.A. 64 F1
Exminster U.K. 19 D8
Exmoor *hills* U.K. 19 D7
Exmoor National Park U.K. 19 D7
Exmore *VA* U.S.A. 64 D4
Exmouth Australia 54 A5
Exmouth U.K. 19 D8
Exmouth, Mount Australia 58 D3
Exmouth Gulf Australia 54 A5
Exmouth Plateau *sea feature* Indian Ocean 73 P7
Expedition National Park Australia 56 E5
Expedition Range *mts* Australia 56 E5
Exton *PA* U.S.A. 64 D2
Extremadura *aut. comm.* Spain 25 D4
Exuma Cays *is* Bahamas 67 I4
Eyasi, Lake *salt l.* Tanz. 48 D4
Eyawadi *r.* Myanmar *see* Irrawaddy
Eye U.K. 19 I6
Eyelenoborsk Rus. Fed. 11 S3
Eyemouth U.K. 20 G5
Eyjafjörður *inlet* Iceland 14 [inset]
Eyl Somalia 48 E3
Eylau Rus. Fed. *see* Bagrationovsk
Eynsham U.K. 19 F7
Eyre (North), Lake Australia 57 B6
Eyre (South), Lake Australia 57 B6
Eyre, Lake Australia 57 B6
Eyre Creek *watercourse* Australia 56 B5
Eyre Mountains N.Z. 59 B7
Eyre Peninsula Australia 57 A7
Eysturoy *i.* Faroe Is 14 [inset]
Ezakheni S. Africa 51 J5
Ezenzeleni S. Africa 51 I4
Ezequiel Ramos Mexía, Embalse *resr* Arg. 70 C5
Ezhou China 43 K6
Ezhva Rus. Fed. 12 K3
Ezine Turkey 27 L5
Ezo *i.* Japan *see* Hokkaidō
Ezousa *r.* Cyprus 39 A2

Faaborg Denmark 15 G9
Faadhippolhu Atoll Maldives 38 B5
Faafxadhuun Somalia 48 E3
Fåborg Denmark *see* Faaborg
Fabriano Italy 26 E3
Fachi Niger 46 E3
Fada Chad 47 F3
Fada-N'Gourma Burkina Faso 46 D3
Fadghāmī Syria 35 F4
Fadiffolu Atoll Maldives *see* Faadhippolhu Atoll
Fadippolu Atoll Maldives *see* Faadhippolhu Atoll
Faenza Italy 26 D2
Færoerne *terr.* N. Atlantic Ocean *see* Faroe Islands
Faeroes *terr.* N. Atlantic Ocean *see* Faroe Islands
Făgăraş Romania 27 K2
Fagatogo American Samoa 53 I3
Fagersta Sweden 15 I7
Fagurhólsmýri Iceland 14 [inset]
Fagwir South Sudan 32 D8
Fahraj Iran 35 I5
Fā'id Egypt 34 D5
Fairbanks U.S.A. 60 D3
Fairchance *PA* U.S.A. 64 B3
Fairfax *VA* U.S.A. 64 C3
Fairfield *CA* U.S.A. 65 A1
Fair Haven *VT* U.S.A. 64 E1
Fair Head *hd* U.K. 21 F2
Fair Isle *i.* U.K. 20 H1
Fairmont *MN* U.S.A. 63 I3
Fairmont *WV* U.S.A. 64 A3
Fairview Australia 56 D2
Fairview *PA* U.S.A. 64 A1
Fairweather, Mount Canada/U.S.A. 60 E4
Faisalabad Pak. 33 L3
Faith U.S.A. 62 G2
Faizabad Afgh. *see* Feyzābād
Faizabad India 37 E4
Fakaofo *atoll* Tokelau 53 I2
Fakaofu *atoll* Tokelau *see* Fakaofo
Fakenham U.K. 19 H6
Fåker Sweden 14 I5
Fakfak Indon. 41 F8
Fakhrābād Iran 35 I5
Fakiragram India 37 G4
Fako *vol.* Cameroon *see* Cameroun, Mont
Fal *r.* U.K. 19 C8
Falaba Sierra Leone 46 B4
Falam Myanmar 37 H5
Falenki Rus. Fed. 12 K4
Falkenberg Sweden 15 H8
Falkirk U.K. 20 F5
Falkland U.K. 20 F4
Falkland Escarpment *sea feature* S. Atlantic Ocean 72 E9
Falkland Islands *terr.* S. Atlantic Ocean 70 E8
Falkland Plateau *sea feature* S. Atlantic Ocean 72 E9
Falkland Sound *sea chan.* Falkland Is 70 D8
Falköping Sweden 15 H7
Fallbrook *CA* U.S.A. 65 D4
Fallieres Coast Antarctica 76 L2
Fallon U.S.A. 62 D4
Fall River *MA* U.S.A. 64 F2
Fall River Pass U.S.A. 62 F3
Falmouth U.K. 19 B8
Falmouth *VA* U.S.A. 64 C3
False Bay S. Africa 50 D8
False Point India 37 F5
Falster *i.* Denmark 15 G9
Fălticeni Romania 13 E7
Falun Sweden 15 I6
Famagusta Cyprus 39 A2
Famagusta Bay Cyprus *see* Ammochostos Bay
Famenin Iran 35 H4
Fame Range *hills* Australia 55 C6
Family Well Australia 54 D5
Fāmūr, Daryācheh-ye *l.* Iran 35 H5
Fana Mali 46 C3
Fanad Head *hd* Ireland 21 E2
Fandriana Madag. 49 E6
Fane *r.* Ireland 21 F4
Fangxian China 43 K6
Fangzheng China 44 C3
Fannich, Loch *l.* U.K. 20 D3
Fano Italy 26 E3
Fanum Fortunae Italy *see* Fano
Faqih Aḩmadān Iran 35 H5
Faraba Mali 46 B3
Faradofay Madag. *see* Tôlañaro
Farafangana Madag. 49 E6
Farāfirah, Wāḩāt al *oasis* Egypt 32 C4
Farafra Oasis *oasis* Egypt *see* Farāfirah, Wāḩāt al
Farāh Afgh. 33 J3
Faraḩābād Iran *see* Khezerābād
Farallones de Cali, Parque Nacional *nat. park* Col. 68 C3
Faranah Guinea 46 B3
Fararah Oman 33 I6
Farasān, Jazā'ir *is* Saudi Arabia 32 F6
Faraulep *atoll* Micronesia 41 G7
Fareham U.K. 19 F8
Farewell, Cape Greenland 61 N3
Farewell, Cape N.Z. 59 D5
Farewell Spit N.Z. 59 D5
Färgelanda Sweden 15 H7
Farghona Uzbek. *see* Farg'ona
Fargo U.S.A. 63 H2
Farg'ona Uzbek. 33 L1
Faribault U.S.A. 63 I3
Faribault, Lac *l.* Canada 61 K4
Faridabad India 36 D3
Faridkot India 36 C3
Faridpur Bangl. 37 G5
Farīmān Iran 33 I2
Farkhar Afgh. *see* Farkhato
Farkhato Afgh. 36 B1
Farmahīn Iran 35 H4
Farmington *ME* U.S.A. 63 M3
Farmington *NH* U.S.A. 64 F1
Farmington *NM* U.S.A. 62 F4
Farmville *VA* U.S.A. 64 B4
Farnborough U.K. 19 G7
Farne Islands U.K. 18 F3
Farnham U.K. 19 G7
Farnham, Lake *imp. l.* Australia 55 D6
Farnham, Mount Canada 62 D1
Faro Brazil 69 G4
Faro Port. 25 C5
Fårö *i.* Sweden 15 K8
Faroe - Iceland Ridge *sea feature* Arctic Ocean 77 I2
Faroe Islands *terr.* N. Atlantic Ocean 14 [inset]
Fårösund Sweden 15 K8
Farquhar Group *is* Seychelles 49 F5
Farquharson Tableland *hills* Australia 55 C6
Farrāshband Iran 35 I5
Farr Bay Antarctica 76 F2
Farrukhabad India *see* Fatehgarh
Farsund Norway 15 E7
Farwell *TX* U.S.A. 62 G5
Fasā Iran 35 I5
Fasano Italy 26 G4
Faşikan Geçidi *pass* Turkey 39 A1
Fastiv Ukr. 13 F6
Fastov Ukr. *see* Fastiv
Fatehabad India 36 C3
Fatehgarh India 36 D4
Fatehpur *Rajasthan* India 36 C4
Fatehpur *Uttar Prad.* India 36 E4
Fatick Senegal 46 B3
Fattoilep *atoll* Micronesia *see* Faraulep
Faughan *r.* U.K. 21 E2
Fauresmith S. Africa 51 G5
Fauske Norway 14 I3
Fawley U.K. 19 F8
Faxaflói *b.* Iceland 14 [inset]
Faxälven *r.* Sweden 14 J5
Faya Chad 47 E3
Fayetteville *AR* U.S.A. 63 I4
Fayetteville *NC* U.S.A. 63 L4
Fayetteville *WV* U.S.A. 64 A3
Fâyid Egypt *see* Fā'id
Faylakah *i.* Kuwait 35 H5
Fazao Malfakassa, Parc National de *nat. park* Togo 46 D4
Fazilka India 36 C3
Fdérik Mauritania 46 B2
Fead Group *is* P.N.G. *see* Nuguria Islands
Feale *r.* Ireland 21 C5
Fear, Cape U.S.A. 63 L5
Featherston N.Z. 59 E5
Feathertop, Mount Australia 58 C6
Fécamp France 24 E2
Federal District *admin. dist.* Brazil *see* Distrito Federal
Federalsburg *MD* U.S.A. 64 D3
Federated Malay States *country* Asia *see* Malaysia
Fedusar India 36 C4
Fehmarn *i.* Germany 17 M3
Feia, Lagoa *lag.* Brazil 71 C3
Feijó Brazil 68 D5
Feilding N.Z. 59 E5
Feio *r.* Brazil *see* Aguapeí
Feira de Santana Brazil 71 D1
Fejd el Abiod *pass* Alg. 26 B6
Feke Turkey 34 D3
Felanitx Spain 25 H4
Felāvarjān Iran 35 H4
Feldberg *mt.* Germany 17 L7
Feldkirch Austria 17 L7
Feldkirchen in Kärnten Austria 17 O7
Felipe C. Puerto Mex. 66 G5
Felixlândia Brazil 71 B2
Felixstowe U.K. 19 I7
Felixton S. Africa 51 J5
Fellowsville *WV* U.S.A. 64 B3
Felsina Italy *see* Bologna
Felton *DE* U.S.A. 64 D3
Feltre Italy 26 D1
Femunden *l.* Norway 14 G5
Femundsmarka Nasjonalpark *nat. park* Norway 14 H5
Fenaio, Punta del *pt* Italy 26 D3
Fener Burnu *hd* Turkey 39 B1
Fénérive Madag. *see* Fenoarivo Atsinanana
Fengari *mt.* Greece 27 K4
Fengguang China 44 B3
Fengman China 44 B4
Fengxian China 42 J6
Fengxiang *Heilong.* China *see* Luobei
Fengxiang *Yunnan* China *see* Lincang
Fengyüan Taiwan 43 M8
Fengzhen China 43 K4
Feni Bangl. 37 G5
Feni Islands P.N.G. 52 F2
Feno, Capo di *c.* *Corse* France 24 I6
Fenoarivo Atsinanana Madag. 49 E5
Fenua Ura *atoll* Fr. Polynesia *see* Manua
Feodosiya Ukr. 34 D1
Fer, Cap de *c.* Alg. 26 B6
Férai Greece *see* Feres
Ferdows Iran 33 I3
Feres Greece 27 L4
Fergus Falls U.S.A. 63 H2
Fergusson Island P.N.G. 52 F2
Fériana Tunisia 26 C7
Ferizaj Kosovo 27 I3
Ferkessédougou Côte d'Ivoire 46 C4
Fermo Italy 26 E3
Fermont Canada 61 L4
Fermoselle Spain 25 C3
Fermoy Ireland 21 D5
Fernandina, Isla *i.* *Galápagos* Ecuador 68 [inset]
Fernandina Beach U.S.A. 63 K5
Fernando de Magallanes, Parque Nacional *nat. park* Chile 70 B8
Fernando de Noronha *i.* Brazil 72 F6
Fernandópolis Brazil 71 A3
Fernando Poó *i.* Equat. Guinea *see* Bioc
Fernão Dias Brazil 71 B2
Ferndown U.K. 19 F8
Fernlee Australia 58 C2
Ferns Ireland 21 F5
Ferozepore India *see* Firozpur
Ferrara Italy 26 D2
Ferreira-Gomes Brazil 69 H3
Ferro, Capo *c.* *Sardegna* Italy 26 C4
Ferrol Spain 25 B2
Ferros Brazil 71 C2
Ferryville Tunisia *see* Menzel Bourguiba
Fertő-tavi *nat. park* Hungary 26 G1
Fès Morocco 22 D5
Feshi Dem. Rep. Congo 49 B4
Fété Bowé Senegal 46 B3
Fethard Ireland 21 E5
Fethiye *Malatya* Turkey *see* Yazıhan
Fethiye *Muğla* Turkey 27 M6
Fethiye Körfezi *b.* Turkey 27 M6
Fetisovo Kazakh. 35 I2
Fetlar *i.* U.K. 20 [inset]
Fettercairn U.K. 20 G4
Feuilles, Rivière aux *r.* Canada 61 K4
Fevral'sk Rus. Fed. 44 C1
Fevzipaşa Turkey 34 E3
Feyzābād Afgh. 36 B1
Fez Morocco *see* Fès
Ffestiniog U.K. 19 D6
Fianarantsoa Madag. 49 E6
Fichē Eth. 48 D3
Fier Albania 27 H4
Fiery Creek *r.* Australia 56 B3
Fife Ness *pt* U.K. 20 G4
Fifield Australia 58 C4
Figeac France 24 F4
Figueira da Foz Port. 25 B3
Figueras Spain *see* Figueres
Figueres Spain 25 H2
Figuig Morocco 22 D5
Figuil Cameroon 47 E4
Fiji *country* S. Pacific Ocean 53 H3
Fīk' Eth. 48 E3
Filadelfia Para. 70 D2
Filchner Ice Shelf Antarctica 76 A1
Filey U.K. 18 G4
Filibe Bulg. *see* Plovdiv
Filingué Niger 46 D3
Filipinas *country* Asia *see* Philippines
Filippiada Greece 27 I5
Filipstad Sweden 15 I7
Fillan Norway 14 F5
Fillmore *CA* U.S.A. 65 C3
Fillmore *UT* U.S.A. 65 F1
Fīltu Eth. 48 E3
Fimbul Ice Shelf Antarctica 76 C2
Fīn Iran 35 H4
Findhorn *r.* U.K. 20 F3
Findlay U.S.A. 63 K3
Finger Lakes *NY* U.S.A. 64 C1
Finike Turkey 27 N6
Finike Körfezi *b.* Turkey 27 N6
Finisterre Spain *see* Fisterra
Finisterre, Cabo *c.* Spain *see* Finisterre, Cape
Finisterre, Cape Spain 25 B2
Finke *watercourse* Australia 56 A5
Finke, Mount *h.* Australia 55 F7
Finke Bay Australia 54 E3
Finke Gorge National Park Australia 55
Finland *country* Europe 14 O5
Finland, Gulf of Europe 15 M7
Finlay *r.* Canada 60 F4
Finn *r.* Ireland 21 E3
Finnigan, Mount Australia 56 D2
Finniss, Cape Australia 55 F8
Finnmarksvidda *reg.* Norway 14 H2
Finnsnes Norway 14 J2
Finspång Sweden 15 I7
Fintona U.K. 21 E3
Finucane Range *hills* Australia 56 C4
Fionn Loch *l.* U.K. 20 D3
Fionnphort U.K. 20 C4
Fiordland National Park N.Z. 59 A7
Fırat *r.* Asia *see* Euphrates
Fırat *r.* Asia 34 E3 *see* Euphrates
Firebaugh *CA* U.S.A. 65 B2
Firenze Italy *see* Florence
Firk, Sha'īb *watercourse* Iraq 35 G5
Firmat Arg. 70 D4
Firminy France 24 G4
Firmum Italy *see* Fermo
Firmum Picenum Italy *see* Fermo
Firovo Rus. Fed. 12 G4
Firozabad India 36 D4
Firozpur India 36 C3
Fīrūzābād Iran 35 I5
Fīrūz Kūh Iran 35 I4
Fischersbrunn Namibia 50 B3
Fish *watercourse* Namibia 50 C5
Fisher (abandoned) Australia 55 E7
Fisher Bay Antarctica 76 G2
Fisher Glacier Antarctica 76 E2
Fishers Island *NY* U.S.A. 64 F2
Fisher Strait Canada 61 J3
Fishguard U.K. 19 C7
Fishing Creek *MD* U.S.A. 64 C3
Fiske, Cape Antarctica 76 L2
Fiskenæsset Greenland *see* Qeqertarsuatsiaat
Fismes France 24 F2
Fisterra Spain 25 B2
Fisterra, Cabo *c.* Spain *see* Finisterre, Ca
Fitri, Lac *l.* Chad 47 E3
Fitzgerald River National Park Australia 55 B8
Fitz Roy Arg. 70 C7
Fitzroy *r.* Australia 54 C4
Fitz Roy, Cerro *mt.* Arg. 70 B7
Fitzroy Crossing Australia 54 D4
Fiume Croatia *see* Rijeka
Fivemiletown U.K. 21 E3
Five Points *CA* U.S.A. 65 B2
Fizi Dem. Rep. Congo 49 C4
Fizuli Azer. *see* Füzuli

Norway 15 F6
gstaff S. Africa 51 I6
gstaff U.S.A. 62 E4
mborough Head hd U.K. 18 G4
minksvlei salt pan S. Africa 50 E6
nnan Isles U.K. 20 B2
sjön l. Sweden 14 I4
thead r. U.S.A. 62 E2
thead Lake U.S.A. 62 E2
ttery, Cape Australia 56 D2
ttery, Cape U.S.A. 62 C2
etwood Australia 56 D4
etwood U.K. 18 D5
etwood PA U.S.A. 64 D2
kkefjord Norway 15 E7
mington NJ U.S.A. 64 D2
n Sweden 15 J7
nsburg Germany 17 L3
rs France 24 D2
tcher Peninsula Antarctica 76 L2
aders r. Australia 56 C3
aders Chase National Park Australia 7 B7
aders Group National Park Australia 6 D2
aders Island Australia 57 [inset]
aders Passage Australia 56 E3
aders Ranges mts Australia 57 B7
aders Ranges National Park Australia 7 B6
aders Reefs Australia 56 E3
Flon Canada 62 G1
t U.K. 18 D5
t U.S.A. 63 K3
t Island Kiribati 75 J6
ton Australia 58 D1
a Norway 15 H6
singskiy, Mys c. Rus. Fed. 28 H2
dden U.K. 18 E3
od Range mts Antarctica 76 J1
ra r. Australia 54 E3
rac France 24 F4
ra Reef Australia 56 D3
rence Italy 26 D3
rence AL U.S.A. 63 J5
rence AZ U.S.A. 62 E5
rence SC U.S.A. 63 L5
rencia Col. 68 C3
rentia Italy see Florence
rentino Ameghino, Embalse resr Arg. 0 C6
res r. Arg. 70 E5
res Guat. 66 G5
res i. Indon. 41 E8
res, Laut sea Indon. 41 D8
resta Brazil 69 K5
riano Brazil 69 J5
rianópolis Brazil 71 A4
rida Uruguay 70 E4
rida state U.S.A. 63 K5
rida, Straits of Bahamas/U.S.A. 67 H4
rida Islands Solomon Is 53 G2
rida Keys is U.S.A. 63 K7
rin CA U.S.A. 65 B1
rina Greece 27 I4
rø Norway 15 D6
yd VA U.S.A. 64 A4
yd, Mount AZ U.S.A. 65 F3
shing Neth. see Vlissingen
r. P.N.G. 52 E2
ng Fish, Cape Antarctica 76 K2
m Lake Canada 62 G1
a Bos.-Herz. 26 H3
a Turkey 27 L5
habers U.K. 20 F3
şani Romania 27 L2
gia Italy 26 F4
o i. Cape Verde 46 [inset]
naven h. U.K. 20 E2
x France 24 E5
da sea chan. Norway 14 I3
dereid Norway 14 H4
dfjorden sea chan. Norway 14 G4
egandros i. Greece 27 K6
eyet Canada 63 K2
ey Island Canada 61 K3
gno Italy 26 E3
kestone U.K. 19 I7
kingham U.K. 19 G6
kston U.S.A. 63 K5
dal Norway 14 G5
onica Italy 26 D3
om Lake CA U.S.A. 65 B1
nboni Comoros 49 E5
nin Rus. Fed. 13 I7
ninskaya Rus. Fed. 12 K2
ninskoye Rus. Fed. 12 I4
da NY U.S.A. 64 D1
d-du-Lac Canada 60 H4
d du Lac U.S.A. 63 J3
devila Spain 25 F3
di Italy 26 E4
ni Sardegna Italy 26 C4
sagrada Spain see A Fonsagrada
seca, Golfo do b. Central America 6 G6
te Boa Brazil 68 E4
tur pt Iceland 14 [inset]
chow China see Fuzhou
aker, Mount U.S.A. 60 C3
aulep atoll Micronesia see Faraulep
bes Australia 58 D4
chheim Germany 17 M6
d City CA U.S.A. 65 C3
de Norway 15 D6
dham U.K. 19 H6
dingbridge U.K. 19 F8
d Range mts Antarctica 76 J1
ds Bridge Australia 58 B2
écariah Guinea 46 B4
el, Mont mt. Greenland 61 O3
eland hd U.K. 19 F8
eland Point U.K. 19 D7
est MS U.S.A. 63 J5
est Creek r. Australia 56 C3
est Hill Australia 58 C5
estville CA U.S.A. 65 A1
far U.K. 20 G4
ges-les-Eaux France 19 I9
ked River NJ U.S.A. 64 D3
k Union VA U.S.A. 64 B4
li Italy 26 E2
mby U.K. 18 D5
mentera i. Spain 25 G4
mentor, Cap de c. Spain 25 H4
mer Yugoslav Republic of Macedonia untry Europe see Macedonia
miga Brazil 71 B3
mosa Arg. 70 E3
mosa country Asia see Taiwan
mosa Brazil 71 B1
mosa, Serra hills Brazil 69 G6
mosa Bay Kenya see Ungwana Bay
mosa Strait China/Taiwan see iwan Strait
moso r. Bahia Brazil 71 B1
moso r. Tocantins Brazil 71 A1
nos Moz. 51 L2
res U.K. 20 F3
rest Vic. Australia 58 A7
rest W.A. Australia 55 E7
Forrestal Range mts Antarctica 76 A1
Forrest City U.S.A. 63 I4
Forrest Lakes imp. l. Australia 55 E7
Fors Sweden 14 J5
Forsayth Australia 56 C3
Forsnäs Sweden 14 M3
Forssa Fin. 15 M6
Forster Australia 58 F4
Forsyth MT U.S.A. 62 F2
Forsyth Range hills Australia 56 C4
Fort Abbas Pak. 36 C3
Fort Albany Canada 63 K1
Fortaleza Brazil 69 K4
Fort Archambault Chad see Sarh
Fort Ashby WV U.S.A. 64 B3
Fort Augustus U.K. 20 E3
Fort Beaufort S. Africa 51 H7
Fort Benton U.S.A. 62 E2
Fort Brabant Canada see Tuktoyaktuk
Fort Bragg U.S.A. 62 C4
Fort Charlet Alg. see Djanet
Fort Chimo Canada see Kuujjuaq
Fort Chipewyan Canada 60 G4
Fort Crampel Cent. Afr. Rep. see Kaga Bandoro
Fort-Dauphin Madag. see Tôlañaro
Fort-de-France Martinique 67 L6
Fort de Kock Indon. see Bukittinggi
Fort de Polignac Alg. see Illizi
Fort Dodge U.S.A. 63 I3
Fort Edward NY U.S.A. 64 E1
Fortescue r. Australia 54 B5
Forte Veneza Brazil 69 H5
Fort Flatters Alg. see Bordj Omer Driss
Fort Foureau Cameroon see Kousséri
Fort Franklin Canada see Déline
Fort Gardel Alg. see Zaouatallaz
Fort George Canada see Chisasibi
Fort Good Hope Canada 60 F3
Fort Gouraud Mauritania see Fdérik
Forth r. U.K. 20 F4
Forth, Firth of est. U.K. 20 F4
Fort Hertz Myanmar see Putao
Fortification Range mts NV U.S.A. 65 E1
Fortín General Mendoza Para. 70 D2
Fortín Leonida Escobar Para. 70 D2
Fortín Madrejón Para. 70 E2
Fortín Pilcomayo Arg. 70 D2
Fortín Ravelo Bol. 68 F7
Fortín Sargento Primero Leyes Arg. 70 E2
Fortín Suárez Arana Bol. 68 F7
Fortín Teniente Juan Echauri López Para. 70 D2
Fort Jameson Zambia see Chipata
Fort Johnston Malawi see Mangochi
Fort Lamy Chad see Ndjamena
Fort Laperrine Alg. see Tamanrasset
Fort Lauderdale U.S.A. 63 K6
Fort Liard Canada 60 F3
Fort Macleod Canada 62 E2
Fort Manning Malawi see Mchinji
Fort McMurray Canada 60 G4
Fort McPherson Canada 60 E3
Fort Munro Pak. 36 B3
Fort Myers U.S.A. 63 K6
Fort Nelson Canada 60 F4
Fort Payne U.S.A. 63 J5
Fort Peck Reservoir U.S.A. 62 F2
Fort Pierce U.S.A. 63 K6
Fort Portal Uganda 48 D3
Fort Providence Canada 60 G3
Fort Randall U.S.A. see Cold Bay
Fort Resolution Canada 60 G3
Fortrose N.Z. 59 B8
Fortrose U.K. 20 E3
Fort Rosebery Zambia see Mansa
Fort Rousset Congo see Owando
Fort Rupert Canada see Waskaganish
Fort Sandeman Pak. see Zhob
Fort Saskatchewan Canada 62 E1
Fort Scott U.S.A. 63 I4
Fort Severn Canada 61 J4
Fort-Shevchenko Kazakh. 30 E2
Fort Simpson Canada 60 F3
Fort Smith Canada 60 G3
Fort Smith U.S.A. 63 I4
Fort Stockton U.S.A. 62 G5
Fort Sumner U.S.A. 62 G5
Fort Trinquet Mauritania see Bîr Mogreïn
Fort Vermilion Canada 77 L3
Fort Victoria Zimbabwe see Masvingo
Fort Ware Canada see Ware
Fort Wayne U.S.A. 63 J3
Fort William U.K. 20 D4
Fort Worth U.S.A. 63 H5
Fort Yukon U.S.A. 60 D3
Forum Iulii France see Fréjus
Forvik Norway 14 H4
Fossano Italy 26 B2
Fossil Downs Australia 54 D4
Foster Australia 58 C7
Fotadrevo Madag. 49 E6
Fotherby U.K. 18 G5
Fotokol Cameroon 47 E3
Fotuna i. Vanuatu see Futuna
Fougères France 24 D2
Foula i. U.K. 20 [inset]
Foulness Point U.K. 19 H7
Foul Point Sri Lanka 38 D4
Foumban Cameroon 46 E4
Foundation Ice Stream glacier Antarctica 76 L1
Fountains Abbey and Royal Water Garden tourist site U.K. 18 F4
Fourches, Mont des h. France 24 G2
Four Corners CA U.S.A. 65 D3
Fouriesburg S. Africa 51 I5
Fournoi i. Greece 27 L6
Fourpeaked Mountain U.S.A. 60 C4
Fouta Djallon reg. Guinea 46 B3
Foveaux Strait N.Z. 59 A8
Fowey r. U.K. 19 C8
Fowler CO U.S.A. 62 G4
Fowler Ice Rise Antarctica 76 L1
Fowlers Bay Australia 52 D5
Fowlers Bay b. Australia 55 F8
Fox Creek Canada 60 G4
Foxdale Isle of Man 18 C4
Foxe Basin g. Canada 61 K3
Foxe Channel Canada 61 J3
Foxe Peninsula Canada 61 K3
Fox Glacier N.Z. 59 C6
Fox Islands U.S.A. 60 B4
Fox Mountain Canada 60 E3
Fox Valley Canada 62 F1
Foyers U.K. 20 E3
Foyle r. Ireland/U.K. 21 E3
Foyle, Lough b. Ireland/U.K. 21 E2
Foynes Ireland 21 C5
Foz de Areia, Represa de resr Brazil 71 A4
Foz do Cunene Angola 49 B5
Foz do Iguaçu Brazil 70 F3
Fraga Spain 25 G3
Frakes, Mount Antarctica 76 K1
Framingham MA U.S.A. 64 F1
Framnes Mountains Antarctica 76 E2
Franca Brazil 71 B3
Français, Récif des rf New Caledonia 53 G3
Francavilla Fontana Italy 26 G4
France country Europe 24 F3
Frances Australia 57 C8
Franceville Gabon 48 B4
Francis atoll Kiribati see Beru
Francisco de Orellana Ecuador see Coca
Francistown Botswana 49 C6
Francois Peron National Park Australia 55 A6
Frankenhöhe hills Germany 17 M6
Frankfort KY U.S.A. 63 K4
Frankfurt Germany see Frankfurt am Main
Frankfurt am Main Germany 17 L5
Frankfurt an der Oder Germany 17 O4
Frank Hann National Park Australia 55 C8
Fränkische Alb hills Germany 17 M6
Fränkische Schweiz reg. Germany 17 M6
Frankland, Cape Australia 57 [inset]
Franklin MA U.S.A. 64 F1
Franklin NH U.S.A. 64 F1
Franklin PA U.S.A. 64 B2
Franklin Bay Canada 77 A2
Franklin D. Roosevelt Lake resr U.S.A. 62 D2
Franklin-Gordon National Park Australia 57 [inset]
Franklin Island Antarctica 76 H1
Franklin Mountains Canada 60 F3
Franklin Strait Canada 61 I2
Franklinville NY U.S.A. 64 B1
Frankston Australia 58 B7
Fränsta Sweden 14 J5
Frantsa-Iosifa, Zemlya is Rus. Fed. 28 G2
Franz Josef Glacier N.Z. 59 C6
Frasca, Capo della c. Sardegna Italy 26 C5
Frascati Italy 26 E4
Fraser r. Australia 54 C4
Fraser r. B.C. Canada 62 C2
Fraser r. Nfld. and Lab. Canada 61 L4
Fraser, Mount h. Australia 55 B6
Fraserburg S. Africa 50 E6
Fraserburgh U.K. 20 G3
Fraserdale Canada 63 K2
Fraser Island Australia 56 F5
Fraser Island National Park Australia 56 F5
Fraser National Park Australia 58 B6
Fraser Range hills Australia 55 C8
Frauenfeld Switz. 24 I3
Fray Bentos Uruguay 70 E4
Freckleton U.K. 18 E5
Frederica DE U.S.A. 64 D3
Fredericia Denmark 15 F9
Frederick MD U.S.A. 64 C3
Frederick Reef Australia 56 F4
Fredericksburg TX U.S.A. 62 H5
Fredericksburg VA U.S.A. 64 C3
Fredericton Canada 61 L5
Frederikshåb Greenland see Paamiut
Frederikshavn Denmark 15 G8
Frederiksværk Denmark 15 H9
Fredonia AZ U.S.A. 65 F2
Fredonia NY U.S.A. 64 B1
Fredrika Sweden 14 K4
Fredrikshamn Fin. see Hamina
Fredrikstad Norway 15 G7
Freehold NJ U.S.A. 64 D2
Freeland PA U.S.A. 64 D2
Freeling Heights h. Australia 57 B6
Freel Peak CA U.S.A. 65 C1
Freeport TX U.S.A. 63 H6
Freeport City Bahamas 67 I3
Free State prov. S. Africa 51 H5
Freetown Sierra Leone 46 B4
Fregenal de la Sierra Spain 25 C4
Fregon Australia 55 F6
Fréhel, Cap c. France 24 C2
Frei (Chile) research stn Antarctica 76 A2
Freiburg Switz. see Fribourg
Freiburg im Breisgau Germany 17 K6
Freising Germany 17 M6
Freistadt Austria 17 O6
Fréjus France 24 H5
Fremantle Australia 55 A8
Fremont CA U.S.A. 65 B2
Fremont NE U.S.A. 63 H3
French Congo country Africa see Congo
French Guiana terr. S. America 69 H3
French Guinea country Africa see Guinea
French Island Australia 58 B7
Frenchman r. U.S.A. 62 F2
Frenchman Lake NV U.S.A. 65 E2
Frenchpark Ireland 21 D4
French Pass N.Z. 59 D5
French Polynesia terr. S. Pacific Ocean 75 K7
French Somaliland country Africa see Djibouti
French Southern and Antarctic Lands terr. Indian Ocean 73 M8
French Sudan country Africa see Mali
French Territory of the Afars and Issas country Africa see Djibouti
Frenda Alg. 25 G6
Fresco r. Brazil 69 H5
Freshford Ireland 21 E5
Fresnillo Mex. 66 D4
Fresno CA U.S.A. 65 C2
Fresno r. CA U.S.A. 65 B2
Freu, Cap des c. Spain 25 H4
Freudenstadt Germany 17 L6
Frew watercourse Australia 56 A4
Frewena Australia 56 A3
Freycinet Estuary inlet Australia 55 A6
Freycinet Peninsula Australia 57 [inset]
Freyming-Merlebach France 24 H2
Fria Guinea 46 B3
Fria, Cape Namibia 49 B5
Friant CA U.S.A. 65 C2
Frias Arg. 70 C3
Fribourg Switz. 24 H3
Friedens PA U.S.A. 64 B2
Friedland Rus. Fed. see Pravdinsk
Friedrichshafen Germany 17 L7
Friendly Islands country S. Pacific Ocean see Tonga
Frinton-on-Sea U.K. 19 I7
Frisco Mountain UT U.S.A. 65 F1
Frissell, Mount h. CT U.S.A. 64 E1
Frobisher Bay Canada see Iqaluit
Frobisher Bay b. Canada 61 L3
Frohavet b. Norway 14 F5
Frolovo Rus. Fed. 13 I6
Frome watercourse Australia 57 B6
Frome U.K. 19 E7
Frome r. U.K. 19 E8
Frome, Lake imp. l. Australia 57 B6
Frome Downs Australia 57 B6
Frontera Tabasco Mex. 66 F5
Fronteras Mex. 66 C2
Front Royal VA U.S.A. 64 B3
Frosinone Italy 26 E4
Frostburg MD U.S.A. 64 B3
Frøya i. Norway 14 F5
Frunze Kyrg. see Bishkek
Frusino Italy see Frosinone
Fruška Gora, Nacionalni Park nat. park Serbia 27 H2
Frýdek-Místek Czech Rep. 17 Q6
Fucheng Shaanxi China see Fuxian
Fuding China 43 M7
Fuenlabrada Spain 25 E3
Fuerte Olimpo Para. 70 E2
Fuerteventura i. Canary Is 46 B2
Fuga i. Phil. 43 M9
Fuhai China 42 F3
Fuḩaymī Iraq 35 F4
Fujairah U.A.E. 33 I4
Fujeira U.A.E. see Fujairah
Fuji Japan 45 E6
Fujian prov. China 43 L7
Fuji-Hakone-Izu Kokuritsu-kōen Japan 45 E6
Fujin China 44 C3
Fujinomiya Japan 45 E6
Fuji-san vol. Japan 45 E6
Fujiyoshida Japan 45 E6
Fûka Egypt see Fūkah
Fūkah Egypt 34 B5
Fukien prov. China see Fujian
Fukuchiyama Japan 45 D6
Fukue-jima i. Japan 45 C6
Fukui Japan 45 E5
Fukuoka Japan 45 C6
Fukushima Japan 45 F5
Fukuyama Japan 45 C7
Fûl, Gebel h. Egypt see Fūl, Jabal
Fūl, Jabal h. Egypt 39 A5
Fulchhari Bangl. 37 G4
Fulda Germany 17 L5
Fulda r. Germany 17 L5
Fulham U.K. 19 G7
Fuli China see Jixian
Fulitun China see Jixian
Fullerton CA U.S.A. 65 D4
Fulton MO U.S.A. 63 I4
Fulton NY U.S.A. 64 C1
Fumane Moz. 51 K3
Fumay France 24 G2
Funabashi Japan 45 E6
Funafuti atoll Tuvalu 53 H2
Funchal Madeira 46 B1
Fundão Brazil 71 C2
Fundão Port. 25 C3
Fundi Italy see Fondi
Fundy, Bay of g. Canada 61 L5
Fünen i. Denmark see Fyn
Funeral Peak CA U.S.A. 65 D2
Fünfkirchen Hungary see Pécs
Funhalouro Moz. 51 L2
Funing Yunnan China 42 J8
Funtua Nigeria 46 D3
Funzie U.K. 20 [inset]
Fürgun, Küh-e mt. Iran 33 I4
Furmanov Rus. Fed. 12 I4
Furmanovka Kazakh. see Moyynkum
Furmanovo Kazakh. see Zhalpaktal
Furnás h. Spain 25 G4
Furnas, Represa resr Brazil 71 B3
Furneaux Group is Australia 57 [inset]
Fürstenwalde Germany 17 O4
Fürth Germany 17 M6
Furukawa Japan 45 F5
Fury and Hecla Strait Canada 61 J3
Fusan S. Korea see Pusan
Fushun China 44 A4
Fushuncheng China see Shuncheng
Fusong China 44 B4
Futuna i. Vanuatu 53 H3
Futuna Islands Wallis and Futuna Is see Hoorn, Îles de
Fuxian Liaoning China see Wafangdian
Fuxian Shaanxi China 43 J5
Fuxin China 43 M4
Fuxing China see Wangmo
Fuxinzhen China see Fuxin
Fuyang Anhui China 43 L6
Fuyu Heilong. China 44 B3
Fuyu Jilin China see Songyuan
Fuyu Jilin China 44 B3
Fuyuan Heilong. China 44 D2
Fuyun China 42 F3
Fuzhou Fujian China 43 L7
Fuzhou Jiangxi China 31 K4
Füzuli Azer. 35 G3
Fyn i. Denmark 15 G9
Fyne, Loch inlet U.K. 20 D5

G

Gaâfour Tunisia 26 C6
Gaalkacyo Somalia 48 E3
Gabakly Turkm. 33 J2
Gabbs NV U.S.A. 65 D1
Gabbs Valley Range mts NV U.S.A. 65 C1
Gabela Angola 49 B5
Gaberones Botswana see Gaborone
Gabès Tunisia 22 G5
Gabès, Golfe de g. Tunisia 22 G5
Gabo Island Australia 58 D6
Gabon country Africa 48 B4
Gaborone Botswana 51 G3
Gabrovo Bulg. 27 K3
Gabú Guinea-Bissau 46 B3
Gadag India 38 B3
Gadaisu P.N.G. 56 E1
Gadap Pak. 36 A4
Gadchiroli India 38 D1
Gäddede Sweden 14 I4
Gades Spain see Cádiz
Gadhra India 36 B5
Gadra Pak. 36 B4
Gadsden U.S.A. 63 J5
Gadwal India 38 C2
Gadyach Ukr. see Hadyach
Gaer U.K. 19 D7
Găeşti Romania 27 K2
Gaeta Italy 26 E4
Gaeta, Golfo di g. Italy 26 E4
Gaferut i. Micronesia 74 F5
Gafsa Tunisia 26 C7
Gagarin Rus. Fed. 13 G5
Gagnoa Côte d'Ivoire 46 C4
Gagnon Canada 63 N1
Gago Coutinho Angola see Lumbala N'guimbo
Gagra Georgia 13 I8
Gaiab watercourse Namibia 50 D5
Gaibanda Bangl. see Gaibandha
Gaibandha Bangl. 37 G4
Gaïdouronisi i. Greece 27 K7
Gaifi, Wâdi el watercourse Egypt see Jayfi, Wādī al
Gaillac France 24 E5
Gaillimh Ireland see Galway
Gaindainqoinkor China 37 G3
Gainesville FL U.S.A. 63 K6
Gainesville GA U.S.A. 63 K5
Gainesville TX U.S.A. 63 H5
Gainsborough U.K. 18 G5
Gairdner, Lake imp. l. Australia 57 A6
Gairloch U.K. 20 D3
Gair Loch b. U.K. 20 D3
Gajipur India see Ghazipur
Gajol India 37 G4
Gakarosa mt. S. Africa 50 F4
Gala China 37 G3
Gala Co l. China 37 G3
Galâla el Baḩarîya, Gebel el plat. Egypt see Jalālah al Baḩrīyah, Jabal
Galana r. Kenya 48 E4
Galanta Slovakia 17 P6
Galápagos, Parque Nacional Ecuador 68 [inset]
Galapagos Islands is Ecuador 75 O6
Galapagos Rise sea feature Pacific Ocean 75 N6
Galashiels U.K. 20 G5
Galați Romania 27 M2
Galatina Italy 26 H4
Gala Water r. U.K. 20 G5
Galaýmor Turkm. 33 J2
Galaymor Turkm. see Galaýmor
Galbally Ireland 21 D5
Galdhøpiggen mt. Norway 15 F6
Galeana Nuevo León Mex. 66 D4
Galena AK U.S.A. 60 C3
Galena MD U.S.A. 64 D3
Galera, Punta pt Chile 70 B6
Galesburg IL U.S.A. 63 I3
Galeshewe S. Africa 50 G5
Galeton PA U.S.A. 64 C2
Galey r. Ireland 21 C5
Galheirão r. Brazil 71 B1
Galich Rus. Fed. 12 I4
Galichskaya Vozvyshennost' hills Rus. Fed. 12 I4
Galicia aut. comm. Spain 25 C2
Galičica nat. park Macedonia 27 I4
Galilee, Lake imp. l. Australia 56 D4
Galilee, Sea of l. Israel 39 B3
Galizia aut. comm. Spain see Galicia
Gallabat Sudan 32 E7
Gallatin TN U.S.A. 63 J4
Galle Sri Lanka 38 D5
Gallego Rise sea feature Pacific Ocean 75 M6
Gallegos r. Arg. 70 C8
Gallia country Europe see France
Gallinas, Punta pt Col. 68 D1
Gallipoli Italy 26 H4
Gallipoli Turkey 27 L4
Gällivare Sweden 14 L3
Gällö Sweden 14 I5
Gallup U.S.A. 62 F4
Galmisdale U.K. 20 C4
Galmudug reg. Somalia 30 D6
Galong Australia 58 D5
Galoya Sri Lanka 38 D4
Gal Oya National Park Sri Lanka 38 D5
Galston U.K. 20 E5
Galt CA U.S.A. 65 B1
Galtat Zemmour W. Sahara 46 B2
Galtee Mountains hills Ireland 21 D5
Galtymore h. Ireland 16 C4
Galveston TX U.S.A. 63 I6
Galveston Bay U.S.A. 63 I6
Galwa Nepal 37 E3
Galway Ireland 21 C4
Galway Bay Ireland 21 C4
Gamalakhe S. Africa 51 J6
Gamba China see Gongbalou
Gamba Gabon 48 A4
Gambēla Eth. 48 D3
Gambēla National Park Eth. 48 D3
Gambell U.S.A. 60 A3
Gambella Eth. see Gambēla
Gambia, The country Africa 46 B3
Gambier, Îles is Fr. Polynesia 75 L7
Gambier Islands Australia 57 B7
Gambier Islands Fr. Polynesia see Gambier, Îles
Gamboma Congo 48 B4
Gamboola Australia 56 C3
Gamboula Cent. Afr. Rep. 48 B3
Gamda China see Zamtang
Gamêtî Canada 60 G3
Gamlakarleby Fin. see Kokkola
Gamleby Sweden 15 J8
Gammelstaden Sweden 14 M4
Gammon Ranges National Park Australia 57 B6
Gamova, Mys pt Rus. Fed. 44 C4
Gamtog China 42 H6
Gamud mt. Eth. 48 D3
Ganado U.S.A. 62 F4
Gäncä Azer. 35 G2
Ganda Angola 49 B5
Gandaingoin China 37 G3
Gandajika Dem. Rep. Congo 49 C4
Gandak Barrage Nepal 37 E4
Gandari Mountain Pak. 36 B3
Gandava Pak. 33 K4
Gander Canada 61 M5
Gandesa Spain 25 G3
Gandhidham India 36 B5
Gandhinagar India 36 C5
Gandhi Sagar resr India 36 C4
Gandia Spain 25 F4
Gandzha Azer. see Gäncä
Ganga r. Bangl./India 37 G5 see Ganges
Ganga Cone sea feature Indian Ocean see Ganges Cone
Gangán Arg. 70 C6
Ganganagar India 36 C3
Gangapur India 36 D4
Ganga Sera India 36 B4
Gangaw Myanmar 37 H5
Gangawati India 38 C3
Gangaw Range mts Myanmar 37 I5
Gangca China 42 I5
Gangdisê Shan mts China 37 E3
Ganges r. Bangl./India 37 G5
Ganges France 24 F5
Ganges, Mouths of the Bangl./India 37 G5
Ganges Cone sea feature Indian Ocean 73 N4
Gangouyi China 42 J5
Gangra Turkey see Çankırı
Gangtok India 37 G4
Ganjig China 43 M4
Ganmain Australia 58 C5
Gannan China 44 A3
Gannat France 24 F3
Gannett Peak U.S.A. 60 H5
Ganq China 42 G5
Gansu prov. China 42 H4
Gantheaume Point Australia 54 C4
Gantsevichi Belarus see Hantsavichy
Ganye Nigeria 46 E4
Ganyushkino Kazakh. 11 P6
Ganzhou China 43 K7
Ganzi South Sudan 47 G4
Gao Mali 46 C3
Gaocheng China see Litang
Gaoleshan China see Xianfeng
Gaotai China 42 H5
Gaoth Dobhair Ireland 21 D2
Gaoua Burkina Faso 46 C3
Gaoual Guinea 46 B3
Gaoxiong Taiwan see Kaohsiung
Gaoyao China see Zhaoqing
Gaoyou Hu l. China 43 L6
Gap France 24 H4
Gapuwiyak Australia 56 A2
Gaqoi China 37 E3
Gar China 36 E2
Gar' r. Rus. Fed. 44 C1
Gara, Lough l. Ireland 21 D4
Garabekewül Turkm. 33 J2
Garabil Belentligi hills Turkm. 33 J2
Garabogaz Turkm. 35 I2
Garabogaz Aylagy b. Turkm. see Garabogazköl Aýlagy
Garabogazköl Aylagy b. Turkm. see Garabogazköl Aýlagy
Garabogazköl Aýlagy b. Turkm. 35 I2
Garabogazköl Bogazy sea chan. Turkm. 35 I2
Garagum des. Turkm. see Karakum Desert
Garagum Kanaly canal Turkm. 33 J2
Garah Australia 58 D2
Garalo Mali 46 C3
Garamba r. Dem. Rep. Congo 48 C3
Garanhuns Brazil 69 K5
Ga-Rankuwa S. Africa 51 H3
Garapuava Brazil 71 B2
Gárasavvon Sweden see Karesuando
Garautha India 36 D4
Garba China see Jiulong
Garbahaarrey Somalia 48 E3
Garba Tula Kenya 48 D3
Garbo China see Lhozhag
Garbsen Germany 17 L4
Garça Brazil 71 A3
Garco China 37 G2
Garda, Lago di Italy see Garda, Lake
Garda, Lake Italy 26 D2
Garde, Cap de c. Alg. 26 B6
Garden City U.S.A. 62 G4
Garden Hill Canada 63 I1
Garden Mountain VA U.S.A. 64 A4
Garden Route National Park S. Africa 50 F7
Gardeyz Afgh. see Gardēz
Gardēz Afgh. 36 B2
Gardinas Belarus see Hrodna
Gardiner, Mount Australia 54 F5
Gardiner Range hills Australia 54 E4
Gardiners Island NY U.S.A. 64 E2
Gardīz Afgh. see Gardēz
Gardner atoll Micronesia see Faraulep
Gardner MA U.S.A. 64 F1
Gardner Inlet Antarctica 76 L1
Gardner Island atoll Kiribati see Nikumaroro
Gardner Pinnacles is U.S.A. 74 I4
Gáregasnjárga Fin. see Karigasniemi
Garelochhead U.K. 20 E4
Garet el Djenoun mt. Alg. 46 D2
Gargano, Parco Nazionale del nat. park Italy 26 F4
Gargunsa China see Gar
Gargždai Lith. 15 L9
Garhchiroli India see Gadchiroli
Garhi Madh. Prad. India 38 C1
Garhi Rajasthan India 36 C5
Garhi Khairo Pak. 36 A3
Garhwa India 37 E4
Gari Rus. Fed. 11 S4
Gariep Dam dam S. Africa 51 G6
Garies S. Africa 50 C6
Garigliano r. Italy 26 E4
Garissa Kenya 48 D4
Garkalne Latvia 15 N8
Garkung Caka l. China 37 F2
Garm Tajik. see Gharm
Garmī Iran 35 H3
Garmsār Iran 35 I4
Garnpung Lake imp. l. Australia 58 A4
Garo Hills India 37 G4
Garonne r. France 24 D4
Garoowe Somalia 48 E3
Garopaba Brazil 71 A5
Garoua Cameroon 46 E4
Garoua Boulai Cameroon 47 E4
Garqêntang China see Sog
Garré Arg. 70 D5
Garruk Pak. 36 A3
Garry r. U.K. 20 E3
Garry Lake Canada 61 H3
Garrynahine U.K. 20 C2
Garsen Kenya 48 E4
Garshy Turkm. see Garşy
Garsila Sudan 47 F3
Garşy Turkm. 35 I2
Garth U.K. 19 D6
Gartok China see Garyarsa
Garub Namibia 50 C4
Garvagh U.K. 21 F3
Garve U.K. 20 E3
Garwa India see Garhwa
Garwha India see Garhwa
Gar Xincun China 36 E2
Gary IN U.S.A. 63 J3
Gary WV U.S.A. 64 A4
Garyarsa China 36 E3
Garyū-zan mt. Japan 45 D6
Garza García Mex. 66 D3
Garzê China 42 H6
Gasan-Kuli Turkm. see Esenguly
Gascogne reg. France see Gascony
Gascogne, Golfe de g. France see Gascony, Gulf of
Gascony reg. France 24 D5
Gascony, Gulf of France 24 C5
Gascoyne r. Australia 55 A6
Gascoyne Junction Australia 55 A6
Gasherbrum I mt. China/Pakistan 36 D2
Gashua Nigeria 46 E3
Gaspé Canada 63 O2
Gaspésie, Péninsule de la pen. Canada 63 N2
Gassan vol. Japan 45 F5
Gassaway WV U.S.A. 64 A3
Gasteiz Spain see Vitoria-Gasteiz
Gastello Rus. Fed. 44 F2
Gastonia U.S.A. 63 K4
Gata, Cabo de c. Spain 25 E5
Gata, Cape Cyprus 39 A2
Gata, Sierra de mts Spain 25 C3
Gatas, Akra c. Cyprus see Gata, Cape
Gatchina Rus. Fed. 12 F4
Gatehouse of Fleet U.K. 20 E6
Gateshead U.K. 18 F4
Gates of the Arctic National Park and Preserve U.S.A. 60 C3
Gatesville U.S.A. 62 H5
Gatineau r. Canada 63 L2
Gatong China see Jomda
Gatooma Zimbabwe see Kadoma
Gatton Australia 58 F1
Gatvand Iran 35 H4
Gatyana S. Africa see Willowvale
Gau i. Fiji 53 H3
Gauhati India see Guwahati
Gaujas nacionālais parks nat. park Latvia 15 N8
Gaul country Europe see France
Gaula r. Norway 14 G5
Gaurama Brazil 71 A4
Gauribidanur India 38 C3
Gauteng prov. S. Africa 51 I4
Gavarr Armenia 35 G2
Gāvbandī Iran 35 I6
Gāvbūs, Kūh-e mts Iran 35 I6
Gavdos i. Greece 27 K7
Gavião r. Brazil 71 C1
Gavīleh Iran 35 G4
Gav Khūnī Iran 35 I4
Gävle Sweden 15 J6
Gavrilovka Vtoraya Rus. Fed. 13 I5
Gavrilov-Yam Rus. Fed. 12 H4

Gawachab Namibia 50 C4
Gawan India 37 F4
Gawilgarh Hills India 36 D5
Gawler Australia 57 B7
Gawler Ranges hills Australia 57 A7
Gaxun Nur salt l. China 42 I4
Gaya India 37 F4
Gaya Niger 46 D3
Gaya He r. China 44 C4
Gayéri Burkina Faso 46 D3
Gaylord U.S.A. 63 K2
Gayndah Australia 57 E5
Gayny Rus. Fed. 12 L3
Gaysin Ukr. see Haysyn
Gayutino Rus. Fed. 12 H4
Gaz Iran 35 H4
Gaza terr. Asia 39 B4
Gaza Gaza 39 B4
Gaza prov. Moz. 51 K2
Gazan Pak. 36 A3
Gazanjyk Turkm. see Bereket
Gaza Strip terr. Asia see Gaza
Gaziantep Turkey 34 E3
Gaziantep prov. Turkey 39 C1
Gazibenli Turkey see Yahyalı
Gazimağusa Cyprus see Famagusta
Gazimurskiy Khrebet mts Rus. Fed. 43 L2
Gazimurskiy Zavod Rus. Fed. 43 L2
Gazipaşa Turkey 39 A1
Gazli Uzbek. 33 J1
Gbarnga Liberia 46 C4
Gboko Nigeria 46 D4
Gcuwa S. Africa see Butterworth
Gdańsk Poland 17 Q3
Gdańsk, Gulf of Poland/Rus. Fed. 17 Q3
Gdańska, Zatoka g. Poland/Rus. Fed. see Gdańsk, Gulf of
Gdingen Poland see Gdynia
Gdov Rus. Fed. 12 E4
Gdynia Poland 17 Q3
Geaidnovuohppi Norway 14 M2
Gearraidh na h-Aibhne U.K. see Garrynahine
Geçitkale Cyprus see Lefkonikon
Gedaref Sudan 32 E7
Gediz r. Turkey 27 L5
Gedney Drove End U.K. 19 H6
Gedser Denmark 15 G9
Geelong Australia 58 B7
Geelvink Channel Australia 55 A7
Geel Vloer salt pan S. Africa 50 E5
Gees Gwardafuy c. Somalia see Gwardafuy, Gees
Gegen Gol China 44 A2
Geidam Nigeria 46 E3
Geikie r. Canada 60 H4
Geilo Norway 15 F6
Geiranger Norway 14 E5
Geisûm, Gezâ'ir is Egypt see Qaysūm, Juzur
Geita Tanz. 48 D4
Gejiu China 42 I8
Gela Sicilia Italy 26 F6
Gêladaindong mt. China 37 G2
Geladī Eth. 48 E3
Gelendzhik Rus. Fed. 34 E1
Gelibolu Turkey see Gallipoli
Gelidonya Burnu pt Turkey see Yardımcı Burnu
Gelincik Dağı mt. Turkey 27 N5
Gemena Dem. Rep. Congo 48 B3
Geminokağı Cyprus see Karavostasi
Gemlik Turkey 27 M4
Gemona del Friuli Italy 26 E1
Gemsa Egypt see Jamsah
Gemsbok National Park Botswana 50 E3
Gemsbokplein well S. Africa 50 E4
Genalē Wenz r. Eth. 48 E3
Genāveh Iran 35 H5
General Acha Arg. 70 D5
General Alvear Arg. 70 C5
General Belgrano II research stn Antarctica see Belgrano II
General Carrera, Lago l. Arg./Chile 70 B7
General Conesa Arg. 70 D6
General Freire Angola see Muxaluando
General Juan Madariaga Arg. 70 E5
General La Madrid Arg. 70 D5
General Machado Angola see Camacupa
General Pico Arg. 70 D5
General Pinedo Arg. 70 D3
General Roca Arg. 70 C5
General Salgado Brazil 71 A3
General San Martín research stn Antarctica see San Martín
General Santos Phil. 41 E7
General Villegas Arg. 70 D5
Genesee PA U.S.A. 64 C2
Geneseo NY U.S.A. 64 C1
Geneva S. Africa 51 H4
Geneva Switz. 24 H3
Geneva NY U.S.A. 64 C1
Geneva OH U.S.A. 64 A2
Geneva, Lake France/Switz. 24 H3
Genève Switz. see Geneva
Genf Switz. see Geneva
Genhe China see Gegen Gol
Genichesk Ukr. see Heniches'k
Genji India 36 C5
Genk Belgium 17 J5
Genoa Australia 58 D6
Genoa Italy 26 C2
Genoa, Gulf of Italy 26 C2
Genova Italy see Genoa
Genova, Golfo di Italy see Genoa, Gulf of
Gent Belgium see Ghent
Gentioux, Plateau de France 24 F4
Genua Italy see Genoa
Geographe Bay Australia 55 A8
Geographical Society Ø i. Greenland 61 P2
Georga, Zemlya i. Rus. Fed. 28 F1
George r. Canada 61 L4
George S. Africa 50 F7
George, Lake Australia 58 D5
George, Lake NY U.S.A. 64 E1
George Land i. Rus. Fed. see Georga, Zemlya
Georges Mills NH U.S.A. 64 E1
George Sound inlet N.Z. 59 A7
Georgetown Australia 56 C3
George Town Cayman Is 67 H5
Georgetown Gambia 46 B3
Georgetown Guyana 69 G2
George Town Malaysia 41 C7
Georgetown DE U.S.A. 64 D3
Georgetown SC U.S.A. 63 L5
Georgetown TX U.S.A. 62 H5
George VI Sound sea chan. Antarctica 76 L2
George V Land reg. Antarctica 76 G2
Georgia country Asia 35 F2
Georgia state U.S.A. 63 K5
Georgian Bay Canada 63 K2
Georgienne, Baie b. Canada see Georgian Bay
Georgina watercourse Australia 56 B5
Georgiu-Dezh Rus. Fed. see Liski
Georgiyevka Vostochnyy Kazakhstan Kazakh. 42 E3
Georgiyevka Zhambylskaya Oblast' Kazakh. see Korday
Georgiyevsk Rus. Fed. 13 I7
Georgiyevskoye Rus. Fed. 12 J4
Georg von Neumayer research stn Antarctica see Neumayer
Gera Germany 17 N5
Geral, Serra mts Brazil 71 A4
Geral de Goiás, Serra hills Brazil 71 B1
Geraldine N.Z. 59 C7
Geral do Paraná, Serra hills Brazil 71 B1
Geraldton Australia 55 A7
Gerar watercourse Israel 39 B4
Gercüş Turkey 35 F3
Gerede Turkey 34 D2
Gereshk Afgh. 33 J3
Gerlachovský štít mt. Slovakia 17 R6
Germania country Europe see Germany
Germanicea Turkey see Kahramanmaraş
German South-West Africa country Africa see Namibia
Germany country Europe 17 L5
Germersheim Germany 17 L6
Gerona Spain see Girona
Gerrit Denys is P.N.G. see Lihir Group
Gers r. France 24 E4
Gersoppa India 38 B3
Géryville Alg. see El Bayadh
Gêrzê China 37 F2
Gerze Turkey 34 D2
Gesoriacum France see Boulogne-sur-Mer
Gettysburg PA U.S.A. 64 C3
Gettysburg SD U.S.A. 62 H2
Gettysburg National Military Park nat. park PA U.S.A. 64 C3
Getz Ice Shelf Antarctica 76 J2
Geurie Australia 58 D4
Gevaş Turkey 35 F3
Gevgelija Macedonia 27 J4
Gexto Spain see Algorta
Gey Iran see Nīkshahr
Geyikli Turkey 27 L5
Geylegphug Bhutan 37 G4
Geysdorp S. Africa 51 G4
Geyserville CA U.S.A. 65 A1
Geyve Turkey 27 N4
Ghaap Plateau S. Africa 50 F4
Ghāb, Wādī al r. Syria 39 C2
Ghabāghib Syria 39 C3
Ghabeish Sudan 32 C7
Ghadaf, Wādī al watercourse Jordan 39 C4
Ghadamés Libya see Ghadāmis
Ghadāmis Libya 46 D1
Gha'em Shahr Iran 35 I3
Ghaghara r. India 37 F4
Ghaibi Dero Pak. 36 A4
Ghana country Africa 46 C4
Ghantila India 36 B5
Ghanwā Saudi Arabia 32 G4
Ghanzi Botswana 49 C6
Ghanzi admin. dist. Botswana 50 F2
Ghap'an Armenia see Kapan
Ghardaïa Alg. 22 E5
Gharghoda India 38 D1
Ghārib, Gebel mt. Egypt see Ghārib, Jabal
Ghārib, Jabal mt. Egypt 34 D5
Gharm Tajik. 33 L2
Gharq Ābād Iran 35 H4
Gharwa India see Garhwa
Gharyān Libya 47 E1
Ghāt Libya 46 E2
Ghatgan India 37 F5
Ghatol India 36 C5
Ghawdex i. Malta see Gozo
Ghazal, Bahr el watercourse Chad 47 E3
Ghazaouet Alg. 25 F6
Ghaziabad India 36 D3
Ghazi Ghat Pak. 36 B3
Ghazipur India 37 E4
Ghazna Afgh. see Ghaznī
Ghaznī Afgh. 36 B2
Ghaznī r. Afgh. 36 A2
Ghazzah Gaza see Gaza
Ghent Belgium 16 I5
Gheorghe Gheorghiu-Dej Romania see Onești
Gheorgheni Romania 27 K1
Gherla Romania 27 J1
Ghijduwon Uzbek. see G'ijduvon
Ghīnah, Wādī al watercourse Saudi Arabia 39 D4
Ghisonaccia Corse France 24 I5
Ghotaru India 36 B4
Ghotki Pak. 33 K4
Ghudamis Libya see Ghadāmis
Ghugri r. India 37 F4
Ghurayfah h. Saudi Arabia 39 C4
Ghūrī Iran 35 I5
Ghūrīan Afgh. 33 J3
Ghuzor Uzbek. see G'uzor
Giaginskaya Rus. Fed. 35 F1
Gialias r. Cyprus 39 A2
Gianisada i. Greece 27 L7
Giannitsa Greece 27 J4
Giant's Castle mt. S. Africa 51 I5
Giant's Causeway lava field U.K. 21 F2
Giarre Sicilia Italy 26 F6
Gibb r. Australia 54 D3
Gibeon Namibia 50 C3
Gibraltar terr. Europe 25 D5
Gibraltar Gibraltar 72 H3
Gibraltar, Strait of Morocco/Spain 25 C6
Gibraltar Range National Park Australia 58 F2
Gibson Australia 55 C8
Gibson Desert Australia 55 C6
Gichgeniyn Nuruu mts Mongolia 42 G3
Giddalur India 38 C3
Giddi, Gebel el h. Egypt see Jiddī, Jabal al
Gīdolē Eth. 47 G4
Gien France 24 F3
Gießen Germany 17 L5
Gifford r. Canada 61 J2
Gifu Japan 45 E6
Gigha i. U.K. 20 D5
Gīgīga Eth. see Jijiga
G'ijduvon Uzbek. 33 J1
Gijón Spain see Gijón-Xixón
Gijón-Xixón Spain 25 D2
Gila r. AZ U.S.A. 65 E4
Gila Bend AZ U.S.A. 65 F4
Gila Bend Mountains AZ U.S.A. 65 F4
Gīlān Afgh. 36 A2
Gīlān-e Gharb Iran 35 G4
Gilbert r. Australia 56 C3
Gilbert Islands Kiribati 74 H5
Gilbert Islands country Pacific Ocean see Kiribati
Gilbert Ridge sea feature Pacific Ocean 74 H6
Gilbert River Australia 56 C3
Gilbués Brazil 69 I5
Gil Chashmeh Iran 35 J4
Gilé Moz. 49 D5
Giles Creek r. Australia 54 E4
Gilgai Australia 58 E2
Gilgandra Australia 58 D3
Gil Gil Creek r. Australia 58 D2
Gilgit Pak. 36 C2
Gilgit r. Pak. 33 L2
Gilgit-Baltistan admin. div. Pak. 36 C1
Gilgunnia Australia 58 C4
Gılındire Turkey see Aydıncık
Gillam Canada 61 I4
Gillen, Lake imp. l. Australia 55 D6
Gilles, Lake imp. l. Australia 57 B7
Gillett PA U.S.A. 64 C2
Gillette U.S.A. 62 F3
Gilliat Australia 56 C4
Gillingham England U.K. 19 E7
Gillingham England U.K. 19 H7
Gilling West U.K. 18 F4
Gilmour Island Canada 61 K4
Gilroy CA U.S.A. 65 B2
Gīmbī Eth. 48 D3
Gimhae S. Korea see Kimhae
Gimli Canada 63 H1
Gimol'skoye, Ozero l. Rus. Fed. 12 G3
Ginebra, Laguna l. Bol. 68 E6
Gineifa Egypt see Junayfah
Gin Gin Australia 56 E5
Gingin Australia 55 A7
Gīnīr Eth. 48 E3
Ginosa Italy 26 G4
Ginzo de Limia Spain see Xinzo de Limia
Gioia del Colle Italy 26 G4
Gippsland reg. Australia 58 B7
Girā, Wādī watercourse Egypt see Jirā', Wādī
Girard PA U.S.A. 64 A1
Giresun Turkey 34 E2
Girgenti Sicilia Italy see Agrigento
Giridh India see Giridih
Giridih India 37 F4
Girilambone Australia 58 C3
Girna r. India 36 C5
Gir National Park India 36 B5
Girne Cyprus see Kyrenia
Girón Ecuador 68 C4
Giron Sweden see Kiruna
Girona Spain 25 H3
Gironde est. France 24 D4
Girot Pak. 36 C2
Girral Australia 58 C4
Girraween National Park Australia 58 E2
Girvan U.K. 20 E5
Girvas Rus. Fed. 12 G3
Gisborne N.Z. 59 G4
Gislaved Sweden 15 H8
Gissar Range mts Tajik./Uzbek. 33 K2
Gissarskiy Khrebet mts Tajik./Uzbek. see Gissar Range
Gitarama Rwanda 48 C4
Gitega Burundi 48 C4
Giuba r. Somalia see Jubba
Giulianova Italy 26 E3
Giurgiu Romania 27 K3
Giuvala, Pasul pass Romania 27 K2
Givar Iran 35 J3
Givors France 24 G4
Giyani S. Africa 51 J2
Giza Egypt 34 C5
Gizhiga Rus. Fed. 29 R3
Gjakovë Kosovo 27 I3
Gjilan Kosovo 27 I3
Gjirokastër Albania 27 I4
Gjirokastra Albania see Gjirokastër
Gjoa Haven Canada 61 I3
Gjøra Norway 14 F5
Gjøvik Norway 15 G6
Gkinas, Akrotirio pt Greece 27 M6
Glace Bay Canada 61 M5
Glacier Bay National Park and Preserve U.S.A. 60 E4
Glacier Peak vol. U.S.A. 62 C2
Gladstad Norway 14 G4
Gladstone Australia 56 E4
Gladys VA U.S.A. 64 B4
Glamis U.K. 20 F4
Glamis CA U.S.A. 65 E4
Glamoč Bos.-Herz. 26 G2
Glanton U.K. 18 F3
Glasgow U.K. 20 E5
Glasgow KY U.S.A. 63 J4
Glasgow MT U.S.A. 62 F2
Glasgow VA U.S.A. 64 B4
Glass, Loch l. U.K. 20 E3
Glass Mountain CA U.S.A. 65 C2
Glastonbury U.K. 19 E7
Glazov Rus. Fed. 12 L4
Gleiwitz Poland see Gliwice
Glen Allen VA U.S.A. 64 C4
Glen Alpine Dam S. Africa 51 I2
Glenamaddy Ireland 21 D4
Glenamoy r. Ireland 21 C3
Glenbawn, Lake Australia 58 E4
Glencoe Ont. Canada 64 A1
Glencoe S. Africa 51 J5
Glendale AZ U.S.A. 62 E5
Glendale CA U.S.A. 65 C3
Glendale UT U.S.A. 65 F2
Glendale Lake PA U.S.A. 64 B2
Glen Davis Australia 58 E4
Glenden Australia 56 E4
Glendive U.S.A. 62 G2
Glenfield NY U.S.A. 64 D1
Glengavlen Ireland 21 E3
Glengyle Australia 56 B5
Glen Innes Australia 58 E2
Glenluce U.K. 20 E6
Glen More val. U.K. 20 E3
Glenmorgan Australia 58 D1
Glennallen U.S.A. 60 D3
Glenore Australia 56 C3
Glenormiston Australia 56 B4
Glenreagh Australia 58 F3
Glenrothes U.K. 20 F4
Glens Falls NY U.S.A. 64 E1
Glen Shee val. U.K. 20 F4
Glenties Ireland 21 D3
Glenveagh National Park Ireland 21 E2
Glenville WV U.S.A. 64 A3
Glenwood Springs U.S.A. 62 F4
Glevum U.K. see Gloucester
Glittertinden mt. Norway 15 F6
Gliwice Poland 17 Q5
Globe U.S.A. 62 E5
Glogau Poland see Głogów
Głogów Poland 17 P5
Glomfjord Norway 14 H3
Glomma r. Norway 14 G7
Glommersträsk Sweden 14 K4
Glorieuses, Îles is Indian Ocean 49 E5
Glorioso Islands Indian Ocean see Glorieuses, Îles
Gloucester Australia 58 E3
Gloucester U.K. 19 E7
Gloucester MA U.S.A. 64 F1
Gloucester VA U.S.A. 64 C4
Gloversville NY U.S.A. 64 D1
Glubinnoye Rus. Fed. 44 D3
Glubokiy Krasnoyarskiy Kray Rus. Fed. 42 H2
Glubokiy Rostovskaya Oblast' Rus. Fed. 13 I6
Glubokoye Belarus see Hlybokaye
Glubokoye Kazakh. 42 E2
Gluggarnir h. Faroe Is 14 [inset]
Glukhov Ukr. see Hlukhiv
Glusburn U.K. 18 F5
Glynebwy U.K. see Ebbw Vale
Gmelinka Rus. Fed. 13 J6
Gmünd Austria 17 O6
Gmunden Austria 17 N7
Gnarp Sweden 15 J5
Gnesen Poland see Gniezno
Gniezno Poland 17 P4
Gnjilane Kosovo see Gjilan
Gnowangerup Australia 55 B8
Gnows Nest Range hills Australia 55 B7
Goa India 38 B3
Goa state India 38 B3
Goageb Namibia 50 C4
Goalen Head hd Australia 58 E6
Goalpara India 37 G4
Goat Fell h. U.K. 20 D5
Goba Eth. 48 E3
Gobabis Namibia 50 D2
Gobannium U.K. see Abergavenny
Gobas Namibia 50 D4
Gobi Desert des. China/Mongolia 42 J4
Gobindpur India 37 F5
Gobō Japan 45 D6
Gochas Namibia 50 D3
Godalming U.K. 19 G7
Godavari r. India 38 D2
Godavari, Cape India 38 D2
Godda India 37 F4
Godē Eth. 48 E3
Godere Eth. 48 E3
Goderich Canada 63 K3
Goderville France 19 H9
Gōd-e Zirah depr. Afgh. 33 J4
Godhavn Greenland see Qeqertarsuaq
Godhra India 36 C5
Gods r. Canada 61 I4
Gods Lake Canada 63 I1
Godthåb Greenland see Nuuk
Godwin-Austen, Mount China/Pakistan see K2
Goedgegun Swaziland see Nhlangano
Goegap Nature Reserve S. Africa 50 D5
Goélands, Lac aux l. Canada 61 L4
Gogra r. India see Ghaghara
Goiana Brazil 69 L5
Goiandira Brazil 71 A2
Goianésia Brazil 71 A1
Goiânia Brazil 71 A2
Goiás Brazil 71 A1
Goiás state Brazil 71 A2
Goio-Erê Brazil 70 F2
Gojra Pak. 36 C3
Gokak India 38 B2
Gokarn India 38 B3
Gök Çay r. Turkey 39 A1
Gökçeada i. Turkey 27 K4
Gökdere r. Turkey 39 A1
Goklenkuy, Solonchak salt l. Turkm. 35 J2
Gökova Körfezi b. Turkey 27 L6
Göksun Turkey 34 E3
Göksu Parkı Turkey 39 A1
Gokwe Zimbabwe 49 C5
Gol Norway 15 F6
Golaghat India 37 H4
Gölbaşı Turkey 34 E3
Gölcük Turkey 27 M4
Gold PA U.S.A. 64 C2
Gołdap Poland 17 S3
Gold Coast country Africa see Ghana
Gold Coast Australia 58 F2
Golden Bay N.Z. 59 D5
Golden Gate Highlands National Park S. Africa 51 I5
Golden Hinde mt. Canada 62 B2
Goldfield NV U.S.A. 65 D2
Goldsboro U.S.A. 63 L4
Goldstone Lake CA U.S.A. 65 D3
Goldsworthy (abandoned) Australia 54 B5
Goldvein VA U.S.A. 64 C3
Göle Turkey 35 F2
Goleta CA U.S.A. 65 C3
Golets-Davydov, Gora mt. Rus. Fed. 43 J2
Golfo di Orosei Gennargentu e Asinara, Parco Nazionale del nat. park Sardegna Italy 26 C4
Gölgeli Dağları mts Turkey 27 M6
Golingka China see Gongbo'gyamda
Gölköy Turkey 34 E2
Gollel Swaziland see Lavumisa
Golmud China 42 G5
Golovnino Rus. Fed. 44 G4
Golpāyegān Iran 35 H3
Gölpazarı Turkey 27 N4
Golspie U.K. 20 F3
Golyama Syutkya mt. Bulg. 27 K4
Golyam Persenk mt. Bulg. 27 K4
Golyshi Rus. Fed. see Vetluzhskiy
Goma Dem. Rep. Congo 48 C4
Gomang Co salt l. China 37 G3
Gomati r. India 33 N4
Gombe Nigeria 46 E3
Gombe r. Tanz. 49 D4
Gombi Nigeria 46 E3
Gombroon Iran see Bandar-e 'Abbās
Gomel' Belarus see Homyel'
Gómez Palacio Mex. 66 D3
Gomīshān Iran 35 I3
Gomo Co salt l. China 37 F2
Gonaïves Haiti 67 J5
Gonarezhou National Park Zimbabwe 49 D6
Gonbad-e Kāvūs Iran 35 I3
Gonda India 37 E4
Gondal India 36 B5
Gondar Eth. see Gonder
Gonder Eth. 48 D2
Gondia India 36 E5
Gondiya India see Gondia
Gönen Turkey 27 L4
Gonfreville-l'Orcher France 19 H9
Gongbalou China 37 G3
Gongbo'gyamda China 37 H3
Gongchang China see Longxi
Gongga Shan mt. China 42 I7
Gonghe Qinghai China 42 I5
Gongogi r. Brazil 71 D1
Gongolgon Australia 58 C3
Gongtang China see Damxung
Gonjog China see Coqên
Gonzales CA U.S.A. 65 B2
Gonzales TX U.S.A. 63 H6
Gonzha Rus. Fed. 44 B1
Goochland VA U.S.A. 64 C4
Goodenough, Cape Antarctica 76 G2
Goodenough Island P.N.G. 52 F2
Good Hope, Cape of S. Africa 50 D8
Goodooga Australia 58 C2
Goodspeed Nunataks Antarctica 76 E2
Goole U.K. 18 G5
Goolgowi Australia 58 B5
Goolma Australia 58 D4
Gooloogong Australia 58 D4
Goomalling Australia 55 B7
Goombalie Australia 58 B2
Goondiwindi Australia 58 E2
Goongarrie, Lake imp. l. Australia 55 C7
Goongarrie National Park Australia 55 C7
Goonyella Australia 56 D4
Goorly, Lake imp. l. Australia 55 B7
Goose Bay Canada see Happy Valley-Goose Bay
Goose Lake U.S.A. 62 C3
Gooty India 38 C3
Gopalganj Bangl. 37 G5
Gopalganj India 37 F4
Gopeshwar India 36 D3
Göppingen Germany 17 L6
Gora Bol'shaya Kovriga h. Rus. Fed. 12 K2
Gorakhpur India 37 E4
Gora Okpan h. Kazakh. 35 H1
Gora Pervyy Brat h. Rus. Fed. 44 F1
Gora Tardoki-Yangi mt. Rus. Fed. 44 E2
Goražde Bos.-Herz. 26 H3
Gorczański Park Narodowy nat. park Poland 17 R6
Gördes Turkey 27 M5
Gordil Cent. Afr. Rep. 48 C3
Gordon U.K. 20 G5
Gordon, Lake Australia 57 [inset]
Gordon Downs (abandoned) Australia 54 E4
Gordon Lake PA U.S.A. 64 B3
Gordonsville VA U.S.A. 64 B3
Goré Chad 47 E4
Gorē Eth. 48 D3
Gore N.Z. 59 B8
Gore VA U.S.A. 64 B3
Gorebridge U.K. 20 F5
Gore Point U.S.A. 60 C4
Gorey Ireland 21 F5
Gorgān Iran 35 I3
Gorgān, Khalīj-e Iran 35 I3
Gorge Range hills Australia 54 B5
Gorgona, Isla i. Col. 68 C3
Gori Georgia 32 F1
Goris Armenia 35 G3
Gorizia Italy 26 E2
Gorki Belarus see Horki
Gor'kiy Rus. Fed. see Nizhniy Novgorod
Gor'kovskoye Vodokhranilishche resr Rus. Fed. 12 I4
Gorlice Poland 13 D6
Görlitz Germany 17 O5
Gorlovka Ukr. see Horlivka
Gorna Dzhumaya Bulg. see Blagoevgrad
Gorna Oryakhovitsa Bulg. 27 K3
Gornji Milanovac Serbia 27 I2
Gornji Vakuf Bos.-Herz. 26 G3
Gorno-Altaysk Rus. Fed. 42 F2
Gornotrakiyska Nizina lowland Bulg. 27 K3
Gornozavodsk Permskiy Kray Rus. Fed. 11 R4
Gornozavodsk Sakhalinskaya Oblast' Rus. Fed. 44 F3
Gornyak Rus. Fed. 42 E2
Gornyy Rus. Fed. 13 K6
Gornyye Klyuchi Rus. Fed. 44 D3
Goro i. Fiji see Koro
Gorodenka Ukr. see Horodenka
Gorodets Rus. Fed. 12 I4
Gorodishche Penzenskaya Oblast' Rus. Fed. 13 J5
Gorodishche Volgogradskaya Oblast' Rus. Fed. 13 J6
Gorodok Belarus see Haradok
Gorodok Rus. Fed. see Zakamensk
Gorodok Khmel'nyts'ka Oblast' Ukr. see Horodok
Gorodok L'vivs'ka Oblast' Ukr. see Horodok
Gorodovikovsk Rus. Fed. 13 I7
Goroka P.N.G. 52 E2
Gorokhovets Rus. Fed. 12 I4
Gorom Gorom Burkina Faso 46 C3
Gorongosa mt. Moz. 49 D5
Gorongosa, Parque Nacional da nat. park Moz. 49 D5
Gorontalo Indon. 41 E7
Gorshechnoye Rus. Fed. 13 H6
Gort Ireland 21 D4
Gort an Choirce Ireland 21 D2
Gorutuba r. Brazil 71 C1
Goryachiy Klyuch Rus. Fed. 35 E1
Gory Shu-Ile mts Kazakh. 42 C4
Gorzów Wielkopolski Poland 17 O4
Gosainthan mt. China see Xixabangma Feng
Gosforth U.K. 18 F3
Goshen CA U.S.A. 65 C2
Goshen NH U.S.A. 64 E1
Goshen NY U.S.A. 64 D2
Goshen VA U.S.A. 64 B4
Goshoba Turkm. see Goşoba
Goşoba Turkm. 35 I2
Gospić Croatia 26 F2
Gosport U.K. 19 F8
Gossi Mali 46 C3
Gostivar Macedonia 27 I4
Göteborg Sweden see Gothenburg
Götene Sweden 15 H7
Gotenhafen Poland see Gdynia
Gotha Germany 17 M5
Gothenburg Sweden 15 G8
Gotland i. Sweden 15 K8
Gotō-rettō is Japan 45 C6
Gotse Delchev Bulg. 27 J4
Gotska Sandön i. Sweden 15 K7
Gōtsu Japan 45 D6
Göttingen Germany 17 L5
Gott Peak Canada 62 C1
Gottwaldow Czech Rep. see Zlín
Gouda Neth. 16 J4
Goudiri Senegal 46 B3
Goudoumaria Niger 46 E3
Goûgaram Niger 46 D3
Gough Island S. Atlantic Ocean 72 H8
Gouin, Réservoir resr Canada 63 M2
Goulburn Australia 58 D5
Goulburn r. N.S.W. Australia 58 E4
Goulburn r. Vic. Australia 58 B6
Goulburn Islands Australia 54 F2
Goulburn River National Park Australia 58 E4
Gould Coast Antarctica 76 J1
Goulou atoll Micronesia see Ngulu
Goundam Mali 46 C3
Goundi Chad 47 E4
Gouraya Alg. 25 G5
Gourcy Burkina Faso 46 C3
Gourdon France 24 E4
Gouré Niger 46 E3
Gouripur Bangl. 37 G4
Gourits r. S. Africa 50 E8
Gourma-Rharous Mali 46 C3
Gournay-en-Bray France 24 E2
Governador Valadares Brazil 71 C2
Governor's Harbour Bahamas 67 I3
Govĭ Altayn Nuruu mts Mongolia 42 H4
Govind Ballash Pant Sagar resr India 37 E4
Gowanda NY U.S.A. 64 B1
Gowan Range hills Australia 56 D5
Gowārān Afgh. 36 A3
Gowd-e Mokh l. Iran 35 I5
Gowmal Kalay Afgh. 36 B2
Gowna, Lough l. Ireland 21 E4
Goya Arg. 70 E3
Göyçay Azer. 35 G2
Goyder watercourse Australia 55 F6
Göygöl Azer. 35 G2
Goymatdag hills Turkm. see Goýmatdag
Goýmatdag hills Turkm. 35 I2
Göynük Turkey 27 N4
Goyoum Cameroon 46 E4
Goz-Beïda Chad 47 F3
Gozha Co salt l. China 36 E2
Gözkaya Turkey 39 C1
Gozo i. Malta 26 F6
Graaff-Reinet S. Africa 50 G7
Grabo Côte d'Ivoire 46 C4
Grabouw S. Africa 50 D8
Gračac Croatia 26 F2
Gradaús, Serra dos hills Brazil 69 H5
Gradiška Bos.-Herz. see Bosanska Gradiška
Grafton Australia 58 F2
Grafton WV U.S.A. 64 A3
Grafton, Cape Australia 56 D3
Grafton, Mount NV U.S.A. 65 E1
Grafton Passage Australia 56 D3
Graham TX U.S.A. 62 H5
Graham Bell Island Rus. Fed. see Greem-Bell, Ostrov
Graham Island B.C. Canada 60 E4
Graham Island Nunavut Canada 61 I2
Graham Land pen. Antarctica 76 L2
Grahamstown S. Africa 51 H7
Grahovo Bos.-Herz. see Bosansko Grahovo
Graigue Ireland 21 F5
Grajaú Brazil 69 I5
Grajaú r. Brazil 69 J4
Grammos mt. Greece 27 I4
Grampian Mountains U.K. 20 E4
Grampians National Park Australia 57
Granada Nicaragua 67 G6
Granada Spain 25 E5
Granard Ireland 21 E4
Granby Canada 63 M2
Gran Canaria i. Canary Is 46 B2
Gran Chaco reg. Arg./Para. 70 D3
Grand r. MO U.S.A. 62 J3
Grand Atlas mts Morocco see Haut Atla
Grand Bahama i. Bahamas 67 I3
Grand Ballon mt. France 17 K7
Grand Bank Canada 61 M5
Grand Banks of Newfoundland sea fea N. Atlantic Ocean 72 E3
Grand-Bassam Côte d'Ivoire 46 C4
Grand Bend Ont. Canada 64 A1
Grand Canal Ireland 21 E4
Grand Canary i. Canary Is see Gran Canaria
Grand Canyon U.S.A. 62 E4
Grand Canyon gorge AZ U.S.A. 65 F2
Grand Canyon National Park AZ U.S.A. 65 F2
Grand Cayman i. Cayman Is 67 H5
Grand Drumont mt. France 17 K7
Grande r. Bahia Brazil 71 B1
Grande r. São Paulo Brazil 71 A3
Grande r. Nicaragua 67 H6
Grande, Bahía b. Arg. 70 C8
Grande, Ilha i. Brazil 71 B3
Grande Comore i. Comoros see Ngazic
Grande Prairie Canada 60 G4
Grand Erg de Bilma des. Niger 46 E3
Grand Erg Occidental des. Alg. 22 D5
Grand Erg Oriental des. Alg. 22 F6
Grande-Rivière Canada 63 O2
Grandes, Salinas salt flat Arg. 70 C4
Grand Falls N.B. Canada 63 N2
Grand Falls-Windsor Nfld. and Lab. Canada 61 M5
Grand Forks U.S.A. 63 H2
Grand Gorge NY U.S.A. 64 D1
Grandioznyy, Pik mt. Rus. Fed. 42 H2
Grand Island U.S.A. 62 H3
Grand Isle U.S.A. 63 I6
Grand Junction U.S.A. 62 F4
Grand-Lahou Côte d'Ivoire 46 C4
Grand Lake N.B. Canada 63 N2
Grand Manan Island Canada 63 N3
Grand Marais MN U.S.A. 63 I2
Grândola Port. 25 B4
Grand Passage New Caledonia 53 G3
Grand Rapids Canada 62 H1
Grand Rapids MI U.S.A. 63 J3
Grand Rapids MN U.S.A. 63 I2
Grand-Sault Canada see Grand Falls
Grand St-Bernard, Col du pass Italy/Sw see Great St Bernard Pass
Grand Teton mt. U.S.A. 62 E3
Grand Turk Turks and Caicos Is 67 J4
Grand Wash Cliffs mts AZ U.S.A. 65 E3
Grange Ireland 21 E6
Grängesberg Sweden 15 I6
Grangeville U.S.A. 62 D2
Granite Mountains CA U.S.A. 65 E3
Granite Mountains CA U.S.A. 65 E4
Granite Peak MT U.S.A. 62 F2
Granitola, Capo c. Sicilia Italy 26 E6
Granja Brazil 69 J4
Gran Laguna Salada l. Arg. 70 C6
Gränna Sweden 15 I7
Gran Paradiso mt. Italy 26 B2
Gran Paradiso, Parco Nazionale del nat. park Italy 26 B2
Gran Pilastro mt. Austria/Italy 17 M7
Gran San Bernardo, Colle del pass Italy Switz. see Great St Bernard Pass
Gran Sasso e Monti della Laga, Parco Nazionale del nat. park Italy 26 E3
Grantham U.K. 19 G6
Grant Island Antarctica 76 J2
Grantown-on-Spey U.K. 20 F3
Grant Range mts NV U.S.A. 65 E1
Grants U.S.A. 62 F4
Grants Pass U.S.A. 62 C3
Grantsville WV U.S.A. 64 A3
Grantville PA U.S.A. 64 C2
Granville France 24 D2
Granville NY U.S.A. 64 E1
Granville Lake Canada 61 H4
Grão Mogol Brazil 71 C2
Grapevine Mountains NV U.S.A. 65 D2
Graskop S. Africa 51 J3
Grasplatz Namibia 50 B4
Grasse France 24 H5
Grassflat PA U.S.A. 64 B2
Grassington U.K. 18 F4
Grass Valley CA U.S.A. 65 B1
Grästorp Sweden 15 H7
Graudenz Poland see Grudziądz
Graus Spain 25 G2
Gravataí Brazil 71 A5
Grave, Pointe de pt France 24 D4
Gravelotte S. Africa 51 J2
Gravesend Australia 58 E2
Gravesend U.K. 19 H7
Gravina in Puglia Italy 26 G4
Gray France 24 G3
Grays U.K. 19 H7
Graz Austria 17 O7
Great Abaco i. Bahamas 67 I3
Great Australian Bight g. Australia 55 E
Great Baddow U.K. 19 H7
Great Bahama Bank sea feature Baham 67 I3
Great Barrier Island N.Z. 59 E3
Great Barrier Reef Australia 56 D1
Great Barrier Reef Marine Park (Cairns Section) Australia 56 D3

eat Barrier Reef Marine Park (Capricorn Section) Australia 56 E4
eat Barrier Reef Marine Park (Central Section) Australia 56 E3
eat Barrier Reef Marine Park (Far North Section) Australia 56 D2
eat Barrington MA U.S.A. 64 E1
eat Basalt Wall National Park Australia 56 D3
eat Basin U.S.A. 62 D4
eat Basin National Park NV U.S.A. 65 E1
eat Bear Lake Canada 60 G3
eat Belt sea chan. Denmark 15 G9
eat Bend U.S.A. 62 H4
eat Bitter Lake Egypt 39 A4
eat Blasket Island Ireland 21 B5
eat Britain i. U.K. 16 G4
eat Clifton U.K. 18 D4
eat Cumbrae i. U.K. 20 E5
eat Dividing Range mts Australia 58 B6
eat Eastern Erg des. Alg. see Grand Erg Oriental
eater Antarctica reg. Antarctica see East Antarctica
eater Antilles is Caribbean Sea 67 H4
eater Khingan Mountains China see Da Hinggan Ling
eat Exuma i. Bahamas 67 I4
eat Falls U.S.A. 62 E2
eat Fish r. S. Africa 51 H7
eat Fish Point S. Africa 51 H7
eat Fish River Reserve Complex nature res. S. Africa 51 H7
eat Gandak r. India 37 F4
eat Ganges atoll Cook Is see Manihiki
eat Inagua i. Bahamas 67 J4
eat Karoo plat. S. Africa 50 F7
eat Kei r. S. Africa 51 I7
eat Lake Australia 57 [inset]
eat Malvern U.K. 19 E6
eat Meteor Tablemount sea feature N. Atlantic Ocean 72 G4
eat Namaqualand reg. Namibia 50 C4
eat Nicobar i. India 31 I6
eat Ormes Head hd U.K. 18 D5
eat Ouse r. U.K. 19 H6
eat Oyster Bay Australia 57 [inset]
eat Palm Islands Australia 56 D3
eat Plain of the Koukdjuak Canada 61 K3
eat Point MA U.S.A. 64 F2
eat Rift Valley Africa 48 D4
eat Ruaha r. Tanz. 49 D4
eat Sacandaga Lake NY U.S.A. 64 D1
eat St Bernard Pass Italy/Switz. 26 B2
eat Salt Lake U.S.A. 62 E3
eat Salt Lake Desert U.S.A. 62 E3
eat Sand Sea des. Egypt/Libya 34 B5
eat Sandy Desert Australia 54 C5
eat Sandy Island Australia see Fraser Island
eat Sea Reef Fiji 53 H3
eat Slave Lake Canada 60 G3
eatstone-on-Sea U.K. 19 H8
eat Stour r. U.K. 19 I7
eat Torrington U.K. 19 C8
eat Victoria Desert Australia 55 E7
eat Wall research stn Antarctica 76 A2
eat Wall tourist site China 43 L4
eat Waltham U.K. 19 H7
eat Western Erg des. Alg. see Grand Erg Occidental
eat Whernside h. U.K. 18 F4
eat Yarmouth U.K. 19 I6
ebenkovskiy Ukr. see Hrebinka
ebyonka Ukr. see Hrebinka
eco, Cape Cyprus see Greko, Cape
edos, Sierra de mts Spain 25 D3
eece country Europe 27 I5
eece NY U.S.A. 64 C1
eeley U.S.A. 62 G3
eem-Bell, Ostrov i. Rus. Fed. 28 H1
een r. WY U.S.A. 62 F4
een Bay U.S.A. 63 J3
een Bay U.S.A. 63 J2
eenbrier r. WV U.S.A. 64 A4
een Cape Australia 58 E6
eencastle U.K. 21 F3
eene NY U.S.A. 64 D1
eeneville U.S.A. 63 K4
eenfield CA U.S.A. 65 B2
eenfield MA U.S.A. 64 E1
een Head hd Australia 55 A7
eenhill Island Australia 54 F2
een Lake Canada 62 F1
eenland terr. N. America 61 N3
eenland Basin sea feature Arctic Ocean 77 I2
eenland Fracture Zone sea feature Arctic Ocean 77 I1
eenland Sea Greenland/Svalbard 28 A2
eenlaw U.K. 20 G5
eenock U.K. 20 E5
eenore Ireland 21 F3
eenport NY U.S.A. 64 E2
een River WY U.S.A. 62 F3
eensburg PA U.S.A. 64 B2
eenstone Point U.K. 20 D3
eenville Liberia 46 C4
eenville AL U.S.A. 63 J5
eenville MS U.S.A. 63 I5
eenville NH U.S.A. 64 F1
eenville PA U.S.A. 64 A2
eenville SC U.S.A. 63 K5
eenville TX U.S.A. 63 H5
eenwich atoll Micronesia see Kapingamarangi
eenwich CT U.S.A. 64 E2
eenwood SC U.S.A. 63 K5
egory r. Australia 56 B3
egory, Lake imp. l. S.A. Australia 57 B6
egory, Lake imp. l. W.A. Australia 54 D5
egory, Lake imp. l. W.A. Australia 55 B6
egory Downs Australia 56 B3
egory National Park Australia 54 E4
egory Range hills Qld Australia 56 C3
egory Range hills W.A. Australia 54 C5
eifswald Germany 17 N3
eko, Cape Cyprus 39 B2
emyachinsk Rus. Fed. 11 R4
enå Denmark 15 G8
enaa Denmark see Grenå
enada U.S.A. 63 J5
enada country West Indies 67 L6
enade France 24 E5
enen spit Denmark 15 G8
enfell Australia 58 D4
enfell Canada 62 G1
enoble France 24 G4
ense-Jakobselv Norway 14 Q2
enville, Cape Australia 56 C1
enville Island Fiji see Rotuma
eshak Pak. 36 A4
essåmoen Nasjonalpark nat. park Norway 14 H4
eta r. U.K. 18 E4
etna U.K. 20 F6
etna VA U.S.A. 64 B4
evena Greece 27 I4
evesmühlen Germany 17 M4
Grey, Cape Australia 56 B2
Greybull U.S.A. 62 F3
Grey Hunter Peak Canada 60 E3
Greylock, Mount MA U.S.A. 64 E1
Greymouth N.Z. 59 C6
Grey Range hills Australia 58 A2
Grey's Plains Australia 55 A6
Greytown S. Africa 51 J5
Gribanovskiy Rus. Fed. 13 I6
Griffin U.S.A. 63 K5
Griffith Australia 58 C5
Grim, Cape Australia 57 [inset]
Grimari Cent. Afr. Rep. 48 C3
Grimmen Germany 17 N3
Grimsby U.K. 18 G5
Grímsey i. Iceland 14 [inset]
Grimshaw Canada 60 G4
Grímsstaðir Iceland 14 [inset]
Grimstad Norway 15 F7
Grindavík Iceland 14 [inset]
Grindsted Denmark 15 F9
Grindul Chituc spit Romania 27 M2
Grinnell Peninsula Canada 61 I2
Griqualand East reg. S. Africa 51 I6
Griqualand West reg. S. Africa 50 F5
Griquatown S. Africa 50 F5
Grise Fiord Canada 61 J2
Grishino Ukr. see Krasnoarmiys'k
Gris Nez, Cap c. France 19 I8
Gritley U.K. 20 G2
Grmeč mts Bos.-Herz. 26 G2
Groblersdal S. Africa 51 I3
Groblershoop S. Africa 50 F5
Grodno Belarus see Hrodna
Groen watercourse S. Africa 50 F6
Groen watercourse S. Africa 50 C6
Groix, Île de i. France 24 C3
Grombalia Tunisia 26 D6
Grong Norway 14 H4
Groningen Neth. 17 K4
Grønland terr. N. America see Greenland
Groom Lake NV U.S.A. 65 E2
Groot-Aar Pan salt pan S. Africa 50 E4
Groot Berg r. S. Africa 50 D7
Groot Brakrivier S. Africa 50 F8
Grootdraaidam dam S. Africa 51 I4
Grootdrink S. Africa 50 E5
Groote Eylandt i. Australia 56 B2
Grootfontein Namibia 49 B5
Groot Karas Berg plat. Namibia 50 D4
Groot Letaba r. S. Africa 51 J2
Groot Marico S. Africa 51 H3
Groot Swartberge mts S. Africa 50 E7
Grootvloer salt pan S. Africa 50 E5
Groot Winterberg mt. S. Africa 51 H7
Gross Barmen Namibia 50 C2
Großer Rachel mt. Germany 17 N6
Grosser Speikkogel mt. Austria 17 O7
Grosseto Italy 26 D3
Grossevichi Rus. Fed. 44 E3
Groß-Gerau Germany 17 L6
Großglockner mt. Austria 17 N7
Gross Ums Namibia 50 D2
Großvenediger mt. Austria 17 N7
Grottoes VA U.S.A. 64 B3
Groundhog r. Canada 63 K2
Grove Mountains Antarctica 76 E2
Grover Beach CA U.S.A. 65 B3
Growler Mountains AZ U.S.A. 65 F4
Groznyy Rus. Fed. 13 J8
Grubišno Polje Croatia 26 G2
Grudovo Bulg. see Sredets
Grudziądz Poland 17 Q4
Grünau Namibia 50 D4
Grünberg Poland see Zielona Góra
Grundarfjörður Iceland 14 [inset]
Gruzinskaya S.S.R. country Asia see Georgia
Gryazi Rus. Fed. 13 H5
Gryazovets Rus. Fed. 12 I4
Gryfice Poland 17 O4
Gryfino Poland 17 O4
Gryfów Śląski Poland 17 O5
Gryllefjord Norway 14 J2
Grytviken S. Georgia 70 I8
Gua India 37 F5
Guacanayabo, Golfo de b. Cuba 67 I4
Guadajoz r. Spain 25 D5
Guadalajara Mex. 66 D4
Guadalajara Spain 25 E3
Guadalcanal i. Solomon Is 53 G2
Guadalete r. Spain 25 C5
Guadalope r. Spain 25 F3
Guadalquivir r. Spain 25 C5
Guadalupe i. Mex. 66 A3
Guadalupe CA U.S.A. 65 B3
Guadalupe, Sierra de mts Spain 25 D4
Guadalupe Peak U.S.A. 62 G5
Guadalupe Victoria Durango Mex. 62 G7
Guadarrama, Sierra de mts Spain 25 D3
Guadeloupe terr. West Indies 67 L5
Guadeloupe Passage Caribbean Sea 67 L5
Guadiana r. Port./Spain 25 C5
Guadix Spain 25 E5
Guafo, Isla i. Chile 70 B6
Guaíba Brazil 71 A5
Guaiçuí Brazil 71 B2
Guaíra Brazil 70 F2
Gualala CA U.S.A. 65 A1
Gualeguay Arg. 70 E4
Gualeguaychu Arg. 70 E4
Gualicho, Salina salt flat Arg. 70 C6
Guam terr. N. Pacific Ocean 41 G6
Guamblin, Isla i. Chile 70 A6
Guampí, Sierra de mts Venez. 68 E2
Guamúchil Mex. 66 C3
Guanajuato Mex. 66 D4
Guanambi Brazil 71 C1
Guanare Venez. 68 E2
Guane Cuba 67 H4
Guangdong prov. China 43 K8
Guanghua China see Laohekou
Guangxi aut. reg. China see Guangxi Zhuangzu Zizhiqu
Guangxi Zhuangzu Zizhiqu aut. reg. China 43 J8
Guangyuan China 42 J6
Guangzhou China 43 K8
Guanhães Brazil 71 C2
Guanipa r. Venez. 68 F2
Guanshui China 44 B4
Guantánamo Cuba 67 I4
Guapé Brazil 71 B3
Guapí Col. 68 C3
Guaporé r. Bol./Brazil 68 E6
Guaporé Brazil 71 A5
Guaqui Bol. 68 E7
Guará r. Brazil 71 B1
Guarabira Brazil 69 K5
Guaranda Ecuador 68 C4
Guarapari Brazil 71 C3
Guarapuava Brazil 71 A4
Guararapes Brazil 71 A3
Guaratinguetá Brazil 71 B3
Guaratuba Brazil 71 A4
Guaratuba, Baía de b. Brazil 71 A4
Guarda Port. 25 C3
Guardafui, Cape Somalia see Gwardafuy, Gees
Guardiagrele Italy 26 F3
Guardo Spain 25 D2
Guárico, del Embalse resr Venez. 68 E2
Guarujá Brazil 71 B3
Guasave Mex. 62 F6
Guasdualito Venez. 68 D2
Guatemala country Central America 66 F5
Guatemala Guat. 66 F6
Guaviare r. Col. 68 E3
Guaxupé Brazil 71 B3
Guayaquil Ecuador 68 C4
Guayaquil, Golfo de g. Ecuador 68 B4
Guaymas Mex. 66 B3
Guba Eth. 48 D2
Guba Buor-Khaya b. Rus. Fed. 29 O2
Gubakha Rus. Fed. 11 R4
Gubbi India 38 C3
Gubio Nigeria 46 E3
Gubkin Rus. Fed. 13 H6
Gucheng China 43 K6
Gudari India 38 D2
Gudbrandsdalen val. Norway 15 F6
Gudermes Rus. Fed. 13 J8
Gudivada India 38 D2
Gudiyattam India 38 C3
Gudur Andhra Prad. India 38 C3
Gudur Andhra Prad. India 38 C3
Gudvangen Norway 15 E6
Gudzhal r. Rus. Fed. 44 D2
Guecho Spain see Algorta
Guéckédou Guinea 46 B4
Guelma Alg. 26 B6
Guelmim Morocco 46 B2
Guelph Ont. Canada 64 A1
Guerara Alg. 22 E5
Guercif Morocco 22 D5
Guéret France 24 E3
Guernsey terr. Channel Is 19 E9
Guérou Mauritania 46 B3
Guerrah Et-Tarf salt pan Alg. 26 B7
Guerrero Negro Mex. 62 E6
Guers, Lac l. Canada 61 L4
Gueugnon France 24 G3
Gugē mt. Eth. 48 D3
Gūgerd, Kūh-e mts Iran 35 I4
Guiana Basin sea feature N. Atlantic Ocean 72 E5
Guiana Highlands mts S. America 68 E2
Guidan-Roumji Niger 46 D3
Guider Cameroon 47 E4
Guidonia-Montecelio Italy 26 E4
Guigang China 43 J8
Guiglo Côte d'Ivoire 46 C4
Guija Moz. 51 K3
Guildford U.K. 19 G7
Guilherme Capelo Angola see Cacongo
Guilin China 43 K7
Guillaume-Delisle, Lac l. Canada 61 K4
Guimarães Brazil 69 J4
Guimarães Port. 25 B3
Guinea country Africa 46 B3
Guinea, Gulf of Africa 46 D4
Guinea Basin sea feature N. Atlantic Ocean 72 H5
Guinea-Bissau country Africa 46 B3
Guinea-Conakry country Africa see Guinea
Guinea Ecuatorial country Africa see Equatorial Guinea
Guiné-Bissau country Africa see Guinea-Bissau
Guinée country Africa see Guinea
Güines Cuba 67 H4
Guingamp France 24 C2
Guipavas France 24 B2
Guiratinga Brazil 69 H7
Guiyang Guizhou China 42 J7
Guizhou prov. China 42 J7
Gujarat state India 36 C5
Gujar Khan Pak. 36 C2
Gujerat state India see Gujarat
Gujranwala Pak. 33 L3
Gujrat Pak. 33 L3
Gukovo Rus. Fed. 13 H6
Gulabgarh India 36 D2
Gulbarga India 38 C2
Gulbene Latvia 15 O8
Gul'cha Kyrg. see Gülchö
Gülchö Kyrg. 42 C4
Gülcihan Turkey 39 B1
Gülek Boğazı pass Turkey 34 D3
Gulfport U.S.A. 63 J5
Gulian China 44 A1
Guliston Uzbek. 33 K1
Guliya Shan mt. China 44 A2
Gulja China see Yining
Gul Kach Pak. 36 B3
Gul'kevichi Rus. Fed. 35 F1
Gull Lake Canada 62 F1
Gullträsk Sweden 14 L3
Güllük Körfezi b. Turkey 27 L6
Gülnar Turkey 39 A1
Gulu Uganda 48 D3
Guluwuru Island Australia 56 B1
Gulyayevskiye Koshki, Ostrova is Rus. Fed. 12 L1
Guma China see Pishan
Gumal r. Pak. 33 L3
Gumare Botswana 49 C5
Gumbaz Pak. 36 B3
Gumbinnen Rus. Fed. see Gusev
Gumdag Turkm. 35 I3
Gumel Nigeria 46 D3
Gümgüm Turkey see Varto
Gumla India 37 F5
Gümüşhacıköy Turkey 34 D2
Gümüşhane Turkey 35 E2
Guna India 36 D4
Guna Terara mt. Eth. 32 E7
Gunbar Australia 58 B5
Gunbower Australia 58 B5
Güncang China 37 H3
Gundabooka National Park Austr. 57 D6
Gundagai Australia 58 D5
Güney Turkey 27 M5
Güneydoğu Toroslar mts Turkey 34 F3
Gungu Dem. Rep. Congo 49 B4
Gunib Rus. Fed. 35 G2
Gunnaur India 36 D3
Gunnbjørn Fjeld nunatak Greenland 61 P3
Gunnedah Australia 58 E3
Gunning Australia 58 D5
Gunnison U.S.A. 62 F4
Güns Hungary see Kőszeg
Guntakal India 38 C3
Guntur India 38 D2
Gunungsitoli Indon. 41 B7
Günyüzü Turkey 34 C3
Gunza Angola see Porto Amboim
Günzburg Germany 17 M6
Guovdageaidnu Norway see Kautokeino
Gupis Pak. 36 C1
Gurbantünggüt Shamo des. China 42 F4
Gurdaspur India 36 C2
Gurdzhaani Georgia see Gurjaani
Güre Turkey 27 M5
Gurgan Iran see Gorgān
Gurgaon India 36 D3
Gurgei, Jebel mt. Sudan 47 F3
Gurha India 36 B4
Guri, Embalse de resr Venez. 68 F2
Gurig National Park Australia 54 F2
Gurinhatã Brazil 71 A2
Gurjaani Georgia 35 G2
Guro Moz. 49 D5
Guru China 37 G3
Gürün Turkey 34 E3
Gurupá Brazil 69 H4
Gurupi Brazil 69 I6
Gurupi r. Brazil 69 I4
Gurupi, Serra do hills Brazil 69 I4
Guru Sikhar mt. India 36 C4
Guruzala India 38 C2
Gur'yev Kazakh. see Atyrau
Gur'yevsk Rus. Fed. 15 L9
Gur'yevskaya Oblast' admin. div. Kazakh. see Atyrauskaya Oblast'
Gusau Nigeria 46 D3
Gusev Rus. Fed. 15 M9
Gushan China 45 A5
Gusino Rus. Fed. 13 F5
Gusinoozersk Rus. Fed. 42 J2
Gus'-Khrustal'nyy Rus. Fed. 12 I5
Guspini Sardegna Italy 26 C5
Gustav Holm, Kap c. Greenland see Tasiilap Karra
Gustine CA U.S.A. 65 B2
Güstrow Germany 17 N4
Gütersloh Germany 17 L5
Gutsuo China 37 F3
Guwahati India 37 G4
Guwēr Iraq 35 F3
Guwlumayak Turkm. see Guwlumaýak
Guwlumaýak Turkm. 35 I2
Guyana country S. America 69 G2
Guyane Française terr. S. America see French Guiana
Guyang Nei Mongol China 43 K4
Guyenne reg. France 24 D4
Guy Fawkes River National Park Australia 58 F3
Guymon U.S.A. 62 G4
Guyra Australia 58 E3
Guyuan Hebei China 43 L4
Guyuan Ningxia China 42 J5
Güzeloluk Turkey 39 B1
Güzelyurt Cyprus see Morfou
Guzmán Mex. 66 C2
Guzmán, Lago de l. Mex. 66 C2
G'uzor Uzbek. 33 K2
Gvardeysk Rus. Fed. 15 L9
Gvasyugi Rus. Fed. 44 E3
Gwa Myanmar 42 G9
Gwabegar Australia 58 D3
Gwaii Haanas Canada 60 E4
Gwaii Haanas National Park Reserve and Haida Heritage Site Canada 60 E4
Gwal Haidarzai Pak. 36 B3
Gwalior India 36 D4
Gwanda Zimbabwe 49 C6
Gwane Dem. Rep. Congo 48 C3
Gwardafuy, Gees c. Somalia 48 F2
Gweebarra Bay Ireland 21 D3
Gwelo Zimbabwe see Gweru
Gweru Zimbabwe 49 C5
Gweta Botswana 49 C6
Gwoza Nigeria 46 E3
Gwydir r. Australia 58 D2
Gyablung China 37 H3
Gyaca China 37 H3
Gya'gya China see Saga
Gyaijêpozhanggê China see Zhidoi
Gyai Qu r. China 37 H3
Gyaisi China see Jiulong
Gyali i. Greece 27 L6
Gyamotang China see Dêngqên
Gyamug China 36 E2
Gyandzha Azer. see Gäncä
Gyangkar China see Dinngyê
Gyangnyi Caka salt l. China 37 F2
Gyangrang China 37 F3
Gyangtse China see Gyangzê
Gyangzê China 37 G3
Gyaring China 37 I2
Gyaring Co l. China 37 G3
Gyaring Hu l. China 42 H6
Gyaros i. Greece 27 K6
Gyarubtang China 42 G6
Gydan, Khrebet mts Rus. Fed. see Kolymskiy, Khrebet
Gydan Peninsula Rus. Fed. 28 I2
Gydanskiy Poluostrov pen. Rus. Fed. see Gydan Peninsula
Gyêgu China see Yushu
Gyêmdong China 37 H3
Gyigang China 42 H7
Gyimda China 37 H3
Gyirong Xizang China 37 F3
Gyirong Xizang China 37 F3
Gyiza China 37 H2
Gyldenløve Fjord inlet Greenland see Umiiviip Kangertiva
Gympie Australia 57 F5
Gyöngyös Hungary 17 Q7
Győr Hungary 26 G1
Gypsumville Canada 62 H1
Gyrfalcon Islands Canada 61 L4
Gytheio Greece 27 J6
Gyula Hungary 27 I1
Gyulafehérvár Romania see Alba Iulia
Gyümai China see Darlag
Gyumri Armenia 35 F2
Gzhatsk Rus. Fed. see Gagarin

H

Ha Bhutan 37 G4
Haa-Alif Atoll Maldives see Ihavandhippolhu Atoll
Ha'apai Group is Tonga 53 I3
Haapajärvi Fin. 14 N5
Haapavesi Fin. 14 N4
Haapsalu Estonia 15 M7
Ha 'Arava watercourse Israel/Jordan see 'Arabah, Wādī al
Ha'Arava, Nahal watercourse Israel/Jordan see Jayb, Wādī al
Haarlem Neth. 16 J4
Haarlem S. Africa 50 F7
Hab r. Pak. 36 A4
Habahe China 42 F3
Habana Cuba see Havana
Habarane Sri Lanka 38 D4
Habaswein Kenya 48 D3
Habbān Yemen 32 G7
Ḥabbānīyah, Hawr al l. Iraq 35 F4
Hab Chauki Pak. 36 A4
Habiganj Bangl. 37 G4
Habra India 37 G5
Hachijō-jima i. Japan 45 E6
Hachinohe Japan 44 F4
Hacıköy Turkey see Çekerek
Hack, Mount Australia 57 B6
Hackberry AZ U.S.A. 65 F3
Hackensack NJ U.S.A. 64 D2
Hacufera Moz. 49 D6
Haḍabat al Jilf al Kabīr plat. Egypt see Jilf al Kabīr, Haḍabat al
Hadagalli India 38 B3
Hadayang China 44 B2
Ḥadd, Ra's al pt Oman 33 I5
Haddington U.K. 20 G5
Hadejia Nigeria 46 E3
Hadera Israel 39 B3
Hadera r. Israel 39 B3
Haderslev Denmark 15 F9
Hadhramaut reg. Yemen see Ḥaḍramawt
Hādī, Jabal al mts Jordan 39 C4
Hadım Turkey 34 D3
Hadleigh U.K. 19 H6
Hadong S. Korea 45 B6
Ḥadraj, Wādī watercourse Saudi Arabia 39 C4
Ḥaḍramawt reg. Yemen 48 E2
Hadranum Sicilia Italy see Adrano
Hadrian's Wall tourist site U.K. 18 E3
Hadrumetum Tunisia see Sousse
Hadsund Denmark 15 G8
Hadyach Ukr. 13 G6
Haeju N. Korea 45 B5
Haeju-man b. N. Korea 45 B5
Haenam S. Korea 45 B6
Haenertsburg S. Africa 51 I2
Ha'erbin China see Harbin
Ḥafar al Bāṭin Saudi Arabia 32 G4
Hafik Turkey 34 E3
Ḥafīrah, Qā' al salt pan Jordan 39 C4
Hafizabad Pak. 36 C2
Haflong India 37 H4
Hafnarfjörður Iceland 14 [inset]
Hafren r. U.K. see Severn
Haft Gel Iran 35 H5
Hafursfjörður b. Iceland 14 [inset]
Haga Myanmar see Haka
Hagar Nish Plateau Eritrea 32 E6
Hagåtña Guam 41 G6
Hagen Germany 17 K5
Hagerstown MD U.S.A. 64 C3
Hagfors Sweden 15 H6
Hagi Japan 45 C6
Ha Giang Vietnam 31 J4
Hagley U.K. 19 E6
Hag's Head hd Ireland 21 C5
Haguenau France 24 H2
Hahajima-rettō is Japan 45 F8
Hai Tanz. 47 G5
Haib watercourse Namibia 50 C5
Haibowan China see Wuhai
Haicheng Liaoning China 43 M4
Haida Gwaii Canada 60 E4
Haifa Israel 39 B3
Haifa, Bay of Israel 39 B3
Haig Australia 55 D7
Haikakan country Asia see Armenia
Haikou China 43 K8
Ḥā'il Saudi Arabia 35 F6
Ḥā'il, Wādī watercourse Saudi Arabia 35 F6
Hailin China 44 C3
Hailong China see Meihekou
Hailsham U.K. 19 H8
Hailun China 44 B3
Hailuoto Fin. 14 N4
Hainan prov. China 43 J9
Hainan Dao i. China 43 K9
Hainan Strait China 43 J9
Haines U.S.A. 60 E4
Haines Junction Canada 60 E3
Haiphong Vietnam see Hai Phong
Hai Phong Vietnam 31 J4
Haiqing China 44 D3
Haitan Dao i. China 43 L7
Haiti country West Indies 67 J5
Haiwee Reservoir CA U.S.A. 65 D2
Haiya Sudan 32 E6
Haiyan Qinghai China 42 I5
Haiyang Dao i. China 45 A5
Haizhou Wan b. China 43 L6
Hajdúböszörmény Hungary 27 I1
Hajeb El Ayoun Tunisia 26 C7
Ḥajhir mt. Yemen 33 H7
Haji Mahesar Pak. 36 A3
Hajipur India 37 F4
Ḥajjah Yemen 32 F6
Ḥājjīābād Fārs Iran 35 I5
Ḥājjīābād Iran 35 I4
Haka Myanmar 37 H5
Hakha Myanmar see Haka
Hakkâri Turkey 35 F3
Hakkas Sweden 14 L3
Hakken-zan mt. Japan 45 D6
Hako-dake mt. Japan 44 F3
Hakodate Japan 44 F4
Hakos Mountains Namibia 50 C2
Hakseen Pan salt pan S. Africa 50 E4
Hakui Japan 45 E5
Haku-san vol. Japan 45 E5
Hala Pak. 36 B4
Ḥalab Syria see Aleppo
Ḥalabja Iraq 35 G4
Halaç Turkm. 33 J2
Halaç Turkm. see Halaç
Halahai China 44 B3
Halaib Sudan 32 E5
Halaib Triangle terr. Egypt/Sudan 32 E5
Halâl, Gebel h. Egypt see Hilāl, Jabal
Ḥalāniyāt, Juzur al is Oman 33 I6
Halba Lebanon 39 C2
Halberstadt Germany 17 M5
Halden Norway 15 G7
Haldensleben Germany 17 M4
Haldwani India 36 D3
Hale watercourse Australia 56 A5
Hāleh Iran 35 I6
Haleparki Deresi r. Syria/Turkey see Quwayq, Nahr
Halesowen U.K. 19 E6
Halesworth U.K. 19 I6
Half Assini Ghana 46 C4
Halfmoon Bay N.Z. 59 B8
Halfway Ireland 21 D6
Halhgol Mongolia 43 L3
Halia India 37 E4
Ḥalībīyah Syria 35 E4
Halicarnassus Turkey see Bodrum
Halifax Canada 61 L5
Halifax U.K. 18 F5
Halifax, Mount Australia 56 D3
Ḥalīmah mt. Lebanon/Syria 39 C2
Halkirk U.K. 20 F2
Hälla Sweden 14 J5
Halladale r. U.K. 20 F2
Halla-san National Park S. Korea 45 B6
Hall Beach Canada 61 J3
Halle (Saale) Germany 17 M5
Hällefors Sweden 15 I7
Hallein Austria 17 N7
Hallett, Cape Antarctica 76 H2
Halley research stn Antarctica 76 B1
Hallgreen, Mount Antarctica 76 B2
Hall Islands Micronesia 74 G5
Hällnäs Sweden 14 K4
Hallock U.S.A. 63 H2
Hall Peninsula Canada 61 L3
Hallsberg Sweden 15 I7
Halls Creek Australia 54 D4
Hallstead PA U.S.A. 64 D2
Hallviken Sweden 14 I5
Halmahera i. Indon. 41 E7
Halmstad Sweden 15 H8
Hals Denmark 15 G8
Hälsingborg Sweden see Helsingborg
Halsua Fin. 14 N5
Haltwhistle U.K. 18 E4
Haly, Mount h. Australia 58 E1
Hamada Japan 45 D6
Ḥamâda El Ḥaricha des. Mali 46 C2
Hamadān Iran 35 H4
Ḥamādat Murzuq plat. Libya 48 B1
Ḥamāh Syria 39 C2
Hamam Turkey 39 C1
Hamamatsu Japan 45 E6
Hamar Norway 15 G6
Hamarøy Norway 14 I2
Ḥamâṭa, Gebel mt. Egypt see Ḥamāṭah, Jabal
Ḥamāṭah, Jabal mt. Egypt 32 D5
Hamatonbetsu Japan 44 F3
Hambantota Sri Lanka 38 D5
Hambleton Hills U.K. 18 F4
Hamburg Germany 17 L4
Hamburg S. Africa 51 H7
Hamburg NY U.S.A. 64 B1
Hamburgisches Wattenmeer, Nationalpark nat. park Germany 17 L4
Ḥamḍ, Wādī al watercourse Saudi Arabia 32 E4
Hamden CT U.S.A. 64 E2
Hämeenlinna Fin. 15 N6
HaMelaḥ, Yam salt l. Asia see Dead Sea
Hamelin Australia 55 A6
Hameln Germany 17 L4
Hamersley Lakes imp. l. Australia 55 B7
Hamersley Range mts Australia 54 B5
Hamhŭng N. Korea 45 B5
Hami China 42 G4
Hamid Sudan 32 D5
Hamilton Qld Australia 56 C4
Hamilton S.A. Australia 57 A5
Hamilton Vic. Australia 57 C8
Hamilton watercourse Qld Australia 56 B4
Hamilton watercourse S.A. Australia 57 A5
Hamilton Bermuda 67 L2
Hamilton Ont. Canada 64 B1
Hamilton r. Canada see Churchill
Hamilton N.Z. 59 E3
Hamilton U.K. 20 E5
Hamilton MT U.S.A. 62 E2
Hamilton NY U.S.A. 64 D1
Hamilton OH U.S.A. 63 K4
Hamilton, Mount CA U.S.A. 65 B2
Hamilton, Mount NV U.S.A. 65 E1
Hamilton Mountain h. NY U.S.A. 64 D1
Hamīm, Wādī al watercourse Libya 23 I5
Hamina Fin. 15 O6
Hamirpur Hima. Prad. India 36 D3
Hamirpur Uttar Prad. India 36 E4
Hamitabat Turkey see Isparta
Hamju N. Korea 45 B5
Hamm Germany 17 K5
Ḥammām al 'Alīl Iraq 35 F3
Hammam Boughrara Alg. 25 F6
Hammamet Tunisia 26 D6
Hammamet, Golfe de g. Tunisia 26 D6
Ḥammār, Hawr al imp. l. Iraq 35 G5
Hammarstrand Sweden 14 J5
Hammerdal Sweden 14 I5
Hammerfest Norway 14 M1
Hammonton NJ U.S.A. 64 D3
Hampden Sydney VA U.S.A. 64 B4
Hampshire Downs hills U.K. 19 F7
Hampton NH U.S.A. 64 F1
Hampton VA U.S.A. 64 C4
Hampton Tableland reg. Australia 55 D8
Ḥamrā, Birkat al well Saudi Arabia 35 F5
Hamra, Vâdii watercourse Syria/Turkey see Ḥimār, Wādī al
Hamrat esh Sheikh Sudan 32 C7
Hamta Pass India 36 D2
Hāmūn-e Jaz Mūrīān imp. l. Iran 33 I4
Hāmūn-e Lowrah dry lake Afgh./Pak. see Hamun-i-Lora
Hāmūn-e Şāberī imp. l. Afgh./Iran 33 J3
Hamun-i-Lora dry lake Afgh./Pak. 33 K4
Hamun-i-Mashkel salt flat Pak. 33 J4
Hamur Turkey 35 F3
Hamwic U.K. see Southampton
Hanahai watercourse Botswana/Namibia 50 F2
Ḥanak Saudi Arabia 32 E4
Hanakpınar Turkey see Çınar
Hanamaki Japan 45 F5
Hanang mt. Tanz. 49 D4
Hanbin China see Ankang
Hancheng China 43 K5
Hancock MD U.S.A. 64 B3
Hancock NY U.S.A. 64 D2
Handa Island U.K. 20 D2
Handan China 43 K5
Handeni Tanz. 49 D4
HaNegev des. Israel see Negev
HaNeqarot watercourse Israel 39 B4
Hanford CA U.S.A. 65 C2
Hangayn Nuruu mts Mongolia 42 H3
Hangchow China see Hangzhou
Hangö Fin. see Hanko
Hangu China 43 L5
Hangya China 42 H5
Hangzhou China 43 M6
Hangzhou Wan b. China 43 M6
Hani Turkey 35 F3
Hanish Kabir i. Yemen see Suyūl Ḥanīsh
Hankey S. Africa 50 G7
Hanko Fin. 15 M7
Hanle India 36 D2
Hann, Mount h. Australia 54 D3
Hanna Canada 62 E1
Hannibal MO U.S.A. 63 I4
Hannibal NY U.S.A. 64 C1
Hannover Germany 17 L4
Hann Range mts Australia 55 F5
Hanöbukten b. Sweden 15 I9
Hanoi Vietnam see Ha Nôi
Ha Nôi Vietnam 31 J4
Hanover Germany see Hannover
Hanover S. Africa 50 G6
Hanover PA U.S.A. 64 C3
Hanover VA U.S.A. 64 C4
Hansen Mountains Antarctica 76 D2
Hansi India 36 D3
Hansnes Norway 14 K2
Hanstholm Denmark 15 F8
Hantsavichy Belarus 15 O10
Hanumangarh India 36 C3
Hanwood Australia 58 C5
Hanzhong China 42 J6
Hao atoll Fr. Polynesia 75 K7
Haora India 37 G5
Haparanda Sweden 14 N4
Happy Valley-Goose Bay Canada 61 L4
Ḥaql Saudi Arabia 39 B5
Ḥarad, Jabal al mt. Jordan 39 B5
Ḥaraḍh Saudi Arabia 32 G5
Haradok Belarus 13 F5
Haramachi Japan 45 F5
Haramukh mt. India 36 C2
Haran Turkey see Harran
Harappa Road Pak. 36 C3
Harar Eth. see Hārer
Harare Zimbabwe 49 D5
Ḥarāsīs, Jiddat al des. Oman 33 I6
Harāt Iran 35 I5
Har-Ayrag Mongolia 43 J3
Haraze-Mangueigne Chad 47 F3

Harb, Jabal mt. Saudi Arabia 34 D6
Harbin China 44 B3
Harboi Hills Pak. 36 A3
Harchoka India 37 E5
Harda India 36 D5
Harda Khas India see Harda
Hardangerfjorden sea chan. Norway 15 D7
Hardangervidda plat. Norway 15 E6
Hardangervidda Nasjonalpark nat. park Norway 15 E6
Hardap admin. reg. Namibia 50 C3
Hardap nature res. Namibia 50 C3
Hardap Dam Namibia 50 C3
Hardeveld mts S. Africa 50 D6
Hardin U.S.A. 62 F2
Harding S. Africa 51 I6
Harding Range hills Australia 55 B6
Hardoi India 36 E4
Hardwar India see Haridwar
Ḩareīdīn, Wādī watercourse Egypt see Ḩuraydīn, Wādī
Hārer Eth. 48 E3
Harf el Mreffi mt. Lebanon 39 B3
Hargeisa Somalia see Hargeysa
Hargele Eth. 48 E3
Hargeysa Somalia 48 E3
Harghita-Mădăraş, Vârful mt. Romania 27 K1
Harhorin Mongolia 42 I3
Har Hu l. China 42 H5
Haridwar India 36 D3
Ḩarif, Har mt. Israel 39 B4
Harihar India 38 B3
Harihari N.Z. 59 C6
Harihharpur India 38 B3
Ḩārim Syria 39 C1
Harima-nada b. Japan 45 D6
Haringhat r. Bangl. 37 G5
Ḩarīr, Wādī adh r. Syria 39 C3
Harī Rūd r. Afgh./Iran 33 J2
Harjavalta Fin. 15 M6
Harlech U.K. 19 C6
Harleston U.K. 19 I6
Harlow U.K. 19 H7
Harlowton U.S.A. 62 F2
Harman WV U.S.A. 64 B3
Harmancık Turkey 27 M5
Harnai India 38 B2
Harnai Pak. 36 A3
Harney Basin U.S.A. 62 C3
Härnösand Sweden 14 J5
Har Nuur l. Mongolia 42 G3
Haroldswick U.K. 20 [inset]
Harper Liberia 46 C4
Harper, Mount U.S.A. 60 D3
Harper Lake CA U.S.A. 65 D3
Harrai India 36 D5
Harran Turkey 39 D1
Harricana, Rivière d' r. Canada 63 L1
Harrington Australia 58 F3
Harrington DE U.S.A. 64 D3
Harris, Lake imp. l. Australia 57 A6
Harris, Mount Australia 55 E6
Harris, Sound of sea chan. U.K. 20 B3
Harrisburg PA U.S.A. 64 C2
Harrismith Australia 55 B8
Harrison AR U.S.A. 63 I4
Harrison, Cape Canada 61 M4
Harrison Bay U.S.A. 60 C2
Harrisonburg VA U.S.A. 64 B3
Harrisonville U.S.A. 63 I4
Harrisville PA U.S.A. 64 A2
Harrisville WV U.S.A. 64 A3
Harrodsville N.Z. see Otorohanga
Harrogate U.K. 18 F5
Har Sai Shan mt. China 37 I2
Harsīn Iran 35 G4
Harşit r. Turkey 34 E2
Hârşova Romania 27 L2
Harstad Norway 14 J2
Harsud India 36 D5
Hart r. Canada 60 E3
Hartbees watercourse S. Africa 50 E5
Hartberg Austria 17 O7
Harteigan mt. Norway 15 E6
Harter Fell h. U.K. 18 E4
Hartford CT U.S.A. 64 E2
Hartland U.K. 19 C8
Hartland Point U.K. 19 C7
Hartlepool U.K. 18 F4
Hartley Zimbabwe see Chegutu
Hartley Bay Canada 60 F4
Hartola Fin. 15 O6
Harts r. S. Africa 51 G5
Härtsfeld hills Germany 17 M6
Harts Range mts Australia 55 F5
Hartswater S. Africa 50 G4
Har Us Nuur l. Mongolia 42 G3
Harvey Australia 55 A8
Harvey U.S.A. 62 G2
Harwich U.K. 19 I7
Haryana state India 36 D3
Harz hills Germany 17 M5
Har Zin Israel 39 B4
Ḩaşāh, Wādī al watercourse Jordan 39 B4
Ḩaşāh, Wādī al watercourse Jordan/Saudi Arabia 39 C4
Ḩasanah, Wādī watercourse Egypt 39 A4
Hasan Dağı mts Turkey 34 D3
Hasan Guli Turkm. see Esenguly
Hasankeyf Turkey 35 F3
Hasanur India 38 C4
Hasardag mt. Turkm. 35 J3
Hasbaïya Lebanon 39 B3
Hasbaya Lebanon see Hasbaïya
HaSharon plain Israel 39 B3
Hashtgerd Iran 35 H4
Hashtpar Iran 35 H3
Hashtrud Iran 35 G3
Haslemere U.K. 19 G7
Hăşmaşul Mare mt. Romania 27 K1
Ḩaşş, Jabal al hills Syria 39 C1
Hassan India 38 C3
Hasselt Belgium 16 J5
Hassi Bel Guebbour Alg. 46 D2
Hassi Messaoud Alg. 22 F5
Hässleholm Sweden 15 H8
Hastings Australia 58 B7
Hastings r. Australia 58 F3
Hastings N.Z. 59 F4
Hastings U.K. 19 H8
Hastings MN U.S.A. 63 I3
Hastings NE U.S.A. 62 H3
Hata India 37 E4
Hatanbulag Mongolia 43 J4
Hatay Turkey see Antakya
Hatay prov. Turkey 39 C1
Hatch UT U.S.A. 65 F2
Hatches Creek (abandoned) Australia 56 A4
Hatfield Australia 58 A4
Hatfield U.K. 18 G5
Hatgal Mongolia 42 I2
Hath India 38 D1
Hat Head National Park Australia 58 F3
Hathras India 36 D4
Hatisar Bhutan see Geylegphug
Hatod India 36 C5
Hato Hud East Timor see Hatudo
Hatra Iraq 35 F4
Hattah Australia 57 C7
Hattah Kulkyne National Park Australia 57 C7
Hatteras, Cape U.S.A. 63 L4
Hatteras Abyssal Plain sea feature S. Atlantic Ocean 72 D4
Hattfjelldal Norway 14 H4
Hattiesburg U.S.A. 63 J5
Hatudo East Timor 54 D2
Hat Yai Thai. 31 J6
Haud reg. Eth. 48 E3
Hauge Norway 15 E7
Haugesund Norway 15 D7
Haukeligrend Norway 15 E7
Haukipudas Fin. 14 N4
Haukivesi l. Fin. 14 P5
Hauraki Gulf N.Z. 59 E3
Haut Atlas mts Morocco 22 C5
Haute-Normandie admin. reg. France 19 I9
Haute-Volta country Africa see Burkina Faso
Haut-Folin h. France 24 G3
Hauts Plateaux Alg. 22 D5
Havana Cuba 67 H4
Havant U.K. 19 G8
Havasu, Lake Arizona/California U.S.A. 65 E3
Havel r. Germany 17 M4
Havelock N.Z. 59 D5
Havelock Swaziland see Bulembu
Havelock Falls Australia 54 F3
Havelock North N.Z. 59 F4
Haverfordwest U.K. 19 C7
Haverhill MA U.S.A. 64 F1
Haveri India 38 B3
Havlíčkův Brod Czech Rep. 17 O6
Havøysund Norway 14 N1
Havran Turkey 27 L5
Havre U.S.A. 62 F2
Havre Rock i. Kermadec Is 53 I5
Havre-St-Pierre Canada 63 O1
Havza Turkey 34 D2
Hawai'i i. U.S.A. 75 J4
Hawai'ian Islands N. Pacific Ocean 74 I4
Hawaiian Ridge sea feature N. Pacific Ocean 74 I4
Ḩawallī Kuwait 35 H5
Hawarden U.K. 18 D5
Hawea, Lake N.Z. 59 B7
Hawera N.Z. 59 E4
Hawes U.K. 18 E4
Hawick U.K. 20 G5
Ḩawīzah, Hawr al imp. l. Iraq 35 G5
Hawkdun Range mts N.Z. 59 B7
Hawke Bay N.Z. 59 F4
Hawkins Peak UT U.S.A. 65 F2
Hawlēr Iraq see Arbīl
Hawley PA U.S.A. 64 D2
Ḩawrān, Wādī watercourse Iraq 35 F4
Hawston S. Africa 50 D8
Hawthorne NV U.S.A. 65 C1
Haxat China 44 B3
Haxby U.K. 18 F4
Hay Australia 58 B5
Hay watercourse Australia 56 B5
Hay r. Canada 60 G3
Hayachine-san mt. Japan 45 F5
Hayastan country Asia see Armenia
Haydān, Wādī al r. Jordan 39 B4
Hayes r. Nunavut Canada 61 I3
Hayes Halvø pen. Greenland 61 L2
Hayfield Reservoir CA U.S.A. 65 E4
Hayl, Wādī watercourse Syria 39 C3
Hayl, Wādī al watercourse Syria 39 D2
Hayle U.K. 19 B8
Haymā' Oman 33 I6
Haymana Turkey 34 D3
Haymarket VA U.S.A. 64 C3
Hay-on-Wye U.K. 19 D6
Hayrabolu Turkey 27 L4
Hay River Canada 60 G3
Hays KS U.S.A. 62 H4
Ḩays Yemen 32 F7
Haysyn Ukr. 13 F6
Ḩayţān, Jabal h. Egypt 39 A4
Hayward CA U.S.A. 65 A2
Haywards Heath U.K. 19 G8
Hazar Turkm. 35 I3
Hazaribag India see Hazaribagh
Hazaribagh India 37 F5
Hazaribagh Range mts India 37 E5
Hazār Masjed, Kūh-e mts Iran 33 I2
Hazelton Canada 60 F4
Hazen Strait Canada 61 G2
Hazleton PA U.S.A. 64 D2
Hazlett, Lake imp. l. Australia 54 E5
Hazrat-e Solţān Afgh. 36 A1
H. Bouchard Arg. 70 D4
Headford Ireland 21 C4
Headingly Australia 56 B4
Head of Bight b. Australia 55 E7
Healdsburg CA U.S.A. 65 A1
Healesville Australia 58 B6
Healy U.S.A. 60 D3
Heanor U.K. 19 F5
Heard Island Indian Ocean 73 M9
Heard and McDonald Islands terr. Indian Ocean 73 M9
Hearst Canada 63 K2
Hearst Island Antarctica 76 L2
Heart of Neolithic Orkney tourist site U.K. 20 F1
Heathcote Australia 58 B6
Heathfield U.K. 19 H8
Heathsville VA U.S.A. 64 C4
Hebei prov. China 43 L5
Hebel Australia 58 C2
Heber City U.S.A. 62 E3
Hebi China 43 K5
Hebron Canada 61 L4
Hebron West Bank 39 B4
Hecate Strait Canada 60 E4
Hechi China 43 J8
Hede Sweden 14 H5
Hedemora Sweden 15 I6
Ḩefa Israel see Haifa
Ḩefa, Mifraz Israel see Haifa, Bay of
Hefei China 43 L6
Hegang China 44 C3
Heidan r. Jordan see Haydān, Wādī al
Heide Germany 17 L3
Heide Namibia 50 C2
Heidelberg Germany 17 L6
Heidelberg S. Africa 51 I4
Heihe China 44 B2
Heilbron S. Africa 51 H4
Heilbronn Germany 17 L6
Heiligenhafen Germany 17 M3
Heilongjiang prov. China 44 C3
Heilong Jiang r. China 44 F1
Heilong Jiang r. China/Rus. Fed. 40 F3
Heilungkiang prov. China see Heilongjiang
Heinola Fin. 15 O6
Heirnkut Myanmar 37 H4
Heishi Beihu l. China 37 E2
Heisker Islands U.K. see Monach Islands
Heiţân, Gebel h. Egypt see Ḩayţān, Jabal
Hejaz reg. Saudi Arabia see Hijaz
Hekimhan Turkey 34 E3
Hekla vol. Iceland 14 [inset]
Hekou Gansu China 42 I5
Helagsfjället mt. Sweden 14 H5
Helan Shan mts China 42 J5
Helena MT U.S.A. 62 E2
Helensburgh U.K. 20 E4
Helen Springs Australia 54 F4
Ḩeleẓ Israel 39 B4
Helgoland i. Germany 17 K3
Helgoländer Bucht g. Germany 17 L3
Heligoland i. Germany see Helgoland
Heligoland Bight g. Germany see Helgoländer Bucht
Heliopolis Lebanon see Ba'albek
Helixi China see Ningguo
Hella Iceland 14 [inset]
Helland Norway 14 J2
Hellas country Europe see Greece
Helleh r. Iran 35 H5
Hellespont str. Turkey see Dardanelles
Hellhole Gorge National Park Australia 56 D5
Hellín Spain 25 F4
Hellinikon tourist site Greece 34 A3
Hell-Ville Madag. see Andoany
Helmand r. Afgh. 36 A2
Helmantica Spain see Salamanca
Helmeringhausen Namibia 50 C3
Helmond Neth. 17 J5
Helmsdale U.K. 20 F2
Helmsdale r. U.K. 20 F2
Helmstedt Germany 17 M4
Helong China 44 C4
Helsingborg Sweden 15 H8
Helsingfors Fin. see Helsinki
Helsingør Denmark 15 H8
Helsinki Fin. 15 N6
Helston U.K. 19 B8
Helvécia Brazil 71 D2
Helvetic Republic country Europe see Switzerland
Ḩelwān Egypt see Ḩulwān
Hemel Hempstead U.K. 19 G7
Hemet CA U.S.A. 65 D4
Hemlock Lake NY U.S.A. 64 C1
Hemsby U.K. 19 I6
Hemse Sweden 15 K8
Henan prov. China 43 K6
Henares r. Spain 25 E3
Henashi-zaki pt Japan 45 E4
Henbury Australia 55 F6
Hendek Turkey 27 N4
Henderson NC U.S.A. 63 L4
Henderson NV U.S.A. 65 E2
Henderson TX U.S.A. 63 I5
Henderson Island Pitcairn Is 75 L7
Henderville atoll Kiribati see Aranuka
Hendon U.K. 19 G7
Hendorābī i. Iran 35 I6
Hendy-Gwyn U.K. see Whitland
Hengduan Shan mts China 42 H7
Hengelo Neth. 17 K4
Hengnan China see Hengyang
Hengshan China 44 C3
Hengshui Hebei China 43 L5
Hengxian China 43 J8
Hengyang Hunan China 43 K7
Hengzhou China see Hengxian
Heniches'k Ukr. 13 G7
Henley N.Z. 59 C7
Henley-on-Thames U.K. 19 G7
Henlopen, Cape DE U.S.A. 64 D3
Hennenman S. Africa 51 H4
Henniker NH U.S.A. 64 F1
Henrietta Maria, Cape Canada 61 J4
Henrique de Carvalho Angola see Saurimo
Henry, Cape NC U.S.A. 64 C4
Henry Ice Rise Antarctica 76 A1
Henryk Arctowski research stn Antarctica see Arctowski
Henry Kater, Cape Canada 61 L3
Hensall Ont. Canada 64 A1
Henshaw, Lake CA U.S.A. 65 D4
Hentiesbaai Namibia 50 B2
Henty Australia 58 C5
Henzada Myanmar see Hinthada
Heptanesus is Greece see Ionian Islands
Heraclea Turkey see Ereğli
Heraclea Pontica Turkey see Ereğli
Heraklion Greece see Iraklion
Herald Cays atolls Australia 56 E3
Herāt Afgh. 33 J3
Hérault r. France 24 F5
Herbert Downs Australia 56 B4
Herbert River Falls National Park Australia 56 D3
Herbert Wash imp. l. Australia 55 D6
Hercules Dome Antarctica 76 K1
Hereford U.K. 19 E6
Hereford U.S.A. 62 G5
Héréhérétué atoll Fr. Polynesia 75 K7
Herford Germany 17 L4
Herīs Iran 35 G3
Herisau Switz. 24 I3
Herkimer NY U.S.A. 64 D1
Herlen Gol r. China/Mongolia 43 L3
Herlen He r. China/Mongolia see Herlen Gol
Herm i. Channel Is 19 E9
Herma Ness hd U.K. 20 [inset]
Hermanus S. Africa 50 D8
Hermel Lebanon 39 C2
Hermes, Cape S. Africa 51 I6
Hermidale Australia 58 C3
Hermite, Islas is Chile 70 C9
Hermit Islands P.N.G. 52 E2
Hermon, Mount Lebanon/Syria 39 B3
Hermonthis Egypt see Armant
Hermopolis Magna Egypt see Al Ashmūnayn
Hermosillo Mex. 66 B3
Hernandarias Para. 70 F3
Herndon CA U.S.A. 65 C2
Herndon PA U.S.A. 64 C2
Herndon WV U.S.A. 64 A4
Herne Bay U.K. 19 I7
Herning Denmark 15 F8
Heroica Nogales Mex. see Nogales
Heroica Puebla de Zaragoza Mex. see Puebla
Hérouville-St-Clair France 19 G9
Herowābād Iran see Khalkhāl
Herrera del Duque Spain 25 D4
Hershey PA U.S.A. 64 C2
Hertford U.K. 19 G7
Hertzogville S. Africa 51 G5
Hervey Islands Cook Is 75 J7
Herzliyya Israel 39 B3
Ḩeşār Iran 35 H5
Heshan China 43 J8
Hesperia CA U.S.A. 65 D3
Hesselberg h. Germany 17 M6
Hetton U.K. 18 E4
Hettstedt Germany 17 M5
Hevron West Bank see Hebron
Hexham U.K. 18 E4
Hexian Guangxi China see Hezhou
Ḩeydarābād Iran 35 G3
Heydebreck Poland see Kędzierzyn-Koźle
Heysham U.K. 18 E4
Heyshope Dam S. Africa 51 J4
Heywood U.K. 18 E5
Heze China 43 L5
Hezhou China 43 K8
Hezuo China 42 I5
Hezuozhen China see Hezuo
Hibbing U.S.A. 63 I2
Hibbs, Point Australia 57 [inset]
Hibernia Reef Australia 54 C3
Hicks, Point Australia 58 D6
Hicks Bay N.Z. 59 G3
Hidaka-sanmyaku mts Japan 44 F4
Hidalgo Mex. 62 H7
Hidalgo del Parral Mex. 66 C3
Hidrolândia Brazil 71 A2
Hierosolyma Israel/West Bank see Jerusalem
Higashi-suidō str. Japan 45 C6
Higgins Bay NY U.S.A. 64 D1
High Atlas mts Morocco see Haut Atlas
High Desert U.S.A. 62 C3
Highland Peak CA U.S.A. 65 C1
Highland Peak NV U.S.A. 65 E2
Highlands NJ U.S.A. 64 E2
Highland Springs VA U.S.A. 64 C4
High Level Canada 60 G4
High Point U.S.A. 63 L4
High Point h. NJ U.S.A. 64 D2
High Prairie Canada 60 G4
High River Canada 62 E1
High Tatras mts Poland/Slovakia see Tatra Mountains
High Wycombe U.K. 19 G7
Higüey Dom. Rep. 67 K5
Hiiumaa i. Estonia 15 M7
Ḩijānah, Buḩayrat al salt pan Syria 39 C3
Hijaz reg. Saudi Arabia 32 E4
Ḩikmah, Ra's al pt Egypt 34 B5
Hiko NV U.S.A. 65 E2
Hikone Japan 45 E6
Hikurangi mt. N.Z. 59 G3
Hila Indon. 54 D1
Hilāl, Jabal h. Egypt 39 A4
Hilāl, Ra's al pt Libya 32 B3
Hilary Coast Antarctica 76 H1
Hildale UT U.S.A. 65 F2
Hildesheim Germany 17 L4
Hillah Iraq 35 G4
Hill End Australia 58 D4
Hillerød Denmark 15 H9
Hillgrove Australia 56 D3
Hillside Australia 54 B5
Hillston Australia 58 B4
Hilton Australia 56 B4
Hilton S. Africa 51 J5
Hilton NY U.S.A. 64 C1
Hilton Head Island U.S.A. 63 K5
Hilvan Turkey 34 E3
Hilversum Neth. 16 J4
Himachal Pradesh state India 36 D3
Himalaya mts Asia 36 D2
Himalchul mt. Nepal 37 F3
Himanka Fin. 14 M4
Ḩimār, Wādī al watercourse Syria/Turkey 39 D1
Himarë Albania 27 H4
Himatnagar India 36 C5
Himeji Japan 45 D6
Ḩimş Syria see Homs
Ḩimş, Baḩrat resr Syria see Qaţţīnah, Buḩayrat
Hinchinbrook Island Australia 56 D3
Hinckley U.K. 19 F6
Hinckley Reservoir NY U.S.A. 64 D1
Hindaun India 36 D4
Hinderwell U.K. 18 G4
Hindley U.K. 18 E5
Hindmarsh, Lake dry lake Australia 57 C8
Hindu Kush mts Afgh./Pak. 33 K3
Hindupur India 38 C3
Hinesville U.S.A. 63 K5
Hinganghat India 38 C1
Hingoli India 38 C2
Hınıs Turkey 35 F3
Hinnøya i. Norway 14 I2
Hinojosa del Duque Spain 25 D4
Hinsdale NH U.S.A. 64 E1
Hinthada Myanmar 42 H9
Hinton WV U.S.A. 64 A4
Hiort i. U.K. see St Kilda
Hipponium Italy see Vibo Valentia
Hippo Regius Alg. see Annaba
Hippo Zarytus Tunisia see Bizerte
Hirabit Dağ mt. Turkey 35 G3
Hirakud Dam India 37 E5
Hirakud Reservoir India 37 E5
Hirapur India 36 D4
Hiriyur India 38 C3
Hirosaki Japan 44 F4
Hiroshima Japan 45 D6
Hirschberg mt. Germany 17 M7
Hirschberg Poland see Jelenia Góra
Hirson France 24 G2
Hîrşova Romania see Hârşova
Hirta is U.K. see St Kilda
Hirtshals Denmark 15 F8
Hisar India 36 C3
Hisar Iran 35 H3
Hisarköy Turkey see Domaniç
Hisarönü Turkey 27 O4
Ḩisb, Sha'īb watercourse Iraq 35 G5
Ḩisbān Jordan 39 B4
Hisor Tizmasi mts Tajik./Uzbek. see Gissar Range
Hispalis Spain see Seville
Hispania country Europe see Spain
Hispaniola i. Caribbean Sea 67 J4
Hispur Glacier Pak. 36 C1
Hissar India see Hisar
Hisua India 37 F4
Ḩisyah Syria 39 C2
Hīt Iraq 35 F4
Hitachi Japan 45 F5
Hitachinaka Japan 45 F5
Hitra i. Norway 14 F5
Hiva Oa i. Fr. Polynesia 75 K6
Hixson Cay rf Australia 56 F4
Hiyyon watercourse Israel 39 B4
Hizan Turkey 35 F3
Hjälmaren l. Sweden 15 I7
Hjerkinn Norway 14 F5
Hjo Sweden 15 I7
Hjørring Denmark 15 G8
Hkakabo Razi mt. China/Myanmar 42 H7
Hlaingdet Myanmar 37 I5
Hlako Kangri mt. China see Lhagoi Kangri
Hlane Royal National Park Swaziland 51 J4
Hlatikulu Swaziland 51 J4
Hlohlowane S. Africa 51 H5
Hlotse Lesotho 51 I5
Hluhluwe-Umfolozi Park nature res. S. Africa 51 J5
Hlukhiv Ukr. 13 G6
Hlusha Belarus 13 F5
Hlybokaye Belarus 15 O9
Ho Ghana 46 D4
Hoachanas Namibia 50 D2
Hoang Sa is S. China Sea see Paracel Islands
Hobart Australia 57 [inset]
Hobart U.S.A. 62 H4
Hobbs U.S.A. 62 G5
Hobbs Coast Antarctica 76 J1
Hobiganj Bangl. see Habiganj
Hobro Denmark 15 F8
Hobyo Somalia 48 E3
Hoceima, Baie d'Al b. Morocco 25 E6
Hochfeiler mt. Austria/Italy see Gran Pilastro
Hochfeld Namibia 49 B6
Hô Chi Minh Vietnam see Ho Chi Minh City
Ho Chi Minh City Vietnam 31 J5
Hochschwab mt. Austria 17 O7
Hochschwab mts Austria 17 O7
Hôd reg. Mauritania 46 C3
Hoddesdon U.K. 19 G7
Hodeidah Yemen 32 F7
Hodgson Downs Australia 54 F3
Hódmezővásárhely Hungary 27 I1
Hodna, Chott el salt l. Alg. 25 I6
Hodo-dan pt N. Korea 45 B5
Hoeryŏng N. Korea 44 C4
Hof Germany 17 M5
Hofmeyr S. Africa 51 G6
Höfn Iceland 14 [inset]
Hofors Sweden 15 J6
Hofsjökull Iceland 14 [inset]
Hofsós Iceland 14 [inset]
Hōfu Japan 45 C6
Hofūf Saudi Arabia 32 G4
Höganäs Sweden 15 H8
Hogan Group is Australia 58 C7
Hoggar plat. Alg. see Ahaggar
Hog Island VA U.S.A. 64 D4
Högsby Sweden 15 J8
Hohensalza Poland see Inowrocław
Hoher Dachstein mt. Austria 17 N7
Hohe Tauern mts Austria 17 N7
Hohhot China 43 K4
Höhmorīt Mongolia 42 G3
Hohneck mt. France 24 H2
Hoh Sai Hu l. China 37 H2
Hoh Xil Hu salt l. China 37 G2
Hoh Xil Shan mts China 37 G2
Hoima Uganda 48 D3
Hojagala Turkm. 35 J3
Hojai India 37 H4
Højryggen mts Greenland 61 M2
Hokitika N.Z. 59 C6
Hokkaidō i. Japan 44 F4
Hokksund Norway 15 F7
Hol Norway 15 F6
Holbæk Denmark 15 G9
Holbeach U.K. 19 H6
Holbrook Australia 58 C5
Holbrook U.S.A. 62 E5
Holden UT U.S.A. 65 F1
Holdrege U.S.A. 62 H3
Holguín Cuba 67 I4
Höljes Sweden 15 H6
Holland country Europe see Netherlands
Holland NY U.S.A. 64 B1
Hollandia Indon. see Jayapura
Hollick-Kenyon Peninsula Antarctica 76 L2
Hollick-Kenyon Plateau Antarctica 76 K1
Hollidaysburg PA U.S.A. 64 B2
Hollister CA U.S.A. 65 B2
Holly Springs U.S.A. 63 J5
Hollywood CA U.S.A. 65 C4
Hollywood FL U.S.A. 63 K6
Holm Norway 14 H4
Holmes Reef Australia 56 D3
Holmestrand Norway 15 G7
Holmgard Rus. Fed. see Velikiy Novgorod
Holm Ø i. Greenland see Kiatassuaq
Holmön i. Sweden 14 L5
Holmsund Sweden 14 L5
Ḩolon Israel 39 B3
Holoog Namibia 50 C4
Holothuria Banks rf Australia 54 D3
Holroyd r. Australia 56 C2
Holstebro Denmark 15 F8
Holsteinsborg Greenland see Sisimiut
Holston r. U.S.A. 63 K4
Holsworthy U.K. 19 C8
Holt U.K. 19 I6
Holycross Ireland 21 E5
Holy Cross U.S.A. 60 C3
Holyhead U.K. 18 C5
Holyhead Bay U.K. 18 C5
Holy Island England U.K. 18 F3
Holy Island Wales U.K. 18 C5
Holyoke U.S.A. 62 G3
Holy See Europe see Vatican City
Holywell U.K. 18 D5
Holzkirchen Germany 17 M7
Homāyūnshahr Iran see Khomeynīshahr
Hombori Mali 46 C3
Home Bay Canada 61 L3
Homer NY U.S.A. 64 C1
Homestead Australia 56 D4
Homnabad India 38 C2
Homoine Moz. 51 L2
Homs Libya see Al Khums
Homs Syria 39 C2
Homyel' Belarus 13 F5
Honan prov. China see Henan
Honavar India 38 B3
Honawad India 38 B2
Honaz Turkey 27 M6
Hondeklipbaai S. Africa 50 C6
Hondo U.S.A. 62 H6
Honduras country Central America 67 G6
Hønefoss Norway 15 G6
Honesdale PA U.S.A. 64 D2
Honey Lake salt l. U.S.A. 62 C3
Honeyoye Lake NY U.S.A. 64 C1
Honfleur France 24 E2
Hông, Sông r. Vietnam see Red
Hongjiang Hunan China 43 J7
Hong Kong H.K. China 43 K8
Hong Kong aut. reg. China 74 D4
Hongliuwan China see Aksay
Hongliuyuan China 42 H4
Hongshilazi China 44 B4
Hongwŏn N. Korea 45 B4
Hongxing China 44 A3
Hongze Hu l. China 43 L6
Honiara Solomon Is 53 F2
Honiton U.K. 19 D8
Honjō Japan 45 F5
Honkajoki Fin. 15 M6
Honningsvåg Norway 14 N1
Honolulu U.S.A. 75 J4
Honshū i. Japan 45 D6
Hood, Mount vol. U.S.A. 62 C2
Hood Point Australia 55 B8
Hood Point P.N.G. 56 D1
Hoogeveen Neth. 17 K4
Hooghly r. mouth India see Hugli
Hook Head hd Ireland 21 F5
Hook Reef Australia 56 E3
Hooper Bay U.S.A. 77 B2
Hooper Island MD U.S.A. 64 C3
Hoopstad S. Africa 51 G4
Höör Sweden 15 H9
Hoorn Neth. 16 J4
Hoorn, Îles de is Wallis and Futuna Is 53 I3
Hoosick NY U.S.A. 64 E1
Hoover Dam Arizona/Nevada U.S.A. 65
Hopa Turkey 35 F2
Hope Canada 62 C2
Hope r. N.Z. 59 D6
Hope AR U.S.A. 63 I5
Hope, Lake imp. l. Australia 55 C8
Hope, Point U.S.A. 60 B3
Hopedale Canada 61 L4
Hopefield S. Africa 50 D7
Hopei prov. China see Hebei
Hope Saddle pass N.Z. 59 D5
Hopes Advance, Cap c. Canada 61 L3
Hopetoun Australia 57 C7
Hopetown S. Africa 50 G5
Hopewell VA U.S.A. 64 C4
Hopewell Islands Canada 61 K4
Hopkins r. Australia 57 C8
Hopkins, Lake imp. l. Australia 55 E6
Hopkinsville U.S.A. 63 J4
Hopland CA U.S.A. 65 A1
Horasan Turkey 35 F2
Hörby Sweden 15 H9
Horizon Deep sea feature S. Pacific Oce 74 I7
Horki Belarus 13 F5
Horlick Mountains Antarctica 76 K1
Horlivka Ukr. 13 H6
Hormuz, Strait of Iran/Oman 33 I4
Horn Austria 17 O6
Horn c. Iceland 14 [inset]
Horn, Cape Chile 70 C9
Hornavan l. Sweden 14 J3
Horncastle U.K. 18 G5
Horndal Sweden 15 J6
Horne, Îles de is Wallis and Futuna Is se Hoorn, Îles de
Hörnefors Sweden 14 K5
Hornell NY U.S.A. 64 C1
Hornepayne Canada 63 K2
Hornisgrinde mt. Germany 17 L6
Hornkranz Namibia 50 C2
Hornos, Cabo de Chile see Horn, Cape
Hornsby Australia 58 E4
Hornsea U.K. 18 G5
Hornslandet pen. Sweden 15 J6
Horodenka Ukr. 13 E6
Horodnya Ukr. 13 F6
Horodok Khmel'nyts'ka Oblast' Ukr. 13
Horodok L'vivs'ka Oblast' Ukr. 13 D6
Horokanai Japan 44 F3
Horoshiri-dake mt. Japan 44 F4
Horqin Youyi Qianqi China see Ulanhot
Horqin Zuoyi Houqi China see Ganjig
Horqin Zuoyi Zhongqi China see Baoka
Horrabridge U.K. 19 C8
Horrocks Australia 55 A7
Horru China 37 G3
Horseheads NY U.S.A. 64 C1
Horseleap Ireland 21 D4
Horsens Denmark 15 F9
Horseshoe Bend Australia 55 F6
Horseshoe Seamounts sea feature N. Atlantic Ocean 72 G3
Horsham Australia 57 C8
Horsham U.K. 19 G7
Horten Norway 15 G7
Hortobágyi nat. park Hungary 27 I1
Horton r. Canada 60 F3
Ḩoseynābād Iran 35 G4
Ḩoseynīyeh Iran 35 H5
Hoshangabad India 36 D5
Hoshiarpur India 36 C3
Hospet India 38 C3
Hospital Ireland 21 D5
Hosséré Vokre mt. Cameroon 46 E4
Hotagen r. Sweden 14 I5
Hotahudo East Timor see Hatudo
Hotan China 36 E1
Hotazel S. Africa 50 F4
Hot Creek Range mts NV U.S.A. 65 D1
Hotgi India 38 C2
Hotham r. Australia 55 B8
Hoting Sweden 14 J4
Hot Springs AR U.S.A. 63 I5
Hot Springs NM U.S.A. see Truth or Consequences
Hot Springs SD U.S.A. 62 G3
Hottah Lake Canada 60 G3
Hottentots Bay Namibia 50 B4
Hottentots Point Namibia 50 B4
Houghton MI U.S.A. 63 J2
Houghton NY U.S.A. 64 B1
Houghton le Spring U.K. 18 F4
Houma China 43 K5
Houma U.S.A. 63 I6
Houston TX U.S.A. 63 H6
Hout r. S. Africa 51 I2
Houtman Abrolhos is Australia 55 A7
Houton U.K. 20 F2
Houwater S. Africa 50 F6
Hovd Hovd Mongolia 42 G3
Hove U.K. 19 G8
Hoveton U.K. 19 I6
Hovmantorp Sweden 15 I8
Hövsgöl Nuur l. Mongolia 42 I2
Howar, Wadi watercourse Sudan 32 C6
Howard Australia 56 F5
Howard PA U.S.A. 64 C2
Howden U.K. 18 G5
Howe, Cape Australia 58 D6
Howe, Mount Antarctica 76 J1
Howick S. Africa 51 J5
Howland Island terr. N. Pacific Ocean 53 I1
Howlong Australia 58 C5
Howrah India see Haora
Howth Ireland 21 F4
Howz i-Mian i-Tak Iran 35 I4
Höxter Germany 17 L5
Hoy i. U.K. 20 F2
Høyanger Norway 15 E6
Hoyerswerda Germany 17 O5
Høylandet Norway 14 H4
Höytiäinen l. Fin. 14 P5
Hpapun Myanmar 42 H9
Hradec Králové Czech Rep. 17 O5
Hrasnica Bos.-Herz. 26 H3
Hrazdan Armenia 35 G2
Hrebinka Ukr. 13 G6
Hrodna Belarus 15 M10
Hrvatska country Europe see Croatia
Hrvatsko Grahovo Bos.-Herz. see Bosansko Grahovo
Hsiang Kang H.K. China see Hong Kon
Hsin-chia-p'o country Asia see Singapor
Hsin-chia-p'o Sing. see Singapore
Hsinchu Taiwan 43 M8
Hsinking China see Changchun
Hsipaw Myanmar 42 H8
Hsi-sha Ch'ün-tao is S. China Sea see Paracel Islands
Huab watercourse Namibia 49 B6
Huacho Peru 68 C6
Huachuan China 44 C3
Huade China 43 K4
Huadian China 44 B4
Huai'an Jiangsu China 43 L6
Huaibei China 43 L6
Huaide China 44 B4
Huaihua China 43 J7
Huailillas mt. Peru 68 C5

ainan China 43 L6
aiyin *Jiangsu* China *see* Huai'an
jialing China 42 I5
juápan de León Mex. 66 E5
lapai Peak *AZ* U.S.A. 65 F3
lian Taiwan *see* Hualien
lien Taiwan 43 M8
allaga *r.* Peru 68 C5
mbo Angola 49 B5
nan China 44 C3
ncane Peru 68 E7
ncavelica Peru 68 C6
ncayo Peru 68 C6
ngcaoba China *see* Xingyi
ngchuan China 43 L6
ng Hai *sea* N. Pacific Ocean *see*
llow Sea
ng He *r.* China *see* Yellow
ngshan China 43 L7
ngshi China 43 L6
ngtu Gaoyuan *plat.* China 43 J5
nren China 44 B4
nuco Peru 68 C5
ráz Peru 68 C5
rmey Peru 68 C6
scarán, Nevado de *mt.* Peru 68 C5
sco Chile 70 B3
shugou China *see* Jingtieshan
shulinzi China 44 B4
tabampo Mex. 66 C3
yxay Laos 42 I8
zangsi China *see* Tianzhu
balli India *see* Hubli
ei *prov.* China 43 K6
li India 38 B3
knall U.K. 19 F5
dersfield U.K. 18 F5
er China 44 A2
iksvall Sweden 15 J6
son *MA* U.S.A. 64 F1
son *MD* U.S.A. 64 C3
son *NY* U.S.A. 64 E1
son *r. NY* U.S.A. 64 E2
son, Baie d' *sea* Canada *see*
udson Bay
son, Détroit d' *str.* Canada *see*
udson Strait
son Bay Canada 62 G1
son Bay *sea* Canada 61 J4
son Falls *NY* U.S.A. 64 E1
son Island Tuvalu *see* Nanumanga
son Mountains Antarctica 76 K2
son Strait Canada 61 K3
Vietnam 31 J5
huetenango Guat. 66 F5
hueto, Cerro *mt.* Mex. 66 C4
lva Spain 25 C5
ntelauquén Chile 70 B4
rcal-Overa Spain 25 F5
sca Spain 25 F2
scar Spain 25 E5
henden Australia 56 D4
hes (abandoned) Australia 55 E7
hson *CA* U.S.A. 65 B2
li *r. mouth* India 37 F5
li-Chinsurah India 37 G5
o *OK* U.S.A. 63 H5
ehot China *see* Hohhot
hot China *see* Hohhot
udi S. Africa 50 G4
anpu China 42 J5
arau Range *mts* N.Z. 59 F4
o-Hoch Plateau Namibia 50 C4
a, Nevado de *vol.* Col. 68 C3
a, Planalto da Angola 49 B5
i China 42 I7
nanguillo Mex. 66 F5
en Nur *l.* China 37 G2
tinen Fin. 15 M6
yang China *see* Huizhou
ze China 42 I7
zhou China 43 K8
Saudi Arabia 32 F4
awng Valley Myanmar 37 I4
untsi Botswana 50 E2
n China 44 B3
n Ergi China 44 A3
ayfah Saudi Arabia 32 F4
ayhilah *well* Syria 39 D2
n China 44 D3
n Gol *r.* China 44 B3
Canada 63 L2
U.K. *see* Kingston upon Hull
Island *atoll* Kiribati *see* Orona
sfred Sweden 15 I8
n China *see* Hulun Buir
n Buir China 43 L3
n Nur *l.* China 43 L3
wān Egypt 34 C5
a China 44 B2
nahuaca Arg. 70 C2
naitá Brazil 68 F5
nber, Mouth of the U.K. 18 H5
nboldt Canada 62 F1
nboldt *r.* U.S.A. 60 G5
neburn Australia 58 B1
ne Reservoir Australia 58 C5
nphrey Island *atoll* Cook Is *see*
anihiki
nphreys, Mount *CA* U.S.A. 65 C2
nphreys Peak U.S.A. 62 E4
Libya 47 E2
aflói *b.* Iceland 14 [inset]
an *prov.* China 43 K7
edoara Romania 27 J2
gary *country* Europe 23 H2
gerford Australia 58 B2
gnam N. Korea 45 B5
jiang China *see* Baishan
s Mountains Namibia 50 C4
stanton U.K. 19 H6
te *r.* Germany 17 L4
ter *r.* Australia 58 E4
ter Island Australia 57 [inset]
ter Island S. Pacific Ocean 53 H4
ter Islands Australia 57 [inset]
tingdon U.K. 19 G6
tingdon *PA* U.S.A. 64 C2
tington *IN* U.S.A. 63 J3
tington *WV* U.S.A. 63 K4
tington Beach *CA* U.S.A. 65 C4
tly N.Z. 59 E3
tly U.K. 20 G3
tsville Canada 63 L2
tsville *AL* U.S.A. 63 J5
tsville *TX* U.S.A. 63 H5
za *reg.* Pak. 36 C1
lin He *r.* China *see* Hulin Gol
longmen China 44 B2
nville Australia 57 [inset]
eh *prov.* China *see* Hubei
nik *r.* Turkey 39 C1
aydin, Wādī *watercourse* Egypt 39 A4
d Island Kiribati *see* Arorae
ghada Egypt *see* Al Ghurdaqah
er's Cross Ireland 21 D5
on *CA* U.S.A. 65 B2
on *SD* U.S.A. 62 H3
on, Lake Canada/U.S.A. 64 A1
ricane *UT* U.S.A. 65 F2
sley U.K. 19 F7
Hurst Green U.K. 19 H7
Husain Nika Pak. 36 B3
Húsavík *Norðurland eystra* Iceland
14 [inset]
Húsavík *Vestfirðir* Iceland 14 [inset]
Huseyinabat Turkey *see* Alaca
Huseyinli Turkey *see* Kızılırmak
Huşi Romania 27 M1
Huskvarna Sweden 15 I8
Husn Jordan *see* Al Ḥiṣn
Ḥusn Āl 'Abr Yemen 32 G6
Husnes Norway 15 D7
Husum Germany 17 L3
Husum Sweden 14 K5
Hutag-Öndör Mongolia 42 I3
Hutchinson *KS* U.S.A. 62 H4
Hutou China 44 D3
Hutton, Mount *h.* Australia 57 E5
Hutton Range *hills* Australia 55 C6
Hüvek Turkey *see* Bozova
Huwaytat *reg.* Saudi Arabia 39 C5
Huzhong China 44 A2
Huzhou China 43 M6
Hvannadalshnúkur *vol.* Iceland 14 [inset]
Hvar *i.* Croatia 26 G3
Hvide Sande Denmark 15 F8
Hvíta *r.* Iceland 14 [inset]
Hwange Zimbabwe 49 C5
Hwange National Park Zimbabwe 49 C5
Hwang Ho *r.* China *see* Yellow
Hwedza Zimbabwe 49 D5
Hwlffordd U.K. *see* Haverfordwest
Hyannis *MA* U.S.A. 64 F2
Hyargas Nuur *salt l.* Mongolia 42 G3
Hyde N.Z. 59 C7
Hyden Australia 55 B8
Hyderabad India 38 C2
Hyderabad Pak. 33 K4
Hydra *i.* Greece *see* Ydra
Hyères France 24 H5
Hyères, Îles d' *is* France 24 H5
Hyesan N. Korea 44 C4
Hyland, Mount Australia 58 F3
Hyland Post Canada 60 F4
Hyllestad Norway 15 D6
Hyltebruk Sweden 15 H8
Hyōno-sen *mt.* Japan 45 D6
Hyrcania Iran *see* Gorgān
Hyrynsalmi Fin. 14 P4
Hythe U.K. 19 I7
Hyūga Japan 45 C6
Hyvinkää Fin. 15 N6

I

Iaciara Brazil 71 B1
Iaco *r.* Brazil 68 E5
Iaçu Brazil 71 C1
Iadera Croatia *see* Zadar
Iakora Madag. 49 E6
Ialomița *r.* Romania 27 L2
Ianca Romania 27 L2
Iași Romania 27 L1
Ibadan Nigeria 46 D4
Ibagué Col. 68 C3
Ibaiti Brazil 71 A3
Ibarra Ecuador 68 C3
Ibb Yemen 32 F7
Iberá, Esteros del *marsh* Arg. 70 E3
Iberia Peru 68 E6
Iberian Peninsula Europe 25
Ibeto Nigeria 46 D3
iBhayi S. Africa *see* Port Elizabeth
Ibi Nigeria 46 D4
Ibiá Brazil 71 B2
Ibiaí Brazil 71 B2
Ibiapaba, Serra da *hills* Brazil 69 J4
Ibiassucê Brazil 71 C1
Ibicaraí Brazil 71 D1
Ibiquera Brazil 71 C1
Ibirama Brazil 71 A4
Ibiranhém Brazil 71 C2
Ibitinga Brazil 71 A3
Ibiza Spain 25 G4
Ibiza *i.* Spain 25 G4
Iblei, Monti *mts Sicilia* Italy 26 F6
Ibn Buşayyiş *well* Saudi Arabia 35 G6
Ibotirama Brazil 69 J6
Iboundji, Mont *h.* Gabon 48 B4
Ibrā' Oman 33 I5
Ibradı Turkey 34 C3
Ibrī Oman 33 I5
Ica *r.* Col. *see* Putumayo
Ica Peru 68 C6
Içana Brazil 68 E3
Içana *r.* Brazil 68 E3
Icaria *i.* Greece *see* Ikaria
Icatu Brazil 69 J4
Iceberg Canyon *gorge NV* U.S.A. 65 E2
İçel *Mersin* Turkey *see* Mersin
Iceland *country* Europe 14 [inset]
Iceland Basin *sea feature* N. Atlantic Ocean
72 G2
Icelandic Plateau *sea feature*
N. Atlantic Ocean 77 I2
Ichalkaranji India 38 B2
Ichifusa-yama *mt.* Japan 45 C6
Ichinomiya Japan 45 E6
Ichinoseki Japan 45 F5
Ichinskaya Sopka, Vulkan *vol.* Rus. Fed.
29 Q4
Ichkeul, Parc National de l' Tunisia 26 C6
Ichnya Ukr. 13 G6
Icó Brazil 69 K5
Iconha Brazil 71 C3
Iconium Turkey *see* Konya
Icosium Alg. *see* Algiers
Iculisma France *see* Angoulême
Icy Cape U.S.A. 60 B2
Id Turkey *see* Narman
Idah Nigeria 46 D4
Idaho *state* U.S.A. 62 E3
Idaho Falls U.S.A. 62 E3
Idalia National Park Australia 56 D5
Idar India 36 C5
Idar-Oberstein Germany 17 K6
Ideriyn Gol *r.* Mongolia 42 I3
Idfū Egypt 32 D5
Idhān Awbārī *des.* Libya 46 E2
Idhān Murzūq *des.* Libya 46 E2
Idhra *i.* Greece *see* Ydra
Idi Amin Dada, Lake Dem. Rep. Congo/
Uganda *see* Edward, Lake
Idiofa Dem. Rep. Congo 49 B4
Idivuoma Sweden 14 M2
Idkū Egypt 34 C5
Idle *r.* U.K. 18 G5
Idlib Syria 39 C2
Idra *i.* Greece *see* Ydra
Idre Sweden 15 H6
Idutywa S. Africa 51 I7
Idzhevan Armenia *see* Ijevan
Iecava Latvia 15 N8
Iepê Brazil 71 A3
Ierapetra Greece 27 K7
Ierissou, Kolpos *b.* Greece 27 J4
Iešjávri *l.* Norway 14 N2
Ifakara Tanz. 49 D4
Ifanadiana Madag. 49 E6
Ife Nigeria 46 D4
Ifenat Chad 47 E3
Iferouâne Niger 46 D3
Iffley Australia 56 C3
Ifjord Norway 14 O1
Ifôghas, Adrar des *hills* Mali 46 D3
Iforas, Adrar des *hills* Mali *see*
Ifôghas, Adrar des
Igan *Sarawak* Malaysia 41 D7
Iganga Uganda 47 G4
Igarapava Brazil 71 B3
Igarka Rus. Fed. 28 J3
Igatpuri India 38 B2
Igbeti Nigeria *see* Igbetti
Igbetti Nigeria 46 D4
Igdir Iran 35 G3
Iğdır Turkey 35 G3
Iggesund Sweden 15 J6
Igikpak, Mount U.S.A. 60 C3
Iglesias *Sardegna* Italy 26 C5
Iglesiente *reg. Sardegna* Italy 26 C5
Igloolik Canada 61 J3
Igluligaarjuk Canada *see* Chesterfield Inlet
Ignace Canada 63 I2
Ignalina Lith. 15 O9
İğneada Turkey 27 L4
İğneada Burnu *pt* Turkey 27 M4
iGoli S. Africa *see* Johannesburg
Igoumenitsa Greece 27 I5
Igra Rus. Fed. 11 Q4
Igrim Rus. Fed. 11 S3
Iguaçu *r.* Brazil 71 A4
Iguaçu, Saltos do *waterfall* Arg./Brazil *see*
Iguaçu Falls
Iguaçu Falls Arg./Brazil 70 F3
Iguaí Brazil 71 C1
Iguala Mex. 66 E5
Igualada Spain 25 G3
Iguape Brazil 71 B4
Iguaraçu Brazil 71 A3
Iguatama Brazil 71 B3
Iguatemi Brazil 70 F2
Iguatu Brazil 69 K5
Iguazú, Cataratas do *waterfall* Arg./Brazil
see Iguaçu Falls
Iguéla Gabon 48 A4
Iguidi, Erg *des.* Alg./Mauritania 46 C2
Igunga Tanz. 49 D4
Iharaña Madag. 49 E5
Ihavandhippolhu Atoll Maldives 38 B5
Ihavandiffulu Atoll Maldives *see*
Ihavandhippolhu Atoll
Ih Bogd Uul *mt.* Mongolia 42 I4
Ihosy Madag. 49 E6
Ih Tol Gol China 44 A2
Iide-san *mt.* Japan 45 E5
Iijärvi *l.* Fin. 14 O2
Iijoki *r.* Fin. 14 N4
Iisalmi Fin. 14 O5
Iizuka Japan 45 C6
Ijebu-Ode Nigeria 46 D4
Ijevan Armenia 35 G2
IJssel *r.* Neth. 17 J4
IJsselmeer *l.* Neth. 22 F1
Ikaahuk Canada *see* Sachs Harbour
Ikaalinen Fin. 15 M6
Ikageleng S. Africa 51 H3
Ikageng S. Africa 51 H4
iKapa S. Africa *see* Cape Town
Ikare Nigeria 46 D4
Ikaria *i.* Greece 27 L6
Ikast Denmark 15 F8
Ikeda Japan 44 F4
Ikela Dem. Rep. Congo 48 C4
Ikhtiman Bulg. 27 J3
Ikhutseng S. Africa 50 G5
Iki-Burul Rus. Fed. 13 J7
Ikom Nigeria 46 D4
Iksan S. Korea 45 B6
Ikungu Tanz. 49 D4
Ilagan Phil. 41 E6
Ilaisamis Kenya 48 D3
Īlām Iran 35 G4
Ilam Nepal 37 F4
Ilan Taiwan 43 M8
Ilave Peru 68 E7
Iława Poland 17 Q4
Ilazārān, Kūh-e *mt.* Iran 33 I4
Ilebo Dem. Rep. Congo 49 C4
Île Europa *i.* Indian Ocean *see* Europa, Île
Ilek Kazakh. 11 Q5
Ilemi Triangle *terr.* Africa 48 D3
Ilen *r.* Ireland 21 C6
Ileret Kenya 48 D3
Ileza Rus. Fed. 12 I3
Ilfeld Germany 17 M5
Ilford U.K. 19 H7
Ilfracombe Australia 56 D4
Ilfracombe U.K. 19 C7
Ilgaz Turkey 34 D2
Ilgın Turkey 34 C3
Ilha Grande, Represa *resr* Brazil 70 F2
Ilha Solteíra, Represa *resr* Brazil 71 A3
Ílhavo Port. 25 B3
Ilhéus Brazil 71 D1
Ili Kazakh. *see* Kapchagay
Iliamna Lake U.S.A. 60 C4
İliç Turkey 34 E3
Il'ichevsk Azer. *see* Şärur
Il'ichevsk Ukr. *see* Illichivs'k
Ilici Spain *see* Elche-Elx
Ilimananngip Nunaa *i.* Greenland 61 P2
Il'inka Rus. Fed. 13 J7
Il'inskiy *Permskiy Kray* Rus. Fed. 11 R4
Il'inskiy *Sakhalinskaya Oblast'* Rus. Fed.
44 F3
Il'insko-Podomskoye Rus. Fed. 12 J3
Ilion *NY* U.S.A. 64 D1
Ilium *tourist site* Turkey *see* Troy
Iliysk Kazakh. *see* Kapchagay
Ilkal India 38 C3
Ilkeston U.K. 19 F6
Ilkley U.K. 18 F5
Illapel Chile 70 B4
Illéla Niger 46 D3
Iller *r.* Germany 17 L6
Illichivs'k Ukr. 27 N1
Illimani, Nevado de *mt.* Bol. 68 E7
Illinois *r.* U.S.A. 63 I4
Illinois *state* U.S.A. 63 J4
Illizi Alg. 46 D2
Illogwa *watercourse* Australia 56 A5
Ilmajoki Fin. 14 M5
Il'men', Ozero *l.* Rus. Fed. 12 F4
Ilminster U.K. 19 E8
Ilo Peru 68 D7
Iloilo Phil. 41 E6
Ilomantsi Fin. 14 Q5
Ilorin Nigeria 46 D4
Ilovlya Rus. Fed. 13 I6
Iluka Australia 58 F2
Ilulissat Greenland 61 M3
Iluppur India 38 C4
Ilva *i.* Italy *see* Elba, Isola d'
Imabari Japan 45 D6
Imaichi Japan 45 E5
Imala Moz. 49 D5
İmamoğlu Turkey 34 D3
Iman Rus. Fed. *see* Dal'nerechensk
Iman *r.* Rus. Fed. 44 D3
Imari Japan 45 C6
Imaruí Brazil 71 A5
Imataca, Serranía de *mts* Venez. 68 F2
Imatra Fin. 15 P6
Imbituva Brazil 71 A4
imeni 26 Bakinskikh Komissarov Azer. *see*
Uzboy
imeni Babushkina Rus. Fed. 12 I4
imeni Kirova Kazakh. *see* Kopbirlik
imeni Petra Stuchki Latvia *see* Aizkraukle
imeni Poliny Osipenko Rus. Fed. 44 E1
imeni Tel'mana Rus. Fed. 44 D2
Īmī Eth. 48 E3
İmishli Azer. *see* İmişli
İmişli Azer. 35 H3
Imit Pak. 36 C1
Imja-do *i.* S. Korea 45 B6
Imola Italy 26 D2
iMonti S. Africa *see* East London
Impendle S. Africa 51 I5
Imperatriz Brazil 69 I5
Imperia Italy 26 C3
Imperial *CA* U.S.A. 65 E4
Imperial Beach *CA* U.S.A. 65 D4
Imperial Dam *Arizona/California* U.S.A.
65 E4
Imperial Valley *plain CA* U.S.A. 65 E4
Imperieuse Reef Australia 54 B4
Impfondo Congo 48 B3
Imphal India 37 H4
İmralı Adası *i.* Turkey 27 M4
İmroz Turkey 27 K4
İmroz *i.* Turkey *see* Gökçeada
Imtān Syria 39 C3
In *r.* Rus. Fed. 44 D2
Ina Japan 45 E6
Inambari *r.* Peru 68 E6
Inari Fin. 14 O2
Inarijärvi *l.* Fin. 14 O2
Inarijoki *r.* Fin./Norway 14 N2
Inca Spain 25 H4
İnce Burnu *pt* Turkey 27 L4
İnce Burun *pt* Turkey 34 D2
Inch Ireland 21 F5
Inchard, Loch *b.* U.K. 20 D2
Incheon S. Korea *see* Inch'ŏn
Inchicronan Lough *l.* Ireland 21 D5
Inch'ŏn S. Korea 45 B5
Incirli Turkey *see* Karasu
Indaal, Loch *b.* U.K. 20 C5
Indalsälven *r.* Sweden 14 J5
Indalstø Norway 15 D6
Inda Silasē Eth. 48 D2
Indé Mex. 66 C3
Indefatigable Island *Galápagos* Ecuador
see Santa Cruz, Isla
Independence *CA* U.S.A. 65 C2
Independence *MO* U.S.A. 63 I4
Inder China 44 A3
Inderbor Kazakh. 30 E2
Indi India 38 C2
India *country* Asia 31 I4
Indiana *PA* U.S.A. 64 B2
Indiana *state* U.S.A. 63 J3
Indian-Antarctic Ridge *sea feature*
Southern Ocean 74 D9
Indianapolis U.S.A. 63 J4
Indian Desert India/Pak. *see* Thar Desert
Indian Head Canada 62 G1
Indian Lake *NY* U.S.A. 64 D1
Indian Lake *PA* U.S.A. 64 B2
Indian Ocean 73
Indianola *IA* U.S.A. 63 I3
Indianola *MS* U.S.A. 63 I5
Indian Peak *UT* U.S.A. 65 F1
Indian Springs *NV* U.S.A. 65 E2
Indiga Rus. Fed. 12 K2
Indigirka *r.* Rus. Fed. 29 P2
Indigskaya Guba *b.* Rus. Fed. 12 K2
Indija Serbia 27 I2
Indio *CA* U.S.A. 65 D4
Indira Priyadarshini Pench National Park
India 36 D5
Indispensable Reefs Solomon Is 53 G3
Indjija Serbia *see* Inđija
Indonesia *country* Asia 41 D8
Indore India 36 C5
Indrapura, Gunung *vol.* Indon. *see*
Kerinci, Gunung
Indravati *r.* India 38 D2
Indre *r.* France 24 E3
Indulkana Australia 55 F6
Indur India *see* Nizamabad
Indus *r.* China/Pakistan 31 F4
Indus, Mouths of the Pak. 33 K5
Indus Cone *sea feature* Indian Ocean
73 M4
Indwe S. Africa 51 H6
İnebolu Turkey 34 D2
İnegöl Turkey 27 M4
Inevi Turkey *see* Cihanbeyli
Infantes Spain *see*
Villanueva de los Infantes
Infiernillo, Presa *resr* Mex. 66 D5
Inga Rus. Fed. 14 S3
Ingersoll *Ont.* Canada 64 A1
Ingham Australia 56 D3
Ingleborough *h.* U.K. 18 E4
Inglefield Land *reg.* Greenland 61 K2
Ingleton U.K. 18 E4
Inglewood *Qld* Australia 58 E2
Inglewood *Vic.* Australia 58 A6
Inglewood *CA* U.S.A. 65 C4
Ingoka Pum *mt.* Myanmar 37 I4
Ingoldmells U.K. 18 H5
Ingolstadt Germany 17 M6
Ingomar Australia 55 F7
Ingraj Bazar India 37 G4
Ingrid Christensen Coast Antarctica 76 E2
Ingwavuma S. Africa 51 K4
Ingwavuma *r.* S. Africa/Swaziland *see*
Ngwavuma
Inhaca Moz. 51 K3
Inhaca, Península *pen.* Moz. 51 K4
Inhambane Moz. 51 L2
Inhambane *prov.* Moz. 51 L2
Inhaminga Moz. 49 D5
Inharrime Moz. 51 L3
Inhassoro Moz. 49 D6
Inhaúmas Brazil 71 B1
Inhobim Brazil 71 C1
Inhumas Brazil 71 A2
Inis Ireland *see* Ennis
Inis Córthaidh Ireland *see* Enniscorthy
Inishark *i.* Ireland 21 B4
Inishbofin *i.* Ireland 21 B4
Inisheer *i.* Ireland 21 C4
Inishkea North *i.* Ireland 21 B3
Inishkea South *i.* Ireland 21 B3
Inishmaan *i.* Ireland 21 C4
Inishmore *i.* Ireland 21 C4
Inishmurray *i.* Ireland 21 D3
Inishowen *pen.* Ireland 21 E2
Inishowen Head *hd* Ireland 21 F2
Inishtrahull *i.* Ireland 21 E2
Inishturk *i.* Ireland 21 B4
Injune Australia 57 E5
Inkerman Australia 56 C3
Inland Kaikoura Range *mts* N.Z. 59 D6
Inland Sea Japan *see* Seto-naikai
Inlet *NY* U.S.A. 64 D1
Inn *r.* Europe 17 M7
Innaanganeq *c.* Greenland 61 L2
Innamincka Australia 57 C5
Innamincka Regional Reserve *nature res.*
Australia 57 C5
Inndyr Norway 14 I3
Inner Sound *sea chan.* U.K. 20 D3
Innes National Park Australia 57 B7
Innisfail Australia 56 D3
Innokent'yevka Rus. Fed. 44 C2
Innoko *r.* U.S.A. 60 C3
Innsbruck Austria 17 M7
Inny *r.* Ireland 21 E4
Inocência Brazil 71 A2
Inongo Dem. Rep. Congo 48 B4
İnönü Turkey 27 N5
Inowrocław Poland 17 Q4
In Salah Alg. 46 D2
Insch U.K. 20 G3
Inscription, Cape Australia 56 B3
Insterburg Rus. Fed. *see* Chernyakhovsk
Inta Rus. Fed. 11 S2
Interamna Italy *see* Teramo
Interlaken Switz. 24 H3
International Falls U.S.A. 63 I2
Intutu Peru 68 D4
Inubō-zaki *pt* Japan 45 F6
Inuvik Canada 60 E3
Inveraray U.K. 20 D4
Inverbervie U.K. 20 G4
Invercargill N.Z. 59 B8
Inverell Australia 58 E2
Invergordon U.K. 20 E3
Inverkeithing U.K. 20 F4
Inverleigh Australia 56 C3
Inverness U.K. 20 E3
Inverness *CA* U.S.A. 65 A1
Inverurie U.K. 20 G3
Investigator Group *is* Australia 55 F8
Investigator Ridge *sea feature*
Indian Ocean 73 O6
Investigator Strait Australia 57 B7
Inwood *WV* U.S.A. 64 B3
Inya Rus. Fed. 42 F2
Inyanga Zimbabwe *see* Nyanga
Inyokern *CA* U.S.A. 65 D3
Inyo Mountains *CA* U.S.A. 65 C2
Inyonga Tanz. 49 D4
Inza Rus. Fed. 13 J5
Inzhavino Rus. Fed. 13 I5
Ioannina Greece 27 I5
Iokan'ga *r.* Rus. Fed. 12 H2
Iola U.S.A. 63 H4
Iolgo, Khrebet *mts* Rus. Fed. 42 F2
Iolotan' Turkm. *see* Ýolöten
Iona *i.* U.K. 20 C4
Iona, Parque Nacional do *nat. park* Angola
49 B5
Ione *NV* U.S.A. 65 D1
Iongo Angola 49 B4
Ionian Islands Greece 27 H5
Ionian Sea Greece/Italy 26 H5
Ionioi Nisoi *is* Greece *see* Ionian Islands
Ios *i.* Greece 27 K6
Iowa *state* U.S.A. 63 I3
Iowa City U.S.A. 63 I3
Ipameri Brazil 71 A2
Ipanema Brazil 71 C2
Iparía Peru 68 D5
Ipatinga Brazil 71 C2
Ipatovo Rus. Fed. 13 I7
Ipelegeng S. Africa 51 G4
Ipiales Col. 68 C3
Ipiaú Brazil 71 D1
Ipirá Brazil 71 D1
Iporá Brazil 71 A4
iPitoli S. Africa *see* Pretoria
Ipixuna *r.* Brazil 68 F5
Ipoh Malaysia 41 C7
Iporá Brazil 71 A2
Ippy Cent. Afr. Rep. 48 C3
Ipsala Turkey 27 L4
Ipswich Australia 58 F1
Ipswich U.K. 19 I6
Ipu Brazil 69 J4
Iqaluit Canada 61 L3
Iquique Chile 70 B2
Iquiri *r.* Brazil *see* Ituxi
Iquitos Peru 68 D4
Irai Brazil 70 F3
Irakleio Greece *see* Iraklion
Iraklion Greece 27 K7
Iramaia Brazil 71 C1
Iran *country* Asia 32 I3
Īrānshahr Iran 33 J4
Irapuato Mex. 66 D4
Iraq *country* Asia 35 F4
Irara Brazil 71 D1
Irati Brazil 71 A4
Irayel' Rus. Fed. 12 L2
Irazú, Volcán *vol.* Costa Rica 67 H7
Irbid Jordan 39 B3
Irbil Iraq *see* Arbil
Irbit Rus. Fed. 28 H4
Irecê Brazil 69 J6
Ireland *country* Europe 21 E4
Ireland *i.* Ireland/U.K. 21
Irema Dem. Rep. Congo 48 C4
Irgiz Kazakh. 42 A3
Iri S. Korea *see* Iksan
Irian, Teluk *b.* Indon. *see*
Cenderawasih, Teluk
Iriba Chad 47 F3
Īrī Dāgh *mt.* Iran 35 G3
Irîgui *reg.* Mali/Mauritania 46 C3
Iringa Tanz. 49 D4
Iriri *r.* Brazil 69 H4
Irish Sea Ireland/U.K. 21 G4
Irituia Brazil 69 I4
Irkutsk Rus. Fed. 42 I2
Irmak Turkey 34 D3
Irminger Basin *sea feature*
N. Atlantic Ocean 72 F2
Iron Baron Australia 57 B7
Irondequoit *NY* U.S.A. 64 C1
Iron Mountain U.S.A. 63 J2
Iron Mountain *UT* U.S.A. 65 F2
Iron Range National Park Australia 56 C2
Irosin Phil. 41 E6
Irpen' Ukr. *see* Irpin'
Irpin' Ukr. 13 F6
Irrawaddy *r.* Myanmar 37 H6
Irrawaddy, Mouths of the Myanmar 31 I5
Irshad Pass Afgh./Pak. 36 C1
Irta Rus. Fed. 12 K3
Irthing *r.* U.K. 18 E4
Irtysh *r.* Kazakh./Rus. Fed. 31 G1
Irun Spain 25 F2
Iruña Spain *see* Pamplona
Iruñea Spain *see* Pamplona
Irvine U.K. 20 E5
Irvine *CA* U.S.A. 65 D4
Irvine Glacier Antarctica 76 L2
Irwin *r.* Australia 55 A7
Isa Nigeria 46 D3
Isaac *r.* Australia 56 E4
Isabela Phil. 41 E7
Isabela, Isla *i. Galápagos* Ecuador
68 [inset]
Isabelia, Cordillera *mts* Nicaragua 67 G6
Isabella Lake *CA* U.S.A. 65 C3
Isachsen, Cape Canada 61 H2
Ísafjarðardjúp *est.* Iceland 14 [inset]
Ísafjörður Iceland 14 [inset]
Isa Khel Pak. 36 B2
Isar *r.* Germany 17 N6
Isbister U.K. 20 [inset]
Ischia, Isola d' *i.* Italy 26 E4
Ise Japan 45 E6
Isère *r.* France 24 G4
Isère, Pointe *pt* Fr. Guiana 69 H2
Isernia Italy 26 F4
Ise-shima Kokuritsu-kōen Japan 45 E6
Ise-wan *b.* Japan 45 E6
Iseyin Nigeria 46 D4
Isfahan Iran *see* Eşfahān
Isheyevka Rus. Fed. 13 K5
Ishigaki Japan 43 M8
Ishikari-wan *b.* Japan 44 F4
Ishim *r.* Kazakh./Rus. Fed. 30 G1
Ishinomaki Japan 45 F5
Ishinomaki-wan *b.* Japan 43 Q5
Ishioka Japan 45 F5
Ishkoshim Tajik. 36 B1
Ishurdi Bangl. 37 G4
Ishwardi Bangl. *see* Ishurdi
Isiboro Sécure, Parque Nacional *nat. park*
Bol. 68 E7
Isigny-sur-Mer France 19 F9
Işıklar Dağı *mts* Turkey 27 L4
Işıklı Turkey 27 M5
Isil'kul' Rus. Fed. 28 I4
iSimangaliso Wetland Park *nature res.*
S. Africa 51 K4
Isipingo S. Africa 51 J5
Isiro Dem. Rep. Congo 48 C3
Isisford Australia 56 D5
Iskateley Rus. Fed. 12 L2
İskenderun Turkey 39 C1
İskenderun Körfezi *b.* Turkey 39 B1
İskilip Turkey 34 D2
Iskitim Rus. Fed. 28 J4
Iskŭr *r.* Bulg. 27 K3
Iskushuban Somalia 48 F2
Isla *r. Scotland* U.K. 20 F4
Isla *r. Scotland* U.K. 20 G3
Isla Gorge National Park Australia 56 E5
İslahiye Turkey 34 E3
Islamabad India *see* Anantnag
Islamabad Pak. 33 L3
Islamgarh Pak. 36 B4
Islamkot Pak. 36 B4
Ísland *country* Europe *see* Iceland
Island Lagoon *imp. l.* Australia 57 B6
Island Lake Canada 63 I1
Island Magee *pen.* U.K. 21 G3
Islands, Bay of N.Z. 59 E2
Islay *i.* U.K. 20 C5
Isle of Man *terr.* Irish Sea 18 C4
Isle of Wight *VA* U.S.A. 64 C4
Ismail Ukr. *see* Izmayil
Ismā'īlīya Egypt *see* Al Ismā'īlīyah
Ismā'īlīya *governorate* Egypt *see*
Al Ismā'īlīyah
İsmailly Azer. *see* İsmayıllı
İsmayıllı Azer. 35 H2
Isojoki Fin. 14 L5
Isoka Zambia 49 D5
Isokylä Fin. 14 O3
Isokyrö Fin. 14 M5
Isola di Capo Rizzuto Italy 26 G5
Ispahan Iran *see* Eşfahān
Isparta Turkey 27 N6
Isperikh Bulg. 27 L3
İspir Turkey 35 F2
Ispisar Tajik. *see* Khŭjand
Isplinji Pak. 36 A3
Israel *country* Asia 39 B4
Israelite Bay Australia 55 C8
Isra'īl *country* Asia *see* Israel
Issia Côte d'Ivoire 46 C4
Issoire France 24 F4
Issyk-Kul' Kyrg. *see* Balykchy
Issyk-Kul', Ozero *salt l.* Kyrg. *see* Ysyk-Köl
İstanbul Turkey 27 M4
İstanbul Boğazı *str.* Turkey *see* Bosporus
Īstgāh-e Eznā Iran 35 H4
Istiaia Greece 27 J5
Istik *r.* Tajik. 36 C1
Istočni Drvar Bos.-Herz. 26 G2
Istra *pen.* Croatia *see* Istria
Istres France 24 G5
Istria *pen.* Croatia 26 E2
Iswardi Bangl. *see* Ishurdi
Itabapoana *r.* Brazil 71 C3
Itaberá Brazil 71 A3
Itaberaba Brazil 71 C1
Itaberaí Brazil 71 A2
Itabira Brazil 71 C2
Itabirito Brazil 71 C3
Itabuna Brazil 71 D1
Itacajá Brazil 69 I5
Itacarambi Brazil 71 B1
Itacoatiara Brazil 69 G4
Itaetê Brazil 71 C1
Itagmatana Iran *see* Hamadān
Itaguaçu Brazil 71 C2
Itaí Brazil 71 A3
Itaiópolis Brazil 71 A4
Itäisen Suomenlahden kansallispuisto
nat. park Fin. 15 O6
Itaituba Brazil 69 G4
Itajaí Brazil 71 A4
Itajubá Brazil 71 B3
Itajuipe Brazil 71 D1
Italia *country* Europe *see* Italy
Italia, Laguna *l.* Bol. 68 F6
Italy *country* Europe 26 E3
Itamarandiba Brazil 71 C2
Itambé Brazil 71 C1
Itambé, Pico de *mt.* Brazil 71 C2
Itampolo Madag. 49 E6
Itanagar India 37 H4
Itanguari *r.* Brazil 71 B1
Itanhaém Brazil 71 B4
Itanhém Brazil 71 C2
Itanhém *r.* Brazil 71 D2
Itaobím Brazil 71 C2
Itapaci Brazil 71 A1
Itapajipe Brazil 71 A2
Itapebi Brazil 71 D1
Itapecerica Brazil 71 B3
Itapecuru Mirim Brazil 69 J4
Itapemirim Brazil 71 C3
Itaperuna Brazil 71 C3
Itapetinga Brazil 71 C1
Itapetininga Brazil 71 A3
Itapeva Brazil 71 A3
Itapeva, Lago *l.* Brazil 71 A5
Itapicuru Brazil 69 J6
Itapicuru, Serra de *hills* Brazil 69 I5
Itapipoca Brazil 69 K4
Itapira Brazil 71 B3
Itaporanga Brazil 71 A3
Itapuã Brazil 71 A5
Itaqui Brazil 70 E3
Itararé Brazil 71 A4
Itarsi India 36 D5
Itarumã Brazil 71 A2
Itatiba Brazil 71 B3
Itatuba Brazil 68 F5
Itaúna Brazil 71 B3
Itaúnas Brazil 71 D2

Itbayat i. Phil. 43 M8
Itea Greece 27 J5
Ithaca NY U.S.A. 64 C1
iThekweni S. Africa see Durban
Ithrah Saudi Arabia 39 C4
Itilleq Greenland 61 M3
Itimbiri r. Dem. Rep. Congo 48 C3
Itinga Brazil 71 C2
Itiquira Brazil 69 H7
Itiruçu Brazil 71 C1
Itiúba, Serra de hills Brazil 69 K6
Itō Japan 45 E6
iTswane S. Africa see Pretoria
Ittiri Sardegna Italy 26 C4
Ittoqqortoormiit Greenland 61 P2
Itu Brazil 71 B3
Ituaçu Brazil 71 C1
Ituberá Brazil 71 D1
Ituí r. Brazil 68 D4
Ituiutaba Brazil 71 A2
Itumbiara Brazil 71 A2
Itumbiara, Barragem resr Brazil 71 A2
Ituni Guyana 69 G2
Itupiranga Brazil 69 I5
Ituporanga Brazil 71 A4
Iturama Brazil 71 A2
Ituri r. Dem. Rep. Congo 48 C3
Iturup, Ostrov i. Rus. Fed. 44 G3
Itutinga Brazil 71 B3
Ituxi r. Brazil 68 F5
Ītyop'ia country Africa see Ethiopia
Itzehoe Germany 17 L4
Iul'tin Rus. Fed. 29 T3
Ivalo Fin. 14 O2
Ivalojoki r. Fin. 14 O2
Ivanava Belarus 15 N10
Ivanhoe Australia 58 B4
Ivankiv Ukr. 13 F6
Ivankovtsy Rus. Fed. 44 D2
Ivano-Frankivs'k Ukr. 13 E6
Ivano-Frankovsk Ukr. see Ivano-Frankivs'k
Ivanovka Rus. Fed. 44 B2
Ivanovo Belarus see Ivanava
Ivanovo tourist site Bulg. 27 K3
Ivanovo Rus. Fed. 12 I4
Ivanteyevka Rus. Fed. 13 K5
Ivantsevichi Belarus see Ivatsevichy
Ivatsevichy Belarus 15 N10
Ivaylovgrad Bulg. 27 L4
Ivdel' Rus. Fed. 11 S3
Ivittuut Greenland 61 N3
Iviza i. Spain see Ibiza
Ivory Coast country Africa see Côte d'Ivoire
Ivrea Italy 26 B2
İvrindi Turkey 27 L5
Ivris Ugheltekhili pass Georgia 35 G2
Ivugivik Canada see Ivujivik
Ivujivik Canada 61 K3
Ivyanyets Belarus 15 O10
Ivydale WV U.S.A. 64 A3
Iwaki Japan 45 F5
Iwaki-san vol. Japan 44 F4
Iwakuni Japan 45 D6
Iwamizawa Japan 44 F4
Iwo Nigeria 46 D4
Iwye Belarus 15 N10
Ixiamas Bol. 68 E6
Ixmiquilpán Mex. 66 E4
Ixopo S. Africa 51 J6
Ixtlán Mex. 66 D4
Ixworth U.K. 19 H6
İyirmi Altı Bakı Komissarı Azer. see Uzboy
Izabal, Lago de l. Guat. 66 G5
Izberbash Rus. Fed. 35 G2
Īzeh Iran 35 H5
Izgal Pak. 36 C2
Izhevsk Rus. Fed. 11 Q4
Izhma Respublika Komi Rus. Fed. 12 L2
Izhma Respublika Komi Rus. Fed. see Sosnogorsk
Izhma r. Rus. Fed. 12 L2
Izmail Ukr. see Izmayil
Izmayil Ukr. 27 M2
İzmir Turkey 27 L5
İzmir Körfezi g. Turkey 27 L5
İzmit Turkey 27 M4
İzmit Körfezi b. Turkey 27 M4
Izozog Bol. 68 F7
Izra' Syria 39 C3
Iztochni Rodopi mts Bulg. 27 K4
Izu-hantō pen. Japan 45 E6
Izuhara Japan 45 C6
Izumo Japan 45 D6
Izu-Ogasawara Trench sea feature N. Pacific Ocean 74 F3
Izu-shotō is Japan 45 E6
Izyaslav Ukr. 13 E6
Iz"yayu Rus. Fed. 12 M2
Izyum Ukr. 13 H6

Jabalón r. Spain 25 D4
Jabalpur India 36 D5
Jabbūl, Sabkhat al salt flat Syria 39 C2
Jabiru Australia 54 F3
Jablah Syria 39 B2
Jablanica Bos.-Herz. 26 G3
Jaboatão dos Guararapes Brazil 69 L5
Jaboticabal Brazil 71 A3
Jacaraci Brazil 71 C1
Jacareacanga Brazil 69 G5
Jacareí Brazil 71 B3
Jacarézinho Brazil 71 A3
Jacinto Brazil 71 C2
Jack r. Australia 56 D2
Jackson Australia 58 D1
Jackson AL U.S.A. 63 J5
Jackson CA U.S.A. 65 B1
Jackson MI U.S.A. 63 K3
Jackson MS U.S.A. 63 I5
Jackson TN U.S.A. 63 J4
Jackson WY U.S.A. 62 E3
Jackson, Mount Antarctica 76 L2
Jackson Head hd N.Z. 59 B6
Jacksonville FL U.S.A. 63 K5
Jacksonville IL U.S.A. 63 I4
Jacksonville NC U.S.A. 63 L5
Jack Wade U.S.A. 60 D3
Jacmel Haiti 67 J5
Jacobabad Pak. 33 K4
Jacobina Brazil 69 J6
Jacobsdal S. Africa 50 G5
Jacques-Cartier, Mont mt. Canada 63 N2
Jacuí Brazil 71 B3
Jacuípe r. Brazil 69 K6
Jacunda Brazil 69 I4
Jaddangi India 38 D2
J. A. D. Jensen Nunatakker nunataks Greenland 61 N3
Jadotville Dem. Rep. Congo see Likasi
Jādū Libya 46 E1
Jaén Spain 25 E5
Jaffa, Cape Australia 57 B8
Jaffna Sri Lanka 38 C4
Jafr, Qā' al imp. l. Jordan 39 C4
Jagadhri India 36 D3
Jagalur India 38 C3
Jagatsinghapur India see Jagatsinghpur
Jagatsinghpur India 37 F5
Jagdalpur India 38 D2
Jagdaqi China 44 B2
Jagersfontein S. Africa 51 G5
Jaggang China 36 D2
Jaggayyapeta India 38 D2
Jaghīn Iran 33 I4
Jagok Tso salt l. China see Urru Co
Jagtial India 38 C2
Jaguariaíva Brazil 71 A4
Jaguaripe Brazil 71 D1
Jahanabad India see Jehanabad
Jahmah well Iraq 35 G5
Jahrom Iran 35 I5
Jaicós Brazil 69 J5
Jaigarh India 38 B2
Jailolo Gilolo i. Indon. see Halmahera
Jainpur India 37 E4
Jaintapur Bangl. see Jaintiapur
Jaintiapur Bangl. 37 H4
Jaipur India 36 C4
Jaipurhat Bangl. see Joypurhat
Jais India 37 E4
Jaisalmer India 36 B4
Jaisamand Lake India 36 C4
Jaitaran India 36 C4
Jaitgarh h. India 38 C1
Jajapur India see Jajpur
Jajarkot Nepal 33 N4
Jajce Bos.-Herz. 26 G2
Jajnagar state India see Orissa
Jajpur India 37 F5
Jakān mt. Afgh. 36 A3
Jakar Bhutan 37 G4
Jakarta Indon. 41 C8
Jakhau India 36 B5
Jäkkvik Sweden 14 J3
Jakliat India 36 C3
Jakobshavn Greenland see Ilulissat
Jakobstad Fin. 14 M5
Jalaid China see Inder
Jalālābād Afgh. 36 B2
Jalal-Abad Kyrg. 42 C4
Jalālah al Baḥrīyah, Jabal plat. Egypt 34 C5
Jalalpur Pirwala Pak. 36 B3
Jalāmid, Ḥazm al ridge Saudi Arabia 35 E5
Jalandhar India 36 C3
Jalapa Mex. see Xalapa
Jalapa Enríquez Mex. see Xalapa
Jalasjärvi Fin. 14 M5
Jalaun India 36 D4
Jalawlā' Iraq 35 G4
Jaldak Afgh. 36 A3
Jaldrug India 38 C2
Jales Brazil 71 A3
Jalesar India 36 D4
Jalgaon India 36 C5
Jalībah Iraq 35 G5
Jalingo Nigeria 46 E4
Jalna India 38 B2
Jalón r. Spain 25 F3
Jalor India see Jalore
Jalore India 36 C4
Jalpa Mex. 66 D4
Jalpaiguri India 37 G4
Jālū Libya 47 F2
Jalūlā Iraq see Jalawlā'
Jamaica country West Indies 67 I5
Jamaica Channel Haiti/Jamaica 67 I5
Jamalpur Bangl. 37 G4
Jamalpur India 37 F4
Jamanxim r. Brazil 69 G4
Jambi Indon. 41 C8
Jambin Australia 56 E5
Jambo India 36 C4
Jamda India 37 F5
Jamekunte India 38 C2
James r. N. Dakota/S. Dakota U.S.A. 62 H3
James r. VA U.S.A. 64 C4
James, Baie b. Canada see James Bay
Jamesabad Pak. 36 B4
James Bay Canada 63 K1
James Island Galápagos Ecuador see San Salvador, Isla
Jameson Land reg. Greenland 61 P2
James Peak N.Z. 59 B7
James Ranges mts Australia 55 F6
James Ross Island Antarctica 76 A2
James Ross Strait Canada 61 I3
Jamestown Australia 57 B7
Jamestown S. Africa 51 H6
Jamestown St Helena 72 H7
Jamestown ND U.S.A. 62 H2
Jamestown NY U.S.A. 64 B1
Jamkhed India 38 B2
Jammu India 36 C2
Jammu and Kashmir terr. Asia 36 D2
Jamnagar India 36 B5
Jampur Pak. 36 B3
Jamrud Pak. 36 B2
Jämsä Fin. 15 N6
Jamsah Egypt 34 D6
Jämsänkoski Fin. 14 N6
Jamshedpur India 37 F5
Jamtari Nigeria 46 E4
Jamui India 37 F4
Jamuna r. Bangl. see Raimangal
Jamuna r. India see Yamuna
Janā i. Saudi Arabia 35 H6
Janāb, Wādī al watercourse Jordan 39 C4
Janakpur India 37 E5
Janaúba Brazil 71 C1
Jandaia Brazil 71 A2
Jandaq Iran 35 I4
Jandola Pak. 36 B2
Jandowae Australia 58 E1
Janesville WI U.S.A. 63 J3
Jangada Brazil 71 A4
Jangamo Moz. 51 L3
Jangaon India 38 C2
Jangipur India 37 G4
Jangngai Turkm. see Jaňňa
Jangngai Ri mts China 37 F2
Jan Mayen terr. Arctic Ocean 77 I2
Jan Mayen Fracture Zone sea feature Arctic Ocean 77 I2
Jaňňa Turkm. 35 I2
Jansenville S. Africa 50 G7
Januária Brazil 71 B1
Janūb Sīnā' governorate Egypt see Janūb Sīnā'
Janūb Sīnā' governorate Egypt 39 A5
Japan country Asia 45 D5
Japan, Sea of N. Pacific Ocean 45 D5
Japan Alps National Park Japan see Chūbu-Sangaku Kokuritsu-kōen
Japan Trench sea feature N. Pacific Ocean 74 F3
Japiim Brazil 68 D5
Japurá r. Brazil 68 F4
Japvo Mount India 37 H4
Jarābulus Syria 39 D1
Jaraguá Brazil 71 A1
Jaraguá, Serra mts Brazil 71 A4
Jaraguá do Sul Brazil 71 A4
Jarash Jordan 39 B3
Jardine River National Park Australia 56 C1
Jardinésia Brazil 71 A2
Jardinópolis Brazil 71 B3
Jargalang China 44 A4
Jargalant Bayanhongor Mongolia 42 H3
Jargalant Hovd Mongolia see Hovd
Jari r. Brazil 69 H4
Järna Sweden 15 J7
Jarocin Poland 17 P5
Jarosław Poland 13 D6
Järpen Sweden 14 H5
Jarrettsville MD U.S.A. 64 C3
Jarú Brazil 68 F6
Jarud China see Lubei
Järvakandi Estonia 15 N7
Järvenpää Fin. 15 N6
Jarvis Island terr. S. Pacific Ocean 74 J6
Jarwa India 37 E4
Jashpurnagar India 37 F5
Jāsk Iran 33 I4
Jasliq Uzbek. 35 J2
Jasło Poland 13 D6
Jasol India 36 C4
Jason Islands Falkland Is 70 D8
Jason Peninsula Antarctica 76 L2
Jasper Canada 62 D1
Jasper NY U.S.A. 64 C1
Jasper TX U.S.A. 63 I5
Jasrasar India 36 C4
Jaşşān Iraq 35 G4
Jassy Romania see Iaşi
Jastrzębie-Zdrój Poland 17 Q6
Jaswantpura India 36 C4
Jászberény Hungary 27 H1
Jataí Brazil 71 A2
Jatapu r. Brazil 69 G4
Jath India 38 B2
Jati Pak. 33 K5
Játiva Spain see Xàtiva
Jatoi Janubi Pak. 36 B3
Jaú Brazil 71 A3
Jaú r. Brazil 68 F4
Jaú, Parque Nacional do nat. park Brazil 68 F4
Jaua-Sarisariñama, Parque Nacional nat. park Venez. 68 F3
Jauja Peru 68 C6
Jaunlutriņi Latvia 15 M8
Jaunpiebalga Latvia 15 O8
Jaunpur India 37 E4
Java Georgia 35 F2
Java i. Indon. 41 C8
Javaés r. Brazil see Formoso
Javari r. Brazil/Peru see Yavari
Java Ridge sea feature Indian Ocean 73 P6
Java Sea Indon. see Jawa, Laut
Java Trench sea feature Indian Ocean 74 C6
Jävre Sweden 14 L4
Jawa i. Indon. see Java
Jawa, Laut sea Indon. 41 D8
Jawhar India 38 B2
Jawhar Somalia 48 E3
Jawor Poland 17 P5
Jaya, Puncak mt. Indon. 41 F8
Jayakusumu mt. Indon. see Jaya, Puncak
Jayakwadi Sagar l. India 38 B2
Jayantiapur Bangl. see Jaintiapur
Jayapura Indon. 41 G8
Jayb, Wādī al watercourse Israel/Jordan 39 B4
Jayfi, Wādī al watercourse Egypt 39 B4
Jaypur India 38 D2
Jayrūd Syria 39 C3
Jazīreh-ye Shīf Iran 35 H5
Jbail Lebanon 39 B2
Jean NV U.S.A. 65 E3
Jeannin, Lac l. Canada 61 L4
Jebel, Bahr el r. Sudan/Uganda see White Nile
Jebel, Bahr el r. Sudan/Uganda see White Nile
Jebel Abyad Plateau Sudan 32 C6
Jebel Bani ridge Morocco 22 C6
Jebel Toubkal mt. Morocco 22 C5
Jech Doab lowland Pak. 36 C3
Jedburgh U.K. 20 G5
Jeddah Saudi Arabia 32 E5
Jedeida Tunisia 26 C6
Jefferson OH U.S.A. 64 A2
Jefferson, Mount NV U.S.A. 65 D1
Jefferson City U.S.A. 63 I4
Jeffreys Bay S. Africa 50 G8
Jehanabad India 37 F4
Jeju S. Korea see Cheju
Jejuí Guazú r. Para. 70 E2
Jēkabpils Latvia 15 N8
Jelbart Ice Shelf Antarctica 76 B2
Jelenia Góra Poland 17 O5
Jelep La pass China/India 37 G4
Jelgava Latvia 15 M8
Jember Indon. 41 D8
Jena Germany 17 M5
Jendouba Tunisia 26 C6
Jengish Chokusu mt. China/Kyrg. see Pobeda Peak
Jenīn West Bank 39 B3
Jenne Mali see Djenné
Jennings U.S.A. 63 I5
Jenolan Caves Australia 58 E4
Jens Munk Island Canada 61 K3
Jeparit Australia 57 C8
Jequié Brazil 71 C1
Jequitaí r. Brazil 71 B2
Jequitinhonha Brazil 71 C2
Jequitinhonha r. Brazil 71 D1
Jerba, Île de i. Tunisia 22 G5
Jerbar South Sudan 47 G4
Jereh Iran 35 H5
Jérémie Haiti 67 J5
Jerez Mex. 66 D4
Jerez de la Frontera Spain 25 C5
Jergol Norway 14 N2
Jergucat Albania 27 I5
Jericho Australia 56 D4
Jericho West Bank 39 B4
Jerid, Chott el salt l. Tunisia 22 F5
Jerilderie Australia 58 B5
Jerimoth Hill RI U.S.A. 64 F2
Jeroaquara Brazil 71 A1
Jerome U.S.A. 62 E3
Jerruck Pak. 36 B4
Jersey terr. Channel Is 19 E9
Jersey City NJ U.S.A. 64 D2
Jersey Shore PA U.S.A. 64 C2
Jerumenha Brazil 69 J5
Jerusalem Israel/West Bank 39 B4
Jervis Bay Australia 58 E5
Jervis Bay b. Australia 58 E5
Jervis Bay Territory admin. div. Australia 58 E5
Jesenice Slovenia 26 F1
Jesi Italy 26 E3
Jesselton Sabah Malaysia see Kota Kinabalu
Jessheim Norway 15 G6
Jessore Bangl. 37 G5
Jesu Maria Island P.N.G. see Rambutyo Island
Jesup U.S.A. 63 K5
Jetpur India 36 B5
Jewish Autonomous Oblast admin. div. Rus. Fed. see Yevreyskaya Avtonomnaya Oblast'
Jeypur India see Jaypur
Jezzine Lebanon 39 B3
Jhabua India 36 C5
Jhajhar India see Jhajjar
Jhajjar India 36 D3
Jhal Pak. 36 A3
Jhalawar India 36 D4
Jhang Pak. 33 L3
Jhansi India 36 D4
Jhapa Nepal 37 F4
Jharia India 37 F5
Jharkhand state India 37 F5
Jharsuguda India 37 F5
Jhawani Nepal 37 F4
Jhelum r. India/Pak. 36 C3
Jhelum Pak. 33 L3
Jhenaidah Bangl. 37 G5
Jhenaidaha Bangl. see Jhenaidah
Jhenida Bangl. see Jhenaidah
Jhimpir Pak. 36 B4
Jhudo Pak. 36 B4
Jhumritilaiya India 37 F4
Jhund India 36 B5
Jhunjhunun India 36 C3
Jiamusi China 44 C3
Ji'an Jiangxi China 43 K7
Ji'an Jilin China 44 B4
Jiangcheng China 42 I8
Jiangjunmiao China 42 F4
Jiangmen China 43 K8
Jiangna China see Yanshan
Jiangsi China see Dejiang
Jiangsu prov. China 43 L6
Jiangxi prov. China 43 L7
Jiangzhesongrong China 37 F3
Jianshui Hu l. China 37 E2
Jianyang Fujian China 43 L7
Jiaohe China 44 B4
Jiaojiang China see Taizhou
Jia Tsuo La pass China 37 F3
Jiaxing China 43 M6
Jiayi Taiwan see Chiai
Jiayin China 44 C2
Jiayuguan China 42 H5
Jībūtī country Africa see Djibouti
Jibuti Djibouti see Djibouti
Jiddah Saudi Arabia see Jeddah
Jiddī, Jabal al h. Egypt 39 A4
Jidong China 44 C3
Jiehkkevárri mt. Norway 14 K2
Jiexiu China 43 K5
Jieyang China 43 L8
Jieznas Lith. 15 N9
Jigzhi China 42 I6
Jiḥār, Wādī al watercourse Syria 39 C2
Jihlava Czech Rep. 17 O6
Jijel Alg. 22 F4
Jijiga Eth. 48 E3
Jījīrud Iran 35 H4
Jil'ād reg. Jordan 39 B3
Jilf al Kabīr, Haḍabat al plat. Egypt 32 C5
Jilib Somalia 48 E3
Jilin China 44 B4
Jilin prov. China 44 B4
Jilin Hada Ling mts China 44 B4
Jiliu He r. China 44 A2
Jilo India 36 C4
Jilong Taiwan see Chilung
Jīma Eth. 48 D3
Jiménez Chihuahua Mex. 66 D3
Jiménez Tamaulipas Mex. 66 E4
Jimía, Cerro mt. Hond. 66 G5
Jimsar China 42 F4
Jim Thorpe PA U.S.A. 64 D2
Jinan China 43 L5
Jin'an China see Songpan
Jinbi China see Dayao
Jinchang China 42 I5
Jincheng Shanxi China 43 K5
Jinchengjiang China see Hechi
Jinchuan Gansu China see Jinchang
Jind India 36 D3
Jindřichův Hradec Czech Rep. 17 O6
Jingbian China 43 J5
Jingdezhen China 43 L7
Jingellic Australia 58 C5
Jinggangshan China see Ciping
Jinghong China 42 I8
Jingle China 43 K5
Jingmen China 43 K6
Jingpo China 44 C4
Jingpo Hu l. China 44 C4
Jingsha China see Jingzhou
Jingtai China 42 I5
Jingtieshan China 42 H5
Jingxian Hunan China see Jingzhou
Jingyu China 44 B4
Jingyuan China 42 I5
Jingzhou Hubei China 43 K6
Jingzhou Hunan China 43 J7
Jinhe Nei Mongol China 44 A2
Jinhua Zhejiang China 43 L7
Jining Nei Mongol China 43 K4
Jining Shandong China 43 L5
Jinja Uganda 48 D3
Jinka Eth. 48 D3
Jinotepe Nicaragua 67 G6
Jinping Guizhou China 43 J7
Jinsen S. Korea see Inch'ŏn
Jinsha Jiang r. China see Yangtze
Jinshan Nei Mongol China see Guyang
Jinshi Hunan China 43 K7
Jintur India 38 C2
Jinxi Jiangxi China 43 L7
Jinxi Liaoning China see Lianshan
Jinxian China 43 L7
Jinz, Qā' al salt flat Jordan 39 C4
Jinzhong China 43 K5
Jinzhou China 43 M4
Ji-Paraná Brazil 68 F6
Jipijapa Ecuador 68 B4
Ji Qu r. China 37 I3
Jiquiriçá Brazil 71 D1
Jiquitaia Brazil 71 D2
Jirā', Wādī watercourse Egypt 39 A5
Jirānīyāt, Shi'bān al watercourse Saudi Arabia 39 D4
Jīroft Iran 33 I4
Jirriiban Somalia 48 E3
Jishou China 43 J7
Jisr ash Shughūr Syria 39 C2
Jiu r. Romania 27 J3
Jiuding Shan mt. China 42 I6
Jiujiang Jiangxi China 43 L7
Jiulian China see Mojiang
Jiulong H.K. China see Kowloon
Jiulong Sichuan China 42 I7
Jiwani Pak. 33 J4
Jiwen China 44 A2
Jixi Heilong. China 44 C3
Jixian China 44 C3
Jīzah, Ahrāmāt al tourist site Egypt see Pyramids of Giza
Jīzān Saudi Arabia 32 F6
Jizzakh Uzbek. see Jizzax
Jizzax Uzbek. 33 K1
Joaçaba Brazil 71 A4
Joaíma Brazil 71 C2
João Belo Moz. see Xai-Xai
João de Almeida Angola see Chibia
João Pessoa Brazil 69 L5
João Pinheiro Brazil 71 B2
Joaquín V. González Arg. 70 D3
Joda India 37 F5
Jodhpur India 36 C4
Jodiya India 36 B5
Joensuu Fin. 14 P5
Jōetsu Japan 45 E5
Jofane Moz. 49 D6
Jogbura Nepal 36 E3
Jõgeva Estonia 15 O7
Jogjakarta Indon. see Yogyakarta
Johannesburg S. Africa 51 H4
Johannesburg CA U.S.A. 65 D3
Johan Peninsula Canada 61 K2
Johi Pak. 36 A4
John Day U.S.A. 62 D3
John Day r. U.S.A. 62 C2
John F. Kennedy airport NY U.S.A. 64 E2
John o' Groats U.K. 20 F2
Johnsonburg PA U.S.A. 64 B2
Johnsondale CA U.S.A. 65 C3
Johnston, Lake imp. l. Australia 55 C8
Johnston and Sand Islands terr. N. Pacific Ocean see Johnston Atoll
Johnston Atoll terr. N. Pacific Ocean 74 I4
Johnstone U.K. 20 E5
Johnstone Lake Canada see Old Wives Lake
Johnston Range hills Australia 55 B7
Johnstown Ireland 21 E5
Johnstown NY U.S.A. 64 D1
Johnstown PA U.S.A. 64 B2
Johor Bahru Malaysia 41 C7
Johore Bahru Malaysia see Johor Bahru
Jõhvi Estonia 15 O7
Joinville Brazil 71 A4
Joinville France 24 G2
Joinville Island Antarctica 76 A2
Jokkmokk Sweden 14 K3
Jökulsá r. Iceland 14 [inset]
Jökulsá á Fjöllum r. Iceland 14 [inset]
Jökulsá í Fljótsdal r. Iceland 14 [inset]
Jolfa Iran 35 G3
Joliet U.S.A. 63 J3
Jolo i. Phil. 41 E7
Jomda China 42 H6
Jonava Lith. 15 N9
Jonesboro AR U.S.A. 63 I4
Jones Sound sea chan. Canada 61 J2
Jonglei Canal South Sudan 32 D8
Jönköping Sweden 15 I8
Jonquière Canada 63 M2
Joplin U.S.A. 63 I4
Joppa Israel see Tel Aviv-Yafo
Jora India 36 D4
Jordan country Asia 39 C4
Jordan r. Asia 39 B4
Jordan U.S.A. 62 F2
Jordânia Brazil 71 C1
Jordet Norway 15 H6
Jorhat India 37 H4
Jorm Afgh. 36 B1
Jörn Sweden 14 L4
Joroinen Fin. 14 O5
Jørpeland Norway 15 E7
Jos Nigeria 46 D4
José de San Martín Arg. 70 B6
Joseph, Lac l. Canada 61 L4
Joseph Bonaparte Gulf Australia 54 E3
Joshimath India 36 D3
Joshipur India 38 E1
Joshua Tree National Park CA U.S.A. 65 E4
Jos Plateau Nigeria 46 D4
Jostedalsbreen Nasjonalpark nat. park Norway 15 E6
Jotunheimen Nasjonalpark nat. park Norway 15 F6
Jouaiya Lebanon 39 B3
Joubertina S. Africa 50 F7
Jouberton S. Africa 51 H4
Jõuga Estonia 15 O7
Joûnié Lebanon 39 B3
Joutsa Fin. 15 O6
Joutseno Fin. 15 P6
Jowai India 37 H4
Jowr Deh Iran 35 H3
Joyce's Country reg. Ireland 21 C4
Joypurhat Bangl. 37 G4
Juan Aldama Mex. 66 D3
Juan de Fuca Strait Canada/U.S.A. 62 C2
Juan Fernández, Archipiélago is S. Pacific Ocean 75 O8
Juan Fernández Islands S. Pacific Ocean see Juan Fernández, Archipiélago
Juanjuí Peru 68 C5
Juankoski Fin. 14 P5
Juazeiro Brazil 69 J5
Juazeiro do Norte Brazil 69 K5
Juba r. Somalia see Jubba
Juba South Sudan 47 G4
Jubany research stn Antarctica 76 A2
Jubba r. Somalia 48 E4
Jubbah Saudi Arabia 35 F5
Jubbulpore India see Jabalpur
Jubilee Lake imp. l. Australia 55 D7
Juby, Cap c. Morocco 46 B2
Juchitán Mex. 66 E5
Jucuruçu Brazil 71 D2
Jucuruçu r. Brazil 71 D2
Judaberg Norway 15 D7
Judaidat al Hamir Iraq 35 F5
Judayyidat 'Ar'ar well Iraq 35 F5
Judenburg Austria 17 O7
Juelsminde Denmark 15 G9
Juerana Brazil 71 D2
Jugar China see Sêrxü
Juigalpa Nicaragua 67 G6
Juina Brazil 69 G6
Juiz de Fora Brazil 71 C3
Julaca Bol. 68 E8
Julia Brazil 68 E4
Juliaca Peru 68 D7
Julia Creek Australia 56 C4
Julian CA U.S.A. 65 D4
Julian Alps mts Slovenia see Julijske Alpe
Julianatop mt. Indon. see Mandala, Puncak
Juliana Top mt. Suriname 69 G3
Julianehåb Greenland see Qaqortoq
Julijske Alpe mts Slovenia 26 E1
Juliomagus France see Angers
Julius, Lake Australia 56 B4
Jullundur India see Jalandhar
Jumbilla Peru 68 C5
Jumilla Spain 25 F4
Jumla Nepal 37 E3
Jumna r. India see Yamuna
Junagadh India 36 B5
Junagarh India 38 D2
Junayfah Egypt 39 A4
Junction TX U.S.A. 62 H5
Junction City KS U.S.A. 63 H4
Jundiaí Brazil 71 B3
Juneau AK U.S.A. 60 E4
Junee Australia 58 C5
Jūn el Khudr b. Lebanon 39 B3
Jungar Qi China see Shagedu
Jungfrau mt. Switz. 24 H3
Junggar Pendi basin China 42 F3
Juniata r. PA U.S.A. 64 C2
Junín Arg. 70 D4
Junín Peru 68 C6
Junior WV U.S.A. 64 B3
Juniper Mountains AZ U.S.A. 65 F3
Junipero Serro Peak CA U.S.A. 65 B2
Junlian China 42 I7
Junsele Sweden 14 J5
Junxian China see Danjiangkou
Juodupė Lith. 15 N8
Jupiá Brazil 71 A3
Jupía, Represa resr Brazil 71 A3
Juquiá r. Brazil 71 B4
Jur r. South Sudan 32 C8
Jura mts France/Switz. 24 G4
Jura i. U.K. 20 D4
Jura, Sound of sea chan. U.K. 20 D5
Juraci Brazil 71 C1
Jurbarkas Lith. 15 M9
Jurf ad Darāwīsh Jordan 39 B4
Jurhen Ul mts China 37 G2
Jūrmala Latvia 15 M8
Jurmu Fin. 14 O4
Juruá r. Brazil 68 E4
Juruena r. Brazil 69 G5
Juruti Brazil 69 G4
Jurva Fin. 14 L5
Jūsīyah Syria 39 C2
Jussara Brazil 71 A1
Jutaí Brazil 68 E5
Jutaí r. Brazil 68 E4
Jutiapa Guat. 66 G6
Juticalpa Hond. 67 G6
Jutis Sweden 14 J3
Jutland pen. Denmark 15 F8
Juuka Fin. 14 P5
Juva Fin. 14 O6
Jūyom Iran 35 I5
Južnoukrajinsk Ukr. see Yuzhnoukrayin
Jwaneng Botswana 50 G3
Jylland pen. Denmark see Jutland
Jyväskylä Fin. 14 N5

K

K2 mt. China/Pakistan 36 D2
Ka r. Nigeria 46 D3
Kaafu Atoll Maldives see Male Atoll
Kaa-Iya del Gran Chaco, Parque Nacio nat. park Bol. 68 F7
Kaakhka Turkm. see Kaka
Kaapstad S. Africa see Cape Town
Kaarina Fin. 15 M6
Kaavi Fin. 14 P5
Kaba China see Habahe
Kabala Sierra Leone 46 B4
Kabale Uganda 48 C4
Kabalega Falls National Park Uganda Murchison Falls National Park
Kabalo Dem. Rep. Congo 49 C4
Kabambare Dem. Rep. Congo 49 C4
Kabangu Dem. Rep. Congo 49 C5
Kabara i. Fiji 53 I3
Kabarega National Park Uganda see Murchison Falls National Park
Kabaw Valley Myanmar 37 H5
Kabbani r. India 38 C3
Kåbdalis Sweden 14 L3
Kabinakagami Lake Canada 63 K2
Kabinda Dem. Rep. Congo 49 C4
Kabīr r. Syria 39 B2
Kabīrkūh mts Iran 35 G4
Kabo Cent. Afr. Rep. 48 B3
Kābol Afgh. see Kābul
Kabompo r. Zambia 49 C5
Kabongo Dem. Rep. Congo 49 C4
Kaboré Tambi, Parc National de nat. p Burkina Faso 46 C3
Kabūd Rāhang Iran 35 H4
Kābul Afgh. 36 B2
Kābul r. Afgh. 36 C2
Kabunda Dem. Rep. Congo 49 C5
Kabunduk Indon. 54 B2
Kabwe Zambia 49 C5
Kachalinskaya Rus. Fed. 13 J6
Kachchh, Great Rann of marsh India s
Kachchh, Rann of
Kachchh, Gulf of India 36 B5
Kachchh, Little Rann of marsh India 3
Kachchh, Rann of marsh India 36 B4
Kachia Nigeria 46 D4
Kachkanar Rus. Fed. 11 R4
Kachret'i Georgia 35 G2
Kachug Rus. Fed. 42 J2
Kaçkar Dağı mt. Turkey 35 F2
Kadaiyanallur India 38 C4
Kadanai r. Afgh./Pak. 36 A3
Kadapa India see Cuddapah
Kadavu i. Fiji 53 H3
Kadavu Passage Fiji 53 H3
Kaddam l. India 38 C2
Kade Ghana 46 C4
Kādhimain Iraq 35 G4
Kadi India 36 C5
Kadıköy Turkey 27 M4
Kadınhanı Turkey 34 D3
Kadiolo Mali 46 C3
Kadiri India 38 C3
Kadirli Turkey 34 E3
Kadiyevka Ukr. see Stakhanov
Kadmat atoll India 38 B4
Ka-do i. N. Korea 45 B5
Kadoma Zimbabwe 49 C5
Kadu Myanmar 37 I4
Kadugli Sudan 32 C7
Kaduna Nigeria 46 D3
Kaduna r. Nigeria 46 D4
Kadusam mt. China/India 37 I3
Kaduy Rus. Fed. 12 H4
Kadyy Rus. Fed. 12 I4
Kadzherom Rus. Fed. 12 L2
Kaédi Mauritania 46 B3
Kaélé Cameroon 47 E3
Kaesŏng N. Korea 45 B5
Kāf Saudi Arabia 39 C4
Kafa Ukr. see Feodosiya
Kafakumba Dem. Rep. Congo 49 C4
Kafan Armenia see Kapan
Kafanchan Nigeria 46 D4
Kafr ash Shaykh Egypt 34 C5
Kafr el Sheikh Egypt see Kafr ash Shay
Kafue Zambia 49 C5
Kafue r. Zambia 49 C5
Kafue National Park Zambia 49 C5
Kaga Japan 45 E5
Kaga Bandoro Cent. Afr. Rep. 48 B3
Kagan Pak. 36 C2
Kaganovichabad Tajik. see Kolkhozobo
Kaganovichi Pervyye Ukr. see Polis'ke
Kagarlyk Ukr. see Kaharlyk
Kåge Sweden 14 L4
Kağızman Turkey 35 F2
Kagmar Sudan 32 D7
Kagoshima Japan 45 C7
Kagoshima pref. Japan 45 C7
Kagul Moldova see Cahul
Kahama Tanz. 48 D4
Kaharlyk Ukr. 13 F6
Kaherekoau Mountains N.Z. 59 A7
Kahperusvaarat mts Fin. 14 L2
Kahramanmaraş Turkey 34 E3
Kahror Pakka Pak. 36 B3
Kâhta Turkey 34 E3

hurangi National Park N.Z. 59 D5
hurangi Point N.Z. 59 D5
huta Pak. 36 C2
huzi-Biega, Parc National du nat. park Dem. Rep. Congo 48 C4
i, Kepulauan is Indon. 41 F8
iapoi N.Z. 59 D6
ibab AZ U.S.A. 65 F2
i Besar i. Indon. 41 F8
ifeng Henan China 43 K6
ihua Yunnan China see Wenshan
iingveld reg. S. Africa 50 E5
i Kecil i. Indon. 41 F8
ikoura N.Z. 59 D6
las mt. China see Kangrinboqê Feng
ilasahar India see Kailashahar
ilashahar India 37 H4
las Range mts China see Gangdisê Shan
ili China 43 J7
ilu China 43 M4
imana Indon. 41 F8
imanawa Mountains N.Z. 59 E4
imar China 37 H2
imur Range hills India 36 E4
ina Estonia 15 M7
inan Japan 45 D6
inji Lake National Park Nigeria 46 D4
ipara Harbour N.Z. 59 E3
irana India 36 D3
irouan Tunisia 26 D7
iser Wilhelm II Land reg. Antarctica 76 E2
itaia N.Z. 59 D2
itangata N.Z. 59 B8
itawa N.Z. 59 F4
ithal India 36 D3
itum Sweden 14 L3
iwatu Indon. 54 D2
iyuan Liaoning China 44 B4
iyuan Yunnan China 42 I8
jaani Fin. 14 O4
jabbi Australia 56 C4
jakī Afgh. 36 A2
arabie, Lake Australia 58 D1
irān Afgh. 36 A2
ka Turkm. 33 I2
kadu National Park Australia 54 F3
kamas S. Africa 50 E5
kamega Kenya 48 D3
kar Pak. 36 A4
kata Liberia 46 B4
kenge Dem. Rep. Congo 49 C4
khi Azer. see Qax
khovka Ukr. 27 O1
khovs'ke Vodoskhovyshche resr Ukr. 3 G7
khul Moldova see Cahul
kī Iran 35 H5
kinada India 38 D2
kogawa Japan 45 D6
kori India 36 E4
kshaal-Too mts China/Kyrg. 42 D4
ktovik U.S.A. 60 D2
a Pak. 36 B3
a Tanz. 49 D4
aâ Kebira Tunisia 26 D7
aallit Nunaat terr. N. America see Greenland
abahi Indon. 41 E8
abáka Greece see Kalampaka
abgur India 38 C2
abo Zambia 49 C5
ach Rus. Fed. 13 I6
acha Dida Kenya 48 D3
ach-na-Donu Rus. Fed. 13 I6
adan r. India/Myanmar 37 H5
agwe Myanmar 37 I5
ahari Desert Africa 50 F2
ahari Gemsbok National Park S. Africa 0 E3
ajoki Fin. 14 M4
am India 38 C1
am Pak. 33 L2
ámai Greece see Kalamata
amare Botswana 51 H2
amaria Greece 27 J4
amata Greece 27 J6
amazoo U.S.A. 63 J3
ampaka Greece 27 I5
anchak Ukr. 27 O1
andula Angola see Calandula
annie Australia 55 B7
anshiyū ar Ramlī al Kabīr, Sarīr des. ibya 32 B3
ār Iraq 35 G4
āt Afgh. 36 A2
āt Sīstān va Balūchestān Iran 33 I4
at Balochistan Pak. 36 A3
aus r. Rus. Fed. 13 J7
aw Myanmar 37 I5
bäcär Azer. 35 G2
barri Australia 55 A6
barri National Park Australia 55 A6
burgi India see Gulbarga
e Turkey 27 M6
ecik Turkey 34 D2
élé Benin 46 D3
emie Dem. Rep. Congo 49 C4
emyo Myanmar 37 H5
-e Namak Iran 35 I4
evala Rus. Fed. 12 F2
ewa Myanmar 37 H5
eybar Iran 35 G3
gan China see Zhangjiakou
ghatgi India 38 B3
goorlie Australia 55 C7
i Croatia 26 F2
i r. India/Nepal 36 E3
iakra, Nos pt Bulg. 27 M3
i Gandaki r. Nepal 37 F4
igiri India 38 C3
ikata India see Kolkata
ima Dem. Rep. Congo 48 C4
imnos i. Greece see Kalymnos
inin Rus. Fed. see Tver'
iningrad Rus. Fed. 12 D5
inino Armenia see Tashir
inino Rus. Fed. 12 I4
ininsk Rus. Fed. 13 J6
ininskaya Rus. Fed. 13 H7
injara India 36 C5
inkavichy Belarus 13 F5
inkovichi Belarus see Kalinkavichy
isch Poland see Kalisz
ispell U.S.A. 62 E2
isz Poland 17 Q5
itva r. Rus. Fed. 13 I6
iua Tanz. 49 D4
iujar India 36 E4
ix Sweden 14 M4
kalighat India 37 H4
kalpen, Nationalpark nat. park Austria 7 O7
kan Turkey 27 M6
kfeld Namibia 49 B6
kfonteindam dam S. Africa 51 G5
kudah Sri Lanka 38 D5
laste Estonia 15 O7
lavesi l. Fin. 14 O5
lsedet Sweden 14 H5
Kallsjön l. Sweden 14 H5
Kallur India 38 C2
Kalmar Sweden 15 J8
Kalmarsund sea chan. Sweden 15 J8
Kalmūkh Qal'eh Iran 35 J3
Kalmunai Sri Lanka 38 D5
Kalmykia aut. rep. Rus. Fed. see Kalmykiya-Khalm'g-Tangch, Respublika
Kalmykiya-Khalm'g-Tangch, Respublika aut. rep. Rus. Fed. 35 G1
Kalmykovo Kazakh. see Taypak
Kalmytskaya Avtonomnaya Oblast' aut. rep. Rus. Fed. see Kalmykiya-Khalm'g-Tangch, Respublika
Kalnai India 37 E5
Kalodnaye Belarus 15 O11
Kalol India 36 C5
Kalomo Zambia 49 C5
Kalone Peak Canada 60 F4
Kalpa India 36 D3
Kalpeni atoll India 38 B4
Kalpetta India 38 C4
Kalpi India 36 D4
Kaltag U.S.A. 60 C3
Kaltukatjara Australia 55 E6
Kaluga Rus. Fed. 13 H5
Kalundborg Denmark 15 G9
Kalush Ukr. 13 E6
Kalvakol India 38 C2
Kälviä Fin. 14 M5
Kal'ya Rus. Fed. 11 R3
Kalyan India 38 B2
Kalyandurg India 33 M7
Kalyansingapuram India 38 D2
Kalyazin Rus. Fed. 12 H4
Kalymnos i. Greece 27 L6
Kama Dem. Rep. Congo 48 C4
Kama r. Rus. Fed. 12 L4
Kamaishi Japan 45 F5
Kamalia Pak. 36 C3
Kaman Turkey 34 D3
Kamanjab Namibia 49 B5
Kamarān i. Yemen 32 F6
Kamaran Island Yemen see Kamarān
Kamaron Sierra Leone 46 B4
Kamasin India 36 E4
Kambalda Australia 55 C7
Kambam India 38 C4
Kambara i. Fiji see Kabara
Kambia Sierra Leone 46 B4
Kambing, Pulau i. East Timor see Ataúro, Ilha de
Kambo-san mt. N. Korea see Kwanmo-bong
Kambove Dem. Rep. Congo 49 C5
Kambūt Libya 34 B5
Kamchatka, Poluostrov pen. Rus. Fed. see Kamchatka Peninsula
Kamchatka Basin sea feature Bering Sea 74 H2
Kamchatka Peninsula Rus. Fed. 29 Q4
Kamchiya r. Bulg. 27 L3
Kameia, Parque Nacional de nat. park Angola see Cameia, Parque Nacional da
Kamelik r. Rus. Fed. 13 K5
Kamen', Gory mts Rus. Fed. 28 K3
Kamenets-Podol'skiy Ukr. see Kam"yanets'-Podil's'kyy
Kamenitsa mt. Bulg. 27 J4
Kamenjak, Rt pt Croatia 26 E2
Kamenka Arkhangel'skaya Oblast' Rus. Fed. 12 J2
Kamenka Penzenskaya Oblast' Rus. Fed. 13 J5
Kamenka Primorskiy Kray Rus. Fed. 44 E3
Kamenka-Bugskaya Ukr. see Kam"yanka-Buz'ka
Kamenka-Strumilovskaya Ukr. see Kam"yanka-Buz'ka
Kamen'-na-Obi Rus. Fed. 42 E2
Kamennogorsk Rus. Fed. 15 P6
Kamennomostskiy Rus. Fed. 35 F1
Kamenolomni Rus. Fed. 13 I7
Kamenongue Angola see Camanongue
Kamen'-Rybolov Rus. Fed. 44 D3
Kamenskoye Rus. Fed. 29 R3
Kamenskoye Ukr. see Dniprodzerzhyns'k
Kamensk-Shakhtinskiy Rus. Fed. 13 I6
Kamensk-Ural'skiy Rus. Fed. 28 H4
Kamet mt. China 36 D3
Kamiesberge mts S. Africa 50 D6
Kamieskroon S. Africa 50 C6
Kamileroi Australia 56 C3
Kamina Dem. Rep. Congo 49 C4
Kaminak Lake Canada 61 I3
Kaminuriak Lake Canada see Qamanirjuaq Lake
Kamishihoro Japan 44 F4
Kamloops Canada 62 C1
Kamo Armenia see Gavarr
Kamoke Pak. 36 C3
Kamonia Dem. Rep. Congo 49 C4
Kampala Uganda 48 D3
Kampara India 38 D1
Kampene Dem. Rep. Congo 48 C4
Kampinoski Park Narodowy nat. park Poland 17 R4
Kâmpóng Cham Cambodia 31 J5
Kâmpóng Saôm Cambodia see Sihanoukville
Kâmpóng Spœ Cambodia 31 J5
Kâmpóng Thum Cambodia 31 J5
Kâmpôt Cambodia 31 J5
Kampuchea country Asia see Cambodia
Kamrau, Teluk b. Indon. 41 F8
Kamsack Canada 62 G1
Kamskoye Vodokhranilishche resr Rus. Fed. 11 R4
Kamsuuma Somalia 48 E3
Kamuli Uganda 48 D3
Kam"yanets'-Podil's'kyy Ukr. 13 E6
Kam"yanka-Buz'ka Ukr. 13 E6
Kamyanyets Belarus 15 M10
Kāmyārān Iran 35 G4
Kamyshin Rus. Fed. 13 J6
Kamyzyak Rus. Fed. 13 K7
Kanab U.S.A. 62 E4
Kanab Creek r. AZ U.S.A. 65 F2
Kanak Pak. 36 A3
Kananga Dem. Rep. Congo 49 C4
Kanangra-Boyd National Park Australia 58 E4
Kanarak India see Konarka
Kanarraville UT U.S.A. 65 F2
Kanas watercourse Namibia 50 C4
Kanash Rus. Fed. 12 J5
Kanauj India see Kannauj
Kanazawa Japan 45 E5
Kanbalu Myanmar 37 H5
Kanchanjanga mt. India/Nepal see Kangchenjunga
Kanchipuram India 38 C3
Kand mt. Pak. 36 A3
Kanda Pak. 36 A3
Kandahār Afgh. 36 A3
Kandalaksha Rus. Fed. 12 G2
Kandalakshskiy Zaliv g. Rus. Fed. 12 G2
Kandar Indon. 54 E2
Kandavu i. Fiji see Kadavu
Kandavu Passage Fiji see Kadavu Passage
Kandé Togo 46 D4
Kandh Kot Pak. 36 B3
Kandi Benin 46 D3
Kandi India 38 C2
Kandiaro Pak. 36 B4
Kandıra Turkey 27 N4
Kandos Australia 58 D4
Kandreho Madag. 49 E5
Kandukur India 38 C3
Kandy Sri Lanka 38 D5
Kandyagash Kazakh. 28 G5
Kane PA U.S.A. 64 B2
Kane Bassin b. Greenland 77 K1
Kaneh watercourse Iran 35 I6
Kaneti Pak. 36 A3
Kanevskaya Rus. Fed. 13 H7
Kang Botswana 50 F2
Kangaamiut Greenland 61 M3
Kangaarsussuaq c. Greenland 61 K2
Kangaba Mali 46 C3
Kangal Turkey 34 E3
Kangān Būshehr Iran 35 I6
Kangandala, Parque Nacional de nat. park Angola see Cangandala, Parque Nacional de
Kangar Malaysia 41 C7
Kangaroo Island Australia 57 B7
Kangaroo Point Australia 56 B3
Kangaslampi Fin. 14 P5
Kangasniemi Fin. 14 O6
Kangāvar Iran 35 G4
Kangchenjunga mt. India/Nepal 37 G4
Kangding China 42 I6
Kangean, Kepulauan is Indon. 41 D8
Kangen r. South Sudan 47 G4
Kangerlussuaq Greenland 61 M3
Kangerlussuaq inlet Greenland 61 M3
Kangerlussuaq inlet Greenland 77 J2
Kangersuatsiaq Greenland 61 M2
Kangertittivaq sea chan. Greenland 61 P2
Kanggye N. Korea 44 B4
Kanghwa S. Korea 45 B5
Kangikajik c. Greenland 61 P2
Kangirsuk Canada 61 K3
Kangmar China 37 F3
Kangnŭng S. Korea 45 C5
Kango Gabon 48 B3
Kangping China 44 A4
Kangri Karpo Pass China/India 37 I3
Kangrinboqê Feng mt. China 36 E3
Kangsangdobdê China see Xainza
Kangto mt. China/India 37 H4
Kangtog China 37 F2
Kanifing Gambia 46 B3
Kanigiri India 38 C3
Kanin, Poluostrov pen. Rus. Fed. 12 J2
Kanin Nos Rus. Fed. 77 G2
Kanin Nos, Mys c. Rus. Fed. 12 I1
Kaninskiy Bereg coastal area Rus. Fed. 12 I2
Kanjiroba mt. Nepal 37 E3
Kankaanpää Fin. 15 M6
Kankakee U.S.A. 63 J3
Kankan Guinea 46 C3
Kanker India 38 D1
Kankesanturai Sri Lanka 38 D4
Kankossa Mauritania 46 B3
Kannauj India 36 D4
Kanniya Kumari c. India see Comorin, Cape
Kannonkoski Fin. 14 N5
Kannur India 38 B4
Kannus Fin. 14 M5
Kano Nigeria 46 D3
Kanonpunt pt S. Africa 50 E8
Kanosh UT U.S.A. 65 F1
Kanovlei Namibia 49 B5
Kanoya Japan 45 C7
Kanpur Orissa India 38 E1
Kanpur Uttar Prad. India 36 E4
Kansai airport Japan 45 D6
Kansas r. U.S.A. 62 I4
Kansas state U.S.A. 62 H4
Kansas City KS U.S.A. 63 I4
Kansk Rus. Fed. 29 K4
Kansu prov. China see Gansu
Kantara h. Cyprus 39 A2
Kantavu i. Fiji see Kadavu
Kantchari Burkina Faso 46 D3
Kantemirovka Rus. Fed. 13 H6
Kanthi India 37 F5
Kantishna r. U.S.A. 60 C3
Kanton atoll Kiribati 53 I2
Kanturk Ireland 21 D5
Kanuku Mountains Guyana 69 G3
Kanur India 38 C3
Kanus Namibia 50 D4
Kanyakubja India see Kannauj
Kanyamazane S. Africa 51 J3
Kanye Botswana 51 G3
Kaohsiung Taiwan 43 M8
Kaokoveld plat. Namibia 49 B5
Kaolack Senegal 46 B3
Kaoma Zambia 49 C5
Kaouadja Cent. Afr. Rep. 48 C3
Kapa S. Africa see Cape Town
Kapan Armenia 35 G3
Kapanga Dem. Rep. Congo 49 C4
Kaparhā Iran 35 H5
Kapatu Zambia 49 D4
Kapchagay Kazakh. 42 D4
Kapchagayskoye Vodokhranilishche resr Kazakh. 42 D4
Kap Dan Greenland see Kulusuk
Kapello, Akrotirio pt Greece 27 J6
Kapellskär Sweden 15 K7
Kapelskär Sweden see Kapellskär
Kapili r. India 37 G4
Kapingamarangi atoll Micronesia 74 G5
Kapingamarangi Rise sea feature N. Pacific Ocean 74 G5
Kapıorman Dağları mts Turkey 27 N4
Kapip Pak. 36 B3
Kapiri Mposhi Zambia 49 C5
Kapisillit Greenland 61 M3
Kapiskau r. Canada 63 K1
Kapiti Island N.Z. 59 E5
Kaplankyr, Chink esc. Asia 35 I2
Kaplankyr Döwlet Gorugy nature res. Turkm. 35 J2
Kapoeta South Sudan 47 G4
Kaposvár Hungary 26 G1
Kappeln Germany 17 L3
Kapsukas Lith. see Marijampolė
Kaptai Bangl. 37 H5
Kapuriya India 36 C4
Kapurthala India 36 C3
Kapuskasing Canada 63 K2
Kapustin Yar Rus. Fed. 13 J6
Kaputar mt. Australia 58 E3
Kaputir Kenya 48 D3
Kapuvár Hungary 26 G1
Kapydzhik, Gora mt. Armenia/Azer. see Qazangödağ
Kapyl' Belarus 15 O10
Kara India 36 E4
Kara Togo 46 D4
Kara r. Turkey 35 F3
Kara-Balta Kyrg. 42 C4
Karabalyk Kazakh. 30 F1
Karabekaul' Turkm. see Garabekewül
Karabiga Turkey 27 L4
Kara-Bogaz-Gol, Proliv sea chan. Turkm. see Garabogazköl Bogazy
Kara-Bogaz-Gol'skiy Zaliv b. Turkm. see Garabogazköl Aýlagy
Karabük Turkey 34 D2
Karaburun Turkey 27 L5
Karabutak Kazakh. 28 H5
Karacabey Turkey 27 M4
Karacaköy Turkey 27 M4
Karacalı Dağ mt. Turkey 35 E3
Karaçal Tepe mt. Turkey 39 A1
Karacasu Turkey 27 M6
Karaca Yarımadası pen. Turkey 27 N6
Karachayevsk Rus. Fed. 13 I8
Karachev Rus. Fed. 13 G5
Karachi Pak. 33 K5
Karacurun Turkey see Hilvan
Karad India 38 B2
Kara Dağ h. Turkey 39 D1
Kara Dağ mt. Turkey 34 D3
Kara Deniz sea Asia/Europe see Black Sea
Karagan Rus. Fed. 44 A1
Karaganda Kazakh. 42 C3
Karagayly Kazakh. 42 D3
Karaginskiy Zaliv b. Rus. Fed. 29 R4
Karagiye, Vpadina depr. Kazakh. 35 H2
Karagola India 37 F4
Karahallı Turkey 27 M5
Karahasanlı Turkey 34 D3
Karaikal India 38 C4
Karaikkudi India 38 C4
Karaisalı Turkey 34 D3
Karaitivu i. Sri Lanka 38 C4
Karaj Iran 35 H4
Karak Jordan see Al Karak
Karakalli Turkey see Özalp
Karakax China see Moyu
Karakax He r. China 36 E1
Karakax Shan mts China 36 E2
Karaki China 36 E1
Karaklis Armenia see Vanadzor
Karakoçan Turkey 35 F3
Kara-Köl Kyrg. 31 G2
Karakol Kyrg. 42 D4
Karakoram Pass China/India 36 D2
Karakoram Range mts Asia 33 M2
Kara K'orē Eth. 48 D2
Karakorum Range mts Asia see Karakoram Range
Karaköse Turkey see Ağrı
Kara Kul' Kyrg. see Kara-Köl
Karakul', Ozero l. Tajik. see Qarokŭl
Kara Kum des. Turkm. see Karakum Desert
Karakum, Peski Kazakh. see Karakum Desert
Karakum Desert Kazakh. 30 E2
Karakum Desert Turkm. 30 F3
Karakurt Turkey 35 F2
Karakuş Dağı ridge Turkey 27 N5
Karal Chad 47 E3
Karala Estonia 15 L7
Karalundi Australia 55 B6
Karaman Turkey 34 D3
Karaman prov. Turkey 39 A1
Karamanlı Turkey 27 M6
Karamay China 42 E3
Karambar Pass Afgh./Pak. 36 C1
Karamea N.Z. 59 D5
Karamea Bight b. N.Z. 59 C5
Karamiran China 37 F1
Karamiran Shankou pass China 37 F1
Karamürsel Turkey 27 M4
Karamyshevo Rus. Fed. 15 P8
Karān i. Saudi Arabia 35 H6
Karangasem Indon. 54 A2
Karanja India 38 C1
Karanjia India 36 E5
Karapınar Gaziantep Turkey 39 C1
Karapınar Konya Turkey 34 D3
Karas admin. reg. Namibia 50 C4
Karasay China 37 E1
Karasburg Namibia 50 D5
Kara Sea Rus. Fed. 28 I2
Kárášjohka Finnmark Norway see Karasjok
Karasjok Norway 14 N2
Kara Strait Rus. Fed. see Karskiye Vorota, Proliv
Karasu r. Syria/Turkey 39 C1
Karasu Bitlis Turkey see Hizan
Karasu Sakarya Turkey 27 N4
Karasu r. Turkey 35 F3
Karasubazar Ukr. see Bilohirs'k
Karasuk Rus. Fed. 28 I4
Karataş Turkey 39 B1
Karataş Burnu hd Turkey see Fener Burnu
Karatau Kazakh. 42 C4
Karatau, Khrebet mts Kazakh. 42 B4
Karatepe Turkey 39 A1
Karatsu Japan 45 C6
Karaudanawa Guyana 69 G3
Karauli India 36 D4
Karavan Kyrg. see Kerben
Karavostasi Cyprus 39 A2
Karayılan Turkey 39 C1
Karayulgan China 42 E4
Karazhal Kazakh. 42 C3
Karbalā' Iraq 35 G4
Karcag Hungary 27 I1
Kardhitsa Greece see Karditsa
Karditsa Greece 27 I5
Kärdla Estonia 15 M7
Karee S. Africa 51 H5
Kareeberge mts S. Africa 50 E6
Kareima Sudan 32 D6
Kareli India 36 D5
Karelia aut. rep. Rus. Fed. see Kareliya, Respublika
Kareliya, Respublika aut. rep. Rus. Fed. 14 R5
Karel'skaya A.S.S.R. aut. rep. Rus. Fed. see Kareliya, Respublika
Karel'skiy Bereg coastal area Rus. Fed. 12 G2
Karema Tanz. 49 D4
Karera India 36 D4
Karesuando Sweden 14 M2
Kargalinskaya Rus. Fed. 35 G2
Kargalinski Rus. Fed. see Kargalinskaya
Kargapazarı Dağları mts Turkey 35 F3
Karghalik China see Yecheng
Kargı Turkey 34 D2
Kargil India 36 D2
Kargilik China see Yecheng
Kargıpınarı Turkey 39 B1
Kargopol' Rus. Fed. 12 H3
Kari Nigeria 46 E3
Kariba Zimbabwe 49 C5
Kariba, Lake resr Zambia/Zimbabwe 49 C5
Kariba Dam Zambia/Zimbabwe 49 C5
Kariba-yama vol. Japan 44 E4
Karibib Namibia 50 B1
Karigasniemi Fin. 14 N2
Karijini National Park Australia 55 B5
Karijoki Fin. 14 L5
Karikachi-tōge pass Japan 44 F4
Karikari, Cape N.Z. 59 D2
Karimata, Selat str. Indon. 41 C8
Karimganj India 37 H4
Karimnagar India 38 C2
Káristos Greece see Karystos
Karjat Mahar. India 38 B2
Karjat Mahar. India 38 B2
Karkaralinsk Kazakh. 42 D3
Karkar Island P.N.G. 52 E2
Karkh Pak. 36 A4
Karkinits'ka Zatoka g. Ukr. 27 O2
Kärkölä Fin. 15 N6
Karkonoski Park Narodowy nat. park Czech Rep./Poland see Krkonošský narodní park
Karksi-Nuia Estonia 15 N7
Karkük Iraq see Kirkuk
Karlachi Pak. 36 B2
Karlholmsbruk Sweden 15 J6
Karlik Shan mts China 42 G4
Karlıova Turkey 35 F3
Karlivka Ukr. 13 G6
Karl Marks, Qullai mt. Tajik. 36 C1
Karl-Marx-Stadt Germany see Chemnitz
Karlovac Croatia 26 F2
Karlovka Ukr. see Karlivka
Karlovo Bulg. 27 K3
Karlovy Vary Czech Rep. 17 N5
Karlsborg Sweden 15 I7
Karlsburg Romania see Alba Iulia
Karlshamn Sweden 15 I8
Karlskoga Sweden 15 I7
Karlskrona Sweden 15 I8
Karlsruhe Germany 17 L6
Karlstad Sweden 15 H7
Karluk U.S.A. 60 C4
Karmala India 38 B2
Karmel, Har h. Israel see Carmel, Mount
Karmona Spain see Córdoba
Karmøy i. Norway 15 D7
Karmpur Pak. 36 C3
Karnafuli Reservoir Bangl. 37 H5
Karnal India 36 D3
Karnataka state India 38 B3
Karnavati India see Ahmadabad
Karnobat Bulg. 27 L3
Karodi Pak. 36 A4
Karoi Zimbabwe 49 C5
Karo La pass China 37 G3
Karong India 37 H4
Karonga Malawi 49 D4
Karonie Australia 55 C7
Karoo National Park S. Africa 50 F7
Karoo Nature Reserve S. Africa 50 G7
Karoonda Australia 57 B7
Karora Eritrea 32 E6
Káros i. Greece see Keros
Karossa, Tanjung pt Indon. 54 B2
Karpasia pen. Cyprus 39 B2
Karpas Peninsula Cyprus see Karpasia
Karpathos i. Greece 27 L7
Karpathou, Steno sea chan. Greece 27 L6
Karpaty mts Europe see Carpathian Mountains
Karpenisi Greece 27 I5
Karpilovka Belarus see Aktsyabrski
Karpinsk Rus. Fed. 11 S4
Karpogory Rus. Fed. 12 J2
Karpuz r. Turkey 39 A1
Karratha Australia 54 B5
Karroo plat. S. Africa see Great Karoo
Kars Turkey 35 F2
Kärsämäki Fin. 14 N5
Kārsava Latvia 15 O8
Karshi Qashqadaryo Uzbek. see Qarshi
Karskiye Vorota, Proliv str. Rus. Fed. 28 G3
Karskoye More sea Rus. Fed. see Kara Sea
Karstula Fin. 14 N5
Karsu Turkey 39 C1
Karsun Rus. Fed. 13 J5
Kartal Turkey 27 M4
Kartaly Rus. Fed. 28 H4
Kartayel' Rus. Fed. 12 L2
Karttula Fin. 14 O5
Karumba Australia 56 C3
Karumbhar Island India 36 B5
Kārūn, Rūd-e r. Iran 35 H5
Karuni Indon. 54 B2
Karur India 38 C4
Karvia Fin. 14 M5
Karviná Czech Rep. 17 Q6
Karwar India 38 B3
Karyagino Azer. see Füzuli
Karymskoye Rus. Fed. 43 K2
Karynzharyk, Peski des. Kazakh. 35 I2
Karystos Greece 27 K5
Kaş Turkey 27 M6
Kasa India 38 B2
Kasaba Turkey see Turgutlu
Kasaï r. Dem. Rep. Congo 48 B4
Kasaï, Plateau du Dem. Rep. Congo 49 C4
Kasaji Dem. Rep. Congo 49 C5
Kasama Zambia 49 D5
Kasane Botswana 49 C5
Kasaragod India 38 B3
Kasargod India see Kasaragod
Kasargode India see Kasaragod
Kasatkino Rus. Fed. 44 C2
Kasba Lake Canada 61 H3
Kasba Tadla Morocco 22 C5
Kasenga Dem. Rep. Congo 49 C5
Kasengu Dem. Rep. Congo 49 C4
Kasese Dem. Rep. Congo 48 C4
Kasese Uganda 48 D3
Kasevo Rus. Fed. see Neftekamsk
Kasganj India 36 D4
Kāshān Iran 35 H4
Kashary Rus. Fed. 13 I6
Kashgar China see Kashi
Kashi China 42 D5
Kashihara Japan 45 D6
Kashima-nada b. Japan 45 F5
Kashin Rus. Fed. 12 H4
Kashipur India 36 D3
Kashira Rus. Fed. 13 H5
Kashiwazaki Japan 45 E5
Kashkarantsy Rus. Fed. 12 H2
Kāshmar Iran 33 I2
Kashmir terr. Asia see Jammu and Kashmir
Kashmir, Vale of reg. India 36 C2
Kashyr Kazakh. 42 D2
Kashyukulu Dem. Rep. Congo 49 C4
Kasi India see Varanasi
Kāsigar Afgh. 36 B2
Kasimov Rus. Fed. 13 I5
Kaskinen Fin. 14 L5
Kaskö Fin. see Kaskinen
Kasongo Dem. Rep. Congo 49 C4
Kasongo-Lunda Dem. Rep. Congo 49 B4
Kasos i. Greece 27 L7
Kaspīy Mangy Oypaty lowland Kazakh./Rus. Fed. see Caspian Lowland
Kaspiysk Rus. Fed. 35 G2
Kaspiyskiy Rus. Fed. see Lagan'
Kaspiyskoye More l. Asia/Europe see Caspian Sea
Kassa Slovakia see Košice
Kassala Sudan 32 E6
Kassandras, Akrotirio pt Greece 27 J5
Kassandras, Kolpos b. Greece 27 J4
Kassel Germany 17 L5
Kasserine Tunisia 26 C7
Kastamonu Turkey 34 D2
Kastéllion Greece see Kissamos
Kastellorizon i. Greece see Megisti
Kastoria Greece 27 I4
Kastornoye Rus. Fed. 13 H6
Kastsyukovichy Belarus 13 G5
Kasulu Tanz. 49 D4
Kasumkent Rus. Fed. 35 H2
Kasungu Malawi 49 D5
Kasungu National Park Malawi 49 D5
Kasur Pak. 36 C3
Katādtlit Nunāt terr. N. America see Greenland
Katahdin, Mount U.S.A. 63 N2
Katahsang Ghar Afgh. 36 A2
Kataklik India 36 D2
Katako-Kombe Dem. Rep. Congo 48 C4
Katakwi Uganda 48 D3
Katana India 36 D5
Katangi India 36 D5
Katanning Australia 55 B8
Katavi National Park Tanz. 49 D4
Katea Dem. Rep. Congo 49 C4
Katerini Greece 27 J4
Katesh Tanz. 49 D4
Kate's Needle mt. Canada/U.S.A. 60 E4
Katete Zambia 49 D5
Katherîna, Gebel mt. Egypt see Kātrīnā, Jabal
Katherine Australia 54 F3
Katherine Gorge National Park Australia see Nitmiluk National Park
Kathiawar pen. India 36 B5
Kathihar India see Katihar
Kathiraveli Sri Lanka 38 D4
Kathiwara India 36 C5
Kathleen Falls Australia 54 E3
Kathmandu Nepal 37 F4
Kathu S. Africa 50 F4
Kathua India 36 C2
Kati Mali 46 C3
Katihar India 37 F4
Katikati S. Africa 51 H7
Katima Mulilo Namibia 49 C5
Katiola Côte d'Ivoire 46 C4
Kā Tiritiri o te Moana mts N.Z. see Southern Alps
Katkop Hills S. Africa 50 E6
Katlehong S. Africa 51 I4
Katmai National Park and Preserve U.S.A. 60 C4
Katmandu Nepal see Kathmandu
Kato Achaïa Greece 27 I5
Katoomba Australia 58 E4
Katowice Poland 17 Q5
Katoya India 37 G5
Katrancık Dağı mts Turkey 27 M6
Kātrīnā, Jabal mt. Egypt 34 D5
Katrine, Loch l. U.K. 20 E4
Katrineholm Sweden 15 J7
Katse Dam Lesotho 51 I5
Katsina Nigeria 46 D3
Katsina-Ala Nigeria 46 D4
Katsuura Japan 45 F6
Kattamudda Well Australia 54 D5
Kattaqo'rg'on Uzbek. 33 K2
Kattaqŭrghon Uzbek. see Kattaqo'rg'on
Kattegat str. Denmark/Sweden 15 G8
Kattowitz Poland see Katowice
Katumbar India 36 D4
Katunino Rus. Fed. 12 J4
Katuri Pak. 36 B3
Katwa India see Katoya
Kaua'i i. U.S.A. 75 J4
Kauhajoki Fin. 14 M5
Kauhava Fin. 14 M5
Kaukkwè Hills Myanmar 37 I4
Kaukonen Fin. 14 N3
Kaunas Lith. 15 M9
Kaunata Latvia 15 O8
Kaundy, Vpadina depr. Kazakh. 35 I2
Kaunia Bangl. 37 G4
Kaura-Namoda Nigeria 46 D3
Kaustinen Fin. 14 M5
Kautokeino Norway 14 M2
Kavadarci Macedonia 27 J4
Kavak Turkey 34 E2
Kavaklıdere Turkey 27 M6
Kavala Greece 27 K4
Kavalas, Kolpos b. Greece 27 K4
Kavalerovo Rus. Fed. 44 D3
Kavali India 38 D3
Kavār Iran 35 I5
Kavaratti India 38 B4
Kavaratti atoll India 38 B4
Kavarna Bulg. 27 M3
Kavendou, Mont mt. Guinea 46 B3
Kaveri r. India 38 C4
Kavīr Iran 35 H4
Kavīr, Dasht-e des. Iran 35 I4
Kavīr-e Ḥājj 'Alī Qolī salt pan Iran 35 I4
Kavīr-e Lūt des. Iran 33 I3
Kavkasioni mts Asia/Europe see Caucasus
Kawagoe Japan 45 E6
Kawaguchi Japan 45 E6
Kawakawa N.Z. 59 E2
Kawambwa Zambia 49 C4
Kawana Zambia 49 C5
Kawardha India 36 E5
Kawasaki Japan 45 E6
Kawau Island N.Z. 59 E3
Kawerau N.Z. 59 F4
Kawhia N.Z. 59 E4
Kawhia Harbour N.Z. 59 E4
Kawich Peak NV U.S.A. 65 D2
Kawich Range mts NV U.S.A. 65 D2
Kawlin Myanmar 37 H5
Kawm Umbū Egypt 32 D5
Kaxgar China see Kashi
Kaxgar He r. China 33 M2
Kax He r. China 42 E4
Kaxtax Shan mts China 37 E1
Kaya Burkina Faso 46 C3
Kayadibi Turkey 34 E3
Kayankulam India 38 C4
Kayar India 38 C2
Kaydak, Sor dry lake Kazakh. 35 I1
Kaydanovo Belarus see Dzyarzhynsk
Kayembe-Mukulu Dem. Rep. Congo 49 C4
Kayenta U.S.A. 62 E4
Kayes Mali 46 B3
Kaymaz Turkey 27 N5
Kaynar Kazakh. 42 D3
Kaynar Turkey 34 E3
Kayseri Turkey 34 D3
Kayuyu Dem. Rep. Congo 48 C4
Kazach'ye Rus. Fed. 29 O2
Kazakh Azer. see Qazax
Kazakhskaya S.S.R. country Asia see Kazakhstan
Kazakhskiy Zaliv b. Kazakh. 35 I2
Kazakhstan country Asia 30 F2
Kazakhstan Kazakh. see Aksay
Kazakh Steppe plain Kazakh. see Saryarka
Kazakstan country Asia see Kazakhstan
Kazan' Rus. Fed. 12 K5
Kazandzhik Turkm. see Bereket
Kazanka r. Rus. Fed. 12 K5
Kazanlı Turkey 39 B1
Kazanlŭk Bulg. 27 K3
Kazan-rettō is Japan see Volcano Islands
Kazatin Ukr. see Kozyatyn
Kazbek mt. Georgia/Rus. Fed. 13 J8
Kaz Dağı mts Turkey 27 L5
Kāzerūn Iran 35 H5

Kazhym Rus. Fed. 12 K3
Kazidi Tajik. see Qozideh
Kazi Magomed Azer. see Qazımämmäd
Kazincbarcika Hungary 13 D6
Kaziranga National Park India 37 H4
Kazret'i Georgia 35 G2
Kaztalovka Kazakh. 11 P6
Kazym r. Rus. Fed. 11 T3
Kazym-Mys Rus. Fed. 11 T3
Keady U.K. 21 F3
Kéamu i. Vanuatu see Anatom
Kearney U.S.A. 62 H3
Keban Turkey 34 E3
Keban Barajı resr Turkey 34 E3
Kébémèr Senegal 46 B3
Kebili Tunisia 22 F5
Kebīr, Nahr al r. Lebanon/Syria 39 B2
Kebkabiya Sudan 47 F3
Kebnekaise mt. Sweden 14 K3
Kebock Head hd U.K. 20 C2
K'ebrī Dehar Eth. 48 E3
Kechika r. Canada 77 A3
Keçiborlu Turkey 27 N6
Kecskemét Hungary 27 H1
K'eda Georgia 35 F2
Kėdainiai Lith. 15 M9
Kedairu Passage Fiji see Kadavu Passage
Kedong China 44 B3
Kedva r. Rus. Fed. 12 L2
Kędzierzyn-Koźle Poland 17 Q5
Keele r. Canada 60 F3
Keele Peak Canada 60 E3
Keeler CA U.S.A. 65 D2
Keeling Islands terr. Indian Ocean see Cocos Islands
Keen, Mount h. U.K. 20 G4
Keene CA U.S.A. 65 C3
Keene NH U.S.A. 64 E1
Keeper Hill h. Ireland 21 D5
Keepit, Lake resr Australia 58 E3
Keep River National Park Australia 54 E3
Keer-weer, Cape Australia 56 C2
Keetmanshoop Namibia 50 D4
Keewatin Canada 63 I2
Kefallinía i. Greece see Cephalonia
Kefallonia i. Greece see Cephalonia
Kefamenanu Indon. 41 E8
Kefe Ukr. see Feodosiya
Keffi Nigeria 46 D4
Keflavík Iceland 14 [inset]
Kegalla Sri Lanka 38 D5
Kegen Kazakh. 42 D4
Keg River Canada 60 G4
Kegul'ta Rus. Fed. 13 J7
Kehra Estonia 15 N7
Keighley U.K. 18 F5
Keila Estonia 15 N7
Keimoes S. Africa 50 E5
Keitele Fin. 14 O5
Keitele l. Fin. 14 O5
Keith Australia 57 C8
Keith U.K. 20 G3
Kékes mt. Hungary 17 R7
Kekri India 36 C4
Kelaa i. Maldives 38 B5
K'elafo Eth. 48 E3
Kelibia Tunisia 26 D6
Kelif Uzboýy marsh Turkm. 33 J2
Kelkit Turkey 35 E2
Kelkit r. Turkey 34 E2
Kéllé Congo 48 B4
Keller Lake Canada 60 F3
Kellett, Cape Canada 60 F2
Kelloselkä Fin. 14 P3
Kells Ireland 21 F4
Kells r. U.K. 21 F3
Kelly Range hills Australia 55 C6
Kelmė Lith. 15 M9
Kélo Chad 47 E4
Kelowna Canada 62 D2
Kelseyville CA U.S.A. 65 A1
Kelso U.K. 20 G5
Kelso CA U.S.A. 65 E3
Keluang Malaysia 41 C7
Kelvington Canada 62 G1
Kem' Rus. Fed. 12 G2
Kem' r. Rus. Fed. 12 G2
Ke Macina Mali see Massina
Kemah Turkey 34 E3
Kemaliye Turkey 34 E3
Kemalpaşa Turkey 27 L5
Kemano (abandoned) Canada 60 F4
Kembé Cent. Afr. Rep. 48 C3
Kemeneshát hills Hungary 26 G1
Kemer Antalya Turkey 27 N6
Kemer Muğla Turkey 27 M6
Kemer Barajı resr Turkey 27 M6
Kemerovo Rus. Fed. 28 J4
Kemi Fin. 14 N4
Kemijärvi Fin. 14 O3
Kemijärvi l. Fin. 14 O3
Kemijoki r. Fin. 14 N4
Kemiö Fin. see Kimito
Kemir Turkm. see Keymir
Kemmerer U.S.A. 62 E3
Kemnay U.K. 20 G3
Kemp Coast reg. Antarctica see Kemp Land
Kempele Fin. 14 N4
Kemp Land reg. Antarctica 76 D2
Kemp Peninsula Antarctica 76 A2
Kempsey Australia 58 F3
Kempt, Lac l. Canada 63 M2
Kempten (Allgäu) Germany 17 M7
Kempton Park S. Africa 51 I4
Ken r. India 36 E4
Kenai U.S.A. 60 C3
Kenai Fiords National Park U.S.A. 60 C4
Kenai Mountains U.S.A. 60 C4
Kenâyis, Râs el pt Egypt see Ḥikmah, Ra's al
Kenbridge VA U.S.A. 64 B4
Kendal U.K. 18 E4
Kendall Australia 58 F3
Kendall, Cape Canada 61 J3
Kendari Indon. 41 E8
Kendawangan Indon. 41 D8
Kendégué Chad 47 E3
Kendrapara India 37 F5
Kendraparha India see Kendrapara
Kendujhar India see Keonjhar
Kendujhargarh India see Keonjhar
Kendyrlisor, Solonchak salt l. Kazakh. 35 I2
Kenebri Australia 58 D3
Kenema Sierra Leone 46 B4
Kenge Dem. Rep. Congo 49 B4
Kenhardt S. Africa 50 E5
Kéniéba Mali 46 B3
Kénitra Morocco 22 C5
Kenmare Ireland 21 C6
Kenmare River inlet Ireland 21 B6
Kenmore NY U.S.A. 64 B1
Kennebunkport ME U.S.A. 64 F1
Kennedy, Cape U.S.A. see Canaveral, Cape
Kennedy Range National Park Australia 55 A6
Kennet r. U.K. 19 G7
Kenneth Range hills Australia 55 B5
Kennewick U.S.A. 62 D2
Kenn Reef Australia 56 F4
Kenora Canada 63 I2
Kenosha U.S.A. 63 J3
Kenozero, Ozero l. Rus. Fed. 12 H3
Kent r. U.K. 18 E4
Kent OH U.S.A. 64 A2
Kent VA U.S.A. 64 A4
Kentani S. Africa 51 I7
Kent Group is Australia 57 [inset]
Kent Peninsula Canada 60 H3
Kentucky state U.S.A. 63 K4
Kenya country Africa 48 D3
Kenya, Mount Kenya 48 D4
Keokuk U.S.A. 63 I3
Keoladeo National Park India 36 D4
Keonjhar India 37 F5
Keonjhargarh India see Keonjhar
Kepina r. Rus. Fed. 12 I2
Keppel Bay Australia 56 E4
Kepsut Turkey 27 M5
Kera India 37 F5
Kerala state India 38 B4
Kerang Australia 58 A5
Kerava Fin. 15 N6
Kerba Alg. 25 G5
Kerbela Iraq see Karbalā'
Kerben Kyrg. 42 C4
Kerbi r. Rus. Fed. 44 E1
Kerch Ukr. 34 E1
Kerchom"ya Rus. Fed. 12 L3
Kerema P.N.G. 52 E2
Kerempe Burun pt Turkey 34 D2
Keren Eritrea 32 E6
Kerewan Gambia 46 B3
Kerguélen, Îles is Indian Ocean 73 M9
Kerguelen Islands Indian Ocean see Kerguélen, Îles
Kerguelen Plateau sea feature Indian Ocean 73 M9
Kericho Kenya 48 D4
Kerikeri N.Z. 59 D2
Kerimäki Fin. 14 P6
Kerinci, Gunung vol. Indon. 41 C8
Kerintji vol. Indon. see Kerinci, Gunung
Keriya He watercourse China 42 E5
Keriya Shankou pass China 37 E2
Kerkennah, Îles is Tunisia 26 D7
Kerkini, Limni l. Greece 27 J4
Kérkira i. Greece see Corfu
Kerkouane tourist site Tunisia 26 D6
Kerkyra Greece 27 H5
Kerkyra i. Greece see Corfu
Kerma Sudan 32 D6
Kermadec Islands S. Pacific Ocean 53 I5
Kermadec Trench sea feature S. Pacific Ocean 74 I8
Kermān Iran 33 I3
Kerman CA U.S.A. 65 B2
Kermānshāh Iran 35 G4
Kermānshāh Iran 35 I5
Kermine Uzbek. see Navoiy
Kermit U.S.A. 62 G5
Kern r. CA U.S.A. 65 C3
Keros i. Greece 27 K6
Keros Rus. Fed. 12 L3
Kérouané Guinea 46 C4
Kerr, Cape Antarctica 76 H1
Kerrville U.S.A. 62 H5
Kerry Head hd Ireland 21 C5
Kerteminde Denmark 15 G9
Kerulen r. China/Mongolia see Herlen Gol
Kerur India 38 B2
Keryneia Cyprus see Kyrenia
Kerzaz Alg. 46 C2
Kerzhenets r. Rus. Fed. 12 J4
Kesagami Lake Canada 63 K1
Kesälahti Fin. 14 P6
Keşan Turkey 27 L4
Keşap Turkey 13 H8
Kesariya India 37 F4
Kesennuma Japan 45 F5
Keshan China 44 B2
Keshem Afgh. 36 B1
Keshendeh-ye Bālā Afgh. 36 A1
Keshod India 36 B5
Keshvar Iran 35 H4
Keskin Turkey 34 D3
Keskozero Rus. Fed. 12 G3
Kesova Gora Rus. Fed. 12 H4
Kestell S. Africa 51 I5
Kesten'ga Rus. Fed. 14 Q4
Kestilä Fin. 14 O4
Keswick U.K. 18 D4
Keszthely Hungary 26 G1
Ketapang Indon. 41 D8
Keti Bandar Pak. 36 A4
Ketmen', Khrebet mts China/Kazakh. 42 E4
Kettering U.K. 19 G6
Kettle Creek r. PA U.S.A. 64 C2
Kettleman City CA U.S.A. 65 C2
Keuka NY U.S.A. 64 C1
Keuka Lake NY U.S.A. 64 C1
Keumgang, Mount N. Korea see Kumgang-san
Keumsang, Mount N. Korea see Kumgang-san
Keuruu Fin. 14 N5
Keweenaw Peninsula U.S.A. 63 J2
Key, Lough l. Ireland 21 D3
Keyala South Sudan 47 G4
Keyihe China 44 A2
Key Largo U.S.A. 63 K6
Keymir Turkm. 35 I3
Keynsham U.K. 19 E7
Keyser WV U.S.A. 64 B3
Keysville VA U.S.A. 64 B4
Keyvy, Vozvyshennost' hills Rus. Fed. 12 H2
Key West U.S.A. 63 K7
Kez Rus. Fed. 11 Q4
Kezi Zimbabwe 49 C6
Kgalagadi admin. dist. Botswana 50 E3
Kgalazadi admin. dist. Botswana see Kgalagadi
Kgatlen admin. dist. Botswana see Kgatleng
Kgatleng admin. dist. Botswana 51 H3
Kgomofatshe Pan salt pan Botswana 50 E2
Kgoro Pan salt pan Botswana 50 G3
Kgotsong S. Africa 51 H4
Khabab Syria 39 C3
Khabar Iran 35 I5
Khabarikha Rus. Fed. 12 L2
Khabarovsk Rus. Fed. 44 D2
Khabarovskiy Kray admin. div. Rus. Fed. 44 D2
Khabarovsk Kray admin. div. Rus. Fed. see Khabarovskiy Kray
Khabary Rus. Fed. 42 D2
Khachmas Azer. see Xaçmaz
Khadro Pak. 36 B4
Khāf Iran 33 J3
Khagaria India 37 F4
Khagrachari Bangl. 37 G5
Khagrachhari Bangl. see Khagrachari
Khairgarh Pak. 36 B3
Khairpur Punjab Pak. 36 C3
Khairpur Sindh Pak. 33 K4
Khaīz, Kūh-e mt. Iran 35 H5
Khajuha India 36 E4
Khāk-e Jabbār Afgh. 36 B2
Khakhea Botswana 50 F3
Khākrīz Afgh. 36 A3
Khalatse India 36 D2
Khalifat mt. Pak. 33 K3
Khalīj Surt g. Libya see Sirte, Gulf of
Khalilabad India 37 E4
Khalīlī Iran 35 I6
Khalkhāl Iran 35 H3
Khálki i. Greece see Chalki
Khalkís Greece see Chalkida
Khallikot India 38 E2
Khalturin Rus. Fed. see Orlov
Khamar-Daban, Khrebet mts Rus. Fed. 42 I2
Khamaria India 38 D1
Khambhat India 36 C5
Khambhat, Gulf of India 38 A2
Khamgaon India 38 C1
Khamir Yemen 32 F6
Khamis Mushayṭ Saudi Arabia 32 F6
Khammam India 38 D2
Khammouan Laos see Thakèk
Khamra Rus. Fed. 29 M3
Khān Afgh. 36 B2
Khānābād Afgh. 36 B1
Khān al Baghdādī Iraq 35 F4
Khān al Mashāhidah Iraq 35 G4
Khān al Muşallá Iraq 35 G4
Khanapur India 38 B2
Khān ar Raḥbah Iraq 35 G5
Khanasur Pass Iran/Turkey 35 G3
Khanbalik China see Beijing
Khānch Iran 35 G3
Khandwa India 36 D5
Khandyga Rus. Fed. 29 O3
Khanewal Pak. 33 L3
Khaniá Greece see Chania
Khānī Yek Iran 35 I5
Khanka, Lake China/Rus. Fed. 44 D3
Khanka, Ozero l. China/Rus. Fed. see Khanka, Lake
Khankendi Azer. see Xankändi
Khanna India 36 D3
Khannā, Qā' salt pan Jordan 39 C3
Khanpur Pak. 36 B3
Khanpur Pak. 33 L4
Khān Ruḥābah Iraq see Khān ar Raḥbah
Khansar Pak. 36 B3
Khān Shaykhūn Syria 39 C2
Khantayskoye, Ozero l. Rus. Fed. 28 K3
Khanthabouli Laos see Savannakhét
Khanty-Mansiysk Rus. Fed. 28 H3
Khān Yūnis Gaza 39 B4
Khanzi admin. dist. Botswana see Ghanzi
Khaplu Pak. 33 M2
Khaptad National Park Nepal 36 E3
Kharabali Rus. Fed. 13 J7
Kharagpur Bihar India 37 F4
Kharagpur W. Bengal India 37 F5
Kharān r. Iran 33 I4
Kharari India see Abu Road
Kharda India 38 B2
Khardi India 36 C6
Khardong La pass India see Khardung La
Khardung La pass India 36 D2
Kharfiyah Iraq 35 G5
Kharga Egypt see Al Khārijah
Kharga r. Rus. Fed. 44 D1
Khârga, El Wâḥât el oasis Egypt see Khārijah, Wāḥāt al
Kharga Oasis Egypt see Khārijah, Wāḥāt al
Kharg Islands Iran 35 H5
Khargon India 36 C5
Khari r. Rajasthan India 36 C4
Khari r. Rajasthan India 36 C4
Kharian Pak. 36 C2
Khariar India 38 D1
Khārijah, Wāḥāt al oasis Egypt 32 D5
Kharīm, Gebel h. Egypt see Kharīm, Jabal
Kharīm, Jabal h. Egypt 39 A4
Kharkhara r. India 36 E5
Kharkiv Ukr. 13 H6
Khar'kov Ukr. see Kharkiv
Khār Kūh mt. Iran 35 I5
Kharlovka Rus. Fed. 12 H1
Kharlu Rus. Fed. 14 Q6
Kharmanli Bulg. 27 K4
Kharovsk Rus. Fed. 12 I4
Kharsia India 37 E5
Khartoum Sudan 32 D6
Khasavyurt Rus. Fed. 13 J8
Khāsh Iran 33 J4
Khashm el Girba Sudan 32 E7
Khashm Şana' Saudi Arabia 34 E6
Khashuri Georgia 35 F2
Khasi Hills India 37 G4
Khaskovo Bulg. 27 K4
Khatanga Rus. Fed. 29 L2
Khatanga, Gulf of Rus. Fed. see Khatangskiy Zaliv
Khatangskiy Zaliv b. Rus. Fed. 29 L2
Khatayakha Rus. Fed. 12 M2
Khatinza Pass Pak. 36 B1
Khatyrka Rus. Fed. 29 S3
Khavda India 36 B5
Khayamnandi S. Africa 51 G6
Khaybar Saudi Arabia 32 E4
Khayelitsha S. Africa 50 D8
Khayrān, Ra's al pt Oman 33 I5
Khefa Israel see Haifa
Khehuene, Ponta pt Moz. 51 L2
Khemis Miliana Alg. 25 H5
Khenchela Alg. 26 B7
Khenifra Morocco 22 C5
Kherāmeh Iran 35 I5
Kherrata Alg. 25 I5
Kherreh Iran 35 I6
Khersan r. Iran 35 H5
Kherson Ukr. 27 O1
Kheta r. Rus. Fed. 29 L2
Khezerābād Iran 35 I3
Khiching India 37 F5
Khilok Rus. Fed. 43 K2
Khilok r. Rus. Fed. 43 J2
Khinganskiy Zapovednik nature res. Rus. Fed. 44 C2
Khíos i. Greece see Chios
Khipro Pak. 33 K4
Khirbat Isrīyah Syria 39 C2
Khitai Dawan Aksai Chin 36 D2
Khīyāv Iran 35 G3
Khiytola Rus. Fed. 15 P6
Khlevnoye Rus. Fed. 13 H5
Khmel'nik Ukr. see Khmil'nyk
Khmel'nitskiy Ukr. see Khmel'nyts'kyy
Khmel'nyts'kyy Ukr. 13 E6
Khmer Republic country Asia see Cambodia
Khmil'nyk Ukr. 13 E6
Khobi Georgia 35 F2
Khodzha-Kala Turkm. see Hojagala
Khodzhent Tajik. see Khŭjand
Khojand Tajik. see Khŭjand
Khokhowe Pan salt pan Botswana 50 E3
Khokhropar Pak. 36 B4
Khoksar India 36 D2
Kholm Afgh. 36 A1
Kholm Poland see Chełm
Kholm Rus. Fed. 12 F4
Kholmsk Rus. Fed. 44 F3
Kholon Israel see Ḥolon
Khomas admin. reg. Namibia 50 C2
Khomas Highland hills Namibia 50 B2
Khomeyn Iran 35 H4
Khomeynīshahr Iran 35 H4
Khong, Mae Nam r. Asia see Mekong
Khong, Mae Nam r. China/Myanmar see Salween
Khonj Iran 35 I6
Khonj, Kūh-e mts Iran 35 I6
Khon Kaen Thai. 31 J5
Khonsa India 37 H4
Khonuu Rus. Fed. 29 P3
Khoper r. Rus. Fed. 13 I6
Khor Rus. Fed. 44 D3
Khor r. Rus. Fed. 44 D3
Khorda India see Khurda
Khordha India see Khurda
Khorey-Ver Rus. Fed. 12 M2
Khorinsk Rus. Fed. 43 J2
Khorixas Namibia 49 B6
Khormūj, Kūh-e mt. Iran 35 H5
Khorog Tajik. see Khorugh
Khorol Rus. Fed. 44 D3
Khorol Ukr. 13 G6
Khoroslū Dāgh hills Iran 35 G3
Khorramābād Iran 35 H4
Khorramshahr Iran 35 H5
Khorugh Tajik. 33 L2
Khosheutovo Rus. Fed. 13 J7
Khoshgort Rus. Fed. 11 T2
Khosrow Afgh. 36 A2
Khosūyeh Iran 35 I5
Khotan China see Hotan
Khouribga Morocco 22 C5
Khowrjān Iran 35 I5
Khrebet Sengilen mts Rus. Fed. 42 H2
Khrebet Shyngystau mts Kazakh. 42 D3
Khreum Myanmar 37 H5
Khroma r. Rus. Fed. 29 P2
Khromtau Kazakh. 28 G4
Khrushchev Ukr. see Svitlovods'k
Khrysokhou Bay Cyprus see Chrysochou Bay
Khrystynivka Ukr. 13 F6
Khudumelapye Botswana 50 G2
Khudzhand Tajik. see Khŭjand
Khuis Botswana 50 E4
Khŭjand Tajik. 33 K1
Khŭjayli Uzbek. see Xoʻjayli
Khulays Saudi Arabia 32 E5
Khulkhuta Rus. Fed. 13 J7
Khulna Bangl. 37 G5
Khulo Georgia 35 F2
Khuma S. Africa 51 H4
Khunayzīr, Jabal al mts Syria 39 C2
Khūnīnshahr Iran see Khorramshahr
Khunjerab Pass China/Pakistan 36 C1
Khurayş Saudi Arabia 32 G4
Khurda India 38 E1
Khurdha India see Khurda
Khurja India 36 D3
Khurmuli Rus. Fed. 44 E2
Khūrrāb Iran 35 I5
Khushab Pak. 33 L3
Khushalgarh Pak. 36 B2
Khushshah, Wādī al watercourse Jordan/Saudi Arabia 39 C5
Khust Ukr. 13 D6
Khutse Game Reserve nature res. Botswana 50 G2
Khutsong S. Africa 51 H4
Khutu r. Rus. Fed. 44 E2
Khuzdar Pak. 33 K4
Khvājeh Iran 35 G3
Khvājeh Do Kūh h. Afgh. 36 A1
Khvalynsk Rus. Fed. 13 K5
Khvānsār Iran 35 H4
Khvormūj Iran 35 H5
Khvors Iran 35 I4
Khvoy Iran 35 G3
Khvoynaya Rus. Fed. 12 G4
Khwaja Amran mt. Pak. 36 A3
Khyber Pakhtunkhwa prov. Pak. 36 B2
Khyber Pass Afgh./Pak. 33 L3
Kiama Australia 58 E5
Kiangsi prov. China see Jiangxi
Kiangsu prov. China see Jiangsu
Kiantajärvi l. Fin. 14 P4
Kīāsar Iran 35 I3
Kiatassuaq i. Greenland 61 M2
Kibaha Tanz. 49 D4
Kibali r. Dem. Rep. Congo 48 C3
Kibangou Congo 48 B4
Kibaya Tanz. 49 D4
Kiboga Uganda 48 D3
Kibombo Dem. Rep. Congo 48 C4
Kibondo Tanz. 48 D4
Kibre Mengist Eth. 47 G4
Kıbrıs country Asia see Cyprus
Kibungo Rwanda 48 D4
Kičevo Macedonia 27 I4
Kichmengskiy Gorodok Rus. Fed. 12 J4
Kiçik Qafqaz mts Asia see Lesser Caucasus
Kicking Horse Pass Canada 62 D1
Kidal Mali 46 D3
Kidderminster U.K. 19 E6
Kidepo Valley National Park Uganda 48 D3
Kidira Senegal 46 B3
Kidmang India 36 D2
Kidnappers, Cape N.Z. 59 F4
Kidsgrove U.K. 19 E5
Kiel Germany 17 M3
Kiel Canal Germany 17 L3
Kielce Poland 17 R5
Kielder Water resr U.K. 18 E3
Kieler Bucht b. Germany 17 M3
Kienge Dem. Rep. Congo 49 C5
Kiev Ukr. 13 F6
Kiffa Mauritania 46 B3
Kifisia Greece 27 J5
Kifrī Iraq 35 G4
Kigali Rwanda 48 D4
Kiğı Turkey 35 F3
Kigoma Tanz. 49 C4
Kihlanki Fin. 14 M3
Kihniö Fin. 14 M5
Kiiminki Fin. 14 N4
Kii-sanchi mts Japan 45 D6
Kii-suidō sea chan. Japan 45 D6
Kikerino Rus. Fed. 15 P7
Kikinda Serbia 27 I2
Kikládhes is Greece see Cyclades
Kiknur Rus. Fed. 12 J4
Kikonai Japan 44 F4
Kikori P.N.G. 52 E2
Kikwit Dem. Rep. Congo 49 B4
Kilafors Sweden 15 J6
Kilar India 36 D2
Kilchu N. Korea 44 C4
Kilcoole Ireland 21 F4
Kilcormac Ireland 21 E4
Kilcoy Australia 58 F1
Kildare Ireland 21 F4
Kil'dinstroy Rus. Fed. 14 R2
Kilemary Rus. Fed. 12 J4
Kilembe Dem. Rep. Congo 49 B4
Kilfinan U.K. 20 D5
Kilham U.K. 18 E3
Kilia Ukr. see Kiliya
Kılıç Dağı mt. Syria/Turkey see Aqra', Jabal al
Kilifi Kenya 48 D4
Kilik Pass China 36 C1
Kilimanjaro vol. Tanz. 48 D4
Kilimanjaro National Park Tanz. 48 D4
Kilinailau Islands P.N.G. 52 F2
Kilindoni Tanz. 49 D4
Kilingi-Nõmme Estonia 15 N7
Kilis Turkey 39 C1
Kilis prov. Turkey 39 C1
Kiliya Ukr. 27 M2
Kilkee Ireland 21 C5
Kilkeel U.K. 21 G3
Kilkenny Ireland 21 E5
Kilkhampton U.K. 19 C8
Kilkis Greece 27 J4
Killala Ireland 21 C3
Killala Bay Ireland 21 C3
Killaloe Ireland 21 D5
Killarney N.T. Australia 54 E4
Killarney Qld Australia 58 F2
Killarney Ireland 21 C5
Killarney National Park Ireland 21 C6
Killary Harbour b. Ireland 21 C4
Killeen U.S.A. 62 H5
Killenaule Ireland 21 E5
Killimor Ireland 21 D4
Killin U.K. 20 E4
Killinchy U.K. 21 G3
Killíni mt. Greece see Kyllini
Killinick Ireland 21 F5
Killorglin Ireland 21 C5
Killurin Ireland 21 F5
Killybegs Ireland 21 D3
Kilmacrenan Ireland 21 E2
Kilmaine Ireland 21 C4
Kilmallock Ireland 21 D5
Kilmaluag U.K. 20 C3
Kilmarnock U.K. 20 E5
Kilmelford U.K. 20 D4
Kil'mez' Rus. Fed. 12 K4
Kil'mez' r. Rus. Fed. 12 K4
Kilmona Ireland 21 D6
Kilmore Australia 58 B6
Kilmore Quay Ireland 21 F5
Kilosa Tanz. 49 D4
Kilpisjärvi Fin. 14 L2
Kilrea U.K. 21 F3
Kilrush Ireland 21 C5
Kilsyth U.K. 20 E5
Kiltan atoll India 38 B4
Kiltullagh Ireland 21 D4
Kilwa Masoko Tanz. 49 D4
Kilwinning U.K. 20 E5
Kimba Australia 55 G8
Kimba Congo 48 B4
Kimball U.S.A. 62 G3
Kimball, Mount U.S.A. 60 D3
Kimbe P.N.G. 52 F2
Kimberley S. Africa 50 G5
Kimberley Plateau Australia 54 D4
Kimberley Range hills Australia 55 B6
Kimch'aek N. Korea 45 C4
Kimch'ŏn S. Korea 45 C5
Kimhae S. Korea 45 C6
Kimhandu mt. Tanz. 49 D4
Kimhwa S. Korea 45 B5
Kími Greece see Kymi
Kimito Fin. 15 M6
Kimmirut Canada 61 L3
Kimolos i. Greece 27 K6
Kimovsk Rus. Fed. 13 H5
Kimpese Dem. Rep. Congo 49 B4
Kimpoku-san mt. Japan see Kinpoku-san
Kimry Rus. Fed. 12 H4
Kimvula Dem. Rep. Congo 49 B4
Kinabalu, Gunung mt. Sabah Malaysia 41 D7
Kinango Kenya 49 D4
Kinbasket Lake Canada 62 D1
Kinbrace U.K. 20 F2
Kincardine U.K. 20 F4
Kinchega National Park Australia 57 C7
Kinda Dem. Rep. Congo 49 C4
Kindat Myanmar 37 H5
Kinder Scout h. U.K. 18 F5
Kindersley Canada 62 F1
Kindia Guinea 46 B3
Kindu Dem. Rep. Congo 48 C4
Kinel' Rus. Fed. 13 K5
Kineshma Rus. Fed. 12 I4
Kingaroy Australia 58 E1
King Christian Island Canada 61 H2
King City CA U.S.A. 65 B2
King Edward VII Land pen. Antarctica see Edward VII Peninsula
King George VA U.S.A. 64 C3
King George Island Antarctica 76 A2
King George Islands Canada 61 K4
King George Islands Fr. Polynesia see Roi Georges, Îles du
King Hill h. Australia 54 C5
Kingisepp Rus. Fed. 12 F4
King Island Australia 57 [inset]
Kingisseppa Estonia see Kuressaare
Kinglake National Park Australia 58 B6
King Leopold and Queen Astrid Coast Antarctica 76 E2
King Leopold Range National Park Australia 54 D4
King Leopold Ranges hills Australia 54 D4
Kingman AZ U.S.A. 65 E3
Kingman Reef terr. N. Pacific Ocean 74 J5
Kingoonya Australia 57 A6
King Peak Antarctica 76 L1
King Peninsula Antarctica 76 K2
Kingri Pak. 36 B3
Kings r. Ireland 21 E5
Kings r. CA U.S.A. 65 B2
King Salmon U.S.A. 60 C4
Kingsbridge U.K. 19 D8
Kingsburg CA U.S.A. 65 C2
Kings Canyon National Park CA U.S.A. 65 C2
Kingscliff Australia 58 F2
Kingscote Australia 57 B7
Kingscourt Ireland 21 F4
King Sejong (South Korea) research stn Antarctica 76 A2
King's Lynn U.K. 19 H6
Kingsmill Group is Kiribati 53 H2
Kingsnorth U.K. 19 H7
King Sound b. Australia 54 C4
Kings Peak U.S.A. 62 E3
Kingsport U.S.A. 63 K4
Kingston Australia 57 [inset]
Kingston Canada 63 L3
Kingston Jamaica 67 I5
Kingston Norfolk I. 53 G4
Kingston NY U.S.A. 64 D2
Kingston PA U.S.A. 64 D2
Kingston Peak CA U.S.A. 65 E3
Kingston South East Australia 57 B8
Kingston upon Hull U.K. 18 G5
Kingstown St Vincent 67 L6
Kingsville U.S.A. 62 H6
Kingswood U.K. 19 E7
Kington U.K. 19 D6
Kingungi Dem. Rep. Congo 49 B4
Kingussie U.K. 20 E3
King William VA U.S.A. 64 C4
King William Island Canada 61 I3
King William's Town S. Africa 51 H7
Kingwood WV U.S.A. 64 B3
Kinloch N.Z. 59 B7
Kinloss U.K. 20 F3
Kinna Sweden 15 H8
Kinnegad Ireland 21 E4
Kinneret, Yam l. Israel see Galilee, Sea of
Kinniyai Sri Lanka 38 D4
Kinnula Fin. 14 N5
Kinpoku-san mt. Japan 45 E5
Kinross U.K. 20 F4
Kinsale Ireland 21 D6
Kinsale VA U.S.A. 64 C3
Kinshasa Dem. Rep. Congo 49 B4
Kinston U.S.A. 63 L4
Kintore U.K. 20 G3
Kintyre pen. U.K. 20 D5
Kin-U Myanmar 37 H5
Kinyeti mt. Sudan 47 G4
Kiparissía Greece see Kyparissia
Kipawa, Lac l. Canada 63 L2
Kipnuk U.S.A. 60 B4
Kiptopeke VA U.S.A. 64 D4
Kipungo Angola see Quipungo
Kipushi Dem. Rep. Congo 49 C5
Kirakira Solomon Is 53 G3
Kirandul India 38 D2
Kircubbin U.K. 21 G3
Kirdimi Chad 47 E3
Kirenga r. Rus. Fed. 43 J1
Kirensk Rus. Fed. 29 L4
Kireyevsk Rus. Fed. 13 H5
Kirghizia country Asia see Kyrgyzstan
Kirghiz Range mts Kazakh./Kyrg. 42 C4
Kirgizskaya S.S.R. country Asia see Kyrgyzstan
Kirgizskiy Khrebet mts Kazakh./Kyrg. see Kirghiz Range
Kirgizstan country Asia see Kyrgyzstan
Kiri Dem. Rep. Congo 48 B4
Kiribati country Pacific Ocean 74 I6
Kırıkhan Turkey 39 C1
Kırıkkale Turkey 34 D3
Kirillov Rus. Fed. 12 H4
Kirillovo (abandoned) Rus. Fed. 44 F3
Kirin China see Jilin
Kirin prov. China see Jilin
Kirinda Sri Lanka 38 D5
Kirinyaga mt. Kenya see Kenya, Mount
Kirishi Rus. Fed. 12 G4
Kirishima-Yaku Kokuritsu-kōen Japan 45 C7
Kirishima-yama vol. Japan 45 C7
Kiritimati atoll Kiribati 75 J5
Kiriwina Islands P.N.G. see Trobriand Islands
Kırkağaç Turkey 27 L5
Kirk Bulāg Dāgi mt. Iran 35 G3
Kirkby U.K. 18 E5
Kirkby in Ashfield U.K. 19 F5
Kirkby Lonsdale U.K. 18 E4
Kirkby Stephen U.K. 18 E4
Kirkcaldy U.K. 20 F4
Kirkcolm U.K. 20 D6
Kirkcudbright U.K. 20 E6
Kirkenær Norway 15 H6
Kirkenes Norway 14 Q2
Kirkintilloch U.K. 20 E5
Kirkkonummi Fin. 15 N6
Kirkland AZ U.S.A. 65 F3
Kirkland Lake Canada 63 K2
Kırklareli Turkey 27 L4
Kirk Michael Isle of Man 18 C4
Kirkoswald U.K. 18 E4
Kirkpatrick, Mount Antarctica 76 H1
Kirksville U.S.A. 63 I3
Kirkūk Iraq 35 G4
Kirkwall U.K. 20 G2
Kirkwood S. Africa 51 G7
Kirman Iran see Kermān
Kirov Kaluzhskaya Oblast' Rus. Fed. 13 G
Kirov Kirovskaya Oblast' Rus. Fed. 12 K4
Kirovabad Azer. see Gäncä
Kirovabad Tajik. see Panj
Kirovakan Armenia see Vanadzor
Kirovo Ukr. see Kirovohrad
Kirovo-Chepetsk Rus. Fed. 12 K4
Kirovo-Chepetskiy Rus. Fed. see Kirovo-Chepetsk
Kirovograd Ukr. see Kirovohrad
Kirovohrad Ukr. 13 G6
Kirovsk Leningradskaya Oblast' Rus. Fed. 12 F4
Kirovsk Murmanskaya Oblast' Rus. Fed. 12 G2
Kirovs'ke Ukr. 34 D1
Kirovskiy Rus. Fed. 44 B1
Kirovskiy Rus. Fed. 44 D3
Kirovskoye Ukr. see Kirovs'ke
Kırpaşa pen. Cyprus see Karpasia
Kirpili Turkm. 35 J3
Kirriemuir U.K. 20 F4
Kirs Rus. Fed. 12 L4
Kirsanov Rus. Fed. 13 I5
Kırşehir Turkey 34 D3
Kirthar National Park Pak. 36 A4
Kirthar Range mts Pak. 33 K4
Kiruna Sweden 14 L3
Kirundu Dem. Rep. Congo 48 C4
Kirwan Escarpment Antarctica 76 B2
Kiryū Japan 45 E5
Kisa Sweden 15 I8
Kisama, Parque Nacional de nat. park Angola see Quiçama, Parque Nacional do
Kisandji Dem. Rep. Congo 49 B4
Kisangani Dem. Rep. Congo 48 C3
Kisantu Dem. Rep. Congo 49 B4
Kisar i. Indon. 54 D2
Kiselevka Rus. Fed. 44 E2
Kiselevsk Rus. Fed. 42 F2
Kishanganj India 37 F4
Kishangarh Madh. Prad. India 36 D4
Kishangarh Rajasthan India 36 B4
Kishangarh Rajasthan India 36 C4
Kishangarh Rajasthan India 36 D4
Kishi Nigeria 46 D4
Kishinev Moldova see Chişinău
Kishkenekol' Kazakh. 31 G1
Kishoreganj Bangl. 37 G4
Kishorganj Bangl. see Kishoreganj
Kisi Nigeria see Kishi
Kisii Kenya 48 D4
Kiska Island U.S.A. 29 S4
Kiskunfélegyháza Hungary 27 H1
Kiskunhalas Hungary 27 H1
Kiskunsági nat. park Hungary 27 H1
Kislovodsk Rus. Fed. 35 F2
Kismaayo Somalia 48 E4
Kismayu Somalia see Kismaayo
Kisoro Uganda 47 F5
Kissamos Greece 27 J7
Kissidougou Guinea 46 B4
Kissimmee U.S.A. 63 K6
Kissimmee, Lake U.S.A. 63 K6
Kistendey Rus. Fed. 13 I5
Kistna r. India see Krishna
Kisumu Kenya 48 D4
Kisykkamys Kazakh. see Dzhangala

a Mali 46 C3
a-Daitō-jima i. Japan 43 O7
aibaraki Japan 45 F5
akami Japan 45 F5
a-Kyūshū Japan 45 C6
ale Kenya 48 D3
ami Japan 44 F4
chener Ont. Canada 64 A1
ee Fin. 14 Q5
gum Uganda 48 D3
hira i. Greece see Kythira
hnos i. Greece see Kythnos
i, Cape Cyprus see Kition, Cape
inen r. Fin. 14 O3
ion, Cape Cyprus 39 A2
iou, Akra c. Cyprus see Kition, Cape
tanning PA U.S.A. 64 B2
tatinny Mountains hills NJ U.S.A. 64 D2
tery ME U.S.A. 64 F1
tilä Fin. 14 N3
tur India 38 B3
ui Kenya 48 D4
we Zambia 49 C5
zbüheler Alpen mts Austria 17 N7
uruvesi Fin. 14 O5
valina U.S.A. 60 B3
vijärvi Fin. 14 N5
viöli Estonia 15 O7
vaba N'zogi Angola 49 B4
ev Ukr. see Kiev
evskoye Vodokhranilishche resr Ukr. see Kyyivs'ke Vodoskhovyshche
ıköy Turkey 27 M4
el Rus. Fed. 11 R4
ema Rus. Fed. 12 J3
ılcadağ Turkey 27 M6
ılca Dağ mt. Turkey 34 C3
ılcahamam Turkey 34 D2
ıldağ mt. Turkey 39 A1
ıldağ mt. Turkey 39 B1
ıl Dağı mt. Turkey 34 E3
ılırmak Turkey 34 D2
ılırmak r. Turkey 34 D2
iltepe Turkey 35 F3
ilyurt Rus. Fed. 13 J8
kalesi Turkey 39 B1
lyar Rus. Fed. 13 J8
lyarskiy Zaliv b. Rus. Fed. 35 G1
ner Rus. Fed. 12 K4
yl-Arbat Turkm. see Serdar
yl-Atrek Turkm. see Etrek
ellefjord Norway 14 O1
epsvik Norway 14 J2
adno Czech Rep. 17 O5
genfurt Austria 17 O7
aipėda Lith. 15 L9
aksvig Faroe Is see Klaksvík
aksvík Faroe Is 14 [inset]
math r. U.S.A. 60 F3
math Falls U.S.A. 62 C3
rälven r. Sweden 15 H7
tovy Czech Rep. 17 N6
wer S. Africa 50 D6
eides Islands Cyprus 39 B2
einbegin S. Africa 50 E5
ein Karas Namibia 50 D4
ein Nama Land reg. S. Africa see Namaqualand
ein Roggeveldberge mts S. Africa 50 E7
einsee S. Africa 50 C5
rksdorp S. Africa 51 H4
etnya Rus. Fed. 13 G5
etsk Belarus see Klyetsk
etskaya Rus. Fed. 13 I6
etskiy Rus. Fed. see Kletskaya
dhes Islands Cyprus see Kleides Islands
mkovka Rus. Fed. 12 K4
movo Rus. Fed. 13 G5
n Rus. Fed. 12 H4
novec mt. Czech Rep. 17 N5
ntehamn Sweden 15 K8
ntsy Rus. Fed. 13 G5
uč Bos.-Herz. 26 G2
dzko Poland 17 P5
osterneuburg Austria 17 P6
ane National Park Canada 60 E3
ang Malaysia see Keluang
uczbork Poland 17 Q5
ekhori Rus. Fed. see Karachayevsk
ekhorskiy, Pereval Georgia/Rus. Fed. 35 F2
kwan U.S.A. 60 E4
etsk Belarus 15 O10
uchevskaya Sopka, Vulkan vol. Rus. Fed. 29 R4
uchi Rus. Fed. 44 B2
äda Sweden 15 I6
aresborough U.K. 18 F4
ighton U.K. 19 D6
ights Landing CA U.S.A. 65 B1
in Croatia 26 G2
ittelfeld Austria 17 O7
jaževac Serbia 27 J3
ob Lake Canada see Schefferville
ob Peak h. Australia 54 E3
ock Ireland 21 D4
ockalongy h. Ireland 21 C3
ockalough Ireland 21 C5
ockanaffrin h. Ireland 21 E5
ockboy h. Ireland 21 C6
ock Hill h. U.K. 20 G3
ockmealdown Mountains hills Ireland 21 D5
ocknaskagh h. Ireland 21 D5
owle U.K. 19 F6
ox PA U.S.A. 64 B2
ox Coast Antarctica 76 F2
oxville TN U.S.A. 63 K4
ud Rasmussen Land reg. Greenland 61 L2
ysna S. Africa 50 F8
, Gora mt. Rus. Fed. 44 E3
artac Canada see Quaqtaq
bbfoss Norway 14 P2
be Japan 45 D6
benhavn Denmark see Copenhagen
benni Mauritania 46 C3
blenz Germany 17 K5
boldo Rus. Fed. 44 D1
brin Belarus see Kobryn
broör i. Indon. 41 F8
bryn Belarus 15 N10
buk Valley National Park U.S.A. 60 C3
bulet'i Georgia 35 F2
caeli Kocaeli Turkey see İzmit
caeli Yarımadası pen. Turkey 27 M4
čani Macedonia 27 J4
casu r. Turkey 27 M4
čevje Slovenia 26 F2
ch Bihar India 37 G4
chevo Rus. Fed. 11 Q4
chi India 38 C4
chi Japan 45 D6
chisar Turkey see Kızıltepe
ch Island Canada 61 K3
chkor Kyrg. 42 D4
chkorka Kyrg. see Kochkor
chkurovo Rus. Fed. 13 J5
chubeyevskoye Rus. Fed. 35 F1
d India 38 B3
dala India 38 E2
Kodarma India 37 F4
Koderma India see Kodarma
Kodiak U.S.A. 60 C4
Kodiak Island U.S.A. 60 C4
Kodibeleng Botswana 51 H2
Kodino Rus. Fed. 12 H3
Kodiyakkarai India 38 C4
Kodok South Sudan 32 D8
Kodyma Ukr. 13 F6
Kodzhaele mt. Bulg./Greece 27 K4
Koedoesberg mts S. Africa 50 E7
Koegrabie S. Africa 50 E5
Koekenaap S. Africa 50 D6
Koës Namibia 50 D3
Kofa Mountains AZ U.S.A. 65 F4
Koffiefontein S. Africa 50 G5
Koforidua Ghana 46 C4
Kōfu Japan 45 E6
Kogaluk r. Canada 61 L4
Kogan Australia 58 E1
Køge Denmark 15 H9
Kogon r. Guinea 46 B3
Kohat Pak. 33 L3
Kohila Estonia 15 N7
Kohima India 37 H4
Kohler Range mts Antarctica 76 K2
Kohlu Pak. 36 B3
Kohtla-Järve Estonia 15 O7
Kohŭng S. Korea 45 B6
Koidu Sierra Leone see Sefadu
Koilkonda India 38 C2
Koin N. Korea 45 B4
Koin r. Rus. Fed. 12 K3
Koi Sanjaq Iraq 35 G3
Kŏje-do i. S. Korea 45 C6
Kojonup Australia 55 B8
Kökar Fin. 15 L7
Kokchetav Kazakh. see Kokshetau
Kokemäenjoki r. Fin. 15 L6
Kokerboom Namibia 50 D5
Kokkilai Sri Lanka 38 D4
Kokkola Fin. 14 M5
Koko Nigeria 46 D3
Kokomo U.S.A. 63 J3
Kokong Botswana 50 F3
Kokosi S. Africa 51 H4
Kokpekty Kazakh. 42 E3
Koksan N. Korea 45 B5
Kokshaal-Tau, Khrebet mts China/Kyrg. see Kakshaal-Too
Koksharka Rus. Fed. 12 J4
Kokshetau Kazakh. 31 F1
Kokstad S. Africa 51 I6
Kokterek Kazakh. 13 K6
Koktokay China see Fuyun
Kola Rus. Fed. 14 R2
Kolachi r. Pak. 36 A4
Kolahoi mt. India 36 C2
Kolaka Indon. 41 E8
Kola Peninsula Rus. Fed. 12 H2
Kolar Chhattisgarh India 38 D2
Kolar Karnataka India 38 C3
Kolaras India 36 D4
Kolar Gold Fields India 38 C3
Kolari Fin. 14 M3
Kolarovgrad Bulg. see Shumen
Kolasib India 37 H4
Kolayat India 36 C4
Kolberg Poland see Kołobrzeg
Kol'chugino Rus. Fed. 12 H4
Kolda Senegal 46 B3
Kolding Denmark 15 F9
Kole Kasaï-Oriental Dem. Rep. Congo 48 C4
Kole Orientale Dem. Rep. Congo 48 C3
Koléa Alg. 25 H5
Koler Sweden 14 L4
Kolguyev, Ostrov i. Rus. Fed. 12 K1
Kolhan reg. India 37 F5
Kolhapur India 38 B2
Kolikata India see Kolkata
Kōljala Estonia 15 M7
Kolkasrags pt Latvia 15 M8
Kolkata India 37 G5
Kolkhozabad Khatlon Tajik. see Kolkhozobod
Kolkhozabad Khatlon Tajik. see Vose
Kolkhozobod Tajik. 33 K2
Kollam India 38 C4
Kolleru Lake India 38 D2
Kolmanskop (abandoned) Namibia 50 B4
Köln Germany see Cologne
Kołobrzeg Poland 17 O3
Kologriv Rus. Fed. 12 J4
Kolokani Mali 46 C3
Kolombangara i. Solomon Is 53 F2
Kolomea Ukr. see Kolomyya
Kolomna Rus. Fed. 13 H5
Kołomyja Ukr. see Kolomyya
Kolomyya Ukr. 13 E6
Kolondiéba Mali 46 C3
Kolonedale Indon. 52 C2
Koloni Cyprus 39 A2
Kolonkwaneng Botswana 50 E4
Kolozsvár Romania see Cluj-Napoca
Kolpashevo Rus. Fed. 28 J4
Kolpos Messaras b. Greece 27 K7
Kol'skiy Poluostrov pen. Rus. Fed. see Kola Peninsula
Kölük Turkey see Kâhta
Koluli Eritrea 32 F7
Kolva r. Rus. Fed. 12 M2
Kolvan India 38 B2
Kolvereid Norway 14 G4
Kolvik Norway 14 N1
Kolvitskoye, Ozero l. Rus. Fed. 14 R3
Kolwezi Dem. Rep. Congo 49 C5
Kolyma r. Rus. Fed. 29 R3
Kolyma Lowland Rus. Fed. see Kolymskaya Nizmennost'
Kolyma Range mts Rus. Fed. see Kolymskiy, Khrebet
Kolymskaya Nizmennost' lowland Rus. Fed. 29 Q3
Kolymskiy, Khrebet mts Rus. Fed. 29 R3
Kolyshley Rus. Fed. 13 J5
Kom mt. Bulg. 27 J3
Komadugu-Gana watercourse Nigeria 46 E3
Komaggas S. Africa 50 C5
Komaki Japan 45 E6
Komandnaya, Gora mt. Rus. Fed. 44 E2
Komandorskiye Ostrova is Rus. Fed. 29 R4
Komárno Slovakia 17 Q7
Komati r. Swaziland 51 J3
Komatipoort S. Africa 51 J3
Komatsu Japan 45 E5
Komba i. Indon. 54 C1
Komga S. Africa 51 H7
Komintern Ukr. see Marhanets'
Kominternivs'ke Ukr. 27 N1
Komiža Croatia 26 G3
Komló Hungary 26 H1
Kommunarsk Ukr. see Alchevs'k
Komodo, Taman Nasional Indon. 54 B2
Kôm Ombo Egypt see Kawm Umbū
Komono Congo 48 B4
Komotini Greece 27 K4
Kompong Cham Cambodia see Kâmpóng Cham
Kompong Som Cambodia see Sihanoukville
Kompong Speu Cambodia see Kâmpóng Spœ
Kompong Thom Cambodia see Kâmpóng Thum
Komrat Moldova see Comrat
Komsberg mts S. Africa 50 E7
Komsomol Kazakh. see Karabalyk
Komsomolets Kazakh. see Karabalyk
Komsomolets, Ostrov i. Rus. Fed. 28 K1
Komsomol's'k Ukr. 13 G6
Komsomol'skiy Chukotskiy Avtonomnyy Okrug Rus. Fed. 77 C2
Komsomol'skiy Khanty-Mansiyskiy Avtonomnyy Okrug-Yugra Rus. Fed. see Yugorsk
Komsomol'skiy Respublika Kalmykiya-Khalm'g-Tangch Rus. Fed. 13 J7
Komsomol'sk-na-Amure Rus. Fed. 44 E2
Komsomol'skoye Rus. Fed. 13 J6
Kömürlü Turkey 35 F2
Kon India 37 E4
Konacık Turkey 39 B1
Konada India 38 D2
Konarak India see Konarka
Konarka India 37 F6
Konch India 36 D4
Kondagaon India 38 D2
Kondinin Australia 55 B8
Kondinskoye Rus. Fed. see Oktyabr'skoye
Kondoa Tanz. 49 D4
Kondol' Rus. Fed. 13 J5
Kondopoga Rus. Fed. 12 G3
Kondoz Afgh. see Kunduz
Kondrovo Rus. Fed. 13 G5
Köneürgenç Turkm. see Köneürgenç
Köneürgenç Turkm. 33 I1
Kong Cameroon 46 E4
Kong Christian IX Land reg. Greenland 61 O3
Kong Christian X Land reg. Greenland 61 P2
Kongelab atoll Marshall Is see Rongelap
Kong Frederik IX Land reg. Greenland 61 M3
Kong Frederik VI Kyst coastal area Greenland 61 N3
Kongolo Dem. Rep. Congo 49 C4
Kongor South Sudan 47 G4
Kong Oscars Fjord inlet Greenland 61 P2
Kongoussi Burkina Faso 46 C3
Kongsberg Norway 15 F7
Kongsvinger Norway 15 H6
Kongur Shan mt. China 42 D5
Königsberg Rus. Fed. see Kaliningrad
Konin Poland 17 Q4
Konjic Bos.-Herz. 26 G3
Konkiep watercourse Namibia 50 C5
Konnevesi Fin. 14 O5
Konosha Rus. Fed. 12 I3
Konotop Ukr. 13 G6
Konpara India 37 E5
Konqi He r. China 42 F4
Konso Eth. 48 D3
Konstantinograd Ukr. see Krasnohrad
Konstantinovka Rus. Fed. 44 B2
Konstantinovka Ukr. see Kostyantynivka
Konstanz Germany 17 L7
Kontiolahti Fin. 14 P5
Konttila Fin. 14 O4
Könugard Ukr. see Kiev
Konushin, Mys pt Rus. Fed. 12 I2
Konya Turkey 34 D3
Konzhakovskiy Kamen', Gora mt. Rus. Fed. 11 R4
Kooch Bihar India see Koch Bihar
Kookynie Australia 55 C7
Koolyanobbing Australia 55 B7
Koondrook Australia 58 B5
Koorawatha Australia 58 D5
Koordarrie Australia 54 A5
Kootenay Lake Canada 62 D2
Kootjieskolk S. Africa 50 E6
Kópasker Iceland 14 [inset]
Kopbirlik Kazakh. 42 D3
Koper Slovenia 26 E2
Kopet Dag mts Iran/Turkm. 33 I2
Kopet-Dag, Khrebet mts Iran/Turkm. see Kopet Dag
Köpetdag Gershi mts Iran/Turkm. see Kopet Dag
Köping Sweden 15 J7
Köpmanholmen Sweden 14 K5
Kopong Botswana 51 G3
Koppal India 38 C3
Koppang Norway 15 G6
Kopparberg Sweden 15 I7
Koppeh Dāgh mts Iran/Turkm. see Kopet Dag
Koppi r. Rus. Fed. 44 F2
Koppies S. Africa 51 H4
Koppieskraal Pan salt pan S. Africa 50 E4
Koprivnica Croatia 26 G1
Köprülü Turkey 39 A1
Köprülü Kanyon Milli Parkı nat. park Turkey 27 N6
Kopyl' Belarus see Kapyl'
Kora India 36 E4
Korablino Rus. Fed. 13 I5
K'orahē Eth. 48 E3
Koramlik China 37 F1
Korangal India 38 C2
Korangi Pak. 36 A4
Korān va Monjān Afgh. 36 B1
Koraput India 38 D2
Korat Thai. see Nakhon Ratchasima
Koratla India 38 C2
Korba India 37 E5
Korçë Albania 27 I4
Korčula Croatia 26 G3
Korčula i. Croatia 26 G3
Korčulanski Kanal sea chan. Croatia 26 G3
Korday Kazakh. 42 C4
Kord Kūy Iran 35 I3
Korea, North country Asia 45 B5
Korea, South country Asia 45 B5
Korea Bay g. China/N. Korea 45 B5
Korea Strait Japan/S. Korea 45 C6
Koregaon India 38 B2
Korenovsk Rus. Fed. 35 E1
Korenovskaya Rus. Fed. see Korenovsk
Korepino Rus. Fed. 11 R3
Korets' Ukr. 13 E6
Körfez Turkey 27 M4
Korff Ice Rise Antarctica 76 L1
Korfovskiy Rus. Fed. 44 D2
Korgalzhyn Kazakh. 42 C2
Korgen Norway 14 H3
Korhogo Côte d'Ivoire 46 C4
Koribundu Sierra Leone 46 B4
Kori Creek inlet India 36 B5
Korinthiakos Kolpos sea chan. Greece see Corinth, Gulf of
Korinthos Greece see Corinth
Kőris-hegy h. Hungary 26 G1
Koritnik mt. Albania 27 I3
Koritsa Albania see Korçë
Kōriyama Japan 45 F5
Korkuteli Turkey 27 N6
Korla China 42 F4
Kormakitis, Cape Cyprus 39 A2
Körmend Hungary 26 G1
Kornati, Nacionalni Park nat. park Croatia 26 F3
Korneyevka Rus. Fed. 13 K6
Koro Côte d'Ivoire 46 C4
Koro i. Fiji 53 H3
Koro Mali 46 C3
Köroğlu Dağları mts Turkey 27 O4
Köroğlu Tepesi mt. Turkey 34 D2
Korogwe Tanz. 49 D4
Koroneia, Limni l. Greece 27 J4
Korong Vale Australia 58 A6
Koror Palau 41 F7
Koro Sea sea Fiji 53 H3
Korosten' Ukr. 13 F6
Korostyshiv Ukr. 13 F6
Koro Toro Chad 47 E3
Korpilahti Fin. 14 N5
Korpo Fin. 15 L6
Korppoo Fin. see Korpo
Korsakov Rus. Fed. 44 F3
Korsnäs Fin. 14 L5
Korsør Denmark 15 G9
Korsun'-Shevchenkivs'kyy Ukr. 13 F6
Korsun'-Shevchenkovskiy Ukr. see Korsun'-Shevchenkivs'kyy
Korsze Poland 17 R3
Kortesjärvi Fin. 14 M5
Korti Sudan 32 D6
Kortkeros Rus. Fed. 12 K3
Kortrijk Belgium 16 I5
Korvala Fin. 14 O3
Koryakskaya, Sopka vol. Rus. Fed. 29 Q4
Koryakskoye Nagor'ye mts Rus. Fed. 29 S3
Koryazhma Rus. Fed. 12 J3
Koryŏng S. Korea 45 C6
Kos i. Greece 27 L6
Kosa Rus. Fed. 11 Q4
Kosam India 36 E4
Kosan N. Korea 45 B5
Kościan Poland 17 P4
Kosciusko, Mount Australia see Kosciuszko, Mount
Kosciuszko, Mount Australia 58 D6
Kosciuszko National Park Australia 58 D6
Köse Turkey 35 E2
Köseçobanlı Turkey 39 A1
Kosgi India 38 C2
Kosh-Agach Rus. Fed. 42 F3
Koshikijima-rettō is Japan 45 C7
Koshki Rus. Fed. 13 K5
Kosi Bay S. Africa 51 K4
Košice Slovakia 13 D6
Kosigi India 38 C3
Koskullskulle Sweden 14 L3
Köslin Poland see Koszalin
Kosma r. Rus. Fed. 12 K2
Kosŏng N. Korea 45 C5
Kosova country Europe see Kosovo
Kosovo country Europe 27 I3
Kosovo-Metohija country Europe see Kosovo
Kosovska Mitrovica Kosovo see Mitrovicë
Kosrae atoll Micronesia 74 G5
Kosta-Khetagurovo Rus. Fed. see Nazran'
Kostanay Kazakh. 30 F1
Kostenets Bulg. 27 J3
Kosti Sudan 32 D7
Kostinbrod Bulg. 27 J3
Kostino Rus. Fed. 28 J3
Kostomuksha Rus. Fed. 12 F2
Kostopil' Ukr. 13 E6
Kostopol' Ukr. see Kostopil'
Kostroma Rus. Fed. 12 I4
Kostrzyn Poland 17 O4
Kostyantynivka Ukr. 13 H6
Kostyukovichi Belarus see Kastsyukovichy
Kos'yu Rus. Fed. 11 R2
Koszalin Poland 17 P3
Kőszeg Hungary 26 G1
Kota Andhra Prad. India 38 D3
Kota Chhattisgarh India 37 E5
Kota Rajasthan India 36 C4
Kota Baharu Malaysia see Kota Bharu
Kotabaru Kalimantan Selatan Indon. 41 D8
Kota Bharu Malaysia 41 C7
Kot Addu Pak. 36 B3
Kota Kinabalu Sabah Malaysia 41 D7
Kotaparh India 38 D2
Kot Diji Pak. 36 B4
Kotel'nich Rus. Fed. 12 K4
Kotel'nikovo Rus. Fed. 13 I7
Kotel'nyy, Ostrov i. Rus. Fed. 29 O2
Kotgar India 38 D2
Kotgarh India 36 D3
Kothagudem India see Kottagudem
Kotido Uganda 47 G4
Kotikovo Rus. Fed. 44 D3
Kotka Fin. 15 O6
Kot Kapura India 36 C3
Kotkino Rus. Fed. 12 K2
Kotlas Rus. Fed. 12 J3
Kotli Pak. 36 C2
Kotlik U.S.A. 60 B3
Kötlutangi pt Iceland 14 [inset]
Kotly Rus. Fed. 15 P7
Kotovo Rus. Fed. 13 J6
Kotovsk Rus. Fed. 13 I5
Kotra India 36 C4
Kotra Pak. 36 A3
Kotri r. India 38 D2
Kot Sarae Pak. 36 A5
Kottagudem India 38 D2
Kottarakara India 38 C4
Kottayam India 38 C4
Kotte Sri Lanka see Sri Jayewardenepura Kotte
Kotto r. Cent. Afr. Rep. 48 C3
Kotturu India 38 C3
Kotuy r. Rus. Fed. 29 L2
Kotzebue U.S.A. 60 B3
Kotzebue Sound sea chan. U.S.A. 60 B3
Kouango Cent. Afr. Rep. 48 C3
Koubia Guinea 46 B3
Koudougou Burkina Faso 46 C3
Kouebokkeveld mts S. Africa 50 D7
Koufey Niger 46 E3
Koufonisi i. Greece 27 L7
Kougaberge mts S. Africa 50 F7
Koukourou r. Cent. Afr. Rep. 48 B3
Koulikoro Mali 46 C3
Koumac New Caledonia 53 G4
Koumpentoum Senegal 46 B3
Koundâra Guinea 46 B3
Koupéla Burkina Faso 46 C3
Kourou Fr. Guiana 69 H2
Kouroussa Guinea 46 C3
Kousséri Cameroon 47 E3
Kouvola Fin. 15 O6
Kovallberget Sweden 14 J4
Kovdor Rus. Fed. 12 F2
Kovel' Ukr. 13 E6
Kovernino Rus. Fed. 12 I4
Kovilpatti India 38 C4
Kovno Lith. see Kaunas
Kovrov Rus. Fed. 12 I4
Kovylkino Rus. Fed. 13 I5
Kovzhskoye, Ozero l. Rus. Fed. 12 H3
Kowanyama Australia 56 C2
Kowloon H.K. China 73 P4
Kowŏn N. Korea 45 B5
Kowtal-e Khāwāk Afgh. 36 B2
Kōyama-misaki pt Japan 45 C6
Koyamutthoor India see Coimbatore
Köyceğiz Turkey 27 M6
Koygorodok Rus. Fed. 12 K3
Koyna Reservoir India 38 B2
Köýtendag Turkm. 33 K2
Koyuk U.S.A. 60 B3
Koyukuk r. U.S.A. 60 C3
Koyulhisar Turkey 34 E2
Kozağacı Turkey see Günyüzü
Kō-zaki pt Japan 45 C6
Kozan Turkey 34 D3
Kozani Greece 27 I4
Kozara mts Bos.-Herz. 26 G2
Kozara, Nacionalni Park nat. park Bos.-Herz. 26 G2
Kozarska Dubica Bos.-Herz. see Bosanska Dubica
Kozelets' Ukr. 13 F6
Kozel'sk Rus. Fed. 13 G5
Kozhikode India 38 B4
Kozhva Rus. Fed. 12 M2
Kozlu Turkey 27 N4
Kozly Rus. Fed. 12 H2
Koz'modem'yansk Rus. Fed. 12 J4
Kožuf mts Greece/Macedonia 27 J4
Kōzu-shima i. Japan 45 E6
Kozyatyn Ukr. 13 F6
Kpalimé Togo 46 D4
Kpandae Ghana 46 C4
Kpungan Pass India/Myanmar 37 I4
Krabi Thai. 31 I6
Krâchéh Cambodia 31 J5
Kraddsele Sweden 14 J4
Kragerø Norway 15 F7
Kragujevac Serbia 27 I2
Krakau Poland see Kraków
Kraków Poland 17 Q5
Kralendijk Bonaire 67 K6
Kramators'k Ukr. 13 H6
Kramfors Sweden 14 J5
Kranidi Greece 27 J6
Kranj Slovenia 26 F1
Kranskop S. Africa 51 J5
Krasavino Rus. Fed. 12 J3
Krasilov Ukr. see Krasyliv
Krasino Rus. Fed. 28 G2
Kraskino Rus. Fed. 44 C4
Krāslava Latvia 15 O9
Krasnaya Gorbatka Rus. Fed. 12 I5
Krasnoarmeysk Rus. Fed. 13 J6
Krasnoarmeysk Ukr. see Krasnoarmiys'k
Krasnoarmiys'k Ukr. 13 H6
Krasnoborsk Rus. Fed. 12 J3
Krasnodar Rus. Fed. 13 H7
Krasnodar Kray admin. div. Rus. Fed. see Krasnodarskiy Kray
Krasnodarskiy Kray admin. div. Rus. Fed. 34 E1
Krasnodon Ukr. 13 H6
Krasnogorodsk Rus. Fed. 15 P8
Krasnogorsk Rus. Fed. 44 F2
Krasnogorskoye Rus. Fed. 12 L4
Krasnograd Ukr. see Krasnohrad
Krasnogvardeysk Uzbek. see Bulung'ur
Krasnogvardeyskoye Rus. Fed. 13 I7
Krasnohrad Ukr. 13 G6
Krasnohvardiys'ke Ukr. 13 G7
Krasnokamsk Rus. Fed. 11 R4
Krasnoperekops'k Ukr. 13 G7
Krasnopol'ye Rus. Fed. 44 F2
Krasnorechenskiy Rus. Fed. 44 D3
Krasnoslobodsk Rus. Fed. 13 I5
Krasnotur'insk Rus. Fed. 11 S4
Krasnoufimsk Rus. Fed. 11 R4
Krasnovishersk Rus. Fed. 11 R3
Krasnovodsk Turkm. see Türkmenbaşy
Krasnovodsk, Mys pt Turkm. 35 I3
Krasnovodskoye Plato plat. Turkm. 35 I2
Krasnowodsk Aylagy b. Turkm. see Türkmenbaşy Aýlagy
Krasnoyarovo Rus. Fed. 44 C2
Krasnoyarsk Rus. Fed. 28 K4
Krasnoyarskoye Vodokhranilishche resr Rus. Fed. 42 G2
Krasnoye Lipetskaya Oblast' Rus. Fed. 13 H5
Krasnoye Respublika Kalmykiya-Khalm'g-Tangch Rus. Fed. see Ulan Erge
Krasnoznamenskiy Kazakh. see Yegindykol'
Krasnoznamenskoye Kazakh. see Yegindykol'
Krasnyy Rus. Fed. 13 F5
Krasnyy Chikoy Rus. Fed. 43 J2
Krasnyye Baki Rus. Fed. 12 J4
Krasnyy Kamyshanik Rus. Fed. see Komsomol'skiy
Krasnyy Kholm Rus. Fed. 12 H4
Krasnyy Kut Rus. Fed. 13 J6
Krasnyy Luch Ukr. 13 H6
Krasnyy Lyman Ukr. 13 H6
Krasnyy Yar Rus. Fed. 13 K7
Krasyliv Ukr. 13 E6
Kratie Cambodia see Krâchéh
Kraulshavn Greenland see Nuussuaq
Kraynovka Rus. Fed. 13 J8
Krefeld Germany 17 K5
Kremenchug Ukr. see Kremenchuk
Kremenchugskoye Vodokhranilishche resr Ukr. see Kremenchuts'ka Vodoskhovyshche
Kremenchuk Ukr. 13 G6
Kremenchuts'ka Vodoskhovyshche resr Ukr. 13 G6
Křemešník h. Czech Rep. 17 O6
Kremges Ukr. see Svitlovods'k
Kremmydi, Akrotirio pt Greece 27 J6
Krems Austria see Krems an der Donau
Krems an der Donau Austria 17 O6
Kresta, Zaliv g. Rus. Fed. 29 T3
Kresttsy Rus. Fed. 12 G4
Kretinga Lith. 15 L9
Kreva Belarus 15 O9
Kribi Cameroon 46 D4
Krichev Belarus see Krychaw
Kriel S. Africa 51 I4
Krikellos Greece 27 I5
Kril'on, Mys c. Rus. Fed. 44 F3
Krishna India 38 C2
Krishna r. India 38 D2
Krishnanagar India 37 G5
Krishnarajanagara India 38 C3
Krishnaraja Sagara l. India 38 C3
Kristiania Norway see Oslo
Kristiansand Norway 15 E7
Kristianstad Sweden 15 I8
Kristiansund Norway 14 E5
Kristiinankaupunki Fin. see Kristinestad
Kristinehamn Sweden 15 I7
Kristinestad Fin. 14 L5
Kristinopol' Ukr. see Chervonohrad
Kriti i. Greece see Crete
Kritiko Pelagos sea Greece 27 K6
Krivoy Rog Ukr. see Kryvyy Rih
Križevci Croatia 26 G1
Krk i. Croatia 26 F2
Krka, Nacionalni Park nat. park Croatia 26 F3
Krkonošský narodní park nat. park Czech Rep./Poland 17 O5
Krokom Sweden 14 I5
Krokstadøra Norway 14 F5
Krokstranda Norway 14 I3
Krolevets' Ukr. 13 G6
Kronoby Fin. 14 M5
Kronprins Christian Land reg. Greenland 77 I1
Kronprins Frederik Bjerge nunataks Greenland 61 O3
Kronshtadt Rus. Fed. 15 P7
Kronstadt Romania see Braşov
Kronstadt Rus. Fed. see Kronshtadt
Kroonstad S. Africa 51 H4
Kropotkin Rus. Fed. 13 I7
Krosno Poland 13 D6
Krotoszyn Poland 17 P5
Kruger National Park S. Africa 51 J2
Kruglikovo Rus. Fed. 44 D2
Kruglyakov Rus. Fed. see Oktyabr'skiy
Krui Indon. 41 C8
Kruisfontein S. Africa 50 G8
Kruja Albania see Krujë
Krujë Albania 27 H4
Krumovgrad Bulg. 27 K4
Krungkao Thai. see Ayutthaya
Krung Thep Thai. see Bangkok
Krupa Bos.-Herz. see Bosanska Krupa
Krupa na Uni Bos.-Herz. see Bosanska Krupa
Krupki Belarus 13 F5
Krusenstern, Cape U.S.A. 60 B3
Kruševac Serbia 27 I3
Krychaw Belarus 13 F5
Krylov Seamount sea feature N. Atlantic Ocean 72 G4
Krym' pen. Ukr. see Crimea
Krymsk Rus. Fed. 13 H7
Krymskaya Rus. Fed. see Krymsk
Kryms'kyy Pivostriv pen. Ukr. see Crimea
Krystynopol Ukr. see Chervonohrad
Kryvyy Rih Ukr. 13 G7
Ksabi Alg. 22 D6
Ksar Chellala Alg. 25 H6
Ksar el Boukhari Alg. 25 H6
Ksar el Kebir Morocco 25 D6
Ksar-es-Souk Morocco see Er Rachidia
Ksenofontova Rus. Fed. 11 R3
Kshirpai India 37 F5
Ksour Essaf Tunisia 26 D7
Kstovo Rus. Fed. 12 J4
Kū', Jabal al h. Saudi Arabia 32 G4
Kuaidamao China see Tonghua
Kuala Dungun Malaysia see Dungun
Kuala Lipis Malaysia 41 C7
Kuala Lumpur Malaysia 41 C7
Kuala Terengganu Malaysia 41 C7
Kuandian China 44 B4
Kuantan Malaysia 41 C7
Kuba Azer. see Quba
Kuban' r. Rus. Fed. 13 H7
Kubār Syria 35 E4
Kubaybāt Syria 39 C2
Kubaysah Iraq 35 F4
Kubenskoye, Ozero l. Rus. Fed. 12 H4
Kubrat Bulg. 27 L3
Kuchema Rus. Fed. 12 I2
Kuching Sarawak Malaysia 41 D7
Kucing Sarawak Malaysia see Kuching
Kuçovë Albania 27 H4
Kuda India 36 B5
Kudal India 38 B3
Kudat Sabah Malaysia 41 D7
Kudligi India 38 C3
Kudremukh mt. India 38 B3
Kudymkar Rus. Fed. 11 Q4
Kufstein Austria 17 N7
Kugaaruk Canada 61 J3
Kugesi Rus. Fed. 12 J4
Kugka Lhai China 37 G3
Kugluktuk Canada 60 G3
Kugmallit Bay Canada 77 A2
Kuhanbokano mt. China 37 E3
Kūhdasht Iran 35 G4
Kūh-e 'Alījūq mt. Iran 35 H5
Kūh-e Band-e Bamposht mts Iran 33 J4
Kūh-e Kalār mt. Iran 35 H5
Kūh-e Kharānaq mt. Iran 35 I4
Kūh-e Khvājeh Moḩammad mts Afgh. 36 B1
Kūh-e Kūrd mt. Afgh. 36 A2
Kūh-e Mazār mt. Afgh. 36 A2
Kūh-e Qarah Taray mt. Afgh. 36 A2
Kūhestānāt Afgh. 36 A2
Kūhhā-ye Zagros mts Iran see Zagros Mountains
Kūhīn Iran 35 H3
Kuhmo Fin. 14 P4
Kuhmoinen Fin. 15 N6
Kūhrān, Kūh-e mt. Iran 33 I4
Kuis Namibia 50 C3
Kuiseb watercourse Namibia 50 B2
Kuito Angola 49 B5
Kuitun China see Kuytun
Kuivaniemi Fin. 14 N4
Kujang N. Korea 45 B5
Kuji Japan 45 F4
Kujū-san vol. Japan 45 C6
Kukan Rus. Fed. 44 D2
Kukës Albania 27 I3
Kukesi Albania see Kukës
Kukmor Rus. Fed. 12 K4
Kukshi India 36 C5
Kukunuru India 38 D2
Kula Turkey 27 M5
Kulaisila India 37 F5
Kula Kangri mt. China/Bhutan 37 G3
Kulandy Kazakh. 28 G5
Kular Rus. Fed. 29 O2
Kuldīga Latvia 15 L8
Kuldja China see Yining
Kul'dur Rus. Fed. 44 C2
Kule Botswana 50 E2
Kulebaki Rus. Fed. 13 I5
Kulgera Australia 55 F6
Kulikovo Rus. Fed. 12 J3
Kulin Australia 55 B8
Kulja Australia 55 B7
Kulkyne watercourse Australia 58 B3
Kullu India 36 D3
Kulmbach Germany 17 M5
Kŭlob Tajik. 33 K2
Kuloy Rus. Fed. 12 I3
Kuloy r. Rus. Fed. 12 I2
Kulp Turkey 35 F3
Kul'sary Kazakh. 30 E2
Kulu India see Kullu
Kulu Turkey 34 D3
Kulunda Rus. Fed. 42 D2
Kulundinskaya Step' plain Kazakh./Rus. Fed. 42 D2
Kulundinskoye, Ozero salt l. Rus. Fed. 42 D2
Kulusuk Greenland 61 O3
Kulwin Australia 57 C7
Kulyab Tajik. see Kŭlob
Kuma r. Rus. Fed. 13 J7

Kumagaya Japan **45** E5
Kumalar Dağı *mts* Turkey **27** N5
Kumamoto Japan **45** C6
Kumano Japan **45** E6
Kumanovo Macedonia **27** I3
Kumara Rus. Fed. **44** B2
Kumasi Ghana **46** C4
Kumayri Armenia *see* **Gyumri**
Kumba Cameroon **46** D4
Kumbakonam India **38** C4
Kümbet Turkey **27** N5
Kumbharli Ghat *mt.* India **38** B2
Kumbla India **38** B3
Kumchuru Botswana **50** F2
Kum-Dag Turkm. *see* **Gumdag**
Kumdah Saudi Arabia **32** G5
Kumel *well* Iran **35** I4
Kumeny Rus. Fed. **12** K4
Kumertau Rus. Fed. **28** G4
Kumgang-san *mt.* N. Korea **45** C5
Kumguri India **37** G4
Kumi S. Korea **45** C5
Kumi Uganda **47** G4
Kumla Sweden **15** I7
Kumlu Turkey **39** C1
Kumo Nigeria **46** E3
Kŭmŏ-do *i.* S. Korea **45** B6
Kumon Range *mts* Myanmar **42** H7
Kums Namibia **50** D5
Kumta India **38** B3
Kumu Dem. Rep. Congo **48** C3
Kumukh Rus. Fed. **35** G2
Kumul China *see* **Hami**
Kumund India **38** D1
Kumylzhenskaya Rus. Fed. *see* **Kumylzhenskiy**
Kumylzhenskiy Rus. Fed. **13** I6
Kunar *r.* Afgh. **36** B2
Kunashir, Ostrov *i.* Rus. Fed. **44** G3
Kunashirskiy Proliv *sea chan.* Japan/Rus. Fed. *see* **Nemuro-kaikyō**
Kunchuk Tso *salt l.* China **37** E2
Kunda Estonia **15** O7
Kunda India **37** E4
Kundapura India **38** B3
Kundelungu, Parc National de *nat. park* Dem. Rep. Congo **49** C5
Kundelungu Ouest, Parc National de *nat. park* Dem. Rep. Congo **49** C5
Kundia India **36** C4
Kunduz Afgh. **36** B1
Kunene *r.* Angola *see* **Cunene**
Kuneneng *admin. dist.* Botswana *see* **Kweneng**
Künes China *see* **Xinyuan**
Kungälv Sweden **15** G8
Kungsbacka Sweden **15** H8
Kungshamn Sweden **15** G7
Kungu Dem. Rep. Congo **48** B3
Kungur *mt.* China *see* **Kongur Shan**
Kungur Rus. Fed. **11** R4
Kuni *r.* India **38** C2
Kunié *i.* New Caledonia *see* **Pins, Île des**
Kunigal India **38** C3
Kunimi-dake *mt.* Japan **45** C6
Kunkavav India **36** B5
Kunlun Shan *mts* China **36** D1
Kunlun Shankou *pass* China **37** H2
Kunming China **42** I7
Kunsan S. Korea **45** B6
Kununurra Australia **54** E3
Kun'ya Rus. Fed. **12** F4
Kunya-Urgench Turkm. *see* **Köneürgenç**
Kuohijärvi *l.* Fin. **15** N6
Kuoloyarvi Rus. Fed. **14** P3
Kuopio Fin. **14** O5
Kuortane Fin. **14** M5
Kupa *r.* Croatia/Slovenia **26** G2
Kupang Indon. **41** E9
Kupari India **37** F5
Kupiškis Lith. **15** N9
Kupreanof Island U.S.A. **60** E4
Kupwara India **36** C2
Kup"yans'k Ukr. **13** H6
Kuqa China **42** E4
Kur *r.* Rus. Fed. **44** D2
Kura *r.* Georgia **35** G2
Kuragino Rus. Fed. **42** G2
Kurakh Rus. Fed. **13** J8
Kurama Range *mts* Asia **33** K1
Kuraminskiy Khrebet *mts* Asia *see* **Kurama Range**
Kurashiki Japan **45** D6
Kurasia India **37** E5
Kurayn *i.* Saudi Arabia **35** H6
Kurayoshi Japan **45** D6
Kurchatov Rus. Fed. **13** G6
Kurchum Kazakh. **42** E3
Kürdämir Azer. **35** H2
Kŭrdzhali Bulg. **27** K4
Kure Japan **45** D6
Küre Turkey **34** D2
Kure Atoll U.S.A. **74** I4
Kuressaare Estonia **15** M7
Kurgal'dzhino Kazakh. *see* **Korgalzhyn**
Kurgal'dzhinskiy Kazakh. *see* **Korgalzhyn**
Kurgan Rus. Fed. **28** H4
Kurganinsk Rus. Fed. **35** F1
Kurgannaya Rus. Fed. *see* **Kurganinsk**
Kurgantyube Tajik. *see* **Qŭrghonteppa**
Kuri India **36** B4
Kuria Muria Islands Oman *see* **Ḥalāniyāt, Juzur al**
Kuridala Australia **56** C4
Kurigram Bangl. **37** G4
Kurikka Fin. **14** M5
Kuril Basin *sea feature* Sea of Okhotsk **74** F2
Kuril Islands Rus. Fed. **44** H3
Kurilovka Rus. Fed. **13** K6
Kuril'sk Rus. Fed. **44** G3
Kuril'skiye Ostrova *is* Rus. Fed. *see* **Kuril Islands**
Kuril Trench *sea feature* N. Pacific Ocean **74** F3
Kurkino Rus. Fed. **13** H5
Kurmashkino Kazakh. *see* **Kurchum**
Kurmuk Sudan **32** D7
Kurnool India **38** C3
Kuroiso Japan **45** F5
Kurow N.Z. **59** C7
Kurram Pak. **36** B2
Kurri Kurri Australia **58** E4
Kursavka Rus. Fed. **35** F1
Kürshim Kazakh. *see* **Kurchum**
Kurshskiy Zaliv *b.* Lith./Rus. Fed. *see* **Courland Lagoon**
Kuršių marios *b.* Lith./Rus. Fed. *see* **Courland Lagoon**
Kursk Rus. Fed. **13** H6
Kurskaya Rus. Fed. **35** G1
Kurskiy Zaliv *b.* Lith./Rus. Fed. *see* **Courland Lagoon**
Kurşunlu Turkey **34** D2
Kurtalan Turkey **35** F3
Kurtoğlu Burnu *pt* Turkey **27** M6
Kurtpınar Turkey **39** B1
Kurucaşile Turkey **34** D2
Kuruçay Turkey **34** E3
Kurukshetra India **36** D3
Kuruktag *mts* China **42** F4
Kuruman S. Africa **50** F4
Kuruman *watercourse* S. Africa **50** E4
Kurume Japan **45** C6
Kurumkan Rus. Fed. **43** K2
Kurunegala Sri Lanka **38** D5
Kurupam India **38** D2
Kurush, Jebel *hills* Sudan **32** D5
Kur'ya Rus. Fed. **11** R3
Kuryk Kazakh. **35** H2
Kuşadası Turkey **27** L6
Kuşadası Körfezi *b.* Turkey **27** L6
Kusaie *atoll* Micronesia *see* **Kosrae**
Kusary Azer. *see* **Qusar**
Kuşcenneti *nat. park* Turkey **39** B1
Kuschke Nature Reserve S. Africa **51** I3
Kuş Gölü *l.* Turkey **27** L4
Kushalgarh India **36** C5
Kushchevskaya Rus. Fed. **13** H7
Kushimoto Japan **45** D6
Kushiro Japan **44** G4
Kushka Turkm. *see* **Serhetabat**
Kushkopala Rus. Fed. **12** J3
Kushmurun Kazakh. **30** F1
Kushtagi India **38** C3
Kushtia Bangl. **37** G5
Kuskan Turkey **39** A1
Kuskokwim *r.* U.S.A. **60** B3
Kuskokwim Bay U.S.A. **60** B4
Kuskokwim Mountains U.S.A. **60** C3
Kuşluyan Turkey *see* **Gölköy**
Kusŏng N. Korea **45** B5
Kustanay Kazakh. *see* **Kostanay**
Küstence Romania *see* **Constanța**
Kustia Bangl. *see* **Kushtia**
Kut Iran **35** H5
Küt 'Abdollāh Iran **35** H5
Kütahya Turkey **27** M5
K'ut'aisi Georgia **35** F2
Kut-al-Imara Iraq *see* **Al Kūt**
Kutan Rus. Fed. **13** J7
Kutaraja Indon. *see* **Banda Aceh**
Kutayfat Turayf *vol.* Saudi Arabia **39** D4
Kutch, Gulf of India *see* **Kachchh, Gulf of**
Kutch, Rann of *marsh* India *see* **Kachchh, Rann of**
Kutchan Japan **44** F4
Kutina Croatia **26** G2
Kutjevo Croatia **26** G2
Kutno Poland **17** Q4
Kutru India **38** D2
Kutu Dem. Rep. Congo **48** B4
Kutubdia Island Bangl. **37** G5
Kutum Sudan **47** F3
Kutztown *PA* U.S.A. **64** D2
Kuujjua *r.* Canada **60** G2
Kuujjuaq Canada **61** L4
Kuujjuarapik Canada **61** K4
Kuusamo Fin. **14** P4
Kuusankoski Fin. **15** O6
Kuvango Angola **49** B5
Kuvshinovo Rus. Fed. **12** G4
Kuwait *country* Asia **32** G4
Kuwait Kuwait **35** G5
Kuwajleen *atoll* Marshall Is *see* **Kwajalein**
Kuybyshev *Novosibirskaya Oblast'* Rus. Fed. **28** I4
Kuybyshev *Respublika Tatarstan* Rus. Fed. *see* **Bolgar**
Kuybyshev *Samarskaya Oblast'* Rus. Fed. *see* **Samara**
Kuybysheve Ukr. **13** H7
Kuybyshevka-Vostochnaya Rus. Fed. *see* **Belogorsk**
Kuybyshevskoye Vodokhranilishche *resr* Rus. Fed. **13** K5
Kuyeda Rus. Fed. **11** R4
Kuygan Kazakh. **42** C3
Kuytun China **42** E4
Kuytun Rus. Fed. **42** I2
Kuyucak Turkey **27** M6
Kuzino Rus. Fed. **11** R4
Kuznechnoye Rus. Fed. **15** P6
Kuznetsk Rus. Fed. **13** J5
Kuznetsovo Rus. Fed. **44** E3
Kuznetsovs'k Ukr. **13** E6
Kuzovatovo Rus. Fed. **13** J5
Kvænangen *sea chan.* Norway **14** L1
Kvaløya *i.* Norway **14** K2
Kvalsund Norway **14** M1
Kvarnerić *sea chan.* Croatia **26** F2
Kvitøya *i.* Svalbard **28** E2
Kwa *r.* Dem. Rep. Congo *see* **Kasaï**
Kwabhaca S. Africa *see* **Mount Frere**
Kwadelen *atoll* Marshall Is *see* **Kwajalein**
Kwajalein *atoll* Marshall Is **74** H5
Kwale Nigeria **46** D4
KwaMashu S. Africa **51** J5
KwaMhlanga S. Africa **51** I3
Kwa Mtoro Tanz. **49** D4
Kwangch'ŏn S. Korea **45** B5
Kwangchow China *see* **Guangzhou**
Kwangju S. Korea **45** B6
Kwangsi Chuang Autonomous Region *aut. reg.* China *see* **Guangxi Zhuangzu Zizhiqu**
Kwangtung *prov.* China *see* **Guangdong**
Kwanmo-bong *mt.* N. Korea **44** C4
Kwanobuhle S. Africa **51** G7
KwaNojoli S. Africa **51** G7
Kwanonqubela S. Africa **51** H7
KwaNonzame S. Africa **50** G6
Kwanza *r.* Angola *see* **Cuanza**
Kwatarkwashi Nigeria **46** D3
Kwatinidubu S. Africa **51** H7
KwaZamokuhle S. Africa **51** I4
KwaZamukucinga S. Africa **50** G7
Kwazamuxolo S. Africa **50** G6
KwaZanele S. Africa **51** I4
KwaZulu-Natal *prov.* S. Africa **51** J5
Kweichow *prov.* China *see* **Guizhou**
Kweiyang China *see* **Guiyang**
Kwekwe Zimbabwe **49** C5
Kweneng *admin. dist.* Botswana **50** G2
Kwenge *r.* Dem. Rep. Congo **49** B4
Kwezi-Naledi S. Africa **51** H6
Kwidzyn Poland **17** Q4
Kwikila P.N.G. **52** E2
Kwilu *r.* Angola/Dem. Rep. Congo **49** B4
Kwoka *mt.* Indon. **41** F8
Kyabra Australia **57** C5
Kyabram Australia **58** B6
Kyakhta Rus. Fed. **42** J2
Kyalite Australia **58** A5
Kyancutta Australia **55** F8
Kyangin Myanmar **37** I6
Kyangngoin China **37** H3
Kyaukpadaung Myanmar **37** H5
Kyaukpyu Myanmar **37** H6
Kyaukse Myanmar **37** I5
Kyauktaw Myanmar **37** H5
Kybartai Lith. **15** M9
Kyêbxang Co *l.* China **37** G2
Kyela Tanz. **49** D4
Kyelang India **36** D2
Kyidaunggan Myanmar **37** I6
Kyiv Ukr. *see* **Kiev**
Kyklades *is* Greece *see* **Cyclades**
Kyle of Lochalsh U.K. **20** D3
Kyllini *mt.* Greece **27** J6
Kymi Greece **27** K5
Kymis, Akrotirio *pt* Greece **27** K5
Kynazhegubskoye Vodokhranilishche *l.* Rus. Fed. **14** R3
Kyneton Australia **58** B6
Kynuna Australia **56** C4
Kyoga, Lake Uganda **48** D3
Kyōga-misaki *pt* Japan **45** D6
Kyogle Australia **58** F2
Kyŏngju S. Korea **45** C6
Kyōto Japan **45** D6
Kyparissia Greece **27** I6
Kypros *country* Asia *see* **Cyprus**
Kypshak, Ozero *salt l.* Kazakh. **31** F1
Kyra Rus. Fed. **43** K3
Kyra Panagia *i.* Greece **27** K5
Kyrenia Cyprus **39** A2
Kyrenia Mountains Cyprus *see* **Pentadaktylos Range**
Kyrgyz Ala-Too *mts* Kazakh./Kyrg. *see* **Kirghiz Range**
Kyrgyzstan *country* Asia **31** G2
Kyrksæterøra Norway **14** F5
Kyrta Rus. Fed. **11** R3
Kyssa Rus. Fed. **12** J2
Kytalyktakh Rus. Fed. **29** O3
Kythira *i.* Greece **27** J6
Kythnos *i.* Greece **27** K6
Kyunglung China **36** E3
Kyunhla Myanmar **37** H5
Kyuquot Canada **62** B1
Kyurdamir Azer. *see* **Kürdämir**
Kyūshū *i.* Japan **45** C7
Kyushu-Palau Ridge *sea feature* N. Pacific Ocean **74** F4
Kyustendil Bulg. **27** J3
Kywong Australia **58** C5
Kyyev Ukr. *see* **Kiev**
Kyyiv Ukr. *see* **Kiev**
Kyyivs'ke Vodoskhovyshche *resr* Ukr. **13** F6
Kyyjärvi Fin. **14** N5
Kyzyl Rus. Fed. **42** G2
Kyzyl-Burun Azer. *see* **Siyäzän**
Kyzyl-Kiya Kyrg. *see* **Kyzyl-Kyya**
Kyzylkum, Peski *des.* Kazakh./Uzbek. *see* **Kyzylkum Desert**
Kyzylkum Desert Kazakh./Uzbek. **30** F2
Kyzyl-Kyya Kyrg. **42** C4
Kyzyl-Mazhalyk Rus. Fed. **42** G2
Kyzylorda Kazakh. **42** B4
Kyzylrabot Tajik. *see* **Qizilrabot**
Kyzylsay Kazakh. **35** I2
Kyzylysor Kazakh. **35** H1
Kyzylzhar Kazakh. **42** B3
Kzyl-Dzhar Kazakh. *see* **Kyzylzhar**
Kzyl-Orda Kazakh. *see* **Kyzylorda**
Kzyltu Kazakh. *see* **Kishkenekol'**

Laagri Estonia **15** N7
La Angostura, Presa de *resr* Mex. **66** F5
Laanila Fin. **14** O2
Laascaanood Somalia **48** E3
La Ascensión, Bahía de *b.* Mex. **67** G5
Laasgoray Somalia **48** E2
Laâyoune W. Sahara **46** B2
La Banda Arg. **70** D3
Labasa Fiji **53** H3
La Baule-Escoublac France **24** C3
Labé Guinea **46** B3
La Bénoué, Parc National de *nat. park* Cameroon **47** E4
Labinsk Rus. Fed. **13** I7
La Boucle du Baoulé, Parc National de *nat. park* Mali **46** C3
Labouheyre France **24** D4
Laboulaye Arg. **70** D4
Labozhskoye Rus. Fed. **12** L2
Labrador *reg.* Canada **61** L4
Labrador City Canada **61** L4
Labrador Sea Canada/Greenland **61** M3
Lábrea Brazil **68** F5
Labudalin China *see* **Ergun**
Labuhanbilik Indon. **41** C7
Labuna Indon. **41** E8
Labyrinth, Lake *imp. l.* Australia **57** A6
Labytnangi Rus. Fed. **28** H3
Laç Albania **27** H4
La Cabrera, Sierra de *mts* Spain **25** C2
La Calle Alg. *see* **El Kala**
La Cañiza Spain *see* **A Cañiza**
La Capelle France **24** F2
La Carlota Arg. **70** D4
La Carolina Spain **25** E4
Lăcăuți, Vârful *mt.* Romania **27** L2
Laccadive, Minicoy and Amindivi Islands *union terr.* India *see* **Lakshadweep**
Laccadive Islands India **38** B4
Lac du Bonnet Canada **63** H1
Lacedaemon Greece *see* **Sparti**
La Ceiba Hond. **67** G5
Lacepede Bay Australia **57** B8
Lacepede Islands Australia **54** C4
Lacha, Ozero *l.* Rus. Fed. **12** H3
Lachlan *r.* Australia **58** A5
La Chorrera Panama **67** I7
Lachute Canada **63** M2
Laçın Azer. **35** G3
La Ciotat France **24** G5
Lac Kivu Dem. Rep. Congo/Rwanda **48** C4
Lac La Biche Canada **60** G4
Lac la Martre Canada *see* **Whatî**
Lacombe Canada **62** E1
La Comoé, Parc National de *nat. park* Côte d'Ivoire **46** C4
Laconi *Sardegna* Italy **26** C5
Laconia *NH* U.S.A. **64** F1
La Coruña Spain *see* **A Coruña**
La Coubre, Pointe de *pt* France **24** D4
La Crosse *WI* U.S.A. **63** I3
La Cruz Mex. **66** C4
La Cuesta Mex. **62** G6
La Culebra, Sierra de *mts* Spain **25** C3
Ladainha Brazil **71** C2
Ladakh *reg.* India/Pak. **36** D2
Ladakh Range *mts* India **36** D2
La Demanda, Sierra de *mts* Spain **25** E2
La Déroute, Passage de *str.* Channel Is/France **19** E9
Ladik Turkey **34** D2
Ladnun India **36** C4
Ladoga, Lake Rus. Fed. **12** F3
Ladozhskoye Ozero *l.* Rus. Fed. *see* **Ladoga, Lake**
Ladrones *terr.* N. Pacific Ocean *see* **Northern Mariana Islands**
Ladu *mt.* India **37** H4
Ladva-Vetka Rus. Fed. **12** G3
Ladybank U.K. **20** F4
Ladybrand S. Africa **51** H5
Lady Frere S. Africa **51** H6
Lady Grey S. Africa **51** H6
Ladysmith S. Africa **51** I5
Ladzhanurges Georgia *see* **Lajanurpekhi**
Lae P.N.G. **52** E2
Lærdalsøyri Norway **15** E6
La Esmeralda Bol. **68** F8
Læsø *i.* Denmark **15** G8
Lafayette Alg. *see* **Bougaa**
Lafayette *IN* U.S.A. **63** J3
Lafayette *LA* U.S.A. **63** I5
Lafia Nigeria **46** D4
Lafiagi Nigeria **46** D4
La Flèche France **24** D3
Laforge Canada **61** K4
Lāft Iran **33** I4
La Galite *i.* Tunisia **26** C6
La Galite, Canal de *sea chan.* Tunisia **26** C6
Lagan' Rus. Fed. **13** J7
Lagan *r.* U.K. **21** G3
La Garamba, Parc National de la *nat. park* Dem. Rep. Congo **48** C3
Lagarto Brazil **69** K6
Lågen *r.* Norway **15** G7
Lagg U.K. **20** D5
Laggan U.K. **20** E3
Lagh Bor *watercourse* Kenya/Somalia **48** E3
Laghouat Alg. **22** E5
Lagkor Co *salt l.* China **37** F2
Lago Agrio Ecuador **68** C3
Lagoa Santa Brazil **71** C2
Lagoa Vermelha Brazil **71** A5
Lagodekhi Georgia **35** G2
Lagolândia Brazil **71** A1
La Gomera *i.* Canary Is **46** B2
La Gonâve, Île de *i.* Haiti **67** J5
Lagos Nigeria **46** D4
Lagos Port. **25** B5
Lagosa Tanz. **49** C4
La Grande *OR* U.S.A. **62** D2
La Grande 3, Réservoir *resr* Canada **61** K4
La Grande 4, Réservoir *resr Que.* Canada **61** K4
La Grange Australia **54** C4
La Grange *CA* U.S.A. **65** B2
La Grange *GA* U.S.A. **63** J5
La Gran Sabana *plat.* Venez. **68** F2
La Grita Venez. **68** D2
La Guajira, Península de *pen.* Col. **68** D1
Laguna Brazil **71** A5
Laguna, Picacho de la *mt.* Mex. **66** B4
Laguna Dam *Arizona/California* U.S.A. **65** E4
Laguna Mountains *CA* U.S.A. **65** D4
Lagunas Chile **70** C2
Laguna San Rafael, Parque Nacional *nat. park* Chile **70** B7
Laha China **44** B2
La Habana Cuba *see* **Havana**
La Habra *CA* U.S.A. **65** D4
Lahad Datu *Sabah* Malaysia **41** D7
La Hague, Cap de *c.* France **24** D2
Laharpur India **36** E4
Lahat Indon. **41** C8
Lahe Myanmar **37** H4
Lahemaa rahvuspark *nat. park* Estonia **15** N7
La Hève, Cap de *c.* France **19** H9
Laḥij Yemen **32** F7
Lāhījān Iran **35** H3
Laholm Sweden **15** H8
Lahore Pak. **33** L3
Lahri Pak. **36** B3
Lahti Fin. **15** N6
Laï Chad **47** E4
Laibach Slovenia *see* **Ljubljana**
Laidley Australia **58** F1
L'Aigle France **24** E2
Laihia Fin. **14** M5
Laimakuri India **37** H4
Laimos, Akrotirio *pt* Greece **27** J5
Laingsburg S. Africa **50** E7
Lainioälven *r.* Sweden **14** M3
L'Aïr, Massif de *mts* Niger **46** D3
Lairg U.K. **20** E2
Laishevo Rus. Fed. **12** K5
Laitila Fin. **15** L6
Laives Italy **26** D1
Laiwu China **43** L5
Laiyang China **43** M5
Laizhou China **43** L5
Laizhou Wan *b.* China **43** L5
Lajamanu Australia **54** E4
Lajanurpekhi Georgia **35** F2
Lajeado Brazil **71** A5
Lajes *Rio Grande do Norte* Brazil **69** K5
Lajes *Santa Catarina* Brazil **71** A4
La Junta Mex. **62** F6
La Junta U.S.A. **62** G4
La Juventud, Isla de la *i.* Cuba **67** H4
Lakadiya India **36** B5
La Kagera, Parc National de la *nat. park* Rwanda *see* **Akagera National Park**
L'Akagera, Parc National de l' *nat. park* Rwanda *see* **Akagera National Park**
Lakeba *i.* Fiji **53** I3
Lake Bardawil Reserve *nature res.* Egypt **39** A4
Lake Bolac Australia **58** A6
Lake Cargelligo Australia **58** C4
Lake Cathie Australia **58** F3
Lake Charles U.S.A. **63** I5
Lake City *FL* U.S.A. **63** K5
Lake Clark National Park and Preserve U.S.A. **60** C3
Lake District National Park U.K. **18** D4
Lake Eyre National Park Australia **57** B6
Lakefield Australia **56** D2
Lakefield National Park Australia **56** D2
Lake Gairdner National Park Australia **57** B7
Lake George *NY* U.S.A. **64** E1
Lake Grace Australia **55** B8
Lake Harbour Canada *see* **Kimmirut**
Lake Havasu City *AZ* U.S.A. **65** E3
Lakehurst *NJ* U.S.A. **64** D2
Lake Isabella *CA* U.S.A. **65** C3
Lake King Australia **55** B8
Lakeland *FL* U.S.A. **63** K6
Lakemba *i.* Fiji *see* **Lakeba**
Lake Nash Australia **56** B4
Lake Paringa N.Z. **59** B6
Lake Pleasant *NY* U.S.A. **64** D1
Lakeport *CA* U.S.A. **65** A1
Lake Providence U.S.A. **63** I5
Lakes Entrance Australia **58** D6
Lakeside *VA* U.S.A. **64** C4
Lake Tabourie Australia **58** E5
Lake Tekapo N.Z. **59** C7
Lake Torrens National Park Australia **57** B6
Lakeview *OR* U.S.A. **62** C3
Lakewood *NJ* U.S.A. **64** D2
Lakewood *NY* U.S.A. **64** B1
Lakha India **36** B4
Lakhdenpokh'ya Rus. Fed. **14** Q6
Lakhimpur *Assam* India *see* **North Lakhimpur**
Lakhimpur *Uttar Prad.* India **36** E4
Lakhisarai India **37** F4
Lakhish *r.* Israel **39** B4
Lakhnadon India **36** D5
Lakhpat India **36** B5
Lakhtar India **36** B5
Lakki Marwat Pak. **33** L3
Lakonikos Kolpos *b.* Greece **27** J6
Lakor *i.* Indon. **54** E2
Lakota Côte d'Ivoire **46** C4
Laksefjorden *sea chan.* Norway **14** O1
Lakselv Norway **14** N1
Lakshadweep *is* India *see* **Laccadive Islands**
Lakshadweep *union terr.* India **38** B4
Lakshettipet India **38** C2
Lakshmipur Bangl. **37** G5
Laksmipur Bangl. *see* **Lakshmipur**
Lalaghat India **37** H4
Lalbara India **38** D1
L'Alcora Spain **25** F3
Lalganj India **37** F4
Lālī Iran **35** H4
La Ligua Chile **70** B4
Laliki Indon. **54** D1
Lalin China **44** B3
Lalín Spain **25** B2
La Línea de la Concepción Spain **25** D5
Lalin He *r.* China **44** B3
Lalitpur India **36** D4
Lalitpur Nepal *see* **Patan**
Lalmanirhat Bangl. *see* **Lalmonirhat**
Lalmonirhat Bangl. **37** G4
La Loche Canada **60** H4
La Louvière Belgium **16** J5
Lal'sk Rus. Fed. **12** J3
Lalung La *pass* China **37** F3
Lama Bangl. **37** H5
La Macarena, Parque Nacional *nat. park* Col. **68** D3
La Maddalena *Sardegna* Italy **26** C4
La Madeleine, Îles de *is* Canada **63** O2
La Madeleine, Monts de *mts* France **24** F3
Lamadian China **44** B3
Lamadianzi China *see* **Lamadian**
La Maiko, Parc National de la *nat. park* Dem. Rep. Congo **48** C4
La Mancha *reg.* Spain **25** E4
La Manche *str.* France/U.K. *see* **English Channel**
Lamar *CO* U.S.A. **62** G4
Lamard Iran **35** I6
La Margeride, Monts de *mts* France **24** F4
La Marmora, Punta *mt. Sardegna* Italy **26** C5
La Martre, Lac *l.* Canada **60** G3
Lamas *r.* Turkey **39** B1
Lambaréné Gabon **48** B4
Lambasa Fiji *see* **Labasa**
Lambayeque Peru **68** C5
Lambay Island Ireland **21** G4
Lambert *atoll* Marshall Is *see* **Ailinglaplap**
Lambert Glacier Antarctica **76** E2
Lambert's Bay S. Africa **50** D7
Lambeth *Ont.* Canada **64** A1
Lambi India **36** C3
Lambourn Downs *hills* U.K. **19** F7
La Medjerda, Monts de *mts* Alg. **26** B6
Lamego Port. **25** C3
La Merced Arg. **70** C3
La Merced Peru **68** C6
Lameroo Australia **57** C7
La Mesa *CA* U.S.A. **65** D4
Lamesa U.S.A. **62** G5
Lamia Greece **27** J5
Lamington National Park Australia **58** F2
Lammerlaw Range *mts* N.Z. **59** B7
Lammermuir Hills U.K. **20** G5
Lammhult Sweden **15** I8
Lammi Fin. **15** N6
Lamont *CA* U.S.A. **65** C3
La Montaña de Covadonga, Parque Nacional de la *nat. park* Spain *see* **Los Picos de Europa, Parque Nacional de**
Lampang Thai. **31** I5
Lampazos Mex. **62** G6
Lampedusa, Isola di *i. Sicilia* Italy **26** E7
Lampeter U.K. **19** C6
Lampsacus Turkey *see* **Lâpseki**
Lamu Kenya **48** E4
Lamu Myanmar **37** H6
La Nao, Cabo de *c.* Spain **25** G4
Lanark U.K. **20** F5
Lancang Jiang *r. Xizang/Yunnan* China *see* **Mekong**
Lancaster U.K. **18** E4
Lancaster *CA* U.S.A. **65** C3
Lancaster *PA* U.S.A. **64** C2
Lancaster *SC* U.S.A. **63** K5
Lancaster *VA* U.S.A. **64** C4
Lancaster Canal U.K. **18** E5
Lancaster Sound *str.* Canada **61** J2
Lanchow China *see* **Lanzhou**
Landana Angola *see* **Cacongo**
Landau an der Isar Germany **17** N6
Landeck Austria **17** M7
Landeh Iran **35** H5
Lander *watercourse* Australia **54** E5
Lander U.S.A. **62** F3
Landhi Pak. **36** A4
Landor Australia **55** B6
Landsberg Poland *see* **Gorzów Wielkopolski**
Landsberg am Lech Germany **17** M6
Land's End *pt* U.K. **19** B8
Landshut Germany **17** N6
Landskrona Sweden **15** H9
Lanesborough Ireland **21** E4
La'nga Co *l.* China **36** E3
Langar Afgh. **36** B2
Langberg *mts* S. Africa **50** F5
Langdon U.S.A. **62** H2
Langeac France **24** F4
Langeberg *mts* S. Africa **50** D7
Langeland *i.* Denmark **15** G9
Längelmäki Fin. **15** N6
Langenthal Switz. **24** H3
Langesund Norway **15** F7
Langfang China **43** L5
Langgar China **37** H3
Langjan Nature Reserve S. Africa **51** I2
Langjökull Iceland **14** [inset]
Langklip S. Africa **50** E5
Langlo Crossing Australia **57** D5
Langøya *i.* Norway **14** I2
Langphu *mt.* China **37** F3
Langport U.K. **19** E7
Langqên Zangbo *r.* China **36** D3
Langres France **24** G3
Langres, Plateau de France **24** G3
Langru China **36** D1
Langsa Indon. **41** B7
Långsele Sweden **14** J5
Lang Sơn Vietnam **42** J8
Langtang National Park Nepal **37** F3
Langting India **37** H4
Langtoft U.K. **18** G4
Languedoc *reg.* France **24** E5
Långvattnet Sweden **14** L4
Lanigan Canada **62** F1
Lanín, Parque Nacional *nat. park* Arg. **70** B5
Lanín, Volcán *vol.* Arg./Chile **70** B5
Lanji India **36** E5
Lanka *country* Asia *see* **Sri Lanka**
Lānkārān Azer. **35** H3
Lannion France **24** C2
Lansån Sweden **14** M3
Lansing U.S.A. **63** K3
Lanxi *Heilong.* China **44** B3
Lan Yü *i.* Taiwan **43** M8
Lanzarote *i.* Canary Is **46** B2
Lanzhou China **42** I5
Lanzijing China **44** A3
Laoag City Phil. **41** E6
Lao Cai Vietnam **31** J4
Laodicea Syria *see* **Latakia**
Laodicea Turkey *see* **Denizli**
Laodicea ad Lycum Turkey *see* **Denizli**
Laodicea ad Mare Syria *see* **Latakia**
Laohekou China **43** K6
Laojunmiao China **42** H5
La Okapi, Parc National de *nat. park* Dem. Rep. Congo **48** C3
Lao Ling *mts* China **44** B4
Lao Mangnai China **42** G5
Laon France **24** F2
La Oroya Peru **68** C6
Laos *country* Asia **41** C6
Laotougou China **44** C4
Laotuding Shan *mt.* China **44** B4
Laowohi *pass* India *see* **Khardung La**
Laoye Ling *mts Heilongjiang/Jilin* China **44** C4
Laoye Ling *mts Jilin* China **44** B4
Lapa Brazil **71** A4
La Palma *i.* Canary Is **46** B2
La Palma Panama **67** I7
La Palma del Condado Spain **25** C5
La Panza Range *mts CA* U.S.A. **65** B3
La Paragua Venez. **68** F2
La Paya, Parque Nacional *nat. park* Col. **68** D3
La Paz Arg. **70** E4
La Paz Bol. **68** E7
La Paz Hond. **66** G6
La Paz Mex. **66** B4
La Pedrera Col. **68** E4
La Pendjari, Parc National de la *nat. park* Benin **46** D3
La Pérouse Strait Japan/Rus. Fed. **44** F3
La Pesca Mex. **66** E4
Lapinlahti Fin. **14** O5
Lapithos Cyprus **39** A2
La Plata Arg. **70** E4
La Plata *MD* U.S.A. **64** C3
La Plata, Isla *i.* Ecuador **68** B4
La Plata, Río de *sea chan.* Arg./Uruguay **70** E4
Lapmežciems Latvia **15** M8
Lapominka Rus. Fed. **12** I2
Laporte *PA* U.S.A. **64** C2
Lappajärvi Fin. **14** M5
Lappajärvi *l.* Fin. **14** M5
Lappeenranta Fin. **15** P6
Lappi Fin. **15** L6
Lappland *reg.* Europe **14** K3
Lâpseki Turkey **27** L4
Laptevo Rus. Fed. *see* **Yasnogorsk**
Laptev Sea Rus. Fed. **29** N2
Lapua Fin. **14** M5
Lapurdum France *see* **Bayonne**
Laqiya Arbain *well* Sudan **32** C5
La Quiaca Arg. **70** C2
L'Aquila Italy **26** E3
La Quinta *CA* U.S.A. **65** D4
Lār Iran **35** I6
Larache Morocco **25** C6
Laramie U.S.A. **62** F3
Laramie Mountains U.S.A. **62** F3
Laranda Turkey *see* **Karaman**
Laranjal Paulista Brazil **71** B3
Laranjeiras do Sul Brazil **70** F3
Laranjinha *r.* Brazil **71** A3
Larantuka Indon. **41** E8
Larat Indon. **54** E1
Larat *i.* Indon. **41** F8
Larba Alg. **25** H5
Lärbro Sweden **15** K8
L'Ardenne, Plateau de *plat.* Belgium *see* **Ardennes**
Laredo Spain **25** E2
Laredo U.S.A. **62** H6
La Reina Adelaida, Archipiélago de *is* Chile **70** B8
Largeau Chad *see* **Faya**
Largs U.K. **20** E5
Lāri Iran **35** G3
L'Ariana Tunisia **26** D6
La Rioja Arg. **70** C3
La Rioja *aut. comm.* Spain **25** E2
Larisa Greece **27** J5
Larissa Greece *see* **Larisa**
Larkana Pak. **33** K4
Lark Passage Australia **56** D2
L'Arli, Parc National de *nat. park* Burkina Faso **46** D3
Larnaca Cyprus **39** A2
Larnaka Cyprus *see* **Larnaca**
Larnaka Bay Cyprus **39** A2
Larnakos, Kolpos *b.* Cyprus *see* **Larnaka Bay**
Larne U.K. **21** G3
La Robla Spain **25** D2
La Rochelle France **24** D3
La Roche-sur-Yon France **24** D3
La Roda Spain **25** E4
La Romana Dom. Rep. **67** K5
La Ronge Canada **60** H4
La Ronge, Lac *l.* Canada **60** H4
Larrey Point Australia **54** B4
Larrimah Australia **54** F3
Lars Christensen Coast Antarctica **76** E2
Larsen Ice Shelf Antarctica **76** L2
Larsmo Fin. **14** M5
Larvik Norway **15** G7
La Salonga Nord, Parc National de *nat. park* Dem. Rep. Congo **48** C4
Las Anod Somalia *see* **Laascaanood**
La Sarre Canada **63** L2
Las Cruces *CA* U.S.A. **65** B3
Las Cruces *NM* U.S.A. **62** F5
La Selle, Pic *mt.* Haiti **67** J5
La Serena Chile **70** B3
La Seu d'Urgell Spain **25** G2
Las Flores Arg. **70** E5
Las Heras Arg. **70** C4
Lashio Myanmar **42** H8
Lashkar India **36** D4
Lashkar Gāh Afgh. **33** J3
Las Juntas Chile **70** C3
Las Lomitas Arg. **70** D2
Las Marismas *marsh* Spain **25** C5
Las Martinetas Arg. **70** C7
Las Minas, Cerro de *mt.* Hond. **66** G6
La Société, Archipel de la *is* Fr. Polynesia *see* **Society Islands**
Las Palmas de Gran Canaria Canary Is **46** B2
Las Petas Bol. **69** G7
La Spezia Italy **26** C2
Las Piedras, Río de *r.* Peru **68** E6
Las Plumas Arg. **70** C6
Laspur Pak. **36** C1
Lassance Brazil **71** B2
Las Tablas Panama **67** H7
Las Tablas de Daimiel, Parque Nacional de *nat. park* Spain **25** E4
Las Termas Arg. **70** D3
Last Mountain Lake Canada **62** F1
Las Tórtolas, Cerro *mt.* Chile **70** C3
Lastoursville Gabon **48** B4

tovo i. Croatia 26 G3
Tres Vírgenes, Volcán vol. Mex. 66 B3
Tunas Cuba 67 I4
Varas Chihuahua Mex. 66 C3
Varas Nayarit Mex. 66 C4
Varillas Arg. 70 D4
Vegas NM U.S.A. 62 F4
Vegas NV U.S.A. 65 E2
Viajas, Isla de i. Peru 68 C6
Villuercas mt. Spain 25 D4
acunga Ecuador 68 C4
ady Island Antarctica 76 L2
akia Syria 39 B2
este-de-Buch France 24 D4
nam Australia 55 B7
neron U.K. 20 F2
ni India 36 B4
no India 36 D2
nrop CA U.S.A. 65 B2
na Italy 26 E4
Tortuga, Isla i. Venez. 68 E1
run West Bank 39 B4
aquié Syria see Latakia
uque Canada 63 M2
ur India 38 C2
via country Europe 15 N8
vija country Europe see Latvia
viyskaya S.S.R. country Europe see atvia
ca, Parque Nacional nat. park Chile 8 E7
chhammer Germany 17 N5
der U.K. 20 G5
dio Spain 25 E2
fen Switz. 24 H3
ge Koch Kyst reg. Greenland 61 L2
ghlen, Mount Australia 55 F5
ka Estonia 15 M7
nceston Australia 57 [inset]
nceston U.K. 19 C8
ne r. Ireland 21 C5
Unión Bol. 68 F7
ra Australia 56 D2
rel DE U.S.A. 64 D3
rel MS U.S.A. 63 J5
reldale PA U.S.A. 64 D2
rel Hill hills PA U.S.A. 64 B3
rencekirk U.K. 20 G4
rieton Australia 58 F3
rinburg U.S.A. 63 L5
ru i. Solomon Is see Choiseul
sanne Switz. 24 H3
t i. Indon. 41 D8
t Bali sea Indon. see Bali, Laut
tem East Timor 54 D2
t Flores sea Indon. see Flores, Laut
toka Fiji 53 H3
t Sawu sea Indon. see Sawu, Laut
vuskylä Fin. 14 P5
al France 24 D2
all d'Uixó Spain 25 F4
in i. Iran 35 I6
anoise, Massif de mts France 24 H4
anoise, Parc National de nat. park ance 24 H4
apié, Punta pt Chile 70 B5
ar Iran 35 H5
eaga Peak CA U.S.A. 65 B2
ega Dom. Rep. 67 J5
erton Australia 55 C7
ila Joiosa Spain see ajoyosa-La Vila Joiosa
iña Peru 68 C5
ongai i. P.N.G. see New Hanover
ras Brazil 71 B3
umisa Swaziland 51 J4
ushi-Manda National Park Zambia 9 D5
a India 36 C4
a Myanmar 37 I4
a Pak. 36 B2
Dome Antarctica 76 F2
n Hill National Park Australia 56 B3
ra Ghana 46 C3
rence KS U.S.A. 63 H4
rence MA U.S.A. 64 F1
renceburg TN U.S.A. 63 J4
rence Wells, Mount h. Australia 55 C6
ton U.S.A. 62 H5
z, Jabal al mt. Saudi Arabia 34 D5
Sweden 15 I7
ey Isle of Man 18 C4
o U.K. 20 [inset]
a r. Rus. Fed. 12 M2
denyy, Mys c. Rus. Fed. 12 J1
á Saudi Arabia 32 G5
a salt pan Saudi Arabia 39 D4
san Island U.S.A. 74 I4
yah Pak. 33 L3
arev Rus. Fed. 44 F1
arevac Serbia 27 I2
aro Cárdenas Mex. 66 D5
cano Uruguay 70 F4
dijai Lith. 15 M9
o Primorskiy Kray Rus. Fed. 44 D4
o Respublika Sakha (Yakutiya) Rus. Fed. 9 O3
d U.S.A. 62 G3
der Water r. U.K. 20 G5
dville Australia 58 D4
f Bay Canada see Tasiujaq
mington Spa, Royal U.K. 19 F6
ne, Lough l. Ireland 21 C5
p Ireland 21 C6
therhead U.K. 19 G7
au Claire, Lac à l. Canada 61 K4
vitt Peak CA U.S.A. 65 C1
anon country Asia 39 B2
anon MO U.S.A. 63 I4
anon NH U.S.A. 64 E1
anon PA U.S.A. 64 C2
anon Mountains Lebanon see ban, Jebel
ec CA U.S.A. 65 C3
edyan' Rus. Fed. 13 H5
Blanc France 24 E3
ork Poland 17 P3
owakgomo S. Africa 51 I3
rija Spain 25 C5
sko, Jezioro lag. Poland 17 P3
u Chile 70 B5
yazh'ye Kazakh. see Akku
yazh'ye Rus. Fed. 12 K4
Caire Egypt see Cairo
ce Italy 26 H4
co Italy 26 C2
h r. Austria/Germany 17 M7
haina Greece 27 I6
hang China 43 K7
Chasseron mt. Switz. 24 H3
htaler Alpen mts Austria 17 M7
k Germany 17 L3
Creusot France 24 G3
Crotoy France 19 I8
toure France 24 E5
lbury U.K. 19 E6
desma Spain 25 D3
lmore U.K. 20 E2
lmozero Rus. Fed. 14 R4
Dorat France 24 E3
r. Ireland 21 D6
Lee MA U.S.A. 64 E1
Leech Lake U.S.A. 63 I2
Leeds U.K. 18 F5
Leedstown U.K. 19 B8
Leek U.K. 19 E5
Leesburg VA U.S.A. 64 C3
Lee Steere Range hills Australia 55 C6
Leesville U.S.A. 63 I5
Leesville Lake OH U.S.A. 64 A2
Leesville Lake VA U.S.A. 64 B4
Leeton Australia 58 C5
Leeu-Gamka S. Africa 50 E7
Leeuwarden Neth. 17 J4
Leeuwin, Cape Australia 55 A8
Leeuwin-Naturaliste National Park Australia 55 A8
Lee Vining CA U.S.A. 65 C2
Leeward Islands Caribbean Sea 67 L5
Lefka Cyprus 39 A2
Lefkada Greece 27 I5
Lefkada i. Greece 27 I5
Lefkás Greece see Lefkada
Lefke Cyprus see Lefka
Lefkimmi Greece 27 I5
Lefkoniko Cyprus see Lefkonikon
Lefkonikon Cyprus 39 A2
Lefkoşa Cyprus see Nicosia
Lefkosia Cyprus see Nicosia
Lefroy, Lake imp. l. Australia 55 C7
Legges Tor mt. Australia 57 [inset]
Leghorn Italy see Livorno
Legnago Italy 26 D2
Legnica Poland 17 P5
Le Grand CA U.S.A. 65 B2
Legune Australia 54 E3
Leh India 36 D2
Le Havre France 24 E2
Lehighton PA U.S.A. 64 D2
Lehmo Fin. 14 P5
Lehtimäki Fin. 14 M5
Lehututu Botswana 50 E2
Leibnitz Austria 17 O7
Leicester U.K. 19 F6
Leichhardt r. Australia 52 B3
Leichhardt Falls Australia 56 B3
Leichhardt Range mts Australia 56 D4
Leiden Neth. 16 J4
Leigh N.Z. 59 E3
Leigh U.K. 18 E5
Leighton Buzzard U.K. 19 G7
Leinster Australia 55 C6
Leinster reg. Ireland 21 F4
Leinster, Mount h. Ireland 21 F5
Leipsoi i. Greece 27 L6
Leipzig Germany 17 N5
Leiranger Norway 14 I3
Leiria Port. 25 B4
Leirvik Norway 15 D7
Leisler, Mount h. Australia 55 E5
Leith Hill h. U.K. 19 G7
Leiva, Cerro mt. Col. 68 D3
Leixlip Ireland 21 F4
Leiyang China 43 K7
Leizhou Bandao pen. China 43 J8
Leka Norway 14 G4
Lékana Congo 48 B4
Le Kef Tunisia 26 C6
Lekhainá Greece see Lechaina
Lekkersing S. Africa 50 C5
Lékoni Gabon 48 B4
Leksand Sweden 15 I6
Leksozero, Ozero l. Rus. Fed. 14 Q5
Lélouma Guinea 46 B3
Lelystad Neth. 16 J4
Le Maire, Estrecho de sea chan. Arg. 70 C9
Léman, Lac l. France/Switz. see Geneva, Lake
Le Mans France 24 E2
Le Mars U.S.A. 63 H3
Lemberg Ukr. see L'viv
Lemdiyya Alg. see Médéa
Leme Brazil 71 B3
Lemesos Cyprus see Limassol
Lemi Fin. 15 O6
Lemieux Islands Canada 61 L3
Lemmenjoen kansallispuisto nat. park Fin. 14 N2
Lemmon U.S.A. 62 G2
Lemnos i. Greece see Limnos
Lemoncove CA U.S.A. 65 C2
Lemoore CA U.S.A. 65 C2
Le Moyne, Lac l. Canada 61 L4
Lemro r. Myanmar 37 H5
Lemtybozh Rus. Fed. 11 R3
Le Murge hills Italy 26 G4
Lemvig Denmark 15 F8
Lem"yu r. Rus. Fed. 12 M3
Lena r. Rus. Fed. 42 J1
Lenadoon Point Ireland 21 C3
Lenchung Tso salt l. China 37 E2
Lençóis Brazil 71 C1
Lençóis Maranhenses, Parque Nacional dos nat. park Brazil 69 J4
Lendery Rus. Fed. 12 F3
Le Neubourg France 19 H9
Lenglong Ling mts China 42 I5
Lengshuijiang China 43 K7
Lenham U.K. 19 H7
Lenhovda Sweden 15 I8
Lenin, Qullai mt. Kyrg./Tajik. see Lenin Peak
Lenina, Pik mt. Kyrg./Tajik. see Lenin Peak
Leninabad Tajik. see Khŭjand
Leninakan Armenia see Gyumri
Lenin Atyndagy Choku mt. Kyrg./Tajik. see Lenin Peak
Lenine Ukr. 34 D1
Leningrad Rus. Fed. see St Petersburg
Leningrad Oblast admin. div. Rus. Fed. see Leningradskaya Oblast'
Leningradskaya Rus. Fed. 13 H7
Leningradskaya Oblast' admin. div. Rus. Fed. 15 R7
Leningradskiy Rus. Fed. 29 S3
Lenino Ukr. see Lenine
Leninobod Tajik. see Khŭjand
Lenin Peak Kyrg./Tajik. 42 C5
Leninsk Kazakh. see Baykonyr
Leninsk Rus. Fed. 13 J6
Leninskiy Rus. Fed. 13 H5
Leninsk-Kuznetskiy Rus. Fed. 28 J4
Leninskoye Kirovskaya Oblast' Rus. Fed. 12 J4
Leninskoye Yevreyskaya Avtonomnaya Oblast' Rus. Fed. 44 D3
Lenkoran' Azer. see Länkäran
Lenox MA U.S.A. 64 E1
Lens France 24 F1
Lensk Rus. Fed. 29 M3
Lenti Hungary 26 G1
Lentini Sicilia Italy 26 F6
Léo Burkina Faso 46 C3
Leoben Austria 17 O7
Leodhais, Eilean i. U.K. see Lewis, Isle of
Leominster U.K. 19 E6
Leominster MA U.S.A. 64 F1
León Mex. 66 D4
León Nicaragua 67 G6
León Spain 25 D2
Leonardtown MD U.S.A. 64 C3
Leonardville Namibia 50 D2
Leongatha Australia 58 B7
Leonidio Greece 27 J6
Leonidovo Rus. Fed. 44 F2
Leonora Australia 55 C7
Leopold WV U.S.A. 64 A3
Leopold and Astrid Coast Antarctica see King Leopold and Queen Astrid Coast
Léopold II, Lac l. Dem. Rep. Congo see Mai-Ndombe, Lac
Leopoldina Brazil 71 C3
Leopoldo de Bulhões Brazil 71 A2
Léopoldville Dem. Rep. Congo see Kinshasa
Lepalale S. Africa see Lephalale
Lepaya Latvia see Liepāja
Lepel' Belarus see Lyepyel'
Lephalala r. S. Africa 51 H2
Lephalale S. Africa 51 H2
Lephepe Botswana 51 G2
Lephoi S. Africa 51 G6
Leping China 43 L7
Lepontine, Alpi mts Italy/Switz. 26 C1
Leppävirta Fin. 14 O5
Lepsa Kazakh. see Lepsy
Lepsy Kazakh. 42 D3
Le Puy France see Le Puy-en-Velay
Le Puy-en-Velay France 24 F4
Lerala Botswana 51 H2
Leratswana S. Africa 51 H5
Léré Mali 46 C3
Leribe Lesotho see Hlotse
Lérida Col. 68 D4
Lérida Spain see Lleida
Lerik Azer. 35 H3
Lerma Spain 25 E2
Lermontov Rus. Fed. 35 F1
Lermontovka Rus. Fed. 44 D3
Lermontovskiy Rus. Fed. see Lermontov
Leros i. Greece 27 L6
Le Roy NY U.S.A. 64 C1
Le Roy, Lac l. Canada 61 K4
Lerum Sweden 15 H8
Lerwick U.K. 20 [inset]
Les Amirantes is Seychelles see Amirante Islands
Lesbos i. Greece 27 K5
Les Cayes Haiti 67 J5
Leshan China 42 I7
Leshukonskoye Rus. Fed. 12 J2
Lesi watercourse South Sudan 47 F4
Leskhimstroy Ukr. see Syeverodonets'k
Leskovac Serbia 27 I3
Lesneven France 24 B2
Lesnoy Kirovskaya Oblast' Rus. Fed. 12 L4
Lesnoy Murmanskaya Oblast' Rus. Fed. see Umba
Lesnoye Rus. Fed. 12 G4
Lesogorskoye Rus. Fed. 44 F2
Lesopil'noye Rus. Fed. 44 D3
Lesosibirsk Rus. Fed. 28 K4
Lesotho country Africa 51 I5
Lesozavodsk Rus. Fed. 44 D3
L'Espérance Rock i. Kermadec Is 53 I5
Les Pieux France 19 F9
Les Sables-d'Olonne France 24 D3
Lesser Antarctica reg. Antarctica see West Antarctica
Lesser Antilles is Caribbean Sea 67 K6
Lesser Caucasus mts Asia 35 F2
Lesser Himalaya mts India/Nepal 36 D3
Lesser Khingan Mountains China see Xiao Hinggan Ling
Lesser Slave Lake Canada 60 G4
Lester WV U.S.A. 64 A4
Lestijärvi Fin. 14 N5
Les Vans France 24 G4
Lesvos i. Greece see Lesbos
Leszno Poland 17 P5
Letaba S. Africa 51 J2
Letchworth Garden City U.K. 19 G7
Le Télégraphe h. France 24 G3
Leteri India 36 D4
Letha Range mts Myanmar 37 H5
Lethbridge Alta Canada 62 E2
Leti i. Indon. 54 D2
Leti, Kepulauan is Indon. 41 E8
Leticia Col. 68 E4
Letlhakeng Botswana 51 G3
Letnerechenskiy Rus. Fed. 12 G2
Letniy Navolok Rus. Fed. 12 H2
Le Touquet-Paris-Plage France 19 I8
Letpadan Myanmar 42 H9
Le Tréport France 24 E1
Letsitele S. Africa 51 J2
Letsopa S. Africa 51 G4
Letterkenny Ireland 21 E3
Lëtzebuerg country Europe see Luxembourg
Léua Angola 49 C5
Leucas Greece see Lefkada
Leucate, Étang de l. France 24 F5
Leuchars U.K. 20 G4
Leukas Greece see Lefkada
Leunovo Rus. Fed. 12 I2
Leura Australia 56 E4
Leuven Belgium 16 J5
Levanger Norway 14 G5
Levante, Riviera di coastal area Italy 26 C2
Levanto Italy 26 C2
Levashi Rus. Fed. 13 J8
Levelland U.S.A. 62 G5
Leven England U.K. 18 G5
Leven Scotland U.K. 20 G4
Leven, Loch l. U.K. 20 F4
Lévêque, Cape Australia 54 C4
Leverkusen Germany 17 K5
Lévézou mts France 24 F4
Levice Slovakia 17 Q6
Levin N.Z. 59 E5
Levitha i. Greece 27 L6
Levittown NY U.S.A. 64 E2
Levittown PA U.S.A. 64 D2
Levkás i. Greece see Lefkada
Levkímmi Greece see Lefkimmi
Levskigrad Bulg. see Karlovo
Lev Tolstoy Rus. Fed. 13 H5
Lévy, Cap c. France 19 F9
Lewe Myanmar 37 I6
Lewerberg mt. S. Africa 50 C5
Lewes U.K. 19 H8
Lewes DE U.S.A. 64 D3
Lewis, Isle of i. U.K. 20 C2
Lewis, Lake imp. l. Australia 54 F5
Lewisburg PA U.S.A. 64 C2
Lewisburg WV U.S.A. 64 A4
Lewis Pass N.Z. 59 D6
Lewis Range hills Australia 54 E5
Lewis Range mts U.S.A. 62 E2
Lewiston ID U.S.A. 62 D2
Lewistown MT U.S.A. 62 F2
Lewistown PA U.S.A. 64 C2
Lexington KY U.S.A. 63 K4
Lexington NE U.S.A. 62 H3
Lexington VA U.S.A. 64 B4
Lexington Park MD U.S.A. 64 C3
Leyden Neth. see Leiden
Leyla Dāgh mt. Iran 35 G3
Lezha Albania see Lezhë
Lezhë Albania 27 H4
L'gov Rus. Fed. 13 G6
Lhagoi Kangri mt. China 37 F3
Lharigarbo China 37 G2
Lhasa China 37 G3
Lhasoi China 37 H3
Lhatog China 37 I3
Lhaviyani Atoll Maldives see Faadhippolhu Atoll
Lhazê Xizang China 37 F3
Lhazê Xizang China 37 H3
Lhazhong China 37 F3
Lhomar China 37 G3
Lhorong China 37 H3
Lhotse mt. China/Nepal 37 F4
Lhozhag China 37 G3
Lhuntshi Bhutan 37 G4
Lhünzê China 42 G7
Lhünzhub China see Gaindainqoinkor
Liakoura mt. Greece 27 J5
Liancourt Rocks i. N. Pacific Ocean 45 C5
Liandu China see Lishui
Liangdaohe China 37 G3
Liangzhou China see Wuwei
Lianhe China see Qianjiang
Lianjiang Jiangxi China see Xingguo
Lianshan Liaoning China 43 M4
Lianyin China 44 A1
Lianyungang China 43 L6
Liaocheng China 43 L5
Liaodong Bandao pen. China 43 M4
Liaodong Wan b. China 43 M4
Liao He r. China 44 A4
Liaoning prov. China 44 A4
Liaoyang China 44 A4
Liaoyuan China 44 B4
Liaozhong China 44 A4
Liapades Greece 27 H5
Liard r. Canada 60 F3
Liari Pak. 36 A4
Liathach mt. U.K. 20 D3
Liban country Asia see Lebanon
Liban, Jebel mts Lebanon 39 C2
Libau Latvia see Liepāja
Libby U.S.A. 62 D2
Libenge Dem. Rep. Congo 48 B3
Liberal U.S.A. 62 G4
Liberdade Brazil 71 B3
Liberec Czech Rep. 17 O5
Liberia country Africa 46 C4
Liberia Costa Rica 67 G6
Liberty NY U.S.A. 64 D2
Liberty Lake MD U.S.A. 64 C3
Libni, Gebel h. Egypt see Libnī, Jabal
Libnī, Jabal h. Egypt 39 A4
Libode S. Africa 51 I6
Libourne France 24 D4
Libral Well Australia 54 D5
Libreville Gabon 48 A3
Libya country Africa 47 E2
Libyan Desert Egypt/Libya 32 C5
Libyan Plateau Egypt 34 B5
Licantén Chile 70 B4
Licata Sicilia Italy 26 E6
Lice Turkey 35 F3
Lichas hills Greece 27 J5
Lichfield U.K. 19 F6
Lichinga Moz. 49 D5
Lichtenburg S. Africa 51 H4
Lida Belarus 15 N10
Liddel Water r. U.K. 20 G5
Lidfontein Namibia 50 D3
Lidköping Sweden 15 H7
Lidsjöberg Sweden 14 I4
Liebig, Mount Australia 55 E5
Liechtenstein country Europe 24 I3
Liège Belgium 17 J5
Liegnitz Poland see Legnica
Lieksa Fin. 14 Q5
Lielupe r. Latvia 15 N8
Lielvārde Latvia 15 N8
Lienart Dem. Rep. Congo 48 C3
Lienchung i. Taiwan see Matsu Tao
Lienz Austria 17 N7
Liepāja Latvia 15 L8
Liepaya Latvia see Liepāja
Lietuva country Europe see Lithuania
Liezen Austria 17 O7
Liffey r. Ireland 21 F4
Lifford Ireland 21 E3
Lifi Mahuida mt. Arg. 70 C6
Lifou i. New Caledonia 53 G4
Lifu i. New Caledonia see Lifou
Līgatne Latvia 15 N8
Lightning Ridge Australia 58 C2
Ligonha r. Moz. 49 D5
Ligure, Mar sea France/Italy see Ligurian Sea
Ligurian Sea France/Italy 26 C3
Ligurienne, Mer sea France/Italy see Ligurian Sea
Ligurta AZ U.S.A. 65 E4
Lihir Group is P.N.G. 52 F2
Lihou Reef and Cays Australia 56 E3
Liivi laht b. Estonia/Latvia see Riga, Gulf of
Lijiang Yunnan China 42 I7
Lika reg. Croatia 26 F2
Likasi Dem. Rep. Congo 49 C5
Likati Dem. Rep. Congo 48 C3
Likhachevo Ukr. see Pervomays'kyy
Likhachyovo Ukr. see Pervomays'kyy
Likhapani India 37 H4
Likhás hills Greece see Lichas
Likhoslavl' Rus. Fed. 12 G4
Liku Indon. 41 C7
Likurga Rus. Fed. 12 I4
L'Île-Rousse Corse France 24 I5
Lilla Pak. 36 C2
Lilla Edet Sweden 15 H7
Lille France 24 F1
Lille Bælt sea chan. Denmark see Little Belt
Lillebonne France 19 H9
Lillehammer Norway 15 G6
Lillesand Norway 15 F7
Lillestrøm Norway 15 G7
Lillholmsjö Sweden 14 I5
Lillian, Point h. Australia 55 D6
Lillooet Canada 62 C1
Lilongwe Malawi 49 D5
Lilydale Australia 57 B7
Lima Peru 68 C6
Lima NY U.S.A. 64 C1
Lima OH U.S.A. 63 K3
Lima Duarte Brazil 71 C3
Liman Rus. Fed. 13 J7
Limar Indon. 54 D1
Limassol Cyprus 39 A2
Limavady U.K. 21 F2
Limay r. Arg. 70 C5
Limbaži Latvia 15 N8
Limbunya Australia 54 E4
Lime Acres S. Africa 50 F5
Limeira Brazil 71 B3
Limerick Ireland 21 D5
Limingen Norway 14 H4
Limingen l. Norway 14 H4
Liminka Fin. 14 N4
Limmen Bight b. Australia 56 B2
Limnos i. Greece 27 K5
Limoeiro Brazil 69 K5
Limoges France 24 E4
Limón Costa Rica see Puerto Limón
Limon U.S.A. 62 G4
Limonlu Turkey 39 B1
Limonum France see Poitiers
Limousin reg. France 24 E4
Limoux France 24 F5
Limpopo prov. S. Africa 51 I2
Limpopo r. S. Africa/Zimbabwe 51 K3
Līnah well Saudi Arabia 35 F5
Linakhamari Rus. Fed. 14 Q2
Linares Chile 70 B5
Linares Mex. 66 E4
Linares Spain 25 E4
Lincang China 42 I8
Linchuan China see Fuzhou
Linck Nunataks nunataks Antarctica 76 K1
Lincoln Arg. 70 D4
Lincoln U.K. 18 G5
Lincoln CA U.S.A. 65 B1
Lincoln IL U.S.A. 63 J3
Lincoln NE U.S.A. 63 H3
Lincoln National Park Australia 57 A7
Lincoln Sea Canada/Greenland 77 J1
Lincolnshire Wolds hills U.K. 18 G5
Linda, Serra hills Brazil 71 C1
Linda Creek watercourse Australia 56 B4
Lindau (Bodensee) Germany 17 L7
Lindeman Group is Australia 56 E4
Linden Guyana 69 G2
Lindesnes c. Norway 15 E7
Lindhos Greece see Lindos
Lindi r. Dem. Rep. Congo 48 C3
Lindi Tanz. 49 D4
Lindian China 44 B3
Lindisfarne i. U.K. see Holy Island
Lindley S. Africa 51 H4
Lindos Greece 23 J4
Lindsay CA U.S.A. 65 C2
Lindside WV U.S.A. 64 A4
Lindum U.K. see Lincoln
Line Islands Kiribati 75 J5
Linesville PA U.S.A. 64 A2
Linfen China 43 K5
Lingampet India 38 C2
Linganamakki Reservoir India 38 B3
Lingcheng Hainan China see Lingshui
Lingelethu S. Africa 51 H7
Lingen (Ems) Germany 17 K4
Lingga, Kepulauan is Indon. 41 C8
Lingomo Dem. Rep. Congo 48 C3
Lingshui China 43 J9
Lingsugur India 38 C2
Linguère Senegal 46 B3
Lingxia China 44 A3
Lingzi Tang plain Aksai Chin 36 D2
Linhai China 43 M7
Linhares Brazil 71 C2
Linhe China 43 J4
Linhpa Myanmar 37 H4
Linjiang China 44 B4
Linköping Sweden 15 I7
Linkou China 44 C3
Linlithgow U.K. 20 F5
Linnansaaren kansallispuisto nat. park Fin. 14 P5
Linnhe, Loch inlet U.K. 20 D4
Linosa, Isola di i. Sicilia Italy 26 E7
Lins Brazil 71 A3
Lintao China 42 I5
Linton ND U.S.A. 62 G2
Linxi China 43 L4
Linxia China 42 I5
Linyi Shandong China 43 L5
Linyi Shandong China 43 L5
Linz Austria 17 O6
Lion, Golfe du g. France 24 F5
Lions, Gulf of France see Lion, Golfe du
Lioua Chad 47 E3
Lipari Sicilia Italy 26 F5
Lipari, Isole is Italy 26 F5
Lipetsk Rus. Fed. 13 H5
Lipin Bor Rus. Fed. 12 H3
Lipova Romania 27 I1
Lipovtsy Rus. Fed. 44 C3
Lipsoi i. Greece see Leipsoi
Lipti Lekh pass Nepal 36 E3
Liptrap, Cape Australia 58 B7
Liqeni i Banjës resr Albania 27 I4
Lira Uganda 48 D3
Liranga Congo 48 B4
Lircay Peru 68 D6
Lisala Dem. Rep. Congo 48 C3
L'Isalo, Massif de mts Madag. 49 E6
L'Isalo, Parc National de nat. park Madag. 49 E6
Lisbellaw U.K. 21 E3
Lisboa Port. see Lisbon
Lisbon Port. 25 B4
Lisbon OH U.S.A. 64 A2
Lisburn U.K. 21 F3
Liscannor Bay Ireland 21 C5
Lisdoonvarna Ireland 21 C4
Lishi Jiangxi China see Dingnan
Lishu China 44 B4
Lishui China 43 L7
Lisichansk Ukr. see Lysychans'k
Lisieux France 24 E2
Liskeard U.K. 19 C8
Liski Rus. Fed. 13 H6
Lismore Australia 58 F2
Lismore Ireland 21 E5
Lisnarrick U.K. 21 E3
Lisnaskea U.K. 21 E3
Liss mt. Saudi Arabia 39 D4
Lissa Poland see Leszno
Lister, Mount Antarctica 76 H1
Listowel Ireland 21 C5
Lit Sweden 14 I5
Litang Guangxi China 43 J8
Litang Sichuan China 42 I6
Lītāni, Nahr el r. Lebanon 39 B3
Litchfield CT U.S.A. 64 E2
Lit-et-Mixe France 24 D4
Lithgow Australia 58 E4
Lithino, Akrotirio pt Greece 27 K7
Lithuania country Europe 15 M9
Lititz PA U.S.A. 64 C2
Litoměřice Czech Rep. 17 O5
Litovko Rus. Fed. 44 D2
Litovskaya S.S.R. country Europe see Lithuania
Little Abaco i. Bahamas 67 I3
Little Andaman i. India 31 I5
Little Barrier i. N.Z. 59 E3
Little Belt sea chan. Denmark 15 F9
Little Bitter Lake Egypt 39 A4
Little Cayman i. Cayman Is 67 H5
Little Creek Peak UT U.S.A. 65 F2
Little Desert National Park Australia 57 C8
Little Egg Harbor inlet NJ U.S.A. 64 D3
Little Falls U.S.A. 63 I2
Littlefield AZ U.S.A. 65 F2
Littlefield TX U.S.A. 62 G5
Littlehampton U.K. 19 G8
Little Karas Berg plat. Namibia 50 D4
Little Karoo plat. S. Africa 50 E7
Little Lake CA U.S.A. 65 D3
Little Mecatina r. Nfld. and Lab./Que. Canada see Petit Mécatina
Little Minch sea chan. U.K. 20 B3
Little Namaqualand reg. S. Africa see Namaqualand
Little Ouse r. U.K. 19 H6
Little Pamir mts Asia 36 C1
Little Rock U.S.A. 63 I5
Littlerock CA U.S.A. 65 D3
Little Salt Lake UT U.S.A. 65 F2
Little Sandy Desert Australia 55 B5
Little Tibet reg. India/Pak. see Ladakh
Little Valley NY U.S.A. 64 B1
Litunde Moz. 49 D5
Liu'an China see Lu'an
Liuchow China see Liuzhou
Liuhe China 44 B4
Liukesong China 44 B3
Liupanshui China 42 I7
Liuwa Plain National Park Zambia 49 C5
Liuzhan China 44 B2
Liuzhou China 43 J8
Livadeia Greece 27 J5
Līvāni Latvia 15 O8
Liveringa Australia 52 C3
Livermore CA U.S.A. 65 B2
Livermore, Mount U.S.A. 62 G5
Liverpool Australia 58 E4
Liverpool Canada 63 O3
Liverpool U.K. 18 E5
Liverpool Bay Canada 60 E3
Liverpool Plains Australia 58 E3
Liverpool Range mts Australia 58 D3
Livingston U.K. 20 F5
Livingston MT U.S.A. 62 E2
Livingston TX U.S.A. 63 I5
Livingston, Lake U.S.A. 63 H5
Livingstone Zambia 49 C5
Livingston Island Antarctica 76 L2
Livingston Manor NY U.S.A. 64 D2
Livno Bos.-Herz. 26 G3
Livny Rus. Fed. 13 H5
Livojoki r. Fin. 14 O4
Livonia NY U.S.A. 64 C1
Livorno Italy 26 D3
Livramento do Brumado Brazil 71 C1
Liwā', Wādī al watercourse Syria 39 C3
Liwale Tanz. 49 D4
Lixus Morocco see Larache
Lizard U.K. 19 B9
Lizarda Brazil 69 I5
Lizard Point U.K. 19 B9
Lizarra Spain see Estella
Lizemores WV U.S.A. 64 A3
Ljouwert Neth. see Leeuwarden
Ljubljana Slovenia 26 F1
Ljugarn Sweden 15 K8
Ljungan r. Sweden 14 J5
Ljungaverk Sweden 14 J5
Ljungby Sweden 15 H8
Ljusdal Sweden 15 J6
Ljusnan r. Sweden 15 J6
Ljusne Sweden 15 J6
Llaima, Volcán vol. Chile 70 B5
Llanandras U.K. see Presteigne
Llanbadarn Fawr U.K. 19 C6
Llanbedr Pont Steffan U.K. see Lampeter
Llanbister U.K. 19 D6
Llandeilo U.K. 19 D7
Llandissilio U.K. 19 C7
Llandovery U.K. 19 D7
Llandrindod Wells U.K. 19 D6
Llandudno U.K. 18 D5
Llandysul U.K. 19 C6
Llanegwad U.K. 19 C7
Llanelli U.K. 19 C7
Llanfair Caereinion U.K. 19 D6
Llanfair-ym-Muallt U.K. see Builth Wells
Llangefni U.K. 18 C5
Llangollen U.K. 19 D6
Llangurig U.K. 19 D6
Llanllyfni U.K. 19 C5
Llannerch-y-medd U.K. 18 C5
Llannor U.K. 19 C6
Llano Estacado plain U.S.A. 62 G5
Llanos plain Col./Venez. 68 E2
Llanquihue, Lago l. Chile 70 B6
Llanrhystud U.K. 19 C6
Llantrisant U.K. 19 D7
Llanuwchllyn U.K. 19 D6
Llanwnog U.K. 19 D6
Llanymddyfri U.K. see Llandovery
Llay U.K. 19 D5
Lleida Spain 25 G3
Llerena Spain 25 C4
Llíria Spain 25 F4
Llodio Spain see Laudio
Lloyd George, Mount Canada 60 F4
Lloyd Lake Canada 60 H4
Lloydminster Canada 62 E1
Llucmayor Spain see Llucmajor
Llucmajor Spain 25 H4
Llullaillaco, Volcán vol. Chile 70 C2
Loa r. Chile 70 B2
Loban' r. Rus. Fed. 12 K4
Lobatejo mt. Spain 25 D5
Lobatse Botswana 51 G3
Lobaye r. Cent. Afr. Rep. 48 B3
Loberia Arg. 70 E5
Lobito Angola 49 B5
Lobos Arg. 70 E5
Lobos de Tierra, Isla i. Peru 68 B5
Lochaline U.K. 20 D4
Loch Baghasdail U.K. see Lochboisdale
Lochboisdale U.K. 20 B3
Lochcarron U.K. 20 D3
Lochearnhead U.K. 20 E4
Lochern National Park Australia 56 C5
Loches France 24 E3
Loch Garman Ireland see Wexford
Lochgelly U.K. 20 F4
Lochgilphead U.K. 20 D4
Lochinver U.K. 20 D2
Loch Lomond and the Trossachs National Park U.K. 20 E4
Lochmaddy U.K. 20 B3
Lochnagar mt. U.K. 20 F4
Loch nam Madadh U.K. see Lochmaddy
Loch Raven Reservoir MD U.S.A. 64 C3
Lochy, Loch l. U.K. 20 E4
Lock Australia 57 A7
Lockerbie U.K. 20 F5
Lockhart Australia 58 C5
Lock Haven PA U.S.A. 64 C2
Lockport NY U.S.A. 64 B1
Lod Israel 39 B4
Loddon r. Australia 58 A5
Lodève France 24 F5
Lodeynoye Pole Rus. Fed. 12 G3
Lodhikheda India 36 D5
Lodhran Pak. 36 B3
Lodi Italy 26 C2
Lodi CA U.S.A. 65 B1
Lødingen Norway 14 I2
Lodja Dem. Rep. Congo 48 C4
Lodomeria Rus. Fed. see Vladimir
Lodrani India 36 B5
Lodwar Kenya 48 D3
Łódź Poland 17 Q5
Loeriesfontein S. Africa 50 D6
Lofoten is Norway 14 H2
Lofusa South Sudan 47 G4
Log Rus. Fed. 13 I6
Loga Niger 46 D3

Logan UT U.S.A. 62 E3
Logan, Mount Canada 60 D3
Logan Creek r. Australia 56 D4
Logatec Slovenia 26 F2
Logroño Spain 25 E2
Logtak Lake India 37 H4
Lohardaga India 37 F5
Loharu India 36 C3
Lohatlha S. Africa 50 F5
Lohawat India 36 C4
Lohit r. China/India see **Zayü Qu**
Lohiniva Fin. 14 N3
Lohjanjärvi l. Fin. 15 M6
Lohtaja Fin. 14 M4
Loi-lem Myanmar 42 H8
Loimaa Fin. 15 M6
Loipyet Hills Myanmar 37 I4
Loire r. France 24 C3
Loja Ecuador 68 C4
Loja Spain 25 D5
Lokan tekojärvi resr Fin. 14 O3
Lokchim r. Rus. Fed. 12 K3
Lokgwabe Botswana 50 E3
Lokichar Kenya 30 C6
Lokichokio Kenya 48 D3
Løkken Denmark 15 F8
Løkken Norway 14 F5
Loknya Rus. Fed. 12 F4
Lokoja Nigeria 46 D4
Lokolama Dem. Rep. Congo 48 B4
Lokossa Benin 46 D4
Lokot' Rus. Fed. 13 G5
Lol South Sudan 47 F4
Lola Guinea 46 C4
Lolland i. Denmark 15 G9
Lollondo Tanz. 48 D4
Lolwane S. Africa 50 F4
Lom Bulg. 27 J3
Lom Norway 15 F6
Lomami r. Dem. Rep. Congo 48 C3
Lomar Pass Afgh. 36 A2
Lomas, Bahía de b. Chile 70 C8
Lomas de Zamora Arg. 70 E4
Lombarda, Serra hills Brazil 69 H3
Lomblen i. Indon. 54 C2
Lombok Indon. 54 B2
Lombok i. Indon. 41 D8
Lombok, Selat sea chan. Indon. 54 A2
Lomé Togo 46 D4
Lomela Dem. Rep. Congo 48 C4
Lomela r. Dem. Rep. Congo 47 F5
Lomond, Loch l. U.K. 20 E4
Lomonosov Rus. Fed. 15 P7
Lomonosov Ridge sea feature Arctic Ocean 77 B1
Lomovoye Rus. Fed. 12 I2
Lompoc CA U.S.A. 65 B3
Łomża Poland 17 S4
Lonar India 38 C2
Londa Bangl. 37 G5
Londa India 38 B3
Londinières France 19 I9
Londinium U.K. see **London**
Londoko Rus. Fed. 44 D2
London Ont. Canada 64 A1
London U.K. 19 G7
London KY U.S.A. 63 K4
Londonderry U.K. 21 E3
Londonderry VT U.S.A. 64 E1
Londonderry, Cape Australia 54 D3
Londrina Brazil 71 A3
Lone Pine CA U.S.A. 65 C2
Longa Angola 49 B5
Longa, Proliv sea chan. Rus. Fed. 29 S2
Long Ashton U.K. 19 E7
Longbao China 37 I2
Long Bay U.S.A. 63 L5
Longbeach N.Z. 59 C7
Long Beach CA U.S.A. 65 C4
Long Branch NJ U.S.A. 64 E2
Long Eaton U.K. 19 F6
Longford Ireland 21 E4
Longhoughton U.K. 18 F3
Longhurst, Mount Antarctica 76 H1
Long Island Bahamas 67 I4
Long Island P.N.G. 52 E2
Long Island NY U.S.A. 64 E2
Long Island Sound sea chan. Connecticut/New York U.S.A. 64 E2
Longjiang China 44 A3
Longlac Canada 63 J2
Longmeadow MA U.S.A. 64 E1
Long Melford U.K. 19 H6
Longmen Heilong. China 44 B2
Longmont U.S.A. 62 F3
Longnan China 42 I6
Long Point Ont. Canada 64 A1
Long Point Ont. Canada 64 A1
Long Point N.Z. 59 B8
Long Point Bay Ont. Canada 64 A1
Long Preston U.K. 18 E4
Long Range Mountains Nfld. and Lab. Canada 61 M5
Longreach Australia 56 D4
Long Stratton U.K. 19 I6
Longtown U.K. 18 E3
Longuyon France 24 G2
Longview TX U.S.A. 63 I5
Longwangmiao China 44 D3
Longwei Co l. China 37 G2
Longxi China 42 I6
Longxingchang China see **Wuyuan**
Long Xuyên Vietnam 31 J5
Longyan China 43 L7
Longyearbyen Svalbard 28 C2
Longzhen China 44 B2
Lönsboda Sweden 15 I8
Lons-le-Saunier France 24 G3
Lonton Myanmar 37 I4
Loochoo Islands Japan see **Ryukyu Islands**
Lookout, Cape U.S.A. 63 L5
Lookout, Point Australia 58 F1
Lookout Point Australia 55 B8
Loolmalasin vol. crater Tanz. 48 D4
Loongana Australia 55 D7
Loop Head hd Ireland 21 C5
Lop China 36 E1
Lopasnya Rus. Fed. see **Chekhov**
Lopatina, Gora mt. Rus. Fed. 44 F2
Lop Buri Thai. 31 J5
Lopez, Cap c. Gabon 48 A4
Lop Nur salt flat China 42 G4
Lopphavet b. Norway 14 L1
Loptyuga Rus. Fed. 12 K3
Lora r. Venez. 68 D2
Lora del Río Spain 25 D5
Lorain U.S.A. 61 J5
Loralai Pak. 33 K3
Loralai r. Pak. 36 B3
Lorca Spain 25 F5
Lordegān Iran 35 H5
Lord Howe Atoll Solomon Is see **Ontong Java Atoll**
Lord Howe Island Australia 53 F5
Lord Howe Rise sea feature S. Pacific Ocean 74 G7
Lordsburg U.S.A. 62 F5
Lore East Timor 54 D2
Lorena Brazil 71 B3
Loreto Brazil 69 I5
Loreto Mex. 62 E6
Lorient France 24 C3
Lorn, Firth of est. U.K. 20 D4
Lorne Australia 56 D5
Lorne watercourse Australia 56 B3
Lorrain, Plateau France 24 H2
Loraine Australia 56 B3
Lorraine reg. France 24 G2
Losal India 36 C4
Los Alamos CA U.S.A. 65 B3
Los Alamos NM U.S.A. 62 F4
Los Alerces, Parque Nacional nat. park Arg. 70 B6
Los Ángeles Chile 70 B5
Los Angeles CA U.S.A. 65 C3
Los Angeles Aqueduct canal CA U.S.A. 65 C3
Los Banos CA U.S.A. 65 B2
Los Blancos Arg. 70 D2
Los Canarreos, Archipiélago de is Cuba 67 H4
Los Chonos, Archipiélago de is Chile 70 A6
Los Estados, Isla de i. Arg. 70 D8
Los Glaciares, Parque Nacional nat. park Arg. 70 B8
Lošinj i. Croatia 26 F2
Los Jardines de la Reina, Archipiélago de is Cuba 67 I4
Los Juríes Arg. 70 D3
Los Katios, Parque Nacional nat. park Col. 67 I7
Loskop Dam S. Africa 51 I3
Los Menucos Arg. 70 C6
Los Mochis Mex. 66 C3
Losombo Dem. Rep. Congo 48 B3
Los Picos de Europa, Parque Nacional de nat. park Spain 25 D2
Los Roques, Islas is Venez. 68 E1
Lossie r. U.K. 20 F3
Lossiemouth U.K. 20 F3
Lost Creek WV U.S.A. 64 A3
Los Teques Venez. 68 E1
Los Testigos is Venez. 68 F1
Lost Hills CA U.S.A. 65 C3
Lostwithiel U.K. 19 C8
Los Vilos Chile 70 B4
Lot r. France 24 E4
Lota Chile 70 B5
Lothringen reg. France see **Lorraine**
Lotikipi Plain Kenya/Sudan 48 D3
Loto Dem. Rep. Congo 48 C4
Lotsane r. Botswana 51 I2
Lot's Wife i. Japan see **Sōfu-gan**
Lotta r. Fin./Rus. Fed. 14 Q2
Louangnamtha Laos 40 C5
Louangphabang Laos 41 C6
Loubomo Congo 49 B4
Loudéac France 24 C2
Louga Senegal 46 B3
Loughborough U.K. 19 F6
Lougheed Island Canada 61 H2
Loughor r. U.K. 19 C7
Loughrea Ireland 21 D4
Loughton U.K. 19 H7
Louhans France 24 G3
Louisa VA U.S.A. 64 C3
Louisburgh Ireland 21 C4
Louis-Gentil Morocco see **Youssoufia**
Louisiade Archipelago is P.N.G. 52 F3
Louisiana state U.S.A. 63 I5
Louis Trichardt S. Africa see **Makhado**
Louisville KY U.S.A. 63 J4
Louisville Ridge sea feature S. Pacific Ocean 74 I8
Loukhi Rus. Fed. 12 G2
Loukoléla Congo 48 B4
Loukouo Congo 47 E5
Loulé Port. 25 B5
Loum Cameroon 46 D4
Louny Czech Rep. 17 N5
Loups Marins, Lacs des lakes Canada 61 K4
Lourdes France 24 D5
Lourenço Marques Moz. see **Maputo**
Lousã Port. 25 B3
Loushan China 44 C3
Louth Australia 58 B3
Louth U.K. 18 G5
Loutra Aidipsou Greece 27 J5
Louvain Belgium see **Leuven**
Louviers France 19 I9
Louwater-Suid Namibia 50 C2
Louwsburg S. Africa 51 J4
Lövånger Sweden 14 L4
Lovat' r. Rus. Fed. 12 F4
Lovech Bulg. 27 K3
Lovelock U.S.A. 62 D3
Loviisa Fin. 15 O6
Lovington U.S.A. 62 G5
Lovozero Rus. Fed. 12 G1
Lóvua Angola 49 C4
Lóvua Angola 49 C5
Low, Cape Canada 61 J3
Lowa Dem. Rep. Congo 48 C4
Lowa r. Dem. Rep. Congo 48 C4
Lowarai Pass Pak. 36 B2
Lowell MA U.S.A. 64 F1
Lower California pen. Mex. see **Baja California**
Lower Glenelg National Park Australia 57 C8
Lower Granite Gorge AZ U.S.A. 65 F3
Lower Hutt N.Z. 59 E5
Lower Lake CA U.S.A. 65 A1
Lower Lough Erne l. U.K. 21 E3
Lower Tunguska r. Rus. Fed. see **Nizhnyaya Tunguska**
Lower Zambezi National Park Zambia 49 C5
Lowestoft U.K. 19 I6
Łowicz Poland 17 Q4
Low Island Kiribati see **Starbuck Island**
Lowther Hills U.K. 20 F5
Lowville U.S.A. 63 L3
Loxton Australia 57 C7
Loyal, Loch l. U.K. 20 E2
Loyalsock Creek r. PA U.S.A. 64 C2
Loyalty Islands New Caledonia see **Loyauté, Îles**
Loyang China see **Luoyang**
Loyauté, Îles is New Caledonia 53 G4
Loyev Belarus see **Loyew**
Loyew Belarus 13 F6
Lozère, Mont mt. France 24 F4
Loznica Serbia 27 H2
Lozova Ukr. 13 H6
Lozovaya Ukr. see **Lozova**
Lua r. Dem. Rep. Congo 48 B3
Luacano Angola 49 C5
Lu'an China 43 L6
Luanda Angola 49 B4
Luang Namtha Laos see **Louangnamtha**
Luang Prabang Laos see **Louangphabang**
Luanhaizi China 37 H2
Luanshya Zambia 49 C5
Luanza Dem. Rep. Congo 49 C4
Luao Angola see **Luau**
Luarca Spain 25 C2
Luashi Dem. Rep. Congo 49 C5
Luau Angola 49 C5
Luba Equat. Guinea 46 D4
Lubaczów Poland 13 D6
Lubalo Angola 49 B4
Lubānas ezers l. Latvia 15 O8
Lubango Angola 49 B5
Lubao Dem. Rep. Congo 49 C4
Lubartów Poland 13 D6
Lubbeskolk salt pan S. Africa 50 D5
Lubbock U.S.A. 62 G5
Lübeck Germany 17 M4
Lubefu Dem. Rep. Congo 49 C4
Lubei China 43 M4
Lüben Poland see **Lubin**
Lubersac France 24 E4
Lubin Poland 17 P5
Lublin Poland 13 D6
Lubnān country Asia see **Lebanon**
Lubnān, Jabal mts Lebanon see **Liban, Jebel**
Lubny Ukr. 13 G6
Lubok Antu Sarawak Malaysia 41 D7
Lubudi Dem. Rep. Congo 49 C4
Lubumbashi Dem. Rep. Congo 49 C5
Lubutu Dem. Rep. Congo 48 C4
Lucala An12gola 49 B4
Lucan Ont. Canada 64 A1
Lucan Ireland 21 F4
Lucapa Angola 49 C4
Lucca Italy 26 D3
Luce Bay U.K. 20 E6
Lucélia Brazil 71 A3
Lucena Phil. 41 E6
Lucena Spain 25 D5
Lučenec Slovakia 17 Q6
Lucera Italy 26 F4
Lucerne Switz. 24 I3
Lucerne Valley CA U.S.A. 65 D3
Luchegorsk Rus. Fed. 44 D3
Lucheng Sichuan China see **Kangding**
Łuck Ukr. see **Luts'k**
Luckeesarai India see **Lakhisarai**
Luckenwalde Germany 17 N4
Luckhoff S. Africa 50 G5
Lucknow India 36 E4
Lucrecia, Cabo c. Cuba 67 I4
Lucusse Angola 49 C5
Lucy Creek Australia 56 B4
Lüda China see **Dalian**
Lüdenscheid Germany 17 K5
Lüderitz Namibia 50 B4
Ludewa Tanz. 49 D4
Ludhiana India 36 C3
Ludlow U.K. 19 E6
Ludlow CA U.S.A. 65 D3
Ludogorie reg. Bulg. 27 L3
Ludvika Sweden 15 I6
Ludwigsburg Germany 17 L6
Ludwigshafen am Rhein Germany 17 L6
Ludwigslust Germany 17 M4
Ludza Latvia 15 O8
Luebo Dem. Rep. Congo 49 C4
Luena Angola 49 B5
Luena Flats plain Zambia 49 C5
Lufeng Guangdong China 43 L8
Lufkin U.S.A. 63 I5
Luga Rus. Fed. 12 F4
Luga r. Rus. Fed. 15 P7
Lugano Switz. 24 I3
Lugansk Ukr. see **Luhans'k**
Lugdunum France see **Lyon**
Lugg r. U.K. 19 E6
Luggudontsen mt. China 37 G3
Lugnaquilla h. Ireland 21 F5
Lugo Italy 26 D2
Lugo Spain 25 C2
Lugoj Romania 27 I2
Luhans'k Ukr. 13 H6
Luḥfī, Wādī watercourse Jordan 39 C3
Luhit r. China/India see **Zayü Qu**
Luhit r. India 37 H4
Luhuo China 42 I6
Luhyny Ukr. 13 F6
Luia Angola 49 C4
Luiana Angola 49 C5
Luichow Peninsula China see **Leizhou Bandao**
Luik Belgium see **Liège**
Luimneach Ireland see **Limerick**
Luiro r. Fin. 14 O3
Luis Echeverría Álvarez Baja California Mex. 65 D4
Luitpold Coast Antarctica 76 A1
Luiza Dem. Rep. Congo 49 C4
Lukachek Rus. Fed. 44 D1
Lukapa Angola see **Lucapa**
Lukavac Bos.-Herz. 26 H2
Lukenga, Lac l. Dem. Rep. Congo 49 C4
Lukenie r. Dem. Rep. Congo 48 B4
Lukh r. Rus. Fed. 12 I4
Lukhovitsy Rus. Fed. 13 H5
Lukovit Bulg. 27 K3
Łuków Poland 13 D6
Lukoyanov Rus. Fed. 13 J5
Lukusuzi National Park Zambia 49 D5
Luleå Sweden 14 M4
Luleälven r. Sweden 14 M4
Lüleburgaz Turkey 27 L4
Lüliang Shanxi China 43 K5
Lüliang Shan mts China 43 K5
Lulimba Dem. Rep. Congo 49 C4
Lulonga r. Dem. Rep. Congo 48 B3
Luluabourg Dem. Rep. Congo see **Kananga**
Lülung China 37 F3
Lumachomo China 37 F3
Lumajangdong Co salt l. China 36 E2
Lumbala Moxico Angola see **Lumbala N'guimbo**
Lumbala Moxico Angola see **Lumbala Kaquengue**
Lumbala Kaquengue Angola 49 C5
Lumbala N'guimbo Angola 49 C5
Lumberton U.S.A. 63 L5
Lumbini Nepal 37 E4
Lumbrales Spain 25 C3
Lumezzane Italy 26 D2
Lumsden Canada 62 G1
Lumsden N.Z. 59 B7
Lunan Bay U.K. 20 G4
Lund Sweden 15 H9
Lund NV U.S.A. 65 E1
Lund UT U.S.A. 65 F1
Lundy U.K. 19 C7
Lune r. U.K. 18 E4
Lüneburg Germany 17 M4
Lunenburg VA U.S.A. 64 B4
Lunéville France 24 H2
Lunga r. Zambia 49 C5
Lungdo China 37 E2
Lunggar China 37 E3
Lungleh India see **Lunglei**
Lunglei India 37 H5
Lungmari mt. China 37 F3
Lungmu Co salt l. China 36 E2
Lungwebungu r. Zambia 49 C5
Lunh Nepal 37 E3
Luni India 36 C4
Luni r. India 36 B4
Luni r. Pak. 36 B3
Luninets Belarus see **Luninyets**
Luning NV U.S.A. 65 C1
Luninyets Belarus 15 O10
Lunkaransar India 36 C3
Lunkha India 36 C3
Lunsar Sierra Leone 46 B4
Lunsklip S. Africa 51 I3
Luntai China 42 E4
Luobei China 44 C3
Luobuzhuang China 42 F5
Luohe China 43 K6
Luoto Fin. see **Larsmo**
Luoyang Henan China 43 K6
Luozigou China 44 C4
Lupane Zimbabwe 49 C5
Lupanshui China see **Liupanshui**
L'Upemba, Parc National de nat. park Dem. Rep. Congo 49 C4
Lupeni Romania 27 J2
Lupilichi Moz. 49 D5
Luray VA U.S.A. 64 B3
Luremo Angola 49 B4
Lurgan U.K. 21 F3
Luring China see **Oma**
Lúrio Moz. 49 E5
Lurio r. Moz. 49 E5
Lusaka Zambia 49 C5
Lusambo Dem. Rep. Congo 49 C4
Lusancay Islands and Reefs P.N.G. 52 F2
Lusangi Dem. Rep. Congo 49 C4
Lush, Mount h. Australia 54 D4
Lushnja Albania see **Lushnjë**
Lushnjë Albania 27 H4
Lushuihe China 44 B4
Lusikisiki S. Africa 51 I6
Lusk U.S.A. 62 G3
Luso Angola see **Luena**
Lussvale Australia 58 C1
Lut, Bahrat salt l. Asia see **Dead Sea**
Lūt, Dasht-e des. Iran see **Kavīr-e Lūt**
Lutetia France see **Paris**
Lutherstadt Wittenberg Germany 17 N5
Luton U.K. 19 G7
Łutselk'e Canada 60 G3
Luts'k Ukr. 13 E6
Lutto r. Fin./Rus. Fed. see **Lotta**
Lützow-Holm Bay Antarctica 76 D2
Lutzputs S. Africa 50 E5
Lutzville S. Africa 50 D6
Luumäki Fin. 15 O6
Luuq Somalia 48 E3
Luverne MN U.S.A. 63 H3
Luvuei Angola 49 C5
Luvuvhu r. S. Africa 51 J2
Luwero Uganda 48 D3
Luwingu Zambia 49 C5
Luwuk Indon. 41 E8
Luxembourg country Europe 17 K6
Luxembourg Lux. 17 K6
Luxemburg country Europe see **Luxembourg**
Luxeuil-les-Bains France 24 H3
Luxi Yunnan China 42 H8
Luxolweni S. Africa 51 G6
Luxor Egypt 32 D4
Luza Rus. Fed. 12 J3
Luza r. Rus. Fed. 12 J3
Luza r. Rus. Fed. 12 M2
Luzern Switz. see **Lucerne**
Luzhou China 42 J7
Luziânia Brazil 71 B2
Luzon i. Phil. 41 E6
Luzon Strait Phil. 41 E5
Luzy France 24 F3
L'viv Ukr. 13 E6
L'vov Ukr. see **L'viv**
Lwów Ukr. see **L'viv**
Lyady Rus. Fed. 15 P7
Lyakhavichy Belarus 15 O10
Lyakhovichi Belarus see **Lyakhavichy**
Lyallpur Pak. see **Faisalabad**
Lyamtsa Rus. Fed. 12 H2
Lycia reg. Turkey 27 M6
Lyck Poland see **Ełk**
Lycksele Sweden 14 K4
Lycopolis Egypt see **Asyūṭ**
Lydd U.K. 19 H8
Lydda Israel see **Lod**
Lyddan Island Antarctica 76 B2
Lydia reg. Turkey 27 L5
Lydney U.K. 19 E7
Lyel'chytsy Belarus 13 F6
Lyell, Mount CA U.S.A. 65 C2
Lyell Brown, Mount h. Australia 55 E5
Lyepyel' Belarus 15 P9
Lykens PA U.S.A. 64 C2
Lyme Bay U.K. 19 E8
Lyme Regis U.K. 19 E8
Lymington U.K. 19 F8
Lynchburg VA U.S.A. 64 B4
Lyndhurst N.S.W. Australia 58 D4
Lyndhurst Qld Australia 56 D3
Lyndhurst S.A. Australia 57 B6
Lyndon Australia 55 A5
Lyndon r. Australia 55 A5
Lyne r. U.K. 18 D4
Lyness U.K. 20 F2
Lyngdal Norway 15 E7
Lynn U.K. see **King's Lynn**
Lynn MA U.S.A. 64 F1
Lynn Lake Canada 61 H4
Lynton U.K. 19 D7
Lynx Lake Canada 60 H3
Lyon France 24 G4
Lyon r. U.K. 20 F4
Lyons Australia 55 F7
Lyons France see **Lyon**
Lyons NY U.S.A. 64 C1
Lyons Falls NY U.S.A. 64 D1
Lyozna Belarus 13 F5
Lyra Reef P.N.G. 52 F2
Lysekil Sweden 15 G7
Lyskovo Rus. Fed. 12 J4
Lys'va Rus. Fed. 11 R4
Lysychans'k Ukr. 13 H6
Lysyye Gory Rus. Fed. 13 J6
Lytham St Anne's U.K. 18 D5
Lyuban' Belarus 15 P10
Lyubertsy Rus. Fed. 11 N4
Lyubeshiv Ukr. 13 E6
Lyubim Rus. Fed. 12 I4
Lyubytino Rus. Fed. 12 G4
Lyudinovo Rus. Fed. 13 G5
Lyunda r. Rus. Fed. 12 J4
Lyzha r. Rus. Fed. 12 M2

M

Ma'agan Israel 39 B3
Maale Maldives see **Male**
Maale Atholhu atoll Maldives see **Male Atoll**
Maalhosmadulu Atholhu Uthuruburi atoll Maldives see **North Maalhosmadulu Atoll**
Maalhosmadulu Atoll Maldives 38 B5
Ma'ān Jordan 39 B4
Maan Turkey see **Nusratiye**
Maaninka Fin. 14 O5
Maaninkavaara Fin. 14 P3
Maardu Estonia 15 N7
Maarianhamina Fin. see **Mariehamn**
Ma'arrat an Nu'mān Syria 39 C2
Maastricht Neth. 17 J5
Maaza Plateau Egypt 34 C6
Mabalane Moz. 51 K2
Mabana Dem. Rep. Congo 48 C3
Mabaruma Guyana 68 G2
Mabein Myanmar 37 I5
Mabel Creek Australia 55 F7
Mabel Downs Australia 54 D4
Mablethorpe U.K. 18 H5
Mabopane S. Africa 51 I3
Mabote Moz. 51 L2
Mabrak, Jabal mt. Jordan 39 B4
Mabuasehube Game Reserve nature res. Botswana 50 F3
Mabule Botswana 50 G3
Mabutsane Botswana 50 F3
Macá, Monte mt. Chile 70 B7
Macadam Plains Australia 55 B6
Macaé Brazil 71 C3
Macajuba Brazil 71 C1
Macaloge Moz. 49 D5
MacAlpine Lake Canada 61 H3
Macandze Moz. 51 K2
Macao China 43 K8
Macapá Brazil 69 H3
Macará Ecuador 68 C4
Macarani Brazil 71 C1
Macas Ecuador 68 C4
Macassar Indon. see **Makassar**
Macau Brazil 69 K5
Macaúba Brazil 69 H6
Macauley Island N.Z. 53 I5
Maccaretane Moz. 51 K3
Macclesfield U.K. 18 E5
Macdonald, Lake imp. l. Australia 55 E5
Macdonald Range hills Australia 54 D3
Macdonnell Ranges mts Australia 55 E5
MacDowell Lake Canada 63 I1
Macduff U.K. 20 G3
Macedo de Cavaleiros Port. 25 C3
Macedon mt. Australia 58 B6
Macedon country Europe see **Macedonia**
Macedonia country Europe 27 I4
Maceió Brazil 69 K5
Macenta Guinea 46 C4
Macerata Italy 26 E3
Macfarlane, Lake imp. l. Australia 57 B7
Macgillycuddy's Reeks mts Ireland 21 C6
Machachi Ecuador 68 C4
Machaila Moz. 51 K2
Machakos Kenya 48 D4
Machala Ecuador 68 C4
Machali China see **Madoi**
Machanga Moz. 49 D6
Machar Marshes South Sudan 32 D8
Machattie, Lake imp. l. Australia 56 B5
Machatuine Moz. 51 K3
Machaze Moz. see **Chitobe**
Macherla India 38 C2
Machhagan India 37 F5
Machias ME U.S.A. 63 N3
Machias NY U.S.A. 64 B1
Machilipatnam India 38 D2
Machiques Venez. 68 D1
Machrihanish U.K. 20 D5
Machu Picchu tourist site Peru 68 D6
Machynlleth U.K. 19 D6
Macia Moz. 51 K3
Macias Nguema i. Equat. Guinea see **Bioco**
Măcin Romania 27 M2
Macintyre r. Australia 58 E2
Macintyre Brook r. Australia 58 E2
Maçka Turkey 35 E2
Mackay Australia 56 E4
Mackay, Lake imp. l. Australia 54 E5
MacKay Lake Canada 60 G3
Mackenzie r. Australia 56 E4
Mackenzie Canada 60 F4
Mackenzie r. Canada 60 E3
Mackenzie Guyana see **Linden**
Mackenzie atoll Micronesia see **Ulithi**
Mackenzie Bay Antarctica 76 E2
Mackenzie Bay Canada 60 E3
Mackenzie King Island Canada 61 G2
Mackenzie Mountains Canada 60 E3
Mackenzie-Peace-Finlay r. Canada 60 E3
Mackillop, Lake imp. l. Australia see **Yamma Yamma, Lake**
Mackintosh Range hills Australia 55 D6
Macklin Canada 62 F1
Macksville Australia 58 F3
Maclean Australia 58 F2
Maclear S. Africa 51 I6
MacLeod Canada see **Fort Macleod**
MacLeod, Lake dry lake Australia 55 A6
Macmillan Pass Canada 60 F3
Macomb U.S.A. 63 I3
Macomer Sardegna Italy 26 C4
Mâcon France 24 G3
Macon GA U.S.A. 63 K5
Macon MO U.S.A. 63 I4
Macondo Angola 49 C5
Macpherson Robertson Land reg. Antarctica see **Mac. Robertson Land**
Macquarie r. Australia 58 C3
Macquarie, Lake b. Australia 58 E4
Macquarie Island S. Pacific Ocean 74 G9
Macquarie Marshes Australia 58 C3
Macquarie Mountain Australia 58 D4
Macquarie Ridge sea feature S. Pacific Ocean 74 G9
Mac. Robertson Land reg. Antarctica 76 E2
Macroom Ireland 21 D6
Macumba Australia 57 A5
Macumba watercourse Australia 57 B5
Macuzari, Presa resr Mex. 66 C3
Mādabā Jordan 39 B4
Madadeni S. Africa 51 J4
Madagascar country Africa 49 E6
Madagascar Basin sea feature Indian Ocean 73 L7
Madagascar Ridge sea feature Indian Ocean 73 K8
Madagasikara country Africa see **Madagascar**
Madakasira India 38 C3
Madama Niger 47 E2
Madan Bulg. 27 K4
Madanapalle India 38 C3
Madang P.N.G. 52 E2
Madaoua Niger 46 D3
Madaripur Bangl. 37 G5
Madau Turkm. see **Madaw**
Madaw Turkm. 35 I3
Madaya Myanmar 37 I5
Madded India 38 D2
Madeira r. Brazil 68 G4
Madeira terr. N. Atlantic Ocean 46 B1
Madeira, Arquipélago da terr. N. Atlantic Ocean see **Madeira**
Maden Turkey 35 E3
Madera Mex. 66 C3
Madera CA U.S.A. 65 B2
Madgaon India 38 B3
Madha India 38 B2
Madhavpur India 36 B5
Madhepura India 37 F4
Madhipura India see **Madhepura**
Madhubani India 37 F4
Madhya Pradesh state India 36 D5
Madibogo S. Africa 51 G4
Madidi r. Bol. 68 E6
Madikeri India 38 B3
Madikwe Game Reserve nature res. S. Africa 51 H3
Madīnat ath Thawrah Syria 39 D2
Madingo-Kayes Congo 49 B4
Madingou Congo 49 B4
Madison IN U.S.A. 63 J4
Madison SD U.S.A. 63 H3
Madison VA U.S.A. 64 B3
Madison WI U.S.A. 63 J3
Madison r. U.S.A. 62 E2
Madison Heights VA U.S.A. 64 B4
Madisonville KY U.S.A. 63 J4
Madley, Mount h. Australia 55 C6
Mado Gashi Kenya 48 D3
Madoi China 37 I2
Madona Latvia 15 O8
Madpura India 36 B4
Madra Dağı mts Turkey 27 L5
Madrakah Saudi Arabia 32 E5
Madrakah, Ra's c. Oman 33 I6
Madras India see **Chennai**
Madras state India see **Tamil Nadu**
Madras U.S.A. 62 C3
Madre, Laguna lag. Mex. 66 E3
Madre de Dios r. Peru 68 E6
Madre de Dios, Isla i. Chile 70 A8
Madre del Sur, Sierra mts Mex. 66 D5
Madre Occidental, Sierra mts Mex. 66 C
Madre Oriental, Sierra mts Mex. 66 D3
Madrid Spain 25 E3
Madridejos Spain 25 E4
Madugula India 38 D2
Madura i. Indon. 41 D8
Madurai India 38 C4
Madurantakam India 38 C3
Madvār, Kūh-e mt. Iran 35 I5
Madwas India 37 E4
Maé i. Vanuatu see **Émaé**
Maebashi Japan 45 E5
Mae Hong Son Thai. 42 H9
Mae Sai Thai. 42 H8
Mae Sariang Thai. 42 H9
Maevatanana Madag. 49 E5
Maéwo i. Vanuatu 53 G3
Mafadi mt. S. Africa 50 I5
Mafeking S. Africa see **Mafikeng**
Mafeteng Lesotho 51 H5
Maffra Australia 58 C6
Mafia Island Tanz. 49 D4
Mafikeng S. Africa 51 G3
Mafinga Tanz. 49 D4
Mafra Brazil 71 A4
Mafraq Jordan see **Al Mafraq**
Magabeni S. Africa 51 J6
Magadan Rus. Fed. 29 Q4
Magadi Kenya 48 D4
Magaiza Moz. 51 K2
Magallanes Chile see **Punta Arenas**
Magallanes, Estrecho de Chile see **Magellan, Strait of**
Magangué Col. 68 D2
Mağara Dağı mt. Turkey 39 A1
Magaramkent Rus. Fed. 35 H2
Magaria Niger 46 D3
Magarida P.N.G. 56 E1
Magas Rus. Fed. 13 J8
Magdagachi Rus. Fed. 44 B1
Magdalena Bol. 68 F6
Magdalena r. Col. 68 D1
Magdalena Sonora Mex. 66 B2
Magdalena, Bahía b. Mex. 66 B4
Magdalena, Isla i. Chile 70 B6
Magdeburg Germany 17 M4
Magdelaine Cays atoll Australia 56 E3
Magellan, Strait of Chile 70 B8
Magellan Seamounts sea feature N. Pacific Ocean 74 F4
Magenta, Lake imp. l. Australia 55 B8
Magerøya i. Norway 14 N1
Maggiorasca, Monte mt. Italy 26 C2
Maggiore, Lago Italy see **Maggiore, Lake**
Maggiore, Lake Italy 26 C2
Maghāgha Egypt see **Maghāghah**
Maghāghah Egypt 34 C5
Maghama Mauritania 46 B3
Maghâra, Gebel h. Egypt see **Maghārah, Jabal**
Maghārah, Jabal h. Egypt 39 A4
Maghera U.K. 21 F3
Magherafelt U.K. 21 F3
Maghnia Alg. 25 F6
Maghull U.K. 18 E5
Magilligan Point U.K. 21 F2
Magna Grande mt. Sicilia Italy 26 F6
Magnetic Island Australia 56 D3
Magnetic Passage Australia 56 D3
Magnetity Rus. Fed. 14 R2
Magnitogorsk Rus. Fed. 28 G4
Magnolia AR U.S.A. 63 I5
Mago Rus. Fed. 44 F1
Màgoé Moz. 49 D5
Mago National Park Eth. 48 D3
Magosa Cyprus see **Famagusta**
Magpie, Lac l. Canada 63 O1
Magta' Lahjar Mauritania 46 B3
Magu Tanz. 48 D4
Magu, Khrebet mts Rus. Fed. 44 E1
Magude Moz. 51 K3
Magura Bangl. 37 G5
Magway Myanmar see **Magwe**
Magwe Myanmar 37 H5
Magyar Köztársaság country Europe see **Hungary**
Magyichaung Myanmar 37 H5
Mahābād Iran 35 G3
Mahabharat Range mts Nepal 37 F4
Mahaboobnagar India see **Mahbubnaga**
Mahad India 38 B2
Mahadeo Hills India 36 D5
Mahaffey PA U.S.A. 64 B2
Mahajan India 36 C3
Mahajanga Madag. 49 E5
Mahalapye Botswana 51 H2
Mahale Mountains National Park Tanz. 49 C4
Mahalevona Madag. 49 E5
Maḩallāt-e Bālā Iran 35 H4
Māhān Iran 33 I3
Mahanadi r. India 38 E1
Mahanoro Madag. 49 E5
Maha Oya Sri Lanka 38 D5
Maharashtra state India 38 B2
Maha Sarakham Thai. 31 J5
Mahasham, Wâdi el watercourse Egypt s **Muhashsham, Wādī al**
Mahbubabad India 38 D2
Mahbubnagar India 38 C2
Mahd adh Dhahab Saudi Arabia 32 F5
Mahdia Alg. 25 G6
Mahdia Guyana 69 G2
Mahdia Tunisia 26 D7
Mahé i. Seychelles 73 L6
Mahendragiri mt. India 38 E2
Mahenge Tanz. 49 D4
Mahesana India 36 C5
Mahi r. India 36 C5
Mahia Peninsula N.Z. 59 F4

Mahilyow Belarus 13 F5
Mahim India 38 B2
Mah Jān Iran 35 I5
Mahlabatini S. Africa 51 J5
Maḩmūdābād Iran 35 I3
Maḩmūd-e 'Erāqī Afgh. see Maḩmūd-e Rāqī
Maḩmūd-e Rāqī Afgh. 36 B2
Maho Sri Lanka 38 D5
Mahoba India 36 D4
Maholi India 36 E4
Mahón Spain 25 I4
Mahrauni India 36 D4
Mahrès Tunisia 26 D7
Mahrūd Iran 33 J3
Mahsana India see Mahesana
Mahudaung mts Myanmar 37 H5
Mahur India 38 C2
Mahuva India 36 B5
Mahwa India 36 D4
Mahya Dağı mt. Turkey 27 L4
Mai i. Vanuatu see Émaé
Maiaia Moz. see Nacala
Maicao Col. 68 D1
Maidenhead U.K. 19 G7
Maidstone Canada 62 F1
Maidstone U.K. 19 H7
Maiduguri Nigeria 46 E3
Maiella, Parco Nazionale della nat. park Italy 26 F3
Mai Gudo mt. Eth. 48 D3
Maigue r. Ireland 21 D5
Maihar India 36 E4
Maikala Range hills India 36 E5
Maiko r. Dem. Rep. Congo 48 C3
Mailan Hill mt. India 37 E5
Mailsi Pak. 36 C3
Main r. U.K. 21 F3
Maindargi India 38 C2
Mai-Ndombe, Lac l. Dem. Rep. Congo 48 B4
Maindong Xizang China see Coqên
Maine state U.S.A. 63 N2
Maine, Gulf of Canada/U.S.A. 61 L5
Mainé Hanari, Cerro h. Col. 68 D4
Maïné-Soroa Niger 46 E3
Maingkaing Myanmar 37 H4
Maingkwan Myanmar 37 I4
Mainland i. Scotland U.K. 20 F1
Mainland i. Scotland U.K. 20 [inset]
Mainoru Australia 54 F3
Mainpat reg. India 37 E5
Mainpuri India 36 D4
Main Range National Park Australia 58 F2
Maintirano Madag. 49 E5
Mainz Germany 17 L5
Maio i. Cape Verde 46 [inset]
Maipú Arg. 70 E5
Maiskhal Island Bangl. 37 G5
Maitengwe Botswana 49 C6
Maitland N.S.W. Australia 58 E4
Maitland S.A. Australia 57 B7
Maitland r. Australia 54 B5
Maitri research stn Antarctica 76 C2
Maiwo i. Vanuatu see Maéwo
Maiyu, Mount h. Australia 54 E4
Maíz, Islas del is Nicaragua 67 H6
Maizar Pak. 36 B2
Maizuru Japan 45 D6
Maja Jezercë mt. Albania 27 H3
Majdel Aanjar tourist site Lebanon 39 B3
Majene Indon. 41 D8
Majī Eth. 48 D3
Majiazi China 44 B2
Majōl country N. Pacific Ocean see Marshall Islands
Major, Puig mt. Spain 25 H4
Majorca i. Spain 25 H4
Majro atoll Marshall Is see Majuro
Majunga Madag. see Mahajanga
Majuro atoll Marshall Is 74 H5
Majwemasweu S. Africa 51 H5
Makabana Congo 48 B4
Makale Indon. 41 D8
Makalu mt. China/Nepal 37 F4
Makalu Barun National Park Nepal 37 F4
Makanchi Kazakh. 42 E3
Makanpur India 36 E4
Makari Mountain National Park Tanz. see Mahale Mountains National Park
Makarov Rus. Fed. 44 F2
Makarov Basin sea feature Arctic Ocean 77 B1
Makarska Croatia 26 G3
Makar'ye Rus. Fed. 12 K4
Makar'yev Rus. Fed. 12 I4
Makasar, Selat str. Indon. see Makassar, Selat
Makassar Indon. 41 D8
Makassar, Selat str. Indon. 41 D8
Makassar Strait Indon. see Makassar, Selat
Makat Kazakh. 30 E2
Makatini Flats lowland S. Africa 51 K4
Makedonija country Europe see Macedonia
Makeni Sierra Leone 46 B4
Makete Tanz. 49 D4
Makeyevka Ukr. see Makiyivka
Makgadikgadi depr. Botswana 49 C6
Makgadikgadi Pans National Park Botswana 49 C6
Makhachkala Rus. Fed. 13 J8
Makhad Pak. 36 B2
Makhado S. Africa 51 I2
Makhāzin, Kathīb al des. Egypt 39 A4
Makhâzin, Kathîb el des. Egypt see Makhāzin, Kathīb al
Makhazine, Barrage El dam Morocco 25 D6
Makhmūr Iraq 35 F4
Makhtal India 38 C2
Makin atoll Kiribati see Butaritari
Makindu Kenya 48 D4
Makinsk Kazakh. 31 G1
Makira i. Solomon Is see San Cristobal
Makiyivka Ukr. 13 H6
Makkah Saudi Arabia see Mecca
Makkovik Canada 61 M4
Makó Hungary 27 I1
Makokou Gabon 48 B3
Makopong Botswana 50 F3
Makotipoko Congo 47 E5
Makran reg. Iran/Pak. 33 J4
Makrana India 36 C4
Makran Coast Range mts Pak. 33 J4
Makri India 38 D2
Maksatikha Rus. Fed. 12 G4
Maksi India 36 D5
Maksimovka Rus. Fed. 44 E3
Maksudangarh India 36 D5
Mākū Iran 35 G3
Makunguwiro Tanz. 49 D5
Makurdi Nigeria 46 D4
Makwassie S. Africa 51 G4
Mal India 37 G4
Mala Ireland see Mallow
Mala i. Solomon Is see Malaita
Malå Sweden 14 K4
Mala, Punta pt Panama 67 H7
Malabar Coast India 38 B3
Malabo Equat. Guinea 46 D4
Malaca Spain see Málaga
Malacca Malaysia see Melaka
Malacca, Strait of Indon./Malaysia 41 B7
Malad City U.S.A. 62 E3
Maladzyechna Belarus 15 O9
Malá Fatra nat. park Slovakia 17 Q6
Málaga Spain 25 D5
Malagasy Republic country Africa see Madagascar
Málainn Mhóir Ireland 21 D3
Malaita i. Solomon Is 53 G2
Malakal South Sudan 32 D8
Malakanagiri India see Malkangiri
Malakheti Nepal 36 E3
Malakula i. Vanuatu 53 G3
Malang Indon. 41 D8
Malangana Nepal see Malangwa
Malange Angola see Malanje
Malangwa Nepal 37 F4
Malanje Angola 49 B4
Malappuram India 38 C4
Mälaren l. Sweden 15 J7
Malargüe Arg. 70 C5
Malatya Turkey 34 E3
Malavalli India 38 C3
Malawi country Africa 49 D5
Malawi, Lake Africa see Nyasa, Lake
Malawi National Park Zambia see Nyika National Park
Malaya pen. Malaysia see Peninsular Malaysia
Malaya Pera Rus. Fed. 12 L2
Malaya Vishera Rus. Fed. 12 G4
Malāyer Iran 35 H4
Malay Reef Australia 56 E3
Malaysia country Asia 41 C7
Malaysia, Semenanjung pen. Malaysia see Peninsular Malaysia
Malazgirt Turkey 35 F3
Malbon Australia 56 C4
Malbork Poland 17 Q3
Malchin Germany 17 N4
Malcolm Australia 55 C7
Malcolm, Point Australia 55 C8
Malden Island Kiribati 75 J6
Maldives country Indian Ocean 38 B5
Maldon Australia 58 B6
Maldon U.K. 19 H7
Maldonado Uruguay 70 F4
Male Maldives 73 M5
Maleas, Akrotirio pt Greece 27 J6
Male Atoll Maldives 31 G6
Malebogo S. Africa 51 G5
Malegaon Mahar. India 38 B1
Malegaon Mahar. India 38 C2
Malé Karpaty hills Slovakia 17 P6
Malele Dem. Rep. Congo 49 B4
Maler Kotla India 36 C3
Maleševske Planine mts Bulg./Macedonia 27 J4
Malgobek Rus. Fed. 35 G2
Malgomaj l. Sweden 14 J4
Malha, Naqb mt. Egypt see Māliḩah, Naqb
Malhada Brazil 71 C1
Malheur Lake U.S.A. 62 D3
Mali country Africa 46 C3
Mali Dem. Rep. Congo 48 C4
Mali Guinea 46 B3
Maliana East Timor 54 D2
Malianjing China 42 H4
Māliḩah, Naqb mt. Egypt 39 A5
Malili Indon. 52 C2
Malin Ukr. see Malyn
Malindi Kenya 48 E4
Malines Belgium see Mechelen
Malin Head hd Ireland 21 E2
Mali Raginac mt. Croatia 26 F2
Malka r. Rus. Fed. 35 G2
Malkangiri India 38 D2
Malkapur India 38 B2
Malkara Turkey 27 L4
Mal'kavichy Belarus 15 O10
Malko Tŭrnovo Bulg. 27 L4
Mallacoota Australia 58 D6
Mallacoota Inlet b. Australia 58 D6
Mallaig U.K. 20 D4
Mallawī Egypt 34 C6
Mallee Cliffs National Park Australia 57 C7
Mallery Lake Canada 61 I3
Mallét Brazil 71 A4
Mallorca i. Spain see Majorca
Mallow Ireland 21 D5
Mallowa Well Australia 54 D5
Mallwyd U.K. 19 D6
Malm Norway 14 G4
Malmberget Sweden 14 L3
Malmédy Belgium 17 K5
Malmesbury S. Africa 50 D7
Malmesbury U.K. 19 E7
Malmö Sweden 15 H9
Malmyzh Rus. Fed. 12 K4
Maloca Brazil 69 G3
Malonje mt. Tanz. 49 D4
Maloshuyka Rus. Fed. 12 H3
Malosmadulu Atoll Maldives see Maalhosmadulu Atoll
Måløy Norway 14 D6
Maloyaroslavets Rus. Fed. 13 H5
Malozemel'skaya Tundra lowland Rus. Fed. 12 K2
Malpelo, Isla de i. N. Pacific Ocean 67 H8
Malprabha r. India 38 C2
Malta country Europe 26 F7
Malta Latvia 15 O8
Malta MT U.S.A. 62 F2
Malta Channel Italy/Malta 26 F6
Maltahöhe Namibia 50 C3
Maltby U.K. 18 F5
Maltby le Marsh U.K. 18 H5
Malton U.K. 18 G4
Malukken is Indon. see Moluccas
Maluku is Indon. see Moluccas
Maluku, Laut sea Indon. 41 E8
Ma'lūlā, Jabal mts Syria 39 C3
Malung Sweden 15 H6
Maluti Mountains Lesotho 51 I5
Malu'u Solomon Is 53 G2
Malvan India 38 B2
Malvasia Greece see Monemvasia
Malvern U.K. see Great Malvern
Malvern U.S.A. 63 I5
Malvérnia Moz. see Chicualacuala
Malvinas, Islas terr. S. Atlantic Ocean see Falkland Islands
Malyn Ukr. 13 F6
Malyy Anyuy r. Rus. Fed. 29 R3
Malyye Derbety Rus. Fed. 13 J7
Malyy Kavkaz mts Asia see Lesser Caucasus
Malyy Lyakhovskiy, Ostrov i. Rus. Fed. 29 P2
Malyy Uzen' r. Kazakh./Rus. Fed. 13 K6
Mamadysh Rus. Fed. 12 K5
Mamafubedu S. Africa 51 I4
Mambai Brazil 71 B1
Mambasa Dem. Rep. Congo 48 C3
Mamelodi S. Africa 51 I3
Mamfe Cameroon 46 D4
Mamison Pass Georgia/Rus. Fed. see Klukhorskiy, Pereval
Mamit India 37 H5
Mammoth Reservoir CA U.S.A. 65 C2
Mamonas Brazil 71 C1
Mamoré r. Bol./Brazil 68 E6
Mamou Guinea 46 B3
Mampikony Madag. 49 E5
Mampong Ghana 46 C4
Mamuju Indon. 52 B2
Mamuno Botswana 50 E2
Man Côte d'Ivoire 46 C4
Man India 38 B2
Man r. India 38 B2
Man, Isle of terr. Irish Sea 18 C4
Manacapuru Brazil 68 F4
Manacor Spain 25 H4
Manado Indon. 41 E7
Managua Nicaragua 67 G6
Manakara Madag. 49 E6
Manakau mt. N.Z. 59 D6
Manākhah Yemen 32 F6
Manama Bahrain 32 H4
Manamadurai India 38 C4
Mana Maroka National Park S. Africa 51 H5
Manamelkudi India 38 C4
Manam Island P.N.G. 52 E2
Mananara Avaratra Madag. 49 E5
Manangoora Australia 56 B3
Mananjary Madag. 49 E6
Manantali, Lac de l. Mali 46 B3
Manantenina Madag. 49 E6
Mana Pass China/India 36 D3
Mana Pools National Park Zimbabwe 49 C5
Manapouri, Lake N.Z. 59 A7
Manasa India 36 C4
Manas He r. China 42 F3
Manas Hu l. China 42 F3
Manaslu mt. Nepal 37 F3
Manassas VA U.S.A. 64 C3
Manastir Macedonia see Bitola
Manas Wildlife Sanctuary nature res. Bhutan 37 G4
Man-aung Kyun Myanmar 42 G9
Manaus Brazil 68 F4
Manavgat Turkey 34 C3
Manbazar India 37 F5
Manbij Syria 39 C1
Manby U.K. 18 H5
Manchar India 38 B2
Manchester U.K. 18 E5
Manchester CT U.S.A. 64 E2
Manchester MD U.S.A. 64 C3
Manchester NH U.S.A. 64 F1
Manchester VT U.S.A. 64 E1
Mancılık Turkey 34 E3
Mand, Rūd-e r. Iran 35 H5
Manda Tanz. 49 D4
Manda, Jebel mt. Sudan 47 F4
Manda, Parc National de nat. park Chad 47 E4
Mandabe Madag. 49 E6
Mandal Norway 15 E7
Mandala, Puncak mt. Indon. 41 G8
Mandalay Myanmar 37 I5
Mandale Myanmar see Mandalay
Mandalgovĭ Mongolia 42 J3
Mandalī Iraq 35 G4
Mandal-Ovoo Mongolia 42 I4
Mandalt China 43 K4
Mandan U.S.A. 62 G2
Mandas Sardegna Italy 26 C5
Mandasa India 38 E2
Mandasor India see Mandsaur
Mandav Hills India 36 B5
Mandel Afgh. 33 J3
Mandera Kenya 48 E3
Manderfield UT U.S.A. 65 F1
Mandeville Jamaica 67 I5
Mandeville N.Z. 59 B7
Mandha India 36 B4
Mandhoúdhion Greece see Mantoudi
Mandi India 36 D3
Mandiana Guinea 46 C3
Mandié Moz. 49 D5
Mandini S. Africa 51 J5
Mandira Dam India 37 F5
Mandla India 36 E5
Mandleshwar India 36 C5
Mandrael India 36 D4
Mandritsara Madag. 49 E5
Mandsaur India 36 C4
Mandurah Australia 55 A8
Manduria Italy 26 G4
Mandvi India 36 B5
Mandya India 38 C3
Manerbio Italy 26 D2
Manevychi Ukr. 13 E6
Manfalūṭ Egypt 34 C6
Manfredonia Italy 26 F4
Manfredonia, Golfo di g. Italy 26 G4
Manga Brazil 71 C1
Manga Burkina Faso 46 C3
Mangabeiras, Serra das hills Brazil 69 I6
Mangai Dem. Rep. Congo 48 B4
Mangaia i. Cook Is 75 J7
Mangakino N.Z. 59 E4
Mangalagiri India 38 D2
Mangalia Romania 27 M3
Mangalmé Chad 47 E3
Mangalore India 38 B3
Mangaon India 38 B2
Mangareva Islands Fr. Polynesia see Gambier, Îles
Mangaung Free State S. Africa 51 H5
Mangaung Free State S. Africa see Bloemfontein
Mangawan India 37 E4
Mangea i. Cook Is see Mangaia
Mangghyshlaq Kazakh. see Mangystau
Mangghystaū Kazakh. see Mangystau
Mangghystaū admin. div. Kazakh. see Mangistauskaya Oblast'
Mangghyt Uzbek. see Mang'it
Manghit Uzbek. see Mang'it
Mangin Range mts Myanmar see Mingin Range
Mangistau Kazakh. see Mangystau
Mangistauskaya Oblast' admin. div. Kazakh. 35 I2
Mang'it Uzbek. 33 J1
Mangla Bangl. see Mongla
Mangla Pak. 36 C3
Mangnai China 42 G5
Mangochi Malawi 49 D5
Mangoky r. Madag. 49 E6
Mangole i. Indon. 41 E8
Mangoli India 38 B2
Mangotsfield U.K. 19 E7
Mangqystaū Shyghanaghy b. Kazakh. see Mangyshlakskiy Zaliv
Mangrol India 36 B5
Mangrul India 38 C1
Mangshi China see Luxi
Mangualde Port. 25 C3
Manguéni, Plateau du Niger 46 E2
Mangui China 44 A2
Mangula Zimbabwe see Mhangura
Mangyshlak Kazakh. see Mangystau
Mangyshlak, Poluostrov pen. Kazakh. 35 H1
Mangyshlak Oblast admin. div. Kazakh. see Mangistauskaya Oblast'
Mangyshlakskaya Oblast' admin. div. Kazakh. see Mangistauskaya Oblast'
Mangyshlakskiy Zaliv b. Kazakh. 35 H1
Mangystau Kazakh. 35 H2
Manhã Brazil 71 B1
Manhattan U.S.A. 63 H4
Manhica Moz. 51 K3
Manhoca Moz. 51 K4
Manhuaçu Brazil 71 C3
Manhuaçu r. Brazil 71 C2
Mani China 37 F2
Mania r. Madag. 49 E5
Maniago Italy 26 E1
Manicouagan r. Canada 63 N2
Manicouagan, Réservoir resr Canada 63 N1
Manīfah Saudi Arabia 35 H6
Manihiki atoll Cook Is 74 J6
Maniitsoq Greenland 61 M3
Manikchhari Bangl. 37 H5
Manikgarh India see Rajura
Manila Phil. 41 E6
Manildra Australia 58 D4
Manilla Australia 58 E3
Maningrida Australia 54 F3
Manipur India see Imphal
Manipur state India 37 H4
Manisa Turkey 27 L5
Manistee U.S.A. 63 J3
Manitoba prov. Canada 62 H1
Manitoba, Lake Canada 61 I4
Manitou Beach NY U.S.A. 64 C1
Manitou Islands U.S.A. 63 J2
Manitoulin Island Canada 63 K2
Manitowoc U.S.A. 63 J3
Maniwaki Canada 63 L2
Manizales Col. 68 C2
Manja Madag. 49 E6
Manjarabad India 38 B3
Manjeri India 38 C4
Manjhand Pak. 36 B4
Manjhi India 37 F4
Manjra r. India 38 C2
Mankaiana Swaziland see Mankayane
Mankato MN U.S.A. 63 I3
Mankayane Swaziland 51 J4
Mankera Pak. 36 B3
Mankono Côte d'Ivoire 46 C4
Manlay Mongolia 42 J4
Manley Hot Springs U.S.A. 60 C3
Manmad India 38 B1
Mann r. Australia 54 F3
Mann, Mount Australia 55 E6
Mannahill Australia 57 B7
Mannar Sri Lanka 38 C4
Mannar, Gulf of India/Sri Lanka 38 C4
Manneru r. India 38 D3
Mannheim Germany 17 L6
Mannicolo Islands Solomon Is see Vanikoro Islands
Manning r. Australia 58 F3
Manning Canada 60 G4
Mannington WV U.S.A. 64 A3
Manningtree U.K. 19 I7
Mann Ranges mts Australia 55 E6
Mannsville NY U.S.A. 64 C1
Mannu, Capo c. Sardegna Italy 26 C4
Man-of-War Rocks is U.S.A. see Gardner Pinnacles
Manoharpur India 36 D4
Manohar Thana India 36 D4
Manokotak U.S.A. 60 C4
Manokwari Indon. 41 F8
Manosque France 24 G5
Manouane, Lac l. Canada 63 M1
Manovo-Gounda Saint Floris, Parc National du nat. park Cent. Afr. Rep. 48 C3
Manp'o N. Korea 44 B4
Manra i. Kiribati 53 I2
Manresa Spain 25 G3
Mansa Gujarat India 36 C5
Mansa Punjab India 36 C3
Mansa Zambia 49 C5
Mansa Konko Gambia 46 B3
Mansehra Pak. 33 L3
Mansel Island Canada 61 K3
Mansfield Australia 58 C6
Mansfield U.K. 19 F5
Mansfield LA U.S.A. 63 I5
Mansfield OH U.S.A. 63 K3
Mansfield PA U.S.A. 64 C2
Mansi Myanmar 37 H4
Manso r. Brazil see Mortes, Rio das
Manta Ecuador 68 B4
Mantaro r. Peru 68 D6
Manteca CA U.S.A. 65 B2
Mantena Brazil 71 C2
Mantes-la-Jolie France 24 E2
Mantiqueira, Serra da mts Brazil 71 B3
Mantoudi Greece 27 J5
Mantova Italy see Mantua
Mäntsälä Fin. 15 N6
Mänttä Fin. 14 N5
Mantua Italy 26 D2
Mantuan Downs Australia 56 D5
Manturovo Rus. Fed. 12 J4
Mäntyharju Fin. 15 O6
Mäntyjärvi Fin. 14 O3
Manú Peru 68 D6
Manu, Parque Nacional nat. park Peru 68 D6
Manuae atoll Fr. Polynesia 75 J7
Manu'a Islands American Samoa 53 I3
Manuel Ribas Brazil 71 A4
Manuel Vitorino Brazil 71 C1
Manuelzinho Brazil 69 H5
Manui i. Indon. 41 E8
Manukau N.Z. 59 E3
Manukau Harbour N.Z. 59 E3
Manunda watercourse Australia 57 B7
Manus Island P.N.G. 52 E2
Manvi India 38 C3
Many U.S.A. 63 I5
Manyana Botswana 51 G3
Manyas Turkey 27 L4
Manyas Gölü l. Turkey see Kuş Gölü
Manyoni Tanz. 49 D4
Manzai Pak. 36 B2
Manzanares Spain 25 E4
Manzanillo Cuba 67 I4
Manzanillo Mex. 66 D5
Manzhouli China 43 L3
Manzini Swaziland 51 J4
Mao Chad 47 E3
Maó Spain see Mahón
Maoke, Pegunungan mts Indon. 41 F8
Maokeng S. Africa 51 H4
Maokui Shan mt. China 44 A4
Maolin China 44 A4
Maoming China 43 K8
Maopora i. Indon. 54 D1
Mapai Moz. 51 J2
Mapam Yumco l. China 37 E3
Mapanza Zambia 49 C5
Maphodi S. Africa 51 G6
Mapinhane Moz. 51 L2
Mapiri Bol. 68 E7
Maple Creek Canada 62 F2
Mapmaker Seamounts sea feature N. Pacific Ocean 74 H4
Mapoon Australia 56 C1
Mapoteng Lesotho 51 H5
Mapuera r. Brazil 69 G4
Mapulanguene Moz. 51 K3
Mapungubwe National Park S. Africa 51 I2
Maputo Moz. 51 K3
Maputo prov. Moz. 51 K3
Maputo r. Moz./S. Africa 51 K4
Maputo, Baía de b. Moz. 51 K4
Maputsoe Lesotho 51 H5
Maqanshy Kazakh. see Makanchi
Maqar an Na'am well Iraq 35 F5
Maqat Kazakh. see Makat
Maqên China 42 I6
Maqên Kangri mt. China 42 H6
Maqnā Saudi Arabia 34 D5
Maqteïr reg. Mauritania 46 B2
Ma Qu r. China see Yellow
Maquan He r. China 37 F3
Maquela do Zombo Angola 49 B4
Maquinchao Arg. 70 C6
Mar, Serra do mts Rio de Janeiro/São Paulo Brazil 71 B3
Mar, Serra do mts Rio Grande do Sul/Santa Catarina Brazil 71 A5
Mara India 37 E5
Mara S. Africa 51 I2
Maraā Brazil 68 E4
Marabá Brazil 69 I5
Maraboon, Lake resr Australia 56 E4
Maracá, Ilha de i. Brazil 69 H3
Maracaibo Venez. 68 D1
Maracaibo, Lago de Venez. see Maracaibo, Lake
Maracaibo, Lake Venez. 68 D2
Maracaju Brazil 70 E2
Maracaju, Serra de hills Brazil 70 E2
Maracanda Uzbek. see Samarqand
Maracás Brazil 71 C1
Maracás, Chapada de hills Brazil 71 C1
Maracay Venez. 68 E1
Marādah Libya 47 E2
Maradi Niger 46 D3
Marāgheh Iran 35 G3
Marahuaca, Cerro mt. Venez. 68 E3
Marajó, Baía de est. Brazil 69 I4
Marajó, Ilha de i. Brazil 69 H4
Marakele National Park S. Africa 51 H3
Maralal Kenya 48 D3
Maralbashi China see Bachu
Maralinga Australia 55 E7
Maralwexi China see Bachu
Maramasike i. Solomon Is 53 G2
Maramba Zambia see Livingstone
Marambio research stn Antarctica 76 A2
Marand Iran 35 G3
Marandellas Zimbabwe see Marondera
Maranhão r. Brazil 71 A1
Maranoa r. Australia 58 D1
Marañón r. Peru 68 D4
Marão Moz. 51 L3
Marão mt. Port. 25 C3
Mara Rosa Brazil 71 A1
Maraş Turkey see Kahramanmaraş
Marathon NY U.S.A. 64 C1
Maraú Brazil 71 D1
Mārāzā Azer. 35 H2
Marbella Spain 25 D5
Marble Bar Australia 54 B5
Marble Hall S. Africa 51 I3
Marbul Pass India 36 C2
Marburg S. Africa 51 J6
Marburg Slovenia see Maribor
Marburg an der Lahn Germany 17 L5
Marca, Ponta do pt Angola 49 B5
Marcali Hungary 26 G1
Marcelino Ramos Brazil 71 A4
March U.K. 19 H6
Marche reg. France 24 E3
Marchena Spain 25 D5
Marchinbar Island Australia 56 B1
Mar Chiquita, Laguna l. Arg. 70 D4
Marchtrenk Austria 17 O6
Marcona Peru 68 C7
Marcus Baker, Mount U.S.A. 60 D3
Marcy, Mount U.S.A. 63 M3
Mardan Pak. 33 L3
Mar del Plata Arg. 70 E5
Mardin Turkey 35 F3
Maré i. New Caledonia 53 G4
Maree, Loch l. U.K. 20 D3
Marevo Rus. Fed. 12 G4
Marfa U.S.A. 62 G5
Marganets Ukr. see Marhanets'
Margao India see Madgaon
Margaret r. Australia 54 D4
Margaret watercourse Australia 57 B6
Margaret, Mount h. Australia 54 B5
Margaret River Australia 55 A8
Margaretville NY U.S.A. 64 D1
Margarita, Isla de i. Venez. 68 F1
Margaritovo Rus. Fed. 44 D4
Margate U.K. 19 I7
Margherita, Lake Eth. see Abaya, Lake
Margherita Peak Dem. Rep. Congo/Uganda 48 C3
Marghilon Uzbek. see Marg'ilon
Marg'ilon Uzbek. 33 L1
Mārgo, Dasht-i des. Afgh. see Mārgow, Dasht-e
Margog Caka l. China 37 F2
Mārgow, Dasht-e des. Afgh. 33 J3
Marguerite, Pic mt. Dem. Rep. Congo/Uganda see Margherita Peak
Marguerite Bay Antarctica 76 L2
Margyang China 37 G3
Marhaj Khalil Iraq 35 G4
Marhanets' Ukr. 13 G7
Marhoum Alg. 22 D5
Maria atoll Fr. Polynesia 75 J7
María Elena Chile 70 C2
Maria Island Australia 56 A2
Maria Island National Park Australia 57 [inset]
Mariala National Park Australia 57 D5
Mariana Brazil 71 C3
Mariana Ridge sea feature N. Pacific Ocean 74 F4
Mariana Trench sea feature N. Pacific Ocean 74 F5
Mariani India 37 H4
Mariánica, Cordillera mts Spain see Morena, Sierra
Marianna AR U.S.A. 63 I5
Marianna FL U.S.A. 63 J5
Mariano Machado Angola see Ganda
Mariánské Lázně Czech Rep. 17 N6
Marías, Islas is Mex. 66 C4
Mariato, Punta pt Panama 67 H7
Maria van Diemen, Cape N.Z. 59 D2
Ma'rib Yemen 32 G6
Maribor Slovenia 26 F1
Marica r. Bulg. see Maritsa
Maricopa CA U.S.A. 65 C3
Maridi South Sudan 47 F4
Marie Byrd Land reg. Antarctica 76 J1
Marie-Galante i. Guadeloupe 67 L5
Mariehamn Fin. 15 K6
Mariembero r. Brazil 71 A1
Marienbad Czech Rep. see Mariánské Lázně
Marienburg Poland see Malbork
Mariental Namibia 50 C3
Marienwerder Poland see Kwidzyn
Mariestad Sweden 15 H7
Marietta GA U.S.A. 63 K5
Marietta OH U.S.A. 64 A3
Marignane France 24 G5
Marii, Mys pt Rus. Fed. 40 G2
Mariinsk Rus. Fed. 28 J4
Mariinskiy Posad Rus. Fed. 12 J4
Marijampolė Lith. 15 M9
Marília Brazil 71 A3
Marillana Australia 54 B5
Marimba Angola 49 B4
Marín Spain 25 B2
Marina CA U.S.A. 65 B2
Marina di Gioiosa Ionica Italy 26 G5
Mar"ina Gorka Belarus see Mar"ina Horka
Mar"ina Horka Belarus 15 P10
Marinette U.S.A. 63 J2
Maringá Brazil 71 A3
Maringa r. Dem. Rep. Congo 48 B3
Marinha Grande Port. 25 B4
Marion IN U.S.A. 63 J3
Marion OH U.S.A. 63 K3
Marion SC U.S.A. 63 L5
Marion VA U.S.A. 63 K4
Marion, Lake U.S.A. 63 K5
Marion Reef Australia 56 F3
Maripa Venez. 68 E2
Mariposa CA U.S.A. 65 C2
Mariscal José Félix Estigarribia Para. 70 D2
Maritime Alps mts France/Italy 24 H4
Maritime Kray admin. div. Rus. Fed. see Primorskiy Kray
Maritimes, Alpes mts France/Italy see Maritime Alps
Maritsa r. Bulg. 27 L4
Marittime, Alpi mts France/Italy see Maritime Alps
Mariupol' Ukr. 13 H7
Mariusa, Parque Nacional nat. park Venez. 68 F2
Marīvān Iran 35 G4
Marjan Afgh. see Wazi Khwa
Marjayoûn Lebanon 39 B3
Marka Somalia 48 E3
Markala Mali 46 C3
Markaryd Sweden 15 H8
Marken S. Africa 51 I2
Markermeer l. Neth. 16 J4
Market Deeping U.K. 19 G6
Market Drayton U.K. 19 E6
Market Harborough U.K. 19 G6
Markethill U.K. 21 F3
Market Weighton U.K. 18 G5
Markha r. Rus. Fed. 29 M3
Markit China 33 M2
Markleeville CA U.S.A. 65 C1
Markounda Cent. Afr. Rep. 48 B3
Markovo Rus. Fed. 29 S3
Marks Rus. Fed. 13 J6
Marla Australia 55 F6
Marlborough Downs hills U.K. 19 F7
Marlinton WV U.S.A. 64 A3
Marlo Australia 58 D6
Marmagao India 38 B3
Marmande France 24 E4
Marmara, Sea of g. Turkey 27 M4
Marmara Denizi g. Turkey see Marmara, Sea of
Marmara Gölü l. Turkey 27 M5
Marmarica reg. Libya 34 B5
Marmaris Turkey 27 M6
Marmion, Lake salt l. Australia 55 C7
Marmolada mt. Italy 26 D1
Marne r. France 24 F2
Marne-la-Vallée France 24 F2
Maroantsetra Madag. 49 E5
Maroc country Africa see Morocco
Marol Pak. 36 D2
Maromokotro mt. Madag. 49 E5
Marondera Zimbabwe 49 D5
Maroochydore Australia 58 F1
Maroonah Australia 55 A5
Marosvás>01a>rhely Romania see Târgu Mureş
Marot Pak. 36 C3
Maroua Cameroon 47 E3
Marovoay Madag. 49 E5
Marqādah Syria 35 F4
Marquard S. Africa 51 H5
Marquesas Islands Fr. Polynesia 75 K6
Marquês de Valença Brazil 71 C3
Marquette U.S.A. 63 J2
Marquise France 19 I8
Marquises, Îles is Fr. Polynesia see Marquesas Islands
Marra Australia 58 A3
Marra r. Australia 58 C3
Marra, Jebel mt. Sudan 47 F3
Marra, Jebel Sudan 47 F3
Marracuene Moz. 51 K3
Marrakech Morocco 22 C5
Marrakesh Morocco see Marrakech
Marrangua, Lagoa l. Moz. 51 L3
Marrar Australia 58 C5
Marrawah Australia 57 [inset]
Marree Australia 57 B6
Marruecos country Africa see Morocco
Marrupa Moz. 49 D5
Marryat Australia 55 F6
Marsá al 'Alam Egypt 32 D4
Marsa 'Alam Egypt see Marsá al 'Alam
Marsa al Burayqah Libya 47 E1
Marsabit Kenya 48 D3
Marsala Sicilia Italy 26 E6
Marsá Maṭrūḩ Egypt 34 B5
Marsciano Italy 26 E3
Marsden Australia 58 C4
Marseille France 24 G5
Marseilles France see Marseille
Marsfjället mt. Sweden 14 I4
Marshall watercourse Australia 56 B4
Marshall MN U.S.A. 63 I3
Marshall MO U.S.A. 63 I4
Marshall TX U.S.A. 63 I5
Marshall Islands country N. Pacific Ocean 74 H5
Marshalltown U.S.A. 63 I3
Marsh Harbour Bahamas 67 I3
Marsh Island U.S.A. 63 I6
Märsta Sweden 15 J7
Marsyaty Rus. Fed. 11 S3
Martapura Indon. 41 D8
Martha's Vineyard i. MA U.S.A. 64 F2
Martigny Switz. 24 H3
Martim Vaz, Ilhas is S. Atlantic Ocean see Martin Vas, Ilhas
Martin Slovakia 17 Q6
Martin SD U.S.A. 62 G3
Martinez Lake AZ U.S.A. 65 E4
Martinho Campos Brazil 71 B2
Martinique terr. West Indies 67 L6
Martinique Passage Dominica/Martinique 67 L5
Martin Peninsula Antarctica 76 K2

Martinsburg *WV* U.S.A. **64** C3
Martins Ferry *OH* U.S.A. **64** A2
Martinsville *VA* U.S.A. **63** L4
Martin Vas, Ilhas *is* S. Atlantic Ocean **72** G7
Martin Vaz Islands S. Atlantic Ocean *see* **Martin Vas, Ilhas**
Martök Kazakh. *see* **Martuk**
Marton N.Z. **59** E5
Martorell Spain **25** G3
Martos Spain **25** E5
Martuk Kazakh. **30** E1
Martuni Armenia **35** G2
Ma'rūf Afgh. **36** A3
Maruim Brazil **69** K6
Marukhis Ugheltekhili *pass* Georgia/Rus. Fed. **35** F2
Marulan Australia **58** D5
Marvast Iran **35** I5
Marv Dasht Iran **35** I5
Marvejols France **24** F4
Mary *r.* Australia **54** E3
Mary Turkm. **30** F3
Maryborough *Qld* Australia **57** F5
Maryborough *Vic.* Australia **58** A6
Marydale S. Africa **50** F5
Maryland *state* U.S.A. **64** C3
Maryport U.K. **18** D4
Marysville *CA* U.S.A. **65** B1
Marysville *KS* U.S.A. **63** H4
Maryvale *N.T.* Australia **55** F6
Maryvale *Qld* Australia **56** D3
Maryville *MO* U.S.A. **63** I3
Maryville *TN* U.S.A. **63** K4
Marzagão Brazil **71** A2
Masada *tourist site* Israel **39** B4
Masai Steppe *plain* Tanz. **49** D4
Masaka Uganda **48** D4
Masakhane S. Africa **51** H6
Masallı Azer. **35** H3
Masan S. Korea **45** C6
Masasi Tanz. **49** D5
Masavi Bol. **68** F7
Masbate *i.* Phil. **41** E6
Mascara Alg. **25** G6
Mascarene Basin *sea feature* Indian Ocean **73** L7
Mascarene Plain *sea feature* Indian Ocean **73** L7
Mascarene Ridge *sea feature* Indian Ocean **73** L6
Mascote Brazil **71** D1
Masela Indon. **54** E2
Masela *i.* Indon. **54** E2
Maseru Lesotho **51** H5
Mashai Lesotho **51** I5
Masherbrum *mt.* Pak. **36** D2
Mashhad Iran **33** I2
Mashishing S. Africa **51** J3
Masi Norway **14** M2
Masibambane S. Africa **51** H6
Masīlah, Wādī al *watercourse* Yemen **32** H6
Masilo S. Africa **51** H5
Masi-Manimba Dem. Rep. Congo **49** B4
Masindi Uganda **48** D3
Masinyusane S. Africa **50** F6
Masira, Gulf of Oman *see* **Maşīrah, Khalīj**
Maşīrah, Jazīrat *i.* Oman **33** I5
Maşīrah, Khalīj *b.* Oman **33** I6
Masira Island Oman *see* **Maşīrah, Jazīrat**
Masjed-e Soleymān Iran **35** H5
Mask, Lough *l.* Ireland **21** C4
Maslovo Rus. Fed. **11** S3
Masoala, Tanjona *c.* Madag. **49** F5
Mason, Lake *imp. l.* Australia **55** B6
Mason Bay N.Z. **59** A8
Mason City U.S.A. **63** I3
Masontown *PA* U.S.A. **64** B3
Masqaţ Oman *see* **Muscat**
Massa Italy **26** D2
Massachusetts *state* U.S.A. **64** E1
Massachusetts Bay *MA* U.S.A. **64** F1
Massafra Italy **26** G4
Massakory Chad **47** E3
Massa Marittimo Italy **26** D3
Massangena Moz. **49** D6
Massango Angola **49** B4
Massawa Eritrea **32** E6
Massenya Chad **47** E3
Masset Canada **60** E4
Massif Central *mts* France **24** F4
Massilia France *see* **Marseille**
Massillon *OH* U.S.A. **64** A2
Massina Mali **46** C3
Massinga Moz. **51** L2
Massingir Moz. **51** K2
Massingir, Barragem de *resr* Moz. **51** K2
Masson Island Antarctica **76** F2
Masteksay Kazakh. **13** K6
Masterton N.Z. **59** E5
Mastung Pak. **30** F4
Mastūrah Saudi Arabia **32** E5
Masty Belarus **15** N10
Masuda Japan **45** C6
Masuku Gabon *see* **Franceville**
Masulipatam India *see* **Machilipatnam**
Masulipatnam India *see* **Machilipatnam**
Masuna *i.* American Samoa *see* **Tutuila**
Masvingo Zimbabwe **49** D6
Masvingo *prov.* Zimbabwe **51** J1
Maswa Tanz. **48** D4
Maşyāf Syria **39** C2
Matabeleland South *prov.* Zimbabwe **51** I1
Matad *Dornod* Mongolia **43** L3
Matadi Dem. Rep. Congo **49** B4
Matagalpa Nicaragua **67** G6
Matagami Canada **63** L2
Matagami, Lac *l.* Canada **63** L2
Matagorda Island U.S.A. **63** H6
Matakana Island N.Z. **59** F3
Matala Angola **49** B5
Maţāli', Jabal *mt.* Saudi Arabia **35** F6
Matam Senegal **46** B3
Matamey Niger **46** D3
Matamoras *PA* U.S.A. **64** D2
Matamoros *Coahuila* Mex. **66** D3
Matamoros *Tamaulipas* Mex. **66** E3
Matandu *r.* Tanz. **49** D4
Matane Canada **63** N2
Matanzas Cuba **67** H4
Matapan, Cape *c.* Greece *see* **Tainaro, Akrotirio**
Matara Sri Lanka **38** D5
Mataram Indon. **41** D8
Matarani Peru **68** D7
Mataranka Australia **54** F3
Matarıpe Brazil **71** D1
Mataró Spain **25** H3
Matatiele S. Africa **51** I6
Matatila Reservoir India **36** D4
Mataura N.Z. **59** B8
Matā'utu Wallis and Futuna Is **53** I3
Mata-Utu Wallis and Futuna Is *see* **Matā'utu**
Matawai N.Z. **59** F4
Matay Kazakh. **42** D3
Mategua Bol. **68** F6
Matehuala Mex. **66** D4
Matemanga Tanz. **49** D5
Matera Italy **26** G4
Mateur Tunisia **26** C6
Mathaji India **36** B4
Mathews *VA* U.S.A. **64** C4
Mathis U.S.A. **62** H6
Mathoura Australia **58** B5
Mathura India **36** D4
Mati Phil. **41** E7
Matiali India **37** G4
Matias Cardoso Brazil **71** C1
Matías Romero Mex. **66** E5
Matin India **37** E5
Matla *r.* India **37** G5
Matlabas *r.* S. Africa **51** H2
Matli Pak. **36** B4
Matlock U.K. **19** F5
Mato, Cerro *mt.* Venez. **68** E2
Matobo Hills Zimbabwe **49** C6
Mato Grosso Brazil **68** G7
Mato Grosso *state* Brazil **71** A1
Mato Grosso, Planalto do *plat.* Brazil **69** H7
Matopo Hills Zimbabwe *see* **Matobo Hills**
Matos Costa Brazil **71** A4
Matosinhos Port. **25** B3
Mato Verde Brazil **71** C1
Matroosberg *mt.* S. Africa **50** D7
Matsesta Rus. Fed. **35** E2
Matsue Japan **45** D6
Matsumoto Japan **45** E5
Matsu Tao *i.* Taiwan **43** M7
Matsuyama Japan **45** D6
Mattagami *r.* Canada **63** K1
Matterhorn *mt.* Italy/Switz. **26** B2
Matterhorn *mt.* U.S.A. **62** D3
Matthew Town Bahamas **67** J4
Matturai Sri Lanka *see* **Matara**
Matuku *i.* Fiji **53** H3
Matumbo Angola **49** B5
Maturín Venez. **68** F2
Matusadona National Park Zimbabwe **49** C5
Matwabeng S. Africa **51** H5
Maty Island P.N.G. *see* **Wuvulu Island**
Mau India *see* **Maunath Bhanjan**
Maúa Moz. **49** D5
Maubourguet France **24** E5
Mauchline U.K. **20** E5
Maudaha India **36** E4
Maude Australia **57** D7
Maud Seamount *sea feature* S. Atlantic Ocean **72** I10
Mau-é-ele Moz. *see* **Marão**
Maués Brazil **69** G4
Maughold Head *hd* Isle of Man **18** C4
Maui *i.* U.S.A. **75** J4
Maukkadaw Myanmar **37** H5
Maule *r.* Chile **70** B5
Maulvi Bazar Bangl. *see* **Moulvibazar**
Maumere Indon. **54** C2
Maumturk Mountains *hills* Ireland **21** C4
Maun Botswana **49** C5
Maunath Bhanjan India **37** E4
Maunatlala Botswana **51** H2
Maungaturoto N.Z. **59** E3
Maungdaw Myanmar **37** H5
Mauriac France **24** F4
Maurice *country* Indian Ocean *see* **Mauritius**
Maurice, Lake *imp. l.* Australia **55** E7
Mauritania *country* Africa **46** B3
Mauritanie *country* Africa *see* **Mauritania**
Mauritius *country* Indian Ocean **73** L7
Maurs France **24** F4
Mava Dem. Rep. Congo **48** C3
Mavago Moz. **49** D5
Mavanza Moz. **51** L2
Mavinga Angola **49** C5
Mavrovo *nat. park* Macedonia **27** I4
Mavume Moz. **51** L2
Mavuya S. Africa **51** H6
Mawana India **36** D3
Mawanga Dem. Rep. Congo **49** B4
Mawjib, Wādī al *r.* Jordan **39** B4
Mawkmai Myanmar **42** H8
Mawlaik Myanmar **37** H5
Mawlamyaing Myanmar **31** I5
Mawlamyine Myanmar *see* **Mawlamyaing**
Mawqaq Saudi Arabia **35** F6
Mawson (Australia) *research stn* Antarctica **76** E2
Mawson Coast Antarctica **76** E2
Mawson Escarpment Antarctica **76** E2
Mawson Peninsula Antarctica **76** H2
Mawza Yemen **32** F7
Maxán Arg. **70** C3
Maxia, Punta *mt. Sardegna* Italy **26** C5
Maxixe Moz. **51** L2
Maxmo Fin. **14** M5
May, Isle of *i.* U.K. **20** G4
Maya *r.* Rus. Fed. **29** O3
Mayaguana *i.* Bahamas **67** J4
Mayagüez Puerto Rico **67** K5
Mayahi Niger **46** D3
Mayak Rus. Fed. **44** E2
Mayakovskiy, Qullai *mt.* Tajik. **36** B1
Mayakovskogo, Pik *mt.* Tajik. *see* **Mayakovskiy, Qullai**
Mayama Congo **48** B4
Maya Mountains Belize/Guat. **66** G5
Mayar *h.* U.K. **20** F4
Maybeury *WV* U.S.A. **64** A4
Maybole U.K. **20** E5
Maych'ew Eth. **48** D2
Maydān Shahr Afgh. *see* **Meydān Shahr**
Maydh Somalia **32** G7
Maydos Turkey *see* **Eceabat**
Mayenne France **24** D2
Mayenne *r.* France **24** D3
Mayêr Kangri *mt.* China **37** F2
Mayfield N.Z. **59** C6
Mayi He *r.* China **44** C3
Maykop Rus. Fed. **13** I7
Maymanah Afgh. **36** A2
Mayna *Respublika Khakasiya* Rus. Fed. **28** K4
Mayna *Ul'yanovskaya Oblast'* Rus. Fed. **13** J5
Mayni India **38** B2
Mayo Alim Cameroon **46** E4
Mayoko Congo **48** B4
Mayor, Puig *mt.* Spain *see* **Major, Puig**
Mayor Island N.Z. **59** F3
Mayor Pablo Lagerenza Para. **70** D1
Mayotte *terr.* Africa **49** E5
Mayskiy *Amurskaya Oblast'* Rus. Fed. **44** C1
Mayskiy *Kabardino-Balkarskaya Respublika* Rus. Fed. **35** G2
Mays Landing *NJ* U.S.A. **64** D3
Mayumba Gabon **48** B4
Mayum La *pass* China **37** E3
Mayuram India **38** C4
Mayville *NY* U.S.A. **64** B1
Mazabuka Zambia **49** C5
Mazaca Turkey *see* **Kayseri**
Mazagan Morocco *see* **El Jadida**
Mazar China **36** D1
Mazara, Val di *reg. Sicilia* Italy **26** E6
Mazara del Vallo *Sicilia* Italy **26** E6
Mazār-e Sharīf Afgh. **36** A1
Mazatán Mex. **62** E6
Mazatlán Mex. **66** C4
Mazdaj Iran **35** H4
Mažeikiai Lith. **15** M8
Mazocruz Peru **68** E7
Mazomora Tanz. **49** D4
Mazu Dao *i.* Taiwan *see* **Matsu Tao**
Mazunga Zimbabwe **49** C6
Mazyr Belarus **13** F5
Mazzouna Tunisia **26** C7
Mbabane Swaziland **51** J4
Mbahiakro Côte d'Ivoire **46** C4
Mbaïki Cent. Afr. Rep. **48** B3
Mbakaou, Lac de *l.* Cameroon **46** E4
Mbala Zambia **49** D4
Mbale Uganda **48** D3
Mbalmayo Cameroon **46** E4
Mbam *r.* Cameroon **46** E4
Mbandaka Dem. Rep. Congo **48** B4
M'banza Congo Angola **49** B4
Mbarara Uganda **47** G5
Mbari *r.* Cent. Afr. Rep. **48** C3
Mbaswana S. Africa **51** K4
Mbemkuru *r.* Tanz. **49** D4
Mbeya Tanz. **49** D4
Mbinga Tanz. **49** D5
Mbini Equat. Guinea **46** D4
Mbizi Zimbabwe **49** D6
Mboki Cent. Afr. Rep. **48** C3
Mbombela S. Africa **51** J3
Mbomo Congo **48** B3
Mbouda Cameroon **46** E4
Mbour Senegal **46** B3
Mbout Mauritania **46** B3
Mbozi Tanz. **49** D4
Mbrès Cent. Afr. Rep. **48** B3
Mbuji-Mayi Dem. Rep. Congo **49** C4
Mbulu Tanz. **48** D4
Mburucuyá Arg. **70** E3
McAlester U.S.A. **63** H5
McAlister *mt.* Australia **58** D5
McAllen U.S.A. **62** H6
McArthur *r.* Australia **56** B2
McCall U.S.A. **62** D3
McClintock, Mount Antarctica **76** H1
McClintock Channel Canada **61** H2
McClintock Range *hills* Australia **54** D4
McClure, Lake *CA* U.S.A. **65** B2
McClure Strait Canada **60** G2
McConaughy, Lake U.S.A. **62** G3
McConnellsburg *PA* U.S.A. **64** C3
McCook U.S.A. **62** G3
McDermitt U.S.A. **62** D3
McDonald Islands Indian Ocean **73** M9
McDonald Peak U.S.A. **62** E2
McDougall's Bay S. Africa **50** C5
McFarland *CA* U.S.A. **65** C3
McGrath *AK* U.S.A. **60** C3
McGraw *NY* U.S.A. **64** C1
McGregor S. Africa **50** D7
McGregor Range *hills* Australia **57** C5
McGuire, Mount U.S.A. **62** E2
Mchinga Tanz. **49** D4
Mchinji Malawi **49** D5
McIlwraith Range *hills* Australia **56** C2
McInnes Lake Canada **63** I1
McKay Range *hills* Australia **54** C5
McKean *i.* Kiribati **53** I2
McKinlay *r.* Australia **56** C4
McKinley, Mount U.S.A. **60** C3
McKittrick *CA* U.S.A. **65** C3
McLennan Canada **60** G4
McMinnville *OR* U.S.A. **62** C2
McMurdo *research stn* Antarctica **76** H1
McMurdo Sound *b.* Antarctica **76** H1
McNaughton Lake Canada *see* **Kinbasket Lake**
McPherson U.S.A. **62** H4
McVeytown *PA* U.S.A. **64** C2
Mdantsane S. Africa **51** H7
M'Daourouch Alg. **26** B6
Mead, Lake *resr NV* U.S.A. **65** E2
Meade *r.* U.S.A. **60** C2
Meadow Australia **55** A6
Meadow *UT* U.S.A. **65** F1
Meadville *PA* U.S.A. **64** A2
Meaken-dake *vol.* Japan **44** G4
Mealhada Port. **25** B3
Mealy Mountains Canada **61** M4
Meandarra Australia **58** D1
Meaux France **24** F2
Mecca Saudi Arabia **32** E5
Mecca *CA* U.S.A. **65** D4
Mechanicsville *VA* U.S.A. **64** C4
Mechelen Belgium **16** J5
Mecheria Alg. **22** D5
Mecitözü Turkey **34** D2
Mecklenburger Bucht *b.* Germany **17** M3
Meda *r.* Australia **54** C4
Meda Port. **25** C3
Medak India **38** C2
Medan Indon. **41** B7
Medanosa, Punta *pt* Arg. **70** C7
Médanos de Coro, Parque Nacional *nat. park* Venez. **68** E1
Medawachchiya Sri Lanka **38** D4
Médéa Alg. **25** H5
Medellín Col. **68** C2
Meden *r.* U.K. **18** G5
Medenine Tunisia **22** G5
Mederdra Mauritania **46** B3
Medford *NY* U.S.A. **64** E2
Medford *OR* U.S.A. **62** C3
Medgidia Romania **27** M2
Media *PA* U.S.A. **64** D3
Mediaş Romania **27** K1
Medicine Bow Mountains U.S.A. **62** F3
Medicine Bow Peak U.S.A. **62** F3
Medicine Hat Canada **62** E1
Medicine Lodge U.S.A. **62** H4
Medina Brazil **71** C2
Medina Saudi Arabia **32** E5
Medina *NY* U.S.A. **64** B1
Medinaceli Spain **25** E3
Medina del Campo Spain **25** D3
Medina de Rioseco Spain **25** D3
Medinipur India **37** F5
Mediolanum Italy *see* **Milan**
Mediterranean Sea **22** K5
Mednyy, Ostrov *i.* Rus. Fed. **74** H2
Médoc *reg.* France **24** D4
Mêdog China **42** H7
Meduro *atoll* Marshall Is *see* **Majuro**
Medvedevo Rus. Fed. **12** J4
Medveditsa *r.* Rus. Fed. **13** I6
Medvednica *mts* Croatia **26** F2
Medvezh'i, Ostrova *is* Rus. Fed. **29** R2
Medvezh'ya *vol.* Rus. Fed. **44** H3
Medvezh'ya, Gora *mt.* Rus. Fed. **44** E3
Medvezh'yegorsk Rus. Fed. **12** G3
Medway *r.* U.K. **19** H7
Meekatharra Australia **55** B6
Meeker *CO* U.S.A. **62** F3
Meerut India **36** D3
Mega Escarpment Eth./Kenya **48** D3
Megalopoli Greece **27** J6
Megara Greece **27** J5
Megezez *mt.* Eth. **48** D3
Meghalaya *state* India **37** G4
Meghasani *mt.* India **37** F5
Meghri Armenia **35** G3
Megisti *i.* Greece **27** M6
Megri Armenia *see* **Meghri**
Mehamn Norway **14** O1
Mehar Pak. **36** A4
Meharry, Mount Australia **55** B5
Mehbubnagar India *see* **Mahbubnagar**
Mehdia Tunisia *see* **Mahdia**
Meherpur Bangl. **37** G5
Meherrin *VA* U.S.A. **64** B4
Meherrin *r. VA* U.S.A. **64** C4
Mehrakān *salt marsh* Iran **35** I6
Mehrān *Hormozgān* Iran **35** I6
Mehrān *Īlām* Iran **35** G4
Mehrīz Iran **35** I5
Mehsana India *see* **Mahesana**
Mehtar Lām Afgh. **36** B2
Meia Ponte *r.* Brazil **71** A2
Meiganga Cameroon **47** E4
Meighen Island Canada **61** I2
Meihekou China **44** B4
Meikle Says Law *h.* U.K. **20** G5
Meiktila Myanmar **37** H5
Meiningen Germany **17** M5
Meißen Germany **17** N5
Meixi China **44** C3
Meixian China *see* **Meizhou**
Meizhou China **43** L8
Mej *r.* India **36** D4
Mejicana *mt.* Arg. **70** C3
Mejillones Chile **70** B2
Mékambo Gabon **48** B3
Mek'elē Eth. **48** D2
Mekelle Eth. *see* **Mek'elē**
Mékhé Senegal **46** B3
Mekhtar Pak. **36** B3
Meknassy Tunisia **26** C7
Meknès Morocco **22** C5
Mekong *r.* Asia **31** J5
Mekong *r. Xizang/Yunnan* China **42** I8
Mekong, Mouths of the Vietnam **31** J6
Mekoryuk U.S.A. **60** B3
Melaka Malaysia **41** C7
Melanesia *is* Pacific Ocean **74** G6
Melanesian Basin *sea feature* Pacific Ocean **74** G5
Melbourne Australia **58** B6
Melbourne U.S.A. **63** K6
Melby U.K. **20** [inset]
Meldorf Germany **17** L3
Melekeok Palau **41** F7
Melekess Rus. Fed. *see* **Dimitrovgrad**
Melenki Rus. Fed. **13** I5
Melet Turkey *see* **Mesudiye**
Mélèzes, Rivière aux *r.* Canada **61** K4
Melfa *VA* U.S.A. **64** D4
Melfi Chad **47** E3
Melfi Italy **26** F4
Melfort Canada **62** G1
Melhus Norway **14** G5
Melide Spain **25** C2
Melilla N. Africa **25** E6
Melimoyu, Monte *mt.* Chile **70** B6
Melitene Turkey *see* **Malatya**
Melitopol' Ukr. **13** G7
Melk Austria **17** O6
Melka Guba Eth. **48** D3
Melksham U.K. **19** E7
Mellakoski Fin. **14** N3
Mellansel Sweden **14** K5
Mellerud Sweden **15** H7
Mellid Spain *see* **Melide**
Mellilla N. Africa *see* **Melilla**
Mellor Glacier Antarctica **76** E2
Melmoth S. Africa **51** J5
Mel'nichoye Rus. Fed. **44** D3
Melo Uruguay **70** F4
Meloco Moz. **49** D5
Melolo Indon. **54** C2
Melozitna *r.* U.S.A. **60** C3
Melrhir, Chott *salt l.* Alg. **22** F5
Melrose Australia **55** C6
Melrose U.K. **20** G5
Melton Australia **58** B6
Melton Mowbray U.K. **19** G6
Melun France **24** F2
Melville Canada **62** G1
Melville, Cape Australia **56** D2
Melville, Lake Canada **61** M4
Melville Bugt *b.* Greenland *see* **Qimusseriarsuaq**
Melville Island Australia **54** E2
Melville Island Canada **61** H2
Melville Peninsula Canada **61** J3
Melvin, Lough *l.* Ireland/U.K. **21** D3
Mêmar Co *salt l.* China **37** E2
Memba Moz. **49** E5
Memberamo *r.* Indon. **52** D2
Memel Lith. *see* **Klaipėda**
Memel S. Africa **51** I4
Memmingen Germany **17** M7
Mempawah Indon. **41** C7
Memphis *tourist site* Egypt **34** C5
Memphis *TN* U.S.A. **63** I4
Memphis *TX* U.S.A. **62** G5
Mena Ukr. **13** G6
Mena U.S.A. **63** I5
Menado Indon. *see* **Manado**
Ménaka Mali **46** D3
Mendanha Brazil **71** C2
Mende France **24** F4
Mendefera Eritrea **32** E7
Mendeleyev Ridge *sea feature* Arctic Ocean **77** B1
Mendeleyevsk Rus. Fed. **12** L5
Mendenhall, Cape U.S.A. **60** B4
Mendī Eth. **48** D3
Mendi P.N.G. **52** E2
Mendip Hills U.K. **19** E7
Mendooran Australia **58** D3
Mendota *CA* U.S.A. **65** B2
Mendoza Arg. **70** C4
Menemen Turkey **27** L5
Ménerville Alg. *see* **Thenia**
Menglie China *see* **Jiangcheng**
Menindee Australia **57** C7
Menindee, Lake Australia **57** C7
Menkere Rus. Fed. **29** N3
Mennecy France **24** F2
Menongue Angola **49** B5
Menorca *i.* Spain *see* **Minorca**
Mentawai, Kepulauan *is* Indon. **41** B8
Menton France **24** H5
Menuf Egypt *see* **Minūf**
Menzel Bourguiba Tunisia **26** C6
Menzelet Barajı *resr* Turkey **34** E3
Menzelinsk Rus. Fed. **11** Q4
Menzel Temime Tunisia **26** D6
Menzies Australia **55** C7
Menzies, Mount Antarctica **76** E2
Meobbaai *b.* Namibia **50** B3
Meppen Germany **17** K4
Mepuze Moz. **51** K2
Meqheleng S. Africa **51** H5
Meråker Norway **14** G5
Merano Italy **26** D1
Meratswe *r.* Botswana **50** G2
Merauke Indon. **41** G8
Merca Somalia *see* **Marka**
Mercantour, Parc National du *nat. park* France **24** H4
Merced *CA* U.S.A. **65** B2
Merced *r. CA* U.S.A. **65** B2
Mercedes Arg. **70** E3
Mercedes Uruguay **70** E4
Mercer *PA* U.S.A. **64** A2
Mercês Brazil **71** C3
Mercury Islands N.Z. **59** E3
Mercy, Cape Canada **61** L3
Merdenik Turkey *see* **Göle**
Mere U.K. **19** E7
Meredith *NH* U.S.A. **64** F1
Merefa Ukr. **13** H6
Merga Oasis Sudan **32** C6
Mergui Archipelago *is* Myanmar **31** I5
Meriç *r.* Turkey **27** L4
Mérida Mex. **66** G4
Mérida Spain **25** C4
Mérida Venez. **68** D2
Mérida, Cordillera de *mts* Venez. **68** D2
Meriden *CT* U.S.A. **64** E2
Meridian *MS* U.S.A. **63** J5
Mérignac France **24** D4
Merijärvi Fin. **14** N4
Merikarvia Fin. **15** L6
Merimbula Australia **58** D6
Merín, Laguna *l.* Brazil/Uruguay *see* **Mirim, Lagoa**
Meringur Australia **57** C7
Merjayoun Lebanon *see* **Marjayoûn**
Merluna Australia **56** C2
Mermaid Reef Australia **54** B4
Meron, Har *mt.* Israel **39** B3
Merowe Sudan **32** D6
Mêrqung Co *l.* China **37** F3
Merredin Australia **55** B7
Merrick *h.* U.K. **20** E5
Merrill *WI* U.S.A. **63** J2
Merriwa Australia **58** E4
Merrygoen Australia **58** D3
Mersa Fatma Eritrea **32** F7
Mersa Maṭrûḥ Egypt *see* **Marsá Maṭrūḥ**
Mersey *est.* U.K. **18** E5
Mersin Turkey **39** B1
Mersin *prov.* Turkey **39** A1
Mērsrags Latvia **15** M8
Merta India **36** C4
Merthyr Tydfil U.K. **19** D7
Mértola Port. **25** C5
Mertz Glacier Antarctica **76** G2
Mertz Glacier Tongue Antarctica **76** G2
Meru *vol.* Tanz. **48** D4
Merv Turkm. *see* **Mary**
Merweville S. Africa **50** E7
Merzifon Turkey **34** D2
Merzig Germany **17** K6
Merz Peninsula Antarctica **76** L2
Mesa *AZ* U.S.A. **62** E5
Mesagne Italy **26** G4
Meselefors Sweden **14** J4
Meshed Iran *see* **Mashhad**
Meshra'er Req South Sudan **32** C8
Mesimeri Greece **27** J4
Mesolongi Greece **27** I5
Mesolóngion Greece *see* **Mesolongi**
Mesopotamia *reg.* Iraq **35** F4
Mesquita Brazil **71** C2
Mesquite *NV* U.S.A. **65** E2
Mesquite Lake *CA* U.S.A. **65** E3
Messaad Alg. **22** E5
Messana *Sicilia* Italy *see* **Messina**
Messina *Sicilia* Italy **26** F5
Messina, Strait of Italy **26** F5
Messina, Stretta di Italy *see* **Messina, Strait of**
Messini Greece **27** J6
Messiniakos Kolpos *g.* Greece **27** J6
Mesta *r.* Bulg. **27** K4
Mesta *r.* Greece *see* **Nestos**
Mesta, Akrotirio *pt* Greece **27** K5
Mestghanem Alg. *see* **Mostaganem**
Mestre Italy **26** E2
Mesudiye Turkey **34** E2
Meta *r.* Col./Venez. **68** E2
Meta Incognita Peninsula Canada **61** L3
Metallifere, Colline *mts* Italy **26** D3
Metán Arg. **70** C3
Meteor Depth *sea feature* S. Atlantic Ocean **72** G9
Methoni Greece **27** I6
Methuen *MA* U.S.A. **64** F1
Methven U.K. **20** F4
Metković Croatia **26** G3
Metlaoui Tunisia **22** F5
Metoro Moz. **49** D5
Metsada *tourist site* Israel *see* **Masada**
Mettler *CA* U.S.A. **65** C3
Mettur India **38** C4
Metu Eth. **48** D3
Metz France **24** H2
Meuse *r.* Belgium/France **16** J5
Mevagissey U.K. **19** C8
Mexia U.S.A. **63** H5
Mexiana, Ilha *i.* Brazil **69** I3
Mexicali Mex. **66** A2
Mexico *country* Central America **66** D4
México Mex. *see* **Mexico City**
Mexico *NY* U.S.A. **64** C1
Mexico, Gulf of Mex./U.S.A. **63** H6
Mexico City Mex. **66** E5
Meybod Iran **35** I4
Meydān Shahr Afgh. **36** B2
Meyersdale *PA* U.S.A. **64** B3
Meymeh Iran **35** H4
Meynypil'gyno Rus. Fed. **77** C2
Mezada *tourist site* Israel *see* **Masada**
Mezdra Bulg. **27** J3
Mezen' Rus. Fed. **12** J2
Mezen' *r.* Rus. Fed. **12** J2
Mézenc, Mont *mt.* France **24** G4
Mezenskaya Guba *b.* Rus. Fed. **12** I2
Mezhdurechensk *Kemerovskaya Oblast'* Rus. Fed. **42** F2
Mezhdurechensk *Respublika Komi* Rus. Fed. **12** K3
Mezhdurechnye Rus. Fed. *see* **Shali**
Mezhdusharskiy, Ostrov *i.* Rus. Fed. **28** G2
Mezitli Turkey **39** B1
Mezőtúr Hungary **27** I1
Mežvidi Latvia **15** O8
Mhàil, Rubh' a' *pt* U.K. **20** C5
Mhangura Zimbabwe **49** D5
Mhlume Swaziland **51** J4
Mhow India **36** C5
Mi *r.* Myanmar **37** H5
Miahuatlán Mex. **66** E5
Miajadas Spain **25** D4
Miaméré Cent. Afr. Rep. **48** B3
Miami *FL* U.S.A. **63** K6
Miami *OK* U.S.A. **63** I4
Miami Beach U.S.A. **63** K6
Mīāndowāb Iran **35** G3
Miandrivazo Madag. **49** E5
Mīāneh Iran **35** G3
Miani Hor *b.* Pak. **36** A4
Mianwali Pak. **33** L3
Mianyang *Hubei* China *see* **Xiantao**
Mianyang *Sichuan* China **42** I6
Miarinarivo Madag. **49** E5
Miarritze France *see* **Biarritz**
Miass Rus. Fed. **28** H4
Mica Creek Canada **62** D1
Michalovce Slovakia **13** D6
Michelson, Mount U.S.A. **60** D3
Michigan *state* U.S.A. **63** J2
Michigan, Lake U.S.A. **63** J3
Michinberi India **38** D2
Michipicoten Island Canada **63** J2
Michipicoten River Canada **63** K2
Michurin Bulg. *see* **Tsarevo**
Michurinsk Rus. Fed. **13** I5
Micronesia *country* N. Pacific Ocean *see* **Micronesia, Federated States of**
Micronesia *is* Pacific Ocean **74** F5
Micronesia, Federated States of *country* N. Pacific Ocean **74** G5
Mid-Atlantic Ridge *sea feature* Atlantic Ocean **72** E4
Mid-Atlantic Ridge *sea feature* Atlantic Ocean **72** G8
Middelburg Neth. **16** I5
Middelburg *E. Cape* S. Africa **51** G6
Middelburg *Mpumalanga* S. Africa **51** I3
Middelfart Denmark **15** F9
Middelwit S. Africa **51** H3
Middle America Trench *sea feature* N. Pacific Ocean **75** N5
Middle Atlas *mts* Morocco *see* **Moyen Atlas**
Middlebourne *WV* U.S.A. **64** A3
Middleburg *PA* U.S.A. **64** C2
Middleburgh *NY* U.S.A. **64** D1
Middle Congo *country* Africa *see* **Congo**
Middlemarch N.Z. **59** C7
Middlemount Australia **56** E4
Middle River *MD* U.S.A. **64** C3
Middlesbrough U.K. **18** F4
Middleton Australia **56** C4
Middleton Island *atoll* American Samoa *see* **Rose Island**
Middletown *CA* U.S.A. **65** A1
Middletown *CT* U.S.A. **64** E2
Middletown *NY* U.S.A. **64** D2
Middletown *VA* U.S.A. **64** B3
Midelt Morocco **22** D5
Midhurst U.K. **19** G8
Midi, Canal du France **24** F5
Mid-Indian Basin *sea feature* Indian Ocean **73** N6
Mid-Indian Ridge *sea feature* Indian Ocean **73** M7
Midland Canada **63** L3
Midland *CA* U.S.A. **65** E4
Midland *TX* U.S.A. **62** G5
Midleton Ireland **21** D6
Midnapore India *see* **Medinipur**
Midnapur India *see* **Medinipur**
Midongy Atsimo Madag. **49** E6
Mid-Pacific Mountains *sea feature* N. Pacific Ocean **74** G4
Miðvágur Faroe Is **14** [inset]
Midway Oman *see* **Thamarīt**
Midway Islands *terr.* N. Pacific Ocean **74** I4
Midway Well Australia **55** C5
Midyat Turkey **35** F3
Midye Turkey *see* **Kıyıköy**
Mid Yell U.K. **20** [inset]
Midzhur *mt.* Bulg./Serbia **34** A2
Miehikkälä Fin. **15** O6
Miekojärvi *l.* Fin. **14** N3
Mielec Poland **13** D6
Mieraslompolo Fin. **14** O2
Mierašluoppal Fin. *see* **Mieraslompolo**
Miercurea-Ciuc Romania **27** K1
Mieres Spain **25** D2
Mieres del Camín Spain *see* **Mieres**
Mī'ēso Eth. **48** E3
Mifflinburg *PA* U.S.A. **64** C2
Mifflintown *PA* U.S.A. **64** C2
Migdol S. Africa **51** G4
Miguel Auza Mex. **62** G7
Mihalıççık Turkey **27** N5
Mihara Japan **45** D6
Mihintale Sri Lanka **38** D4
Mihmandar Turkey **39** B1
Mijares *r.* Spain *see* **Millárs**
Mikhaylov Rus. Fed. **13** H5
Mikhaylovgrad Bulg. *see* **Montana**
Mikhaylov Island Antarctica **76** E2
Mikhaylovka *Amurskaya Oblast'* Rus. Fed. **44** C2
Mikhaylovka *Primorskiy Kray* Rus. Fed. **44** D4
Mikhaylovka *Tul'skaya Oblast'* Rus. Fed. *see* **Kimovsk**
Mikhaylovka *Volgogradskaya Oblast'* Rus. Fed. **13** I6
Mikhaylovskoye Rus. Fed. **42** D2
Mikhaylovskoye Rus. Fed. *see* **Shpakovskoye**
Mikhrot Timna Israel **39** B5
Mikir Hills India **37** H4
Mikkeli Fin. **15** O6
Mikkelin mlk Fin. **15** O6
Mikonos *i.* Greece *see* **Mykonos**
Mikoyan Armenia *see* **Yeghegnadzor**
Mikulkin, Mys *c.* Rus. Fed. **12** J2
Mikumi National Park Tanz. **49** D4
Mikun' Rus. Fed. **12** K3
Mikuni-sanmyaku *mts* Japan **45** E5
Mikura-jima *i.* Japan **45** E6
Miladhunmadulu Atoll Maldives **38** B5
Miladummadulu Atoll Maldives *see* **Miladhunmadulu Atoll**
Milan Italy **26** C2
Milange Moz. **49** D5
Milano Italy *see* **Milan**
Milas Turkey **27** L6
Milazzo *Sicilia* Italy **26** F5
Milazzo, Capo di *c. Sicilia* Italy **26** F5
Mildenhall U.K. **19** H6
Mildura Australia **57** C7
Mile China **42** I8
Mileiz, Wādi el *watercourse* Egypt *see* **Mulayz, Wādī al**
Miles Australia **58** E1
Miles City U.S.A. **62** F2
Milestone Ireland **21** D5
Miletto, Monte *mt.* Italy **26** F4
Mileura Australia **55** B6
Milford Ireland **21** E2
Milford *DE* U.S.A. **64** D3
Milford *MA* U.S.A. **64** F1
Milford *NH* U.S.A. **64** F1
Milford *PA* U.S.A. **64** D2
Milford *UT* U.S.A. **65** F1
Milford *VA* U.S.A. **64** C3
Milford Haven U.K. **19** B7
Milford Sound N.Z. **59** A7
Milford Sound *inlet* N.Z. **59** A7
Milgarra Australia **56** C3
Milḥ, Baḥr al *l.* Iraq *see* **Razāzah, Buḥayrat ar**
Miliana Alg. **25** H5
Milid Turkey *see* **Malatya**
Milikapiti Australia **54** E2
Miling Australia **55** B7
Milk *r.* U.S.A. **62** F2
Milk, Wadi el *watercourse* Sudan **32** D6
Mil'kovo Rus. Fed. **29** Q4
Millaa Millaa Australia **56** D3
Millárs *r.* Spain **25** F4

llau France 24 F4
lledgeville U.S.A. 63 K5
lle Lacs lakes U.S.A. 63 I2
lle Lacs, Lac des l. Canada 61 I5
llennium Island atoll Kiribati see
Caroline Island
ller U.S.A. 62 H3
llerovo Rus. Fed. 13 I6
llersburg PA U.S.A. 64 C2
llerton Lake CA U.S.A. 65 C2
lleur Point U.K. 20 D5
ll Hall PA U.S.A. 64 C2
llicent Australia 57 C8
ll Island Canada 61 K3
llmerran Australia 58 E1
llom U.K. 18 D4
llport U.K. 20 E5
llsboro DE U.S.A. 64 D3
lls Creek watercourse Australia 56 C4
lls Lake Canada 60 G3
llstone WV U.S.A. 64 A3
llstream-Chichester National Park
Australia 54 B5
llthorpe Australia 58 D4
llungera Australia 56 C3
llville NJ U.S.A. 64 D3
lly Milly Australia 55 B6
lne Land i. Greenland see
Imannangip Nunaa
lo r. Guinea 46 C3
logradovo Rus. Fed. 44 D4
los i. Greece 27 K6
lparinka Australia 57 C6
lpitas CA U.S.A. 65 B2
lroy PA U.S.A. 64 C2
lton N.Z. 59 B8
lton DE U.S.A. 64 D3
lton NH U.S.A. 64 F1
lton Keynes U.K. 19 G6
ltown Malbay Ireland 21 C5
lverton Ont. Canada 64 A1
lwaukee U.S.A. 63 J3
lwaukee Deep sea feature
Caribbean Sea 72 D4
mili Australia 55 F6
misal India 38 C4
mizan France 24 D4
mongo Gabon 48 B4
mosa Rocks National Park Australia
58 E6
na NV U.S.A. 65 C1
nāb Iran 33 I4
naçu Brazil 71 A1
nahasa, Semenanjung pen. Indon.
41 E7
nahassa Peninsula Indon. see
Minahasa, Semenanjung
naker Canada see Prophet River
nakh Syria 39 C1
namia Australia 54 F3
nami-Daitō-jima i. Japan 43 O7
nas Indon. 41 C7
nas Uruguay 70 E4
nas Gerais state Brazil 71 B2
nas Novas Brazil 71 C2
natitlán Mex. 66 F5
nbu Myanmar 37 H5
nbya Myanmar 37 H5
nchinmávida vol. Chile 70 B6
ndanao i. Phil. 41 E6
ndanao Trench sea feature
N. Pacific Ocean see Philippine Trench
ndelo Cape Verde 46 [inset]
nden Germany 17 L4
nden LA U.S.A. 63 I5
nden NE U.S.A. 62 H3
nden NV U.S.A. 65 C1
ndon Myanmar 37 H6
ndoro i. Phil. 41 E6
ndouli Congo 48 B4
ne Head hd Ireland 21 E6
nehead U.K. 19 D7
neola NY U.S.A. 64 E2
neral VA U.S.A. 64 C3
neral'nyye Vody Rus. Fed. 35 F1
neral Wells U.S.A. 62 H5
neralwells WV U.S.A. 64 A3
nersville UT U.S.A. 65 F1
nerva OH U.S.A. 64 A2
nerva Reefs Fiji 53 I4
nfeng China 37 E1
nga Dem. Rep. Congo 49 C5
ngäçevir Azer. 35 G2
ngäçevir Su Anbarı resr Azer. 35 G2
ngala Cent. Afr. Rep. 48 C3
ngbuloq Uzbek. 33 J1
ngechaur Azer. see Mingäçevir
ngechaurskoye Vodokhranilishche resr
Azer. see Mingäçevir Su Anbarı
ngenew Australia 55 A7
ngin Range mts Myanmar 37 H5
nglanilla Spain 25 F4
ngoyo Tanz. 49 D5
ngshui Gansu China 42 H4
ngshui Heilong. China 44 B3
ngteke China 36 C1
ngulay i. U.K. 20 B4
ngzhou China see Suide
nhe China see Jinxian
nhla Magwe Myanmar 37 H6
nho r. Port./Spain see Miño
nicoy atoll India 38 B4
nigwal, Lake imp. l. Australia 55 C7
nilya Australia 55 A5
nilya r. Australia 55 A5
nipi Lake Canada 61 L4
nna Nigeria 46 D4
nna Bluff pt Antarctica 76 H1
nne Sweden 14 I5
nneapolis MN U.S.A. 63 I3
nnedosa Canada 62 H1
nnehaha Springs WV U.S.A. 64 B3
nnesota r. U.S.A. 63 I3
nnesota state U.S.A. 63 I2
nnitaki Lake Canada 63 I1
ño r. Port./Spain 25 B3
norca i. Spain 25 H3
not U.S.A. 62 G2
nqār, Ghadīr imp. l. Syria 39 C3
nsin Myanmar 37 H4
nsk Belarus 15 O10
ńsk Mazowiecki Poland 17 R4
nsterley U.K. 19 E6
ntaka Pass China/Pakistan 36 C1
nto, Lac l. Canada 61 K4
nto, Mount Antarctica 76 H2
nto Inlet Canada 60 G2
nū Dasht Iran 35 I3
nūf Egypt 34 C5
nusinsk Rus. Fed. 42 G2
nvoul Gabon 48 B3
nxian China 42 I6
nya Konka mt. China see Gongga Shan
nywa Myanmar 37 H5
nzong India 37 I4
rabela Brazil 71 B2
rador, Parque Nacional de nat. park
Brazil 69 I5
rah, Wādī al watercourse Iraq/
Saudi Arabia 35 F4
raí Brazil 71 C3
Miraj India 38 B2
Miramar Arg. 70 E5
Miramichi Canada 63 N2
Miramichi Bay Canada 63 N2
Mirampellou, Kolpos b. Greece 27 K7
Miranda Brazil 70 E2
Miranda Moz. see Macaloge
Miranda, Lake imp. l. Australia 55 C6
Miranda de Ebro Spain 25 E2
Mirandela Port. 25 C3
Mirandola Italy 26 D2
Mirante Brazil 71 C1
Mirante, Serra do hills Brazil 71 A3
Mirassol Brazil 71 A3
Mir-Bashir Azer. see Tärtär
Mirbāṭ Oman 33 H6
Mirboo North Australia 58 C7
Mirepoix France 24 E5
Mirgorod Ukr. see Myrhorod
Miri Sarawak Malaysia 41 D7
Mirialguda India 38 C2
Miri Hills India 37 H4
Mirim, Lagoa l. Brazil/Uruguay 70 F4
Mirim, Lagoa do l. Brazil 71 A5
Mirintu watercourse Australia 58 A2
Mirjan India 38 B3
Mirny (Russian Federation) research stn
Antarctica 76 F2
Mirnyy Arkhangel'skaya Oblast' Rus. Fed.
12 I3
Mirnyy Respublika Sakha (Yakutiya)
Rus. Fed. 29 M3
Mironovka Ukr. see Myronivka
Mirpur Khas Pak. 33 K4
Mirpur Sakro Pak. 36 A4
Mirtoan Sea Greece see Myrtoo Pelagos
Miryalaguda India see Mirialguda
Miryang S. Korea 45 C6
Mirzachul Uzbek. see Guliston
Mirzapur India 37 E4
Mirzawal India 36 C3
Mi-shima i. Japan 45 C6
Mishmi Hills India 37 H3
Mishvan' Rus. Fed. 12 L2
Misima Island P.N.G. 56 F1
Misis Dağ hills Turkey 39 B1
Miskitos, Cayos is Nicaragua 67 H6
Miskolc Hungary 13 D6
Mismā, Tall al h. Jordan 39 C3
Misoöl i. Indon. 41 F8
Misr country Africa see Egypt
Misraç Turkey see Kurtalan
Mişrātah Libya 47 E1
Missinaibi r. Canada 63 K1
Mission Beach Australia 56 D3
Mission Viejo CA U.S.A. 65 D4
Missisa Lake Canada 63 J1
Mississauga Ont. Canada 64 B1
Mississippi r. U.S.A. 63 J6
Mississippi state U.S.A. 63 J5
Mississippi Delta U.S.A. 63 J6
Mississippi-Missouri r. U.S.A. 63 I4
Missolonghi Greece see Mesolongi
Missoula U.S.A. 62 E2
Missouri r. U.S.A. 63 I4
Missouri state U.S.A. 63 I4
Mistassibi r. Canada 61 K5
Mistassini, Lac l. Canada 63 M1
Mistelbach Austria 17 P6
Mistinibi, Lac l. Canada 61 L4
Mistissini Canada 63 M1
Misty Fiords National Monument
Wilderness nat. park U.S.A. 60 E4
Misumba Dem. Rep. Congo 49 C4
Misuratah Libya see Mişrātah
Mitchell Australia 57 D5
Mitchell r. N.S.W. Australia 58 F2
Mitchell r. Qld Australia 56 C2
Mitchell r. Vic. Australia 58 C6
Mitchell Ont. Canada 64 A1
Mitchell SD U.S.A. 62 H3
Mitchell, Lake Australia 56 D3
Mitchell, Mount U.S.A. 63 K4
Mitchell and Alice Rivers National Park
Australia 56 C2
Mitchell Island Cook Is see Nassau
Mitchell Island atoll Tuvalu see
Nukulaelae
Mitchell Point Australia 54 E2
Mitchelstown Ireland 21 D5
Mît Ghamr Egypt see Mīt Ghamr
Mīt Ghamr Egypt 34 C5
Mithi Pak. 36 B4
Mithri Pak. 36 A3
Mitilíni Greece see Mytilini
Mito Japan 45 F5
Mitole Tanz. 49 D4
Mitre mt. N.Z. 59 E5
Mitre Island Solomon Is 53 H3
Mitrofanovka Rus. Fed. 13 H6
Mitrovica Kosovo see Mitrovicë
Mitrovicë Kosovo 27 I3
Mitsinjo Madag. 49 E5
Mits'iwa Eritrea see Massawa
Mitta Mitta Australia 58 C6
Mittellandkanal canal Germany 17 L4
Mittimatalik Canada see Pond Inlet
Mitú Col. 68 D3
Mitumba, Chaîne des mts
Dem. Rep. Congo 49 C5
Mitzic Gabon 48 B3
Miughalaigh i. U.K. see Mingulay
Miura Japan 45 E6
Miyake-jima i. Japan 45 E6
Miyako Japan 45 F5
Miyakonojō Japan 45 C7
Miyang China see Mile
Miyani India 36 B5
Miyazaki Japan 45 C7
Miyazu Japan 45 D6
Miyoshi Japan 45 D6
Mīzān 'Alāqadārī Afgh. 36 A2
Mīzan Teferī Eth. 48 D3
Mizdah Libya 47 E1
Mizen Head hd Ireland 21 C6
Mizhhir"ya Ukr. 13 D6
Mizo Hills state India see Mizoram
Mizoram state India 37 H5
Mizpé Ramon Israel 39 B4
Mizusawa Japan 45 F5
Mjölby Sweden 15 I7
Mkata Tanz. 49 D4
Mkushi Zambia 49 C5
Mladá Boleslav Czech Rep. 17 O5
Mladenovac Serbia 27 I2
Mława Poland 17 R4
Mlilwane Nature Reserve Swaziland 51 J4
Mljet i. Croatia 26 G3
Mlungisi S. Africa 51 H6
Mmabatho S. Africa 51 G3
Mmamabula Botswana 51 H2
Mmathethe Botswana 51 G3
Mo Norway 15 D6
Moa i. Indon. 54 E2
Moab reg. Jordan 39 B4
Moab U.S.A. 62 F4
Moa Island Australia 56 C1
Moala i. Fiji 53 H3
Mo'alla Iran 35 I4
Moamba Moz. 51 K3
Moanda Gabon 48 B4
Moapa NV U.S.A. 65 E2
Moate Ireland 21 E4
Mobārakeh Iran 35 H4
Mobayembongo Dem. Rep. Congo see
Mobayi-Mbongo
Mobayi-Mbongo Dem. Rep. Congo 48 C3
Moberly U.S.A. 63 I4
Mobha India 36 C5
Mobile AL U.S.A. 63 J5
Mobile Bay U.S.A. 63 J5
Moble watercourse Australia 58 B1
Mobridge U.S.A. 62 G2
Mobutu, Lake Dem. Rep. Congo/Uganda
see Albert, Lake
Mobutu Sese Seko, Lake
Dem. Rep. Congo/Uganda see Albert, Lake
Moca Geçidi pass Turkey 39 A1
Moçambique country Africa see
Mozambique
Moçambique Moz. 49 E5
Moçâmedes Angola see Namibe
Mocha Yemen 32 F7
Mocha, Isla i. Chile 70 B5
Mochirma, Parque Nacional nat. park
Venez. 68 F1
Mochudi Botswana 51 H3
Mochudi admin. dist. Botswana see
Kgatleng
Mocimboa da Praia Moz. 49 E5
Mockträsk Sweden 14 L4
Mocoa Col. 68 C3
Mococa Brazil 71 B3
Mocoduene Moz. 51 L2
Moctezuma Chihuahua Mex. 66 C2
Moctezuma San Luis Potosí Mex. 66 D4
Mocuba Moz. 49 D5
Modane France 24 H4
Modder r. S. Africa 51 G5
Modena Italy 26 D2
Modena UT U.S.A. 65 F2
Modesto CA U.S.A. 65 B2
Modesto Lake CA U.S.A. 65 B2
Modimolle S. Africa 51 I3
Modung China 37 I3
Moe Australia 58 C7
Moel Sych h. U.K. 19 D6
Moelv Norway 15 G6
Moen Norway 14 K2
Moeraki Point N.Z. 59 C7
Moero, Lake Dem. Rep. Congo/Zambia
see Mweru, Lake
Moffat U.K. 20 F5
Moga India 36 C3
Mogadishu Somalia 48 E3
Mogador Morocco see Essaouira
Mogadore Reservoir OH U.S.A. 64 A2
Moganyaka S. Africa 51 I3
Mogaung Myanmar 42 H7
Mogdy Rus. Fed. 44 D2
Mogilev Belarus see Mahilyow
Mogilev Podol'skiy Ukr. see
Mohyliv Podil's'kyy
Mogi-Mirim Brazil 71 B3
Mogiquiçaba Brazil 71 D2
Mogocha Rus. Fed. 43 L2
Mogod mts Tunisia 26 C6
Mogoditshane Botswana 51 G3
Mogontiacum Germany see Mainz
Mogroum Chad 47 E3
Moguqi China 44 A3
Mogwadi S. Africa 51 I2
Mogwase S. Africa 51 H3
Mogzon Rus. Fed. 43 K2
Mohács Hungary 26 H2
Mohaka r. N.Z. 59 F4
Mohala India 38 D1
Mohale Dam Lesotho 51 I5
Mohale's Hoek Lesotho 51 H6
Mohali India 36 D3
Mohammadia Alg. 25 G6
Mohan r. India/Nepal 36 E3
Mohana India 36 D4
Mohave, Lake NV U.S.A. 65 E3
Mohawk r. NY U.S.A. 64 E1
Mohawk Mountains AZ U.S.A. 65 F4
Mohenjo Daro tourist site Pak. 36 B4
Moher, Cliffs of Ireland 21 C5
Mohill Ireland 21 E4
Mohon Peak AZ U.S.A. 65 F3
Mohoro Tanz. 49 D4
Mohyliv Podil's'kyy Ukr. 13 E6
Moi Norway 15 E7
Moijabana Botswana 51 H2
Moincêr China 36 E3
Moinda China 37 G3
Moine Moz. 51 K3
Moineşti Romania 27 L1
Mointy Kazakh. see Moyynty
Mo i Rana Norway 14 I3
Mõisaküla Estonia 15 N7
Moissac France 24 E4
Mojave CA U.S.A. 65 C3
Mojave r. CA U.S.A. 65 D3
Mojave Desert CA U.S.A. 65 D3
Mojiang China 42 I8
Moji das Cruzes Brazil 71 B3
Mojos, Llanos de plain Bol. 68 E6
Moju r. Brazil 69 I4
Mokama India 37 F4
Mokau N.Z. 59 E4
Mokau r. N.Z. 59 E4
Mokelumne r. CA U.S.A. 65 B1
Mokelumne Aqueduct canal CA U.S.A.
65 B1
Mokhoabong Pass Lesotho 51 I5
Mokhotlong Lesotho 51 I5
Moknine Tunisia 26 D7
Mokohinau Islands N.Z. 59 E2
Mokokchung India 37 H4
Mokolo Cameroon 47 E3
Mokolo r. S. Africa 51 H2
Mokopane S. Africa 51 I3
Mokp'o S. Korea 45 B6
Mokrous Rus. Fed. 13 J6
Moksha r. Rus. Fed. 13 I5
Mokshan Rus. Fed. 13 J5
Möksy Fin. 14 N5
Môktama, Gulf of Myanmar see
Mottama, Gulf of
Mokundurra India see Mukandwara
Mokwa Nigeria 46 D4
Molatón mt. Spain 25 F4
Moldavia country Europe see Moldova
Moldavskaya S.S.R. country Europe see
Moldova
Molde Norway 14 E5
Moldjord Norway 14 I3
Moldova country Europe 13 F7
Moldoveanu, Vârful mt. Romania 27 K2
Moldovei de Sud, Cîmpia plain Moldova
27 M1
Molen r. S. Africa 51 I4
Mole National Park Ghana 46 C4
Molepolole Botswana 51 G3
Molėtai Lith. 15 N9
Molfetta Italy 26 G4
Molière Alg. see Bordj Bounaama
Molihong Shan mt. China see
Morihong Shan
Molina de Aragón Spain 25 F3
Molkom Sweden 15 H7
Mollagara Turkm. 35 I3
Mollakara Turkm. see Mollagara
Mol Len mt. India 37 H4
Mollendo Peru 68 D7
Mölnlycke Sweden 15 H8
Molochnyy Rus. Fed. 14 R2
Molodechno Belarus see Maladzyechna
Molodezhnaya (Russian Federation)
research stn Antarctica 76 D2
Moloma r. Rus. Fed. 12 K4
Molong Australia 58 D4
Molopo watercourse Botswana/S. Africa
50 E5
Molotov Rus. Fed. see Perm'
Molotovsk Arkhangel'skaya Oblast'
Rus. Fed. see Severodvinsk
Molotovsk Kirovskaya Oblast' Rus. Fed. see
Nolinsk
Moloundou Cameroon 47 E4
Molson Lake Canada 63 H1
Moluccas is Indon. 41 E8
Molucca Sea sea Indon. see Maluku, Laut
Moma Moz. 49 D5
Moma r. Rus. Fed. 29 P3
Momba Australia 58 A3
Mombaça Brazil 69 K5
Mombasa Kenya 48 D4
Mombetsu Hokkaidō Japan see Monbetsu
Mombetsu Hokkaidō Japan see Monbetsu
Mombi New India 37 H4
Momchilgrad Bulg. 27 K4
Momi, Ra's pt Yemen 33 H7
Mompós Col. 68 D2
Møn i. Denmark 15 H9
Mon India 37 H4
Mona terr. Irish Sea see Isle of Man
Monach, Sound of sea chan. U.K. 20 B3
Monach Islands U.K. 20 B3
Monaco country Europe 24 H5
Monaco Basin sea feature
N. Atlantic Ocean 72 G4
Monadhliath Mountains U.K. 20 E3
Monaghan Ireland 21 F3
Mona Passage Dom. Rep./Puerto Rico
67 K5
Monapo Moz. 49 E5
Monar, Loch l. U.K. 20 D3
Monarch Pass U.S.A. 62 F4
Monastir Macedonia see Bitola
Monastir Tunisia 26 D7
Monastyrishche Ukr. see Monastyryshche
Monastyryshche Ukr. 13 F6
Monbetsu Hokkaidō Japan 44 F3
Monbetsu Hokkaidō Japan 44 F4
Moncalieri Italy 26 B2
Moncayo mt. Spain 25 F3
Monchegorsk Rus. Fed. 12 G2
Mönchengladbach Germany 17 K5
Monchique Port. 25 B5
Monclova Mex. 66 D3
Moncton Canada 63 O2
Mondego r. Port. 25 B3
Mondlo S. Africa 51 J4
Mondo Chad 47 E3
Mondovì Italy 26 B2
Mondragone Italy 26 E4
Mondy Rus. Fed. 42 I2
Monemvasia Greece 27 J6
Monessen PA U.S.A. 64 B2
Moneygall Ireland 21 E5
Moneymore U.K. 21 F3
Monfalcone Italy 26 E2
Monfalut Egypt see Manfalūṭ
Monforte de Lemos Spain 25 C2
Monga Dem. Rep. Congo 48 C3
Mongala r. Dem. Rep. Congo 48 B3
Mongar Bhutan 37 G4
Mongbwalu Dem. Rep. Congo 48 D3
Móng Cai Vietnam 31 J4
Mongers Lake imp. l. Australia 55 B7
Monghyr India see Munger
Mongla Bangl. 37 G5
Mong Loi Myanmar 42 I8
Mong Long Myanmar 37 I5
Mongo Chad 47 E3
Mongolia country Asia 42 I3
Mongol Uls country Asia see Mongolia
Mongora Pak. 33 L3
Mongour h. U.K. 20 G4
Mongu Zambia 49 C5
Monguno Nigeria 46 E3
Mönhaan Mongolia 43 K3
Mönh Hayrhan Uul mt. Mongolia 42 G3
Moniaive U.K. 20 F5
Monitor Mountain NV U.S.A. 65 D1
Monitor Range mts NV U.S.A. 65 D1
Monivea Ireland 21 D4
Monkey Bay Malawi 49 D5
Monkira Australia 56 C5
Monkton Ont. Canada 64 A1
Monmouth U.K. 19 E7
Monnow r. U.K. 19 E7
Mono, Punta del pt Nicaragua 67 H6
Mono Lake CA U.S.A. 65 C1
Monolithos Greece 27 L6
Monomoy Point MA U.S.A. 64 F2
Monopoli Italy 26 G4
Monreal del Campo Spain 25 F3
Monreale Sicilia Italy 26 E5
Monroe LA U.S.A. 63 I5
Monroeton PA U.S.A. 64 C2
Monrovia Liberia 46 B4
Mons Belgium 16 I5
Monselice Italy 26 D2
Montagne d'Ambre, Parc National de
nat. park Madag. 49 E5
Montagu S. Africa 50 E7
Montague Range hills Australia 55 B6
Montalto mt. Italy 26 F5
Montalto Uffugo Italy 26 G5
Montana Bulg. 27 J3
Montana state U.S.A. 62 F2
Montanhas do Tumucumaque, Parque
Nacional nat. park 69 H3
Montargis France 24 F3
Montauban France 24 E4
Montauk NY U.S.A. 64 F2
Montauk Point NY U.S.A. 64 F2
Mont-aux-Sources mt. Lesotho 51 I5
Montbard France 24 G3
Mont Blanc mt. France/Italy 24 H4
Montblanc Spain 25 G3
Montblanch Spain see Montblanc
Montbrison France 24 G4
Montceau-les-Mines France 24 G3
Mont-de-Marsan France 24 D5
Montdidier France 24 F2
Monte Alegre Brazil 69 H4
Monte Alegre de Goiás Brazil 71 B1
Monte Alegre de Minas Brazil 71 A2
Monte Azul Brazil 71 C1
Monte Azul Paulista Brazil 71 A3
Montebello Islands Australia 54 A5
Montebelluna Italy 26 E2
Monte-Carlo Monaco 24 H5
Monte Cristi Dom. Rep. 67 J5
Monte Cristo S. Africa 51 H2
Monte Dourado Brazil 69 H4
Monte Falterona, Campigna e delle
Foreste Casentinesi, Parco Nazionale del
nat. park Italy 26 D3
Montego Bay Jamaica 67 I5
Montélimar France 24 G4
Monte Lindo r. Para. 70 E2
Montemorelos Mex. 66 E3
Montemor-o-Novo Port. 25 B4
Montenegro country Europe 26 H3
Montepulciano Italy 26 D3
Monte Quemado Arg. 70 D3
Montereau-Fault-Yonne France 24 F2
Monterey Mex. see Monterrey
Monterey CA U.S.A. 65 B2
Monterey VA U.S.A. 64 B3
Monterey Bay CA U.S.A. 65 A2
Montería Col. 68 C2
Monteros Arg. 70 C3
Monterrey Baja California Mex. 65 E4
Monterrey Nuevo León Mex. 66 D3
Montesano sulla Marcellana Italy 26 F4
Monte Santo Brazil 69 K6
Monte Santu, Capo di c. Sardegna Italy
26 C4
Montes Claros Brazil 71 C2
Montesilvano Italy 26 F3
Montevarchi Italy 26 D3
Montevideo Uruguay 70 E4
Montezuma Peak NV U.S.A. 65 D2
Montgomery U.K. 19 D6
Montgomery AL U.S.A. 63 J5
Montgomery WV U.S.A. 64 A3
Montgomery Islands Australia 54 C3
Monthey Switz. 24 H3
Monticello NY U.S.A. 64 D2
Monticello UT U.S.A. 62 F4
Montignac France 24 E4
Montilla Spain 25 D5
Monti Sibillini, Parco Nazionale dei
nat. park Italy 26 E3
Montividiu Brazil 71 A2
Montivilliers France 19 H9
Mont-Laurier Canada 63 L2
Montluçon France 24 F3
Montmagny Canada 63 M2
Montmorillon France 24 E3
Monto Australia 56 E5
Montour Falls NY U.S.A. 64 C1
Montpelier VT U.S.A. 63 M3
Montpellier France 24 F5
Montréal Canada 61 K5
Montreal Lake Canada 62 F1
Montreuil France 19 I8
Montreux Switz. 24 H3
Montrose well S. Africa 50 E4
Montrose U.K. 20 G4
Montrose CO U.S.A. 62 F4
Montrose PA U.S.A. 64 D2
Montross VA U.S.A. 64 C3
Mont-St-Aignan France 19 I9
Montserrat terr. West Indies 67 L5
Monywa Myanmar 37 H5
Monza Italy 26 C2
Monze, Cape c. Pak. see Muari, Ras
Monzón Spain 25 G3
Mooi r. S. Africa 51 J5
Mooifontein Namibia 50 C4
Mookane Botswana 51 H2
Mookgopong S. Africa see Naboomspruit
Moolawatana Australia 57 B6
Moomba Australia 57 C6
Moomin Creek r. Australia 58 D2
Moonaree Australia 57 A6
Moonbi Range mts Australia 58 E3
Moonda Lake imp. l. Australia 57 C5
Moonie Australia 58 E1
Moonie r. Australia 58 D2
Moora Australia 55 B7
Mooraberree Australia 56 C5
Moore r. Australia 55 A7
Moore, Lake imp. l. Australia 55 B7
Moore Embayment b. Antarctica 76 H1
Moorefield WV U.S.A. 64 B3
Moore Reef Australia 56 E3
Moore River National Park Australia 55 A7
Moorfoot Hills U.K. 20 F5
Moorhead U.S.A. 63 H2
Moornanyah Lake imp. l. Australia 58 A4
Mooroopna Australia 58 B6
Moorreesburg S. Africa 50 D7
Moorrinya National Park Australia 56 D4
Moose r. Canada 63 K1
Moose Factory Canada 63 K1
Moosehead Lake U.S.A. 63 N2
Moose Jaw Canada 62 F1
Moosomin Canada 62 G1
Moosonee Canada 63 K1
Mootwingee National Park Australia
57 C6
Mopane S. Africa 51 I2
Mopeia Moz. 49 D5
Mopipi Botswana 49 C6
Mopti Mali 46 C3
Moqor Afgh. 36 A2
Moquegua Peru 68 D7
Mora Cameroon 47 E3
Mora Spain 25 E4
Mora Sweden 15 I6
Moradabad India 36 D3
Morada Nova Brazil 69 K5
Moraleda, Canal sea chan. Chile 70 B6
Moram India 38 C2
Moramanga Madag. 49 E5
Moranbah Australia 56 E4
Morang Nepal see Biratnagar
Morar, Loch l. U.K. 20 D4
Morari, Tso l. India 36 D2
Moratuwa Sri Lanka 38 C5
Morava reg. Czech Rep. 17 P6
Moravia NY U.S.A. 64 C1
Morawa Australia 55 A7
Moray Firth b. U.K. 20 E3
Moray Range hills Australia 54 E3
Morbeng S. Africa see Soekmekaar
Morbi India 36 B5
Morcenx France 24 D4
Mordaga China 43 M2
Mor Dağı mt. Turkey 35 G3
Morden Canada 62 H2
Mordovo Rus. Fed. 13 I5
Morecambe U.K. 18 E4
Morecambe Bay U.K. 18 D4
Moree Australia 58 D2
Morehead P.N.G. 52 E2
Morehead City U.S.A. 67 I2
More Laptevykh sea Rus. Fed. see
Laptev Sea
Morelia Mex. 66 D5
Morella Australia 56 C4
Morella Spain 25 F3
Morena India 36 D4
Morena, Sierra mts Spain 25 C5
Moreni Romania 27 K2
Moreno Valley CA U.S.A. 65 D4
Moresby, Mount Canada 60 E4
Moresby Island Canada see Gwaii Haanas
Moreswe Pan salt pan Botswana 50 G2
Moreton Bay Australia 58 F1
Moreton-in-Marsh U.K. 19 F7
Moreton Island Australia 58 F1
Moreton Island National Park Australia
58 F1
Morez France 24 H3
Morfou Cyprus 39 A2
Morfou Bay Cyprus 39 A2
Morgan Hill CA U.S.A. 65 B2
Morganton U.S.A. 63 K4
Morgantown WV U.S.A. 64 B3
Morgenzon S. Africa 51 I4
Morges Switz. 24 H3
Morgh, Kowtal-e Afgh. 36 B2
Morghāb r. Afgh. 36 A2
Morhar r. India 37 F4
Mori China 42 G4
Mori Japan 44 F4
Moriah, Mount NV U.S.A. 65 E1
Moriarty's Range hills Australia 58 B2
Morice Lake Canada 60 F4
Morichal Col. 68 D3
Morihong Shan mt. China 44 B4
Morija Lesotho 51 H5
Morin Dawa China see Nirji
Morioka Japan 45 F5
Morisset Australia 58 E4
Moriyoshi-zan vol. Japan 45 F5
Morjärv Sweden 14 M3
Morki Rus. Fed. 12 K4
Morlaix France 24 C2
Morley U.K. 18 F5
Mormugao India see Marmagao
Morne Diablotins vol. Dominica 67 L5
Morney watercourse Australia 56 C5
Mornington, Isla i. Chile 70 A7
Mornington Abyssal Plain sea feature
S. Atlantic Ocean 72 C9
Mornington Island Australia 56 B3
Mornington Peninsula National Park
Australia 58 B7
Moro Pak. 36 A4
Morobe P.N.G. 52 E2
Morocco country Africa 46 C1
Morococala mt. Bol. 68 E7
Morogoro Tanz. 49 D4
Moro Gulf Phil. 41 E7
Morojaneng S. Africa 51 H5
Morokweng S. Africa 50 F4
Morombe Madag. 49 E6
Mörön Mongolia 42 I3
Morondava Madag. 49 E6
Morón de la Frontera Spain 25 D5
Moroni Comoros 49 E5
Moron Us He r. China see Tongtian He
Morotai i. Indon. 41 E7
Moroto Uganda 48 D3
Morozovsk Rus. Fed. 13 I6
Morpeth Ont. Canada 64 A1
Morpeth U.K. 18 F3
Morphou Cyprus see Morfou
Morrinhos Brazil 71 A2
Morris Canada 63 H2
Morris PA U.S.A. 64 C2
Morristown AZ U.S.A. 65 F4
Morristown NJ U.S.A. 64 D2
Morristown TN U.S.A. 63 K4
Morrisville NY U.S.A. 64 D1
Morro Brazil 71 B2
Morro Bay CA U.S.A. 65 B3
Morro d'Anta Brazil 71 D2
Morro do Chapéu Brazil 69 J6
Morro Grande h. Brazil 69 H4
Morrosquillo, Golfo de b. Col. 68 C2
Morrumbene Moz. 51 L2
Morse, Cape Antarctica 76 G2
Morshanka Rus. Fed. 13 I5
Morshansk Rus. Fed. see Morshanka
Morsott Alg. 26 C7
Mort watercourse Australia 56 C4
Mortagne-au-Perche France 24 E2
Mortagne-sur-Sèvre France 24 D3
Mortara Italy 26 C2
Mortehoe U.K. 19 C7
Morteros Arg. 70 D4
Mortes, Rio das r. Brazil 71 A1
Mortlake Australia 58 A7
Mortlock Islands Micronesia 74 G5
Mortlock Islands P.N.G. see Takuu Islands
Morton U.K. 19 G6
Morton National Park Australia 58 E5
Morundah Australia 58 C5
Morupule Botswana 51 H2
Moruroa atoll Fr. Polynesia see Mururoa
Moruya Australia 58 E5
Morven Australia 57 D5
Morven h. U.K. 20 F2
Morvern reg. U.K. 20 D4
Morvi India see Morbi
Morwell Australia 58 C7
Morzhovets, Ostrov i. Rus. Fed. 12 I2
Mosbach Germany 17 L6
Mosborough U.K. 18 F5
Moscow Rus. Fed. 12 H5
Moscow ID U.S.A. 62 D2
Moscow PA U.S.A. 64 D2
Moscow University Ice Shelf Antarctica
76 G2
Moselebe watercourse Botswana 50 F3
Moselle r. France 24 H2
Moses Lake U.S.A. 62 D2
Mosgiel N.Z. 59 C7
Moshaweng watercourse S. Africa 50 F4
Moshchnyy, Ostrov i. Rus. Fed. 15 O7
Moshi Tanz. 48 D4
Mosh'yuga Rus. Fed. 12 L2
Mosi-oa-Tunya waterfall Zambia/
Zimbabwe see Victoria Falls
Mosjøen Norway 14 H4
Moskal'vo Rus. Fed. 44 F1
Moskenesøy i. Norway 14 H3
Moskva Rus. Fed. see Moscow
Mosonmagyaróvár Hungary 17 P7
Mosquera Col. 68 C3
Mosquito r. Brazil 71 C1
Mosquito Creek Lake OH U.S.A. 64 A2
Mosquitos, Golfo de los b. Panama 67 H7
Moss Norway 15 G7
Mossâmedes Angola see Namibe
Mossat U.K. 20 G3
Mossburn N.Z. 59 B7
Mosselbaai S. Africa see Mossel Bay
Mossel Bay S. Africa 50 F8
Mossel Bay b. S. Africa 50 F8
Mossgiel Australia 58 B4
Mossman Australia 56 D3
Mossoró Brazil 69 K5
Moss Vale Australia 58 E5
Most Czech Rep. 17 N5
Mostaganem Alg. 25 G6
Mostar Bos.-Herz. 26 G3
Mostovskoy Rus. Fed. 35 F1
Mosty Belarus see Masty
Mosul Iraq 35 F3
Møsvatnet l. Norway 15 F7
Motala Sweden 15 I7
Motaze Moz. 51 K3
Motetema S. Africa 51 I3
Moth India 36 D4
Motherwell U.K. 20 F5
Motian Ling h. China 44 A4
Motihari India 37 F4
Motilla del Palancar Spain 25 F4
Motiti Island N.Z. 59 F3
Motokwe Botswana 50 F3
Motril Spain 25 E5
Motru Romania 27 J2
Mottama, Gulf of Myanmar 31 I5

Motu Ihupuku i. N.Z. see Campbell Island
Motul Mex. 66 G4
Mouaskar Alg. see Mascara
Moudjéria Mauritania 46 B3
Moudros Greece 27 K5
Mouhijärvi Fin. 15 M6
Mouila Gabon 48 B4
Moulamein Australia 58 B5
Moulamein Creek r. Australia 58 A5
Moulavibazar Bangl. see Moulvibazar
Mould Bay Canada 60 G2
Moulèngui Binza Gabon 48 B4
Moulins France 24 F3
Moulouya r. Morocco 22 D4
Moultrie U.S.A. 63 K5
Moultrie, Lake U.S.A. 63 L5
Moulvibazar Bangl. 37 G4
Moundou Chad 47 E4
Moundsville WV U.S.A. 64 A3
Mount Abu India 36 C4
Mountain Home AR U.S.A. 63 I4
Mountain Home ID U.S.A. 62 D3
Mountain Lake Park MD U.S.A. 64 B3
Mountain Zebra National Park S. Africa 51 G7
Mount Aspiring National Park N.Z. 59 B7
Mount Ayliff S. Africa 51 I6
Mountbellew Ireland 21 D4
Mount Buffalo National Park Australia 58 C6
Mount Carmel Junction UT U.S.A. 65 F2
Mount Coolon Australia 56 D4
Mount Darwin Zimbabwe 49 D5
Mount Denison Australia 54 F5
Mount Dutton Australia 57 A5
Mount Eba Australia 57 A6
Mount Elgon National Park Uganda 48 D3
Mount Fletcher S. Africa 51 I6
Mount Frankland National Park Australia 55 B8
Mount Frere S. Africa 51 I6
Mount Gambier Australia 57 C8
Mount Hagen P.N.G. 52 E2
Mount Holly NJ U.S.A. 64 D3
Mount Hope Australia 58 B4
Mount Hope WV U.S.A. 64 A4
Mount Howitt Australia 57 C5
Mount Isa Australia 56 B4
Mount Jackson VA U.S.A. 64 B3
Mount Jewett PA U.S.A. 64 B2
Mount Joy PA U.S.A. 64 C2
Mount Kaputar National Park Australia 58 E3
Mount Keith Australia 55 C6
Mount Lofty Range mts Australia 57 B7
Mount Magnet Australia 55 B7
Mount Manara Australia 58 A4
Mount McKinley National Park U.S.A. see Denali National Park and Preserve
Mountmellick Ireland 21 E4
Mount Moorosi Lesotho 51 H6
Mount Morgan Australia 56 E4
Mount Morris NY U.S.A. 64 C1
Mount Murchison Australia 58 A3
Mount Nebo WV U.S.A. 64 A3
Mount Pleasant IA U.S.A. 63 I3
Mount Pleasant MI U.S.A. 63 K3
Mount Pleasant TX U.S.A. 63 I5
Mount Remarkable National Park Australia 57 B7
Mount Rogers National Recreation Area park VA U.S.A. 64 A4
Mount Sanford Australia 54 E4
Mount's Bay U.K. 19 B8
Mount Shasta U.S.A. 62 C3
Mountsorrel U.K. 19 F6
Mount Swan Australia 56 A4
Mount Union PA U.S.A. 64 C2
Mount Vernon Australia 55 B6
Mount Vernon IL U.S.A. 63 J4
Mount Vernon WA U.S.A. 62 C2
Mount William National Park Australia 57 [inset]
Mount Willoughby Australia 55 F6
Moura Australia 56 E5
Moura Brazil 68 F4
Moura Port. 25 C4
Mourdi, Dépression du depr. Chad 47 F3
Mourdiah Mali 46 C3
Mourne r. U.K. 21 E3
Mourne Mountains hills U.K. 21 F3
Mousa i. U.K. 20 [inset]
Mouscron Belgium 16 I5
Mousgougou Chad 48 B2
Moussafoyo Chad 47 E4
Moussoro Chad 47 E3
Moutamba Congo 48 B4
Moutong Indon. 41 E7
Mouydir, Monts du plat. Alg. 46 D2
Mowbullan, Mount Australia 58 E1
Moy r. Ireland 21 C3
Moyale Eth. 48 D3
Moyen Atlas mts Morocco 22 C5
Moyen Congo country Africa see Congo
Moyeni Lesotho 51 H6
Moynalyk Rus. Fed. 42 H2
Moyo i. Indon. 54 B2
Moyobamba Peru 68 C5
Moyola r. U.K. 21 F3
Moyu China 36 D1
Moyynkum Kazakh. 42 C4
Moyynkum, Peski des. Kazakh. 42 B4
Moyynty Kazakh. 28 I5
Mozambique country Africa 49 D6
Mozambique Channel Africa 49 E6
Mozambique Ridge sea feature Indian Ocean 73 K7
Mozdok Rus. Fed. 13 J8
Mozhaysk Rus. Fed. 13 H5
Mozhga Rus. Fed. 12 L4
Mozyr' Belarus see Mazyr
Mpaathutlwa Pan salt pan Botswana 50 E3
Mpanda Tanz. 49 D4
Mpen India 37 I4
Mpika Zambia 49 D5
Mpolweni S. Africa 51 J5
Mporokoso Zambia 49 D4
Mpulungu Zambia 49 D4
Mpumalanga prov. S. Africa 51 I4
Mpunde mt. Tanz. 49 D4
Mpwapwa Tanz. 49 D4
Mqanduli S. Africa 51 I6
Mqinvartsveri mt. Georgia/Rus. Fed. see Kazbek
Mrauk-U Myanmar 37 H5
Mrewa Zimbabwe see Murehwa
Mrkonjić-Grad Bos.-Herz. 26 G2
M'Saken Tunisia 26 D7
Mshinskaya Rus. Fed. 15 P7
M'Sila Alg. 25 I6
Msta r. Rus. Fed. 12 F4
Mstislavl' Belarus see Mstsislaw
Mstsislaw Belarus 13 F5
Mtelo Kenya 48 D3
Mthatha S. Africa 51 I6
Mtoko Zimbabwe see Mutoko
Mtorwi Tanz. 49 D4
Mtsensk Rus. Fed. 13 H5
Mts'ire Kavkasioni Asia see Lesser Caucasus
Mtubatuba S. Africa 51 K5
Mtunzini S. Africa 51 J5
Mtwara Tanz. 49 E5
Mu r. Myanmar 37 H5
Mu'āb, Jibāl reg. Jordan see Moab
Muanda Dem. Rep. Congo 49 B4
Muang Khoua Laos 42 I8
Muang Pakxan Laos see Pakxan
Muang Sing Laos 42 I8
Muang Thai country Asia see Thailand
Muari, Ras c. Pak. 36 A4
Mu'aygil, Khashm al h. Saudi Arabia 35 H6
Mubarek Uzbek. see Muborak
Mubarraz well Saudi Arabia 35 F5
Mubende Uganda 48 D3
Mubi Nigeria 46 E3
Muborak Uzbek. 33 K2
Mucajaí, Serra do mts Brazil 68 F3
Muccan Australia 54 C5
Muchinga Escarpment Zambia 49 D5
Muck i. U.K. 20 C4
Mucojo Moz. 49 E5
Muconda Angola 49 C5
Mucubela Moz. 49 D5
Mucugê Brazil 71 C1
Mucur Turkey 34 D3
Mucuri Brazil 71 D2
Mucuri r. Brazil 71 D2
Mudabidri India 38 B3
Mudan China see Heze
Mudanjiang China 44 C3
Mudan Jiang r. China 44 C3
Mudan Ling mts China 44 B4
Mudanya Turkey 27 M4
Mudaysīsāt, Jabal al h. Jordan 39 C4
Muddus nationalpark nat. park Sweden 14 K3
Muddy r. NV U.S.A. 65 E2
Muddy Peak NV U.S.A. 65 E2
Mudgal India 38 C3
Mudgee Australia 58 D4
Mudhol India 38 B2
Mudigere India 38 B3
Mud Lake NV U.S.A. 65 D2
Mudraya country Africa see Egypt
Mudurnu Turkey 27 N4
Mud'yuga Rus. Fed. 12 H3
Mueda Moz. 49 D5
Mueller Range hills Australia 54 D4
Muftyuga Rus. Fed. 12 J2
Mufulira Zambia 49 C5
Mufumbwe Zambia 49 C5
Muğan Düzü lowland Azer. 35 H3
Mugarripug China 37 F2
Mughalbhin Pak. see Jati
Mughal Kot Pak. 36 B3
Mughal Sarai India 37 E4
Mūghār Iran 35 I4
Mughayrā' Saudi Arabia 39 C5
Muğla Turkey 27 M6
Mugxung China 37 H2
Mūḩ, Sabkhat imp. l. Syria 39 D2
Muhammad Ashraf Pak. 36 B4
Muhammad Qol Sudan 32 E5
Muhammarah Iran see Khorramshahr
Muhashsham, Wādī al watercourse Egypt 39 B4
Muḩaysh, Wādī al watercourse Jordan 39 C5
Muhaysin Syria 39 D1
Mühlhausen (Thüringen) Germany 17 M5
Mühlig-Hofmann Mountains Antarctica 76 C2
Muhos Fin. 14 N4
Muḩradah Syria 39 C2
Mui Bai Bung c. Vietnam see Mui Ca Mau
Mui Ca Mau c. Vietnam 31 J6
Muié Angola 49 C5
Muineachán Ireland see Monaghan
Muirkirk U.K. 20 E5
Muir of Ord U.K. 20 E3
Muite Moz. 49 D5
Muji China 36 D1
Muju S. Korea 45 B5
Mukacheve Ukr. 13 D6
Mukachevo Ukr. see Mukacheve
Mukalla Yemen 32 G7
Mukandwara India 36 D4
Mukden China see Shenyang
Mukhen Rus. Fed. 44 E2
Mukhino Rus. Fed. 44 B1
Mukhtuya Rus. Fed. see Lensk
Mukinbudin Australia 55 B7
Mukojima-rettō is Japan 45 F8
Muktsar India 36 C3
Mula r. India 38 B2
Mulan China 44 C3
Mulanje, Mount Malawi 49 D5
Mulapula, Lake imp. l. Australia 57 B6
Mulayz, Wādī al watercourse Egypt 39 A4
Mulchatna r. U.S.A. 60 C3
Mulde r. Germany 17 N5
Mulegé Mex. 66 B3
Mules i. Indon. 54 C2
Muleshoe U.S.A. 62 G5
Mulga Park Australia 55 E6
Mulgathing Australia 55 F7
Mulhacén mt. Spain 25 E5
Mülhausen France see Mulhouse
Mulhouse France 24 H3
Muli China 42 I7
Muli Rus. Fed. see Vysokogorniy
Muling Heilong. China 44 C3
Muling Heilong. China 44 C3
Muling He r. China 44 D3
Mull i. U.K. 20 D4
Mull, Sound of sea chan. U.K. 20 C4
Mullaghcleevaun h. Ireland 21 F4
Mullaittivu Sri Lanka 38 D4
Mullaley Australia 58 D3
Mullengudgery Australia 58 C3
Mullens WV U.S.A. 64 A4
Muller watercourse Australia 54 F5
Mullewa Australia 55 A7
Mullica r. NJ U.S.A. 64 D3
Mullingar Ireland 21 E4
Mullion Creek Australia 58 D4
Mull of Galloway c. U.K. 20 E6
Mull of Kintyre hd U.K. 20 D5
Mull of Oa hd U.K. 20 C5
Mullumbimby Australia 58 F2
Mulobezi Zambia 49 C5
Mulshi Lake India 38 B2
Multai India 36 D5
Multan Pak. 33 L3
Multia Fin. 14 N5
Mulug India 38 C2
Mumbai India 38 B2
Mumbil Australia 58 D4
Mumbwa Zambia 49 C5
Muna Mex. 66 G4
Muna r. Rus. Fed. 29 N3
Munabao Pak. 36 B4
Munaðarnes Iceland 14 [inset]
München Germany see Munich
München-Gladbach Germany see Mönchengladbach
Muncoonie West, Lake imp. l. Australia 56 B5
Muncy PA U.S.A. 64 C2
Munda Pak. 36 B3
Mundel Lake Sri Lanka 38 C5
Mundesley U.K. 19 I6
Mundford U.K. 19 H6
Mundiwindi Australia 55 C5
Mundra India 36 B5
Mundrabilla Australia 52 C5
Mundubbera Australia 57 E5
Mundwa India 36 C4
Mungallala Australia 57 D5
Mungana Australia 56 D3
Mungāri Moz. 49 D5
Mungbere Dem. Rep. Congo 48 C3
Mungeli India 37 E5
Munger India 37 F4
Mu Nggava i. Solomon Is see Rennell
Mungindi Australia 58 D2
Mungla Bangl. see Mongla
Mungo Angola 49 B5
Mungo, Lake Australia 58 A4
Mungo National Park Australia 58 A4
Munich Germany 17 M6
Munjpur India 36 B5
Munkács Ukr. see Mukacheve
Munkebakken Norway 14 P2
Munkedal Sweden 15 G7
Munkfors Sweden 15 H7
Munkhafaḍ al Qaṭṭārah depr. Egypt see Qattara Depression
Munku-Sardyk, Gora mt. Mongolia/Rus. Fed. 42 I2
Munnik S. Africa 51 I2
Munsan S. Korea 45 B5
Münster Niedersachsen Germany 17 M4
Münster Nordrhein-Westfalen Germany 17 K5
Munster reg. Ireland 21 D5
Muntadgin Australia 55 B7
Muntervary hd Ireland 21 C6
Munyal-Par sea feature India see Bassas de Pedro Padua Bank
Munzur Vadisi Milli Parkı nat. park Turkey 23 L4
Muojärvi l. Fin. 14 P4
Muonio Fin. 14 M3
Muonioälven r. Fin./Sweden 14 M3
Muonionjoki r. Fin./Sweden see Muonioälven
Mupa, Parque Nacional da nat. park Angola 49 B5
Muqdisho Somalia see Mogadishu
Muquem Brazil 71 A1
Muqui Brazil 71 C3
Mur r. Austria 17 P7
Mura r. Austria see Mur
Murakami Japan 45 E5
Murallón, Cerro mt. Chile 70 B7
Muramvya Burundi 48 C4
Murashi Rus. Fed. 12 K4
Murat r. Turkey 35 E3
Muratlı Turkey 27 L4
Muraysah, Ra's al pt Libya 34 B5
Murchison watercourse Australia 55 A6
Murchison, Mount Antarctica 76 H2
Murchison, Mount h. Australia 55 B6
Murchison Falls National Park Uganda 48 D3
Murcia Spain 25 F5
Murcia aut. comm. Spain 25 F5
Murehwa Zimbabwe 49 D5
Mureşul r. Romania 27 I1
Muret France 24 E5
Murewa Zimbabwe see Murehwa
Murfreesboro TN U.S.A. 63 J4
Murgab Tajik. see Murghob
Murgab r. Turkm. see Murgap
Murgap r. Turkm. 33 J2
Murgha Kibzai Pak. 36 B3
Murghob Tajik. 33 L2
Murgon Australia 57 E5
Murgoo Australia 55 B6
Muri India 37 F5
Muriaé Brazil 71 C3
Muriege Angola 49 C4
Müritz l. Germany 17 N4
Müritz, Nationalpark nat. park Germany 17 N4
Murmansk Rus. Fed. 12 G1
Murmanskaya Oblast' admin. div. Rus. Fed. 14 S2
Murmanskiy Bereg coastal area Rus. Fed. 12 G1
Murmansk Oblast admin. div. Rus. Fed. see Murmanskaya Oblast'
Muro, Capo di c. Corse France 24 I6
Murom Rus. Fed. 12 I5
Muroran Japan 44 F4
Muros Spain 25 B2
Muroto Japan 45 D6
Muroto-zaki pt Japan 45 D6
Murrah al Kubrá, Al Buḩayrah al l. Egypt see Great Bitter Lake
Murrah aş Şughrá, Al Buḩayrah al l. Egypt see Little Bitter Lake
Murra Murra Australia 58 C2
Murrat el Kubra, Buheirat l. Egypt see Great Bitter Lake
Murrat el Sughra, Buheirat l. Egypt see Little Bitter Lake
Murray r. S.A. Australia 57 B7
Murray r. W.A. Australia 55 A8
Murray KY U.S.A. 63 J4
Murray, Lake P.N.G. 52 E2
Murray Bridge Australia 57 B7
Murray-Darling r. Australia 52 E5
Murray Downs Australia 54 F5
Murray Range hills Australia 55 E6
Murraysburg S. Africa 50 F6
Murray Sunset National Park Australia 57 C7
Murrieta CA U.S.A. 65 D4
Murringo Australia 58 D5
Murrisk reg. Ireland 21 C4
Murroogh Ireland 21 C4
Murrumbidgee r. Australia 58 A5
Murrumburrah Australia 58 D5
Murrurundi Australia 58 E3
Mursan India 36 D4
Murshidabad India 37 G4
Murska Sobota Slovenia 26 G1
Murtoa Australia 57 C8
Murua i. P.N.G. see Woodlark Island
Murud India 38 B2
Murunkan Sri Lanka 38 D4
Murupara N.Z. 59 F4
Mururoa atoll Fr. Polynesia 75 K7
Murviedro Spain see Sagunto
Murwara India 36 E5
Murwillumbah Australia 58 F2
Murzūq Libya 47 E2
Mürzzuschlag Austria 17 O7
Muş Turkey 35 F3
Mūsā, Khowr-e b. Iran 35 H5
Musakhel Pak. 36 B3
Musala mt. Bulg. 27 J3
Musan N. Korea 44 C4
Muscat Oman 33 I5
Muscat and Oman country Asia see Oman
Muscatine U.S.A. 63 I3
Musgrave Australia 56 C2
Musgrave Ranges mts Australia 55 E6
Mushāsh al Kabid well Jordan 39 C5
Mushayyish, Wādī al watercourse Jordan 39 C4
Mushie Dem. Rep. Congo 48 B4
Music Mountain AZ U.S.A. 65 F3
Musina S. Africa 51 J2
Muskeget Channel MA U.S.A. 64 F2
Muskogee U.S.A. 63 H4
Musmar Sudan 32 E6
Musoma Tanz. 48 D4
Mussau Island P.N.G. 52 E2
Musselburgh U.K. 20 F5
Mussende Angola 49 B5
Mustafakemalpaşa Turkey 27 M4
Mustjala Estonia 15 M7
Mustvee Estonia 15 O7
Musu-dan pt N. Korea 44 C4
Muswellbrook Australia 58 E4
Mūţ Egypt 32 C4
Mut Turkey 39 A1
Mutá, Ponta do pt Brazil 71 D1
Mutare Zimbabwe 49 D5
Mutina Italy see Modena
Mutis Col. 68 C2
Mutnyy Materik Rus. Fed. 12 L2
Mutoko Zimbabwe 49 D5
Mutsamudu Comoros 49 E5
Mutsu Japan 44 F4
Muttaburra Australia 56 D4
Mutton Island Ireland 21 C5
Muttukuru India 38 D3
Muttupet India 38 C4
Mutum Brazil 71 C2
Mutunópolis Brazil 71 A1
Mutur Sri Lanka 38 D4
Mutusjärvi l. Fin. 14 O2
Muurola Fin. 14 N3
Mu Us Shamo des. China 43 J5
Muxaluando Angola 49 B4
Muxima Angola 49 B4
Muyezerskiy Rus. Fed. 12 G3
Muyinga Burundi 48 D4
Muyumba Dem. Rep. Congo 49 C4
Muyunkum, Peski des. Kazakh. see Moyynkum, Peski
Muzaffarabad Pak. 36 C2
Muzaffargarh Pak. 33 L3
Muzaffarnagar India 36 D3
Muzaffarpur India 37 F4
Muzamane Moz. 51 K2
Muzhi Rus. Fed. 11 S2
Múzquiz Mex. 66 D3
Muztag mt. China 36 E2
Muz Tag mt. China 37 F1
Muztor Kyrg. see Toktogul
Mvadi Gabon 48 B3
Mvolo South Sudan 47 F4
Mvuma Zimbabwe 49 D5
Mwanza Malawi 49 D5
Mwanza Tanz. 48 D4
Mweelrea h. Ireland 21 C4
Mweka Dem. Rep. Congo 49 C4
Mwene-Ditu Dem. Rep. Congo 49 C4
Mwenezi Zimbabwe 49 D6
Mwenga Dem. Rep. Congo 48 C4
Mweru, Lake Dem. Rep. Congo/Zambia 49 C4
Mweru Wantipa National Park Zambia 49 C4
Mwimba Dem. Rep. Congo 49 C4
Mwinilunga Zambia 49 C5
Myadaung Myanmar 37 I5
Myadzyel Belarus 15 O9
Myajlar India 36 B4
Myall Lakes National Park Australia 58 F4
Myanmar country Asia 31 I4
Myauk-U Myanmar see Mrauk-U
Mybster U.K. 20 F2
Myeik Myanmar 31 I5
Myingyan Myanmar 37 H5
Myitkyina Myanmar 42 H7
Myitson Myanmar 37 I5
Myittha Myanmar 37 I5
Mykolayiv Ukr. 27 O1
Mykonos i. Greece 27 K6
Myla Rus. Fed. 12 K2
Myla r. Rus. Fed. 12 K2
Mylae Sicilia Italy see Milazzo
Mylasa Turkey see Milas
Mymensing Bangl. see Mymensingh
Mymensingh Bangl. 37 G4
Mynämäki Fin. 15 M6
Myŏnggan N. Korea 44 C4
Myory Belarus 15 O9
Mýrdalsjökull Iceland 14 [inset]
Myre Norway 14 I2
Myrheden Sweden 14 L4
Myrhorod Ukr. 13 G6
Myronivka Ukr. 13 F6
Myrtle Beach U.S.A. 63 L5
Myrtleford Australia 58 C6
Myrtoo Pelagos sea Greece 27 J6
Mys Abramovskiy pt Rus. Fed. 12 I2
Mys Articheskiy c. Rus. Fed. 77 E1
Mys Chelyuskin Rus. Fed. 77 E1
Mysia reg. Turkey 27 L5
Mys Lazareva Rus. Fed. see Lazarev
Myślibórz Poland 17 O4
Mysore India 38 C3
Mysore state India see Karnataka
Mys Shmidta Rus. Fed. 29 T3
Mysuru India see Mysore
Mysy Rus. Fed. 12 L3
My Tho Vietnam 31 J5
Mytilene i. Greece see Lesbos
Mytilini Greece 27 L5
Mytilini Strait Greece/Turkey 27 L5
Mytishchi Rus. Fed. 12 H5
Myyeldino Rus. Fed. 12 L3
Mzamomhle S. Africa 51 H6
Mzimba Malawi 49 D5
Mzuzu Malawi 49 D5

N

Naantali Fin. 15 M6
Naas Ireland 21 F4
Naba Myanmar 37 I4
Nababeep S. Africa 50 C5
Nababganj Bangl. see Nawabganj
Nabadwip India see Navadwip
Nabarangapur India 38 D2
Nabarangpur India see Nabarangapur
Nabari Japan 45 E6
Nabatîyé et Tahta Lebanon 39 B3
Nabatiyet et Tahta Lebanon see Nabatîyé et Tahta
Nabberu, Lake imp. l. Australia 55 C6
Naberera Tanz. 49 D4
Naberezhnyye Chelny Rus. Fed. 11 Q4
Nabeul Tunisia 26 D6
Nabha India 36 D3
Nabil'skiy Zaliv b. Rus. Fed. 44 F2
Nabire Indon. 41 F8
Nabi Younés, Ras en pt Lebanon 39 B3
Nāblus West Bank 39 B3
Naboomspruit S. Africa 51 I3
Nabq Reserve nature res. Egypt 34 D5
Nābulus West Bank see Nāblus
Nacala Moz. 49 E5
Nachalovo Rus. Fed. 13 K7
Nachingwea Tanz. 49 D5
Nachna India 36 B4
Nacimiento Reservoir CA U.S.A. 65 B3
Nacogdoches U.S.A. 63 I5
Nådendal Fin. see Naantali
Nadezhdinskoye Rus. Fed. 44 D2
Nadiad India 36 C5
Nadol India 36 C4
Nador Morocco 25 E6
Nadūshan Iran 35 I4
Nadvirna Ukr. 13 E6
Nadvoitsy Rus. Fed. 12 G3
Nadvornaya Ukr. see Nadvirna
Nadym Rus. Fed. 28 I3
Næstved Denmark 15 G9
Nafarroa aut. comm. Spain see Navarra
Nafas, Ra's an mt. Egypt 39 B5
Nafḩa, Har h. Israel 39 B4
Nafpaktos Greece 27 I5
Nafplio Greece 27 J6
Naftalan Azer. 35 G2
Naft-e Sefīd Iran 35 H5
Naft-e Shāh Iran see Naft Shahr
Naft Shahr Iran 35 G4
Nafūd al Ghuwayṭah des. Saudi Arabia 39 D5
Nafūd al Jur'ā des. Saudi Arabia 35 G6
Nafūsah, Jabal hills Libya 46 E1
Nafy Saudi Arabia 32 F4
Nag, Co l. China 37 G2
Naga Phil. 41 E6
Nagahama Japan 45 D6
Naga Hills India 37 H4
Naga Hills state India see Nagaland
Nagaland state India 37 H4
Nagamangala India 38 C3
Nagambie Australia 58 B6
Nagano Japan 45 E5
Nagaoka Japan 45 E5
Nagaon India 37 H4
Nagapatam India see Nagappattinam
Nagappattinam India 38 C4
Nagar Hima. Prad. India 33 M3
Nagar Karnataka India 38 B3
Nagaram India 38 D2
Nagari Hills India 38 C3
Nagarjuna Sagar Reservoir India 38 C2
Nagar Parkar Pak. 33 L5
Nagar Untari India 37 E4
Nagasaki Japan 45 C6
Nagato Japan 45 C6
Nagaur India 36 C4
Nagbhir India 38 C1
Nagda India 36 C5
Nagercoil India 38 C4
Nag' Hammâdi Egypt see Naj' Ḩammādī
Nagina India 36 D3
Nagong Chu r. China see Parlung Zangbo
Nagorno-Karabakh aut. reg. Azer. see Dağlıq Qarabağ
Nagornyy Karabakh aut. reg. Azer. see Dağlıq Qarabağ
Nagorsk Rus. Fed. 12 K4
Nagoya Japan 45 E6
Nagpur India 36 D5
Nagqu China 42 G6
Nag Qu r. China 37 H3
Nagurskoye Rus. Fed. 28 F1
Nagyatád Hungary 26 G1
Nagybecskerek Serbia see Zrenjanin
Nagyenyed Romania see Aiud
Nagykanizsa Hungary 26 G1
Nagyvárad Romania see Oradea
Naha Japan 43 N7
Nahan India 36 D3
Nahanni Butte Canada 60 F3
Nahanni National Park Reserve Canada 60 F3
Naharāyim Jordan 39 B3
Nahariyya Israel 39 B3
Nahāvand Iran 35 H4
Nahr Dijlah r. Asia 35 G5 see Tigris
Nahrīn Afgh. 36 B1
Nahuel Huapi, Parque Nacional nat. park Arg. 70 B6
Naij Tal China 37 H2
Naikliu Indon. 54 C2
Nain Canada 61 L4
Nā'īn Iran 35 I4
Naini Tal India see Nainital
Nainital India 36 D3
Nairn U.K. 20 F3
Nairn r. U.K. 20 F3
Nairobi Kenya 48 D4
Naissus Serbia see Niš
Naivasha Kenya 48 D4
Najafābād Iran 35 H4
Najd reg. Saudi Arabia 32 F4
Nájera Spain 25 E2
Naj' Ḩammādī Egypt 32 D4
Naji China 44 A2
Najibabad India 36 D3
Najin N. Korea 44 C4
Najitun China see Naji
Najrān Saudi Arabia 32 F6
Nakadōri-shima i. Japan 45 C6
Nakambé r. Burkina Faso/Ghana see White Volta
Nakanbe r. Burkina Faso/Ghana see White Volta
Nakanno Rus. Fed. 29 L3
Nakano-shima i. Japan 45 D5
Nakasongola Uganda 47 G4
Nakatsu Japan 45 C6
Nakatsugawa Japan 45 E6
Nakfa Eritrea 32 E6
Nakhichevan' Azer. see Naxçıvan
Nakhl Egypt 39 A5
Nakhodka Rus. Fed. 44 D4
Nakhola India 37 H4
Nakhon Pathom Thai. 31 J5
Nakhon Ratchasima Thai. 31 J5
Nakhon Sawan Thai. 31 J5
Nakhon Si Thammarat Thai. 31 I6
Nakhtarana India 36 B5
Nakina Canada 63 J1
Naknek U.S.A. 60 C4
Nakonde Zambia 49 D4
Nakskov Denmark 15 G9
Naktong-gang r. S. Korea 45 C6
Nakuru Kenya 48 D4
Nal Pak. 36 A4
Nalázi Moz. 51 K3
Nalbari India 37 G4
Nal'chik Rus. Fed. 13 I8
Naldurg India 38 C2
Nalgonda India 38 C2
Naliya India 36 B5
Nallamala Hills India 38 C3
Nallıhan Turkey 27 N4
Nālūt Libya 46 E1
Namaacha Moz. 51 K3
Namacurra Moz. 49 D5
Namadgi National Park Australia 58 D5
Namahadi S. Africa 51 I4
Namak, Daryācheh-ye imp. l. Iran 35 H4
Namakkal India 38 C4
Namakwaland reg. Namibia see Great Namaqualand
Namakzar-e Shadad salt flat Iran 33 I3
Namaland reg. Namibia see Great Namaqualand
Namangan Uzbek. 33 L1
Namaqualand reg. Namibia see Great Namaqualand
Namaqualand reg. S. Africa 50 C5
Namaqua National Park S. Africa 50 C6
Namatanai P.N.G. 52 F2
Nambour Australia 58 F1
Nambucca Heads Australia 58 F3
Nambung National Park Australia 55 A7
Namcha Barwa mt. China see Namjagbarwa Feng
Namche Bazar Nepal 37 F4
Nam Co salt l. China 37 G3
Namdalen val. Norway 14 H4
Namdalseid Norway 14 G4
Nam Định Vietnam 31 J4
Namen Belgium see Namur
Nam-gang r. N. Korea 45 B5
Namhae-do i. S. Korea 45 B6
Namib Desert Namibia 50 B3
Namibe Angola 49 B5
Namibia country Africa 49 B6
Namibia Abyssal Plain sea feature N. Atlantic Ocean 72 I8
Namib-Naukluft Game Park nature res. Namibia 50 B3
Namie Japan 45 F5
Namīn Iran 35 H3
Namjagbarwa Feng mt. China 42 G7
Namoi r. Australia 58 D3
Nampa mt. Nepal 36 E3
Nampa U.S.A. 62 D3
Nampala Mali 46 C3
Namp'o N. Korea 45 B5
Nampula Moz. 49 D5
Namsen r. Norway 14 G4
Namsos Norway 14 G4
Namti Myanmar 37 I4
Namtsy Rus. Fed. 29 N3
Namtu Myanmar 42 H8
Namuli, Monte mt. Moz. 49 D5
Namuno Moz. 49 D5
Namur Belgium 16 J5
Namutoni Namibia 49 B5
Namwŏn S. Korea 45 B6
Namya Ra Myanmar 37 I4
Nan Thai. 31 J5
Nana Bakassa Cent. Afr. Rep. 48 B3
Nanaimo Canada 62 C2
Nanam N. Korea 44 C4
Nan'an China 43 L8
Nanango Australia 58 F1
Nananib Plateau Namibia 50 C3
Nanao Japan 45 E5
Nanatsu-shima i. Japan 45 E5
Nancha China 44 C3
Nanchang Jiangxi China 43 L7
Nanchong China 42 J6
Nancy France 24 H2
Nanda Devi mt. India 36 E3
Nanda Kot mt. India 36 E3
Nandapur India 38 D2
Nanded India 38 C2
Nander India see Nanded
Nandewar Range mts Australia 58 E3
Nandod India 38 B1
Nandurbar India 36 C5
Nandyal India 38 C3
Nang China 42 G7
Nanga Eboko Cameroon 46 E4
Nanga Parbat mt. Pak. 36 C2
Nangar National Park Australia 58 D4
Nangnim-sanmaek mts N. Korea 45 B4
Nangqên China 42 H6
Nangulangwa Tanz. 49 D4
Nanguneri India 38 C4
Nanjing China 43 L6
Nanking China see Nanjing
Nankova Angola 49 B5
Nan Ling mts China 43 K8
Nannilam India 38 C4
Nannine Australia 55 B6
Nanning China 43 J8
Nannup Australia 55 A8
Nanortalik Greenland 61 N3
Nanouki atoll Kiribati see Nonouti
Nanouti atoll Kiribati see Nonouti
Nanping China 43 L7
Nansei-shotō is Japan see Ryukyu Islands
Nansen Basin sea feature Arctic Ocean 77 H1
Nansen Sound sea chan. Canada 61 I1
Nan-sha Ch'ün-tao is S. China Sea see Spratly Islands
Nansha Qundao is S. China Sea see Spratly Islands
Nansio Tanz. 48 D4
Nantes France 24 D3
Nantes à Brest, Canal de France 24 C3
Nanthi Kadal Lagoon lag. Sri Lanka 38 D4
Nanticoke Ont. Canada 64 A1
Nanticoke MD U.S.A. 64 D3
Nanticoke PA U.S.A. 64 D2
Nantong China 43 M6
Nantucket MA U.S.A. 64 F2
Nantucket Island MA U.S.A. 64 G2
Nantucket Sound g. MA U.S.A. 64 F2
Nantwich U.K. 19 E5
Nanumaga i. Tuvalu see Nanumanga
Nanumanga i. Tuvalu 53 H2
Nanumea atoll Tuvalu 53 H2
Nanuque Brazil 71 C2
Nanxiong China 43 K7
Nanyang China 43 K6
Nanyuki Kenya 48 D4
Naococane, Lac l. Canada 61 K4
Naoero country S. Pacific Ocean see Nau
Naogaon Bangl. 37 G4
Naoli He r. China 44 D3
Naoshera India 36 C2
Napa CA U.S.A. 65 A1
Napaktulik Lake Canada 60 G3
Napasoq Greenland 61 M3
Napier N.Z. 59 F4
Napier Range hills Australia 54 D4
Naples Italy 26 F4
Naples FL U.S.A. 63 K6
Napoli Italy see Naples
Naqadeh Iran 35 G3
Nara India 36 B5
Nara Japan 45 D6
Nara Mali 46 C3
Narach Belarus 15 O9
Naracoorte Australia 57 C8
Naradhan Australia 58 C4
Narainpur India 38 D2
Naralua India 37 F4
Naranjal Ecuador 68 C4
Narasapur India 38 D2
Narasaraopet India 38 D2
Narasinghapur India 38 E1
Narayanganj Bangl. 37 G5
Narayanganj India 36 E5
Narayangarh India 36 C4
Narbada r. India see Narmada
Narberth U.K. 19 C7
Narbo France see Narbonne

rbonne France 24 F5
rborough Island Galápagos Ecuador see
ernandina, Isla
rcea r. Spain 25 C2
rdò Italy 26 H4
rechi r. Pak. 36 B3
rembeen Australia 55 B8
res Abyssal Plain sea feature
. Atlantic Ocean 72 D4
res Deep sea feature N. Atlantic Ocean
2 D4
res Strait Canada/Greenland 61 K2
retha Australia 55 D7
rew r. Poland 17 R4
rib Namibia 50 C3
rimanov Rus. Fed. 13 J7
rimskiy Khrebet mts Kazakh. see
larymskiy Khrebet
rince Turkey 34 E3
rin Gol watercourse China 37 H1
rkher India 36 D5
rmada r. India 36 C5
rman Turkey 35 F2
rnaul India 36 D3
rni Italy 26 E3
rnia Italy see Narni
rodnaya, Gora mt. Rus. Fed. 11 S3
ro-Fominsk Rus. Fed. 13 H5
rok Kenya 48 D4
rooma Australia 58 E6
rovchat Rus. Fed. 13 I5
rowlya Belarus 13 F6
rpes Fin. 14 L5
rrabri Australia 58 D3
rragansett Bay RI U.S.A. 64 F2
rran r. Australia 58 C2
rrandera Australia 58 C5
rran Lake Australia 58 C2
rrogin Australia 55 B8
rromine Australia 58 D4
rrows VA U.S.A. 64 A4
rrowsburg NY U.S.A. 64 D2
rsapur India 38 C2
rsaq Greenland 61 N3
rshingdi Bangl. see Narsingdi
rsimhapur India see Narsinghpur
rsingdi Bangl. 37 G5
rsinghpur India 36 D5
rsipatnam India 38 D2
rtkala Rus. Fed. 35 F2
ruto Japan 45 D6
rva Estonia 15 P7
rva Bay Estonia/Rus. Fed. 15 O7
rva laht b. Estonia/Rus. Fed. see
larva Bay
rva Reservoir resr Estonia/Rus. Fed. see
larvskoye Vodokhranilishche
rva veehoidla resr Estonia/Rus. Fed. see
larvskoye Vodokhranilishche
rvik Norway 14 J2
rvskiy Zaliv b. Estonia/Rus. Fed. see
larva Bay
rvskoye Vodokhranilishche resr
stonia/Rus. Fed. 15 P7
rwana India 36 D3
r'yan-Mar Rus. Fed. 12 L2
rymskiy Khrebet mts Kazakh. 42 E3
ryn Kyrg. 42 D4
säker Sweden 14 J5
shik India 38 B1
shua NH U.S.A. 64 F1
shville TN U.S.A. 63 J4
sib Syria 39 C3
sijärvi l. Fin. 15 M6
sik India see Nashik
sinu Fiji 53 H3
sir Pak. 36 B3
sir South Sudan 32 D8
sirabad Bangl. see Mymensingh
sirabad India 36 C4
sr Egypt 34 C5
sratabad Iran see Zābol
srīan-e Pā'īn Iran 35 G4
ssau r. Australia 56 C2
ssau Bahamas 67 I3
ssau i. Cook Is 53 J3
ssau NY U.S.A. 64 E1
ssawadox VA U.S.A. 64 D4
sser, Lake resr Egypt 32 D5
ssjö Sweden 15 I8
ssuttooq inlet Greenland 61 M3
sva Rus. Fed. 12 F4
a Botswana 49 C6
al Brazil 69 K5
al prov. S. Africa see KwaZulu-Natal
al Basin sea feature Indian Ocean
3 K8
anz Iran 35 H4
ashquan r. Canada 61 L4
chez U.S.A. 63 I5
chitoches U.S.A. 63 I5
halia Australia 58 B6
hia Gali Pak. 36 C2
i, Punta pt Spain 25 H3
ional City CA U.S.A. 65 D4
ionalpark Bayerischer Wald nat. park
ermany 17 N6
ional West Coast Tourist Recreation
a park Namibia 50 B2
itingou Benin 46 D3
ividade Brazil 69 I6
mauk Myanmar 37 H5
or Bangl. see Natore
ore Bangl. 37 G4
ori Japan 45 F5
ron, Lake salt l. Tanz. 48 D4
sionalen Park Pirin nat. park Bulg.
7 J4
tai National Park Australia 58 E5
una, Kepulauan is Indon. 41 C7
una Besar i. Indon. 41 C7
uraliste, Cape Australia 55 A8
uraliste Plateau sea feature
ndian Ocean 73 P8
urpark Maas-Schwalm-Nette nat. park
ermany/Neth. 17 J5
uchas Namibia 50 C2
Co l. China 37 E2
ufragados, Ponta dos pt Brazil 71 A4
ujoji Akmenė Lith. 15 M8
ukh India 36 C4
ukot Pak. 33 K5
upada India 38 E2
ūr Jordan 39 B4
uroz Kalat Pak. 36 A3
urskaya Rus. Fed. 13 J8
uru i. Nauru 53 G2
uru country S. Pacific Ocean 53 G2
ustdal Norway 15 D6
uta Peru 68 D4
utaca Uzbek. see Qarshi
ute Dam Namibia 50 C4
vadwip India 37 G5
vahrudak Belarus 15 N10
valmoral de la Mata Spain 25 D4
valvillar de Pela Spain 25 D4
van Ireland 21 F4
vangar India see Jamnagar
vapolatsk Belarus 15 P9
var, Dasht-e depr. Afgh. 36 A2
varin, Mys c. Rus. Fed. 29 S3
Navarra aut. comm. Spain 25 F2
Navarra, Comunidad Foral de aut. comm.
Spain see Navarra
Navarre Australia 58 A6
Navarre aut. comm. Spain see Navarra
Navarro r. CA U.S.A. 65 A1
Navashino Rus. Fed. 12 I5
Navassa Island terr. West Indies 67 I5
Naver r. U.K. 20 E2
Näverede Sweden 14 I5
Navi Mumbai Mahar. 38 B2
Navlakhi India 36 B5
Navlya Rus. Fed. 13 G5
Năvodari Romania 27 M2
Navoiy Uzbek. 33 K1
Navojoa Mex. 66 C3
Navolato Mex. 66 C4
Návpaktos Greece see Nafpaktos
Návplion Greece see Nafplio
Navşar Turkey see Şemdinli
Navsari India 38 B1
Nawá Syria 39 C3
Nawabganj Bangl. 37 G4
Nawabshah Pak. 33 K4
Nawada India 37 F4
Nāwah Afgh. 36 A2
Nawalgarh India 36 C4
Nawanshahr India 36 D3
Nawan Shehar India see Nawanshahr
Nawar, Dasht-i depr. Afgh. see
Nāvar, Dasht-e
Nawarangpur India see Nabarangapur
Nawngleng Myanmar 42 H8
Nawoiy Uzbek. see Navoiy
Naxçıvan Azer. 35 G3
Naxos i. Greece 27 K6
Nayagarh India 38 E1
Nayak Afgh. 36 A2
Nayar Mex. 66 D4
Nāy Band, Kūh-e mt. Iran 33 I3
Nayoro Japan 44 F3
Nay Pyi Taw Mandalay Myanmar 31 I5
Nazaré Brazil 71 D1
Nazareth Israel 39 B3
Nazário Brazil 71 A2
Nazas r. Mex. 62 G6
Nazca Peru 68 D6
Nazca Ridge sea feature S. Pacific Ocean
75 O7
Nazilli Turkey 27 M6
Nazimabad Pak. 36 A4
Nazımiye Turkey 35 E3
Nazir Hat Bangl. 37 G5
Nazran' Rus. Fed. 35 G2
Nazrēt Eth. 48 D3
Nazwá Oman 33 I5
Ncojane Botswana 50 E2
N'dalatando Angola 49 B4
Ndélé Cent. Afr. Rep. 48 C3
Ndendé Gabon 48 B4
Ndende i. Solomon Is see Ndeni
Ndeni i. Solomon Is 53 G3
Ndjamena Chad 47 E3
N'Djamena Chad see Ndjamena
Ndjouani i. Comoros see Nzwani
Ndoi i. Fiji see Doi
Ndola Zambia 49 C5
Nduke i. Solomon Is see Kolombangara
Ndwedwe S. Africa 51 J5
Neabul Creek r. Australia 58 C1
Neagh, Lough l. U.K. 21 F3
Neale, Lake imp. l. Australia 55 E6
Nea Liosia Greece 27 J5
Neapoli Greece 27 J6
Neapolis Italy see Naples
Nea Roda Greece 27 J4
Neath U.K. 19 D7
Neath r. U.K. 19 D7
Nebbi Uganda 48 D3
Nebine Creek r. Australia 58 C2
Neblina, Pico da mt. Brazil 68 E3
Nebo Australia 56 E4
Nebolchi Rus. Fed. 12 G4
Nebraska state U.S.A. 62 G3
Nebraska City U.S.A. 63 H3
Nebrodi, Monti mts Sicilia Italy 26 F6
Nechisar National Park Eth. 48 D3
Necker Island U.S.A. 74 J4
Necochea Arg. 70 E5
Nederland country Europe see
Netherlands
Nedlouc, Lac l. Canada 61 K4
Nedluk Lake Canada see Nedlouc, Lac
Nêdong China see Zêtang
Nedre Soppero Sweden 14 L2
Nédroma Alg. 25 F6
Needles CA U.S.A. 65 E3
Neemach India see Neemuch
Neemuch India 36 C4
Neepawa Canada 62 H1
Neergaard Lake Canada 61 J2
Neftçala Azer. 35 H3
Neftechala Azer. see Neftçala
Neftechala Azer. see Uzboy
Neftegorsk Sakhalinskaya Oblast'
Rus. Fed. 44 F1
Neftegorsk Samarskaya Oblast' Rus. Fed.
13 K5
Neftekamsk Rus. Fed. 11 Q4
Neftekumsk Rus. Fed. 35 G1
Nefteyugansk Rus. Fed. 28 I3
Neftezavodsk Turkm. see Seýdi
Neftezawodsk Turkm. see Seýdi
Nefyn U.K. 19 C6
Nefza Tunisia 26 C6
Negage Angola 49 B4
Negara Indon. 54 A2
Negēlē Eth. 48 D3
Negev des. Israel 39 B4
Negomane Moz. 49 D5
Negombo Sri Lanka 38 C5
Negotino Macedonia 27 J4
Negra, Cordillera mts Peru 68 C5
Negra, Punta pt Peru 68 B5
Negra, Serra mts Brazil 71 C2
Négrine Alg. 26 B7
Negro r. Arg. 70 D6
Negro r. Brazil 69 G7
Negro r. Brazil 71 A4
Negro r. S. America 68 G4
Negro, Cabo c. Morocco 25 D6
Negroponte i. Greece see Evvoia
Negros i. Phil. 41 E7
Negru Vodă, Podișul plat. Romania 27 M3
Nehbandān Iran 33 J3
Nehe China 44 B2
Neijiang China 42 J7
Neimenggu aut. reg. China see
Nei Mongol Zizhiqu
Nei Mongol Zizhiqu aut. reg. China 44 A2
Neiva Col. 68 C3
Nejanilini Lake Canada 61 I4
Nejd reg. Saudi Arabia see Najd
Nekā Iran 35 I3
Nek'emtē Eth. 48 D3
Nekrasovskoye Rus. Fed. 12 I4
Neksø Denmark 15 I9
Nelang India 36 D3
Nelia Australia 56 C4
Nelidovo Rus. Fed. 12 G4
Nel'kan Rus. Fed. 29 P3
Nelligere India 38 C3
Nellore India 38 C3
Nelluz watercourse Turkey 39 D1
Nel'ma Rus. Fed. 44 E3
Nelson Canada 62 D2
Nelson r. Canada 61 I4
Nelson N.Z. 59 D5
Nelson U.K. 18 E5
Nelson AZ U.S.A. 65 F3
Nelson, Cape Australia 57 C8
Nelson, Estrecho str. Chile 70 A8
Nelson Bay Australia 58 F4
Nelsonia VA U.S.A. 64 D4
Nelson Lakes National Park N.Z. 59 D6
Nelspruit S. Africa see Mbombela
Néma Mauritania 46 C3
Nema Rus. Fed. 12 K4
Neman r. Belarus/Lith. see Nyoman
Neman Rus. Fed. 15 M9
Nemausus France see Nîmes
Nemawar India 36 D5
Nemed Rus. Fed. 12 L3
Nementcha, Monts des mts Alg. 26 B7
Nemetocenna France see Arras
Nemetskiy, Mys c. Rus. Fed. 14 Q2
Nemirov Ukr. see Nemyriv
Nemor He r. China 44 B2
Nemours Alg. see Ghazaouet
Nemours France 24 F2
Nemrut Dağı mt. Turkey 35 F3
Nemunas r. Belarus/Lith. see Nyoman
Nemuro Japan 44 G4
Nemuro-kaikyō sea chan. Japan/Rus. Fed.
44 G4
Nemyriv Ukr. 13 F6
Nenagh Ireland 21 D5
Nenana U.S.A. 60 D3
Nene r. U.K. 19 H6
Nenjiang China 44 B2
Nen Jiang r. China 44 B3
Neosho U.S.A. 63 I4
Nepal country Asia 37 E3
Nepalganj Nepal 37 E3
Nepean, Point Australia 58 B7
Nephi U.S.A. 62 E4
Nephin h. Ireland 21 C3
Nephin Beg Range hills Ireland 21 C3
Nepisiguit r. Canada 63 N2
Nepoko r. Dem. Rep. Congo 48 C3
Nérac France 24 E4
Nerang Australia 58 F1
Nera Tso l. China 37 H3
Nerchinsk Rus. Fed. 43 L2
Nerekhta Rus. Fed. 12 I4
Neretva r. Bos.-Herz./Croatia 26 G3
Nêri Pünco l. China 37 G3
Neriquinha Angola 49 C5
Neris r. Lith. 15 M9
Nerl' r. Rus. Fed. 12 H4
Nerópolis Brazil 71 A2
Neryungri Rus. Fed. 29 N4
Nes Norway 15 F6
Nes' Rus. Fed. 12 J2
Nesbyen Norway 15 F6
Neskaupstaður Iceland 14 [inset]
Nesna Norway 14 H3
Nesri India 38 B2
Ness r. U.K. 20 E3
Ness, Loch l. U.K. 20 E3
Nestos r. Greece 27 K4
Nesvizh Belarus see Nyasvizh
Netanya Israel 39 B3
Netherlands country Europe 16 J4
Netrakona Bangl. 37 G4
Netrokona Bangl. see Netrakona
Nettilling Lake Canada 61 K3
Neubrandenburg Germany 17 N4
Neuchâtel Switz. 24 H3
Neuchâtel, Lac de l. Switz. 24 H3
Neufchâteau France 24 G2
Neufchâtel-en-Bray France 24 E2
Neufchâtel-Hardelot France 19 I8
Neuhausen Rus. Fed. see Gur'yevsk
Neukuhren Rus. Fed. see Pionerskiy
Neumayer research stn Antarctica 76 B2
Neumünster Germany 17 L3
Neunkirchen Austria 17 P7
Neunkirchen Germany 17 K6
Neuquén Arg. 70 C5
Neuruppin Germany 17 N4
Neu Sandez Poland see Nowy Sącz
Neusiedler See l. Austria/Hungary 17 P7
Neusiedler See Seewinkel, Nationalpark
nat. park Austria 17 P7
Neustrelitz Germany 17 N4
Neuville-lès-Dieppe France 19 I9
Neuwied Germany 17 K5
Nevada MO U.S.A. 63 I4
Nevada state U.S.A. 62 D4
Nevada, Sierra mts Spain 25 E5
Nevada, Sierra mts U.S.A. 62 C3
Nevado, Cerro mt. Arg. 70 C5
Nevado, Sierra del mts Arg. 70 C5
Nevasa India 38 B2
Nevatim Israel 39 B4
Nevdubstroy Rus. Fed. see Kirovsk
Nevel' Rus. Fed. 12 F4
Nevel'sk Rus. Fed. 44 F3
Never Rus. Fed. 44 B1
Nevers France 24 F3
Nevertire Australia 58 C3
Nevesinje Bos.-Herz. 26 H3
Nevinnomyssk Rus. Fed. 13 I7
Nevşehir Turkey 34 D3
Nevskoye Rus. Fed. 44 D3
New r. CA U.S.A. 65 E4
New r. WV U.S.A. 64 A4
Newala Tanz. 49 D5
New Albany IN U.S.A. 63 J4
New Amsterdam Guyana 69 G2
New Angledool Australia 58 C2
Newark DE U.S.A. 64 D3
Newark NJ U.S.A. 64 D2
Newark NY U.S.A. 64 C1
Newark-on-Trent U.K. 19 G5
New Bedford MA U.S.A. 64 F2
New Berlin NY U.S.A. 64 D1
New Bern U.S.A. 63 L4
Newberry SC U.S.A. 63 K5
Newberry Springs CA U.S.A. 65 D3
New Bethlehem PA U.S.A. 64 B2
Newbiggin-by-the-Sea U.K. 18 F3
New Bloomfield PA U.S.A. 64 C2
New Boston TX U.S.A. 63 I5
New Braunfels U.S.A. 62 H6
Newbridge Ireland 21 F4
New Britain i. P.N.G. 52 E2
New Britain CT U.S.A. 64 E2
New Britain Trench sea feature
S. Pacific Ocean 74 G6
New Brunswick prov. Canada 63 N2
New Brunswick NJ U.S.A. 64 D2
Newburgh U.K. 20 G3
Newburgh NY U.S.A. 64 D2
Newbury U.K. 19 F7
Newburyport MA U.S.A. 64 F1
Newby Bridge U.K. 18 E4
New Caledonia terr. S. Pacific Ocean
53 G4
New Caledonia Trough sea feature
Tasman Sea 74 G7
Newcastle Australia 58 E4
Newcastle Ireland 21 F4
Newcastle S. Africa 51 I4
Newcastle U.K. 21 G3
New Castle PA U.S.A. 64 A2
Newcastle UT U.S.A. 65 F2
New Castle VA U.S.A. 64 A4
Newcastle WY U.S.A. 62 G3
Newcastle Emlyn U.K. 19 C6
Newcastle-under-Lyme U.K. 19 E5
Newcastle upon Tyne U.K. 18 F4
Newcastle Waters Australia 54 F4
Newcastle West Ireland 21 C5
Newchwang China see Yingkou
New City NY U.S.A. 64 E2
New Cumberland WV U.S.A. 64 A2
New Cumnock U.K. 20 E5
New Deer U.K. 20 G3
New Delhi India 36 D3
New Don Pedro Reservoir CA U.S.A.
65 B2
Newell, Lake imp. l. Australia 55 D6
New England National Park Australia
58 F3
New England Range mts Australia 58 E3
New England Seamounts sea feature
N. Atlantic Ocean 72 E3
Newenham, Cape U.S.A. 60 B4
Newent U.K. 19 E7
Newfane NY U.S.A. 64 B1
Newfane VT U.S.A. 64 E1
Newfoundland i. Canada 61 M5
Newfoundland prov. Canada see
Newfoundland and Labrador
Newfoundland and Labrador prov. Canada
61 M4
New Galloway U.K. 20 E5
New Georgia i. Solomon Is 53 F2
New Georgia Islands Solomon Is 53 F2
New Georgia Sound sea chan. Solomon Is
53 F2
New Guinea i. Indon./P.N.G. 41 G8
New Halfa Sudan 32 E6
New Hampshire state U.S.A. 64 F1
New Hanover i. P.N.G. 52 F2
Newhaven U.K. 19 H8
New Haven CT U.S.A. 64 E2
New Hebrides country S. Pacific Ocean see
Vanuatu
New Hebrides Trench sea feature
S. Pacific Ocean 74 H7
New Iberia U.S.A. 63 I5
Newington S. Africa 51 J3
Newinn Ireland 21 E5
New Ireland i. P.N.G. 52 F2
New Jersey state U.S.A. 64 D3
New Kensington PA U.S.A. 64 B2
New Kent VA U.S.A. 64 C4
New Lanark U.K. 20 F5
Newland Range hills Australia 55 C7
New Liskeard Canada 63 L2
New London CT U.S.A. 64 E2
Newman Australia 55 B5
Newman CA U.S.A. 65 B2
Newmarket Ireland 21 C5
Newmarket U.K. 19 H6
New Market VA U.S.A. 64 B3
Newmarket-on-Fergus Ireland 21 D5
New Martinsville WV U.S.A. 64 A3
New Mexico state U.S.A. 62 F5
New Milford PA U.S.A. 64 D2
New Orleans U.S.A. 63 I6
New Philadelphia OH U.S.A. 64 A2
New Pitsligo U.K. 20 G3
New Plymouth N.Z. 59 E4
Newport Mayo Ireland 21 C4
Newport North Tipperary Ireland 21 D5
Newport England U.K. 19 E6
Newport England U.K. 19 F8
Newport Wales U.K. 19 D7
Newport AR U.S.A. 63 I4
Newport NH U.S.A. 64 E1
Newport NJ U.S.A. 64 D3
Newport OR U.S.A. 62 C3
Newport RI U.S.A. 64 F2
Newport VT U.S.A. 63 M3
Newport WA U.S.A. 62 D2
Newport Beach CA U.S.A. 65 D4
Newport News VA U.S.A. 64 C4
Newport Pagnell U.K. 19 G6
Newquay U.K. 19 B8
New Roads U.S.A. 63 I5
New Rochelle NY U.S.A. 64 E2
New Romney U.K. 19 H8
New Ross Ireland 21 F5
Newry Australia 54 E4
Newry U.K. 21 F3
New Siberia Islands Rus. Fed. 29 P2
New South Wales state Australia 58 C4
New Stanton PA U.S.A. 64 B2
Newton U.K. 18 E5
Newton IA U.S.A. 63 I3
Newton KS U.S.A. 63 H4
Newton MA U.S.A. 64 F1
Newton NJ U.S.A. 64 D2
Newton Abbot U.K. 19 D8
Newton Mearns U.K. 20 E5
Newton Stewart U.K. 20 E6
Newtown Ireland 21 D5
Newtown England U.K. 19 E6
Newtown Wales U.K. 19 D6
New Town U.S.A. 62 G2
Newtownabbey U.K. 21 G3
Newtownards U.K. 21 G3
Newtownbarry Ireland see Bunclody
Newtownbutler U.K. 21 E3
Newtown Mount Kennedy Ireland 21 F4
Newtown St Boswells U.K. 20 G5
Newtownstewart U.K. 21 E3
New Ulm U.S.A. 63 I3
Newville PA U.S.A. 64 C2
New York NY U.S.A. 64 E2
New York state U.S.A. 64 D1
New Zealand country Oceania 59 D5
Neya Rus. Fed. 12 I4
Neyrīz Iran 35 I5
Neyshābūr Iran 33 I2
Nezhin Ukr. see Nizhyn
Ngabé Congo 48 B4
Ngagahtawng Myanmar 37 I4
Ngagau mt. Tanz. 49 D4
Ngalu Indon. 54 C2
Ngamring China 37 F3
Ngangla Ringco salt l. China 37 E3
Nganglong Kangri mt. China 36 E2
Nganglong Kangri mts China 36 E2
Ngangzê Co salt l. China 37 F3
Ngangzê Shan mts China 37 F3
Ngaoundal Cameroon 46 E4
Ngaoundéré Cameroon 47 E4
Ngape Myanmar 37 H5
Ngarrab China see Gyaca
Ngau i. Fiji see Gau
Ngawa China see Aba
Ngazidja i. Comoros 49 E5
Ngga Pulu mt. Indon. see Jaya, Puncak
Ngilmina Indon. 54 D2
Ngiva Angola see Ondjiva
Ngo Congo 48 B4
Ngoako Ramalepe S. Africa see
Duiwelskloof
Ngoin, Co salt l. China 37 G3
Ngoko r. Cameroon/Congo 47 E4
Ngom Qu r. China see Ji Qu
Ngoqumaima China 37 F2
Ngoring China 37 I2
Ngoring Hu l. China 37 I2
Ngourti Niger 46 E3
Nguigmi Niger 46 E3
Nguiu Australia 54 E2
Ngukurr Australia 54 F3
Ngulu atoll Micronesia 41 F7
Ngunza Angola see Sumbe
Ngunza-Kabolu Angola see Sumbe
Nguru Nigeria 46 E3
Ngwaketse admin. dist. Botswana see
Southern
Ngwane country Africa see Swaziland
Ngwathe S. Africa 51 H4
Ngwavuma r. S. Africa/Swaziland 51 K4
Ngwelezana S. Africa 51 J5
Nhachengue Moz. 51 L2
Nhamalabué Moz. 49 D5
Nha Trang Vietnam 31 J5
Nhecolândia Brazil 69 G7
Nhill Australia 57 C8
Nhlangano Swaziland 51 J4
Nhow i. Fiji see Gau
Nhulunbuy Australia 56 B2
Niafounké Mali 46 C3
Niagara Falls Ont. Canada 64 B1
Niagara Falls NY U.S.A. 64 B1
Niagara-on-the-Lake Ont. Canada 64 B1
Niagzu Aksai Chin 36 D2
Niakaramandougou Côte d'Ivoire 46 C4
Niamey Niger 46 D3
Niangara Dem. Rep. Congo 48 C3
Niangay, Lac l. Mali 46 C3
Nianzishan China 44 A3
Nias i. Indon. 41 B7
Niassa, Lago l. Africa see Nyasa, Lake
Nibil Well Australia 54 D5
Nīca Latvia 15 L8
Nicaragua country Central America 67 G6
Nicaragua, Lago de Nicaragua 67 G6
Nicastro Italy 26 G5
Nice France 24 H5
Nice CA U.S.A. 65 A1
Nicephorium Syria see Ar Raqqah
Nicholson r. Australia 56 B3
Nicholson Range hills Australia 55 B6
Nicobar Islands India 31 I6
Nicolaus CA U.S.A. 65 B1
Nicomedia Kocaeli Turkey see İzmit
Nicosia Cyprus 39 A2
Nicoya, Península de pen. Costa Rica
67 G7
Nida Lith. 15 L9
Nidagunda India 38 C2
Nidd r. U.K. 18 F4
Nidzica Poland 17 R4
Niebüll Germany 17 L3
Niedere Tauern mts Austria 17 N7
Niedersächsisches Wattenmeer,
Nationalpark nat. park Germany 16 K4
Niefang Equat. Guinea 46 E4
Niellé Côte d'Ivoire 46 C3
Nienburg (Weser) Germany 17 L4
Nieuw Nickerie Suriname 69 G2
Nieuwoudtville S. Africa 50 D6
Niğde Turkey 34 D3
Niger country Africa 46 D3
Niger r. Africa 46 D4
Niger, Mouths of the Nigeria 46 D4
Niger Cone sea feature S. Atlantic Ocean
72 I5
Nigeria country Africa 46 D4
Nighthawk Lake Canada 63 K2
Nigrita Greece 27 J4
Nihing Pak. 36 A3
Nihon country Asia see Japan
Niigata Japan 45 E5
Niihama Japan 45 D6
Nii-jima i. Japan 45 E6
Niimi Japan 45 D6
Niitsu Japan 45 E5
Nijil, Wādī watercourse Jordan 39 B4
Nijmegen Neth. 17 J5
Nikel' Rus. Fed. 12 F1
Nikki Benin 46 D4
Nikkō Kokuritsu-kōen Japan 45 E5
Nikolayev Ukr. see Mykolayiv
Nikolayevka Rus. Fed. 13 J5
Nikolayevsk Rus. Fed. 13 J6
Nikolayevskiy Rus. Fed. see Nikolayevsk
Nikolayevsk-na-Amure Rus. Fed. 44 F1
Nikol'sk Rus. Fed. 12 J4
Nikol'skiy Kazakh. see Satpayev
Nikol'skoye Kamchatskiy Kray Rus. Fed.
29 R4
Nikol'skoye Vologod. Obl. Rus. Fed. see
Sheksna
Nikopol' Ukr. 13 G7
Niksar Turkey 34 E2
Nīkshahr Iran 33 J4
Nikšić Montenegro 26 H3
Nikumaroro atoll Kiribati 53 I2
Nikunau i. Kiribati 53 H2
Nīl, Bahr el r. Africa see Nile
Nilagiri India 37 F5
Niland CA U.S.A. 65 E4
Nilang India see Nelang
Nilanga India 38 C2
Nilaveli Sri Lanka 38 D4
Nile r. Africa 34 C5
Niles OH U.S.A. 64 A2
Nilgiri Hills India 38 C4
Nīl Kowtal Afgh. 36 A2
Nilphamari Bangl. 37 G4
Nilsiä Fin. 14 P5
Nimach India see Neemuch
Niman r. Rus. Fed. 44 D2
Nimba, Monts mt. Africa see
Nimba, Mount
Nimba, Mount Africa 46 C4
Nimbal India 38 B2
Nimberra Well Australia 55 C5
Nimelen r. Rus. Fed. 44 E1
Nîmes France 24 G5
Nimmitabel Australia 57 E8
Nimrod Glacier Antarctica 76 H1
Nimu India 36 D2
Nimule Sudan 47 G4
Nimwegen Neth. see Nijmegen
Nindigully Australia 58 D2
Nine Degree Channel India 38 B4
Nine Islands P.N.G. see Kilinailau Islands
Ninetyeast Ridge sea feature Indian Ocean
73 N8
Ninety Mile Beach Australia 58 C7
Ninety Mile Beach N.Z. 59 D2
Nineveh NY U.S.A. 64 D1
Ning'an China 44 C3
Ningbo China 43 M7
Ningde China 43 L7
Ningguo China 43 L6
Ninghsia Hui Autonomous Region
aut. reg. China see Ningxia Huizu
Zizhiqu
Ningjiang China see Songyuan
Ningjing Shan mts China 42 H6
Ningnan China 42 I7
Ningwu China 43 K5
Ningxia aut. reg. China see
Ningxia Huizu Zizhiqu
Ningxia Huizu Zizhiqu aut. reg. China
42 J5
Ningxian China 43 J5
Ninh Binh Vietnam 42 J8
Ninnis Glacier Antarctica 76 G2
Ninnis Glacier Tongue Antarctica 76 H2
Ninohe Japan 45 F4
Niobrara r. U.S.A. 62 H3
Niokolo Koba, Parc National du nat. park
Senegal 46 B3
Niono Mali 46 C3
Nioro Mali 46 C3
Niort France 24 D3
Nipani India 38 B2
Niphad India 38 B1
Nipigon Canada 61 J5
Nipigon, Lake Canada 61 J5
Nipissing, Lake Canada 63 L2
Nippon country Asia see Japan
Nippon Hai sea N. Pacific Ocean see
Japan, Sea of
Nipton CA U.S.A. 65 E3
Niquelândia Brazil 71 A1
Nir Ardabīl Iran 35 G3
Nir Yazd Iran 35 I5
Nira r. India 38 B2
Nirji China 44 B2
Nirmal India 38 C2
Nirmali India 37 F4
Nirmal Range hills India 38 C2
Niš Serbia 27 I3
Nisa Port. 25 C4
Nisarpur India 38 B1
Niscemi Sicilia Italy 26 F6
Nīshāpūr Iran see Neyshābūr
Nishino-shima vol. Japan 45 F8
Nishi-Sonogi-hantō pen. Japan 45 C6
Nisibis Turkey see Nusaybin
Nísiros i. Greece see Nisyros
Nissan r. Sweden 15 H8
Nistru r. Moldova see Dniester
Nistru r. Ukr. 27 N1 see Dniester
Nisyros i. Greece 27 L6
Niṭā Saudi Arabia 35 H6
Nitendi i. Solomon Is see Ndeni
Niterói Brazil 71 C3
Nith r. U.K. 20 F5
Nitibe East Timor 54 D2
Niti Pass China/India 36 D3
Niti Shankou pass China/India see
Niti Pass
Nitmiluk National Park Australia 54 F3
Nitra Slovakia 17 Q6
Niuafo'ou i. Tonga 53 I3
Niuatoputopu i. Tonga 53 I3
Niue terr. S. Pacific Ocean 53 J3
Niulakita i. Tuvalu 53 H3
Niutao i. Tuvalu 53 H2
Nivala Fin. 14 N5
Nive watercourse Australia 56 D5
Niwai India 36 C4
Niwas India 36 E5
Nixia China see Sêrxü
Niya China see Minfeng
Niya He r. China 37 E1
Nizamabad India 38 C2
Nizam Sagar l. India 38 C2
Nizhnedevitsk Rus. Fed. 13 H6
Nizhnekamsk Rus. Fed. 12 K5
Nizhnekamskoye Vodokhranilishche resr
Rus. Fed. 11 Q4
Nizhnekolymsk Rus. Fed. 29 R3
Nizhnetambovskoye Rus. Fed. 44 E2
Nizhneudinsk Rus. Fed. 42 H2
Nizhnevartovsk Rus. Fed. 28 I3
Nizhnevolzhsk Rus. Fed. see Narimanov
Nizhneyansk Rus. Fed. 29 O2
Nizhniy Baskunchak Rus. Fed. 13 J6
Nizhniye Kresty Rus. Fed. see Cherskiy
Nizhniy Lomov Rus. Fed. 13 I5
Nizhniy Novgorod Rus. Fed. 12 I4
Nizhniy Odes Rus. Fed. 12 L3
Nizhniy Tagil Rus. Fed. 11 R4
Nizhnyaya Mola Rus. Fed. 12 J2
Nizhnyaya Omra Rus. Fed. 12 L3
Nizhnyaya Pirenga, Ozero l. Rus. Fed.
14 R3
Nizhnyaya Tunguska r. Rus. Fed. 28 J3
Nizhnyaya Tura Rus. Fed. 11 R4
Nizhyn Ukr. 13 F6
Nizina Mazowiecka reg. Pol. 17 R4
Nizip Turkey 39 C1
Nízke Tatry nat. park Slovakia 17 Q6
Nizwā Oman see Nazwá
Nizza France see Nice
Njallavarri mt. Norway 14 L2
Njavve Sweden 14 K3
Njombe Tanz. 49 D4
Njurundabommen Sweden 14 J5
Nkambe Cameroon 46 E4
Nkandla S. Africa 51 J5
Nkawkaw Ghana 46 C4
Nkhata Bay Malawi 49 D5
Nkhotakota Malawi 49 D5
Nkondwe Tanz. 49 D4
Nkongsamba Cameroon 46 D4
Nkululeko S. Africa 51 H6
Nkwenkwezi S. Africa 51 H7
Noakhali Bangl. 37 G5
Noatak r. U.S.A. 60 B3
Nobber Ireland 21 F4
Nobeoka Japan 45 C6
Noboribetsu Japan 44 F4
Noccundra Australia 57 C5
Nockatunga Australia 57 C5
Noel Kempff Mercado, Parque Nacional
nat. park Bol. 68 F6
Nogales Mex. 66 B2
Nogales U.S.A. 62 E5
Nōgata Japan 45 C6
Nogent-le-Rotrou France 24 E2
Noginsk Rus. Fed. 12 H5
Nogliki Rus. Fed. 44 F2
Nogoa r. Australia 56 E4
Nohar India 36 C3
Noheji Japan 44 F4
Noida India 36 D3
Noirmoutier, Île de i. France 24 C3
Noirmoutier-en-l'Île France 24 C3
Nōkis Uzbek. see Nukus
Nok Kundi Pak. 33 J4
Nokou Chad 47 E3
Nokrek Peak India 37 G4
Nola Cent. Afr. Rep. 48 B3
Nolinsk Rus. Fed. 12 K4
No Mans Land i. MA U.S.A. 64 F2
Nome U.S.A. 60 B3
Nomgon Mongolia 42 J4
Nomhon China 37 I1
Nomoi Islands Micronesia see
Mortlock Islands
Nomonde S. Africa 51 H6
Nomzha Rus. Fed. 12 I4
Nondweni S. Africa 51 J5
Nong'an China 44 B3
Nongoma S. Africa 51 J4
Nongstoin India 37 G4
Nonidas Namibia 50 B2

Nonni r. China see Nen Jiang
Nonning Australia 57 B7
Nonouti atoll Kiribati 53 H2
Nonzwakazi S. Africa 50 G6
Noolyeanna Lake imp. l. Australia 57 B5
Noondie, Lake imp. l. Australia 55 B7
Noonkanbah Australia 54 D4
Noonthorangee Range hills Australia 57 C6
Noorama Creek watercourse Australia 58 B1
Nora r. Rus. Fed. 44 C2
Noranda Canada 63 L2
Nor-Bayazet Armenia see Gavarr
Norberg Sweden 15 I6
Nord Greenland see Station Nord
Nordaustlandet i. Svalbard 28 D2
Norden Germany 17 K4
Nordenshel'da, Arkhipelag is Rus. Fed. 28 K2
Nordenskjold Archipelago is Rus. Fed. see Nordenshel'da, Arkhipelag
Norderstedt Germany 17 M4
Nordfjordeid Norway 14 D6
Nordfold Norway 14 I3
Nordfriesische Inseln Germany see North Frisian Islands
Nordhausen Germany 17 M5
Nordkapp c. Norway see North Cape
Nordkinnhalvøya i. Norway 14 O1
Nordkjosbotn Norway 14 K2
Nordli Norway 14 H4
Nordmaling Sweden 14 K5
Nord- og Østgrønland, Nationalparken i nat. park Greenland 61 O2
Nordøstrundingen c. Greenland 77 I1
Nord-Ostsee-Kanal Germany see Kiel Canal
Norðoyar is Faroe Is 10 E3
Nordre Strømfjord inlet Greenland see Nassuttooq
Nordvik Rus. Fed. 29 M2
Nore r. Ireland 21 F5
Nore, Pic de mt. France 24 F5
Noreg country Europe see Norway
Norfolk NE U.S.A. 63 H3
Norfolk VA U.S.A. 64 C4
Norfolk Island terr. S. Pacific Ocean 53 G4
Norfolk Island Ridge sea feature Tasman Sea 74 H7
Norge country Europe see Norway
Norheimsund Norway 15 E6
Noril'sk Rus. Fed. 28 J3
Norkyung China 37 G3
Norma Co l. China 37 G2
Norman U.S.A. 63 H4
Normanby Island P.N.G. 56 E1
Normandes, Îles is English Chan. see Channel Islands
Normandia Brazil 69 G3
Normandie reg. France see Normandy
Normandie, Collines de hills France 24 D2
Normandy reg. France 24 D2
Normanton Australia 56 C3
Ñorquinco Arg. 70 B6
Norra Kvarken str. Fin./Sweden 14 L5
Norra Storfjället mts Sweden 14 I4
Norristown PA U.S.A. 64 D2
Norrköping Sweden 15 J7
Norrtälje Sweden 15 K7
Norseman Australia 55 C8
Norsjö Sweden 14 K4
Norsk Rus. Fed. 44 C1
Norsup Vanuatu 53 G3
Norte, Punta pt Arg. 70 E5
Norte, Serra do hills Brazil 69 G6
Nortelândia Brazil 69 G6
North, Cape Antarctica 76 H2
Northallerton U.K. 18 F4
Northam Australia 55 B7
Northampton Australia 52 B4
Northampton U.K. 19 G6
Northampton MA U.S.A. 64 E1
Northampton PA U.S.A. 64 D2
North Anna r. VA U.S.A. 64 C4
North Atlantic Ocean Atlantic Ocean 63 O4
North Australian Basin sea feature Indian Ocean 73 P6
North Battleford Canada 62 F1
North Bay Canada 63 L2
North Berwick U.K. 20 G4
North Berwick ME U.S.A. 64 F1
North Bourke Australia 58 B3
North Canton OH U.S.A. 64 A2
North Cape Norway 14 N1
North Cape N.Z. 59 D2
North Cape U.S.A. 60 A4
North Caribou Lake Canada 63 I1
North Carolina state U.S.A. 63 L4
North Channel lake channel Canada 63 K2
North Channel U.K. 21 G2
Northcliffe Glacier Antarctica 76 F2
North Collins NY U.S.A. 64 B1
North Dakota state U.S.A. 62 G2
North Downs hills U.K. 19 G7
North East PA U.S.A. 64 B1
Northeast Foreland c. Greenland see Nordøstrundingen
North-East Frontier Agency state India see Arunachal Pradesh
Northeast Pacific Basin sea feature N. Pacific Ocean 75 J4
North Edwards CA U.S.A. 65 D3
Northern prov. S. Africa see Limpopo
Northern Areas admin. div. Pak. see Gilgit-Baltistan
Northern Cape prov. S. Africa 50 D5
Northern Donets r. Rus. Fed./Ukr. see Severskiy Donets
Northern Dvina r. Rus. Fed. see Severnaya Dvina
Northern Ireland prov. U.K. 21 F3
Northern Lau Group is Fiji 53 I3
Northern Mariana Islands terr. N. Pacific Ocean 41 G6
Northern Rhodesia country Africa see Zambia
Northern Sporades is Greece see Voreies Sporades
Northern Territory admin. div. Australia 52 D3
Northern Transvaal prov. S. Africa see Limpopo
North Esk r. U.K. 20 G4
North Foreland c. U.K. 19 I7
North Fork CA U.S.A. 65 C2
North Fork Pass Canada 60 E3
North Frisian Islands Germany 17 L3
North Geomagnetic Pole Arctic Ocean 61 K2
North Grimston U.K. 18 G4
North Haven CT U.S.A. 64 E2
North Head hd N.Z. 59 E3
North Horr Kenya 48 D3
North Island India 38 B4
North Island N.Z. 59 D4
North Kingsville OH U.S.A. 64 A2
North Knife Lake Canada 61 I4
North Korea country Asia 45 B5
North Lakhimpur India 37 H4
North Las Vegas NV U.S.A. 65 E2
North Luangwa National Park Zambia 49 D5
North Maalhosmadulu Atoll Maldives 38 B5
North Magnetic Pole Canada 77 A1
North Malosmadulu Atoll Maldives see North Maalhosmadulu Atoll
North Palisade mt. CA U.S.A. 65 C2
North Perry OH U.S.A. 64 A2
North Platte U.S.A. 62 G3
North Platte r. U.S.A. 62 G3
North Pole Arctic Ocean 77 I1
North Rona i. U.K. see Rona
North Ronaldsay i. U.K. 20 G1
North Ronaldsay Firth sea chan. U.K. 20 G1
North Saskatchewan r. Canada 62 F1
North Sea Europe 16 H2
North Shields U.K. 18 F3
North Shoshone Peak NV U.S.A. 65 D1
North Siberian Lowland Rus. Fed. 28 L2
North Siberian Lowland Rus. Fed. 77 E2
North Sinai governorate Egypt see Shamāl Sīnā'
North Slope plain U.S.A. 60 D3
North Somercotes U.K. 18 H5
North Spirit Lake Canada 63 I1
North Stradbroke Island Australia 58 F1
North Sunderland U.K. 18 F3
North Syracuse NY U.S.A. 64 C1
North Taranaki Bight b. N.Z. 59 E4
Northton U.K. 20 B3
North Tonawanda NY U.S.A. 64 B1
North Trap rf N.Z. 59 A8
North Tyne r. U.K. 18 E4
North Uist i. U.K. 20 B3
Northumberland National Park U.K. 18 E3
Northumberland Strait Canada 63 O2
Northville NY U.S.A. 64 D1
North Walsham U.K. 19 I6
North West prov. S. Africa 50 G4
Northwest Atlantic Mid-Ocean Channel N. Atlantic Ocean 72 E1
North West Cape Australia 54 A5
North West Frontier prov. Pak. see Khyber Pakhtunkhwa
North West Nelson Forest Park nat. park N.Z. see Kahurangi National Park
Northwest Pacific Basin sea feature N. Pacific Ocean 74 G3
Northwest Territories admin. div. Canada 60 H3
Northwich U.K. 18 E5
North Wildwood NJ U.S.A. 64 D3
Northwind Ridge sea feature Arctic Ocean 77 B1
Northwood NH U.S.A. 64 F1
North York Moors moorland U.K. 18 G4
North York Moors National Park U.K. 18 G4
Norton U.K. 18 G4
Norton de Matos Angola see Balombo
Norton Sound sea chan. U.S.A. 60 B3
Norvegia, Cape Antarctica 76 B2
Norwalk CT U.S.A. 64 E2
Norwalk OH U.S.A. 63 K3
Norway country Europe 14 E6
Norway House Canada 62 H1
Norwegian Basin sea feature N. Atlantic Ocean 72 H1
Norwegian Bay Canada 61 I2
Norwegian Sea N. Atlantic Ocean 77 H2
Norwich Ont. Canada 64 A1
Norwich U.K. 19 I6
Norwich CT U.S.A. 64 E2
Norwich NY U.S.A. 64 D1
Noshiro Japan 45 F4
Nosovaya Rus. Fed. 12 L1
Noşratābād Iran 33 I4
Noss, Isle of i. U.K. 20 [inset]
Nossebro Sweden 15 H7
Nossob watercourse Africa 50 D2
Nossob watercourse Africa 50 D2
Notch Peak UT U.S.A. 65 F1
Noteć r. Poland 17 O4
Noto, Golfo di g. Sicilia Italy 26 F6
Notodden Norway 15 F7
Noto-hantō pen. Japan 45 E5
Notre-Dame, Monts mts Canada 63 N2
Notre Dame Bay Canada 61 M5
Notre-Dame-de-Koartac Canada see Quaqtaq
Nottaway r. Canada 63 L1
Nottingham U.K. 19 F6
Nottingham Island Canada 61 K3
Nottoway r. VA U.S.A. 64 C4
Nouabalé-Ndoki, Parc National de nat. park Congo 48 B3
Nouâdhibou Mauritania 46 B2
Nouâdhibou, Râs c. Mauritania 46 B2
Nouakchott Mauritania 46 B3
Nouâmghâr Mauritania 46 B3
Nouméa New Caledonia 53 G4
Nouna Burkina Faso 46 C3
Noupoort S. Africa 50 G6
Nousu Fin. 14 P3
Nouveau-Brunswick prov. Canada see New Brunswick
Nouveau-Comptoir Canada see Wemindji
Nouvelle Calédonie i. S. Pacific Ocean 53 G4
Nouvelle Calédonie terr. S. Pacific Ocean see New Caledonia
Nouvelle-France, Cap de c. Canada 61 K3
Nouvelles Hébrides country S. Pacific Ocean see Vanuatu
Nova América Brazil 71 A1
Nova Chaves Angola see Muconda
Nova Freixa Moz. see Cuamba
Nova Friburgo Brazil 71 C3
Nova Gaia Angola see Cambundi-Catembo
Nova Goa India see Panaji
Nova Gradiška Croatia 26 G2
Nova Iguaçu Brazil 71 C3
Nova Kakhovka Ukr. 27 O1
Nova Lima Brazil 71 C2
Nova Lisboa Angola see Huambo
Novalukoml' Belarus 13 F5
Nova Mambone Moz. 49 D6
Nova Nabúri Moz. 49 D5
Nova Odesa Ukr. 13 F7
Nova Pilão Arcado Brazil 69 J5
Nova Ponte Brazil 71 B2
Nova Ponte, Represa resr Brazil 71 B2
Novara Italy 26 C2
Nova Roma Brazil 71 B1
Nova Scotia prov. Canada 63 N3
Nova Sento Sé Brazil 69 J5
Novato CA U.S.A. 65 A1
Nova Trento Brazil 71 A4
Nova Venécia Brazil 71 C2
Nova Xavantino Brazil 69 H6
Novaya Kakhovka Ukr. see Nova Kakhovka
Novaya Kazanka Kazakh. 11 P6
Novaya Ladoga Rus. Fed. 12 G3
Novaya Lyalya Rus. Fed. 11 S4
Novaya Odessa Ukr. see Nova Odesa
Novaya Sibir', Ostrov i. Rus. Fed. 29 P2
Novaya Zemlya is Rus. Fed. 28 G2
Nova Zagora Bulg. 27 L3
Novelda Spain 25 F4
Nové Zámky Slovakia 17 Q7
Novgorod Rus. Fed. see Velikiy Novgorod
Novgorod-Severskiy Ukr. see Novhorod-Sivers'kyy
Novgorod-Volynskiy Ukr. see Novohrad-Volyns'kyy
Novhorod-Sivers'kyy Ukr. 13 G6
Novi Grad Bos.-Herz. see Bosanski Novi
Novi Iskŭr Bulg. 27 J3
Novikovo Rus. Fed. 44 F3
Novi Kritsim Bulg. see Stamboliyski
Novi Ligure Italy 26 C2
Novi Pazar Bulg. 27 L3
Novi Pazar Serbia 27 I3
Novi Sad Serbia 27 H2
Novo Acre Brazil 71 C1
Novoaltaysk Rus. Fed. 42 E2
Novoanninskiy Rus. Fed. 13 I6
Novo Aripuanã Brazil 68 F5
Novoazovs'k Ukr. 13 H7
Novocheboksarsk Rus. Fed. 12 J4
Novocherkassk Rus. Fed. 13 I7
Novo Cruzeiro Brazil 71 C2
Novodugino Rus. Fed. 12 G5
Novodvinsk Rus. Fed. 12 I2
Novoekonomicheskoye Ukr. see Dymytrov
Novogeorgiyevka Rus. Fed. 44 B2
Novogrudok Belarus see Navahrudak
Novo Hamburgo Brazil 71 A5
Novohradské hory mts Czech Rep. 17 O6
Novohrad-Volyns'kyy Ukr. 13 E6
Novokhopersk Rus. Fed. 13 I6
Novokiyevskiy Uval Rus. Fed. 44 C2
Novokubansk Rus. Fed. 35 F1
Novokubanskiy Rus. Fed. see Novokubansk
Novokuybyshevsk Rus. Fed. 13 K5
Novokuznetsk Rus. Fed. 42 F2
Novolazarevskaya (Russian Federation) research stn Antarctica 76 C2
Novolukoml' Belarus see Novalukoml'
Novo Mesto Slovenia 26 F2
Novomikhaylovskiy Rus. Fed. 34 E1
Novomoskovsk Rus. Fed. 13 H5
Novomoskovs'k Ukr. 13 G6
Novonikolayevsk Rus. Fed. see Novosibirsk
Novonikolayevskiy Rus. Fed. 13 I6
Novooleksiyivka Ukr. 13 G7
Novo Paraíso Brazil 68 F3
Novopashiyskiy Rus. Fed. see Gornozavodsk
Novopokrovka Rus. Fed. 44 D3
Novopokrovskaya Rus. Fed. 13 I7
Novopolotsk Belarus see Navapolatsk
Novopskov Ukr. 13 H6
Novo Redondo Angola see Sumbe
Novorossiyka Rus. Fed. 44 C1
Novorossiysk Rus. Fed. 13 H7
Novorybnaya Rus. Fed. 29 L2
Novorzhev Rus. Fed. 12 F4
Novoselovo Rus. Fed. 42 G1
Novoselskoye Rus. Fed. see Achkhoy-Martan
Novosel'ye Rus. Fed. 15 P7
Novosergiyevka Rus. Fed. 11 Q5
Novoshakhtinsk Rus. Fed. 13 H7
Novosheshminsk Rus. Fed. 12 K5
Novosibirsk Rus. Fed. 28 J4
Novosibirskiye Ostrova is Rus. Fed. see New Siberia Islands
Novosil' Rus. Fed. 13 H5
Novosokol'niki Rus. Fed. 12 F4
Novospasskoye Rus. Fed. 13 J5
Novotroyits'ke Ukr. 13 G7
Novoukrainka Ukr. see Novoukrayinka
Novoukrayinka Ukr. 13 F6
Novoussuri (abandoned) Rus. Fed. 44 E2
Novouzensk Rus. Fed. 13 K6
Novovolyns'k Ukr. 13 E6
Novovoronezh Rus. Fed. 13 H6
Novovoronezhskiy Rus. Fed. see Novovoronezh
Novovoskresenovka Rus. Fed. 44 B1
Novozybkov Rus. Fed. 13 F5
Nový Jičín Czech Rep. 17 P6
Novyy Afon Georgia see Akhali Ap'oni
Novyy Bor Rus. Fed. 12 L2
Novyy Donbass Ukr. see Dymytrov
Novyye Petushki Rus. Fed. see Petushki
Novyy Kholmogory Rus. Fed. see Archangel
Novyy Margelan Uzbek. see Farg'ona
Novyy Nekouz Rus. Fed. 12 H4
Novyy Oskol Rus. Fed. 13 H6
Novyy Port Rus. Fed. 28 I3
Novyy Urengoy Rus. Fed. 28 I3
Novyy Urgal Rus. Fed. 44 D2
Novyy Uzen' Kazakh. see Zhanaozen
Novyy Zay Rus. Fed. 12 L5
Now Iran 35 I5
Nowabganj Bangl. see Nawabganj
Nowdī Iran 35 H3
Nowgong India see Nagaon
Now Kharegan Iran 35 I3
Nowogard Poland 17 O4
Noworadomsk Poland see Radomsko
Nowra Australia 58 E5
Nowrangapur India see Nabarangapur
Nowshera Pak. 33 L3
Nowy Sącz Poland 17 R6
Nowy Targ Poland 17 R6
Noxen PA U.S.A. 64 C2
Noyabr'sk Rus. Fed. 28 I3
Noyon Mongolia 42 I4
Nozizwe S. Africa 51 G6
Nqamakwe S. Africa 51 H7
Nqutu S. Africa 51 J5
Nsanje Malawi 49 D5
Nsombo Zambia 49 C5
Nsukka Nigeria 46 D4
Nsumbu National Park Zambia see Sumbu National Park
Ntambu Zambia 49 C5
Ntha S. Africa 51 H4
Ntoro, Kavo pt Greece 27 K5
Ntoum Gabon 48 A3
Ntungamo Uganda 48 D4
Nuanetsi Zimbabwe see Mwenezi
Nuba Mountains Sudan 32 D7
Nubian Desert Sudan 32 D5
Nudo Coropuna mt. Peru 68 D7
Nueltin Lake Canada 61 I3
Nueva Gerona Cuba 67 H4
Nueva Harberton Arg. 70 C8
Nueva Imperial Chile 70 B5
Nueva Loja Ecuador see Lago Agrio
Nueva Rosita Mex. 62 G6
Nueva San Salvador El Salvador 66 G6
Nueve de Julio Arg. see 9 de Julio
Nuevitas Cuba 67 I4
Nuevo, Golfo g. Arg. 70 D6
Nuevo Casas Grandes Mex. 66 C2
Nuevo Ideal Mex. 66 C4
Nuevo Laredo Mex. 66 E3
Nuevo Rocatuerte Ecuador 68 C4
Nugaal watercourse Somalia 48 E3
Nugget Point N.Z. 59 B8
Nugur India 38 D2
Nuguria Islands P.N.G. 52 F2
Nuhaka N.Z. 59 F4
Nui atoll Tuvalu 53 H2
Nui Con Voi r. Vietnam see Red
Nuiqsut U.S.A. 60 C2
Nujiang China 37 I3
Nu Jiang r. China/Myanmar see Salween
Nu Jiang r. China/Myanmar see Salween
Nukey Bluff h. Australia 57 A7
Nukha Azer. see Şäki
Nuku'alofa Tonga 53 I4
Nukufetau atoll Tuvalu 53 H2
Nukuhiva i. Fr. Polynesia see Nuku Hiva
Nuku Hiva i. Fr. Polynesia 75 K6
Nukulaelae atoll Tuvalu 53 H2
Nukulailai atoll Tuvalu see Nukulaelae
Nukumanu Islands P.N.G. 53 F2
Nukunau i. Kiribati see Nikunau
Nukunono atoll Tokelau see Nukunonu
Nukunonu atoll Tokelau 53 I2
Nukus Uzbek. 33 I1
Nulato U.S.A. 60 C3
Nullagine Australia 54 C5
Nullarbor Australia 55 E7
Nullarbor National Park Australia 55 E7
Nullarbor Plain Australia 55 E7
Nullarbor Regional Reserve nature res. Australia 55 E7
Nulu'erhu Shan mts China 43 L4
Numalla, Lake imp. l. Australia 58 B2
Numan Nigeria 48 B3
Numanuma P.N.G. 56 E1
Numazu Japan 45 E6
Numbulwar Australia 56 A2
Numedal val. Norway 15 F6
Numfoor i. Indon. 41 F8
Numin He r. China 44 B3
Numurkah Australia 58 B6
Nunakuluut i. Greenland 61 N3
Nunap Isua c. Greenland see Farewell, Cape
Nunarsuit i. Greenland see Nunakuluut
Nunavut admin. div. Canada 61 J2
Nunda NY U.S.A. 64 C1
Nundle Australia 58 E3
Nuneaton U.K. 19 F6
Nungba India 37 H4
Nungnain Sum China 43 L3
Nunivak Island U.S.A. 60 B4
Nunkapasi India 38 E1
Nunkun mt. India 36 D2
Nunligran Rus. Fed. 29 T3
Nuñomoral Spain 25 C3
Nuoro Sardegna Italy 26 C4
Nupani i. Solomon Is 53 G3
Nuqrah Saudi Arabia 32 F4
Nūrābād Iran 35 H5
Nurakita i. Tuvalu see Niulakita
Nur Dağları mts Turkey 39 B1
Nuremberg Germany 17 M6
Nurla India 36 D2
Nurlat Rus. Fed. 13 K5
Nurmes Fin. 14 P5
Nurmo Fin. 14 M5
Nürnberg Germany see Nuremberg
Nurota Uzbek. 33 K1
Nurri, Mount h. Australia 58 C3
Nusaybin Turkey 35 F3
Nu Shan mts China 42 H7
Nushki Pak. 36 A3
Nusratiye Turkey 39 D1
Nuttal Pak. 36 B3
Nutwood Downs Australia 54 F3
Nuuk Greenland 61 M3
Nuupas Fin. 14 O3
Nuussuaq Greenland 61 M2
Nuussuaq pen. Greenland 61 M2
Nuwaybi' al Muzayyinah Egypt 34 D5
Nuweiba el Muzeina Egypt see Nuwaybi' al Muzayyinah
Nuwerus S. Africa 50 D6
Nuweveldberge mts S. Africa 50 E7
Nuyts, Point Australia 55 B8
Nuyts Archipelago is Australia 55 F8
Nuzvid India 38 D2
Nwanedi Nature Reserve S. Africa 51 J2
Nxai Pan National Park Botswana 49 C5
Nyagan' Rus. Fed. 11 T3
Nyahururu Kenya 48 D3
Nyah West Australia 58 A5
Nyainqêntanglha Feng mt. China 37 G3
Nyainqêntanglha Shan mts China 37 G3
Nyainrong China 42 G6
Nyainronglung China see Nyainrong
Nyåker Sweden 14 K5
Nyakh Rus. Fed. see Nyagan'
Nyaksimvol' Rus. Fed. 11 S3
Nyala Sudan 47 F3
Nyalam China see Congdü
Nyalikungu Tanz. see Maswa
Nyamandhlovu Zimbabwe 49 C5
Nyamtumbo Tanz. 49 D5
Nyande Zimbabwe see Masvingo
Nyandoma Rus. Fed. 12 I3
Nyandomskaya Vozvyshennost' hills Rus. Fed. 12 H3
Nyanga Congo 48 B4
Nyanga Zimbabwe 49 D5
Nyangani mt. Zimbabwe 49 D5
Nyasa, Lake Africa 49 D4
Nyasaland country Africa see Malawi
Nyashabozh Rus. Fed. 12 L2
Nyasvizh Belarus 15 O10
Nyborg Denmark 15 G9
Nyborg Norway 14 P1
Nybro Sweden 15 I8
Nyeboe Land reg. Greenland 61 M1
Nyêmo China 37 G3
Nyenchen Tanglha Range mts China see Nyainqêntanglha Shan
Nyeri Kenya 48 D4
Nyi, Co l. China 37 F2
Nyika National Park Zambia 49 D5
Nyima China 37 F3
Nyimba Zambia 49 D5
Nyingchi China 37 H3
Nyingchi China 37 H3
Nyíregyháza Hungary 13 D7
Nyiru, Mount Kenya 48 D3
Nykarleby Fin. 14 M5
Nykøbing Denmark 15 G9
Nykøbing Sjælland Denmark 15 G9
Nyköping Sweden 15 J7
Nyland Sweden 14 J5
Nylsvley nature res. S. Africa 51 I3
Nymagee Australia 58 C4
Nymboida National Park Australia 58 F2
Nynäshamn Sweden 15 J7
Nyngan Australia 58 C3
Nyogzê China 37 E3
Nyoman r. Belarus/Lith. 15 M10
Nyon Switz. 24 H3
Nyons France 24 G4
Nyrob Rus. Fed. 11 R3
Nysa Poland 17 P5
Nysh Rus. Fed. 44 F2
Nystad Fin. see Uusikaupunki
Nytva Rus. Fed. 11 R4
Nyuksenitsa Rus. Fed. 12 J3
Nyunzu Dem. Rep. Congo 49 C4
Nyurba Rus. Fed. 29 M3
Nyyskiy Zaliv b. Rus. Fed. 44 F1
Nzambi Congo 48 B4
Nzega Tanz. 49 D4
Nzérékoré Guinea 46 C4
N'zeto Angola 49 B4
Nzwani i. Comoros 49 E5

Oahe, Lake U.S.A. 62 G3
O'ahu i. U.S.A. 75 J4
Oaitupu i. Tuvalu see Vaitupu
Oak Bluffs MA U.S.A. 64 F2
Oakey Australia 58 E1
Oakham U.K. 19 G6
Oak Hill WV U.S.A. 64 A4
Oakhurst CA U.S.A. 65 C2
Oakland CA U.S.A. 65 A2
Oakland MD U.S.A. 64 B3
Oakland airport CA U.S.A. 65 A2
Oaklands Australia 58 C5
Oakover r. Australia 54 C5
Oakridge U.S.A. 62 C3
Oakvale Australia 57 C7
Oak View CA U.S.A. 65 C3
Oakville Ont. Canada 64 B1
Oamaru N.Z. 59 C7
Oaro N.Z. 59 D6
Oasis CA U.S.A. 65 D2
Oates Coast reg. Antarctica see Oates Land
Oates Land reg. Antarctica 76 H2
Oaxaca Mex. 66 E5
Oaxaca de Juárez Mex. see Oaxaca
Ob' r. Rus. Fed. 42 E2
Ob, Gulf of sea chan. Rus. Fed. see Obskaya Guba
Oba i. Vanuatu see Aoba
Obala Cameroon 46 E4
Obama Japan 45 D6
Oban U.K. 20 D4
O Barco Spain 25 C2
Obbia Somalia see Hobyo
Obdorsk Rus. Fed. see Salekhard
Óbecse Serbia see Bečej
Oberon Australia 58 D4
Oberpfälzer Wald mts Germany 17 N6
Obi i. Indon. 41 E8
Óbidos Brazil 69 G4
Obihiro Japan 44 F4
Obil'noye Rus. Fed. 13 J7
Ob'-Irtysh r. Rus. Fed. 28 H3
Obluch'ye Rus. Fed. 44 C2
Obninsk Rus. Fed. 13 H5
Obo Cent. Afr. Rep. 48 C3
Obock Djibouti 32 F7
Ŏbŏk N. Korea 44 C4
Obokote Dem. Rep. Congo 48 C4
Obo Liang China 42 G5
Obouya Congo 48 B4
Oboyan' Rus. Fed. 13 H6
Obozerskiy Rus. Fed. 12 I3
Obrenovac Serbia 27 I2
Obruk Turkey 34 D3
Observatory Hill h. Australia 55 F7
Obshchiy Syrt hills Rus. Fed. 11 Q5
Obskaya Guba sea chan. Rus. Fed. 28 I3
Obuasi Ghana 46 C4
Ob"yachevo Rus. Fed. 12 K3
Ocala U.S.A. 63 K6
Ocaña Col. 68 D2
Ocaña Spain 25 E4
Occidental, Cordillera mts Chile 68 E7
Occidental, Cordillera mts Col. 68 C3
Occidental, Cordillera mts Peru 68 D7
Ocean City MD U.S.A. 64 D3
Ocean City NJ U.S.A. 64 D3
Ocean Falls Canada 60 F4
Ocean Island Kiribati see Banaba
Ocean Island atoll U.S.A. see Kure Atoll
Oceanside CA U.S.A. 65 D4
Ochakiv Ukr. 27 N1
Och'amch'ire Georgia 35 F2
Ocher Rus. Fed. 11 Q4
Ochiishi-misaki pt Japan 44 G4
Ochil Hills U.K. 20 F4
Ochito r. Pak. 36 A4
Ochrida, Lake Albania/Macedonia see Ohrid, Lake
Ochsenfurt Germany 17 M6
Ockelbo Sweden 15 J6
Ocolaşul Mare, Vârful mt. Romania 27 K1
Octeville-sur-Mer France 19 H9
October Revolution Island Rus. Fed. see Oktyabr'skoy Revolyutsii, Ostrov
Ocussi enclave East Timor 54 D2
Ocussi-Ambeno enclave East Timor see Ocussi
Oda, Jebel mt. Sudan 32 E5
Ódáðahraun lava field Iceland 14 [inset]
Ŏdaejin N. Korea 44 C4
Odae-san National Park S. Korea 45 C5
Ōdate Japan 45 F4
Odawara Japan 45 E6
Odda Norway 15 E6
Odemira Port. 25 B5
Ödemiş Turkey 27 L5
Ödenburg Hungary see Sopron
Odense Denmark 15 G9
Odenwald reg. Germany 17 L6
Oderbucht b. Germany 17 O3
Ödeshog Sweden 15 I7
Odessa Ukr. 27 N1
Odessa TX U.S.A. 62 G5
Odessus Bulg. see Varna
Odiel r. Spain 25 C5
Odienné Côte d'Ivoire 46 C4
Odintsovo Rus. Fed. 12 H5
Odra r. Germany/Pol. 17 Q6
Odzala, Parc National d' nat. park Congo 48 B3
Oea Libya see Tripoli
Oé-Cusse enclave East Timor see Ocussi
Oecussi enclave East Timor see Ocussi
Oeiras Brazil 69 J5
Oekussi enclave East Timor see Ocussi
Oenpelli Australia 54 F3
Oesel i. Estonia see Hiiumaa
Of Turkey 35 F2
Ofanto r. Italy 26 G4
Ofaqim Israel 39 B4
Offa Nigeria 46 D4
Offenbach am Main Germany 17 L5
Offenburg Germany 17 K6
Oga Japan 45 E5
Ogadēn reg. Eth. 48 E3
Oga-hantō pen. Japan 45 E5
Ōgaki Japan 45 E6
Ogallala U.S.A. 62 G3
Ogasawara-shotō is Japan see Bonin Islands
Ogbomosho Nigeria 46 D4
Ogbomoso Nigeria see Ogbomosho
Ogden UT U.S.A. 62 E3
Ogilvie r. Canada 60 E3
Ogilvie Mountains Canada 60 D3
Oglio r. Italy 26 D2
Oglongi Rus. Fed. 44 E1
Ogmore Australia 56 E4
Ogodzha Rus. Fed. 44 D1
Ogoja Nigeria 46 D4
Ogoki Reservoir Canada 63 J1
Ogoron Rus. Fed. 44 C1
Ogosta r. Bulg. 27 J3
Ogre Latvia 15 N8
Ogulin Croatia 26 F2
Ogurchinskiy, Ostrov i. Turkm. see Ogurjaly Adasy
Ogurjaly Adasy i. Turkm. 35 I3
Oğuzeli Turkey 39 C1
Ohai N.Z. 59 A7
Ohakune N.Z. 59 E4
Ohanet Alg. 46 D2
Ōhata Japan 44 F4
Ohcejohka Fin. see Utsjoki
O'Higgins (Chile) research stn Antarctica 76 A2
O'Higgins, Lago l. Chile 70 B7
Ohio r. Ohio/West Virginia U.S.A. 64 A3
Ohio state U.S.A. 64 A2
Ohrid Macedonia 27 I4
Ohrid, Lake Albania/Macedonia 27 I4
Ohridsko Ezero l. Albania/Macedonia see Ohrid, Lake
Ohrigstad S. Africa 51 J3
Ohrit, Liqeni i l. Albania/Macedonia see Ohrid, Lake
Ohura N.Z. 59 E4
Oich r. U.K. 20 E3
Oiga China 37 H3
Oil City PA U.S.A. 64 B2
Oise r. France 24 F2
Ōita Japan 45 C6
Oiti mt. Greece 27 J5
Ojai CA U.S.A. 65 C3
Ojalava i. Samoa see 'Upolu
Ojinaga Mex. 66 D3
Ojiya Japan 45 E5
Ojos del Salado, Nevado mt. Arg./Chile 70 C3
Oka r. Rus. Fed. 13 I4
Oka r. Rus. Fed. 42 I1
Okahandja Namibia 50 C1
Okahukura N.Z. 59 E4
Okakarara Namibia 49 B6
Okanagan Lake Canada 62 D2
Okanda Sri Lanka 38 D5
Okano r. Gabon 48 B4
Okanogan U.S.A. 62 D2
Okanogan r. U.S.A. 62 D2
Okara Pak. 33 L3
Okarem Turkm. see Ekerem
Okataina vol. N.Z. see Tarawera, Mount
Okaukuejo Namibia 49 B5
Okavango r. Africa 49 C5
Okavango Delta swamp Botswana 49 C
Okavango Swamps Botswana see Okavango Delta
Okaya Japan 45 E5
Okayama Japan 45 D6
Okazaki Japan 45 E6
Okeechobee, Lake U.S.A. 63 K6
Okefenokee Swamp U.S.A. 63 K5
Okehampton U.K. 19 C8
Okha India 36 B5
Okha Rus. Fed. 44 F1
Okha Rann marsh India 36 B5
Okhotsk Rus. Fed. 29 P4
Okhotsk, Sea of Japan/Rus. Fed. 44 G3
Okhotskoye More sea Japan/Rus. Fed. see Okhotsk, Sea of
Okhtyrka Ukr. 13 G6
Okinawa i. Japan 45 B8
Okinawa-guntō is Japan see Okinawa-shotō
Okinawa-shotō is Japan 45 B8
Okino-Daitō-jima i. Japan 43 O8
Okino-Tori-shima i. Japan 43 P8
Oki-shotō is Japan 43 O5
Oki-shotō is Japan 45 D5
Oklahoma state U.S.A. 62 H4
Oklahoma City U.S.A. 63 H4
Okmulgee U.S.A. 63 H4
Okondja Gabon 48 B4
Okovskiy Les for. Rus. Fed. 12 G5
Okoyo Congo 48 B4
Øksfjord Norway 14 M1
Oktemberyan Armenia see Armavir
Oktyabr' Kazakh. see Kandyagash
Oktyabr'sk Kazakh. see Kandyagash
Oktyabr'skiy Belarus see Aktsyabrski
Oktyabr'skiy Amurskaya Oblast' Rus. Fed. 44 C1
Oktyabr'skiy Arkhangel'skaya Oblast' Rus. Fed. 12 I3
Oktyabr'skiy Kamchatskiy Kray Rus. Fed. 29 Q4
Oktyabr'skiy Respublika Bashkortostan Rus. Fed. 11 Q5
Oktyabr'skiy Volgogradskaya Oblast' Rus. Fed. 13 I7
Oktyabr'skoye Rus. Fed. 11 T3
Oktyabr'skoy Revolyutsii, Ostrov i. Rus. Fed. 29 K2
Okulovka Rus. Fed. 12 G4
Okushiri-tō i. Japan 44 E4
Okusi enclave East Timor see Ocussi
Okuta Nigeria 46 D4
Okwa watercourse Botswana 50 G1
Ólafsvík Iceland 14 [inset]
Olakkur India 38 C3
Olancha CA U.S.A. 65 C2
Olancha Peak CA U.S.A. 65 C2
Öland i. Sweden 15 J8
Olary Australia 57 C7
Olavarría Arg. 70 D5
Oława Poland 17 P5
Olbia Sardegna Italy 26 C4
Old Bastar India 38 D2
Oldcastle Ireland 21 E4
Old Cork Australia 56 C4
Old Crow Canada 60 E3
Oldenburg Germany 17 L4
Oldenburg in Holstein Germany 17 M
Olderdalen Norway 14 L2
Old Gidgee Australia 55 B6
Oldham U.K. 18 E5
Old Harbor U.S.A. 60 C4
Old Head of Kinsale hd Ireland 21 D6
Oldmeldrum U.K. 20 G3
Old River CA U.S.A. 65 C3
Olds Canada 62 E1
Old Wives Lake Canada 62 F1
Olean NY U.S.A. 64 B1
Olecko Poland 17 S3
Olekma r. Rus. Fed. 29 N3
Olekminsk Rus. Fed. 29 N3
Olekminskiy Stanovik mts Rus. Fed. 43 M2
Oleksandrivs'k Ukr. see Zaporizhzhya
Oleksandriya Ukr. 13 G6
Ølen Norway 15 D7
Olenegorsk Rus. Fed. 12 G1
Olenek Rus. Fed. 29 M3
Olenek r. Rus. Fed. 29 M2
Olenek Bay Rus. Fed. see Olenekskiy Z
Olenekskiy Zaliv b. Rus. Fed. 29 N2
Olenino Rus. Fed. 12 G4
Olenitsa Rus. Fed. 12 G2

lenivs'ki Kar"yery Ukr. see Dokuchayevs'k
lenya Rus. Fed. see Olenegorsk
leshky Ukr. see Tsyurupyns'k
levs'k Ukr. 13 E6
l'ga Rus. Fed. 44 D4
lga, Mount Australia 55 E6
l'ginsk Rus. Fed. 44 D1
lginskoye Rus. Fed. see Kochubeyevskoye
lgiy Mongolia 42 F3
lhão Port. 25 C5
lia Chain mts Australia 55 E6
lifants r. Moz./S. Africa 51 J3
lifants watercourse Namibia 50 D3
lifants S. Africa 51 J2
lifants r. W. Cape S. Africa 50 D6
lifants r. W. Cape S. Africa 50 E7
lifantshoek S. Africa 50 F4
lifantsrivierberge mts S. Africa 50 D7
limbos h. Cyprus see Olympos
limbos mt. Greece see Olympus, Mount
limpos Beydağları Milli Parkı nat. park Turkey 27 N6
linda Brazil 69 L5
linga Moz. 49 D5
lio Australia 56 C4
liphants Drift S. Africa 51 H3
lisipo Port. see Lisbon
liva Spain 25 F4
liva, Cordillera de mts Arg./Chile 70 C3
livares, Cerro de mt. Arg./Chile 70 C4
livehurst CA U.S.A. 65 B1
liveira dos Brejinhos Brazil 71 C1
livença Moz. see Lupilichi
livenza Spain 25 C4
l'khovka Rus. Fed. 13 J6
llagüe Chile 70 C2
llombo Congo 48 B4
lmaliq Uzbek. 42 B4
lmos Peru 68 C5
lmütz Czech Rep. see Olomouc
lney U.K. 19 G6
ney MD U.S.A. 64 C3
lofström Sweden 15 I8
lomouc Czech Rep. 17 P6
lonets Rus. Fed. 12 G3
loron-Ste-Marie France 24 D5
losenga atoll American Samoa see Swains Island
lot Spain 25 H2
lovyannaya Rus. Fed. 43 L2
loy r. Rus. Fed. 29 Q3
loy, Qatorkŭhi mts Asia see Alai Range
lsztyn Poland 17 R4
lt r. Romania 27 K3
lten Switz. 24 H3
ltenița Romania 27 L2
ltu Turkey 35 F2
l'viopol' Ukr. see Pervomays'k
ymbos h. Cyprus see Olympos
lympia U.S.A. 62 C2
lympos h. Cyprus 39 A2
lympos Greece see Olympus, Mount
lympou, Ethnikos Drymos nat. park Greece 27 J4
lympus, Mount Greece 27 J4
lympus, Mount U.S.A. 62 C2
lyutorskiy, Mys c. Rus. Fed. 29 S4
lyutorskiy Zaliv b. Rus. Fed. 29 R4
lzheras Rus. Fed. see Mezhdurechensk
ma China 37 E2
ma r. Rus. Fed. 12 J2
magh U.K. 21 E3
maha U.S.A. 63 H3
maheke admin. reg. Namibia 50 D2
mal'skiy Khrebet mts Rus. Fed. 44 E1
man country Asia 33 I6
man, Gulf of Asia 33 I4
maruru Namibia 49 B6
mate Peru 68 D7
maweneno Botswana 50 F3
mba i. Vanuatu see Aoba
mbai, Selat sea chan. Indon. 54 D2
mbalantu Namibia see Uutapi
mboué Gabon 48 A4
mbu China 37 F3
mdraaisvlei S. Africa 50 F6
mdurman Sudan 32 D6
meo Australia 58 C6
metepec Mex. 66 E5
m Hajër Eritrea 32 E7
midīyeh Iran 35 H5
mineca Mountains Canada 60 F4
mirzak Kazakh. 35 H2
mitara Namibia 50 C2
miya Japan 45 E6
molon Rus. Fed. 29 R3
mo National Park Eth. 48 D3
msk Rus. Fed. 28 I4
msukchan Rus. Fed. 29 Q3
mū Japan 44 F3
mu, Vârful mt. Romania 27 K2
mura Japan 45 C6
mutninsk Rus. Fed. 12 L4
nancock VA U.S.A. 64 D4
nangué, Lac l. Gabon 48 B4
naping Lake Canada 63 K2
ncativo Arg. 70 D4
nchan Isle of Man 18 C4
ncócua Angola 49 B5
ncül Turkey 39 D1
ndal India see Andal
ndangwa Namibia 49 B5
nderstedorings S. Africa 50 E6
ndjiva Angola 49 B5
ndo Nigeria 46 D4
ndörhaan Mongolia 43 K3
ndörshil Mongolia 43 J3
ndozero Rus. Fed. 12 G3
ne Botswana 50 E2
nega Rus. Fed. 12 H3
nega r. Rus. Fed. 12 H3
nega, Lake l. Rus. Fed. see Onezhskoye Ozero
nega, Lake Rus. Fed. 12 G3
nega Bay g. Rus. Fed. see Onezhskaya Guba
ne Hundred Mile House Canada see 100 Mile House
neida NY U.S.A. 64 D1
neida Lake NY U.S.A. 64 D1
'Neill U.S.A. 62 H3
nekotan, Ostrov i. Rus. Fed. 29 Q5
neonta NY U.S.A. 64 D1
nești Romania 27 L1
nezhskaya Guba g. Rus. Fed. 12 G2
nezhskoye Ozero Rus. Fed. 11 N3
nezhskoye Ozero l. Rus. Fed. see Onega, Lake
ng r. India 38 D1
nga Gabon 48 B4
ngers watercourse S. Africa 50 F5
ngjin N. Korea 45 B5
ngole India 38 D3
nilahy r. Madag. 49 E6
nitsha Nigeria 46 D4
njati Mountain Namibia 50 C2
njiva Angola see Ondjiva
no-i-Lau i. Fiji 53 I4
nomichi Japan 45 D6
Onor, Gora mt. Rus. Fed. 44 F2
Onotoa atoll Kiribati 53 H2
Onseepkans S. Africa 50 D5
Onslow Australia 54 A5
Ontake-san vol. Japan 45 E6
Ontario prov. Canada 64 A1
Ontario CA U.S.A. 65 D3
Ontario, Lake Canada/U.S.A. 64 C1
Ontong Java Atoll Solomon Is 53 F2
Onutu atoll Kiribati see Onotoa
Onverwacht Suriname 69 G2
Onyx CA U.S.A. 65 C3
Oodnadatta Australia 57 A5
Oodweyne Somalia 48 E3
Ooldea Australia 55 E7
Ooldea Range hills Australia 55 E7
Ooratippra r. Australia 56 B4
Oos-Londen S. Africa see East London
Oostende Belgium see Ostend
Ootacamund India see Udagamandalam
Ootsa Lake Canada 60 F4
Opala Dem. Rep. Congo 48 C4
Oparino Rus. Fed. 12 K4
Oparo i. Fr. Polynesia see Rapa
Opataca, Lac l. Canada 63 M1
Opava Czech Rep. 17 P6
Opelika U.S.A. 63 J5
Opelousas U.S.A. 63 I5
Opienge Dem. Rep. Congo 48 C3
Opinaca, Réservoir resr Canada 63 L1
Opiscotéo, Lac l. Canada 61 L4
Opochka Rus. Fed. 12 F4
Opodepe Mex. 66 B3
Opole Poland 17 P5
Oporto Port. 25 B3
Opotiki N.Z. 59 F4
Oppdal Norway 14 F5
Oppeln Poland see Opole
Opunake N.Z. 59 D4
Opuwo Namibia 49 B5
Oradea Romania 27 I1
Orahovac Kosovo see Rahovec
Orai India 36 D4
Oral Kazakh. see Ural'sk
Oran Alg. 25 F6
Orán Arg. 70 D2
Orang India 37 H4
Ŏrang N. Korea 44 C4
Orange Australia 58 D4
Orange France 24 G4
Orange r. Namibia/S. Africa 50 C5
Orange CA U.S.A. 65 D4
Orange MA U.S.A. 64 E1
Orange TX U.S.A. 63 I5
Orange VA U.S.A. 64 B3
Orange, Cabo c. Brazil 69 H3
Orangeburg U.S.A. 63 K5
Orange Cone sea feature S. Atlantic Ocean 72 I8
Orange Free State prov. S. Africa see Free State
Orange Walk Belize 66 G5
Oranje r. Namibia/S. Africa see Orange
Oranje Gebergte hills Suriname 69 G3
Oranjemund Namibia 50 C5
Oranjestad Aruba 67 J6
Oranmore Ireland 21 D4
Orapa Botswana 49 C6
Orăștie Romania 27 J2
Orașul Stalin Romania see Brașov
Oravais Fin. 14 M5
Orba Co l. China 36 E2
Orbetello Italy 26 D3
Orbost Australia 58 D6
Orcadas (Argentina) research stn S. Atlantic Ocean 76 A2
Orchha India 36 D4
Orchila, Isla i. Venez. 68 E1
Orchy r. U.K. 20 D4
Orcutt CA U.S.A. 65 B3
Ord r. Australia 54 E3
Ord, Mount h. Australia 54 D4
Órdenes Spain see Ordes
Orderville UT U.S.A. 65 F2
Ordes Spain 25 B2
Ordesa y Monte Perdido, Parque Nacional de nat. park Spain 25 G2
Ord Mountain CA U.S.A. 65 D3
Ordos Nei Mongol China 43 K5
Ord River Dam Australia 54 E4
Ordu Hatay Turkey see Yayladağı
Ordu Ordu Turkey 34 E2
Ordubad Azer. 35 G3
Ordzhonikidze Rus. Fed. see Vladikavkaz
Ore Nigeria 46 D4
Örebro Sweden 15 I7
Oregon state U.S.A. 62 C3
Oregon City U.S.A. 62 C2
Orekhov Ukr. see Orikhiv
Orekhovo-Zuyevo Rus. Fed. 12 H5
Orel Rus. Fed. 13 H5
Orel, Gora mt. Rus. Fed. 44 E1
Orel', Ozero l. Rus. Fed. 44 E1
Ore Mountains Czech Rep./Germany see Erzgebirge
Orenburg Rus. Fed. 28 G4
Orense Spain see Ourense
Oreor Palau see Koror
Orepuki N.Z. 59 A8
Öresund str. Denmark/Sweden 15 H9
Oretana, Cordillera mts Spain see Toledo, Montes de
Orewa N.Z. 59 E3
Orfanou, Kolpos b. Greece 27 J4
Orford Australia 57 [inset]
Orford U.K. 19 I6
Orford Ness hd U.K. 19 I6
Organabo Fr. Guiana 69 H2
Orgūn Afgh. 36 B2
Orhaneli Turkey 27 M5
Orhangazi Turkey 27 M4
Orhon Gol r. Mongolia 42 J2
Orichi Rus. Fed. 12 K4
Oriental, Cordillera mts Bol. 68 E7
Oriental, Cordillera mts Col. 68 D2
Oriental, Cordillera mts Peru 68 E6
Orihuela Spain 25 F4
Orikhiv Ukr. 13 G7
Orimattila Fin. 15 N6
Orinoco r. Col./Venez. 68 F2
Orissa state India 38 E1
Orissaare Estonia 15 M7
Oristano Sardegna Italy 26 C5
Orivesi Fin. 15 N6
Orivesi l. Fin. 14 P5
Oriximiná Brazil 69 G4
Orizaba Mex. 66 E5
Orizaba, Pico de vol. Mex. 66 E5
Orizona Brazil 71 A2
Orkanger Norway 14 F5
Örkelljunga Sweden 15 H8
Orkla r. Norway 14 F5
Orkney S. Africa 51 H4
Orkney Islands is U.K. 20 F1
Orlândia Brazil 71 B3
Orlando U.S.A. 63 K6
Orleaes Brazil 71 A5
Orléans France 24 E3
Orléansville Alg. see Chlef
Orlik Rus. Fed. 42 H2
Orlov Rus. Fed. 12 K4
Orlov Gay Rus. Fed. 13 K6
Orlovskiy Rus. Fed. 13 I7
Ormara Pak. 33 J4
Ormskirk U.K. 18 E5
Orne r. France 24 D2
Ørnes Norway 14 H3
Örnsköldsvik Sweden 14 K5
Orobie, Alpi mts Italy 26 C1
Orobo, Serra do hills Brazil 71 C1
Orodara Burkina Faso 46 C3
Orofino U.S.A. 62 D2
Oro Grande CA U.S.A. 65 D3
Orol Dengizi salt l. Kazakh./Uzbek. see Aral Sea
Oron Israel 39 B4
Orona atoll Kiribati 53 I2
Orontes r. Asia 34 E3 see 'Āṣī, Nahr al
Orontes r. Lebanon/Syria 39 C2
Oroqen Zizhiqi China see Alihe
Oroquieta Phil. 41 E7
Orós, Açude resr Brazil 69 K5
Orosei, Golfo di b. Sardegna Italy 26 C4
Orosháza Hungary 27 I1
Oroville U.S.A. 62 C4
Orqohan China 44 A2
Orsa Sweden 15 I6
Orsha Belarus 13 F5
Orshanka Rus. Fed. 12 J4
Orsk Rus. Fed. 28 G4
Ørsta Norway 14 E5
Orta Toroslar plat. Turkey 39 A1
Ortegal, Cabo c. Spain 25 C2
Orthez France 24 D5
Ortigueira Spain 25 C2
Ortles mt. Italy 26 D1
Orton U.K. 18 E4
Ortona Italy 26 F3
Ortonville U.S.A. 63 H2
Ortospana Afgh. see Kābul
Orulgan, Khrebet mts Rus. Fed. 29 N3
Orumbo Namibia 50 C2
Orūmīyeh Iran see Urmia
Oruro Bol. 68 E7
Orūzgān Afgh. 36 A2
Orvieto Italy 26 E3
Orville Coast Antarctica 76 L1
Orwell OH U.S.A. 64 A2
Oryol Rus. Fed. see Orel
Os Norway 14 G5
Osa Rus. Fed. 11 R4
Osa, Península de pen. Costa Rica 67 H7
Ōsaka Japan 45 D6
Osakarovka Kazakh. 42 C2
Osby Sweden 15 H8
Oschiri Sardegna Italy 26 C4
Ösel i. Estonia see Hiiumaa
Osetr r. Rus. Fed. 13 H5
Ōse-zaki pt Japan 45 C6
Osh Kyrg. 42 C4
Oshakati Namibia 49 B5
Oshawa Canada 63 L3
Oshika-hantō pen. Japan 45 F5
Ō-shima i. Japan 44 E4
Ō-shima i. Japan 45 E6
Oshkosh WI U.S.A. 63 J3
Oshmyany Belarus see Ashmyany
Oshnovīyeh Iran 35 G3
Oshogbo Nigeria 46 D4
Oshtorān Kūh mt. Iran 35 H4
Oshwe Dem. Rep. Congo 48 B4
Osijek Croatia 26 H2
Osimo Italy 26 E3
Osipenko Ukr. see Berdyans'k
Osipovichi Belarus see Asipovichy
Osiyan India 36 C4
Osizweni S. Africa 51 J4
Osječenica mts Bos.-Herz. 26 G2
Ösjön l. Sweden 14 I5
Oskarshamn Sweden 15 J8
Öskemen Kazakh. see Ust'-Kamenogorsk
Oslo Norway 15 G7
Oslofjorden sea chan. Norway 15 G7
Osmanabad India 38 C2
Osmancık Turkey 34 D2
Osmaneli Turkey 27 M4
Osmaniye Turkey 34 E3
Osmannagar India 38 C2
Os'mino Rus. Fed. 15 P7
Osnabrück Germany 17 L4
Osnaburg atoll Fr. Polynesia see Mururoa
Osogbo Nigeria see Oshogbo
Osogovska Planina mts Bulg./Macedonia 27 J3
Osogovske Planine mts Bulg./Macedonia see Osogovska Planina
Osogovski Planini mts Bulg./Macedonia see Osogovska Planina
Osorno Chile 70 B6
Osorno Spain 25 D2
Osoyoos Canada 62 D2
Osøyri Norway 15 D6
Osprey Reef Australia 56 D2
Oss Neth. 17 J5
Ossa, Mount Australia 57 [inset]
Ossining NY U.S.A. 64 E2
Ossora Rus. Fed. 29 R4
Ostashkov Rus. Fed. 12 G4
Ostend Belgium 16 I5
Ostende Belgium see Ostend
Österbymo Sweden 15 I8
Österdalälven r. Sweden 15 H6
Østerdalen val. Norway 15 G5
Österreich country Europe see Austria
Östersund Sweden 14 I5
Ostfriesische Inseln Germany see East Frisian Islands
Östhammar Sweden 15 K6
Ostrava Czech Rep. 17 Q6
Ostróda Poland 17 Q4
Ostrogozhsk Rus. Fed. 13 H6
Ostrov Rus. Fed. 12 F4
Ostrova Tyulen'i is Kazakh. 35 H1
Ostrovets Poland see Ostrowiec Świętokrzyski
Ostrovnoy Rus. Fed. 77 G2
Ostrovskoye Rus. Fed. 12 I4
Ostrov Vrangelya i. Rus. Fed. see Wrangel Island
Ostrów Poland see Ostrów Wielkopolski
Ostrowiec Poland see Ostrowiec Świętokrzyski
Ostrowiec Świętokrzyski Poland 13 D6
Ostrów Mazowiecka Poland 17 R4
Ostrowo Poland see Ostrów Wielkopolski
Ostrów Wielkopolski Poland 17 P5
Osŭm r. Bulg. 27 K3
Ōsumi-shotō is Japan 45 C7
Osuna Spain 25 D5
Oswego NY U.S.A. 64 C1
Oswestry U.K. 19 D6
Otago Peninsula N.Z. 59 C7
Otahiti i. Fr. Polynesia see Tahiti
Otaki N.Z. 59 E5
Otanmäki Fin. 14 O4
Otaru Japan 44 F4
Otavi Namibia 49 B5
Ōtawara Japan 45 F5
Otdia atoll Marshall Is see Wotje
Otematata N.Z. 59 C7
Otepää Estonia 15 O7
Otgon Tenger Uul mt. Mongolia 42 H3
Otira N.Z. 59 C6
Otjinene Namibia 49 B6
Otjiwarongo Namibia 49 B6
Otjozondjupa admin. reg. Namibia 50 C1
Otley U.K. 18 F5
Otorohanga N.Z. 59 E4
Otpor Rus. Fed. see Zabaykal'sk
Otradnoye Rus. Fed. see Otradnyy
Otradnyy Rus. Fed. 13 K5
Otranto Italy 26 H4
Otranto, Strait of Albania/Italy 26 H4
Otrogovo Rus. Fed. see Stepnoye
Otrozhnyy Rus. Fed. 29 S3
Otsego Lake NY U.S.A. 64 D1
Ōtsu Japan 45 D6
Otta Norway 15 F6
Ottawa Canada 61 K5
Ottawa r. Canada 63 M2
Ottawa IL U.S.A. 63 J3
Ottawa KS U.S.A. 63 H4
Ottawa Islands Canada 61 J4
Otter r. U.K. 19 D8
Otterburn U.K. 18 E3
Otter Rapids Canada 63 K1
Ottumwa U.S.A. 63 I3
Otukpo Nigeria 46 D4
Oturkpo Nigeria see Otukpo
Otuzco Peru 68 C5
Otway, Cape Australia 58 A7
Otway National Park Australia 58 A7
Ouachita Mountains Arkansas/Oklahoma U.S.A. 63 I5
Ouadda Cent. Afr. Rep. 48 C3
Ouaddaï reg. Chad 47 F3
Ouagadougou Burkina Faso 46 C3
Ouahigouya Burkina Faso 46 C3
Ouahran Alg. see Oran
Ouaka r. Cent. Afr. Rep. 48 B3
Oualâta Mauritania 46 C3
Ouallam Niger 46 D3
Ouanda-Djailé Cent. Afr. Rep. 48 C3
Ouando Cent. Afr. Rep. 48 C3
Ouango Cent. Afr. Rep. 48 C3
Ouara r. Cent. Afr. Rep. 48 C3
Ouarâne reg. Mauritania 46 C2
Ouargaye Burkina Faso 46 D3
Ouargla Alg. 22 F5
Ouarogou Burkina Faso see Ouargaye
Ouarzazate Morocco 22 C5
Oubangui r. Cent. Afr. Rep./Dem. Rep. Congo see Ubangi
Oubergpas pass S. Africa 50 G7
Oudtshoorn S. Africa 50 F7
Oued Tlélat Alg. 25 F6
Oued Zem Morocco 22 C5
Oued Zénati Alg. 26 B6
Ouessant, Île d' i. France 24 B2
Ouesso Congo 48 B3
Ouezzane Morocco 25 D6
Oughter, Lough l. Ireland 21 E3
Ouguati Namibia 50 B1
Ouistreham France 19 G9
Oujda Morocco 25 F6
Oujeft Mauritania 46 B3
Oulainen Fin. 14 N4
Oulangan kansallispuisto nat. park Fin. 14 P3
Ouled Djellal Alg. 25 I6
Ouled Farès Alg. 25 G5
Ouled Naïl, Monts des mts Alg. 25 H6
Oulu Fin. 14 N4
Oulujärvi l. Fin. 14 O4
Oulujoki r. Fin. 14 N4
Oulunsalo Fin. 14 N4
Oulx Italy 26 B2
Oum-Chalouba Chad 47 F3
Oum el Bouaghi Alg. 26 B7
Oum-Hadjer Chad 47 E3
Ounasjoki r. Fin. 14 N3
Oundle U.K. 19 G6
Ounianga Kébir Chad 47 F3
Oura, Akrotirio pt Greece 27 L5
Ourense Spain 25 C2
Ouricurí Brazil 69 J5
Ourinhos Brazil 71 A3
Ouro r. Brazil 71 A1
Ouro Preto Brazil 71 C3
Ous Rus. Fed. 11 S3
Ouse r. England U.K. 18 G5
Ouse r. England U.K. 19 H8
Outaouais, Rivière des r. Canada see Ottawa
Outardes Quatre, Réservoir resr Canada 63 N1
Outer Hebrides is U.K. 20 B3
Outer Mongolia country Asia see Mongolia
Outer Santa Barbara Channel CA U.S.A. 65 C4
Outjo Namibia 49 B6
Outlook Canada 62 F1
Outokumpu Fin. 14 P5
Out Skerries is U.K. 20 [inset]
Ouvéa atoll New Caledonia 53 G4
Ouyen Australia 57 C7
Ouzel r. U.K. 19 G6
Ovace, Punta d' mt. Corse France 24 I6
Ovacık Turkey 39 A1
Ovada Italy 26 C2
Ovalle Chile 70 B4
Ovamboland reg. Namibia 49 B5
Ovan Gabon 48 B3
Ovar Port. 25 B3
Överkalix Sweden 14 M3
Overlander Roadhouse Australia 55 A6
Overton NV U.S.A. 65 E2
Övertorneå Sweden 14 M3
Överum Sweden 15 J8
Ovid NY U.S.A. 64 C1
Oviedo Spain 25 D2
Øvre Anárjohka Nasjonalpark nat. park Norway 14 N2
Øvre Dividal Nasjonalpark nat. park Norway 14 K2
Øvre Rendal Norway 15 G6
Ovruch Ukr. 13 F6
Ovsyanka Rus. Fed. 44 B1
Owando Congo 48 B4
Owa Rafa i. Solomon Is see Santa Ana
Owasco Lake NY U.S.A. 64 C1
Owase Japan 45 E6
Owatonna U.S.A. 63 I3
Owbeh Afgh. 33 J3
Owego NY U.S.A. 64 C1
Owel, Lough l. Ireland 21 E4
Owenmore r. Ireland 21 C3
Owenmore r. Ireland 21 D3
Owenreagh r. U.K. 21 E3
Owen River N.Z. 59 D5
Owens r. CA U.S.A. 65 D2
Owensboro U.S.A. 63 J4
Owens Lake CA U.S.A. 65 D2
Owen Sound Canada 63 K3
Owen Stanley Range mts P.N.G. 52 E2
Owerri Nigeria 46 D4
Owo Nigeria 46 D4
Owyhee U.S.A. 62 D3
Owyhee r. U.S.A. 62 D3
Öxarfjörður b. Iceland 14 [inset]
Oxelösund Sweden 15 J7
Oxford N.Z. 59 D6
Oxford U.K. 19 F7
Oxford MD U.S.A. 64 C3
Oxford MS U.S.A. 63 J5
Oxford NY U.S.A. 64 D1
Oxford Lake Canada 63 H1
Oxley Australia 58 B5
Oxleys Peak Australia 58 E3
Oxley Wild Rivers National Park Australia 58 F3
Ox Mountains hills Ireland 21 C4
Oxnard CA U.S.A. 65 C3
Oxus r. Asia see Amudar'ya
Øya Norway 14 H3
Oyama Japan 45 E5
Oyapock r. Brazil/Fr. Guiana 69 H3
Oyem Gabon 48 B3
Oyen Canada 62 E1
Oykel r. U.K. 20 E3
Oyo Nigeria 46 D4
Oyonnax France 24 G3
Oyster Rocks is India 38 B3
Oytograk China 37 E1
Oyukludağı mt. Turkey 39 A1
Özalp Turkey 35 G3
Ozark AL U.S.A. 63 J5
Ozark Plateau U.S.A. 63 I4
Ozarks, Lake of the U.S.A. 63 I4
O'zbekiston country Asia see Uzbekistan
Özen Kazakh. see Kyzylsay
Ozernovskiy Rus. Fed. 29 Q4
Ozernyy Rus. Fed. 13 G5
Ozerpakh Rus. Fed. 44 F1
Ozersk Rus. Fed. 15 M9
Ozerskiy Rus. Fed. 44 F3
Ozery Rus. Fed. 13 H5
Ozeryane Rus. Fed. 44 C2
Ozieri Sardegna Italy 26 C4
Ozinki Rus. Fed. 13 K6
Oznachennoye Rus. Fed. see Sayanogorsk
Ozuki Japan 45 C6

P

Paamiut Greenland 61 N3
Paanayarvi Natsional'nyy Park nat. park Rus. Fed. 14 Q3
Paanopa i. Kiribati see Banaba
Paarl S. Africa 50 D7
Paatsjoki r. Europe see Patsoyoki
Paballelo S. Africa 50 E5
P'abal-li N. Korea 44 C4
Pabbay i. U.K. 20 B3
Pabianice Poland 17 Q5
Pabianitz Poland see Pabianice
Pabna Bangl. 37 G4
Pabradė Lith. 15 N9
Pab Range mts Pak. 33 K4
Pacaás Novos, Parque Nacional nat. park Brazil 68 F6
Pacasmayo Peru 68 C5
Pachagarh Bangl. see Panchagarh
Pachikha Rus. Fed. 12 J3
Pachino Sicilia Italy 26 F6
Pachmarhi India 36 D5
Pachor India 36 D5
Pachora India 38 B1
Pachpadra India 36 C4
Pachuca Mex. 66 E4
Pachuca de Soto Mex. see Pachuca
Pacific-Antarctic Ridge sea feature S. Pacific Ocean 75 J9
Pacific Grove CA U.S.A. 65 B2
Pacific Ocean 74
Packsaddle Australia 57 C6
Pacoval Brazil 69 H4
Pacuí r. Brazil 71 B2
Paczków Poland 17 P5
Padali Rus. Fed. see Amursk
Padampur India 36 C3
Padang Indon. 41 C8
Padany Rus. Fed. 12 G3
Padatha, Kūh-e mt. Iran 35 H4
Padcaya Bol. 68 F8
Paddington Australia 58 B4
Paden City WV U.S.A. 64 A3
Paderborn Germany 17 L5
Padeșu, Vârful mt. Romania 27 J2
Padibyu Myanmar 37 I5
Padilla Bol. 68 F7
Padjelanta nationalpark nat. park Sweden 14 J3
Padova Italy see Padua
Padrão, Ponta pt Angola 49 B4
Padrauna India 37 F4
Padre Island U.S.A. 63 H6
Padstow U.K. 19 C8
Padsvillye Belarus 15 O9
Padua India 38 D2
Padua Italy 26 D2
Paducah KY U.S.A. 63 J4
Paducah TX U.S.A. 62 G5
Padum India 36 D2
Paegam N. Korea 44 C4
Paektu-san mt. China/N. Korea see Baitou Shan
Paengnyŏng-do i. S. Korea 45 B5
Pafos Cyprus see Paphos
Pafuri Moz. 51 J2
Pag Croatia 26 F2
Pag i. Croatia 26 F2
Paga Indon. 54 C2
Pagadian Phil. 41 E7
Pagai Selatan i. Indon. 41 C8
Pagalu i. Equat. Guinea see Annobón
Pagan i. N. Mariana Is 41 G6
Pagasitikos Kolpos b. Greece 27 J5
Paget, Mount S. Georgia 70 I8
Paget Cay rf Australia 56 F3
Pagon i. N. Mariana Is see Pagan
Pagosa Springs U.S.A. 62 F4
Pahlgam India 36 C2
Pahranagat Range mts NV U.S.A. 65 E2
Pahrump NV U.S.A. 65 E2
Pahuj r. India 36 D4
Pahute Mesa plat. NV U.S.A. 65 D2
Paicines CA U.S.A. 65 B2
Paide Estonia 15 N7
Paignton U.K. 19 D8
Päijänne l. Fin. 15 N6
Paikü Co l. China 37 F3
Paimio Fin. 15 M6
Painel Brazil 71 A4
Painesville OH U.S.A. 64 A2
Pains Brazil 71 B3
Painted Rock Dam AZ U.S.A. 65 F4
Paint Hills Canada see Wemindji
Paisley U.K. 20 E5
Paita Peru 68 B5
Paiva Couceiro Angola see Quipungo
Pajala Sweden 14 M3
Pakala India 38 C3
Pakangyi Myanmar 37 H5
Pakaraima Mountains Guyana 67 M8
Pakaraima Mountains S. America 68 F3
Pakaur India 37 F4
Pakhachi Rus. Fed. 29 R3
Pakhoi China see Beihai
Paki Nigeria 46 D3
Pakistan country Asia 33 J4
Paknampho Thai. see Nakhon Sawan
Pakokku Myanmar 37 H5
Pakpattan Pak. 36 C3
Pakruojis Lith. 15 M9
Paks Hungary 26 H1
Pakse Laos see Pakxé
Pakur India see Pakaur
Pakxan Laos 42 I9
Pakxé Laos 41 C6
Pala Chad 47 E4
Palaestinia reg. Asia see Palestine
Palaiochora Greece 27 J7
Palaiseau France 24 F2
Palakkad India 38 C4
Palakkat India see Palakkad
Palamakoloi Botswana 50 F2
Palamau India see Palamu
Palamós Spain 25 H3
Palamu India 37 F5
Palana Rus. Fed. 29 Q4
Palandur India 38 D1
Palani India 38 C4
Palanpur India 36 C4
Palapye Botswana 51 H2
Palatka Rus. Fed. 29 Q3
Palatka U.S.A. 63 K6
Palau country N. Pacific Ocean 41 F7
Palau Islands Palau 41 F7
Palawan i. Phil. 41 D7
Palawan Trough sea feature N. Pacific Ocean 74 D5
Palayankottai India 38 C4
Palchal Lake India 38 D2
Paldiski Estonia 15 N7
Palekh Rus. Fed. 12 I4
Palembang Indon. 41 C8
Palena Chile 70 B6
Palencia Spain 25 D2
Palermo Sicilia Italy 26 E5
Palestine reg. Asia 39 B3
Palestine U.S.A. 63 H5
Paletwa Myanmar 37 H5
Palezgir Chauki Pak. 36 B3
Palghat India see Palakkad
Palgrave, Mount h. Australia 55 A5
Palhoca Brazil 71 A4
Pali Chhattisgarh India 38 D1
Pali Mahar. India 38 B2
Pali Rajasthan India 36 C4
Palikir Micronesia 74 G5
Palinuro, Capo c. Italy 26 F4
Paliouri, Akrotirio pt Greece 27 J5
Palitana India 36 B5
Palivere Estonia 15 M7
Palk Bay Sri Lanka 38 C4
Palkino Rus. Fed. 15 P8
Palkonda Range hills India 38 C3
Palk Strait India/Sri Lanka 38 C4
Palla Bianca mt. Austria/Italy see Weißkugel
Pallamallawa Australia 58 E2
Pallas Green New Ireland 21 D5
Pallasovka Rus. Fed. 13 J6
Pallas-Yllästunturin kansallispuisto nat. park Fin. 14 M2
Pallavaram India 38 D3
Palliser, Cape N.Z. 59 E5
Palliser, Îles is Fr. Polynesia 75 K7
Palliser Bay N.Z. 59 E5
Pallu India 36 C3
Palma r. Brazil 71 B1
Palma del Río Spain 25 D5
Palma de Mallorca Spain 25 H4
Palmaner India 38 C3
Palmares Brazil 69 K5
Palmares do Sul Brazil 71 A5
Palmas Brazil 71 A4
Palmas Tocantins 68 I6
Palmas, Cape Liberia 46 C4
Palmdale CA U.S.A. 65 C3
Palmeira Brazil 71 A4
Palmeira das Missões Brazil 70 F3
Palmeira dos Índios Brazil 69 K5
Palmeirais Brazil 69 J5
Palmeiras Brazil 71 C1
Palmeirinhas, Ponta das pt Angola 49 B4
Palmer research stn Antarctica 76 L2
Palmer r. Australia 56 C3
Palmer watercourse Australia 55 F6
Palmer U.S.A. 60 D3
Palmer Land reg. Antarctica 76 L2
Palmerston N.T. Australia see Darwin
Palmerston N.T. Australia 54 E3
Palmerston atoll Cook Is 53 J3
Palmerston N.Z. 59 C7
Palmerston North N.Z. 59 E5
Palmerton PA U.S.A. 64 D2
Palmerville Australia 56 D2
Palmi Italy 26 F5
Palmira Col. 68 C3
Palm Springs CA U.S.A. 65 D4
Palmyra Syria see Tadmur
Palmyra VA U.S.A. 64 B4
Palmyra Atoll terr. N. Pacific Ocean 74 J5
Palmyras Point India 37 F5
Palni Hills India 38 C4
Palo Alto CA U.S.A. 65 A2
Paloich South Sudan 32 D7
Palojärvi Fin. 14 M2
Palojoensuu Fin. 14 M2
Palomaa Fin. 14 O2
Palomar Mountain CA U.S.A. 65 D4
Paloncha India 38 D2
Palopo Indon. 41 E8
Palos, Cabo de c. Spain 25 F5
Palo Verde CA U.S.A. 65 E4
Paltamo Fin. 14 O4
Palu Indon. 41 D8
Palu i. Indon. 54 C2
Palu Turkey 35 E3
Palwal India 36 D3
Palwancha India see Paloncha
Palyeskaya Nizina marsh Belarus/Ukr. see Pripet Marshes
Pamana i. Indon. 54 C2
Pambarra Moz. 51 L1
Pambula Australia 58 D6
Pamidi India 38 C3
Pamiers France 24 E5
Pamir mts Asia 33 L2
Pamlico Sound sea chan. U.S.A. 63 L4
Pampa de Infierno Arg. 70 D3
Pampas reg. Arg. 70 D5
Pampeluna Spain see Pamplona
Pamphylia reg. Turkey 27 N6
Pamplin VA U.S.A. 64 B4
Pamplona Col. 68 D2
Pamplona Spain 25 F2
Pamukova Turkey 27 N4
Pamzal India 36 D2
Panaca NV U.S.A. 65 E2
Panagyurishte Bulg. 27 K3
Panaji India 38 B3
Panama country Central America 67 H7
Panamá Panama see Panama City
Panamá, Canal de Panama 67 I7
Panamá, Golfo de Panama see Panama, Gulf of
Panama, Gulf of Panama 67 I7
Panama, Isthmus of Panama 67 I7
Panamá, Istmo de Panama see Panama, Isthmus of

Panama City Panama 67 I7
Panama City U.S.A. 63 J5
Panamint Range mts CA U.S.A. 65 D2
Panamint Valley val. CA U.S.A. 65 D2
Panao Peru 68 C5
Panarea, Isola i. Italy 26 F5
Panay i. Phil. 41 E6
Pancake Range mts NV U.S.A. 65 E1
Pančevo Serbia 27 I2
Panchagarh Bangl. 37 G4
Pancsova Serbia see Pančevo
Panda Moz. 51 L3
Pandeiros r. Brazil 71 B1
Pandharpur India 38 B2
Pandy U.K. 19 E7
Paneas Syria see Bāniyās
Panevėžys Lith. 15 N9
Panfilov Kazakh. see Zharkent
Pangi Range mts Pak. 36 C2
Pangkalanbuun Indon. 41 D8
Pangkalpinang Indon. 41 C8
Pangnirtung Canada 61 L3
Pangody Rus. Fed. 28 I3
Pangong Tso salt l. China/India see Bangong Co
Pangu He r. China 44 B1
Panguitch UT U.S.A. 65 F2
Panipat India 36 D3
Panir Pak. 36 A3
Panj Tajik. 36 B1
Panjāb Afgh. 36 A2
Panjim India see Panaji
Panjnad r. Pak. 36 B3
Pankakoski Fin. 14 Q5
Panna India 36 E4
Panna reg. India 36 D4
Pannawonica Australia 54 B5
Pano Lefkara Cyprus 39 A2
Panorama Brazil 71 A3
Panormus Sicilia Italy see Palermo
Panshi China 44 B4
Pantanal marsh Brazil 69 G7
Pantanal Matogrossense, Parque Nacional do nat. park Brazil 69 G7
Pantar i. Indon. 54 D2
Pantelaria Sicilia Italy see Pantelleria
Pantelleria Sicilia Italy 26 D6
Pantelleria, Isola di i. Sicilia Italy 26 E6
Pantha Myanmar 37 H5
Panth Piploda India 36 C5
Panticapaeum Ukr. see Kerch
Pánuco Veracruz Mex. 66 E4
Panwari India 36 D4
Panzhihua China 42 I7
Panzi Dem. Rep. Congo 49 B4
Paola Italy 26 G5
Paoua Cent. Afr. Rep. 48 B3
Pápa Hungary 26 G1
Papa, Monte del mt. Italy 26 F4
Papagni r. India 38 C3
Papakura N.Z. 59 E3
Papanasam India 38 C4
Papantla Mex. 66 E4
Paparoa National Park N.Z. 59 C6
Papa Stour i. U.K. 20 [inset]
Papa Westray i. U.K. 20 G1
Papay i. U.K. see Papa Westray
Papeete Fr. Polynesia 75 K7
Papenburg Germany 17 K4
Paphos Cyprus 39 A2
Paphus Cyprus see Paphos
Papoose Lake NV U.S.A. 65 E2
Papua, Gulf of P.N.G. 52 E2
Papua New Guinea country Oceania 52 E2
Par U.K. 19 C8
Pará r. Brazil 71 B2
Pará, Rio do r. Brazil 69 I4
Paraburdoo Australia 55 B5
Paracatu Brazil 71 B2
Paracatu r. Brazil 71 B2
Paracel Islands S. China Sea 41 D6
Parachilna Australia 57 B6
Paraćin Serbia 27 I3
Paracuru Brazil 69 K4
Pará de Minas Brazil 71 B2
Paradise U.S.A. 62 C4
Paradise Peak NV U.S.A. 65 D1
Paradwip India 37 F5
Paraetonium Egypt see Marsá Maṭrūḩ
Paragominas Brazil 69 I4
Paragould U.S.A. 63 I4
Paragua i. Phil. see Palawan
Paraguaçu Paulista Brazil 71 A3
Paraguay r. Arg./Para. 70 E3
Paraguay country S. America 70 E2
Paraíba do Sul r. Brazil 71 C3
Parainen Fin. see Pargas
Paraíso do Norte Brazil 69 I6
Paraisópolis Brazil 71 B3
Parak Iran 35 I6
Parakou Benin 46 D4
Paralakhemundi India 38 E2
Paralkot India 38 D2
Paramagudi India see Paramakkudi
Paramakkudi India 38 C4
Paramaribo Suriname 69 G2
Paramillo, Parque Nacional nat. park Col. 68 C2
Paramirim Brazil 71 C1
Paramo Frontino mt. Col. 68 C2
Paramus NJ U.S.A. 64 D2
Paramushir, Ostrov i. Rus. Fed. 29 Q4
Paran watercourse Israel 39 B4
Paraná Arg. 70 D4
Paranã Brazil 71 B1
Paraná r. Brazil 71 A1
Paraná state Brazil 71 A4
Paraná r. S. America 70 E4
Paranã, Serra do hills Brazil 71 B1
Paranaguá Brazil 71 A4
Paranaíba Brazil 71 A2
Paranaíba r. Brazil 71 A3
Paranapiacaba, Serra mts Brazil 71 A4
Paranavaí Brazil 70 F2
Parangi Aru r. Sri Lanka 38 D4
Parang Pass India 36 D2
Parângul Mare, Vârful mt. Romania 27 J2
Parantan Sri Lanka 38 D4
Paraopeba Brazil 71 B2
Pārapāra Iraq 35 G4
Paraparaumu N.Z. 59 E5
Paras Pak. 36 C2
Paraspori, Akrotirio pt Greece 27 L7
Parateca Brazil 71 C1
Paratinga Brazil 71 C1
Parāū, Kūh-e mt. Iraq 35 G4
Paraúna Brazil 71 A2
Parbhani India 38 C2
Parchim Germany 17 M4
Parding China 37 G2
Pardo r. Bahia Brazil 71 D1
Pardo r. Mato Grosso do Sul Brazil 70 F2
Pardo r. São Paulo Brazil 71 A3
Pardoo Australia 54 B5
Pardubice Czech Rep. 17 O5
Parece Vela i. Japan see Okino-Tori-shima
Parecis, Serra dos hills Brazil 68 F6
Pareh Iran 35 G3
Parenda India 38 B2
Parent, Lac l. Canada 63 L2
Pareora N.Z. 59 C7
Parepare Indon. 41 D8
Parga Greece 27 I5
Pargas Fin. 15 M6
Parghelia Italy 26 F5
Pargi India 38 C2
Paria, Gulf of Trin. and Tob./Venez. 67 L6
Paria, Península de pen. Venez. 68 F1
Parikkala Fin. 15 P6
Parikud Islands India 38 E2
Parima, Serra mts Brazil 68 F3
Parima-Tapirapecó, Parque Nacional nat. park Venez. 68 F3
Parintins Brazil 69 G4
Paris Ont. Canada 64 A1
Paris France 24 F2
Paris TX U.S.A. 63 H5
Park U.K. 21 E3
Parkano Fin. 15 M5
Parker AZ U.S.A. 65 E3
Parker Dam CA U.S.A. 65 E3
Parker Range hills Australia 55 B8
Parkersburg WV U.S.A. 64 A3
Parkes Australia 58 D4
Park Falls U.S.A. 63 I2
Parkhill Ont. Canada 64 A1
Parkutta Pak. 36 D2
Parla Kimedi India see Paralakhemundi
Parlakimidi India see Paralakhemundi
Parli Vaijnath India 38 C2
Parlung Zangbo r. China 37 H3
Parma Italy 26 D2
Parma OH U.S.A. 64 A2
Parnaíba Brazil 69 J4
Parnaíba r. Brazil 69 J4
Parnassus N.Z. 59 D6
Parner India 38 B2
Parnonas mts Greece 27 J6
Pärnu Estonia 15 N7
Pärnu-Jaagupi Estonia 15 N7
Paro Bhutan 37 G4
Paroikia Greece 27 K6
Paroo watercourse Australia 58 A3
Paroo Channel watercourse Australia 58 A3
Paros i. Greece 27 K6
Parowan UT U.S.A. 65 F2
Parral Chile 70 B5
Parramatta Australia 58 E4
Parramore Island VA U.S.A. 64 D4
Parras Mex. 66 D3
Parrett r. U.K. 19 D7
Parry, Cape Canada 77 A2
Parry, Kap c. Greenland see Kangaarsussuaq
Parry Bay Canada 61 J3
Parry Channel Canada 61 G2
Parry Islands Canada 61 G2
Parry Range hills Australia 54 A5
Parry Sound Canada 63 K2
Parsnip Peak NV U.S.A. 65 E1
Parsons KS U.S.A. 63 H4
Parsons WV U.S.A. 64 B3
Parsons Range hills Australia 54 F3
Partabgarh India 38 B2
Partabpur India 37 E5
Parthenay France 24 D3
Partizansk Rus. Fed. 44 D4
Partney U.K. 18 H5
Partry Ireland 21 C4
Partry Mountains hills Ireland 21 C4
Paru r. Brazil 69 H4
Paryang China 37 E3
Parys S. Africa 51 H4
Pasa Dağı mt. Turkey 34 D3
Pasadena CA U.S.A. 65 C3
Pasado, Cabo c. Ecuador 68 B4
Pascagoula U.S.A. 63 J5
Paşcani Romania 27 L1
Pascoal, Monte h. Brazil 71 D2
Pascua, Isla de i. S. Pacific Ocean see Easter Island
Pas de Calais str. France/U.K. see Dover, Strait of
Pasewalk Germany 17 O4
Pasha Rus. Fed. 12 G3
Pashih Haihsia sea chan. Phil./Taiwan see Bashi Channel
Pashkovo Rus. Fed. 44 C2
Pashkovskiy Rus. Fed. 13 H7
Pasighat India 37 H3
Pasinler Turkey 35 F3
Pasni Pak. 73 M4
Paso de los Toros Uruguay 70 E4
Pasok Myanmar 37 H5
Paso Robles CA U.S.A. 65 B3
Passaic NJ U.S.A. 64 D2
Passa Tempo Brazil 71 B3
Passau Germany 17 N6
Passo del San Gottardo Switz. see St Gotthard Pass
Passo Fundo Brazil 70 F3
Passos Brazil 71 B3
Passur r. Bangl. see Pusur
Passuri Nadi r. Bangl. see Pusur
Pastavy Belarus 15 O9
Pastaza r. Peru 68 C4
Pasto Col. 68 C3
Pastos Bons Brazil 69 J5
Pasu Pak. 36 C1
Pasur Turkey see Kulp
Pasvalys Lith. 15 N8
Pasvikelva r. Europe see Patsoyoki
Patache, Punta pt Chile 70 B2
Patagonia reg. Arg. 70 B8
Pataliputra India see Patna
Patan Gujarat India see Somnath
Patan Gujarat India 36 C5
Patan Mahar. India 38 B2
Patan Nepal 37 F4
Patan Pak. 36 C2
Patavium Italy see Padua
Patea N.Z. 59 E4
Patea inlet N.Z. see Doubtful Sound
Pate Island Kenya 48 E4
Pateley Bridge U.K. 18 F4
Patensie S. Africa 50 G7
Patera India 36 D4
Paterson Australia 58 E4
Paterson r. Australia 58 C2
Paterson NJ U.S.A. 64 D2
Paterson Range hills Australia 54 C5
Pathanamthitta India 38 C4
Pathankot India 36 C2
Pathari India 36 D5
Pathein Myanmar see Bassein
Patía r. Col. 68 C3
Patiala India 36 D3
Patkai Bum mts India/Myanmar 37 H4
Patkaklik China 37 F1
Patmos i. Greece 27 L6
Patna India 37 F4
Patna Orissa India 37 F5
Patnagarh India 37 E5
Patnos Turkey 35 F3
Pato Branco Brazil 70 F3
Patoda India 38 B2
Patos Albania 27 H4
Patos Brazil 69 K5
Patos, Lagoa dos l. Brazil 70 F4
Patos de Minas Brazil 71 B2
Patquía Arg. 70 C4
Patra Greece see Patras
Patrae Greece see Patras
Pátrai Greece see Patras
Patras Greece 27 I5
Patreksfjörður Iceland 14 [inset]
Patricio Lynch, Isla i. Chile 70 A7
Patrick Creek watercourse Australia 56 D4
Patrimônio Brazil 71 A2
Patrocínio Brazil 71 B2
Patsoyoki r. Europe 14 Q2
Pattadakal tourist site India 38 B2
Patterson CA U.S.A. 65 B2
Patti India 37 E4
Pattijoki Fin. 14 N4
Pättikkä Fin. 14 L2
Patton PA U.S.A. 64 B2
Pattullo, Mount Canada 60 F4
Patu Brazil 69 K5
Patuakhali Bangl. 37 G5
Patuca, Punta pt Hond. 67 H5
Patur India 38 C1
Patuxent r. MD U.S.A. 64 C3
Patuxent Range mts Antarctica 76 L1
Patvinsuon kansallispuisto nat. park Fin. 14 Q5
Pau France 24 D5
Pauhunri mt. China/India 37 G4
Pauillac France 24 D4
Pauini Brazil 68 E5
Pauini r. Brazil 68 E5
Pauk Myanmar 37 H5
Paulatuk Canada 77 A2
Paulicéia Brazil 71 A3
Paulis Dem. Rep. Congo see Isiro
Paulo Afonso Brazil 69 K5
Paulo de Faria Brazil 71 A3
Paulpietersburg S. Africa 51 J4
Paul Roux S. Africa 51 H5
Paumotu, Îles is Fr. Polynesia see Tuamotu Islands
Paungbyin Myanmar 37 H4
Pauni India 38 C1
Pauri India 36 D3
Pavagada India 38 C3
Pavão Brazil 71 C2
Pāveh Iran 35 G4
Pavia Italy 26 C2
Pāvilosta Latvia 15 L8
Pavino Rus. Fed. 12 J4
Pavlikeni Bulg. 27 K3
Pavlodar Kazakh. 42 D2
Pavlof Volcano U.S.A. 60 B4
Pavlograd Ukr. see Pavlohrad
Pavlohrad Ukr. 13 G6
Pavlovka Rus. Fed. 13 J5
Pavlovo Rus. Fed. 12 I5
Pavlovsk Altayskiy Kray Rus. Fed. 42 E2
Pavlovsk Voronezhskaya Oblast' Rus. Fed. 13 I6
Pavlovskaya Rus. Fed. 13 H7
Pawai India 36 E4
Paw Paw WV U.S.A. 64 B3
Pawtucket RI U.S.A. 64 F2
Paxson U.S.A. 60 D3
Payakumbuh Indon. 41 C8
Pāy Bowldak Ghar mt. Afgh. 36 A3
Payette U.S.A. 62 D3
Pay-Khoy, Khrebet hills Rus. Fed. 28 H3
Payne Canada see Kangirsuk
Payne, Lac l. Canada 61 K4
Payne's Find Australia 55 B7
Paysandú Uruguay 70 E4
Pazar Turkey 35 F2
Pazarcık Turkey 34 E3
Pazardzhik Bulg. 27 K3
Pazin Croatia 26 E2
Peabody MA U.S.A. 64 F1
Peace r. Canada 60 G4
Peace River Canada 77 L3
Peach Springs AZ U.S.A. 65 F3
Peak Charles h. Australia 55 C8
Peak Charles National Park Australia 55 C8
Peak District National Park U.K. 18 F5
Peake watercourse Australia 57 B6
Peak Hill N.S.W. Australia 58 D4
Peak Hill W.A. Australia 55 B6
Peale, Mount U.S.A. 62 F4
Pearce Point Australia 54 E3
Pearisburg VA U.S.A. 64 A4
Pearl r. U.S.A. 63 J5
Pearsall U.S.A. 62 H6
Pearston S. Africa 51 G7
Peary Channel Canada 61 I2
Peary Land reg. Greenland 77 J1
Pebane Moz. 49 D5
Pebas Peru 68 D4
Peć Kosovo see Pejë
Peçanha Brazil 71 C2
Peças, Ilha das i. Brazil 71 A4
Pechenga Rus. Fed. 14 Q2
Pechora Rus. Fed. 12 M2
Pechora r. Rus. Fed. 12 L1
Pechora Sea Rus. Fed. see Pechorskoye More
Pechorskaya Guba b. Rus. Fed. 12 L1
Pechorskoye More sea Rus. Fed. 77 G2
Pechory Rus. Fed. 15 O8
Pecos U.S.A. 62 G5
Pecos r. U.S.A. 62 G6
Pécs Hungary 26 H1
Pedda Vagu r. India 38 C2
Pedder, Lake Australia 57 [inset]
Peddie S. Africa 51 H7
Pedernales Dom. Rep. 67 J5
Pedersöre Fin. 14 M5
Pediaios r. Cyprus 39 A2
Pediva Angola 49 B5
Pedra Azul Brazil 71 C1
Pedra Preta, Serra da mts Brazil 71 A1
Pedras de Maria da Cruz Brazil 71 B1
Pedregulho Brazil 71 B3
Pedreiras Brazil 69 J4
Pedro, Point Sri Lanka 38 D4
Pedro II, Ilha reg. Brazil/Venez. 68 E3
Pedro Juan Caballero Para. 70 E2
Peebles U.K. 20 F5
Pee Dee r. U.S.A. 63 L5
Peekskill NY U.S.A. 64 E2
Peel r. Australia 58 E3
Peel r. Canada 60 E3
Peel Isle of Man 18 C4
Peera Peera Poolanna Lake imp. l. Australia 57 B5
Peery Lake salt l. Australia 58 A3
Pegasus Bay N.Z. 59 D6
Pegu Myanmar 42 H9
Pegu Yoma mts Myanmar 37 H6
Pegysh Rus. Fed. 12 K3
Pehuajó Arg. 70 D5
Peine Chile 70 C2
Peint India 38 B1
Peipsi järv l. Estonia/Rus. Fed. see Peipus, Lake
Peipus, Lake Estonia/Rus. Fed. 15 O7
Peiraias Greece see Piraeus
Pei Shan mts China see Bei Shan
Peixe Brazil 69 I6
Peixe r. Brazil 71 A1
Peixoto de Azevedo Brazil 69 H6
Pejë Kosovo 27 I3
Pèk Laos see Phônsavan
Peka Lesotho 51 H5
Peking China see Beijing
Pekinga Benin 46 D3
Pelagie, Isole is Sicilia Italy 26 E7
Peleaga, Vârful mt. Romania 27 J2
Peles Rus. Fed. 12 K3
Pelkosenniemi Fin. 14 O3
Pella S. Africa 50 D5
Pello Fin. 14 M3
Pelly Crossing Canada 60 E3
Pelly Mountains Canada 60 E3
Peloponnese admin. reg. Greece 27 J6
Pelopónnesos admin. reg. Greece see Peloponnese
Peloponnisos admin. reg. Greece see Peloponnese
Pelotas Brazil 70 F4
Pelotas, Rio das r. Brazil 71 A4
Pelusium tourist site Egypt 39 A4
Pelusium, Bay of Egypt see Ţīnah, Khalīj aţ
Pemba Moz. 49 E5
Pemba Island Tanz. 49 D4
Pemberton Canada 62 C1
Pembina r. Canada 62 D1
Pembroke Canada 63 L2
Pembroke U.K. 19 C7
Pembrokeshire Coast National Park U.K. 19 B7
Pen India 38 B2
Peña Cerredo mt. Spain see Torrecerredo
Peñalara mt. Spain 25 E3
Penamar Brazil 71 C1
Peña Nevada, Cerro mt. Mex. 66 E4
Penang Malaysia see George Town
Penápolis Brazil 71 A3
Peñaranda de Bracamonte Spain 25 D3
Penarie Australia 58 A5
Penarlâg U.K. see Hawarden
Peñarroya mt. Spain 25 F3
Peñarroya-Pueblonuevo Spain 25 D4
Penarth U.K. 19 D7
Peñas, Cabo de c. Spain 25 D2
Penas, Golfo de g. Chile 70 A7
Peña Ubiña mt. Spain 25 D2
Pendle Hill h. U.K. 18 E5
Pendleton U.S.A. 62 D2
Pend Oreille Lake U.S.A. 62 D2
Pendra India 37 E5
Penduv India 38 B2
Peneda Gerês, Parque Nacional da nat. park Port. 25 B3
Penfro U.K. see Pembroke
Penganga r. India 38 C2
Penge Dem. Rep. Congo 49 C4
Penge S. Africa 51 J3
P'enghu Ch'üntao is Taiwan 43 L8
P'enghu Liehtao is Taiwan see P'enghu Ch'üntao
Penha Brazil 71 A4
Penhoek Pass S. Africa 51 H6
Penhook VA U.S.A. 64 B4
Peniche Port. 25 B4
Penicuik U.K. 20 F5
Peninga Rus. Fed. 14 R5
Peninsular Malaysia Malaysia 41 C7
Penitente, Serra do hills Brazil 69 I5
Pennell Coast Antarctica 76 H2
Penn Hills PA U.S.A. 64 B2
Pennine, Alpi mts Italy/Switz. 26 B2
Pennine Alps mts Italy/Switz. see Pennine, Alpi
Pennines hills U.K. 18 E4
Pennsburg PA U.S.A. 64 D2
Penns Grove NJ U.S.A. 64 D3
Pennsville NJ U.S.A. 64 D3
Pennsylvania state U.S.A. 64 B2
Penn Yan NY U.S.A. 64 C1
Penny Icecap Canada 61 L3
Penny Point Antarctica 76 H1
Penola Australia 57 C8
Penong Australia 55 F7
Penonomé Panama 67 H7
Penrhyn atoll Cook Is 75 J6
Penrhyn Basin sea feature S. Pacific Ocean 75 J6
Penrith Australia 58 E4
Penrith U.K. 18 E4
Pensacola U.S.A. 63 J5
Pensacola Mountains Antarctica 76 L1
Pensi La pass India 36 D2
Pentadaktylos Range mts Cyprus 39 A2
Pentakota India 38 D2
Pentecost Island Vanuatu 53 G3
Pentecôte, Île i. Vanuatu see Pentecost Island
Penticton Canada 62 D2
Pentire Point U.K. 19 B8
Pentland Australia 56 D4
Pentland Firth sea chan. U.K. 20 F2
Pentland Hills U.K. 20 F5
Pen-y-Bont ar Ogwr U.K. see Bridgend
Penygadair h. U.K. 19 D6
Penza Rus. Fed. 13 J5
Penzance U.K. 19 B8
Penzhinskaya Guba b. Rus. Fed. 29 R3
Peoria IL U.S.A. 63 J3
Peradeniya Sri Lanka 38 D5
Pera Head hd Australia 56 C2
Perales del Alfambra Spain 25 F3
Perambalur India 38 C4
Perämeren kansallispuisto nat. park Fin. 14 N4
Peräseinäjoki Fin. 14 M5
Percival Lakes imp. l. Australia 54 D5
Percy Isles Australia 56 E4
Perdizes Brazil 71 B2
Peregrebnoye Rus. Fed. 11 T3
Pereira Col. 68 C3
Pereira Barreto Brazil 71 A3
Pereira de Eça Angola see Ondjiva
Peremul Par rf India 38 B4
Peremyshlyany Ukr. 13 E6
Perenjori Australia 55 B7
Pereslavl'-Zalesskiy Rus. Fed. 12 H4
Pereslavskiy Natsional'nyy Park nat. park Rus. Fed. 12 H4
Pereyaslavka Rus. Fed. 44 D3
Pereyaslav-Khmel'nitskiy Ukr. see Pereyaslav-Khmel'nyts'kyy
Pereyaslav-Khmel'nyts'kyy Ukr. 13 F6
Pergamino Arg. 70 D4
Perho Fin. 14 N5
Perico Arg. 70 C2
Pericos Mex. 66 C3
Périgueux France 24 E4
Perijá, Parque Nacional nat. park Venez. 68 D2
Periyar India see Erode
Perlas, Punta de pt Nicaragua 67 H6
Perleberg Germany 17 M4
Perm' Rus. Fed. 11 R4
Permas Rus. Fed. 12 J4
Pernambuco Brazil see Recife
Pernambuco Plain sea feature S. Atlantic Ocean 72 G6
Pernatty Lagoon imp. l. Australia 57 B6
Pernem India 38 B3
Pernik Bulg. 27 J3
Pernov Estonia see Pärnu
Perojpur Bangl. see Pirojpur
Peron Islands Australia 54 E3
Péronne France 24 F2
Perpignan France 24 F5
Perranporth U.K. 19 B8
Perrégaux Alg. see Mohammadia
Perris CA U.S.A. 65 D4
Perros-Guirec France 24 C2
Perry FL U.S.A. 63 K5
Perryton U.S.A. 62 G4
Perryville AK U.S.A. 60 C4
Perryville MO U.S.A. 63 J4
Perseverancia Bol. 68 F6
Pershore U.K. 19 E6
Persia country Asia see Iran
Persian Gulf Asia see The Gulf
Pertek Turkey 35 E3
Perth Australia 55 A7
Perth Canada 63 L3
Perth U.K. 20 F4
Perth Amboy NJ U.S.A. 64 D2
Perth Basin sea feature Indian Ocean 73 P7
Pertominsk Rus. Fed. 12 H2
Pertunmaa Fin. 15 O6
Pertusato, Capo c. Corse France 24 I6
Peru atoll Kiribati see Beru
Peru country S. America 68 D6
Peru-Chile Trench sea feature S. Pacific Ocean 75 O6
Perugia Italy 26 E3
Peruru India 38 C3
Perusia Italy see Perugia
Pervomaysk Rus. Fed. 13 I5
Pervomays'k Ukr. 13 F6
Pervomayskiy Kazakh. 42 E2
Pervomayskiy Arkhangel'skaya Oblast' Rus. Fed. see Novodvinsk
Pervomayskiy Tambovskaya Oblast' Rus. Fed. 13 I5
Pervomays'kyy Ukr. 13 H6
Pervorechenskiy Rus. Fed. 29 R3
Pesaro Italy 26 E3
Pescadores is Taiwan see P'enghu Ch'üntao
Pescara Italy 26 F3
Pescara r. Italy 26 F3
Peschanokopskoye Rus. Fed. 13 I7
Peschanoye Rus. Fed. see Yashkul'
Peschanyy, Mys pt Kazakh. 35 H2
Pesha r. Rus. Fed. 12 J2
Peshawar Pak. 33 L3
Peshkopi Albania 27 I4
Peshtera Bulg. 27 K3
Peski Karakumy des. Turkm. see Karakum Desert
Peskovka Rus. Fed. 12 L4
Pesnica Slovenia 26 F1
Pessac France 24 D4
Pestovo Rus. Fed. 12 G4
Pestravka Rus. Fed. 13 K5
Petah Tiqwa Israel 39 B3
Petäjävesi Fin. 14 N5
Petalion, Kolpos sea chan. Greece 27 K5
Petaluma CA U.S.A. 65 A1
Petatlán Mex. 66 D5
Petauke Zambia 49 D5
Peterborough Australia 57 B7
Peterborough Canada 63 L3
Peterborough U.K. 19 G6
Peterborough NH U.S.A. 64 F1
Peterculter U.K. 20 G3
Peterhead U.K. 20 H3
Peter I Island Antarctica 76 K2
Peter I Øy i. Antarctica see Peter I Island
Peterlee U.K. 18 F4
Petermann Bjerg nunatak Greenland 61 P2
Petermann Ranges mts Australia 55 E6
Peter Pond Lake Canada 60 H4
Petersburg NY U.S.A. 64 E1
Petersburg VA U.S.A. 64 C4
Petersburg WV U.S.A. 64 B3
Petersfield U.K. 19 G7
Petersville U.S.A. 60 C3
Peter the Great Bay Rus. Fed. see Petra Velikogo, Zaliv
Peth India 38 B2
Petilia Policastro Italy 26 G5
Petit Atlas mts Morocco see Anti Atlas
Petitjean Morocco see Sidi Kacem
Petit Lac Manicouagan l. Canada 61 L4
Petit Mécatina r. Nfld. and Lab./Que. Canada 61 M4
Petit St-Bernard, Col du pass France 24 H4
Petit Saut, Barrage du dam Fr. Guiana 69 H3
Peto Mex. 66 G4
Petoskey U.S.A. 63 K2
Petra tourist site Jordan 39 B4
Petra Velikogo, Zaliv b. Rus. Fed. 44 C4
Petre, Point Ont. Canada 64 C1
Petrich Bulg. 27 J4
Petrikau Poland see Piotrków Trybunalski
Petrikov Belarus see Pyetrykaw
Petrinja Croatia 26 G2
Petroaleksandrovsk Uzbek. see To'rtko'l
Petrograd Rus. Fed. see St Petersburg
Petrokhanski Prokhod pass Bulg. 27 J3
Petrokov Poland see Piotrków Trybunalski
Petrolina Brazil 69 J5
Petrolina de Goiás Brazil 71 A2
Petropavl Kazakh. see Petropavlovsk
Petropavlovsk Kazakh. 31 F1
Petropavlovsk Rus. Fed. see Petropavlovsk-Kamchatskiy
Petropavlovsk-Kamchatskiy Rus. Fed. 29 Q4
Petrópolis Brazil 71 C3
Petroşani Romania 27 J2
Petrovsk Rus. Fed. 13 J5
Petrovskoye Rus. Fed. see Svetlograd
Petrovsk-Zabaykal'skiy Rus. Fed. 43 J2
Petrozavodsk Rus. Fed. 12 G3
Petrus Steyn S. Africa 51 I4
Petrusville S. Africa 50 G6
Petsamo Rus. Fed. see Pechenga
Pettau Slovenia see Ptuj
Pettigo U.K. 21 E3
Petukhovo Rus. Fed. 28 H4
Petushki Rus. Fed. 12 H5
Petzeck mt. Austria 17 N7
Pevek Rus. Fed. 29 S3
Pêxung China 37 H2
Peza r. Rus. Fed. 12 J2
Pezinok Slovakia 17 P6
Pezu Pak. 36 B2
Pforzheim Germany 17 L6
Phagameng Limpopo S. Africa 51 I3
Phagwara India 36 C3
Phahameng Free State S. Africa 51 H5
Phalaborwa S. Africa 51 J2
Phalodi India 36 C4
Phalsund India 36 B4
Phaltan India 38 B2
Phalut Peak India/Nepal 37 G4
Phangnga Thai. 31 I6
Phăng Xi Păng mt. Vietnam 42 I8
Phan Rang-Thap Cham Vietnam 31 J5
Phan Thiêt Vietnam 31 J5
Phatthalung Thai. 31 J6
Phayao Thai. 42 H9
Phek India 37 H4
Phenix VA U.S.A. 64 B4
Phenix City U.S.A. 63 J5
Phet Buri Thai. 31 I5
Philadelphia Jordan see 'Ammān
Philadelphia Turkey see Alaşehir
Philadelphia PA U.S.A. 64 D3
Philip U.S.A. 62 G3
Philip Atoll Micronesia see Sorol
Philippeville Alg. see Skikda
Philippi WV U.S.A. 64 A3
Philippi, Lake imp. l. Australia 56 B5
Philippine Basin sea feature N. Pacific Ocean 74 E4
Philippines country Asia 41 E6
Philippine Sea N. Pacific Ocean 41 E6
Philippine Trench sea feature N. Pacific Ocean 74 E4
Philippolis S. Africa 51 G6
Philippopolis Bulg. see Plovdiv
Philipsburg PA U.S.A. 64 B2
Philip Smith Mountains U.S.A. 60 D3
Philipstown S. Africa 50 G6
Phillip Island Australia 58 B7
Phillipsburg U.S.A. 62 H4
Phillips Range hills Australia 54 D4
Philmont NY U.S.A. 64 E1
Philomelium Turkey see Akşehir
Phiritona S. Africa 51 H4
Phitsanulok Thai. 31 J5
Phnom Penh Cambodia 31 J5
Phnum Pénh Cambodia see Phnom Penh
Phoenicia NY U.S.A. 64 D1
Phoenix U.S.A. 62 E5
Phoenix Island Kiribati see Rawaki
Phoenix Islands Kiribati 53 I2
Phôngsali Laos 40 C5
Phong Saly Laos see Phôngsali
Phônsavan Laos 41 C6
Phosphate Hill Australia 56 C4
Phrae Thai. 31 J5
Phra Nakhon Si Ayutthaya Thai. see Ayutthaya
Phuket Thai. 31 I6
Phulabani India see Phulbani
Phulbani India 38 E1
Phulchhari Ghat Bangl. see Fulchhari
Phulji Pak. 36 A4
Phyu Myanmar 42 H9
Piaca Brazil 69 I5
Piacenza Italy 26 C2
Pian r. Australia 58 D3
Pianosa, Isola i. Italy 26 D3
Piatra Neamţ Romania 27 L1
Piave r. Italy 26 E2
Pibor Post South Sudan 47 G4
Picardie admin. reg. France 19 J9
Picardie reg. France see Picardy
Picardy admin. reg. France see Picardie
Picardy reg. France 24 E2
Picauville France 19 F9
Picayune U.S.A. 63 J5
Pichanal Arg. 70 D2
Pichhor India 36 D4
Pichilemu Chile 70 B4
Pichilingue Mex. 66 B4
Pickens WV U.S.A. 64 A3
Pickering U.K. 18 G4
Pickering, Vale of val. U.K. 18 G4
Pickle Lake Canada 61 I4
Pico da Neblina, Parque Nacional do nat. park Brazil 68 E3
Picos Brazil 69 J5
Pico Truncado Arg. 70 C7
Picton Australia 58 E5
Picton N.Z. 59 E5
Pidurutalagala mt. Sri Lanka 38 D5
Piedade Brazil 71 B3
Piedra de Águila Arg. 70 B6
Piedras, Punta pt Arg. 70 E5
Piedras Blancas Point CA U.S.A. 65 B3
Piedras Negras Mex. 66 D3
Pieksämäki Fin. 14 O5
Pielavesi Fin. 14 O5
Pielinen l. Fin. 14 P5
Pieljekaise nationalpark nat. park Sweden 14 J3
Pienaarsrivier S. Africa 51 I3
Pieniński Park Narodowy nat. park Poland 17 R6
Pieninský nat. park Slovakia 17 R6
Pieria mts Greece 27 J4
Pierowall U.K. 20 G1
Pierpont OH U.S.A. 64 A2
Pierre U.S.A. 62 G3
Pierrelatte France 24 G4
Pietermaritzburg S. Africa 51 J5
Pietersaari Fin. see Jakobstad
Pietra Spada, Passo di pass Italy 26 G5
Piet Retief S. Africa see eMkhondo
Pietrosa mt. Romania 27 K1
Pigeon Lake Canada 62 E1
Pigg's Peak Swaziland 51 J3
Pihij India 36 C5
Pihkva järv l. Estonia/Rus. Fed. see Pskov, Lake
Pihlajavesi l. Fin. 14 P6
Pihlava Fin. 15 L6
Pihtipudas Fin. 14 N5
Piippola Fin. 14 N4
Piispajärvi Fin. 14 P4
Pikalevo Rus. Fed. 12 G4
Pike WV U.S.A. 64 A3
Pikelot i. Micronesia 74 F5
Pikes Peak U.S.A. 62 G4
Pikeville KY U.S.A. 63 K4
Pikinni atoll Marshall Is see Bikini
Piła Poland 17 P4
Pilanesberg National Park S. Africa 51 H3
Pilar Arg. 70 E4
Pilar Para. 70 E3
Pilar de Goiás Brazil 71 A1
Pilaya r. Bol. 68 F8
Pilcomayo r. Bol./Para. 68 F8
Piler India 38 C3
Pili, Cerro mt. Chile 70 C2
Pilibangan India 36 C3
Pilibhit India 36 D3
Pilipinas country Asia see Philippines
Pillau Rus. Fed. see Baltiysk
Pillcopata Peru 68 D6
Pilliga Australia 58 D3
Pil'na Rus. Fed. 12 J5
Pil'nya, Ozero l. Rus. Fed. 12 M1
Pilões, Serra dos mts Brazil 71 B2
Pílos Greece see Pylos
Pilot Peak NV U.S.A. 65 D1
Pilot Station U.S.A. 60 B3
Pilsen Czech Rep. see Plzeň
Piltene Latvia 15 L8
Pil'tun, Zaliv b. Rus. Fed. 44 F1
Pilu Pak. 36 B4
Pimenta Bueno Brazil 68 F6
Pimpalner India 38 B1
Pin r. India 36 D2
Pin r. Myanmar 37 H5
Pinahat India 36 D4
Pinamar Arg. 70 E5
Pinang Malaysia see George Town
Pınarbaşı Turkey 34 E3

ar del Río Cuba 67 H4
arhisar Turkey 27 L4
as Ecuador 68 C4
czów Poland 17 R5
daí Brazil 71 C1
damonhangaba Brazil 71 B3
dar Australia 55 A7
daré r. Brazil 69 J4
dhos Óros mts Greece see
ndus Mountains
dos mts Greece see Pindus Mountains
drei India 36 E5
dus Mountains Greece 27 I5
e watercourse Australia 57 C7
e Bluff U.S.A. 63 I5
e Creek Australia 54 E3
e Creek r. PA U.S.A. 64 C2
ecrest CA U.S.A. 65 B1
edale WY U.S.A. 62 F3
e Flat Lake CA U.S.A. 65 C2
ega Rus. Fed. 12 I2
ega r. Rus. Fed. 12 I2
egrove Australia 55 A6
e Grove PA U.S.A. 64 C2
eios r. Greece 27 J5
e Island Bay Antarctica 75 N10
e Island Glacier Antarctica 76 K1
e Mountain CA U.S.A. 65 B3
e Peak AZ U.S.A. 65 F3
e Point (abandoned) Canada 60 G3
eridge CA U.S.A. 65 C2
e Ridge U.S.A. 62 G3
erolo Italy 26 B2
es, Akrotirio pt Greece 27 K4
es, Isle of i. Cuba see
Juventud, Isla de
es, Isle of i. New Caledonia see
ns, Île des
etown S. Africa 51 J5
e Valley NY U.S.A. 64 C1
g'an China 42 I5
g'anyi China see Ping'an
gdingbu China see Guyuan
gdingshan China 43 K6
gdong Taiwan see P'ingtung
gdu Shandong China 43 L5
ggang China 44 B4
gliang China 42 J5
gtan Dao i. China see Haitan Dao
gtung Taiwan 43 M8
gxiang Guangxi China 42 J8
gxiang Jiangxi China 43 K7
gyang Heilong. China 44 B2
al Brazil 71 B3
eiro Brazil 69 I4
ee U.K. 19 D8
ós r. Greece see Pineios
n Australia 55 C7
ebu Myanmar 37 H4
nacle h. Virginia/West Virginia U.S.A.
B3
nacles National Monument nat. park
U.S.A. 65 B2
n Hills CA U.S.A. 65 D3
es, Isla de i. Cuba see La Juventud,
a de
es, Mount CA U.S.A. 65 C3
tepa Nacional Mex. 66 E5
, Île des i. New Caledonia 53 G4
, Pointe aux pt Ont. Canada 64 A1
k Belarus 15 O10
ados Chile 70 C2
ura UT U.S.A. 65 F2
he NV U.S.A. 65 E2
i Dem. Rep. Congo 49 C4
er, Ostrov i. Rus. Fed. 28 K2
erskiy Kaliningradskaya Oblast'
s. Fed. 15 L9
erskiy Khanty-Mansiyskiy Avtonomnyy
krug-Yugra Rus. Fed. 11 S3
ki Poland 17 R5
io N.Z. 59 E4
iotahi inlet N.Z. see Milford Sound
ni, Lago l. Brazil 68 F4
rków Trybunalski Poland 17 Q5
Dingzi mt. China 44 C4
r India 36 C4
r Road India 36 C4
ri i. Greece 27 K5
r Peak NV U.S.A. 65 D2
India 36 C3
nuacan, Réservoir resr Canada 63 M2
iri r. Brazil 71 A4
Benin 46 D4
canjuba Brazil 71 A2
cicaba Brazil 71 B3
cicaba r. Brazil 71 C2
gununga Brazil 71 B3
curuca Brazil 69 J4
eus Greece 27 J6
do Sul Brazil 71 A4
evs Greece see Piraeus
u Brazil 71 A3
uí Brazil 71 A3
lahı Adası Azer. 35 H2
nhas Bahia Brazil 71 C1
nhas Goiás Brazil 69 H7
nhas r. Brazil 71 A2
nhas r. Rio Grande do Norte Brazil
K5
pora Brazil 71 B2
va India 36 D4
nópolis Brazil 71 A1
s do Rio Brazil 71 A2
os Greece see Pyrgos
eos mts Europe see Pyrenees
iri Brazil 69 J4
rkondu Turkey see Taşkent
pur Bangl. 37 G5
anjal Pass India 36 C2
anjal Range mts India/Pak. 36 C2
atin Ukr. see Pyryatyn
Italy 26 D3
e Italy see Pisa
gua Chile 68 D7
urum Italy see Pesaro
Peru 68 C6
k Czech Rep. 17 O6
China see Ningnan
an China 36 D1
n Iran 33 J4
n Pak. 36 A3
pek Kyrg. see Bishkek
ia reg. Turkey 34 C3
s, Cerro mt. Arg. 70 C3
Mex. 66 G4
cci Italy 26 G4
oia Italy 26 D3
oriae Italy see Pistoia
erga r. Spain 25 D3
Guinea 46 B3
aga Brazil 71 A4
agui Brazil 71 B2
India 36 B5
jungja Lake imp. l. Australia 58 A5
irn, Henderson, Ducie and Oeno
ands terr. S. Pacific Ocean see
cairn Islands
irn Island Pitcairn Is 75 L7
irn Islands terr. S. Pacific Ocean 75 L7
Sweden 14 L4

Piteälven r. Sweden 14 L4
Pitelino Rus. Fed. 13 I5
Piterka Rus. Fed. 13 J6
Piteşti Romania 27 K2
Pithoragarh India 36 E3
Pitihra India 36 D5
Pitkyaranta Rus. Fed. 12 F3
Pitlochry U.K. 20 F4
Pitsane Siding Botswana 51 G3
Pitti i. India 38 B4
Pitt Island N.Z. 53 I6
Pitt Islands Solomon Is see
Vanikoro Islands
Pittsburgh PA U.S.A. 64 B2
Pittsfield MA U.S.A. 64 E1
Pittsworth Australia 58 E1
Piumhí Brazil 71 B3
Piura Peru 68 B5
Piute Mountains CA U.S.A. 65 E3
Piute Peak CA U.S.A. 65 C3
Piuthan Nepal 37 E3
Pivka Slovenia 26 F2
Pixariá mt. Greece see Pyxaria
Pixley CA U.S.A. 65 C3
Piz Bernina mt. Italy/Switz. 26 C1
Piz Buin mt. Austria/Switz. 17 M7
Pizhanka Rus. Fed. 12 K4
Pizhi Nigeria 46 D4
Pizhma Rus. Fed. 12 J4
Pizhma r. Rus. Fed. 12 K4
Pizhma r. Rus. Fed. 12 L2
Placentia Italy see Piacenza
Placerville CA U.S.A. 65 B1
Plácido de Castro Brazil 68 E6
Plainfield CT U.S.A. 64 F2
Plainview U.S.A. 62 G5
Plaka, Akrotirio pt Greece 27 L7
Plakoti, Cape Cyprus 39 B2
Planada CA U.S.A. 65 B2
Planaltina Brazil 71 B1
Planalto do Brasil plat. Brazil 69 J7
Planura Brazil 71 A3
Plaquemine U.S.A. 63 I5
Plasencia Spain 25 C3
Plaster City CA U.S.A. 65 E4
Plastun Rus. Fed. 44 E3
Platani r. Sicilia Italy 26 E6
Platberg mt. S. Africa 51 I5
Platinum U.S.A. 77 B3
Plato Col. 68 D2
Plato Kendirli-Kiyasan plat. Kazakh. 35 I2
Plato Putorana mts Rus. Fed. 77 E2
Platte r. U.S.A. 62 H3
Plattsburgh U.S.A. 63 M3
Plauen Germany 17 N5
Plavsk Rus. Fed. 13 H5
Playas Ecuador 68 B4
Pleasant Bay MA U.S.A. 64 G2
Pleasanton U.S.A. 66 E3
Pleasant Point N.Z. 59 C7
Pleasantville NJ U.S.A. 64 D3
Pleaux France 24 F4
Pledger Lake Canada 63 K1
Plenty watercourse Australia 56 B5
Plenty, Bay of g. N.Z. 59 F3
Plentywood U.S.A. 62 G2
Plesetsk Rus. Fed. 12 I3
Pleshchentsy Belarus see Plyeshchanitsy
Plétipi, Lac l. Canada 63 M1
Plettenberg Bay S. Africa 50 F8
Pleven Bulg. 27 K3
Plevna Bulg. see Pleven
Pljevlja Montenegro 27 H3
Płock Poland 17 Q4
Pločno mt. Bos.-Herz. 26 G3
Plodovoye Rus. Fed. 12 F3
Ploemeur France 24 C3
Ploeşti Romania see Ploieşti
Ploieşti Romania 27 L2
Plomb du Cantal mt. France 24 F4
Ploskoye Rus. Fed. see Stanovoye
Płoty Poland 17 O4
Ploudalmézeau France 24 B2
Plouzané France 24 B2
Plovdiv Bulg. 27 K3
Plozk Poland see Płock
Plumridge Lakes imp. l. Australia 55 D7
Plungė Lith. 15 L9
Plyeshchanitsy Belarus 15 O9
Plymouth U.K. 19 C8
Plymouth CA U.S.A. 65 B1
Plymouth IN U.S.A. 63 J3
Plymouth MA U.S.A. 64 F2
Plymouth (abandoned) Montserrat 67 L5
Plymouth Bay MA U.S.A. 64 F2
Plynlimon h. U.K. 19 D6
Plyussa Rus. Fed. 15 P7
Plzeň Czech Rep. 17 N6
Pô Burkina Faso 46 C3
Po r. Italy 26 E2
Pobeda Peak China/Kyrg. 31 H2
Pobedy, Pik mt. China/Kyrg. see
Pobeda Peak
Pocatello U.S.A. 62 E3
Pochala South Sudan 47 G4
Pochayiv Ukr. 13 E6
Pochep Rus. Fed. 13 G5
Pochinki Rus. Fed. 13 J5
Pochinok Rus. Fed. 13 G5
Pochutla Mex. 66 E5
Pocking Germany 17 N6
Pocklington U.K. 18 G5
Poções Brazil 71 C1
Pocomoke City MD U.S.A. 64 D3
Pocomoke Sound b. Maryland/Virginia
U.S.A. 64 D4
Poconé Brazil 69 G7
Pocono Mountains hills PA U.S.A. 64 D2
Poços de Caldas Brazil 71 B3
Podanur India 38 C4
Poddor'ye Rus. Fed. 12 F4
Podgorenskiy Rus. Fed. 13 H6
Podgorica Montenegro 27 H3
Podgornoye Rus. Fed. 28 J4
Podile India 38 C3
Podişul Transilvaniei plat. Romania see
Transylvanian Basin
Podkamennaya Tunguska r. Rus. Fed.
29 K3
Podocarpus, Parque Nacional nat. park
Ecuador 68 C4
Podol'sk Rus. Fed. 13 H5
Podporozh'ye Rus. Fed. 12 G3
Podujevë Kosovo 27 I3
Podujevo Kosovo see Podujevë
Podz' Rus. Fed. 12 K3
Poelela, Lagoa l. Moz. 51 L3
Poeppel Corner imp. l. Australia 57 B5
Poetovio Slovenia see Ptuj
Pofadder S. Africa 50 D5
Pogar Rus. Fed. 13 G5
Poggibonsi Italy 26 D3
Poggio di Montieri mt. Italy 26 D3
Pogradec Albania 27 I4
Pogranichnyy Rus. Fed. 44 C3
Po Hai g. China see Bo Hai
P'ohang S. Korea 45 C5
Pohnpei atoll Micronesia 74 G5
Pohri India 36 D4
Poi India 37 H4
Poiana Mare Romania 27 J3

Poinsett, Cape Antarctica 76 F2
Point Arena CA U.S.A. 65 A1
Pointe-à-Pitre Guadeloupe 67 L5
Pointe-Noire Congo 49 B4
Point Hope U.S.A. 60 B3
Point Lake Canada 60 G3
Point Pleasant NJ U.S.A. 64 D2
Poitiers France 24 E3
Poitou reg. France 24 E3
Pojuca r. Brazil 71 D1
Pokaran India 36 B4
Pokataroo Australia 58 D2
Pokcha Rus. Fed. 11 R3
Pokhara Nepal 37 E3
Pokhran Landi Pak. 36 A4
Pokhvistnevo Rus. Fed. 11 Q5
Poko Dem. Rep. Congo 48 C3
Pokosnoye Rus. Fed. 42 I1
P'ok'r Kovkas mts Asia see
Lesser Caucasus
Pokrovka Primorskiy Kray Rus. Fed. 44 C4
Pokrovka Zabaykal'skiy Kray Rus. Fed.
44 A1
Pokrovsk Respublika Sakha (Yakutiya)
Rus. Fed. 29 N3
Pokrovsk Saratovskaya Oblast' Rus. Fed.
see Engel's
Pokrovskoye Rus. Fed. 13 H7
Pokshen'ga r. Rus. Fed. 12 J3
Pol India 36 C5
Pola Croatia see Pula
Pola de Lena Spain 25 D2
Pola de Siero Spain 25 D2
Poland country Europe 10 J5
Poland NY U.S.A. 64 D1
Polar Plateau Antarctica 76 A1
Polatlı Turkey 34 D3
Polatsk Belarus 15 P9
Polavaram India 38 D2
Polcirkeln Sweden 14 L3
Pol-e Fāsā Iran 35 I5
Pol-e Khomrī Afgh. 36 B2
Polessk Rus. Fed. 15 L9
Poles'ye marsh Belarus/Ukr. see
Pripet Marshes
Polgahawela Sri Lanka 38 D5
Poli Cyprus see Polis
Políaigos i. Greece see Polyaigos
Police Poland 17 O4
Policoro Italy 26 G4
Poligny France 24 G3
Polikastron Greece see Polykastro
Polillo Islands Phil. 41 E6
Polis Cyprus 39 A2
Polis'ke Ukr. 13 F6
Polis'kyy Zapovidnyk nat. park Ukr. 13 F6
Politovo Rus. Fed. 12 K2
Políyiros Greece see Polygyros
Polkowice Poland 17 P5
Pollachi India 38 C4
Pollard Islands U.S.A. see
Gardner Pinnacles
Pollino, Monte mt. Italy 26 G5
Pollino, Parco Nazionale del nat. park Italy
26 G5
Pollock Pines CA U.S.A. 65 B1
Pollock Reef Australia 55 C8
Polmak Norway 14 O1
Polnovat Rus. Fed. 11 T3
Polo Fin. 14 P4
Pologi Ukr. see Polohy
Polohy Ukr. 13 H7
Polokwane S. Africa 51 I2
Polonne Ukr. 13 E6
Polonnoye Ukr. see Polonne
Polotsk Belarus see Polatsk
Polperro U.K. 19 C8
Polska country Europe see Poland
Polson U.S.A. 62 E2
Polta r. Rus. Fed. 12 I2
Poltava Ukr. 13 G6
Poltoratsk Turkm. see Aşgabat
Põltsamaa Estonia 15 N7
Polunochnoye Rus. Fed. 11 S3
Põlva Estonia 15 O7
Polvijärvi Fin. 14 P5
Polyaigos i. Greece 27 K6
Polyanovgrad Bulg. see Karnobat
Polyarnyy Chukotskiy Avtonomnyy Okrug
Rus. Fed. 29 S3
Polyarnyy Murmanskaya Oblast' Rus. Fed.
12 G1
Polyarnyye Zori Rus. Fed. 12 G2
Polyarnyy Ural mts Rus. Fed. 11 S2
Polygyros Greece 27 J4
Polykastro Greece 27 J4
Polynesia is Pacific Ocean 74 I6
Polynésie Française terr. S. Pacific Ocean
see French Polynesia
Pomarkku Fin. 15 M6
Pombal Pará Brazil 69 H4
Pombal Paraíba Brazil 69 K5
Pombal Port. 25 B4
Pomene Moz. 51 L2
Pomeranian Bay Poland 17 O3
Pomeroy S. Africa 51 J5
Pomeroy U.K. 21 F3
Pomezia Italy 26 E4
Pomfret S. Africa 50 F3
Pomona Namibia 50 B4
Pomona CA U.S.A. 65 D3
Pomorie Bulg. 27 L3
Pomorskie, Pojezierze reg. Poland 17 O4
Pomorskiy Bereg coastal area Rus. Fed.
12 G2
Pomorskiy Proliv sea chan. Rus. Fed. 12 K1
Pomos Point Cyprus 39 A2
Pomo Tso l. China see Puma Yumco
Pomou, Akra pt Cyprus see Pomos Point
Pomozdino Rus. Fed. 12 L3
Pompain China 37 H3
Pompei Italy 26 F4
Pompéia Brazil 71 A3
Pompeyevka Rus. Fed. 44 C2
Ponape atoll Micronesia see Pohnpei
Ponazyrevo Rus. Fed. 12 J4
Ponca City U.S.A. 63 H4
Ponce Puerto Rico 67 K5
Pondicherry India see Puducherry
Pond Inlet Canada 77 K2
Ponds Bay Canada see Pond Inlet
Ponente, Riviera di coastal area Italy 26 B3
Ponferrada Spain 25 C2
Pongara, Pointe pt Gabon 48 A3
Pongaroa N.Z. 59 F5
Pongo watercourse South Sudan 47 F4
Pongola r. S. Africa 51 K4
Pongolapoort Dam dam S. Africa 51 J4
Ponnagyun Myanmar 37 H5
Ponnaivar r. India 38 C4
Ponnampet India 38 B3
Ponnani India 38 B4
Ponnyadaung Range mts Myanmar 37 H5
Ponoka Canada 62 E1
Ponoy r. Rus. Fed. 12 I2
Ponta Delgada Arquipélago dos Açores
72 G3
Ponta Grossa Brazil 71 A4
Pontal Brazil 71 A3
Pontalina Brazil 71 A2
Pont-à-Mousson France 24 H2

Ponta Porã Brazil 70 E2
Pontarfynach U.K. see Devil's Bridge
Pont-Audemer France 19 H9
Ponte Alta do Norte Brazil 69 I6
Ponte de Sor Port. 25 B4
Ponte Firme Brazil 71 B2
Pontefract U.K. 18 F5
Ponteix Canada 62 F2
Ponteland U.K. 18 F3
Ponte Nova Brazil 71 C3
Pontes-e-Lacerda Brazil 69 G7
Pontevedra Spain 25 B2
Ponthierville Dem. Rep. Congo see
Ubundu
Pontiac IL U.S.A. 63 J3
Pontiac MI U.S.A. 63 K3
Pontiae is Italy see Ponziane, Isole
Pontianak Indon. 41 C8
Pontine Islands is Italy see Ponziane, Isole
Pont-l'Abbé France 24 B3
Pontoise France 24 F2
Ponton watercourse Australia 55 C7
Ponton Canada 62 H1
Pontypool U.K. 19 D7
Pontypridd U.K. 19 D7
Ponza, Isola di i. Italy 26 E4
Ponziane, Isole is Italy 26 E4
Poochera Australia 55 F8
Poole U.K. 19 F8
Poolowanna Lake imp. l. Australia 57 B5
Poona India see Pune
Pooncarie Australia 57 C7
Poonch India see Punch
Poopelloe Lake salt l. Australia 58 B3
Poopó, Lago de l. Bol. 68 E7
Poor Knights Islands N.Z. 59 E2
Popayán Col. 68 C3
Popigay r. Rus. Fed. 29 L2
Popiltah Australia 57 C7
Popilta Lake imp. l. Australia 57 C7
Poplar r. Canada 63 H1
Poplar Bluff U.S.A. 63 I4
Poplar Camp VA U.S.A. 64 A4
Popocatépetl, Volcán vol. Mex. 66 E5
Popokabaka Dem. Rep. Congo 49 B4
Popovichskaya Rus. Fed. see Kalininskaya
Popovo Bulg. 27 L3
Popovo Polje plain Bos.-Herz. 26 G3
Poprad Slovakia 17 R6
Poquoson VA U.S.A. 64 C4
Porangahau N.Z. 59 F5
Porangatu Brazil 71 A1
Porbandar India 36 B5
Porcos r. Brazil 71 B1
Porcupine Abyssal Plain sea feature
N. Atlantic Ocean 72 G2
Porcupine Gorge National Park Australia
56 D4
Poreč Croatia 26 E2
Porecatu Brazil 71 A3
Poretskoye Rus. Fed. 13 J5
Pori Fin. 15 L6
Porirua N.Z. 59 E5
Porkhov Rus. Fed. 12 F4
Porlamar Venez. 68 F1
Pormpuraaw Australia 56 C2
Pornic France 24 C3
Poronaysk Rus. Fed. 44 F2
Porong China 37 G3
Poros Greece 27 J6
Porosozero Rus. Fed. 12 G3
Porpoise Bay Antarctica 76 G2
Porsangerfjorden sea chan. Norway 14 N1
Porsangerhalvøya pen. Norway 14 N1
Porsgrunn Norway 15 F7
Porsuk r. Turkey 27 N5
Portadown U.K. 21 F3
Portaferry U.K. 21 G3
Portage PA U.S.A. 64 B2
Portage WI U.S.A. 63 J3
Portage la Prairie Canada 62 H2
Port Alberni Canada 62 C2
Port Albert Australia 58 C7
Portalegre Port. 25 C4
Portales U.S.A. 62 G5
Port Alfred S. Africa 51 H7
Port Allegany PA U.S.A. 64 B2
Port Alma Australia 56 E4
Port Angeles U.S.A. 62 C2
Port Antonio Jamaica 67 I5
Portarlington Ireland 21 E4
Port Arthur Australia 57 [inset]
Port Arthur U.S.A. 63 I6
Port Askaig U.K. 20 C5
Port Augusta Australia 57 B7
Port-au-Prince Haiti 67 J5
Portavogie U.K. 21 G3
Port Beaufort S. Africa 50 E8
Port Blair India 31 I5
Portbou Spain 25 H2
Port Burwell Ont. Canada 64 A1
Port Campbell Australia 58 A7
Port Campbell National Park Australia
58 A7
Port Chalmers N.Z. 59 C7
Port Charlotte U.S.A. 63 K6
Port-de-Paix Haiti 67 J5
Port Douglas Australia 56 D3
Port Edward S. Africa 51 J6
Porteira Brazil 69 G4
Porteirinha Brazil 71 C1
Portel Brazil 69 H4
Port Elizabeth S. Africa 51 G7
Port Ellen U.K. 20 C5
Port Erin Isle of Man 18 C4
Porterville S. Africa 50 D7
Porterville CA U.S.A. 65 C2
Port Étienne Mauritania see Nouâdhibou
Port Everglades U.S.A. see Fort Lauderdale
Port Fitzroy N.Z. 59 E3
Port Francqui Dem. Rep. Congo see Ilebo
Port-Gentil Gabon 48 A4
Port Glasgow U.K. 20 E5
Port Harcourt Nigeria 46 D4
Porthcawl U.K. 19 D7
Port Hedland Australia 54 B5
Port Herald Malawi see Nsanje
Porthleven U.K. 19 B8
Porthmadog U.K. 19 C6
Port Hope Simpson Canada 61 M4
Port Hueneme CA U.S.A. 65 C3
Port Huron U.S.A. 63 K3
Portimão Port. 25 B5
Port Jackson Australia see Sydney
Port Jackson inlet Australia 58 E4
Port Keats Australia see Wadeye
Port Láirge Ireland see Waterford
Portland N.S.W. Australia 58 D4
Portland Vic. Australia 57 C8
Portland ME U.S.A. 63 M3
Portland OR U.S.A. 62 C2
Portland, Isle of pen. U.K. 19 E8
Portland Bill hd U.K. see Bill of Portland
Portland Roads Australia 56 C2
Port-la-Nouvelle France 24 F5
Portlaoise Ireland 21 E4
Port Lavaca U.S.A. 63 H6
Portlaw Ireland 21 E5
Portlethen U.K. 20 G3
Port Lincoln Australia 57 A7
Port Loko Sierra Leone 46 B4

Port Louis Mauritius 73 L7
Port-Lyautrey Morocco see Kénitra
Port Macquarie Australia 58 F3
Portmadoc U.K. see Porthmadog
Port McNeill Canada 62 B1
Port-Menier Canada 63 O2
Port Moresby P.N.G. 52 E2
Portnaguran U.K. 20 C2
Portnahaven U.K. 20 C5
Port nan Giúran U.K. see Portnaguran
Port Neill Australia 57 B7
Port Nis U.K. see Port of Ness
Port Noarlunga Australia 57 B7
Port Nolloth S. Africa 50 C5
Port Norris NJ U.S.A. 64 D3
Porto Port. see Oporto
Porto Acre Brazil 68 E5
Porto Alegre Brazil 71 A5
Porto Alexandre Angola see Tombua
Porto Amboim Angola 49 B5
Porto Amélia Moz. see Pemba
Porto Artur Brazil 69 G6
Porto Belo Brazil 71 A4
Porto de Moz Brazil 69 H4
Porto de Santa Cruz Brazil 71 C1
Porto dos Gaúchos Óbidos Brazil 69 G6
Porto Esperança Brazil 69 G7
Porto Esperidião Brazil 69 G7
Portoferraio Italy 26 D3
Port of Ness U.K. 20 C2
Porto Franco Brazil 69 I5
Port of Spain Trin. and Tob. 67 L6
Porto Grande Brazil 69 H3
Portogruaro Italy 26 E2
Porto Jofre Brazil 69 G7
Portomaggiore Italy 26 D2
Porto Mendes Brazil 70 F2
Porto Murtinho Brazil 70 E2
Porto Nacional Brazil 69 I6
Porto-Novo Benin 46 D4
Porto Novo Cape Verde 46 [inset]
Porto Primavera, Represa resr Brazil 70 F2
Porto Rico Angola 49 B4
Porto Santo, Ilha de i. Madeira 46 B1
Porto Seguro Brazil 71 D2
Porto Tolle Italy 26 E2
Porto Torres Sardegna Italy 26 C4
Porto União Brazil 71 A4
Porto-Vecchio Corse France 24 I6
Porto Velho Brazil 68 F5
Portoviejo Ecuador 68 B4
Porto Walter Brazil 68 D5
Portpatrick U.K. 20 D6
Port Phillip Bay Australia 58 B7
Port Pirie Australia 57 B7
Portreath U.K. 19 B8
Portree U.K. 20 C3
Port Royal VA U.S.A. 64 C3
Portrush U.K. 21 F2
Port Safaga Egypt see Būr Safājah
Port Said Egypt 39 A4
Port St Joe U.S.A. 63 J6
Port St Mary Isle of Man 18 C4
Portsalon Ireland 21 E2
Port Shepstone S. Africa 51 J6
Portsmouth U.K. 19 F8
Portsmouth NH U.S.A. 64 F1
Portsmouth OH U.S.A. 63 K4
Portsmouth VA U.S.A. 64 C4
Portsoy U.K. 20 G3
Port Stanley Falkland Is see Stanley
Port Stephens b. Australia 58 F4
Portstewart U.K. 21 F2
Port Sudan Sudan 32 E6
Port Talbot U.K. 19 D7
Porttipahdan tekojärvi resr Fin. 14 O2
Portugal country Europe 25 C4
Portugália Angola see Chitato
Portuguese East Africa country Africa see
Mozambique
Portuguese Guinea country Africa see
Guinea-Bissau
Portuguese Timor country Asia see
East Timor
Portuguese West Africa country Africa see
Angola
Portumna Ireland 21 D4
Portus Herculis Monoeci country Europe
see Monaco
Port-Vendres France 24 F5
Port Vila Vanuatu 53 G3
Portville NY U.S.A. 64 B1
Port-Vladimir Rus. Fed. 14 R2
Port Waikato N.Z. 59 E3
Port William U.K. 20 E6
Porvenir Bol. 68 E6
Porvenir Chile 70 B8
Porvoo Fin. 15 N6
Posada Spain 25 D2
Posada de Llanera Spain see Posada
Posadas Arg. 70 E3
Posen Poland see Poznań
Poshekhon'ye Rus. Fed. 12 H4
Poshekon'ye-Volodarsk Rus. Fed. see
Poshekhon'ye
Posht-e Kūh mts Iran 35 G4
Posht Kūh h. Iran 35 H3
Posio Fin. 14 P3
Poso Indon. 41 E8
Posof Turkey 35 F2
Posŏng S. Korea 45 B6
Possession Island Namibia 50 B4
Post U.S.A. 62 G5
Postavy Belarus see Pastavy
Poste-de-la-Baleine Canada see
Kuujjuarapik
Poste Weygand Alg. 46 D2
Postmasburg S. Africa 50 F5
Poston AZ U.S.A. 65 E3
Postysheve Ukr. see Krasnoarmiys'k
Pota Indon. 54 C2
Poté Brazil 71 C2
Potegaon India 38 D2
Potentia Italy see Potenza
Potenza Italy 26 F4
P'ot'i Georgia 35 F2
Potikal India 38 D2
Potiraguá Brazil 71 D1
Potiskum Nigeria 46 E3
Potomac r. Maryland/Virginia U.S.A. 64 C3
Potosí Bol. 68 E7
Potosi Mountain NV U.S.A. 65 E3
Potrerillos Chile 70 C3
Potsdam Germany 17 N4
Potterne U.K. 19 E7
Potters Bar U.K. 19 G7
Pottstown PA U.S.A. 64 D2
Pottsville PA U.S.A. 64 C2
Pottuvil Sri Lanka 38 D5
Poughkeepsie NY U.S.A. 64 E2
Poulton-le-Fylde U.K. 18 E5
Pouso Alegre Brazil 71 B3
Poŭthĭsăt Cambodia 31 J5
Považská Bystrica Slovakia 17 Q6
Povenets Rus. Fed. 12 G3
Poverty Bay N.Z. 59 F4
Povlen mt. Serbia 27 H2
Póvoa de Varzim Port. 25 B3
Povorino Rus. Fed. 13 I6
Povorotnyy, Mys hd Rus. Fed. 44 D4
Poway CA U.S.A. 65 D4

Powell, Lake resr U.S.A. 62 E4
Powell Mountain NV U.S.A. 65 C1
Powell River Canada 62 C2
Powhatan VA U.S.A. 64 C4
Poxoréu Brazil 69 H7
Poyang Hu l. China 43 L7
Poyarkovo Rus. Fed. 44 C2
Pozantı Turkey 34 D3
Požarevac Serbia 27 I2
Poza Rica Mex. 66 E4
Pozdeyevka Rus. Fed. 44 C2
Požega Croatia 26 G2
Požega Serbia 27 I3
Pozharskoye Rus. Fed. 44 D3
Poznań Poland 17 P4
Pozoblanco Spain 25 D4
Pozo Colorado Para. 70 E2
Pozsony Slovakia see Bratislava
Pozzuoli Italy 26 F4
Prachatice Czech Rep. 17 O6
Prachi r. India 37 F6
Prachuap Khiri Khan Thai. 31 I5
Prades France 24 F5
Prado Brazil 71 D2
Prague Czech Rep. 17 O5
Praha Czech Rep. see Prague
Praia Cape Verde 46 [inset]
Praia do Bilene Moz. 51 K3
Prainha Brazil 69 H4
Prairie Australia 56 D4
Prairie du Chien U.S.A. 63 I3
Prasonisi, Akrotirio pt Greece 27 L7
Prata Brazil 71 A2
Prata r. Brazil 71 A2
Prat de Llobregat Spain see El Prat
de Llobregat
Prathes Thai country Asia see Thailand
Prato Italy 26 D3
Pratt U.S.A. 62 H4
Pravdinsk Rus. Fed. 15 L9
Praya Indon. 54 B2
Preble NY U.S.A. 64 C1
Prechistoye Smolenskaya Oblast' Rus. Fed.
13 G5
Prechistoye Yaroslavskaya Oblast'
Rus. Fed. 12 I4
Precipice National Park Australia 56 E5
Preeceville Canada 62 G1
Pregolya r. Rus. Fed. 15 L9
Preiļi Latvia 15 O8
Premer Australia 58 D3
Prémery France 24 F3
Prenzlau Germany 17 N4
Přerov Czech Rep. 17 P6
Prescelly Mts hills U.K. see
Preseli, Mynydd
Prescott AZ U.S.A. 62 E5
Preseli, Mynydd hills U.K. 19 C7
Preševo Serbia 27 I3
Presidencia Roque Sáenz Peña Arg. 70 D3
Presidente Dutra Brazil 69 J5
Presidente Hermes Brazil 68 F6
Presidente Olegário Brazil 71 B2
Presidente Prudente Brazil 71 A3
Presidente Venceslau Brazil 71 A3
Presidio U.S.A. 62 G6
Preslav Bulg. see Veliki Preslav
Prešov Slovakia 13 D6
Prespa, Lake Europe 27 I4
Prespansko Ezero l. Europe see
Prespa, Lake
Prespës, Liqeni i l. Europe see
Prespa, Lake
Presque Isle ME U.S.A. 63 N2
Pressburg Slovakia see Bratislava
Presteigne U.K. 19 D6
Preston U.K. 18 E5
Preston ID U.S.A. 62 E3
Preston, Cape Australia 54 B5
Prestonpans U.K. 20 G5
Prestwick U.K. 20 E5
Preto r. Bahia Brazil 69 J6
Preto r. Minas Gerais Brazil 71 B2
Preto r. Brazil 71 D1
Pretoria S. Africa 51 I3
Pretoria-Witwatersrand-Vereeniging prov.
S. Africa see Gauteng
Preussisch-Eylau Rus. Fed. see
Bagrationovsk
Preußisch Stargard Poland see
Starogard Gdański
Preveza Greece 27 I5
Priargunsk Rus. Fed. 43 L2
Pribilof Islands U.S.A. 60 A4
Priboj Serbia 27 H3
Price r. Australia 54 E3
Price UT U.S.A. 62 E4
Pridorozhnoye Rus. Fed. see Khulkhuta
Priekule Latvia 15 L8
Priekuļi Latvia 15 N8
Priel'brus'ye, Natsional'nyy Park nat. park
Rus. Fed. 13 I8
Prienai Lith. 15 M9
Prieska S. Africa 50 F5
Prievidza Slovakia 17 Q6
Prijedor Bos.-Herz. 26 G2
Prijepolje Serbia 27 H3
Prikaspiyskaya Nizmennost' lowland
Kazakh./Rus. Fed. see Caspian Lowland
Prilep Macedonia 27 I4
Priluki Ukr. see Pryluky
Primorsk Rus. Fed. 15 P6
Primorsk Ukr. see Prymors'k
Primorskiy Kray admin. div. Rus. Fed.
44 D3
Primorsko-Akhtarsk Rus. Fed. 13 H7
Prince Albert Canada 60 H4
Prince Albert S. Africa 50 F7
Prince Albert Mountains Antarctica 76 H1
Prince Albert Peninsula Canada 60 G2
Prince Albert Road S. Africa 50 E7
Prince Alfred, Cape Canada 60 F2
Prince Alfred Hamlet S. Africa 50 D7
Prince Charles Island Canada 61 K3
Prince Charles Mountains Antarctica
76 E2
Prince Edward Island prov. Canada 63 O2
Prince Edward Islands Indian Ocean
73 K9
Prince Frederick MD U.S.A. 64 C3
Prince George Canada 60 F4
Prince Harald Coast Antarctica 76 D2
Prince of Wales, Cape U.S.A. 60 B3
Prince of Wales Island Australia 56 C1
Prince of Wales Island Canada 61 I2
Prince of Wales Island U.S.A. 60 E4
Prince of Wales Strait Canada 60 G2
Prince Patrick Island Canada 60 G2
Prince Regent Inlet sea chan. Canada
61 I2
Prince Rupert Canada 60 E4
Princess Anne MD U.S.A. 64 D3
Princess Astrid Coast Antarctica 76 C2
Princess Charlotte Bay Australia 56 C2
Princess Elisabeth (Belgium) research stn
Antarctica 76 C21685673
Princess Elizabeth Land reg. Antarctica
76 E2
Princess Ragnhild Coast Antarctica 76 C2
Princess Royal Island Canada 60 F4
Princeton Canada 62 C2

Princeton NJ U.S.A. 64 D2
Princeton WV U.S.A. 64 A4
Prince William Sound b. U.S.A. 60 D3
Príncipe i. São Tomé and Príncipe 46 D4
Prins Harald Kyst coastal area Antarctica see Prince Harald Coast
Prinzapolca Nicaragua 67 H6
Priozersk Rus. Fed. 12 F3
Priozyorsk Rus. Fed. see Priozersk
Pripet r. Belarus 23 J1
Pripet r. Belarus/Ukr. 13 F6
Pripet Marshes Belarus/Ukr. 13 E6
Pirechnyy Rus. Fed. 14 Q2
Prishtinë Kosovo 27 I3
Priština Kosovo see Prishtinë
Pritzwalk Germany 17 N4
Privas France 24 G4
Privlaka Croatia 26 F2
Privolzhsk Rus. Fed. 12 I4
Privolzhskaya Vozvyshennost' hills Rus. Fed. 13 J6
Privolzhskiy Rus. Fed. 13 J6
Privolzh'ye Rus. Fed. 13 K5
Priyutnoye Rus. Fed. 13 I7
Prizren Kosovo 27 I3
Probus U.K. 19 C8
Proddatur India 38 C3
Professor van Blommestein Meer resr Suriname 69 G3
Progreso Hond. see El Progreso
Progress Rus. Fed. 44 C2
Prokhladnyy Rus. Fed. 13 J8
Prokop'yevsk Rus. Fed. 42 F2
Prokuplje Serbia 27 I3
Proletarsk Rus. Fed. 13 I7
Proletarskaya Rus. Fed. see Proletarsk
Proletarskoye Vodokhranilishche l. Rus. Fed. 13 I7
Prome Myanmar see Pyè
Promissão Brazil 71 A3
Promissão, Represa resr Brazil 71 A3
Prophet r. Canada 77 A3
Prophet River Canada 60 F4
Propriá Brazil 69 K6
Proskurov Ukr. see Khmel'nyts'kyy
Protem S. Africa 50 E8
Provadiya Bulg. 27 L3
Prøven Greenland see Kangersuatsiaq
Provence reg. France 24 G5
Providence RI U.S.A. 64 F2
Providence, Cape N.Z. 59 A8
Providencia, Isla de i. Caribbean Sea 67 H6
Provideniya Rus. Fed. 29 T3
Provincetown MA U.S.A. 64 F1
Provo U.S.A. 62 E3
Prudentópolis Brazil 71 A4
Prudhoe Bay U.S.A. 60 D2
Prunelli-di-Fiumorbo Corse France 24 I5
Prusa Turkey see Bursa
Prushkov Poland see Pruszków
Pruszków Poland 17 R4
Prut r. Europe 13 F7
Prydz Bay Antarctica 76 E2
Pryluky Ukr. 13 G6
Prymors'k Ukr. 13 H7
Prymors'ke Ukr. see Sartana
Pryp"yat' r. Belarus/Ukr. 13 F6 see Pripet
Prypyats' r. Belarus see Pripet
Prypyats' r. Belarus/Ukr. 11 L5 see Pripet
Przemyśl Poland 13 D6
Przheval'sk Kyrg. see Karakol
Psara i. Greece 27 K5
Pskov Rus. Fed. 12 F4
Pskov, Lake Estonia/Rus. Fed. 15 O7
Pskov Oblast admin. div. Rus. Fed. see Pskovskaya Oblast'
Pskovskaya Oblast' admin. div. Rus. Fed. 15 P8
Pskovskoye Ozero l. Estonia/Rus. Fed. see Pskov, Lake
Ptolemaïda Greece 27 I4
Ptolemais Israel see 'Akko
Ptuj Slovenia 26 F1
Puan S. Korea 45 B6
Pucallpa Peru 68 D5
Puchezh Rus. Fed. 12 I4
Puch'ŏn S. Korea 45 B5
Puck Poland 17 Q3
Pudasjärvi Fin. 14 O4
Pudimoe S. Africa 50 G4
Pudozh Rus. Fed. 12 H3
Pudsey U.K. 18 F5
Pudu China see Suizhou
Puduchcheri India see Puducherry
Puducherry India 38 C4
Puducherry union terr. India 38 C4
Pudukkottai India 38 C4
Puebla Baja California Mex. 65 E4
Puebla Puebla Mex. 66 E5
Puebla de Sanabria Spain 25 C2
Puebla de Zaragoza Mex. see Puebla
Pueblo U.S.A. 62 G4
Puelches Arg. 70 C5
Puelén Arg. 70 C5
Puente Genil Spain 25 D5
Puerto Acosta Bol. 68 E7
Puerto Alegre Bol. 68 F6
Puerto Ángel Mex. 66 E5
Puerto Armuelles Panama 67 H7
Puerto Ayacucho Venez. 68 E2
Puerto Bahía Negra Para. see Bahía Negra
Puerto Baquerizo Moreno Galápagos Ecuador 68 [inset]
Puerto Barrios Guat. 66 G5
Puerto Cabello Venez. 68 E1
Puerto Cabezas Nicaragua 67 H6
Puerto Carreño Col. 68 E2
Puerto Casado Para. 70 E2
Puerto Cavinas Bol. 68 E6
Puerto Coig Arg. 70 C8
Puerto Cortés Mex. 66 B4
Puerto Escondido Mex. 66 E5
Puerto Francisco de Orellana Ecuador see Coca
Puerto Frey Bol. 68 F6
Puerto Génova Bol. 68 E6
Puerto Guarani Para. 70 E2
Puerto Heath Bol. 68 E6
Puerto Huitoto Col. 68 D3
Puerto Inírida Col. 68 E3
Puerto Isabel Bol. 69 G7
Puerto Leguízamo Col. 68 D4
Puerto Lempira Hond. 67 H5
Puerto Libertad Mex. 66 B3
Puerto Limón Costa Rica 67 H6
Puertollano Spain 25 D4
Puerto Lobos Arg. 70 C6
Puerto Madryn Arg. 70 C6
Puerto Maldonado Peru 68 E6
Puerto Máncora Peru 68 B4
Puerto México Mex. see Coatzacoalcos
Puerto Montt Chile 70 B6
Puerto Natales Chile 70 B8
Puerto Nuevo Col. 68 E2
Puerto Peñasco Mex. 66 B2
Puerto Pirámides Arg. 70 D6
Puerto Plata Dom. Rep. 67 J5
Puerto Portillo Peru 68 D5
Puerto Prado Peru 68 D6
Puerto Princesa Phil. 41 D7
Puerto Rico Arg. 70 E3
Puerto Rico Bol. 68 E6
Puerto Rico terr. West Indies 67 K5
Puerto Rico Trench sea feature Caribbean Sea 72 D4
Puerto Santa Cruz Arg. 70 C8
Puerto Sastre Para. 70 E2
Puerto Saucedo Bol. 68 F6
Puerto Suárez Bol. 69 G7
Puerto Supe Peru 68 C6
Puerto Vallarta Mex. 66 C4
Puerto Victoria Peru 68 D5
Puerto Visser Arg. 70 C7
Puerto Yartou Chile 70 B8
Puerto Ybapobó Para. 70 E2
Pugachev Rus. Fed. 13 K5
Pugal India 36 C3
Puhiwaero c. N.Z. see South West Cape
Puigmal mt. France/Spain 24 F5
Pukaki, Lake N.Z. 59 C7
Pukapuka atoll Cook Is 53 J3
Pukchin N. Korea 45 B4
Pukch'ŏng N. Korea 45 C4
Pukekohe N.Z. 59 E3
Puketeraki Range mts N.Z. 59 D6
Pukeuri Junction N.Z. 59 C7
Puksubaek-san mt. N. Korea 44 B4
Pula China see Nyingchi
Pula Croatia 26 E2
Pula Sardegna Italy 26 C5
Pulaski NY U.S.A. 64 C1
Pulaski VA U.S.A. 64 A4
Pulicat Lake inlet India 38 D3
Pulivendla India 38 C3
Pulkkila Fin. 14 N4
Pullman U.S.A. 62 D2
Pulozero Rus. Fed. 14 R2
Pulu China 36 E1
Pülümür Turkey 35 E3
Pumasillo, Cerro mt. Peru 68 D6
Puma Yumco l. China 37 G3
Puná, Isla i. Ecuador 68 B4
Punakha Bhutan 37 G4
Punch India 36 C2
Punda Maria S. Africa 51 J2
Pundri India 36 D3
Pune India 38 B2
P'ungsan N. Korea 44 C4
Punjab state India 36 C3
Punjab prov. Pak. 36 B3
Punmah Glacier China/Pakistan 36 D2
Puno Peru 68 D7
Punta, Cerro de mt. Puerto Rico 67 K5
Punta Alta Arg. 70 D5
Punta Arenas Chile 70 B8
Punta Balestrieri mt. Italy 26 C4
Punta del Este Uruguay 70 F5
Punta Delgada Arg. 70 D6
Punta Gorda Belize 66 G5
Punta Norte Arg. 70 D6
Puntarenas Costa Rica 67 H6
Puntland reg. Somalia 48 E3
Punxsutawney PA U.S.A. 64 B2
Puokio Fin. 14 O4
Puolanka Fin. 14 O4
Puqi China see Chibi
Pur r. Rus. Fed. 28 I3
Puracé, Volcán de vol. Col. 68 C3
Puri India 38 E2
Purna r. Mahar. India 36 D5
Purna r. Mahar. India 38 C2
Purnea India see Purnia
Purnia India 37 F4
Purnululu National Park Australia 54 E4
Pursat Cambodia see Poŭthĭsăt
Puruliya India 37 F5
Purus r. Peru 68 F4
Puruvesi l. Fin. 14 P6
Puryŏng N. Korea 44 C4
Pusad India 38 C2
Pusan S. Korea 45 C6
Pusatlı Dağı mt. Turkey 39 A1
Pushchino Rus. Fed. 13 H5
Pushkin Rus. Fed. 12 F4
Pushkino Azer. see Biläsuvar
Pushkinskaya, Gora mt. Rus. Fed. 44 F3
Pushkinskiye Gory Rus. Fed. 15 P8
Pushma Rus. Fed. 12 J3
Pustoshka Rus. Fed. 12 F4
Pusur r. Bangl. 37 G5
Putain Indon. 54 D2
Putao Myanmar 42 H7
Puteoli Italy see Pozzuoli
Puthein Myanmar see Bassein
Putian China 43 L7
Putna r. Romania 27 L2
Putney VT U.S.A. 64 E1
Putrajaya Malaysia 41 C7
Putre Chile 68 E7
Putsonderwater S. Africa 50 E5
Puttalam Sri Lanka 38 C4
Puttalam Lagoon Sri Lanka 38 C4
Puttgarden Germany 17 M3
Putumayo r. Col. 68 D4
Puumala Fin. 15 P6
Puvirnituq Canada 61 K3
Puyang China 43 L5
Puy de Sancy mt. France 24 F4
Puyehue, Parque Nacional nat. park Chile 70 B6
Puysegur Point N.Z. 59 A8
Puzla Rus. Fed. 12 L3
Pweto Dem. Rep. Congo 49 C4
Pwinbyu Myanmar 37 H5
Pwllheli U.K. 19 C6
Pyal'ma Rus. Fed. 12 G3
Pyalo Myanmar 37 H6
Pyandzh Tajik. see Panj
Pyaozero, Ozero l. Rus. Fed. 14 Q3
Pyaozerskiy Rus. Fed. 14 Q4
Pyapali India 38 C3
Pyasina r. Rus. Fed. 28 J2
Pyatigorsk Rus. Fed. 13 I7
Pyatikhatki Ukr. see P"yatykhatky
P"yatykhatky Ukr. 13 G6
Pyay Myanmar see Pyè
Pychas Rus. Fed. 12 L4
Pyè Myanmar 42 H9
Pye, Mount h. N.Z. 59 B8
Pyetrykaw Belarus 13 F5
Pyhäjoki Fin. 14 N4
Pyhäjoki r. Fin. 14 N4
Pyhäntä Fin. 14 O4
Pyhäsalmi Fin. 14 N5
Pyhäselkä l. Fin. 14 P5
Pyi Myanmar see Pyè
Pyin Myanmar see Pyè
Pyingaing Myanmar 37 H5
Pyinmana Myanmar 37 I6
Pyin-U-Lwin Myanmar 37 I5
Pyle U.K. 19 D7
Pyl'karamo Rus. Fed. 28 J3
Pylos Greece 27 I6
Pymatuning Reservoir PA U.S.A. 64 A2
Pyŏktong N. Korea 44 B4
P'yŏnggang N. Korea 45 B5
P'yŏnghae S. Korea 45 C5
P'yŏngsong N. Korea 45 B5
P'yŏngt'aek S. Korea 45 B5
P'yŏngyang N. Korea 45 B5
Pyramid Hill Australia 58 B6
Pyramid Lake U.S.A. 62 D3
Pyramids of Giza tourist site Egypt 34 C5
Pyrénées mts Europe see Pyrenees
Pyrenees mts Europe 25 H2
Pyrénées, Parc National des nat. park France/Spain 24 D5
Pyrgos Greece 27 I6
Pyryatyn Ukr. 13 G6
Pyrzyce Poland 17 O4
Pyshchug Rus. Fed. 12 J4
Pytalovo Rus. Fed. 15 O8
Pyxaria mt. Greece 27 J5

Q

Qaa Lebanon 39 C2
Qaanaaq Greenland see Thule
Qabātiya West Bank 39 B3
Qabnag China 37 H3
Qabqa China see Gonghe
Qacentina Alg. see Constantine
Qacha's Nek Lesotho 51 I6
Qāder Ābād Iran 35 I5
Qadirpur Pak. 36 C3
Qādisīyah, Sadd dam Iraq 35 F4
Qadisiyah Dam Iraq see Qādisīyah, Sadd
Qā'en Iran 33 I3
Qagan China 43 L3
Qagan Nur China 43 K4
Qagan Nur salt l. China 44 B3
Qagan Us Nei Mongol China 43 K4
Qagan Us Qinghai China see Dulan
Qagbasêrag China 37 H3
Qagcaka China 37 E2
Qahremānshahr Iran see Kermānshāh
Qaidabad Pak. 36 B2
Qaidam He r. China 37 H1
Qaidam Pendi basin China 42 G5
Qainaqangma China 37 G2
Qaisar, Koh-i- mt. Afgh. see Qeyşār, Kūh-e
Qakar China 36 E1
Qalā Diza Iraq 35 G3
Qalāgay (abandoned) Afgh. 36 B2
Qalansīyah Yemen 33 H7
Qalāt Afgh. see Kalāt
Qal'at al Ḥiṣn Syria 39 C2
Qal'at al Mu'aẓẓam Saudi Arabia 34 E6
Qal'at Bīshah Saudi Arabia 32 F5
Qal'at Muqaybirah, Jabal mt. Syria 39 D2
Qal'eh Dāgh mt. Iran 35 G3
Qal'eh-ye Now Afgh. 33 J3
Qal'eh-ye Sang-e Takht Afgh. 36 A2
Qalīb Bāqūr well Iraq 35 G5
Qalyūb Egypt 34 C5
Qalyûb Egypt see Qalyūb
Qamanirjuaq Lake Canada 61 I3
Qamanittuaq Canada see Baker Lake
Qamashi Uzbek. 33 K2
Qamata S. Africa 51 H6
Qamdo China 42 H6
Qandahar Afgh. see Kandahār
Qandarānbāshī, Kūh-e mt. Iran 35 G3
Qandyaghash Kazakh. see Kandyagash
Qangzê China 36 D3
Qapan Iran 35 I3
Qapshagay Kazakh. see Kapchagay
Qapshagay Bögeni resr Kazakh. see Kapchagayskoye Vodokhranilishche
Qapugtang China see Zadoi
Qaqortoq Greenland 61 N3
Qara Āghach r. Iran see Mand, Rūd-e
Qarabutaq Kazakh. see Karabutak
Qara Ertis r. China/Kazakh. see Ertix He
Qaraghandy Kazakh. see Karaganda
Qaraghayly Kazakh. see Karagayly
Qārah Egypt 34 B5
Qārah Saudi Arabia 35 F5
Qaraqum des. Turkm. see Karakum Desert
Qarasu Azer. 35 H2
Qara Şū Chāy r. Syria/Turkey see Karasu
Qarataū Kazakh. see Karatau
Qarataū Zhotasy mts Kazakh. see Karatau, Khrebet
Qara Tikan Iran 35 H3
Qarazhal Kazakh. see Karazhal
Qardho Somalia 48 E3
Qareh Bāgh Afgh. 36 B2
Qareh Chāy r. Iran 35 H4
Qareh Qūzī-ye Bālā Iran 35 I3
Qareh Sū r. Iran 35 G3
Qarhan China 37 H1
Qarkilik China see Ruoqiang
Qarn al Kabsh, Jabal mt. Egypt 34 D5
Qarn el Kabsh, Gebel mt. Egypt see Qarn al Kabsh, Jabal
Qarokŭl l. Tajik. 33 L2
Qarqan China see Qiemo
Qarqan He r. China 42 F5
Qarqaraly Kazakh. see Karkaralinsk
Qarshi Uzbek. 33 K2
Qartaba Lebanon 39 B2
Qārūh, Jazīrat i. Kuwait 35 H5
Qārūn, Birkat salt l. Egypt 34 C5
Qârûn, Birket salt l. Egypt see Qārūn, Birkat
Qaryat al Gharab Iraq 35 G5
Qaryat al Ulyā Saudi Arabia 35 G6
Qasigiannguit Greenland 61 M3
Qaşr al Azraq tourist site Jordan 39 C4
Qaşr al Farāfirah Egypt 34 B6
Qaşr al Kharānah Jordan 39 C4
Qaşr al Khubbāz Iraq 35 F4
Qaşr 'Amrah tourist site Jordan 39 C4
Qaşr Burqu' tourist site Jordan 39 C3
Qaşr-e Shīrīn Iran 35 G4
Qaşr Farâfra Egypt see Qaşr al Farāfirah
Qassimiut Greenland 61 N3
Qaţanā Syria 39 C3
Qatar country Asia 32 H4
Qaţmah Syria 39 C1
Qaţrūyeh Iran 35 I5
Qaţţāfī, Wādī al watercourse Jordan 39 C4
Qaţţâra, Râs esc. Egypt see Qaţţārah, Ra's
Qattara Depression Egypt 34 B5
Qaţţārah, Ra's esc. Egypt 34 B5
Qaţţīnah, Buḩayrat resr Syria 39 C2
Qax Azer. 35 G2
Qaynar Kazakh. see Kaynar
Qaysīyah, Qā' al marsh Jordan 39 C4
Qaysūm, Juzur is Egypt 34 D6
Qayyārah Iraq 35 F4
Qazangödağ mt. Armenia/Azer. 35 G3
Qazaq Shyghanaghy b. Kazakh. see Kazakhskiy Zaliv
Qazaqstan country Asia see Kazakhstan
Qazax Azer. 32 G1
Qazi Ahmad Pak. 36 B4
Qazımämmäd Azer. 35 H2
Qazvīn Iran 35 H3
Qeisûm, Gezâ'ir is Egypt see Qaysūm, Juzur
Qeisum Islands Egypt see Qaysūm, Juzur
Qena Egypt see Qinā
Qeqertarsuaq Greenland 61 M3
Qeqertarsuaq i. Greenland 61 M3
Qeqertarsuatsiaat Greenland 61 M3
Qeqertarsuup Tunua b. Greenland 61 M3
Qeshm Iran 33 I4
Qeydār Iran 35 H3
Qeydū Iran 35 H4
Qeyşār, Kūh-e mt. Afgh. 33 J3
Qezel Owzan, Rūdkhāneh-ye r. Iran 35 H3
Qezi'ot Israel 39 B4
Qian'an China 44 B3
Qian Gorlos China see Qianguozhen
Qianguozhen China 44 B3
Qianjiang Chongqing China 43 J7
Qianjin Heilong. China 44 D3
Qianjin Jilin China 44 C3
Qianqihao China 44 A3
Qian Shan mts China 44 A4
Qiaowa China see Muli
Qiaowan China 42 H4
Qībā' Saudi Arabia 35 G6
Qibing S. Africa 51 H5
Qidukou China 37 H2
Qiemo China 42 F5
Qijiaojing China 42 G4
Qikiqtarjuaq Canada 61 L3
Qila Ladgasht Pak. 33 J4
Qila Saifullah Pak. 33 K3
Qilian Shan mts China 42 H5
Qillak i. Greenland 61 O3
Qiman Tag mts China 37 G1
Qimusseriarsuaq b. Greenland 61 L2
Qinā Egypt 32 D4
Qing'an China 44 B3
Qingdao China 43 M5
Qinggang China 44 B3
Qinggil China see Qinghe
Qinghai prov. China 42 H5
Qinghai Hu salt l. China 42 I5
Qinghai Nanshan mts China 31 I3
Qinghe Heilong. China 44 C3
Qinghe Xinjiang China 42 G3
Qinghecheng China 44 B4
Qingjiang Jiangsu China see Huai'an
Qingjiang Jiangxi China see Zhangshu
Qinglung China 37 G3
Qingshan China see Wudalianchi
Qingshuihe Nei Mongol China 43 K5
Qingshuihe Qinghai China 42 H6
Qingyuan Guangdong China 43 K8
Qingyuan Guangxi China see Yizhou
Qingyuan Liaoning China 44 B4
Qingzang Gaoyuan plat. China see Tibet, Plateau of
Qinhuangdao China 43 L5
Qin Ling mts China 43 J6
Qinzhou China 43 J8
Qionghai China 43 K9
Qiongjiexue China see Qonggyai
Qionglai China 42 I6
Qionglai Shan mts China 42 I6
Qiongzhou Haixia str. China see Hainan Strait
Qiqian China 44 A1
Qiqihar China 44 A3
Qīr Iran 35 I5
Qira China 36 E1
Qīrāïya, Wâdi watercourse Egypt see Qurayyah, Wādī
Qiryat Israel 39 B3
Qiryat Shemona Israel 39 B3
Qishon r. Israel 39 B3
Qitab ash Shāmah vol. crater Saudi Arabia 39 C4
Qitaihe China 44 C3
Qixing He r. China 44 D3
Qizilqum des. Kazakh./Uzbek. see Kyzylkum Desert
Qizilrabot Tajik. 36 C1
Qogir Feng mt. China/Pakistan see K2
Qog Qi China see Sain Us
Qom Iran 35 H4
Qomdo China see Qumdo
Qomīsheh Iran see Shahrezā
Qomolangma Feng mt. China/Nepal see Everest, Mount
Qomsheh Iran see Shahrezā
Qonāq, Kūh-e h. Iran 35 H4
Qondūz Afgh. see Kunduz
Qonggyai China 37 G3
Qo'ng'irot Uzbek. 33 I1
Qong Muztag mt. China 37 E2
Qongrat Uzbek. see Qo'ng'irot
Qoornoq Greenland 61 M3
Qoqek China see Tacheng
Qo'qon Uzbek. 33 L1
Qorako'l Uzbek. 33 J2
Qorghalzhyn Kazakh. see Korgalzhyn
Qornet es Saouda mt. Lebanon 39 C2
Qorveh Iran 35 G4
Qosh Tepe Iraq 35 F3
Qostanay Kazakh. see Kostanay
Qoubaiyat Lebanon 39 C2
Qowowuyag mt. China/Nepal see Cho Oyu
Qozideh Tajik. 36 B1
Quabbin Reservoir MA U.S.A. 64 E1
Quadros, Lago dos l. Brazil 71 A5
Quail Mountains CA U.S.A. 65 D3
Quairading Australia 55 B8
Quambatook Australia 58 A5
Quambone Australia 58 C3
Quamby Australia 56 C4
Quan Dao Hoang Sa is S. China Sea see Paracel Islands
Quan Dao Truong Sa is S. China Sea see Spratly Islands
Quang Ngai Vietnam 31 J5
Quan Long Vietnam see Ca Mau
Quantock Hills U.K. 19 D7
Quanzhou Fujian China 43 L8
Quanzhou Guangxi China 43 K7
Qu'Appelle r. Canada 62 G1
Quaqtaq Canada 61 L3
Quartu Sant'Elena Sardegna Italy 26 C5
Quartzite Mountain NV U.S.A. 65 D2
Quartzsite AZ U.S.A. 65 E4
Quba Azer. 35 H2
Qūchān Iran 33 I2
Qudaym Syria 39 D2
Queanbeyan Australia 58 D5
Québec Canada 61 K5
Québec prov. Canada 63 L2
Quebra Anzol r. Brazil 71 B2
Queen Adelaide Islands Chile see La Reina Adelaida, Archipiélago de
Queen Anne MD U.S.A. 64 D3
Queen Charlotte Islands Canada see Haida Gwaii
Queen Charlotte Sound sea chan. Canada 60 F4
Queen Charlotte Strait Canada 62 B1
Queen Elizabeth Islands Canada 61 H2
Queen Elizabeth National Park Uganda 48 C4
Queen Mary Land reg. Antarctica 76 F2
Queen Maud Gulf Canada 61 H3
Queen Maud Land Antarctica 72 G10
Queen Maud Land reg. Antarctica 76 C2
Queen Maud Mountains Antarctica 76 J1
Queenscliff Australia 58 B7
Queensland state Australia 58 B1
Queenstown Australia 57 [inset]
Queenstown Ireland see Cobh
Queenstown N.Z. 59 B7
Queenstown S. Africa 51 H6
Queimada, Ilha i. Brazil 69 H4
Quelimane Moz. 49 D5
Quellón Chile 70 B6
Quelpart Island S. Korea see Cheju-do
Que Que Zimbabwe see Kwekwe
Querétaro Mex. 66 D4
Querétaro de Arteaga Mex. see Querétaro
Quesnel Lake Canada 62 C1
Quetta Pak. 33 K3
Quetzaltenango Guat. 66 F6
Queuco Chile 70 B5
Quezaltenango Guat. see Quetzaltenango
Quezon City Phil. 41 E6
Quibala Angola 49 B5
Quibaxe Angola 49 B4
Quibdó Col. 68 C2
Quiberon France 24 C3
Quiçama, Parque Nacional do nat. park Angola 49 B4
Quilengues Angola 49 B5
Quillabamba Peru 68 D6
Quillacollo Bol. 68 E7
Quillan France 24 F5
Quill Lakes Canada 62 G1
Quilmes Arg. 70 E4
Quilon India see Kollam
Quilpie Australia 58 B1
Quilpué Chile 70 B4
Quimbele Angola 49 B4
Quimilí Arg. 70 D3
Quimper France 24 B3
Quimperlé France 24 C3
Quinag h. U.K. 20 D2
Quincy IL U.S.A. 63 I4
Quincy MA U.S.A. 64 F1
Quines Arg. 70 C4
Quinga Moz. 49 E5
Quinn Canyon Range mts NV U.S.A. 65 E2
Quinto Spain 25 F3
Quionga Moz. 49 E5
Quipungo Angola 49 B5
Quirima Angola 49 B5
Quirindi Australia 58 E3
Quirinópolis Brazil 71 A2
Quissanga Moz. 49 E5
Quissico Moz. 51 L3
Quitapa Angola 49 B5
Quitilipi Arg. 70 D3
Quito Ecuador 68 C4
Quixadá Brazil 69 K4
Quixeramobim Brazil 69 K5
Qujing China 42 I7
Qulandy Kazakh. see Kulandy
Qulin Gol r. China 44 A3
Qulsary Kazakh. see Kul'sary
Qulyndy Zhazyghy plain Kazakh./Rus. Fed. see Kulundinskaya Step'
Qulzum, Baḩr al Egypt see Suez Bay
Qumar He r. China 42 G6
Qumarheyan China 37 H2
Qumarlêb China see Sêrwolungwa
Qumarrabdün China 37 H2
Qumbu S. Africa 51 I6
Qumdo China 37 H3
Qumrha S. Africa 51 H7
Qumulangma mt. China/Nepal see Everest, Mount
Qundūz Afgh. see Kunduz
Qŭnghirot Uzbek. see Qo'ng'irot
Quntamari China 37 G2
Quoich, Loch l. U.K. 20 D3
Quoile r. U.K. 21 G3
Quoin Point S. Africa 50 D8
Quoxo r. Botswana 50 G2
Qŭqon Uzbek. see Qo'qon
Qurama, Qatorkŭhi mts Asia see Kurama Range
Qurama Tizmasi mts Asia see Kurama Range
Qurayyah, Wādī watercourse Egypt 39 B4
Qurayyāt al Milḩ l. Jordan 39 C4
Qŭrghonteppa Tajik. 33 K2
Qusar Azer. 35 H2
Qusmuryn Kazakh. see Kushmurun
Qusum China 36 D2
Quthing Lesotho see Moyeni
Quttinirpaaq National Park Canada 61 K1
Quwayq, Nahr r. Syria/Turkey 39 C2
Quxar China see Lhazê
Quxian Zhejiang China see Quzhou
Quyang China see Jingzhou
Quyghan Kazakh. see Kuygan
Quy Nhơn Vietnam 31 J5
Qūyūn Eshek i. Iran 35 G3
Quzhou China 43 L7
Qypshaq Köli salt l. Kazakh. see Kypshak, Ozero
Qyrghyz Zhotasy mts Kazakh./Kyrg. see Kirghiz Range
Qyteti Stalin Albania see Kuçovë
Qyzylorda Kazakh. see Kyzylorda
Qyzylqum des. Kazakh./Uzbek. see Kyzylkum Desert
Qyzyltū Kazakh. see Kishkenekol'
Qyzylzhar Kazakh. see Kyzylzhar

R

Raa Atoll Maldives see North Maalhosmadulu Atoll
Raab r. Austria 17 P7
Raab Hungary see Győr
Raahe Fin. 14 N4
Rääkkylä Fin. 14 P5
Raanujärvi Fin. 14 N3
Raasay i. U.K. 20 C3
Raasay, Sound of sea chan. U.K. 20 C3
Raba Indon. 41 D8
Rabang China 36 E2
Rabat Malta see Victoria
Rabat Malta 26 F7
Rabat Morocco 22 C5
Rabaul P.N.G. 52 F2
Rabbath Ammon Jordan see 'Ammān
Rabbit Flat Australia 54 E5
Rābigh Saudi Arabia 32 E5
Rabnabad Islands Bangl. 37 G5
Râbniţa Moldova see Rîbniţa
Rabocheostrovsk Rus. Fed. 12 G2
Racaka China 37 I3
Race, Cape Canada 72 E2
Race Point MA U.S.A. 64 F1
Rachaïya Lebanon 39 B3
Rachaya Lebanon see Rachaïya
Rachel NV U.S.A. 65 E2
Rach Gia Vietnam 31 J6
Racibórz Poland 17 Q5
Racine WI U.S.A. 63 J3
Rădăuţi Romania 13 E7
Radcliff U.S.A. 63 J4
Radde Rus. Fed. 44 C2
Radford VA U.S.A. 64 A4
Radisson Que. Canada 61 K4
Radlinski, Mount Antarctica 76 K1
Radnevo Bulg. 27 K3
Radom Poland 17 R5
Radom Sudan 47 F4
Radomir Bulg. 27 J3
Radom National Park Sudan 47 F4
Radomsko Poland 17 Q5
Radoviš Macedonia 34 A2
Radstock U.K. 19 E7
Radstock, Cape Australia 55 F8
Radun' Belarus 15 N9
Radviliškis Lith. 15 M9
Radyvyliv Ukr. 13 E6
Rae Bareli India 36 E4
Rae-Edzo Canada see Behchokǫ̀
Raeside, Lake salt flat Australia 55 C7
Raetihi N.Z. 59 E4
Rāf h. Saudi Arabia 35 E5
Rafaela Arg. 70 D4
Rafaḩ Gaza see Rafiah
Rafaï Cent. Afr. Rep. 48 C3
Rafḩā' Saudi Arabia 35 F5
Rafiah Gaza 39 B4
Rafsanjān Iran 33 I3
Raga South Sudan 47 F4
Ragged, Mount h. Australia 55 C8
Rago Nasjonalpark nat. park Norway 14 J3
Ragusa Croatia see Dubrovnik
Ragusa Sicilia Italy 26 F6
Raha Indon. 41 E8
Rahachow Belarus 13 F5
Rahad r. Sudan 32 D7
Rahaeng Thai. see Tak
Rahimyar Khan Pak. 33 L4
Rahovec Kosovo 27 I3
Rahuri India 38 B2
Raiatea i. Fr. Polynesia 75 J7
Raichur India 38 C2
Raiganj India 37 G4
Raigarh Chhattisgarh India 37 E5
Raigarh Orissa India 38 D2
Raijua i. Indon. 54 C2
Railroad Valley val. NV U.S.A. 65 E1
Raimangal r. Bangl. 37 G5
Rainbow Lake Canada 60 G4
Raine Island Australia 56 D1
Rainelle WV U.S.A. 64 A4
Raini r. Pak. 36 B3
Rainier, Mount vol. U.S.A. 62 C2
Rainy Lake Canada/U.S.A. 63 I2
Rainy River Canada 63 I2
Raipur Chhattisgarh India 37 E5
Raipur W. Bengal India 37 F5
Raisen India 36 D5
Raisio Fin. 15 M6
Raitalai India 36 D5
Raivavae i. Fr. Polynesia 75 K7
Raiwind Pak. 36 C3
Rajahmundry India 38 D2
Raja-Jooseppi Fin. 14 P2
Rajanpur Pak. 36 B3
Rajapalayam India 38 C4
Rajapur India 38 B2
Rajasthan state India 36 C4
Rajasthan Canal India 36 C3
Rajauri India see Rajouri
Rajevadi India 38 B2
Rajgarh India 36 D4
Rājijovsset Fin. see Raja-Jooseppi
Rajkot India 36 B5
Raj Mahal India 36 C4
Rajmahal Hills India 37 F4
Raj Nandgaon India 36 E5
Rajouri India 36 C2
Rajpipla India 36 C5
Rajpur India 36 C5
Rajpura India 36 D3
Rajputana Agency state India see Rajasthan
Rajsamand India 36 C4
Rajshahi Bangl. 37 G4
Rājū Syria 39 C1
Rajula India 38 A1
Rajur India 38 C1
Rajura India 38 C2
Raka China 37 F3
Rakaposhi mt. Pak. 36 C1
Raka Zangbo r. China see Dogxung Zangbo
Rakhiv Ukr. 13 E6
Rakhni Pak. 33 K3
Rakhni r. Pak. 36 B3
Rakitnoye Belgorodskaya Oblast' Rus. Fed. 13 G6
Rakitnoye Primorskiy Kray Rus. Fed. 44 D3
Rakiura i. N.Z. see Stewart Island
Rakiura National Park N.Z. 59 A8
Rakke Estonia 15 O7
Rakkestad Norway 15 G7
Rakovski Bulg. 27 K3
Rakushechnyy, Mys pt Kazakh. 35 H2
Rakvere Estonia 15 O7
Raleigh U.S.A. 63 L4
Ralston PA U.S.A. 64 C2
Ramagiri India 38 E2
Ramalho, Serra do hills Brazil 71 B1
Rāmallāh West Bank 39 B4
Ramanagaram India 38 C3
Ramanathapuram India 38 C4
Ramapo Deep sea feature N. Pacific O. 74 F3
Ramapur India 38 D1
Ramas, Cape India 38 B3
Ramatlabama S. Africa 51 G3
Rambhapur India 36 C5
Rambutyo Island P.N.G. 52 E2
Rame Head hd Australia 58 D6
Rame Head hd U.K. 19 C8
Rameshki Rus. Fed. 12 H4
Ramgarh Jharkhand India 37 F5
Ramgarh Rajasthan India 36 B4
Ramgarh Rajasthan India 36 C3
Rāmhormoz Iran 35 H5
Ramingining Australia 54 F3
Ramla Israel 39 B4
Ramlat Rabyānah des. Libya see Rebiana Sand Sea
Ramm, Jabal mts Jordan 39 B5
Ramnad India see Ramanathapuram
Râmnicu Sărat Romania 27 L2
Râmnicu Vâlcea Romania 27 K2
Ramon' Rus. Fed. 13 H6
Ramona CA U.S.A. 65 D4
Ramotswa Botswana 51 G3
Rampart of Genghis Khan tourist site 43 K3
Rampur India 36 D3
Rampur Boalia Bangl. see Rajshahi
Ramree Island Myanmar 37 H6
Rāmsar Iran 35 H3
Ramsele Sweden 14 J5
Ramsey Isle of Man 18 C4
Ramsey U.K. 19 G6
Ramsey NJ U.S.A. 64 D2
Ramsey Bay Isle of Man 18 C4
Ramsey Island U.K. 19 B7
Ramsgate U.K. 19 I7
Rāmshir Iran 35 H5
Ramsing mt. India 37 H3
Ramu Bangl. 37 H5
Ramygala Lith. 15 N9
Ranaghat India 37 G5
Rana Pratap Sagar resr India 36 C4
Ranapur India 36 C5
Ranasar India 36 B4
Rancagua Chile 70 B4
Rancharia Brazil 71 A3

nchi India 37 F5
nco, Lago l. Chile 70 B6
nd Australia 58 C5
ndalstown U.K. 21 F3
nders Denmark 15 G8
ndijaure l. Sweden 14 K3
ndsjö Sweden 14 H5
neå Sweden 14 M4
nérou Senegal 46 B3
nfurly N.Z. 59 C7
ngamati Bangl. 37 H5
ngapara India 37 H4
ngiora N.Z. 59 D6
ngitata r. N.Z. 59 C7
ngitikei r. N.Z. 59 E5
ngke China see Zamtang
ngôn Myanmar see Rangoon
ngoon Myanmar 31 I5
ngpur Bangl. 37 G4
ngse Myanmar 37 H4
nibennur India 38 B3
niganj India 37 F5
nipur Pak. 36 B4
niwara India 36 C4
nkin's Springs Australia 58 C4
nna Estonia 15 O7
nnes Australia 56 E5
nnoch, Loch U.K. 20 E4
nong Thai. 31 I6
npur India 36 B5
nsa Iran 35 H4
nsby Sweden 15 H6
ntasalmi Fin. 14 P5
ntauprapat Indon. 41 B7
ntsila Fin. 14 N4
nua Fin. 14 O4
nya Iraq 35 G3
nyah, Wādī watercourse Saudi Arabia 32 F5
o Go mt. Laos/Vietnam 42 J9
ohe China 44 D3
oul Island Kermadec Is 53 I4
pa i. Fr. Polynesia 75 K7
pa-iti i. Fr. Polynesia see Rapa
pallo Italy 26 C2
par India 36 B5
phoe Ireland 21 E3
pidan r. VA U.S.A. 64 C3
pid City U.S.A. 62 G3
pla Estonia 15 N7
pur Andhra Prad. India 38 C3
pur Gujarat India 36 B5
qqa Syria see Ar Raqqah
ra National Park Nepal 37 E3
ritan Bay NJ U.S.A. 64 D2
rkan Pak. 36 B3
roia atoll Fr. Polynesia 75 K7
rotonga i. Cook Is 75 J7
s India 36 C4
sa, Punta pt Arg. 70 D6
s ad Daqm Oman 33 I6
s al Ḥikmah Egypt 34 B5
s al Khaimah U.A.E. see Ra's al Khaymah
's al Khaymah U.A.E. 33 I4
's an Naqb Jordan 39 B4
s Dashen mt. Eth. see Ras Dejen
s Dejen mt. Eth. 48 D2
seiniai Lith. 15 M9
s el Hikma Egypt see Ra's al Ḥikmah
's Ghārib Egypt 34 D5
shad Sudan 32 D7
shīd Egypt see Rashīd
shīd Egypt 34 C5
shīd Qal'eh Afgh. 36 A3
sht Iran 35 H3
skam mts China 36 C1
skoh mts Pak. 33 K4
so, Cabo c. Arg. 70 C6
so da Catarina hills Brazil 69 K5
son Lake imp. l. Australia 55 D7
sony Belarus 15 P9
sra India 37 E4
sshua, Ostrov i. Rus. Fed. 43 S3
ss Jebel Tunisia 26 D6
sskazovo Rus. Fed. 13 I5
sul Pak. 36 C2
tae U.K. see Leicester
tan Sweden 14 I5
tanda S. Africa 51 I4
tangarh India 36 C3
tansbyn Sweden 14 I5
t Buri Thai. 31 I5
thangan Ireland 21 F4
thdowney Ireland 21 E5
thdrum Ireland 21 F5
thedaung Myanmar 37 H5
thenow Germany 17 N4
thfriland U.K. 21 F3
thkeale Ireland 21 D5
thlin Island U.K. 21 F2
tibor Poland see Racibórz
tisbon Germany see Regensburg
tiya India 36 C3
tlam India 36 D5
tnagiri India 38 B2
tnapura Sri Lanka 38 D5
tne Ukr. 13 E6
tno Ukr. see Ratne
ton U.S.A. 62 G4
tray Head hd U.K. 20 H3
ttvik Sweden 15 I6
uðamýri Iceland 14 [inset]
udhatain Kuwait 35 G5
ufarhöfn Iceland 14 [inset]
ukumara Range mts N.Z. 59 F4
ul Soares Brazil 71 C3
uma Fin. 15 L6
urkela India 37 F5
uschen Rus. Fed. see Svetlogorsk
usu Japan 44 G3
utavaara Fin. 14 P5
utjärvi Fin. 15 P6
vānsar Iran 35 G4
var Iran 33 I3
vena NY U.S.A. 64 E1
venglass U.K. 18 D4
venna Italy 26 E2
venna OH U.S.A. 64 A2
vensburg Germany 17 L7
venshoe Australia 56 D3
venswood Australia 56 D4
vi r. Pak. 36 B3
wah Iraq 35 F4
waki i. Kiribati 53 I2
walpindi Pak. 33 L3
wāndiz Iraq 35 G3
wicz Poland 17 P5
wlinna Australia 55 D7
wlins U.S.A. 62 F3
wlinson Range hills Australia 55 E6
wson Arg. 70 C6
wu China 42 H7
xón, Cerro mt. Guat. 66 G5
yachoti India 38 C3
yadurg India 38 C3
yagada India 38 D2
yagarha India see Rayagada
yak Lebanon 39 C3
ychikhinsk Rus. Fed. 44 C2
ydah Yemen 32 F6
yes Peak CA U.S.A. 65 C3
Rayevskiy Rus. Fed. 11 Q5
Rayleigh U.K. 19 H7
Raymond NH U.S.A. 64 F1
Raymond Terrace Australia 58 E4
Raymondville U.S.A. 62 H6
Rayner Glacier Antarctica 76 D2
Raystown Lake PA U.S.A. 64 B2
Raz, Pointe du pt France 24 B2
Razan Iran 35 H4
Rāzān Iran 35 H4
Razani Pak. 36 B2
Razāzah, Buḥayrat ar l. Iraq 35 F4
Razdan Armenia see Hrazdan
Razdel'naya Ukr. see Rozdil'na
Razdol'noye Rus. Fed. 44 C4
Razeh Iran 35 H4
Razgrad Bulg. 27 L3
Razim, Lacul lag. Romania 27 M2
Razlog Bulg. 27 J4
Razmak Pak. 36 B2
Raz"yezd 3km Rus. Fed. see Novyy Urgal
Ré, Île de i. France 24 D3
Reading U.K. 19 G7
Reading PA U.S.A. 64 D2
Reagile S. Africa 51 H3
Realicó Arg. 70 D5
Réalmont France 24 F5
Reate Italy see Rieti
Rebecca, Lake imp. l. Australia 55 C7
Rebiana Sand Sea des. Libya 47 F2
Reboly Rus. Fed. 14 Q5
Rebrikha Rus. Fed. 42 E2
Rebun-tō i. Japan 44 F3
Recherche, Archipelago of the is Australia 55 C8
Rechitsa Belarus see Rechytsa
Rechna Doab lowland Pak. 36 C3
Rechytsa Belarus 13 F5
Recife Brazil 69 L5
Recife, Cape S. Africa 51 G8
Recklinghausen Germany 17 K5
Reconquista Arg. 70 E3
Recreo Arg. 70 C3
Red r. Australia 56 C3
Red r. U.S.A. 63 I5
Red r. Vietnam 42 J8
Red Bank NJ U.S.A. 64 D2
Red Basin China see Sichuan Pendi
Red Bluff U.S.A. 62 C3
Redcar U.K. 18 F4
Redcliffe, Mount h. Australia 55 C7
Red Cliffs Australia 57 C7
Red Deer Canada 62 E1
Red Deer r. Alberta/Saskatchewan Canada 62 E1
Red Deer Lake Canada 62 G1
Reddersburg S. Africa 51 H5
Redding U.S.A. 62 C3
Redditch U.K. 19 F6
Rede r. U.K. 18 E3
Redenção Brazil 69 H5
Redeyef Tunisia 26 C7
Redfield U.S.A. 62 H3
Red Hook NY U.S.A. 64 E2
Redkino Rus. Fed. 12 H4
Red Lake Canada 63 I1
Red Lakes U.S.A. 63 I2
Redlands CA U.S.A. 65 D3
Red Lion PA U.S.A. 64 C3
Red Oak U.S.A. 63 H3
Redondo Port. 25 C4
Redondo Beach CA U.S.A. 65 C4
Red Rock PA U.S.A. 64 C2
Red Sea Africa/Asia 32 D4
Red Wing U.S.A. 63 I3
Redwood City CA U.S.A. 65 A2
Redwood Falls U.S.A. 63 H3
Ree, Lough l. Ireland 21 E4
Reedley CA U.S.A. 65 C2
Reedville VA U.S.A. 64 C4
Reedy WV U.S.A. 64 A3
Reedy Glacier Antarctica 76 J1
Reefton N.Z. 59 C6
Refahiye Turkey 34 E3
Regen Germany 17 N6
Regência Brazil 71 D2
Regensburg Germany 17 N6
Reggane Alg. 46 D2
Reggio Calabria Italy see Reggio di Calabria
Reggio Emilia-Romagna Italy see Reggio nell'Emilia
Reggio di Calabria Italy 26 F5
Reggio Emilia Italy see Reggio nell'Emilia
Reggio nell'Emilia Italy 26 D2
Reghin Romania 27 K1
Regina Canada 62 G1
Régina Fr. Guiana 69 H3
Registro Brazil 70 G2
Registro do Araguaia Brazil 71 A1
Regium Lepidum Italy see Reggio nell'Emilia
Regozero Rus. Fed. 14 Q4
Rehli India 36 D5
Rehoboth Namibia 50 C2
Rehoboth Bay DE U.S.A. 64 D3
Rehovot Israel 39 B4
Reibell Alg. see Ksar Chellala
Reichshoffen France 24 H2
Reid Australia 55 E7
Reidh, Rubha pt U.K. 20 D3
Reigate U.K. 19 G7
Reims France 24 G2
Reinbek Germany 17 M4
Reindeer r. Canada 60 H4
Reindeer Island Canada 62 H1
Reindeer Lake Canada 61 H4
Reine Norway 14 H3
Reinosa Spain 25 D2
Reipholsfjöll h. Iceland 14 [inset]
Reisaelva r. Norway 14 L2
Reisa Nasjonalpark nat. park Norway 14 M2
Reisjärvi Fin. 14 N5
Reitz S. Africa 51 I4
Rekapalle India 38 D2
Relizane Alg. 25 G6
Remarkable, Mount h. Australia 57 B7
Remeshk Iran 33 I4
Remhoogte Pass Namibia 50 C2
Remi France see Reims
Rena Norway 15 G6
Renapur India 38 C2
Rendsburg Germany 17 L3
Renfrew U.K. 20 E5
Rengali Reservoir India 37 F5
Rengo Chile 70 B4
Reni Ukr. 27 M2
Renick WV U.S.A. 64 A4
Renland reg. Greenland see Tuttut Nunaat
Rennell i. Solomon Is 53 G3
Rennes France 24 D2
Rennick Glacier Antarctica 76 H2
Reno r. Italy 26 E2
Reno U.S.A. 62 D4
Renovo PA U.S.A. 64 C2
Réo Burkina Faso 46 C3
Reo Indon. 54 C2
Repalle India 38 D2
Repolka Rus. Fed. 15 P7
Republican r. U.S.A. 62 H4
Republic of South Africa country Africa 50 F5
Repulse Bay b. Australia 56 E4
Repulse Bay Canada 61 J3
Requena Peru 68 D5
Requena Spain 25 F4
Reşadiye Turkey 34 E2
Reserva Brazil 71 A4
Reshm Iran 35 I4
Reshteh-ye Alborz mts Iran see Elburz Mountains
Resistencia Arg. 70 E3
Reşita Romania 27 I2
Resolute Canada 61 I2
Resolution Island Canada 61 L3
Resolution Island N.Z. 59 A7
Resplendor Brazil 71 C2
Resülayn Turkey see Ceylanpınar
Retalhuleu Guat. 66 F6
Retezat, Parcul Național nat. park Romania 27 J2
Retford U.K. 18 G5
Rethel France 24 G2
Réthimnon Greece see Rethymno
Rethymno Greece 27 K7
Retreat Australia 56 C5
Réunion terr. Indian Ocean 73 L7
Reus Spain 25 G3
Reutlingen Germany 17 L6
Reval Estonia see Tallinn
Revda Rus. Fed. 12 G2
Revel Estonia see Tallinn
Revel France 24 F5
Revillagigedo, Islas is Mex. 66 B5
Revillagigedo Island U.S.A. 60 E4
Revivim Israel 39 B4
Rewa India 36 E4
Rewari India 36 D3
Rexburg U.S.A. 62 E3
Reyes, Point CA U.S.A. 65 A1
Reyhanlı Turkey 39 C1
Reykir Iceland 14 [inset]
Reykjanes Ridge sea feature N. Atlantic Ocean 72 F2
Reykjanestá pt Iceland 14 [inset]
Reykjavík Iceland 14 [inset]
Reyneke, Ostrov i. Rus. Fed. 44 E1
Reynolds Range mts Australia 54 F5
Reynosa Mex. 66 E3
Rezā Iran 35 I4
Reẓā'īyeh Iran see Urmia
Reẓā'īyeh, Daryācheh-ye salt l. Iran see Urmia, Lake
Rēzekne Latvia 15 O8
Rezvāndeh Iran see Rezvānshahr
Rezvānshahr Iran 35 H3
Rhaeader Gwy U.K. see Rhayader
Rhayader U.K. 19 D6
Rhegium Italy see Reggio di Calabria
Rheims France see Reims
Rhein r. Germany see Rhine
Rheine Germany 17 K4
Rhemilès well Alg. 46 C2
Rhin r. France see Rhine
Rhine r. France 24 I2
Rhine r. Germany 17 K5
Rhinebeck NY U.S.A. 64 E2
Rhinelander U.S.A. 63 J2
Rhiwabon U.K. see Ruabon
Rho Italy 26 C2
Rhode Island state U.S.A. 64 F2
Rhodes Greece 27 M6
Rhodes i. Greece 27 M6
Rhodesia country Africa see Zimbabwe
Rhodope Mountains Bulg./Greece 27 J4
Rhodus i. Greece see Rhodes
Rhône r. France/Switz. 24 G5
Rhum i. U.K. see Rum
Rhuthun U.K. see Ruthin
Rhydaman U.K. see Ammanford
Rhyl U.K. 18 D5
Riachão Brazil 69 I5
Riacho Brazil 71 C2
Riacho de Santana Brazil 71 C1
Riacho dos Machados Brazil 71 C1
Rialma Brazil 71 A1
Rialto CA U.S.A. 65 D3
Riasi India 36 C2
Riau, Kepulauan is Indon. 41 C7
Ribadeo Spain 25 C2
Ribadesella Spain 25 D2
Ribas do Rio Pardo Brazil 70 F2
Ribáuè Moz. 49 D5
Ribble r. U.K. 18 E5
Ribblesdale val. U.K. 18 E4
Ribe Denmark 15 F9
Ribeira r. Brazil 71 B4
Ribeirão Preto Brazil 71 B3
Ribérac France 24 E4
Riberalta Bol. 68 E6
Rîbnița Moldova 13 F7
Ribnitz-Damgarten Germany 17 N3
Říčany Czech Rep. 17 O6
Rice VA U.S.A. 64 B4
Richards Bay S. Africa 51 K5
Richards Inlet Antarctica 76 H1
Richards Island Canada 60 E3
Richardson Mountains Canada 60 E3
Richardson Mountains N.Z. 59 B7
Richfield U.S.A. 62 E4
Richfield Springs NY U.S.A. 64 D1
Richford NY U.S.A. 64 C1
Richgrove CA U.S.A. 65 C3
Richland U.S.A. 62 D2
Richmond N.S.W. Australia 58 E4
Richmond Qld Australia 56 C4
Richmond N.Z. 59 D5
Richmond KwaZulu-Natal S. Africa 51 J5
Richmond N. Cape S. Africa 50 F6
Richmond U.K. 18 F4
Richmond CA U.S.A. 65 A2
Richmond IN U.S.A. 63 K4
Richmond KY U.S.A. 63 K4
Richmond VA U.S.A. 64 C4
Richmond Range hills Australia 58 F2
Richtersveld National Park S. Africa 50 C5
Richwood WV U.S.A. 64 A3
Ricomagus France see Riom
Riddell Nunataks Antarctica 76 E2
Rideau Lakes Canada 63 L3
Ridgecrest CA U.S.A. 65 D3
Ridgway PA U.S.A. 64 B2
Riecito Venez. 68 E1
Riesa Germany 17 N5
Riesco, Isla i. Chile 70 B8
Riet watercourse S. Africa 50 E6
Rietavas Lith. 15 L9
Rietfontein S. Africa 50 E4
Rieti Italy 26 E3
Rifā'ī, Tall mt. Jordan/Syria 39 C3
Rifstangi pt Iceland 14 [inset]
Rift Valley Lakes National Park Eth. see Abijatta-Shalla National Park
Riga Latvia 15 N8
Riga, Gulf of Estonia/Latvia 15 M8
Rigain Pünco l. China 37 F2
Rīgān Iran 33 I4
Rigas jūras līcis b. Estonia/Latvia see Riga, Gulf of
Rigby U.S.A. 62 E3
Rigside U.K. 20 F5
Riia laht b. Estonia/Latvia see Riga, Gulf of
Riihimäki Fin. 15 N6
Riiser-Larsen Ice Shelf Antarctica 76 B2
Rijau Nigeria 46 D3
Rijeka Croatia 26 F2
Rikuchū-kaigan Kokuritsu-kōen Japan 45 F5
Rikuzen-takata Japan 45 F5
Rila mts Bulg. 27 J3
Rila China 37 F3
Rileyville VA U.S.A. 64 B3
Rillieux-la-Pape France 24 G4
Rimah, Wādī al watercourse Saudi Arabia 32 F4
Rimavská Sobota Slovakia 17 R6
Rimini Italy 26 E2
Râmnicu Sărat Romania see Râmnicu Sărat
Râmnicu Vâlcea Romania see Râmnicu Vâlcea
Rimouski Canada 63 N2
Rimsdale, Loch l. U.K. 20 E2
Rinbung China 37 G3
Rincão Brazil 71 A3
Rindal Norway 14 F5
Ringarooma Bay Australia 57 [inset]
Ringas India 36 C4
Ringebu Norway 15 G6
Ringkøbing Denmark 15 F8
Ringsend U.K. 21 F2
Ringsted Denmark 15 G9
Ringtor China 37 E3
Ringvassøya i. Norway 14 K2
Ringwood Australia 58 B6
Ringwood U.K. 19 F8
Rinns Point U.K. 20 C5
Rinqênzê China 37 G3
Río Abiseo, Parque Nacional nat. park Peru 68 C5
Rio Azul Brazil 71 A4
Riobamba Ecuador 68 C4
Rio Bonito Brazil 71 C3
Rio Branco Brazil 68 E6
Rio Branco, Parque Nacional do nat. park Brazil 68 F3
Rio Brilhante Brazil 70 F2
Rio Casca Brazil 71 C3
Rio Claro Brazil 71 B3
Río Colorado Arg. 70 D5
Río Cuarto Arg. 70 D4
Rio das Pedras Moz. 51 L2
Rio de Contas Brazil 71 C1
Rio de Janeiro Brazil 71 C3
Rio de Janeiro state Brazil 71 C3
Río de la Plata-Paraná r. S. America 70 E4
Rio do Sul Brazil 71 A4
Río Gallegos Arg. 70 C8
Río Grande Arg. 70 C8
Rio Grande Brazil 70 F4
Río Grande Mex. 66 D4
Rio Grande r. Mex./U.S.A. 62 H6
Rio Grande City U.S.A. 62 H6
Rio Grande do Sul state Brazil 71 A5
Rio Grande Rise sea feature S. Atlantic Ocean 72 F8
Ríohacha Col. 68 D1
Río Hondo, Embalse resr Arg. 70 C3
Rioja Peru 68 C5
Río Lagartos Mex. 66 G4
Rio Largo Brazil 69 K5
Riom France 24 F4
Rio Manso, Represa do resr Brazil 69 G6
Rio Mulatos Bol. 68 E7
Río Muni reg. Equat. Guinea 46 E4
Río Negro, Embalse del resr Uruguay 70 E4
Rioni r. Georgia 35 F2
Rio Novo Brazil 71 C3
Rio Pardo de Minas Brazil 71 C1
Rio Preto Brazil 71 C3
Rio Preto, Serra do hills Brazil 71 B2
Rio Rancho U.S.A. 62 F4
Río Tigre Ecuador 68 C4
Rio Verde Brazil 71 A2
Rio Verde de Mato Grosso Brazil 69 H7
Rio Vista CA U.S.A. 65 B1
Ripky Ukr. 13 F6
Ripley England U.K. 18 F4
Ripley England U.K. 19 F5
Ripley NY U.S.A. 64 B1
Ripoll Spain 25 H2
Ripon U.K. 18 F4
Ripon CA U.S.A. 65 B2
Ripu India 37 G4
Risca U.K. 19 D7
Rishiri-tō i. Japan 44 F3
Rishon LeẔiyyon Israel 39 B4
Rising Sun MD U.S.A. 64 C3
Risle r. France 19 H9
Risør Norway 15 F7
Rissa Norway 14 F5
Ristiina Fin. 15 O6
Ristijärvi Fin. 14 P4
Ristikent Rus. Fed. 12 F1
Risum China 36 D2
Ritchie S. Africa 50 G5
Ritscher Upland mts Antarctica 76 B2
Ritsem Sweden 14 J3
Ritter, Mount CA U.S.A. 65 C2
Ritzville U.S.A. 62 D2
Riva del Garda Italy 26 D2
Rivas Nicaragua 67 G6
Rivera Arg. 70 D5
Rivera Uruguay 70 E4
River Cess Liberia 46 C4
Riverhead NY U.S.A. 64 E2
Riverhurst Canada 62 F1
Riverina Australia 55 C7
Riverina reg. Australia 58 B5
Riversdale S. Africa 50 E8
Riverside S. Africa 51 I6
Riverside CA U.S.A. 65 D4
Riversleigh Australia 56 B3
Riverton N.Z. 59 B8
Riverton VA U.S.A. 64 B3
Riverton WY U.S.A. 62 F3
Riverview Canada 63 O2
Rivesaltes France 24 F5
Rivière-du-Loup Canada 63 N2
Rivne Ukr. 13 E6
Rivungo Angola 49 C5
Riwaka N.Z. 59 D5
Riwoqê China see Racaka
Riyadh Saudi Arabia 32 G5
Riza well Iran 35 I4
Rize Turkey 35 F2
Rizokarpaso Cyprus see Rizokarpason
Rizokarpason Cyprus 39 B2
Rjukan Norway 15 F7
Rjuvbrokkene mt. Norway 15 E7
Rkîz Mauritania 46 B3
Roa Norway 15 G6
Roach Lake NV U.S.A. 65 E3
Roade U.K. 19 G6
Road Town Virgin Is (U.K.) 67 L5
Roan Norway 14 G4
Roan Fell h. U.K. 20 G5
Roanne France 24 G3
Roanoke VA U.S.A. 64 B4
Roanoke Rapids U.S.A. 63 L4
Roaring Spring PA U.S.A. 64 B2
Roaringwater Bay Ireland 21 C6
Roatán Hond. 67 G5
Röbäck Sweden 14 L5
Robāṭak Afgh. 36 B1
Robāṭ-e Posht-e Bādām Iran 35 I4
Robāṭ Karīm Iran 35 H4
Robāṭ Tork Iran 35 H4
Robbins Island Australia 57 [inset]
Robe Australia 57 B8
Robe r. Australia 54 A5
Robe r. Ireland 21 C4
Robert-Bourassa, Réservoir resr Canada 61 K4
Robert Glacier Antarctica 76 D2
Roberts, Mount Australia 58 F2
Roberts Butte mt. Antarctica 76 H2
Robertsfors Sweden 14 L4
Robertsganj India 37 E4
Robertson S. Africa 50 D7
Robertson Bay Antarctica 76 H2
Robertson Island Antarctica 76 A2
Robertson Range hills Australia 55 C5
Robertsport Liberia 46 B4
Roberval Canada 63 M2
Robhanais, Rubha hd U.K. see Butt of Lewis
Robin Hood's Bay U.K. 18 G4
Robinson Ranges hills Australia 55 B6
Robinson River Australia 56 B3
Robson, Mount Canada 62 D1
Roçadas Angola see Xangongo
Rocca Busambra mt. Sicilia Italy 26 E6
Rocha Uruguay 70 F4
Rochdale U.K. 18 E5
Rochechouart France 24 E4
Rochefort France 24 D4
Rochegda Rus. Fed. 12 I3
Rochester Australia 58 B6
Rochester U.K. 19 H7
Rochester MN U.S.A. 63 I3
Rochester NH U.S.A. 64 F1
Rochester NY U.S.A. 64 C1
Rochford U.K. 19 H7
Roc'h Trévezel h. France 24 C2
Rockall i. N. Atlantic Ocean 10 D4
Rockall Bank sea feature N. Atlantic Ocean 72 G2
Rock Creek OH U.S.A. 64 A2
Rockefeller Plateau Antarctica 76 J1
Rockford IL U.S.A. 63 J3
Rockhampton Australia 56 E4
Rockhampton Downs Australia 54 F4
Rockingham Australia 55 A8
Rockingham Bay Australia 56 D3
Rock Island U.S.A. 63 I3
Rockland MA U.S.A. 64 F1
Rocksprings U.S.A. 62 G6
Rockstone Guyana 69 G2
Rockville CT U.S.A. 64 E2
Rockville MD U.S.A. 64 C3
Rockwood PA U.S.A. 64 B3
Rocky Mount VA U.S.A. 64 B4
Rocky Mountains Canada/U.S.A. 62 F3
Rodberg Norway 15 F6
Rødbyhavn Denmark 15 G9
Rodeio Brazil 71 A4
Rodel U.K. 20 C3
Rodeo Arg. 70 C4
Rodez France 24 F4
Ródhos i. Greece see Rhodes
Rodi i. Greece see Rhodes
Rodney, Cape U.S.A. 60 B3
Rodniki Rus. Fed. 12 I4
Rodopi Planina mts Bulg./Greece see Rhodope Mountains
Rodos Greece see Rhodes
Rodos i. Greece see Rhodes
Rodosto Turkey see Tekirdağ
Rodrigues Island Mauritius 73 M7
Roe r. U.K. 21 F2
Roebourne Australia 54 B5
Roebuck Bay Australia 54 C4
Roedtan S. Africa 51 I3
Roe Plains Australia 55 D7
Roeselare Belgium 16 I5
Roes Welcome Sound sea chan. Canada 61 J3
Rogachev Belarus see Rahachow
Rogers Lake CA U.S.A. 65 D3
Roggeveen Basin sea feature S. Pacific Ocean 75 O8
Roggeveld plat. S. Africa 50 E7
Roggeveldberge esc. S. Africa 50 E7
Roghadal U.K. see Rodel
Rognan Norway 14 I3
Roha India 38 B2
Rohnert Park CA U.S.A. 65 A1
Rohrbach in Oberösterreich Austria 17 N6
Rohri Sangar Pak. 36 B4
Rohtak India 36 D3
Roi Georges, Îles du is Fr. Polynesia 75 K6
Rois-Bheinn h. U.K. 20 D4
Roja Latvia 15 M8
Rojas Arg. 70 D4
Rokeby Australia 56 C2
Rokeby National Park Australia 56 C2
Rokiškis Lith. 15 N9
Roknäs Sweden 14 L4
Rokytne Ukr. 13 E6
Rolagang China 37 G2
Rola Kangri mt. China 37 G2
Rolândia Brazil 71 A3
Rolim de Moura Brazil 68 F6
Roll AZ U.S.A. 65 F4
Rolla MO U.S.A. 63 I4
Rollag Norway 15 F6
Rolleston Australia 56 E5
Roma Australia 57 E5
Roma Italy see Rome
Roma Lesotho 51 H5
Roma Sweden 15 K8
Romain, Cape U.S.A. 63 L5
Roman Romania 27 L1
Română, Câmpia plain Romania 27 J2
Romanche Gap sea feature S. Atlantic Ocean 72 G6
Romang, Pulau i. Indon. 41 E8
Romania country Europe 27 K2
Roman-Kosh mt. Ukr. 34 D1
Romanovka Rus. Fed. 43 K2
Romans-sur-Isère France 24 G4
Romanzof, Cape U.S.A. 60 B3
Rombas France 24 H2
Romblon Phil. 41 E6
Rome Italy 26 E4
Rome GA U.S.A. 63 J5
Rome NY U.S.A. 64 D1
Romford U.K. 19 H7
Romilly-sur-Seine France 24 F2
Romney WV U.S.A. 64 B3
Romney Marsh reg. U.K. 19 H7
Romny Ukr. 13 G6
Rømø i. Denmark 15 F9
Romodanovo Rus. Fed. 13 J5
Romorantin-Lanthenay France 24 E3
Romsey U.K. 19 F8
Ron India 38 B3
Rona i. U.K. 20 D1
Ronas Hill h. U.K. 20 [inset]
Roncador, Serra do hills Brazil 69 H6
Roncador Reef Solomon Is 53 F2
Ronda Spain 25 D5
Ronda, Serranía de mts Spain 25 D5
Rondane Nasjonalpark nat. park Norway 15 F6
Rondon Brazil 70 F2
Rondonópolis Brazil 69 H7
Rondout Reservoir NY U.S.A. 64 D2
Rong Chu r. China 37 G3
Rongelap atoll Marshall Is 74 H5
Rongklang Range mts Myanmar 37 H5
Rongwo China see Tongren
Rongyul China 37 I3
Rōnlap atoll Marshall Is see Rongelap
Rønne Denmark 15 I9
Ronneby Sweden 15 I8
Ronne Entrance str. Antarctica 76 L2
Ronne Ice Shelf Antarctica 76 L1
Rooke Island P.N.G. see Umboi
Roorkee India 36 D3
Roosendaal Neth. 16 J5
Roosevelt, Mount Canada 60 F4
Roosevelt Island Antarctica 76 I1
Ropar India see Rupnagar
Roper r. Australia 56 A2
Roper Bar Australia 54 F3
Roquefort France 24 D4
Roraima, Mount Guyana 68 F2
Rori India 36 C3
Røros Norway 14 G5
Rørvik Norway 14 G4
Rosamond CA U.S.A. 65 C3
Rosamond Lake CA U.S.A. 65 C3
Rosario Arg. 70 D4
Rosário Brazil 69 J4
Rosario Baja California Mex. 66 A2
Rosario Sinaloa Mex. 66 C4
Rosario Sonora Mex. 62 F6
Rosario Venez. 68 D1
Rosário do Sul Brazil 70 F4
Rosário Oeste Brazil 69 G6
Rosarito Baja California Mex. 62 E6
Rosarno Italy 26 F5
Roscoff France 24 C2
Roscommon Ireland 21 D4
Roscrea Ireland 21 E5
Rose r. Australia 56 A2
Rose Atoll American Samoa see Rose Island
Roseau Dominica 67 L5
Roseberth Australia 57 B5
Roseburg U.S.A. 62 C3
Rosedale Abbey U.K. 18 G4
Roseires Reservoir Sudan 32 D7
Rose Island atoll American Samoa 53 J3
Rosenberg U.S.A. 63 H6
Rosendal Norway 15 E7
Rosendal S. Africa 51 H5
Rosenheim Germany 17 N7
Roseto degli Abruzzi Italy 26 F3
Rosetown Canada 62 F1
Rosetta Egypt see Rashīd
Roseville CA U.S.A. 65 B1
Rosewood Australia 58 F1
Roshchino Rus. Fed. 15 P6
Rosh Pinah Namibia 50 C4
Roshtkala Tajik. see Roshtqal'a
Roshtqal'a Tajik. 36 B1
Rosignano Marittimo Italy 26 D3
Roșiori de Vede Romania 27 K2
Roskilde Denmark 15 H9
Roslavl' Rus. Fed. 13 G5
Roslyakovo Rus. Fed. 14 R2
Roslyatino Rus. Fed. 12 J4
Ross N.Z. 59 C6
Ross, Mount h. N.Z. 59 E5
Rossano Italy 26 G5
Rossan Point Ireland 21 D3
Rosscarbery Ireland 21 C6
Ross Dependency Antarctica 76 I2
Rossel Island P.N.G. 56 F1
Ross Ice Shelf Antarctica 76 I1
Rössing Namibia 50 B2
Ross Island Antarctica 76 H1
Rossiyskaya Sovetskaya Federativnaya Sotsialisticheskaya Respublika country Asia/Europe see Russian Federation
Rosslare Ireland 21 F5
Rosslare Harbour Ireland 21 F5
Rosso Mauritania 46 B3
Ross-on-Wye U.K. 19 E7
Rossony Belarus see Rasony
Rossosh' Rus. Fed. 13 H6
Ross River Canada 60 E3
Ross Sea Antarctica 76 H1
Røssvatnet l. Norway 14 I4
Rostāq Afgh. 36 B1
Rostāq Iran 35 I6
Rosthern Canada 62 F1
Rostock Germany 17 N3
Rostov Rus. Fed. 12 H4
Rostov-na-Donu Rus. Fed. 13 H7
Rostov-on-Don Rus. Fed. see Rostov-na-Donu
Rosvik Sweden 14 L4
Roswell U.S.A. 62 G5
Rota i. N. Mariana Is 41 G6
Rotch Island Kiribati see Tamana
Rote i. Indon. 41 E9
Roth Germany 17 M6
Rothbury U.K. 18 F3
Rothenburg ob der Tauber Germany 17 M6
Rother r. U.K. 19 G8
Rothera research stn Antarctica 76 L2
Rotherham U.K. 18 F5
Rothes U.K. 20 F3
Rothesay U.K. 20 D5
Rothwell U.K. 19 G6
Roti Indon. 54 C2
Roti i. Indon. see Rote
Roto Australia 58 B4
Rotomagus France see Rouen
Rotomanu N.Z. 59 C6
Rotondo, Monte mt. Corse France 24 I5
Rotorua N.Z. 59 F4
Rotorua, Lake N.Z. 59 F4
Rottenmann Austria 17 O7
Rotterdam Neth. 16 J5
Rottnest Island Australia 55 A8
Rottweil Germany 17 L6
Rotuma i. Fiji 53 H3
Rötviken Sweden 14 I5
Roubaix France 24 F1
Rouen France 24 E2
Roulers Belgium see Roeselare
Roumania country Europe see Romania
Round Hill h. U.K. 18 F4
Round Mountain Australia 58 F3
Roundup U.S.A. 62 F2
Rousay i. U.K. 20 F1
Rouxville S. Africa 51 H6
Rouyn-Noranda Canada 63 L2
Rovaniemi Fin. 14 N3
Roven'ki Rus. Fed. 13 H6
Rovereto Italy 26 D2
Rovigo Italy 26 D2
Rovinj Croatia 26 E2
Rovno Ukr. see Rivne
Rovnoye Rus. Fed. 13 J6
Rovuma r. Moz./Tanz. see Ruvuma
Rowena Australia 58 D2

Rowley Island Canada 61 K3
Rowley Shoals *sea feature* Australia 54 B4
Równe Ukr. *see* Rivne
Roxburgh N.Z. 59 B7
Roxburgh Island Cook Is *see* Rarotonga
Roxby Downs Australia 57 B6
Roxo, Cabo *c.* Senegal 46 B3
Royal Canal Ireland 21 E4
Royal Chitwan National Park Nepal 37 F4
Royale, Île *i.* Canada *see* Cape Breton Island
Royale, Isle *i.* U.S.A. 63 J2
Royal Natal National Park S. Africa 51 I5
Royal National Park Australia 58 E5
Royal Sukla Phanta Wildlife Reserve Nepal 36 E3
Royan France 24 D4
Roy Hill Australia 54 B5
Royston U.K. 19 G6
Rozdil'na Ukr. 27 N1
Rozivka Ukr. 13 H7
Rtishchevo Rus. Fed. 13 I5
Ruabon U.K. 19 D6
Ruaha National Park Tanz. 49 D4
Ruahine Range *mts* N.Z. 59 F5
Ruanda *country* Africa *see* Rwanda
Ruapehu, Mount *vol.* N.Z. 59 E4
Ruapuke Island N.Z. 59 B8
Ruatoria N.Z. 59 G3
Ruba Belarus 13 F5
Rub' al Khālī *des.* Saudi Arabia 32 G6
Rubtsovsk Rus. Fed. 42 E2
Ruby U.S.A. 60 C3
Ruckersville *VA* U.S.A. 64 B3
Rudall River National Park Australia 54 C5
Rudarpur India 37 E4
Ruda Śląska Poland 17 Q5
Rudauli India 37 E4
Rūdbār Iran 35 H3
Rūd-e Arghandāb *r.* Afgh. 36 A3
Rudkøbing Denmark 15 G9
Rudnaya Pristan' Rus. Fed. 44 D3
Rudnichnyy Rus. Fed. 12 L4
Rudnya *Smolenskaya Oblast'* Rus. Fed. 13 F5
Rudnya *Volgogradskaya Oblast'* Rus. Fed. 13 J6
Rudnyy Kazakh. 30 F1
Rudolf, Lake *salt l.* Eth./Kenya *see* Turkana, Lake
Rudol'fa, Ostrov *i.* Rus. Fed. 28 G1
Rudolph Island Rus. Fed. *see* Rudol'fa, Ostrov
Rūdsar Iran 35 H3
Rue France 19 I8
Rufiji *r.* Tanz. 49 D4
Rufino Arg. 70 D4
Rufisque Senegal 46 B3
Rufunsa Zambia 49 C5
Rugao China 43 M6
Rugby U.K. 19 F6
Rugby U.S.A. 62 G2
Rugeley U.K. 19 F6
Rügen *i.* Germany 17 N3
Ruhengeri Rwanda 48 C4
Ruhnu *i.* Estonia 15 M8
Ruhuna National Park Sri Lanka 38 D5
Rui Barbosa Brazil 71 C1
Ruijin China 43 L7
Ruipa Tanz. 49 D4
Ruiz Mex. 66 C4
Ruiz, Nevado del *vol.* Col. 68 C3
Rujaylah, Ḥarrat ar *lava field* Jordan 39 C3
Rūjiena Latvia 15 N8
Ruk *is* Micronesia *see* Chuuk
Rukumkot Nepal 37 E3
Rukwa, Lake Tanz. 49 D4
Rum *i.* U.K. 20 C4
Rum, Jebel *mts* Jordan *see* Ramm, Jabal
Ruma Serbia 27 H2
Rumāh Saudi Arabia 32 G4
Rumania *country* Europe *see* Romania
Rumbek South Sudan 47 F4
Rum Cay *i.* Bahamas 67 J4
Rum Jungle Australia 54 E3
Rummānā *h.* Syria 39 D3
Rumphi Malawi 49 D5
Runaway, Cape N.Z. 59 F3
Runcorn U.K. 18 E5
Rundu Namibia 49 B5
Rundvik Sweden 14 K5
Rungwa Tanz. 49 D4
Rungwa *r.* Tanz. 49 D4
Runton Range *hills* Australia 55 C5
Ruokolahti Fin. 15 P6
Ruoqiang China 42 F5
Rupa India 37 H4
Rupert *r.* Canada 63 L1
Rupert *WV* U.S.A. 64 A4
Rupert Bay Canada 63 L1
Rupert Coast Antarctica 76 J1
Rupert House Canada *see* Waskaganish
Rupnagar India 36 D3
Rupshu *reg.* India 36 D2
Ruqqād, Wādī ar *watercourse* Israel 39 B3
Rural Retreat *VA* U.S.A. 64 A4
Rusaddir N. Africa *see* Melilla
Rusape Zimbabwe 49 D5
Ruschuk Bulg. *see* Ruse
Ruse Bulg. 27 K3
Rusera India 37 F4
Rushden U.K. 19 G6
Rushinga Zimbabwe 49 D5
Rushville *NE* U.S.A. 62 G3
Rushworth Australia 58 B6
Russell N.Z. 59 E2
Russell *PA* U.S.A. 64 B2
Russell Bay Antarctica 76 J2
Russell Range *hills* Australia 55 C8
Russellville *AR* U.S.A. 63 I4
Rüsselsheim Germany 17 L5
Russia *country* Asia/Europe *see* Russian Federation
Russian *r.* *CA* U.S.A. 65 A1
Russian Federation *country* Asia/Europe 28 I3
Russian Soviet Federal Socialist Republic *country* Asia/Europe *see* Russian Federation
Russkiy, Ostrov *i.* Rus. Fed. 44 C4
Russkiy Kameshkir Rus. Fed. 13 J5
Rust'avi Georgia 35 G2
Rustburg *VA* U.S.A. 64 B4
Rustenburg S. Africa 51 H3
Ruston U.S.A. 63 I5
Rutanzige, Lake Dem. Rep. Congo/Uganda *see* Edward, Lake
Ruteng Indon. 41 E8
Rutherglen Australia 58 C6
Ruther Glen *VA* U.S.A. 64 C4
Ruthin U.K. 19 D5
Ruthiyai India 36 D4
Rutka *r.* Rus. Fed. 12 J4
Rutland *VT* U.S.A. 64 E1
Rutland Water *resr* U.K. 19 G6
Rutög China *see* Dêrub
Rutog *Xizang* China 37 F3
Rutog *Xizang* China 42 G7
Rutul Rus. Fed. 35 G2
Ruukki Fin. 14 N4
Ruvuma *r.* Moz./Tanz. 49 E5
Ruwayshid, Wādī *watercourse* Jordan 39 C3
Ruwayṭah, Wādī *watercourse* Jordan 39 C5
Ruwenzori National Park Uganda *see* Queen Elizabeth National Park
Ruza Rus. Fed. 12 H5
Ruzayevka Kazakh. 30 F1
Ruzayevka Rus. Fed. 13 J5
Ružomberok Slovakia 17 Q6
Rwanda *country* Africa 48 C4
Ryābād Iran 35 I3
Ryan, Loch *b.* U.K. 20 D5
Ryazan' Rus. Fed. 13 H5
Ryazhsk Rus. Fed. 13 I5
Rybachiy, Poluostrov *pen.* Rus. Fed. 12 G1
Rybach'ye Kyrg. *see* Balykchy
Rybinsk Rus. Fed. 12 H4
Rybinskoye Vodokhranilishche *resr* Rus. Fed. 12 H4
Rybnik Poland 17 Q5
Rybnitsa Moldova *see* Rîbnița
Rybnoye Rus. Fed. 13 H5
Rybreka Rus. Fed. 12 G3
Ryd Sweden 15 I8
Rydberg Peninsula Antarctica 76 L2
Ryde U.K. 19 F8
Rye U.K. 19 H8
Rye *r.* U.K. 18 G4
Rye Bay U.K. 19 H8
Rykovo Ukr. *see* Yenakiyeve
Ryl'sk Rus. Fed. 13 G6
Rylstone Australia 58 D4
Ryn-Peski *des.* Kazakh. 11 P6
Ryukyu Islands Japan 45 B8
Ryūkyū-rettō *is* Japan *see* Ryukyu Islands
Ryukyu Trench *sea feature* N. Pacific Ocean 74 E4
Rzeszów Poland 13 D6
Rzhaksa Rus. Fed. 13 I5
Rzhev Rus. Fed. 12 G4

S

Sa'ādah al Barṣā' *pass* Saudi Arabia 39 C5
Saale *r.* Germany 17 M5
Saalfeld Germany 17 M5
Saarbrücken Germany 17 K6
Saaremaa *i.* Estonia 15 M7
Saarenkylä Fin. 14 N3
Saarijärvi Fin. 14 N5
Saari-Kämä Fin. 14 O3
Saarikoski Fin. 14 L2
Saaristomeren kansallispuisto *nat. park* Fin. *see* Skärgårdshavets nationalpark
Saarlouis Germany 17 K6
Saatlı Azer. 35 H3
Saatly Azer. *see* Saatlı
Sab'a Egypt *see* Saba'ah
Saba'ah Egypt 39 A4
Sab' Ābār Syria 39 C3
Šabac Serbia 27 H2
Sabadell Spain 25 H3
Sabae Japan 45 E6
Sabana, Archipiélago de *is* Cuba 67 H4
Şabanözü Turkey 34 D2
Sabará Brazil 71 C2
Sabastiya West Bank 39 B3
Sab'atayn, Ramlat as *des.* Yemen 32 G6
Sabaudia Italy 26 E4
Sabaya Bol. 68 E7
Sabelo S. Africa 50 F6
Şabḥā Jordan 39 C3
Sabhā Libya 47 E2
Sabhrai India 36 B5
Sabi *r.* India 36 D3
Sabi *r.* Moz./Zimbabwe *see* Save
Sabie Moz. 51 K3
Sabie *r.* Moz./S. Africa 51 K3
Sabie S. Africa 51 J3
Sabinas Mex. 66 D3
Sabinas Hidalgo Mex. 66 D3
Sabini, Monti *mts* Italy 26 E3
Sabirabad Azer. 35 H2
Sabkhat al Bardawīl Reserve *nature res.* Egypt *see* Lake Bardawil Reserve
Sable, Cape Canada 61 L5
Sable, Cape U.S.A. 63 K6
Sable Island Canada 61 M5
Sabon Kafi Niger 46 D3
Sabrina Coast Antarctica 76 F2
Sabugal Port. 25 C3
Sabzawar Afgh. *see* Shīndand
Sabzevār Iran 33 I2
Sabzvārān Iran *see* Jīroft
Sacalinul Mare, Insula *i.* Romania 27 M2
Săcele Romania 27 K2
Sachigo Lake Canada 63 I1
Sachin India 36 C5
Sach'on S. Korea 45 C6
Sach Pass India 36 D2
Sachs Harbour Canada 60 F2
Sacirsuyu *r.* Syria/Turkey *see* Sājūr, Nahr
Saco *ME* U.S.A. 64 F1
Sacramento Brazil 71 B2
Sacramento *CA* U.S.A. 65 B1
Sacramento *r.* *CA* U.S.A. 65 B1
Sacramento Mountains U.S.A. 62 F5
Sada S. Africa 51 H7
Sádaba Spain 25 F2
Sá da Bandeira Angola *see* Lubango
Şadad Syria 39 C2
Ṣa'dah Yemen 32 F6
Saddat al Hindīyah Iraq 35 G4
Saddle Hill *h.* Australia 56 D2
Sadêng China 37 H3
Sadiola Mali 46 B3
Sadiqabad Pak. 36 B3
Sad Ishtragh *mt.* Afgh./Pak. 36 C1
Sa'dīyah, Hawr as *imp. l.* Iraq 35 G4
Sado *r.* Port. 25 B4
Sadoga-shima *i.* Japan 45 E5
Sadot Egypt *see* Sadūt
Sadovoye Rus. Fed. 13 J7
Sa Dragonera *i.* Spain 25 H4
Sadras India 38 D3
Sadūt Egypt *see* Sadūt
Sadūt Egypt 39 B4
Sæby Denmark 15 G8
Saena Julia Italy *see* Siena
Safad Israel *see* Zefat
Safāshahr Iran 35 I5
Safayal Maqūf *well* Iraq 35 G5
Safed Koh *mts* Afgh./Pak. 36 B2
Saffānīyah, Ra's as *hd* Saudi Arabia 35 H5
Säffle Sweden 15 H7
Safford U.S.A. 62 F5
Saffron Walden U.K. 19 H6
Safi Morocco 22 C5
Safīdār, Kūh-e *mt.* Iran 35 I5
Safid Kūh *mts* Afgh. *see* Safīd Kūh
Safīd Kūh *mts* Afgh. 33 J2
Safiras, Serra das *mts* Brazil 71 C2
Şāfītā Syria 39 C2
Safonovo *Arkhangel'skaya Oblast'* Rus. Fed. 12 K2
Safonovo *Smolenskaya Oblast'* Rus. Fed. 13 G5
Safrā' as Sark *esc.* Saudi Arabia 32 F4
Safranbolu Turkey 34 D2
Saga China 37 F3
Saga Japan 45 C6
Sagaing Myanmar 37 H5
Sagami-nada *g.* Japan 45 E6
Sagamore *PA* U.S.A. 64 B2
Sagar *Karnataka* India 38 B3
Sagar *Karnataka* India 38 C2
Sagar *Madh. Prad.* India 36 D5
Sagaredzho Georgia *see* Sagarejo
Sagarejo Georgia 35 G2
Sagar Island India 37 G5
Sagarmatha National Park Nepal 37 F4
Sagastyr Rus. Fed. 29 N2
Sagavanirktok *r.* U.S.A. 60 D2
Saggi, Har *mt.* Israel 39 B4
Sāghand Iran 35 I4
Saginaw U.S.A. 63 K3
Saginaw Bay U.S.A. 63 K3
Saglouc Canada *see* Salluit
Sagone, Golfe de *b.* *Corse* France 24 I5
Sagres Port. 25 B5
Sagthale India 36 C5
Sagua la Grande Cuba 67 H4
Saguenay *r.* Canada 63 N2
Sagunt Spain *see* Sagunto
Sagunto Spain 25 F4
Saguntum Spain *see* Sagunto
Sahagún Spain 25 D2
Sahand, Kūh-e *mt.* Iran 35 G3
Sahara *des.* Africa 46 D3
Şaḥara el Gharbîya *des.* Egypt *see* Western Desert
Şaḥara el Sharqîya *des.* Egypt *see* Eastern Desert
Saharan Atlas *mts* Alg. *see* Atlas Saharien
Saharanpur India 36 D3
Sahara Well Australia 54 C5
Saharsa India 37 F4
Sahaswan India 36 D3
Sahat, Kūh-e *h.* Iran 35 I4
Sahatwar India 37 F4
Şahbuz Azer. 35 G3
Sahdol India *see* Shahdol
Sahebganj India *see* Sahibganj
Sahebgunj India *see* Sahibganj
Saheira, Wâdi el *watercourse* Egypt *see* Suhaymī, Wādī as
Sahel *reg.* Africa 46 C3
Sahibganj India 37 F4
Sahiwal Pak. 36 C3
Sahlābād Iran 33 I3
Ṣaḥneh Iran 35 G4
Ṣaḥrā al Ḥijārah *reg.* Iraq 35 G5
Sahuayo Mex. 66 D4
Sahuteng China *see* Zadoi
Sahyadri *mts* India *see* Western Ghats
Sahyadriparvat Range *hills* India 38 B1
Sai *r.* India 37 E4
Saïda Alg. 25 G6
Saïda Lebanon *see* Sidon
Saïda Morocco 25 E6
Sa'īdīyeh Iran *see* Solṭānīyeh
Saidpur Bangl. 37 G4
Saiha India 37 H5
Saihan Tal China 43 K4
Saijō Japan 45 D6
Saikai Kokuritsu-kōen Japan 45 C6
Saiki Japan 45 C6
Sailana India 36 C5
Saimaa *l.* Fin. 15 P6
Saimbeyli Turkey 34 E3
Sa'īndezh Iran 35 G3
Sa'īn Qal'eh Iran *see* Sa'īndezh
St Abb's Head *hd* U.K. 20 G5
St Agnes U.K. 19 B8
St Agnes *i.* U.K. 19 A9
St Albans U.K. 19 G7
St Aldhelm's Head *hd* U.K. 19 E8
St-Amand-Montrond France 24 F3
St-Amour France 24 G3
St-André, Cap *c.* Madag. *see* Vilanandro, Tanjona
St Andrews U.K. 20 G4
St Ann's Bay Jamaica 67 I5
St Anthony Canada 61 M4
St Anthony U.S.A. 62 E3
St-Arnaud Alg. *see* El Eulma
St Arnaud Australia 58 A6
St Arnaud Range *mts* N.Z. 59 D6
St Augustin *r.* Canada 61 M4
St Augustine U.S.A. 63 K6
St Austell U.K. 19 C8
St-Avertin France 24 E3
St-Barthélemy *terr.* West Indies 67 L5
St Bees U.K. 18 D4
St Bees Head *hd* U.K. 18 D4
St Bride's Bay U.K. 19 B7
St-Brieuc France 24 C2
St Catharines *Ont.* Canada 64 B1
St Catherine's Point U.K. 19 F8
St-Céré France 24 E4
St-Chamond France 24 G4
St Charles *MD* U.S.A. 64 C3
St Charles *MO* U.S.A. 63 I4
St-Chély-d'Apcher France 24 F4
St Christopher and Nevis *country* West Indies *see* St Kitts and Nevis
St Clair, Lake Canada/U.S.A. 63 K3
St-Claude France 24 G3
St Clears U.K. 19 C7
St Cloud U.S.A. 63 I2
St Croix *r.* U.S.A. 63 I2
St David's Head *hd* U.K. 19 B7
St-Denis Réunion 73 L7
St-Denis-du-Sig Alg. *see* Sig
St-Dié France 24 H2
St-Dizier France 24 G2
St-Domingue *country* West Indies *see* Haiti
St Elias, Cape U.S.A. 60 D4
St Elias Mountains Canada 60 D3
Ste-Marie, Cap *c.* Madag. *see* Vohimena, Tanjona
Sainte-Marie, Île *i.* Madag. *see* Boraha, Nosy
Ste-Maxime France 24 H5
Sainte Rose du Lac Canada 62 H1
Saintes France 24 D4
St-Étienne France 24 G4
St-Étienne-du-Rouvray France 19 I9
Saintfield U.K. 21 G3
St-Florent *Corse* France 24 I5
St-Florent-sur-Cher France 24 F3
St-Flour France 24 F4
St Francis U.S.A. 62 G4
St Francis Isles Australia 55 F8
St-Gaudens France 24 E5
St George Australia 58 D2
St George *r.* Australia 56 D3
St George *AK* U.S.A. 60 B4
St George *UT* U.S.A. 65 F2
St George Island U.S.A. 60 B4
St George Range *hills* Australia 54 D4
St George's Grenada 67 L6
St George's Channel Ireland/U.K. 21 F6
St George's Channel P.N.G. 52 F2
St George's Head *hd* Australia 58 E5
St Gotthard Hungary *see* Szentgotthárd
St Gotthard Pass Switz. 24 I3
St Govan's Head *hd* U.K. 19 C7
St Helena *i.* S. Atlantic Ocean 72 H7
St Helena *CA* U.S.A. 65 A1
St Helena, Ascension and Tristan da Cunha *terr.* S. Atlantic Ocean 72 H7
St Helena Bay S. Africa 50 D7
St Helens Australia 57 [inset]
St Helens U.K. 18 E5
St Helens, Mount *vol.* U.S.A. 62 C2
St Helens Point Australia 57 [inset]
St Helier Channel Is 19 E9
Sainthiya India 37 F5
St Ignace U.S.A. 63 K2
St Ignace Island Canada 63 J2
St Ishmael U.K. 19 C7
St Ives *England* U.K. 19 B8
St Ives *England* U.K. 19 G6
St James, Cape Canada 60 E4
St-Jean, Lac *l.* Canada 63 M2
St-Jean-d'Acre Israel *see* 'Akko
St-Jean-d'Angély France 24 D4
St-Jean-de-Monts France 24 C3
St-Jérôme Canada 63 M2
St Joe *r.* U.S.A. 62 D2
Saint John Canada 63 N2
St John *r.* U.S.A. 63 N2
St John's Antigua and Barbuda 67 L5
St John's Canada 61 M5
St Johns *AZ* U.S.A. 62 F5
St Johnsbury U.S.A. 63 M3
St John's Chapel U.K. 18 E4
St Joseph *MO* U.S.A. 63 I4
St Joseph, Lake Canada 63 I1
St-Joseph-d'Alma Canada *see* Alma
St-Junien France 24 E4
St Just U.K. 19 B8
St Keverne U.K. 19 B8
St Kilda *i.* U.K. 10 E4
St Kilda *is* U.K. 16 C2
St Kitts and Nevis *country* West Indies 67 L5
St-Laurent *inlet* Canada *see* St Lawrence
St-Laurent, Golfe du *g.* Canada *see* St Lawrence, Gulf of
St-Laurent-du-Maroni Fr. Guiana 69 H2
St Lawrence *inlet* Canada 61 L5
St Lawrence, Gulf of Canada 61 L5
St Lawrence Island U.S.A. 60 B3
St Leonard *MD* U.S.A. 64 C3
St-Lô France 24 D2
St-Louis Senegal 46 B3
St Louis *MO* U.S.A. 63 I4
St Louis *r.* U.S.A. 63 I2
St Lucia *country* West Indies 67 L6
St Lucia, Lake S. Africa 51 K5
St Lucia Estuary S. Africa 51 K5
St Magnus Bay U.K. 20 [inset]
St-Maixent-l'École France 24 D3
St-Malo France 24 C2
St-Malo, Golfe de *g.* France 24 C2
St-Marc Haiti 67 J5
St Marks S. Africa 51 H7
St Mark's S. Africa *see* Cofimvaba
St-Martin *terr.* West Indies 67 L5
St Martin, Cape S. Africa 50 C7
St Martin's *i.* U.K. 19 A9
St Mary Peak Australia 57 B6
St Mary's *Ont.* Canada 64 A1
St Mary's U.K. 20 G2
St Mary's *i.* U.K. 19 A9
St Marys *PA* U.S.A. 64 B2
St Marys *WV* U.S.A. 64 A3
St Marys City *MD* U.S.A. 64 C3
St Matthew Island U.S.A. 60 A3
St Matthias Group *is* P.N.G. 52 E2
St-Maurice *r.* Canada 63 M2
St Mawes U.K. 19 B8
St-Médard-en-Jalles France 24 D4
St Michaels *MD* U.S.A. 64 C3
St-Nazaire France 24 C3
St Neots U.K. 19 G6
St-Nicolas Belgium *see* Sint-Niklaas
St-Nicolas-de-Port France 24 H2
St-Omer France 24 F1
Saintonge *reg.* France 24 D4
St-Palais France 24 D5
St-Paul *atoll* Fr. Polynesia *see* Héréhérétué
St Paul *AK* U.S.A. 60 A4
St Paul *MN* U.S.A. 63 I3
St-Paul, Île *i.* Indian Ocean 73 N8
St Paul Island U.S.A. 60 A4
St Peter and St Paul Rocks *is* N. Atlantic Ocean *see* São Pedro e São Paulo
St Peter Port Channel Is 19 E9
St Petersburg Rus. Fed. 12 F4
St Petersburg U.S.A. 63 K6
St-Pierre *mt.* France 24 G5
St-Pierre St Pierre and Miquelon 61 M5
St Pierre and Miquelon *terr.* N. America 61 M5
St-Pierre-d'Oléron France 24 D4
St-Pierre-le-Moûtier France 24 F3
St-Pourçain-sur-Sioule France 24 F3
St-Quentin France 24 F2
St-Saëns France 19 I9
St Sebastian Bay S. Africa 50 E8
St-Siméon Canada 63 N2
St Theresa Point Canada 63 I1
St Thomas *Ont.* Canada 64 A1
St-Tropez France 24 H5
St-Tropez, Cap de *c.* France 24 H5
St-Vaast-la-Hougue France 19 F9
St-Valery-en-Caux France 19 H9
St-Véran France 24 H4
St Vincent *country* West Indies *see* St Vincent and the Grenadines
St Vincent, Cape Australia 57 [inset]
St Vincent, Cape Port. *see* São Vicente, Cabo de
St Vincent, Gulf Australia 57 B7
St Vincent and the Grenadines *country* West Indies 67 L6
St Vincent Passage St Lucia/St Vincent 67 L6
St Williams *Ont.* Canada 64 A1
St-Yrieix-la-Perche France 24 E4
Sain Us China 42 J4
Saioa *mt.* Spain 25 F2
Saipal *mt.* Nepal 36 E3
Saipan *i.* N. Mariana Is 41 G6
Saiteli Turkey *see* Kadınhanı
Saittanulkki *h.* Fin. 14 N3
Sajama, Nevado *mt.* Bol. 68 E7
Sājūr, Nahr *r.* Syria/Turkey 39 D1
Sak *watercourse* S. Africa 50 E5
Sakaide Japan 45 D6
Sakākah Saudi Arabia 35 F5
Sakakawea, Lake U.S.A. 62 G2
Sakar *mts* Bulg. 27 L4
Sakaraha Madag. 49 E6
Sak'art'velo *country* Asia *see* Georgia
Sakarya *r.* Turkey 27 N4
Sakassou Côte d'Ivoire 46 C4
Sakata Japan 45 E5
Sakchu N. Korea 45 B4
Sakhalin *i.* Rus. Fed. 44 F2
Sakhalin Oblast *admin. div.* Rus. Fed. *see* Sakhalinskaya Oblast'
Sakhalinskaya Oblast' *admin. div.* Rus. Fed. 44 F2
Sakhalinskiy Zaliv *b.* Rus. Fed. 44 F1
Sakhi India 36 C3
Sakhile S. Africa 51 I4
Şäki Azer. 35 G2
Saki Nigeria *see* Shaki
Saki Ukr. *see* Saky
Šakiai Lith. 15 M9
Sakir *mt.* Pak. 36 A3
Sakishima-shotō *is* Japan 43 M8
Sakoli India 36 D5
Sakon Nakhon Thai. 31 J5
Sakrivier S. Africa 50 E6
Sakura Japan 45 F6
Saky Ukr. 34 D1
Säkylä Fin. 15 M6
Sal *i.* Cape Verde 46 [inset]
Sal *r.* Rus. Fed. 13 I7
Sala Sweden 15 J7
Salacgrīva Latvia 15 N8
Sala Consilina Italy 26 F4
Saladas Arg. 70 E3
Salado *r.* *Buenos Aires* Arg. 70 E5
Salado *r.* *Santa Fé* Arg. 70 D4
Salado *r.* Arg. 70 C5
Salaga Ghana 46 C4
Salairskiy Kryazh *ridge* Rus. Fed. 42 E2
Salajwe Botswana 50 G2
Şalālah Oman 33 H6
Salamanca Mex. 66 D4
Salamanca Spain 25 D3
Salamanca *NY* U.S.A. 64 B1
Salamanga Moz. 51 K4
Salamantica Spain *see* Salamanca
Salamat, Bahr *r.* Chad 47 E4
Salamina *i.* Greece 27 J6
Salamis *tourist site* Cyprus 39 A2
Salamis *i.* Greece *see* Salamina
Salamīyah Syria 39 C2
Sālang, Tūnel-e Afgh. 36 B2
Salantai Lith. 15 L8
Salar de Pocitos Arg. 70 C2
Salas Spain 25 C2
Salaspils Latvia 15 N8
Salawati *i.* Indon. 41 F8
Salawin, Mae Nam *r.* China/Myanmar *see* Salween
Salaya India 36 B5
Salayar *i.* Indon. 41 E8
Sala y Gómez, Isla *i.* S. Pacific Ocean 75 M7
Salazar Angola *see* N'dalatando
Salbris France 24 F3
Šalčininkai Lith. 15 N9
Salcombe U.K. 19 D8
Saldae Alg. *see* Bejaïa
Saldaña Spain 25 D2
Saldanha S. Africa 50 C7
Saldanha Bay S. Africa 50 C7
Saldus Latvia 15 M8
Sale Australia 58 C7
Şāleḥābād Iran 35 H4
Salekhard Rus. Fed. 28 H3
Salem India 38 C4
Salem *MA* U.S.A. 64 F1
Salem *NJ* U.S.A. 64 D3
Salem *NY* U.S.A. 64 E1
Salem *OH* U.S.A. 64 A2
Salem *OR* U.S.A. 62 C3
Salem *VA* U.S.A. 64 A4
Salen *Scotland* U.K. 20 D4
Salen *Scotland* U.K. 20 D4
Salerno Italy 26 F4
Salerno, Golfo di *g.* Italy 26 F4
Salernum Italy *see* Salerno
Salford U.K. 18 E5
Salgótarján Hungary 17 Q6
Salgueiro Brazil 69 K5
Salida U.S.A. 62 F4
Salies-de-Béarn France 24 D5
Salihli Turkey 27 M5
Salihorsk Belarus 15 O10
Salima Malawi 49 D5
Salina *KS* U.S.A. 62 H4
Salina, Isola *i.* Italy 26 F5
Salina Cruz Mex. 66 E5
Salinas Brazil 71 C2
Salinas Ecuador 68 B4
Salinas Mex. 66 D4
Salinas *CA* U.S.A. 65 B2
Salinas *r.* *CA* U.S.A. 65 B2
Salinas, Cabo de *c.* Spain *see* Ses Salines, Cap de
Salinas, Ponta das *pt* Angola 49 B5
Saline Valley *depr.* *CA* U.S.A. 65 D2
Salinópolis Brazil 69 I4
Salinosó Lachay, Punta *pt* Peru 68 C6
Salisbury U.K. 19 F7
Salisbury *MD* U.S.A. 64 D3
Salisbury Zimbabwe *see* Harare
Salisbury Plain U.K. 19 E7
Şalkhad Syria 39 C3
Salla Fin. 14 P3
Salluit Canada 77 K2
Sallum, Khalīj as *b.* Egypt 34 B5
Sallyana Nepal 37 E3
Salmās Iran 35 G3
Salmi Rus. Fed. 12 F3
Salmon U.S.A. 62 E2
Salmon *r.* U.S.A. 62 D2
Salmon Arm Canada 62 D1
Salmon Gums Australia 55 C8
Salmon Reservoir *NY* U.S.A. 64 D1
Salmon River Mountains U.S.A. 62 D3
Salo Fin. 15 M6
Salome *AZ* U.S.A. 65 F4
Salon India 36 E4
Salon-de-Provence France 24 G5
Salonica Greece *see* Thessaloniki
Salonika Greece *see* Thessaloniki
Salpausselkä *reg.* Fin. 15 N6
Salqīn Syria 39 C1
Salses, Étang de *l.* France *see* Leucate, Étang de
Sal'sk Rus. Fed. 13 I7
Salsomaggiore Terme Italy 26 C2
Salt Jordan *see* As Salṭ
Salt *watercourse* S. Africa 50 F7
Salt *r.* U.S.A. 62 E5
Salta Arg. 70 C2
Saltaire U.K. 18 F5
Saltash U.K. 19 C8
Saltcoats U.K. 20 E5
Saltee Islands Ireland 21 F5
Saltfjellet Svartisen Nasjonalpark *nat. park* Norway 14 I3
Saltfjorden *sea chan.* Norway 14 H3
Salt Fork Lake *OH* U.S.A. 64 A2
Saltillo Mex. 66 D3
Salt Lake City U.S.A. 62 E3
Salto Brazil 71 B3
Salto Uruguay 70 E4
Salto da Divisa Brazil 71 D2
Salto Grande Brazil 71 A3
Salton Sea *salt l.* *CA* U.S.A. 65 E4
Salto Santiago, Represa de *resr* Brazil 70 F3
Salt Range *hills* Pak. 36 C2
Saluda *VA* U.S.A. 64 C4
Salūm Egypt *see* As Sallūm
Salûm, Khalîg el *b.* Egypt *see* Sallum, Khalīj as
Salur India 38 D2
Saluzzo Italy 26 B2
Salvador Brazil 71 D1
Salvador *country* Central America *see* El Salvador
Salvaleón de Higüey Dom. Rep. *see* Higüey
Salwah Saudi Arabia 48 F1
Salween *r.* China/Myanmar 31 I4
Salween *r.* China/Myanmar 42 H9
Salyan Azer. 35 H3
Salyan Nepal *see* Sallyana
Sal'yany Azer. *see* Salyan
Salzbrunn Namibia 50 C3
Salzburg Austria 17 N7
Salzgitter Germany 17 M4
Sam India 36 B4
Samagaltay Rus. Fed. 42 G2
Samāh *well* Saudi Arabia 35 G5
Samaida Iran *see* Someydeh
Samaixung China 37 E2
Samalayuca Mex. 62 F5
Samalkot India 38 D2
Samālūṭ Egypt 34 C5
Samālūṭ Egypt *see* Samālūṭ
Samanala *mt.* Sri Lanka *see* Adam's Peak
Samandağı Turkey 39 B1
Samangān Afgh. *see* Āybak
Samani Japan 44 F4
Samanlı Dağları *mts* Turkey 27 M4
Samar Kazakh. *see* Samarskoye
Samar *i.* Phil. 41 E6
Samara Rus. Fed. 13 K5
Samara *r.* Rus. Fed. 11 Q5
Samarga Rus. Fed. 44 E3
Samarinda Indon. 41 D8
Samarka Rus. Fed. 44 D3
Samarobriva France *see* Amiens
Samarqand Uzbek. 33 K2
Sāmarrā' Iraq 35 F4
Samarskoye Kazakh. 42 E3
Samasata Pak. 36 B3
Samastipur India 37 F4
Şamaxı Azer. 35 H2
Samba India 36 C2
Sambaliung *mts* Indon. 41 D7
Sambalpur India 37 E5
Sambar, Tanjung *pt* Indon. 41 D8
Sambat Ukr. *see* Kiev
Sambava Madag. 49 F5
Sambha India 37 G4
Sambhajinagar India *see* Aurangabad
Sambhal India 36 D3
Sambhar Lake India 36 C4
Sambir Ukr. 13 D6
Sambito *r.* Brazil 69 J5
Sambor Ukr. *see* Sambir
Samborombón, Bahía *b.* Arg. 70 E5
Samch'ŏk S. Korea 45 C5
Samch'ŏnp'o S. Korea *see* Sach'on
Same Tanz. 48 D4
Samer France 19 I8
Sami India 36 B5
Samīrah Saudi Arabia 32 F4
Samirum Iran *see* Yazd-e Khvāst
Samjiyŏn N. Korea 44 C4
Şämkir Azer. 35 G2
Sam Neua Laos *see* Xam Nua
Samoa *country* S. Pacific Ocean 53 I3
Samoa Basin *sea feature* S. Pacific Ocean 74 I7
Samoa i Sisifo *country* S. Pacific Ocean *see* Samoa
Samobor Croatia 26 F2
Samoded Rus. Fed. 12 I3
Samokov Bulg. 27 J3
Šamorín Slovakia 17 P6
Samos *i.* Greece 27 L6
Samothrace *i.* Greece *see* Samothraki
Samothraki *i.* Greece 27 K4
Samoylovka Rus. Fed. 13 I6
Sampit Indon. 41 D8
Sam Rayburn Reservoir U.S.A. 63 I5
Samsang China 37 E3
Samsun Turkey 34 E2
Samyai China 37 G3
San Mali 46 C3
Şan'ā' Yemen 32 F6
Sanaa Yemen *see* Şan'ā'
SANAE IV *research stn* Antarctica 76 B2
San Agostín U.S.A. *see* St Augustine
Sanak Island U.S.A. 60 B4
Sanandaj Iran 35 G4
San Andreas *CA* U.S.A. 65 B1
San Andrés, Isla de *i.* Caribbean Sea 67 H6
San Angelo U.S.A. 62 G5
San Antonio Chile 70 B4
San Antonio *TX* U.S.A. 62 H6
San Antonio, Cabo *c.* Cuba 67 H4
San Antonio Oeste Arg. 70 D6
San Antonio Reservoir *CA* U.S.A. 65 B3
San Augustín de Valle Fértil Arg. 70 C4
San Benedetto del Tronto Italy 26 E3
San Benedicto, Isla *i.* Mex. 66 B5
San Benito *r.* *CA* U.S.A. 65 B2
San Benito Mountain *CA* U.S.A. 65 B2
San Bernardino *CA* U.S.A. 65 D3
San Bernardino Mountains *CA* U.S.A. 65 D3
San Bernardo Chile 70 B4
San Blas, Cape U.S.A. 63 J6
San Borja Bol. 68 E6
Sanbornville *NH* U.S.A. 64 F1
San Buenaventura Mex. 66 D3
San Carlos Chile 70 B5
San Carlos Equat. Guinea *see* Luba
San Carlos Venez. 68 E2
San Carlos de Bariloche Arg. 70 B6
San Carlos de Bolívar Arg. 70 D5
Sanchahe China *see* Fuyu
Sanchi India 36 D5
Sanchor India 36 B4
San Clemente *CA* U.S.A. 65 D4
San Clemente Island *CA* U.S.A. 65 C4
Sanclêr U.K. *see* St Clears
San Cristóbal Arg. 70 D4
San Cristóbal *i.* Solomon Is 53 G3
San Cristóbal Venez. 68 D2
San Cristóbal, Isla *i.* *Galápagos* Ecuador 68 [inset]
San Cristóbal de las Casas Mex. 66 F5
Sancti Spíritus Cuba 67 I4
Sand *r.* S. Africa 51 J2
Sandagou Rus. Fed. 44 D4
Sanda Island U.K. 20 D5
Sandakan *Sabah* Malaysia 41 D7
Sandane Norway 14 E6
Sandanski Bulg. 27 J4
Sandaré Mali 46 B3
Sanday *i.* U.K. 20 G1
Sandbach U.K. 19 E5
Sand Cay *rf* India 38 B4
Sandefjord Norway 15 G7
Sandercock Nunataks Antarctica 76 D2
Sanderson U.S.A. 62 G5
Sandfire Roadhouse Australia 54 C4
Sand Fork *WV* U.S.A. 64 A3
Sandgate Australia 58 F1
Sandhead U.K. 20 E6
Sandia Peru 68 E6
San Diego *CA* U.S.A. 65 D4
Sandıklı Turkey 27 N5
Sandila India 36 E4

ndnes Norway 15 D7
ndnessjøen Norway 14 H3
ndø i. Faroe Is see Sandoy
ndoa Dem. Rep. Congo 49 C4
ndomierz Poland 13 D6
n Donà di Piave Italy 26 E2
ndover watercourse Australia 56 B4
ndovo Rus. Fed. 12 H4
ndown U.K. 19 F8
ndoy i. Faroe Is 14 [inset]
nd Point U.S.A. 60 B4
ndpoint U.S.A. 62 D2
ndray i. U.K. 20 B4
ndringham Australia 56 B5
ndrul Mare, Vârful mt. Romania 27 L1
ndsjö Sweden 15 I6
ndstone Australia 55 B6
ndur Faroe Is 14 [inset]
ndusky OH U.S.A. 63 K3
ndveld mts S. Africa 50 D6
ndverhaar Namibia 50 C4
ndvika Akershus Norway 15 G7
ndvika Nord-Trøndelag Norway 14 H5
ndviken Sweden 15 J6
ndwich Island Vanuatu see Éfaté
ndwich Islands N. Pacific Ocean see Hawai'ian Islands
ndwick U.K. 20 [inset]
ndwip Bangl. 37 G5
ndy Cape Qld Australia 56 F5
ndy Cape Tas. Australia 57 [inset]
ndy Hook pt NJ U.S.A. 64 D2
ndy Island Australia 54 C3
ndy Lake Ont. Canada 63 I1
ndy Lake Ont. Canada 63 I1
n Estanislao Para. 70 E2
n Felipe Chile 70 B4
n Felipe Baja California Mex. 66 B2
n Felipe Venez. 68 E1
n Felipe de Puerto Plata Dom. Rep. see Puerto Plata
n Fernando Chile 70 B4
n Fernando Mex. 66 E4
n Fernando Phil. 41 E6
n Fernando Spain 25 C5
n Fernando Trin. and Tob. 67 L6
n Fernando CA U.S.A. 65 C3
n Fernando de Apure Venez. 68 E2
n Fernando de Atabapo Venez. 68 E3
n Fernando de Monte Cristi Dom. Rep. see Monte Cristi
nford FL U.S.A. 63 K6
nford ME U.S.A. 64 F1
nford NC U.S.A. 63 L4
nford, Mount U.S.A. 60 D3
n Francisco Arg. 70 D4
n Francisco CA U.S.A. 65 A2
n Francisco, Cabo de c. Ecuador 68 B3
n Francisco, Passo de pass Arg./Chile 70 C3
n Francisco Bay inlet CA U.S.A. 65 A2
n Francisco de Paula, Cabo c. Arg. 70 C7
n Gabriel Mountains CA U.S.A. 65 C3
ngachaly Azer. see Sanqaçal
ngameshwar India 38 B2
ngan, Koh-i- mt. Afgh. see Sangān, Kūh-e
ngān, Kūh-e mt. Afgh. 36 A2
ngar Rus. Fed. 29 N3
ngareddi India 38 C2
ngareddy India see Sangareddi
n Gavino Monreale Sardegna Italy 26 C5
ngay, Parque Nacional nat. park Ecuador 68 C4
ngeang i. Indon. 54 B2
nger CA U.S.A. 65 C2
ng-e Surakh Iran 35 J3
n Giovanni in Fiore Italy 26 G5
ngir India 36 C5
ngir, Kepulauan is Indon. 41 E7
ngkulirang Indon. 41 D7
ngli India 38 B2
ngmai China see Dêrong
ngmélima Cameroon 46 E4
ngo Zimbabwe 49 D6
ngole India 38 B2
n Gorgonio Mountain CA U.S.A. 65 D3
ngre de Cristo Range mts U.S.A. 62 F4
ngrur India 36 C3
ngu r. Bangl. 37 G5
nguem India 38 B3
ngutane r. Moz. 51 K3
n Hipólito, Punta pt Mex. 66 B3
nhûr Egypt 34 C5
nhûr Egypt see Sanhūr
n Ignacio Beni Bol. 68 E6
n Ignacio Santa Cruz Bol. 68 F7
n Ignacio Santa Cruz Bol. 68 F7
n Ignacio Para. 70 E3
nikiluaq Canada 61 K4
nin-kaigan Kokuritsu-kōen Japan 45 D6
n Jacinto CA U.S.A. 65 D4
n Jacinto Peak CA U.S.A. 65 D4
n Javier Bol. 68 F7
njiang Guizhou China see Jinping
njiangkou China 44 A4
njiaocheng China see Haiyan
njō Japan 45 E5
n Joaquin r. CA U.S.A. 65 B1
n Joaquin Valley val. CA U.S.A. 65 B2
njoli India 36 C5
n Jorge, Golfo de g. Arg. 70 C7
n Jorge, Golfo de g. Spain see Sant Jordi, Golf de
n José Costa Rica 67 H7
n Jose Phil. 41 E6
n Jose CA U.S.A. 65 B2
n José, Isla i. Mex. 66 B4
n José de Amacuro Venez. 68 F2
n José de Bavicora Mex. 62 F6
n Jose de Buenavista Phil. 41 E6
n José de Chiquitos Bol. 68 F7
n José de Comondú Mex. 66 B3
n José del Cabo Mex. 66 C4
n José del Guaviare Col. 68 D3
n José de Mayo Uruguay 70 E4
n Juan Arg. 70 C4
n Juan r. Costa Rica/Nicaragua 67 H6
n Juan Puerto Rico 67 K5
n Juan r. U.S.A. 62 E4
n Juan, Cabo c. Arg. 70 D8
n Juan, Cabo c. Equat. Guinea 46 D4
n Juan Bautista Para. 70 E3
n Juan Bautista de las Misiones Para. see San Juan Bautista
n Juan de los Morros Venez. 68 E2
n Juan Mountains U.S.A. 62 F4
n Julián Arg. 70 C7
n Justo Arg. 70 D4
nkari Drug India 38 C4
nkhu India 36 C3
nkra Chhattisgarh India 38 D1
nkra Rajasthan India 36 B4
nkt Gallen Switz. 24 D3
nkt-Peterburg Rus. Fed. see St Petersburg
nkt Pölten Austria 17 O6
nkt Veit an der Glan Austria 17 O7
nku India 36 D2
nliurfa Turkey 34 E3
Şanlıurfa prov. Turkey 39 D1
San Lorenzo Arg. 70 D4
San Lorenzo Beni Bol. 68 E7
San Lorenzo Tarija Bol. 68 F8
San Lorenzo Ecuador 68 C3
San Lorenzo mt. Spain 25 E2
San Lorenzo, Monte mt. Arg./Chile 70 B7
Sanlúcar de Barrameda Spain 25 C5
San Lucas, Serranía de mts Col. 68 D2
San Luis Arg. 70 C4
San Luis AZ U.S.A. 65 E4
San Luis Obispo CA U.S.A. 65 B3
San Luis Obispo Bay CA U.S.A. 65 B3
San Luis Potosí Mex. 66 D4
San Luis Reservoir CA U.S.A. 65 B2
San Luis Río Colorado Mex. 62 E5
San Marcos U.S.A. 62 H6
San Marino country Europe 26 E3
San Marino San Marino 26 E3
San Martín research stn Antarctica 76 L2
San Martín Catamarca Arg. 70 C3
San Martín Mendoza Arg. 70 C4
San Martín, Lago l. Arg./Chile 70 B7
San Martín de los Andes Arg. 70 B6
San Mateo CA U.S.A. 65 A2
San Matías Bol. 69 G7
San Matías, Golfo g. Arg. 70 D6
Sanmenxia China 43 K6
San Miguel El Salvador 66 G6
San Miguel CA U.S.A. 65 B3
San Miguel de Huachi Bol. 68 E7
San Miguel de Tucumán Arg. 70 C3
San Miguel do Araguaia Brazil 71 A1
San Miguel Island CA U.S.A. 65 B3
Sanming China 43 L7
Sanndraigh i. U.K. see Sandray
Sannicandro Garganico Italy 26 F4
San Nicolas Island CA U.S.A. 65 C4
Sannieshof S. Africa 51 G4
Sanniquellie Liberia 46 C4
Sanok Poland 13 D6
San Pablo Bol. 68 E8
San Pablo de Manta Ecuador see Manta
San Pedro Arg. 70 D2
San Pedro Bol. 68 F7
San Pedro Chile 70 C2
San-Pédro Côte d'Ivoire 46 C4
San Pedro Baja California Sur Mex. 62 E7
San Pedro Para. see San Pedrode Ycuamandyyú
San Pedro watercourse U.S.A. 62 E5
San Pedro, Sierra de mts Spain 25 C4
San Pedro Channel CA U.S.A. 65 C4
San Pedro de Arimena Col. 68 D3
San Pedro de Atacama Chile 70 C2
San Pedro de las Colonias Mex. 66 D3
San Pedro de Macorís Dom. Rep. 67 K5
San Pedro de Ycuamandyyú Para. 70 E2
San Pedro Sula Hond. 66 G5
San Pietro, Isola di i. Sardegna Italy 26 C5
Sanqaçal Azer. 35 H2
Sanquhar U.K. 20 F5
Sanquianga, Parque Nacional nat. park Col. 68 C3
San Quintín, Cabo c. Mex. 66 A2
San Rafael Arg. 70 C4
San Rafael CA U.S.A. 65 A2
San Rafael Mountains CA U.S.A. 65 B3
San Ramón Bol. 68 F6
San Remo Italy 26 B3
San Roque Spain 25 B2
San Salvador i. Bahamas 67 J4
San Salvador El Salvador 66 G6
San Salvador, Isla i. Galápagos Ecuador 68 [inset]
San Salvador de Jujuy Arg. 70 C2
Sansanné-Mango Togo 46 D3
San Sebastián Arg. 70 C8
San Sebastián Spain 25 F2
San Sebastián de los Reyes Spain 25 E3
Sansepolcro Italy 26 E3
San Severo Italy 26 F4
Sanski Most Bos.-Herz. 26 G2
Santa r. Peru 68 C5
Santa Ana Bol. 68 E7
Santa Ana El Salvador 66 G6
Santa Ana Mex. 66 B2
Santa Ana i. Solomon Is 53 G3
Santa Ana CA U.S.A. 65 D4
Santa Ana de Yacuma Bol. 68 E6
Santa Bárbara Brazil 71 C2
Santa Bárbara Mex. 66 C3
Santa Barbara CA U.S.A. 65 C3
Santa Bárbara, Ilha i. Brazil 71 D2
Santa Barbara Channel CA U.S.A. 65 B3
Santa Bárbara d'Oeste Brazil 71 B3
Santa Barbara Island CA U.S.A. 65 C4
Santa Catalina, Gulf of CA U.S.A. 65 D4
Santa Catalina de Armada Spain 25 B2
Santa Catalina Island CA U.S.A. 65 C4
Santa Catarina state Brazil 71 A4
Santa Catarina, Ilha de i. Brazil 71 A4
Santa Clara Col. 68 E4
Santa Clara Cuba 67 I4
Santa Clara CA U.S.A. 65 B2
Santa Clara UT U.S.A. 65 F2
Santa Clarita CA U.S.A. 65 C3
Santa Clotilde Peru 68 D4
Santa Comba Angola see Waku-Kungo
Santa Croce, Capo c. Sicilia Italy 26 F6
Santa Cruz Bol. 68 F7
Santa Cruz Brazil 69 K5
Santa Cruz Costa Rica 68 A1
Santa Cruz CA U.S.A. 65 A2
Santa Cruz, Isla i. Galápagos Ecuador 68 [inset]
Santa Cruz Cabrália Brazil 71 D2
Santa Cruz de Goiás Brazil 71 A2
Santa Cruz de la Palma Canary Is 46 B2
Santa Cruz del Sur Cuba 67 I4
Santa Cruz de Moya Spain 25 F4
Santa Cruz de Tenerife Canary Is 46 B2
Santa Cruz do Sul Brazil 70 F3
Santa Cruz Island CA U.S.A. 65 C3
Santa Cruz Islands Solomon Is 53 G3
Santa Elena, Bahía de b. Ecuador 68 B4
Santa Elena, Cabo c. Costa Rica 67 G6
Santa Elena, Punta pt Ecuador 68 B4
Santa Eudóxia Brazil 71 B3
Santa Eufemia, Golfo di g. Italy 26 G5
Santa Fe Arg. 70 D4
Santa Fe U.S.A. 62 F4
Santa Fé de Bogotá Col. see Bogotá
Santa Fé de Minas Brazil 71 B2
Santa Fé do Sul Brazil 71 A3
Santa Helena Brazil 69 I4
Santa Helena de Goiás Brazil 71 A2
Santa Inês Brazil 69 I4
Santa Inés, Isla i. Chile 76 L3
Santa Isabel Arg. 70 C5
Santa Isabel Equat. Guinea see Malabo
Santa Isabel i. Solomon Is 53 F2
Santa Juliana Brazil 71 B2
Santalpur India 36 B5
Santa Lucia Range mts CA U.S.A. 65 B2
Santa Margarita CA U.S.A. 65 B3
Santa Margarita, Isla i. Mex. 66 B4
Santa María Arg. 70 C3
Santa Maria Amazonas Brazil 69 G4
Santa Maria Rio Grande do Sul Brazil 70 F3
Santa Maria Cape Verde 46 [inset]
Santa María r. Mex. 66 C2
Santa María Peru 68 D4
Santa Maria CA U.S.A. 65 B3
Santa Maria r. AZ U.S.A. 65 F3
Santa Maria, Cabo de c. Moz. 51 K4
Santa Maria, Cabo de c. Port. 25 C5
Santa Maria, Chapadão de hills Brazil 71 B1
Santa María, Isla i. Galápagos Ecuador 68 [inset]
Santa Maria, Serra de hills Brazil 71 B1
Santa Maria da Vitória Brazil 71 B1
Santa Maria do Suaçuí Brazil 71 C2
Santa Maria Island Vanuatu 53 G3
Santa Maria Madalena Brazil 71 C3
Santa Maria Mountains AZ U.S.A. 65 F3
Santa Marta Col. 68 D1
Santa Marta, Cabo de c. Angola 49 B5
Santa Marta Grande, Cabo de c. Brazil 71 A5
Santa Maura i. Greece see Lefkada
Santa Monica CA U.S.A. 65 C3
Santa Monica Bay CA U.S.A. 65 C4
Santana Brazil 71 C1
Santana r. Brazil 71 A2
Santana do Araguaia Brazil 69 H5
Santander Spain 25 E2
Santa Nella CA U.S.A. 65 B2
Santanilla, Islas is Caribbean Sea see Cisne, Islas del
Sant'Antioco Sardegna Italy 26 C5
Sant'Antioco, Isola di i. Sardegna Italy 26 C5
Sant Antoni de Portmany Spain 25 G4
Santapilly India 38 D2
Santa Quitéria Brazil 69 J4
Santarém Brazil 69 H4
Santarém Port. 25 B4
Santa Rosa Arg. 70 D5
Santa Rosa Rio Grande do Sul Brazil 70 F3
Santa Rosa CA U.S.A. 65 A1
Santa Rosa NM U.S.A. 62 G5
Santa Rosa de Copán Hond. 66 G6
Santa Rosa de la Roca Bol. 68 F7
Santa Rosa do Purus Acre Brazil 68 D5
Santa Rosa Island CA U.S.A. 65 B4
Santa Rosalía Mex. 66 B3
Santa Sylvina Arg. 70 D3
Santa Teresa Australia 55 F6
Santa Teresa r. Brazil 71 A1
Santa Vitória Brazil 71 A2
Santa Ynez r. CA U.S.A. 65 B3
Santa Ysabel i. Solomon Is see Santa Isabel
Santee CA U.S.A. 65 D4
Sant Francesc de Formentera Spain 25 G4
Santiago Brazil 70 F3
Santiago i. Cape Verde 46 [inset]
Santiago Chile 70 B4
Santiago Dom. Rep. 67 J5
Santiago Panama 67 H7
Santiago de Compostela Spain 25 B2
Santiago de Cuba Cuba 67 I4
Santiago del Estero Arg. 70 D3
Santiago de los Caballeros Dom. Rep. see Santiago
Santiago de Veraguas Panama see Santiago
Santipur India see Shantipur
Sant Jordi, Golf de g. Spain 25 G3
Santo Amaro Brazil 71 D1
Santo Amaro de Campos Brazil 71 C3
Santo Anastácio Brazil 71 A3
Santo André Brazil 71 B3
Santo Angelo Brazil 70 F3
Santo Antão i. Cape Verde 46 [inset]
Santo Antônio Brazil 68 F4
Santo Antônio r. Brazil 71 C2
Santo António São Tomé and Príncipe 46 D4
Santo Antônio, Cabo c. Brazil 71 D1
Santo Antônio da Platina Brazil 71 A3
Santo Antônio de Jesus Brazil 71 D1
Santo Antônio do Içá Brazil 68 E4
Santo Corazón Bol. 69 G7
Santo Domingo Dom. Rep. 67 K5
Santo Domingo country West Indies see Dominican Republic
Santo Domingo de Guzmán Dom. Rep. see Santo Domingo
Santo Hipólito Brazil 71 B2
Santorini i. Greece 27 K6
Santos Brazil 71 B3
Santos Dumont Brazil 71 C3
Santos Luzardo, Parque Nacional nat. park Venez. 68 E2
Santos Plateau sea feature S. Atlantic Ocean 72 E7
Santo Tomás Peru 68 D6
Santo Tomé Arg. 70 E3
Sanup Plateau AZ U.S.A. 65 F2
San Valentín, Cerro mt. Chile 70 B7
San Vicente El Salvador 66 G6
San Vicente Mex. 66 A2
San Vicente de Baracaldo Spain see Barakaldo
San Vicente de Cañete Peru 68 C6
San Vincenzo Italy 26 D3
San Vito, Capo c. Sicilia Italy 26 E5
Sanwer India 36 C5
Sanya China 43 J9
Sanza Pombo Angola 49 B4
São Bernardo do Campo Brazil 71 B3
São Borja Brazil 70 E3
São Carlos Brazil 71 B3
São Domingos Brazil 71 B1
São Felipe, Serra de hills Brazil 71 B1
São Félix Bahia Brazil 71 D1
São Félix Mato Grosso Brazil 69 H6
São Félix Pará Brazil 69 H5
São Fidélis Brazil 71 C3
São Francisco Brazil 71 B1
São Francisco r. Brazil 71 C1
São Francisco, Ilha de i. Brazil 71 A4
São Francisco de Paula Brazil 71 A5
São Francisco de Sales Brazil 71 A2
São Francisco do Sul Brazil 71 A4
São Gabriel Brazil 70 F4
São Gonçalo Brazil 71 C3
São Gonçalo do Abaeté Brazil 71 B2
São Gonçalo do Sapucaí Brazil 71 B3
São Gotardo Brazil 71 B2
São João, Ilhas de is Brazil 69 J4
São João da Barra Brazil 71 C3
São João da Boa Vista Brazil 71 B3
São João da Madeira Port. 25 B3
São João da Ponte Brazil 71 B1
São João del Rei Brazil 71 B3
São João do Paraíso Brazil 71 C1
São Joaquim Brazil 71 A5
São Joaquim da Barra Brazil 71 B3
São José Amazonas Brazil 68 E4
São José Santa Catarina Brazil 71 A4
São José do Rio Preto Brazil 71 A3
São José dos Campos Brazil 71 B3
São José dos Pinhais Brazil 71 A4
São Leopoldo Brazil 71 A5
São Lourenço Brazil 71 B3
São Lourenço r. Brazil 69 G7
São Luís Brazil 69 J4
São Luís Brazil 69 G4
São Luís de Montes Belos Brazil 71 A2
São Manuel Brazil 71 A3
São Marcos r. Brazil 71 B2
São Mateus Brazil 71 D2
São Mateus do Sul Brazil 71 A4
São Miguel i. Arquipélago dos Açores 72 G3
São Miguel r. Brazil 71 B2
São Miguel do Tapuio Brazil 69 J5
Saône r. France 24 G4
Saoner India 36 D5
São Nicolau i. Cape Verde 46 [inset]
São Paulo Brazil 71 B3
São Paulo state Brazil 71 A3
São Paulo de Olivença Brazil 68 E4
São Pedro da Aldeia Brazil 71 C3
São Pedro e São Paulo is N. Atlantic Ocean 72 G5
São Pires r. Brazil see Teles Pires
São Raimundo Nonato Brazil 69 J5
São Romão Amazonas Brazil 68 E5
São Romão Minas Gerais Brazil 71 B2
São Roque Brazil 71 B3
São Roque, Cabo de c. Brazil 69 K5
São Salvador Angola see M'banza Congo
São Salvador do Congo Angola see M'banza Congo
São Sebastião Brazil 71 B3
São Sebastião, Ilha do i. Brazil 71 B3
São Sebastião do Paraíso Brazil 71 B3
São Sebastião dos Poções Brazil 71 B1
São Simão Goiás Brazil 69 H7
São Simão São Paulo Brazil 71 B3
São Simão, Barragem de resr Brazil 71 A2
São Tiago i. Cape Verde see Santiago
São Tomé São Tomé and Príncipe 46 D4
São Tomé i. São Tomé and Príncipe 46 D4
São Tomé, Cabo de c. Brazil 71 C3
São Tomé, Pico de mt. São Tomé and Príncipe 46 D4
São Tomé and Príncipe country Africa 46 D4
Saoura, Oued watercourse Alg. 22 D6
São Vicente Brazil 71 B3
São Vicente i. Cape Verde 46 [inset]
São Vicente, Cabo de c. Port. 25 B5
Sapanca Turkey 27 N4
Sapaul India see Supaul
Şaphane Dağı mt. Turkey 27 N5
Sapo National Park Liberia 46 C4
Sapouy Burkina Faso 46 C3
Sapozhok Rus. Fed. 13 I5
Sapporo Japan 44 F4
Saputang China see Zadoi
Saqqez Iran 35 G3
Sarā Iran 35 G3
Sarāb Iran 35 G3
Sara Buri Thai. 31 J5
Saradiya India 36 B5
Saragossa Spain see Zaragoza
Saragt Turkm. 33 J2
Saraguro Ecuador 68 C4
Sarahs Turkm. see Saragt
Sarai Sidhu Pak. 36 C3
Sarajevo Bos.-Herz. 26 H3
Saraktash Rus. Fed. 28 G4
Saramati mt. India/Myanmar 37 H4
Saran' Kazakh. 42 C3
Saranda Albania see Sarandë
Sarandë Albania 27 I5
Sarandib country Asia see Sri Lanka
Sarangpur India 36 D5
Saransk Rus. Fed. 13 J5
Sara Peak Nigeria 46 D4
Sarapul Rus. Fed. 11 Q4
Sarāqib Syria 39 C2
Sarasota U.S.A. 63 K6
Sarata Ukr. 27 M1
Saratoga CA U.S.A. 65 A2
Saratoga WY U.S.A. 62 F3
Saratok Sarawak Malaysia 41 D7
Saratov Rus. Fed. 13 J6
Saratovskoye Vodokhranilishche resr Rus. Fed. 13 J5
Saratsina, Akrotirio pt Greece 27 K5
Sarāvān Iran 33 J4
Saray Turkey 27 L4
Sarayköy Turkey 27 M6
Sarayönü Turkey 34 D3
Sarbāz Iran 33 J4
Sarbhang Bhutan 37 G4
Sarbīsheh Iran 33 I3
Sarda r. Nepal 37 E3
Sard Āb Afgh. 36 B1
Sardarshahr India 36 C3
Sar Dasht Iran 35 G3
Sardegna i. Sardegna Italy see Sardinia
Sardica Bulg. see Sofia
Sardinia i. Sardegna Italy 26 C4
Sardis WV U.S.A. 64 A3
Sar-e Būm Afgh. 36 A2
Sareks nationalpark nat. park Sweden 14 J3
Sarektjåkkå mt. Sweden 14 J3
Sar-e Pol Afgh. 36 A1
Sar-e Pol-e Ẕahāb Iran 35 G4
Sar Eskandar Iran see Hashtrud
Sargasso Sea N. Atlantic Ocean 75 P4
Sargodha Pak. 33 L3
Sarh Chad 47 E4
Sārī Iran 35 I3
Saria i. Greece 27 L7
Sar-i-Bum Afgh. see Sar-e Būm
Sarigh Jilganang Kol salt l. Aksai Chin 36 D2
Sarıgöl Turkey 27 M5
Sarıkamış Turkey 35 F2
Sarila India 36 D4
Sarina Australia 56 E4
Sarıoğlan Kayseri Turkey 34 D3
Sarıoğlan Konya Turkey see Belören
Sariqamish Kuli salt l. Turkm./Uzbek. see Sarykamyshskoye Ozero
Sarīr Tibesti des. Libya 47 E2
Sarıveliler Turkey 39 A1
Sariwŏn N. Korea 45 B5
Sarıyar Barajı resr Turkey 27 N5
Sarıyer Turkey 27 M4
Sarız Turkey 34 E3
Sark i. Channel Is 19 E9
Sarkand Kazakh. 42 D3
Şarkikaraağaç Turkey 27 N5
Şarkışla Turkey 34 E3
Şarköy Turkey 27 L4
Sarlath Range mts Afgh./Pak. 36 A3
Sarmi Indon. 41 F8
Särna Sweden 15 H6
Sarneh Iran 35 G4
Sarnen Switz. 24 I3
Sarni India see Amla
Sarnia Canada 63 K3
Sarny Ukr. 13 E6
Saroma-ko l. Japan 44 F3
Saronikos Kolpos g. Greece 27 J6
Saros Körfezi b. Turkey 27 L4
Sarov Rus. Fed. 13 I5
Sarpa, Ozero l. Rus. Fed. 13 J6
Sarpan i. N. Mariana Is see Rota
Sarpsborg Norway 15 G7
Sarqant Kazakh. see Sarkand
Sarrebourg France 24 H2
Sarria Spain 25 C2
Sartana Ukr. 13 H7
Sartène Corse France 24 I6
Sarthe r. France 24 D3
Sartu China see Daqing
Saruna Pak. 36 A4
Sarupsar India 36 C3
Şărur Azer. 35 G3
Sarvābād Iran 35 G4
Sárvár Hungary 26 G1
Sarv-e Bālā Iran 35 I4
Sarwar India 36 C4
Saryarka plain Kazakh. 42 C2
Sar Yazd Iran 35 I5
Sarygamysh Köli salt l. Turkm./Uzbek. see Sarykamyshskoye Ozero
Sary-Ishikotrau, Peski des. Kazakh. see Saryyesik-Atyrau, Peski
Sarykamyshskoye Ozero salt l. Turkm./Uzbek. 35 J2
Saryozek Kazakh. 42 D4
Saryshagan Kazakh. 42 C3
Sarysu watercourse Kazakh. 42 B3
Sarytash Kazakh. 35 H1
Sary-Tash Kyrg. 33 L2
Saryyesik-Atyrau, Peski des. Kazakh. 42 D3
Sarzha Kazakh. 35 H2
Sasar, Tanjung pt Indon. 54 B2
Sasaram India 37 F4
Sasebo Japan 45 C6
Saskatchewan prov. Canada 62 F1
Saskatchewan r. Canada 60 H4
Saskatoon Canada 60 H4
Saskylakh Rus. Fed. 29 M2
Saslaya mt. Nicaragua 67 H6
Sasoi r. India 36 B5
Sasolburg S. Africa 51 H4
Sasovo Rus. Fed. 13 I5
Sassandra Côte d'Ivoire 46 C4
Sassari Sardegna Italy 26 C4
Sassnitz Germany 17 N3
Sass Town Liberia 46 C4
Sasykkol', Ozero l. Kazakh. 42 E3
Sasykoli Rus. Fed. 13 J7
Sasyqköl l. Kazakh. see Sasykkol', Ozero
Sata-misaki c. Japan 45 C7
Satana India 38 B1
Satara India 38 B2
Satara S. Africa 51 J3
Sātbaev Kazakh. see Satpayev
Satırlar Turkey see Yeşilova
Satkania Bangl. 37 H5
Satkhira Bangl. 37 G5
Satluj r. India/Pak. see Sutlej
Satmala Range hills India 38 C2
Satna India 36 E4
Satpayev Kazakh. 42 B3
Satpura Range mts India 36 C5
Satsuma-hantō pen. Japan 45 C7
Satthwa Myanmar 37 H6
Satu Mare Romania 13 D7
Satwas India 36 D5
Saucillo Mex. 66 C3
Sauda Norway 15 E7
Sauðárkrókur Iceland 14 [inset]
Saudi Arabia country Asia 32 F4
Sāūjbolāgh Iran see Mahābād
Saulieu France 24 G3
Sault Sainte Marie Canada 63 K2
Sault Sainte Marie U.S.A. 63 K2
Saumalkol' Kazakh. 30 F1
Saumarez Reef Australia 56 F4
Saumlakki Indon. 54 E2
Saumur France 24 D3
Saundatti India 38 B3
Saunders, Mount h. Australia 54 E3
Saunders Coast Antarctica 76 J1
Saurimo Angola 49 C4
Sautar Angola 49 B5
Sava r. Europe 26 I2
Savage River Australia 57 [inset]
Savai'i i. Samoa 53 I3
Savala r. Rus. Fed. 13 I6
Savalou Benin 46 D4
Savanat Iran see Eşṭahbān
Savannah GA U.S.A. 63 K5
Savannah r. U.S.A. 63 K5
Savannakhét Laos 41 C6
Savanna-la-Mar Jamaica 67 I5
Savant Lake Canada 63 I1
Savanur India 38 B3
Sävar Sweden 14 L5
Savaştepe Turkey 27 L5
Savè Benin 46 D4
Save r. Moz./Zimbabwe 49 D6
Sāveh Iran 35 H4
Saviaho Fin. 14 P5
Savinskiy Rus. Fed. 12 I3
Savitri r. India 38 B2
Savli India 36 C5
Savoie reg. France see Savoy
Savona Italy 26 C2
Savonlinna Fin. 14 P6
Savonranta Fin. 14 P5
Savoy reg. France 24 H3
Şavşat Turkey 35 F2
Sävsjö Sweden 15 I8
Savu i. Indon. 54 C2
Savukoski Fin. 14 P3
Savur Turkey 35 F3
Saw Myanmar 37 H5
Sawai Madhopur India 36 D4
Sawar India 36 C4
Sawel Mountain h. U.K. 21 E3
Sawtell Australia 58 F3
Sawu Indon. 54 C2
Sawu i. Indon. see Savu
Sawu, Laut sea Indon. 41 E8
Sawye Myanmar 37 I5
Saxilby U.K. 18 G5
Saxmundham U.K. 19 I6
Saxnäs Sweden 14 I4
Saxton PA U.S.A. 64 B2
Say Niger 46 D3
Sayabouri Laos see Xaignabouli
Sayak Kazakh. 42 D3
Sayanogorsk Rus. Fed. 42 G2
Sayano-Shushenskoye Vodokhranilishche resr Rus. Fed. 42 G2
Sayansk Rus. Fed. 42 I2
Sayaq Kazakh. see Sayak
Şaydā Lebanon see Sidon
Sāyen Iran 35 I5
Sayghān Afgh. 36 A2
Sayhan-Ovoo Mongolia 42 I3
Sayḩūt Yemen 32 H6
Saykhin Kazakh. 11 P6
Saylac Somalia 47 H3
Saylan country Asia see Sri Lanka
Saynshand Mongolia 43 K4
Sayoa mt. Spain see Saioa
Şayqal, Baḩr imp. l. Syria 39 C3
Sayqyn Kazakh. see Saykhin
Sayre PA U.S.A. 64 C2
Sayreville NJ U.S.A. 64 D2
Sayula Mex. 66 F5
Sazdy Kazakh. 13 K7
Sazin Pak. 36 C2
Sbaa Alg. 22 D6
Sbeitla Tunisia 26 C7
Scaddan Australia 55 C8
Scafell Pike h. U.K. 18 D4
Scalasaig U.K. 20 C4
Scalea Italy 26 F5
Scalloway U.K. 20 [inset]
Scalpaigh, Eilean i. U.K. see Scalpay
Scalpay i. U.K. 20 C3
Scapa Flow inlet U.K. 20 F2
Scarba i. U.K. 20 D4
Scarborough Canada 63 L3
Scarborough Trin. and Tob. 67 L6
Scarborough U.K. 18 G4
Scariff Island Ireland 21 B6
Scarp i. U.K. 20 B2
Scarpanto i. Greece see Karpathos
Schaffhausen Switz. 24 I3
Schakalskuppe Namibia 50 C4
Schärding Austria 17 N6
Schefferville Canada 61 L4
Scheibbs Austria 17 O6
Schellville CA U.S.A. 65 A1
Schenectady NY U.S.A. 64 E1
Schio Italy 26 D2
Schleswig Germany 17 L3
Schleswig-Holsteinisches Wattenmeer, Nationalpark nat. park Germany 17 L3
Schmidt Island Rus. Fed. see Shmidta, Ostrov
Schmidt Peninsula Rus. Fed. see Shmidta, Poluostrov
Schneidemühl Poland see Piła
Schneverdingen Germany 17 L4
Schoharie NY U.S.A. 64 D1
Schönebeck (Elbe) Germany 17 M4
Schouten Island Australia 57 [inset]
Schouten Islands P.N.G. 52 E2
Schrankogel mt. Austria 17 M7
Schröttersburg Poland see Płock
Schull Ireland 21 C6
Schuyler Lake NY U.S.A. 64 D1
Schwäbische Alb mts Germany 17 L7
Schwäbisch Hall Germany 17 L6
Schwandorf Germany 17 N6
Schwarzenberg Germany 17 N5
Schwarzrand mts Namibia 50 C3
Schwarzwald mts Germany see Black Forest
Schwatka Mountains U.S.A. 60 C3
Schwaz Austria 17 M7
Schwedt an der Oder Germany 17 O4
Schweinfurt Germany 17 M5
Schweiz country Europe see Switzerland
Schweizer-Reneke S. Africa 51 G4
Schwerin Germany 17 M4
Schwyz Switz. 24 I3
Sciacca Sicilia Italy 26 E6
Scicli Sicilia Italy 26 F6
Scilly, Île atoll Fr. Polynesia see Manuae
Scilly, Isles of U.K. 19 A9
Scodra Albania see Shkodër
Scole U.K. 19 I6
Scone Australia 58 E4
Scone U.K. 20 F4
Scoresby Land reg. Greenland 61 P2
Scoresbysund Greenland see Ittoqqortoormiit
Scoresby Sund sea chan. Greenland see Kangertittivaq
Scorno, Punta dello pt Sardegna Italy see Caprara, Punta
Scorpion Bight b. Australia 55 D8
Scotia Ridge sea feature S. Atlantic Ocean 72 E9
Scotia Ridge sea feature S. Atlantic Ocean 76 A2
Scotia Sea S. Atlantic Ocean 72 F9
Scotland Ont. Canada 64 A1
Scotland admin. div. U.K. 20 F3
Scotland MD U.S.A. 64 C3
Scott, Cape Australia 54 E3
Scott, Cape Canada 62 B1
Scott Base (New Zealand) research stn Antarctica 76 H1
Scottburgh S. Africa 51 J6
Scott City U.S.A. 62 G4
Scott Coast Antarctica 76 H1
Scott Glacier Antarctica 76 I1
Scott Island Antarctica 76 H2
Scott Mountains Antarctica 76 D2
Scott Reef Australia 54 C3
Scottsbluff U.S.A. 62 G3
Scottsboro U.S.A. 63 J5
Scottsville VA U.S.A. 64 B4
Scourie U.K. 20 D2
Scousburgh U.K. 20 [inset]
Scrabster U.K. 20 F2
Scranton PA U.S.A. 64 D2
Scunthorpe U.K. 18 G5
Scuol Switz. 24 J3
Scupi Macedonia see Skopje
Scutari Albania see Shkodër
Scutari, Lake Albania/Montenegro 27 H3
Seabrook, Lake imp. l. Australia 55 B7
Seaford U.K. 19 H8
Seaforth Ont. Canada 64 A1
Seal r. Canada 61 I4
Seal, Cape S. Africa 50 F8
Sea Lake Australia 57 C7
Sealy U.S.A. 63 H6
Seaman Range mts NV U.S.A. 65 E2
Seamer U.K. 18 G4
Searchlight NV U.S.A. 65 E3
Searcy U.S.A. 63 I4
Searles Lake CA U.S.A. 65 D3
Seaside Park NJ U.S.A. 64 D3
Seattle U.S.A. 62 C2
Seaview Range mts Australia 56 D3
Seba Indon. 54 C2
Sebastea Turkey see Sivas
Sebastián Vizcaíno, Bahía b. Mex. 66 B3
Sebastopol Ukr. see Sevastopol'
Sebastopol CA U.S.A. 65 A1
Sebba Burkina Faso 46 D3
Seben Turkey 27 N4
Sebenico Croatia see Šibenik
Sebeş Romania 27 J2
Sebezh Rus. Fed. 15 P8
Şebinkarahisar Turkey 34 E2
Sebring U.S.A. 63 K6
Sebrovo Rus. Fed. 13 I6
Sebta N. Africa see Ceuta
Sechelt Canada 62 C2
Sechenovo Rus. Fed. 13 J5
Sechura Peru 68 B5
Sechura, Bahía de b. Peru 68 B5
Secretary Island N.Z. 59 A7
Secunda S. Africa 51 I4
Secunderabad India 38 C2
Sedalia U.S.A. 63 I4
Sedam India 38 C2
Sedan France 24 G2
Sedan Dip Australia 56 C3
Seddon N.Z. 59 E5
Seddonville N.Z. 59 C5
Sedeh Iran 33 I3
Sederot Israel 39 B4
Sedlčany Czech Rep. 17 O6
Sedlets Poland see Siedlce
Sedom Israel 39 B4
Sédrata Alg. 26 B6
Šeduva Lith. 15 M9

Seeheim Namibia 50 C4
Seelig, Mount Antarctica 76 K1
Sées France 24 E2
Sefadu Sierra Leone 46 B4
Sefare Botswana 51 H2
Seferihisar Turkey 27 L5
Sefīd Kūh mts Afgh. 36 A2
Sefophe Botswana 51 H2
Segalstad Norway 15 G6
Segamat Malaysia 31 J6
Ségbana Benin 46 D3
Segezha Rus. Fed. 12 G3
Segontia U.K. see Caernarfon
Segontium U.K. see Caernarfon
Segorbe Spain 25 F4
Ségou Mali 46 C3
Segovia r. Hond./Nicaragua see Coco
Segovia Spain 25 D3
Segozerskoye Vodokhranilishche resr Rus. Fed. 12 G3
Seguam Island U.S.A. 60 A4
Séguédine Niger 46 E2
Séguéla Côte d'Ivoire 46 C4
Seguin U.S.A. 62 H6
Segura r. Spain 25 F4
Segura, Sierra de mts Spain 25 E5
Segzī Iran 35 I4
Sehithwa Botswana 49 C6
Sehlabathebe National Park Lesotho 51 I5
Sehore India 36 D5
Sehwan Pak. 36 A4
Seikphyu Myanmar 37 H5
Seiland i. Norway 14 M1
Seinäjoki Fin. 14 M5
Seine r. France 24 E2
Seine, Baie de b. France 24 D2
Seine, Val de val. France 24 F2
Sejny Poland 15 M9
Sekayu Indon. 41 C8
Sekoma Botswana 50 F3
Sekondi Ghana 46 C4
Sek'ot'a Eth. 48 D2
Şela Rus. Fed. see Shali
Selaru i. Indon. 41 F8
Selatan, Tanjung pt Indon. 41 D8
Selat Makassar str. Indon. see Makassar, Selat
Selawik U.S.A. 60 B3
Selbekken Norway 14 F5
Selbu Norway 14 G5
Selby U.K. 18 F5
Selbyville DE U.S.A. 64 D3
Selebi-Phikwe Botswana 49 C6
Selebi-Pikwe Botswana see Selebi-Phikwe
Selemdzha r. Rus. Fed. 44 C1
Selemdzhinsk Rus. Fed. 44 C1
Selemdzhinskiy Khrebet mts Rus. Fed. 44 D1
Selendi Turkey 27 M5
Selenga r. Mongolia/Rus. Fed. 42 J2
Selenga Mörön r. Mongolia/Rus. Fed. see Selenga
Selety r. Kazakh. see Sileti
Seletyteniz, Ozero salt l. Kazakh. see Siletiteniz, Ozero
Seleucia Turkey see Silifke
Seleucia Pieria Turkey see Samandağı
Sel'gon Rus. Fed. 44 D2
Sélibabi Mauritania 46 B3
Selibe-Phikwe Botswana see Selebi-Phikwe
Seliger, Ozero l. Rus. Fed. 12 G4
Seligman AZ U.S.A. 65 F3
Selikhino Rus. Fed. 44 E2
Selîma Oasis Sudan 32 C5
Selimiye Turkey 27 L6
Selinsgrove PA U.S.A. 64 C2
Selizharovo Rus. Fed. 12 G4
Seljord Norway 15 F7
Selkirk Canada 63 H1
Selkirk U.K. 20 G5
Selkirk Mountains Canada 62 D1
Sellafield U.K. 18 D4
Selma AL U.S.A. 63 J5
Selma CA U.S.A. 65 C2
Selsey Bill hd U.K. 19 G8
Sel'tso Rus. Fed. 13 G5
Selty Rus. Fed. 12 L4
Selu i. Indon. 54 E1
Selvas reg. Brazil 68 D5
Selway r. U.S.A. 62 D2
Selwyn Lake Canada 60 H3
Selwyn Mountains Canada 60 E3
Selwyn Range hills Australia 56 B4
Sel'yb Rus. Fed. 12 K3
Semarang Indon. 41 D8
Semau i. Indon. 54 C2
Sembé Congo 48 B3
Şemdinli Turkey 35 G3
Semendire Serbia see Smederevo
Semenivka Ukr. 13 G5
Semenov Rus. Fed. 12 J4
Semenovka Ukr. see Semenivka
Semey Kazakh. 42 E2
Semidi Islands U.S.A. 60 C4
Semikarakorsk Rus. Fed. 13 I7
Semiluki Rus. Fed. 13 H6
Seminole U.S.A. 62 G5
Semipalatinsk Kazakh. see Semey
Semīrom Iran 35 H5
Sem Kolodezey Ukr. see Lenine
Semnān Iran 35 I4
Semyonovskoye Arkhangel'skaya Oblast' Rus. Fed. see Bereznik
Semyonovskoye Kostromskaya Oblast' Rus. Fed. see Ostrovskoye
Sena Bol. 68 E6
Sena Madureira Brazil 68 E5
Senanga Zambia 49 C5
Sendai Kagoshima Japan 45 C7
Sendai Miyagi Japan 45 F5
Sêndo China 37 H3
Seneca Lake NY U.S.A. 64 C1
Seneca Rocks WV U.S.A. 64 B3
Senecaville Lake OH U.S.A. 64 A3
Senegal country Africa 46 B3
Sénégal r. Mauritania/Senegal 46 B3
Senftenberg Germany 17 O5
Senga Hill Zambia 49 D4
Sengerema Tanz. 48 D4
Sengeyskiy, Ostrov i. Rus. Fed. 12 K1
Sengiley Rus. Fed. 13 K5
Sengirli, Mys pt Kazakh. see Syngyrli, Mys
Senhor do Bonfim Brazil 69 J6
Senigallia Italy 26 E3
Senj Croatia 26 F2
Senja i. Norway 14 J2
Sen'kina Rus. Fed. 12 K2
Şenkōy Turkey 39 C1
Senlac S. Africa 50 F3
Senlin Shan mt. China 44 C4
Senlis France 24 F2
Sennar Sudan 32 D7
Sennen U.K. 19 B8
Senneterre Canada 63 L2
Senqu r. Lesotho 51 H6
Sens France 24 F2
Sensuntepeque El Salvador 66 G6
Senta Serbia 27 I2
Senthal India 36 D3
Sentinel AZ U.S.A. 65 F4
Sentinel Peak Canada 60 F4
Senwabarwana S. Africa 51 I2
Şenyurt Turkey 35 F3
Seo de Urgell Spain see La Seu d'Urgell
Seonath r. India 38 D1
Seoni India 36 D5
Seorinarayan India 37 E5
Seoul S. Korea 45 B5
Separation Well Australia 54 C5
Sepik r. P.N.G. 52 E2
Sep'o N. Korea 45 B5
Sepon India 37 H4
Seppa India 37 H4
Sept-Îles Canada 63 N1
Sequoia National Park CA U.S.A. 65 C2
Serafimovich Rus. Fed. 13 I6
Sêraitang China see Baima
Serakhs Ahal Turkm. see Saragt
Seram i. Indon. 41 E8
Seram, Laut sea Indon. 41 F8
Serbāl, Gebel mt. Egypt see Sirbāl, Jabal
Serbia country Europe 27 I3
Sêrbug Co l. China 37 G2
Sêrca China 37 H3
Serchhip India 37 H5
Serdar Turkm. 35 J3
Serder Turkm. see Serdar
Serdica Bulg. see Sofia
Serdo Eth. 48 E2
Serdoba r. Rus. Fed. 13 J5
Serdobsk Rus. Fed. 13 J5
Seredka Rus. Fed. 15 P7
Şereflikoçhisar Turkey 34 D3
Seremban Malaysia 41 C7
Serengeti National Park Tanz. 48 D4
Serenje Zambia 49 D5
Serezha r. Rus. Fed. 12 I5
Sergach Rus. Fed. 12 J5
Sergeyevka Rus. Fed. 44 B2
Sergiyev Posad Rus. Fed. 12 H4
Sergo Ukr. see Stakhanov
Serh China 37 I1
Serhetabat Turkm. 33 J2
Serifos i. Greece 27 K6
Serik Turkey 34 C3
Seringapatam Reef Australia 54 C3
Sêrkang China see Nyainrong
Sermata, Kepulauan is Indon. 54 E2
Sermersuaq glacier Greenland 61 M2
Sermilik inlet Greenland 61 O3
Sernovodsk Rus. Fed. 13 K5
Sernur Rus. Fed. 12 K4
Seronga Botswana 49 C5
Serov Rus. Fed. 11 S4
Serowe Botswana 51 H2
Serpa Port. 25 C5
Serpa Pinto Angola see Menongue
Serpentine Lakes imp. l. Australia 55 E7
Serpukhov Rus. Fed. 13 H5
Serra Brazil 71 C3
Serra Alta Brazil 71 A4
Serrachis r. Cyprus 39 A2
Serra da Bocaina, Parque Nacional da nat. park Brazil 71 B3
Serra da Canastra, Parque Nacional da nat. park Brazil 71 B3
Serra da Mesa, Represa resr Brazil 71 A1
Serra das Araras Brazil 71 B1
Serra do Divisor, Parque Nacional da nat. park Brazil 68 D5
Sérrai Greece see Serres
Serranía de la Neblina, Parque Nacional nat. park Venez. 68 E3
Serraria, Ilha i. Brazil see Queimada, Ilha
Serra Talhada Brazil 69 K5
Serres Greece 27 J4
Serrinha Brazil 69 K6
Sêrro Brazil 71 C2
Sers Tunisia 26 C6
Sertanópolis Brazil 71 A3
Sertãozinho Brazil 71 B3
Sertavul Geçidi pass Turkey 39 A1
Sertolovo Rus. Fed. 15 Q6
Serule Botswana 49 C6
Seruna India 36 C3
Sêrwolungwa China 37 H2
Sêrxü China 42 H6
Seryshevo Rus. Fed. 44 C2
Sese Islands Uganda 48 D4
Sesel country Indian Ocean see Seychelles
Sesfontein Namibia 49 B5
Seshachalam Hills India 38 C3
Sesheke Zambia 49 C5
Sesostris Bank sea feature India 38 A3
Ses Salines, Cap de c. Spain 25 H4
Sestri Levante Italy 26 C2
Sestroretsk Rus. Fed. 15 P6
Sète France 24 F5
Sete Lagoas Brazil 71 B2
Setermoen Norway 14 K2
Setesdal val. Norway 15 E7
Seti r. Nepal 36 E3
Sétif Alg. 22 F4
Seto Japan 45 E6
Seto-naikai sea Japan 43 O6
Seto-naikai Kokuritsu-kōen Japan 45 D6
Settat Morocco 22 C5
Settepani, Monte mt. Italy 26 C2
Settle U.K. 18 E4
Setúbal Port. 25 B4
Setúbal, Baía de b. Port. 25 B4
Seul, Lac l. Canada 63 I1
Sevan Armenia 35 G2
Sevan, Lake Armenia 35 G2
Sevan, Ozero l. Armenia see Sevan, Lake
Sevana Lich l. Armenia see Sevan, Lake
Sevastopol' Ukr. 34 D1
Seven Islands Canada see Sept-Îles
Sevenoaks U.K. 19 H7
Sévérac-le-Château France 24 F4
Severn r. Australia 58 E2
Severn r. Canada 63 I1
Severn S. Africa 50 F4
Severn r. U.K. 19 E7
Severnaya Dvina r. Rus. Fed. 12 I2
Severnaya Sos'va r. Rus. Fed. 11 T3
Severnaya Zemlya is Rus. Fed. 29 L1
Severnoye Rus. Fed. 11 Q5
Severnyy Nenetskiy Avtonomnyy Okrug Rus. Fed. 12 K1
Severnyy Respublika Komi Rus. Fed. 28 H3
Severobaykal'sk Rus. Fed. 43 J1
Severo-Baykal'skoye Nagor'ye mts Rus. Fed. 29 M4
Severodonetsk Ukr. see Syeverodonets'k
Severodvinsk Rus. Fed. 12 H2
Severo-Kuril'sk Rus. Fed. 29 Q4
Severomorsk Rus. Fed. 12 G1
Severoonezhsk Rus. Fed. 12 H3
Severo-Sibirskaya Nizmennost' lowland Rus. Fed. see North Siberian Lowland
Severoural'sk Rus. Fed. 11 R3
Severo-Yeniseyskiy Rus. Fed. 28 K3
Severskaya Rus. Fed. 34 E1
Severskiy Donets r. Rus. Fed./Ukr. 13 I7
Sevier Lake UT U.S.A. 65 F1
Sevilla Col. 68 C3
Sevilla Spain see Seville
Seville Spain 25 D5
Sevlyush Ukr. see Vynohradiv
Sewani India 36 C3
Seward AK U.S.A. 60 D3
Seward Mountains Antarctica 76 L2
Seward Peninsula U.S.A. 60 B3
Sexi Spain see Almuñécar
Seyakha Rus. Fed. 77 F2
Seychelles country Indian Ocean 73 L6
Seydi Turkm. 33 J2
Seydişehir Turkey 34 C3
Seyðisfjörður Iceland 14 [inset]
Seyhan Turkey see Adana
Seyhan r. Turkey 39 B1
Seyitgazi Turkey 27 N5
Seym r. Rus. Fed./Ukr. 13 G6
Seymchan Rus. Fed. 29 Q3
Seymour Australia 58 B6
Seymour S. Africa 51 H7
Seymour IN U.S.A. 63 J4
Seymour TX U.S.A. 62 H5
Seymour Range mts Australia 55 F6
Seypan i. N. Mariana Is see Saipan
Seyyedābād Afgh. 36 B2
Sézanne France 24 F2
Sfântu Gheorghe Romania 27 K2
Sfax Tunisia 26 D7
Sfikias, Limni resr Greece 27 J4
Sfîntu Gheorghe Romania see Sfântu Gheorghe
Sgiersch Poland see Zgierz
's-Gravenhage Neth. see The Hague
Sgurr Alasdair h. U.K. 20 C3
Sgurr Dhomhnuill h. U.K. 20 D4
Sgurr Mòr mt. U.K. 20 D3
Sgurr na Ciche mt. U.K. 20 D3
Shaanxi prov. China 43 J6
Shaartuz Tajik. see Shahrtuz
Shaban Pak. 36 A3
Shabani Zimbabwe see Zvishavane
Shabestar Iran 35 G3
Shabībī, Jabal ash mt. Jordan 39 B5
Shabla, Nos pt Bulg. 27 M3
Shabunda Dem. Rep. Congo 48 C4
Shache China 42 D5
Shackleton Coast Antarctica 76 H1
Shackleton Glacier Antarctica 76 I1
Shackleton Ice Shelf Antarctica 76 F2
Shackleton Range mts Antarctica 76 A1
Shādegān Iran 35 H5
Shady Spring WV U.S.A. 64 A4
Shafer Peak Antarctica 76 H2
Shafter CA U.S.A. 65 C3
Shaftesbury U.K. 19 E7
Shagedu China 43 K5
Shageluk U.S.A. 60 C3
Shagonar Rus. Fed. 42 G2
Shag Point N.Z. 59 C7
Shag Rocks is S. Georgia 70 H8
Shahabad Karnataka India 38 C2
Shahabad Rajasthan India 36 D4
Shahabad Uttar Prad. India 36 E4
Shāhābād Iran see Eslāmābād-e Gharb
Shah Bandar Pak. 36 A4
Shahdol India 36 E5
Shāh Fūlādī mt. Afgh. 36 A2
Shāhīn Dezh Iran see Sa'īndezh
Shahjahanpur India 36 D4
Shāh Jūy Afgh. 36 A2
Shāhpūr Iran see Salmās
Shahrak Afgh. 33 J3
Shahr-e Bābak Iran 35 I5
Shahr-e Kord Iran 35 H4
Shahr-e Şafā Afgh. 36 A3
Shahrezā Iran 35 H4
Shahrig Pak. 36 A3
Shahrisabz Uzbek. 33 K2
Shahr Rey Iran 35 H4
Shahr Sultan Pak. 36 B3
Shahrtuz Tajik. 36 B1
Shāhrūd Iran see Emāmrūd
Shāhrūd, Rūdkhāneh-ye r. Iran 35 H3
Shaikh Husain mt. Pak. 36 A3
Shaikhpura India see Sheikhpura
Shā'īr, Jabal mts Syria 39 C2
Sha'īra, Gebel mt. Egypt see Sha'īrah, Jabal
Sha'īrah, Jabal mt. Egypt 39 B5
Shajapur India 36 D5
Shajianzi China 44 B4
Shakaville S. Africa 51 J5
Shakh Tajik. see Shoh
Shakhbuz Azer. see Şahbuz
Shakhovskaya Rus. Fed. 12 G4
Shakhrisabz Uzbek. see Shahrisabz
Shakhtinsk Kazakh. 42 C3
Shakhty Respublika Buryatiya Rus. Fed. see Gusinoozersk
Shakhty Rostovskaya Oblast' Rus. Fed. 13 I7
Shakhun'ya Rus. Fed. 12 J4
Shaki Nigeria 46 D4
Shakotan-hantō pen. Japan 44 F4
Shalakusha Rus. Fed. 12 I3
Shali Rus. Fed. 35 G2
Shaliuhe China see Gangca
Shalkar India 36 D3
Shalkar Kazakh. 28 G5
Shalkarteniz, Solonchak salt marsh Kazakh. 42 A3
Shalqar Kazakh. see Shalkar
Shaluli Shan mts China 42 H6
Shaluni mt. India 37 I3
Shama r. Tanz. 49 D4
Shamāl Sīnā' governorate Egypt 39 A4
Shamâl Sînâ' governorate Egypt see Shamāl Sīnā'
Shāmat al Akbād des. Saudi Arabia 35 F5
Shamattawa Canada 61 I4
Shambār Iran 35 H4
Shamgong Bhutan see Shemgang
Shāmīyah des. Iraq/Syria 39 D2
Shamkhor Azer. see Şämkir
Shamrock U.S.A. 62 G4
Shamūlzā'ī Afgh. 36 A3
Shandong prov. China 43 L5
Shandong Bandao pen. China 43 M5
Shandur Pass Pak. 36 C1
Shangdu China 43 K4
Shangganling China 44 C3
Shanghai China 43 M6
Shanghai mun. China 43 M6
Shangluo China 43 J6
Shangnan China 43 K6
Shangrao China 43 L7
Shangyou Shuiku resr China 42 E4
Shangzhi China 44 B3
Shanhe China 44 B3
Shanlaragh Ireland 21 C6
Shannon airport Ireland 21 D5
Shannon est. Ireland 21 D5
Shannon r. Ireland 21 D5
Shannon, Mouth of the Ireland 21 C5
Shannon National Park Australia 55 B8
Shannon Ø i. Greenland 77 I1
Shansi prov. China see Shanxi
Shantipur India 37 G5
Shantou China 43 L8
Shantung prov. China see Shandong
Shanxi prov. China 43 K5
Shaoguan China 43 K8
Shaowu China 43 L7
Shaoxing China 43 M6
Shaoyang China 43 K7
Shap U.K. 18 E4
Shapinsay i. U.K. 20 G1
Shapkina r. Rus. Fed. 12 L2
Shapshal'skiy Khrebet mts Rus. Fed. 42 F2
Shaqrā' Saudi Arabia 32 G4
Shār, Jabal mt. Saudi Arabia 34 D6
Sharaf well Iraq 35 F5
Sharan Jogizai Pak. 36 B3
Shardara Kazakh. 33 K1
Shardara, Step' plain Kazakh. see Chardara, Step'
Shari r. Cameroon/Chad see Chari
Shārī, Buḩayrat imp. l. Iraq 35 G4
Shari-dake vol. Japan 44 G4
Sharīfah Syria 39 C2
Sharjah U.A.E. 33 I4
Sharka-leb La pass China 37 G3
Sharkawshchyna Belarus 15 O9
Shark Bay Australia 55 A6
Shark Reef Australia 56 D2
Sharlyk Rus. Fed. 11 Q5
Sharm ash Shaykh Egypt 34 D6
Sharm el Sheikh Egypt see Sharm ash Shaykh
Sharon PA U.S.A. 64 A2
Sharqat Iraq see Ash Sharqāţ
Sharqī, Jabal ash mts Lebanon/Syria 39 B3
Sharur Azer. see Şärur
Shar'ya Rus. Fed. 12 J4
Shashe r. Botswana/Zimbabwe 49 C6
Shashemenē Eth. 48 D3
Shashi China see Jingzhou
Shasta, Mount vol. U.S.A. 62 C3
Shatilki Belarus see Svyetlahorsk
Shatki Rus. Fed. 13 J5
Shaţnat as Salmās, Wādī watercourse Syria 39 D2
Shatoy Rus. Fed. 35 G2
Shatsk Rus. Fed. 13 I5
Shaţţ al 'Arab r. Iran/Iraq 35 H5
Shatura Rus. Fed. 13 H5
Shaubak Jordan see Ash Shawbak
Shaunavon Canada 62 F2
Shaver Lake CA U.S.A. 65 C2
Shaw r. Australia 54 B5
Shawangunk Mountains hills NY U.S.A. 64 D2
Shawano U.S.A. 63 J3
Shawnee OK U.S.A. 63 H4
Shay Gap (abandoned) Australia 54 C5
Shaykh, Jabal ash mt. Lebanon/Syria see Hermon, Mount
Shaykh Miskīn Syria 39 C3
Shāzand Iran 35 H4
Shaẕāẕ, Jabal mt. Saudi Arabia 35 F6
Shchekino Rus. Fed. 13 H5
Shchel'yayur Rus. Fed. 12 L2
Shcherbakov Rus. Fed. see Rybinsk
Shchigry Rus. Fed. 13 H6
Shchors Ukr. 13 F6
Shchuchin Belarus see Shchuchyn
Shchuchyn Belarus 15 N10
Shebalino Rus. Fed. 42 F2
Shebekino Rus. Fed. 13 H6
Sheberghān Afgh. 36 A1
Sheboygan U.S.A. 63 J3
Shebshi Mountains Nigeria 46 E4
Shebunino Rus. Fed. 44 F3
Shedok Rus. Fed. 35 F1
Sheelin, Lough l. Ireland 21 E4
Sheep Haven b. Ireland 21 E2
Sheepmoor S. Africa 51 J4
Sheep Peak NV U.S.A. 65 E2
Sheep's Head hd Ireland see Muntervary
Sheerness U.K. 19 H7
Shefar'am Israel 39 B3
Sheffield N.Z. 59 D6
Sheffield U.K. 18 F5
Sheffield PA U.S.A. 64 B2
Sheghnān Afgh. 36 B1
Shegmas Rus. Fed. 12 K2
Sheikh, Jebel esh mt. Lebanon/Syria see Hermon, Mount
Sheikhpura India 37 F4
Sheikhupura Pak. 33 L3
Shekār Āb Iran 35 I4
Shekhem West Bank see Nāblus
Shekhpura India see Sheikhpura
Sheki Azer. see Şäki
Sheksna Rus. Fed. 12 H4
Sheksninskoye Vodokhranilishche resr Rus. Fed. 12 H4
Shela China 37 H3
Shelagskiy, Mys pt Rus. Fed. 29 S2
Shelburne N.S. Canada 63 N3
Shelburne Bay Australia 56 C1
Shelby MT U.S.A. 62 E2
Shelbyville TN U.S.A. 63 J4
Shelikhova, Zaliv g. Rus. Fed. 29 Q3
Shelikof Strait U.S.A. 60 C4
Shellbrook Canada 62 F1
Shellharbour Australia 58 E5
Shelter Island NY U.S.A. 64 E2
Shelter Point N.Z. 59 B8
Shemakha Azer. see Şamaxı
Shemgang Bhutan 37 G4
Shemordan Rus. Fed. 12 K4
Shenandoah PA U.S.A. 64 C2
Shenandoah Mountains Virginia/West Virginia U.S.A. 64 B3
Shenandoah National Park VA U.S.A. 64 B3
Shendam Nigeria 46 D4
Shending Shan h. China 44 D3
Shengena mt. Tanz. 49 D4
Shengli Feng mt. China/Kyrg. see Pobeda Peak
Shengping China 44 B3
Shenkursk Rus. Fed. 12 I3
Shenmu China 43 K5
Shenshu China 44 C3
Shensi prov. China see Shaanxi
Shentala Rus. Fed. 13 K5
Shenton, Mount h. Australia 55 C7
Shenyang China 44 A4
Shenzhen China 43 K8
Sheopur India 36 D4
Shepetivka Ukr. 13 E6
Shepetovka Ukr. see Shepetivka
Shepherd Islands Vanuatu 53 G3
Shepparton Australia 58 B6
Sheppey, Isle of i. U.K. 19 H7
Sherabad Uzbek. see Sherobod
Sherborne U.K. 19 E8
Sherbro Island Sierra Leone 46 B4
Sherbrooke Canada 63 M2
Sherburne NY U.S.A. 64 D1
Shercock Ireland 21 F4
Shereiq Sudan 32 D6
Shergaon India 37 H4
Shergarh India 36 C4
Sheridan WY U.S.A. 62 F3
Sheringham U.K. 19 I6
Sherman U.S.A. 63 H5
Sherobod Uzbek. 33 K2
Sherpur Dhaka Bangl. 37 G4
Sherpur Rajshahi Bangl. 37 G4
's-Hertogenbosch Neth. 16 J5
Sherwood Forest reg. U.K. 19 F5
Shetland Islands is U.K. 20 [inset]
Shetpe Kazakh. 30 E2
Shevchenko Kazakh. see Aktau
Shevli r. Rus. Fed. 44 D1
Sheyenne r. U.S.A. 62 H2
Shey Phoksundo National Park Nepal 37 E3
Shezhin II Kazakh. 13 K6
Shiant Islands U.K. 20 C3
Shiashkotan, Ostrov i. Rus. Fed. 29 Q5
Shibām Yemen 32 G6
Shībar, Kowtal-e Afgh. 36 B2
Shibata Japan 45 E5
Shibazhan China 44 B1
Shibh Jazīrat Sīnā' pen. Egypt see Sinai
Shibīn al Kawm Egypt 34 C5
Shibîn el Kôm Egypt see Shibīn al Kawm
Shibotsu-jima i. Rus. Fed. see Zelenyy, Ostrov
Shidād al Mismā' h. Saudi Arabia 39 D4
Shidao China 43 M5
Shiel, Loch l. U.K. 20 D4
Shield, Cape Australia 56 B2
Shīeli Kazakh. see Chiili
Shifa, Jabal ash mts Saudi Arabia 34 D5
Shigatse China see Xigazê
Shīḩān mt. Jordan 39 B4
Shihezi China 42 F4
Shihkiachwang China see Shijiazhuang
Shijiazhuang China 43 K5
Shikarpur Pak. 33 K4
Shikhany Rus. Fed. 13 J5
Shikohabad India 36 D4
Shikoku i. Japan 45 D6
Shikoku-sanchi mts Japan 45 D6
Shikotan, Ostrov i. Rus. Fed. 44 G4
Shikotan-tō i. Rus. Fed. see Shikotan, Ostrov
Shikotsu-Tōya Kokuritsu-kōen Japan 44 F4
Shildon U.K. 18 F4
Shilega Rus. Fed. 12 J2
Shiliguri India 37 G4
Shilla mt. India 36 D2
Shillelagh Ireland 21 F5
Shillo r. Israel 39 B3
Shillong India 37 G4
Shilovo Rus. Fed. 13 I5
Shimada Japan 45 E6
Shimanovsk Rus. Fed. 44 B1
Shimbiris mt. Somalia 48 E2
Shimen Hunan China 43 K7
Shimla India 36 D3
Shimoga India 38 B3
Shimokita-hantō pen. Japan 44 F4
Shimoni Kenya 49 D4
Shimonoseki Japan 45 C6
Shimsk Rus. Fed. 12 F4
Shin, Loch l. U.K. 20 E2
Shināfīyah Iraq see Ash Shanāfīyah
Shīndand Afgh. 33 J3
Shingbwiyang Myanmar 37 I4
Shinghshal Pass Pak. 36 C1
Shingū Japan 45 E6
Shingwedzi S. Africa 51 J2
Shingwedzi r. S. Africa 51 J2
Shīnkay Afgh. 36 A3
Shīnkay Ghar Afgh. 36 B2
Shinnston WV U.S.A. 64 A3
Shinshār Syria 39 C2
Shinyanga Tanz. 48 D4
Shiogama Japan 45 F5
Shiono-misaki c. Japan 45 D6
Shipki Pass China/India 36 D3
Shipman VA U.S.A. 64 B4
Shippensburg PA U.S.A. 64 C2
Shipunovo Rus. Fed. 42 E2
Shiquanhe Xizang China see Ali
Shiquanhe Xizang China see Gar
Shira Rus. Fed. 42 F2
Shīrābād Iran 35 H3
Shirakawa-go and Gokayama tourist site Japan 45 E5
Shirane-san vol. Japan 45 E5
Shirase Coast Antarctica 76 J1
Shirase Glacier Antarctica 76 D2
Shīrāz Iran 35 I5
Shire r. Malawi 49 D5
Shīrīn Tagāb Afgh. 36 A1
Shiriya-zaki c. Japan 44 F4
Shīr Khān Afgh. 36 B1
Shīr Kūh mt. Iran 35 I5
Shiroro Reservoir Nigeria 46 D3
Shirpur India 36 C5
Shirten Holoy Gobi des. China 42 H4
Shīrvān Iran 33 I2
Shisanzhan China 44 B2
Shishaldin Volcano U.S.A. 60 B4
Shisha Pangma mt. China see Xixabangma Feng
Shithāthah Iraq 35 F4
Shiv India 36 B4
Shivamogga India see Shimoga
Shiveluch, Vulkan vol. Rus. Fed. 29 R4
Shivpuri India 36 D4
Shivwits UT U.S.A. 65 F2
Shivwits Plateau AZ U.S.A. 65 F2
Shiwan'gandu Zambia 49 D5
Shiyan China 43 K6
Shizuishan China 42 J5
Shizuoka Japan 45 E6
Shkhara mt. Georgia/Rus. Fed. 35 F2
Shklov Belarus see Shklow
Shklow Belarus 13 F5
Shkodër Albania 27 H3
Shkodra Albania see Shkodër
Shkodrës, Liqeni i l. Albania/Montenegro see Scutari, Lake
Shmidta, Ostrov i. Rus. Fed. 28 K1
Shmidta, Poluostrov pen. Rus. Fed. 44 F1
Shōbara Japan 45 D6
Shoh Tajik. 36 B1
Shohi Pass Pak. see Tal Pass
Shokanbetsu-dake mt. Japan 44 F4
Sholakkorgan Kazakh. 42 B4
Sholapur India see Solapur
Sholaqorghan Kazakh. see Sholakkorgan
Shomba r. Rus. Fed. 14 R4
Shomvukovo Rus. Fed. 12 K3
Shona Ridge sea feature S. Atlantic Ocean 72 I9
Shonzha Kazakh. see Chundzha
Shor India 36 D2
Shorapur India 38 C2
Sho'rchi Uzbek. 36 A1
Shorkot Pak. 36 C3
Shorkozakhly, Solonchak salt flat Turkm. 35 J2
Shoshone CA U.S.A. 65 D3
Shoshone Peak NV U.S.A. 65 D2
Shoshong Botswana 51 H2
Shostka Ukr. 13 G6
Shotor Khūn Afgh. 36 A2
Showak Sudan 32 E7
Show Low U.S.A. 62 E5
Shoyna Rus. Fed. 12 J2
Shpakovskoye Rus. Fed. 35 F1
Shpola Ukr. 13 F6
Shqipëria country Europe see Albania
Shreveport U.S.A. 63 I5
Shrewsbury U.K. 19 E6
Shri Lanka country Asia see Sri Lanka
Shri Mohangarh India 36 B4
Shrirampur India 37 G5
Shu Kazakh. 42 C4
Shū r. Kazakh./Kyrg. see Chu
Shu'ab, Ra's pt Yemen 33 H7
Shuangcheng Heilong. China 44 B3
Shuanghedagang China 44 C2
Shuangliao China 44 A4
Shuangshipu China see Fengxian
Shuangyang China 44 B4
Shuangyashan China 44 C3
Shubarkudyk Kazakh. 28 G5
Shubayh well Saudi Arabia 39 D4
Shugozero Rus. Fed. 12 G4
Shuicheng China see Liupanshui
Shulan China 44 B3
Shumagin Islands U.S.A. 60 B4
Shumba Zimbabwe 49 C5
Shumen Bulg. 27 L3
Shumerlya Rus. Fed. 12 J5
Shumilina Belarus 13 F5
Shumyachi Rus. Fed. 13 G5
Shuncheng China 44 A4
Shuoxian China see Shuozhou
Shuozhou China 43 K5
Shuqrah Yemen 32 G7
Shūr r. Iran 35 I5
Shūrjestān Iran 35 I5
Shuryshkarskiy Sor, Ozero l. Rus. Fed. 11 T2
Shūsh Iran 35 H4
Shusha Azer. see Şuşa
Shushtar Iran 35 H4
Shuwaysh, Tall ash h. Jordan 39 C4
Shuya Ivanovskaya Oblast' Rus. Fed. 12 I
Shuya Respublika Kareliya Rus. Fed. 12 G
Shuyskoye Rus. Fed. 12 I4
Shwebo Myanmar 37 H5
Shwedwin Myanmar 37 H4
Shweudaung mt. Myanmar 37 I5
Shyghanaq Kazakh. see Chiganak
Shymkent Kazakh. 42 B4
Shyok India 36 D2
Shypuvate Ukr. 13 H6
Shyroke Ukr. 13 G7
Sia Indon. 41 F8
Siahan Range mts Pak. 33 J4
Sīāh Chashmeh Iran 35 G3
Sīāhgerd Afgh. 36 A1
Sialkot Pak. 33 L3
Siam country Asia see Thailand
Sian China see Xi'an
Sian Rus. Fed. 44 B1
Siang r. India see Brahmaputra
Šiauliai Lith. 15 M9
Siazan' Azer. see Siyäzän
Sibasa S. Africa 51 J2
Şibayi, Lake S. Africa 51 K4
Šibenik Croatia 26 F3
Siberia reg. Rus. Fed. 29 M3
Siberut i. Indon. 41 B8
Sibi Pak. 33 K4
Sibiloi National Park Kenya 48 D3
Sibir' reg. Rus. Fed. see Siberia
Sibiti Congo 48 B4
Sibiu Romania 27 K2
Sibolga Indon. 41 B7
Sibsagar India 37 H4
Sibu Sarawak Malaysia 41 D7
Sibut Cent. Afr. Rep. 48 B3
Sicca Veneria Tunisia see Le Kef
Siccus watercourse Australia 57 B6
Sicheng Anhui China see Sixian
Sichuan prov. China 42 I6
Sichuan Pendi basin China 42 I7
Sicié, Cap c. France 24 G5
Sicilia i. Italy see Sicily
Sicilian Channel Italy/Tunisia 26 E6
Sicily i. Italy 26 F5
Sicuani Peru 68 D6
Siddhapur India 36 C5
Siddipet India 38 C2
Sideros, Akrotirio pt Greece 27 L7
Sidesaviwa S. Africa 50 F7
Sidhauli India 36 E4
Sidhi India 37 E4
Sidhpur India see Siddhapur
Sidi Aïssa Alg. 25 H6
Sidi Ali Alg. 25 G5
Sīdī Barrānī Egypt 34 B5
Sidi Bel Abbès Alg. 25 F6
Sidi Bennour Morocco 22 C5
Sidi Bou Sa'īd Tunisia see Sidi Bouzid
Sidi Bouzid Tunisia 26 C7
Sidi el Barrâni Egypt see Sīdī Barrānī
Sidi El Hani, Sebkhet de salt pan Tunisia 26 D7
Sidi Ifni Morocco 46 B2
Sidi Kacem Morocco 22 C5
Sidi Khaled Alg. 22 E5
Sidlaw Hills U.K. 20 F4
Sidley, Mount Antarctica 76 J1
Sidli India 37 G4
Sidmouth U.K. 19 D8
Sidney MT U.S.A. 62 G2
Sidney NE U.S.A. 62 G3
Sidney OH U.S.A. 63 K3
Sidoktaya Myanmar 37 H5
Sidon Lebanon 39 B3
Sidr Egypt see Sudr
Siedlce Poland 13 D5
Siegen Germany 17 L5
Siena Italy 26 D3
Sieradz Poland 17 Q5
Si'erdingka China 37 H3
Sierra Colorada Arg. 70 C6
Sierra de Perijá mts Venez. 68 D2
Sierra Grande Arg. 70 C6
Sierra Leone country Africa 46 B4
Sierra Leone Basin sea feature N. Atlantic Ocean 72 G5
Sierra Leone Rise sea feature N. Atlantic Ocean 72 G5
Sierra Madre Mountains CA U.S.A. 65 B
Sierra Nevada, Parque Nacional nat. par Venez. 68 D2
Sierra Nevada de Santa Marta, Parque Nacional nat. park Col. 68 D1
Sierra Vista U.S.A. 62 E5
Sierre Switz. 24 H3
Sievi Fin. 14 N5
Sifangtai China 44 B3
Sīfenī Eth. 48 E2
Sifnos i. Greece 27 K6
Sig Alg. 25 F6
Sigguup Nunaa pen. Greenland 61 M2
Sighetu Marmației Romania 13 D7
Sighișoara Romania 27 K1
Sigli Indon. 41 B7
Siglufjörður Iceland 14 [inset]
Signal de la Ste-Baume mt. France 24 G
Signal Peak AZ U.S.A. 65 E4
Sigsbee Deep sea feature G. of Mexico 75 N4
Sigüenza Spain 25 E3
Siguiri Guinea 46 C3
Sigulda Latvia 15 N8
Sihanoukville Cambodia 31 J5
Sihaung Myauk Myanmar 37 H5
Sihawa India 38 D1
Sihora India 36 E5
Siikajoki Fin. 14 N4

njärvi Fin. 14 O5
: Turkey 35 F3
wal Pak. 36 B4
aka Saudi Arabia see Sakākah
andra Rao India 36 D4
ar India 36 C4
ärām Sar mt. Afgh. 36 B2
asso Mali 46 C3
eston U.S.A. 63 J4
note-Alin' mts Rus. Fed. 44 D4
note-Alinskiy Zapovednik nature res.
us. Fed. 44 E3
nos i. Greece 27 K6
ka India 36 B5
kim state India 37 G4
sjö Sweden 14 J4
r. Spain 25 C2
' i. Saudi Arabia 34 D6
alė Lith. 15 M9
vatturai Sri Lanka 38 C4
har India 37 H4
Turkey 27 M4
ru r. India 38 D2
esia reg. Czech Rep./Pol. 17 P5
eti r. Kazakh. 42 C2
etiteniz, Ozero salt l. Kazakh. 31 G1
ghat India 37 H4
ana Tunisia 26 C6
fke Turkey 39 A1
guri India see Shiliguri
ng Co salt l. China 37 G3
our India 36 D4
stra Bulg. 27 L2
stria Bulg. see Silistra
vri Turkey 27 M4
an l. Sweden 15 I6
eborg Denmark 15 F8
ajhuay mt. Chile 68 E7
amäe Estonia 15 O7
e Turkey 34 D3
India 37 F5
od India 38 B1
bela S. Africa 51 J4
aharju Fin. 14 O3
atė Lith. 15 L9
an Turkey 35 F3
ânia Brazil 71 A2
assa India 38 B1
er Bank Passage Turks and Caicos Is
7 J4
er City NM U.S.A. 62 F5
er City NV U.S.A. 65 C1
er Lake CA U.S.A. 65 D3
ermine Mountains hills Ireland
1 D5
er Peak Range mts NV U.S.A. 65 D2
er Spring MD U.S.A. 64 C3
erton U.K. 19 D8
erton CO U.S.A. 62 F4
anggang Sarawak Malaysia see
ri Aman
a China 37 G3
nao China 42 I8
ard, Lac l. Canada 63 L2
aria India 37 F4
av Turkey 27 M5
av Dağları mts Turkey 27 M5
nba Dem. Rep. Congo 48 C3
nbirsk Rus. Fed. see Ul'yanovsk
ncoe Ont. Canada 64 A1
ncoe, Lake Canada 63 L3
ndega India 38 E1
nēn mts Eth. 48 D2
nēn Mountains Eth. see Simēn
neulue i. Indon. 41 B7
nferopol' Ukr. 34 D1
ni i. Greece see Symi
nikot Nepal 37 E3
ni Valley CA U.S.A. 65 C3
nla India see Shimla
nleu Silvaniei Romania 27 J1
nlipal National Park India 37 F5
nojärvi l. Fin. 14 O3
nplício Mendes Brazil 69 J5
nplon Pass Switz. 24 I3
npson Desert Australia 56 B5
npson Desert National Park Australia
6 B5
npson Desert Regional Reserve
nature res. Australia 57 B5
npson Peninsula Canada 61 J3
nrishamn Sweden 15 I9
nushir, Ostrov i. Rus. Fed. 43 S3
a r. India 38 B2
ai pen. Egypt 39 A5
ai al Janūbīya governorate Egypt see
anūb Sīnā'
ai ash Shamālīya governorate Egypt see
hamāl Sīnā'
alunga Italy 26 D3
ancha Rus. Fed. see Cheremshany
byugyun Myanmar 37 H5
can Turkey 35 F3
celejo Col. 68 C2
cora, Serra do hills Brazil 71 C1
d r. India 36 D4
d Pak. see Thul
d prov. Pak. see Sindh
da Rus. Fed. 44 E2
dari India 36 B4
delfingen Germany 17 L6
dh prov. Pak. 36 B4
dhuli Garhi Nepal 37 F4
dhulimadi Nepal see Sindhuli Garhi
dırgı Turkey 27 M5
dor Rus. Fed. 12 K3
dou Burkina Faso 46 C3
dri India 37 F5
d Sagar Doab lowland Pak. 36 B3
nel'nikovo Ukr. see Synel'nykove
es Port. 25 B5
es, Cabo de c. Port. 25 B5
että Fin. 14 N3
fra Côte d'Ivoire 46 C4
ga Sudan 32 D7
ganallur India 38 C4
gapore country Asia 41 C7
gapore Sing. 41 C7
gapura country Asia see Singapore
gapura Sing. see Singapore
gapuru India 38 D2
garaja Indon. 54 A2
ghana India 36 C3
gida Tanz. 49 D4
gidunum Serbia see Belgrade
gkaling Hkamti Myanmar 37 H4
gkawang Indon. 41 C7
gleton Australia 58 E4
gleton, Mount h. N.T. Australia 54 E5
gleton, Mount h. W.A. Australia 55 B7
gora Thai. see Songkhla
n'gosan N. Korea see Kosan
gra India 37 G4
gri India 37 H4
gwara India 38 D1
n'gye N. Korea 45 B5
hala country Asia see Sri Lanka
ing China see Xining
niscola Sardegna Italy 26 C4
j Croatia 26 G3
jai Indon. 41 E8

Sinjār, Jabal mt. Iraq 35 F3
Sinkat Sudan 32 E6
Sinkiang aut. reg. China see Xinjiang Uygur Zizhiqu
Sinkiang Uighur Autonomous Region aut. reg. China see Xinjiang Uygur Zizhiqu
Sinmi-do i. N. Korea 45 B5
Sinnamary Fr. Guiana 69 H2
Sinn Bishr, Gebel h. Egypt see Sinn Bishr, Jabal
Sinn Bishr, Jabal h. Egypt 39 A5
Sinneh Iran see Sanandaj
Sinoia Zimbabwe see Chinhoyi
Sinop Brazil 69 G6
Sinop Turkey 34 D2
Sinope Turkey see Sinop
Sinp'a N. Korea 44 B4
Sinp'o N. Korea 45 C4
Sinsang N. Korea 45 B5
Sint Eustatius i. West Indies 67 L5
Sint Maarten i. West Indies 67 L5
Sint-Niklaas Belgium 16 J5
Sintra Port. 25 B4
Sinŭiju N. Korea 45 B4
Siófok Hungary 26 H1
Sioma Ngwezi National Park Zambia 49 C5
Sion Switz. 24 H3
Sion Mills U.K. 21 E3
Siorapaluk Greenland 61 K2
Sioux Center U.S.A. 63 H3
Sioux City U.S.A. 63 H3
Sioux Falls U.S.A. 63 H3
Sioux Lookout Canada 63 I1
Siphaqeni S. Africa see Flagstaff
Siping China 44 B4
Siple, Mount Antarctica 76 J2
Siple Coast Antarctica 76 I1
Siple Island Antarctica 76 J2
Sipura i. Indon. 41 B8
Sīq, Wādī as watercourse Egypt 39 A5
Sir r. Pak. 36 B5
Sir, Dar"yoi r. Asia see Syrdar'ya
Sira India 38 C3
Sira r. Norway 15 E7
Siracusa Sicilia Italy see Syracuse
Siraha Nepal see Sirha
Sirajganj Bangl. 37 G4
Şiran Turkey 35 E2
Sirbāl, Jabal mt. Egypt 34 D5
Sircilla India see Sirsilla
Sirdaryo r. Asia see Syrdar'ya
Sirdaryo Uzbek. 42 B4
Sirdingka China see Si'erdingka
Sir Edward Pellew Group is Australia 56 B2
Sirha Nepal 37 F4
Sirḥān, Wādī as watercourse Jordan/Saudi Arabia 39 C4
Sirína i. Greece see Syrna
Sir James MacBrien, Mount Canada 60 F3
Sīrjān Iran 33 I4
Sirkazhi India 38 C4
Sirmilik National Park Canada 61 K2
Şırnak Turkey 35 F3
Sirohi India 36 C4
Sironj India 36 D4
Síros i. Greece see Syros
Sirpur India 38 C2
Sirretta Peak CA U.S.A. 65 C3
Sirsa India 36 C3
Sirsi Karnataka India 38 B3
Sirsi Madh. Prad. India 36 D4
Sirsi Uttar Prad. India 36 D3
Sirsilla India 38 C2
Sirte Libya 47 E1
Sirte, Gulf of Libya 47 E1
Sir Thomas, Mount h. Australia 55 E6
Siruguppa India 38 C3
Sirur India 38 B2
Şirvan Azer. 35 H3
Şirvan Turkey 35 F3
Sirvel India 38 C3
Širvintai Lith. see Širvintos
Širvintos Lith. 15 N9
Sīrwān r. Iraq 35 G4
Sis Turkey see Kozan
Sisak Croatia 26 G2
Siscia Croatia see Sisak
Sishen S. Africa 50 F4
Sisian Armenia 35 G3
Sisimiut Greenland 61 M3
Sisteron France 24 G4
Sitamarhi India 37 F4
Sitapur India 36 E4
Siteia Greece 27 L7
Siteki Swaziland 51 J4
Sithonias, Chersonisos pen. Greece 27 J4
Sitía Greece see Siteia
Sitidgi Lake Canada 60 E3
Sitila Moz. 51 L2
Sítio do Mato Brazil 71 C1
Sitka U.S.A. 60 E4
Sitra oasis Egypt see Sitrah
Sitrah oasis Egypt 34 B5
Sittaung Myanmar 37 H4
Sittingbourne U.K. 19 H7
Sittwe Myanmar 37 H5
Siuri India 37 F5
Sivaganga India 38 C4
Sivakasi India 38 C4
Sivaki Rus. Fed. 44 B1
Sivan India see Siwan
Sivas Turkey 34 E3
Sivaslı Turkey 27 M5
Siverek Turkey 35 E3
Siverskiy Rus. Fed. 15 Q7
Sivers'kyy Donets' r. Rus. Fed./Ukr. see Severskiy Donets
Sivomaskinskiy Rus. Fed. 11 S2
Sivrice Turkey 35 E3
Sivrihisar Turkey 27 N5
Sivukile S. Africa 51 I4
Sīwa Egypt see Sīwah
Sīwah Egypt 34 B5
Sīwah, Wāḥāt oasis Egypt 34 B5
Siwalik Range mts India/Nepal 36 D3
Siwan India 37 F4
Siwana India 36 C4
Siwa Oasis oasis Egypt see Sīwah, Wāḥāt
Sixian China 43 L6
Sixmilecross U.K. 21 E3
Siyabuswa S. Africa 51 I3
Siyäzän Azer. 35 H2
Siyunī Iran 35 I4
Siziwang Qi China see Ulan Hua
Sjælland i. Denmark see Zealand
Sjenica Serbia 27 I3
Sjöbo Sweden 15 H9
Sjøvegan Norway 14 J2
Skadarsko Jezero, Nacionalni Park nat. park Montenegro 27 H3
Skadovs'k Ukr. 27 O1
Skaftárós r. mouth Iceland 14 [inset]
Skagafjörður inlet Iceland 14 [inset]
Skagen Denmark 15 G8
Skagerrak str. Denmark/Norway 15 F8
Skagit r. U.S.A. 62 C2
Skagway U.S.A. 77 A3
Skaidi Norway 14 N1

Skaland Norway 14 J2
Skalmodal Sweden 14 I4
Skanderborg Denmark 15 F8
Skaneateles Lake NY U.S.A. 64 C1
Skara Sweden 15 H7
Skardarsko Jezero l. Albania/Montenegro see Scutari, Lake
Skardu Pak. 36 C2
Skärgårdshavets nationalpark nat. park Fin. 15 L7
Skarnes Norway 15 G6
Skarżysko-Kamienna Poland 17 R5
Skaulo Sweden 14 L3
Skawina Poland 17 Q6
Skeena Mountains Canada 60 F4
Skegness U.K. 18 H5
Skellefteå Sweden 14 L4
Skellefteälven r. Sweden 14 L4
Skelleftehamn Sweden 14 L4
Skelmersdale U.K. 18 E5
Skerries Ireland 21 F4
Ski Norway 15 G7
Skiathos i. Greece 27 J5
Skibbereen Ireland 21 C6
Skibotn Norway 14 L2
Skiddaw h. U.K. 18 D4
Skien Norway 15 F7
Skierniewice Poland 17 R5
Skikda Alg. 26 B6
Skipsea U.K. 18 G5
Skipton Australia 58 A6
Skipton U.K. 18 E5
Skirlaugh U.K. 18 G5
Skíros i. Greece see Skyros
Skive Denmark 15 F8
Skjern Denmark 15 F9
Skjolden Norway 15 E6
Skobelev Uzbek. see Farg'ona
Skodje Norway 14 E5
Skoganvarri Norway 14 N2
Skomer Island U.K. 19 B7
Skopelos i. Greece 27 J5
Skopin Rus. Fed. 13 H5
Skopje Macedonia 27 I4
Skoplje Macedonia see Skopje
Skövde Sweden 15 H7
Skovorodino Rus. Fed. 44 A1
Skowhegan U.S.A. 63 N3
Skrunda Latvia 15 M8
Skukum, Mount Canada 60 E3
Skukuza S. Africa 51 J3
Skull Valley AZ U.S.A. 65 F3
Skuodas Lith. 15 L8
Skurup Sweden 15 H9
Skutskär Sweden 15 J6
Skvyra Ukr. 13 F6
Skye i. U.K. 20 C3
Skyring, Seno b. Chile 70 B8
Skyros Greece 27 K5
Skyros i. Greece 27 K5
Skytrain Ice Rise Antarctica 76 L1
Slættaratindur h. Faroe Is 14 [inset]
Slagelse Denmark 15 G9
Slagnäs Sweden 14 K4
Slane Ireland 21 F4
Slaney r. Ireland 21 F5
Slantsy Rus. Fed. 15 P7
Slashers Reefs Australia 56 D3
Slatina Croatia 26 G2
Slatina Romania 27 K2
Slaty Fork WV U.S.A. 64 A3
Slava Rus. Fed. 44 C1
Slave r. Canada 77 L2
Slave Coast Africa 46 D4
Slave Lake Canada 77 L3
Slavgorod Belarus see Slawharad
Slavgorod Rus. Fed. 42 D2
Slavkovichi Rus. Fed. 15 P8
Slavonska Požega Croatia see Požega
Slavonski Brod Croatia 26 H2
Slavuta Ukr. 13 E6
Slavutych Ukr. 13 F6
Slavyanka Rus. Fed. 44 C4
Slavyansk Ukr. see Slov"yans'k
Slavyanskaya Rus. Fed. see Slavyansk-na-Kubani
Slavyansk-na-Kubani Rus. Fed. 34 E1
Slawharad Belarus 13 F5
Sławno Poland 17 P3
Sleaford U.K. 19 G5
Slea Head hd Ireland 21 B5
Sleat, Sound of sea chan. U.K. 20 D3
Sleeper Islands Canada 61 K4
Slessor Glacier Antarctica 76 B1
Slide Mountain NY U.S.A. 64 D2
Slieve Bloom Mountains hills Ireland 21 E5
Slieve Car h. Ireland 21 C3
Slieve Donard h. U.K. 21 G3
Slieve Mish Mountains hills Ireland 21 B5
Slieve Snaght h. Ireland 21 E2
Sligachan U.K. 20 C3
Sligeach Ireland see Sligo
Sligo Ireland 21 D3
Sligo PA U.S.A. 64 B2
Sligo Bay Ireland 21 D3
Slippery Rock PA U.S.A. 64 A2
Slite Sweden 15 K8
Sliven Bulg. 27 L3
Sloan NV U.S.A. 65 E3
Sloboda Rus. Fed. see Ezhva
Slobodchikovo Rus. Fed. 12 K3
Slobodskoy Rus. Fed. 12 K4
Slobozia Romania 27 L2
Slonim Belarus 15 N10
Slough U.K. 19 G7
Slovakia country Europe 10 J6
Slovenia country Europe 26 F2
Slovenija country Europe see Slovenia
Slovenj Gradec Slovenia 26 F1
Slovensko country Europe see Slovakia
Slovenský raj nat. park Slovakia 17 R6
Slov"yans'k Ukr. 13 H6
Słowiński Park Narodowy nat. park Poland 17 P3
Sluch r. Ukr. 13 E6
Słupsk Poland 17 P3
Slussfors Sweden 14 J4
Slutsk Belarus 15 O10
Slyne Head hd Ireland 21 B4
Slyudyanka Rus. Fed. 42 I2
Smallwood Reservoir Canada 61 L4
Smalyavichy Belarus 15 P9
Smalyenskaya Wzwyshsha hills Belarus/Rus. Fed. see Smolensko-Moskovskaya Vozvyshennost'
Smarhon' Belarus 15 O9
Smeaton Canada 62 G1
Smederevo Serbia 27 I2
Smederevska Palanka Serbia 27 I2
Smela Ukr. see Smila
Smethport PA U.S.A. 64 B2
Smidovich Rus. Fed. 44 D2
Smila Ukr. 13 F6
Smiltene Latvia 15 N8
Smirnykh Rus. Fed. 44 F2
Smithfield S. Africa 51 H6
Smith Glacier Antarctica 76 K1
Smith Island MD U.S.A. 64 C3
Smith Island VA U.S.A. 64 D4
Smith Mountain Lake VA U.S.A. 64 B4

Smiths Falls Canada 63 L3
Smithton Australia 57 [inset]
Smithtown Australia 58 F3
Smithville WV U.S.A. 64 A3
Smoky Bay Australia 55 F8
Smoky Cape Australia 58 F3
Smoky Hills KS U.S.A. 62 H4
Smøla i. Norway 14 E5
Smolenka Rus. Fed. 13 K6
Smolensk Rus. Fed. 13 G5
Smolensk-Moscow Upland hills Belarus/Rus. Fed. see Smolensko-Moskovskaya Vozvyshennost'
Smolensko-Moskovskaya Vozvyshennost' hills Belarus/Rus. Fed. 13 G5
Smolevichi Belarus see Smalyavichy
Smolyan Bulg. 27 K4
Smørfjord Norway 14 N1
Smorgon' Belarus see Smarhon'
Smyley Island Antarctica 76 L2
Smyrna Turkey see İzmir
Smyrna DE U.S.A. 64 D3
Smyth Island atoll Marshall Is see Taongi
Snæfell mt. Iceland 14 [inset]
Snaefell h. Isle of Man 18 C4
Snake r. U.S.A. 62 D2
Snake Island Australia 58 C7
Snake River Plain U.S.A. 62 E3
Snare Lakes Canada see Wekweètì
Snares Islands N.Z. 53 G6
Snåsa Norway 14 H4
Sneek Neth. 17 J4
Sneem Ireland 21 C6
Sneeuberge mts S. Africa 50 G6
Snegurovka Ukr. see Tetiyiv
Snelling CA U.S.A. 65 B2
Snettisham U.K. 19 H6
Snezhnogorsk Rus. Fed. 28 J3
Snežnik mt. Slovenia 26 F2
Sniečkus Lith. see Visaginas
Snihurivka Ukr. 13 G7
Snits Neth. see Sneek
Snizort, Loch b. U.K. 20 C3
Snøtinden mt. Norway 14 H3
Snovsk Ukr. see Shchors
Snowbird Lake Canada 61 H3
Snowdon mt. U.K. 19 C5
Snowdonia National Park U.K. 19 D6
Snowdrift Canada see Łutselk'e
Snowdrift r. Canada 60 G3
Snow Hill MD U.S.A. 64 D3
Snow Lake Canada 61 H4
Snowy r. Australia 58 D6
Snowy Mountain NY U.S.A. 64 D1
Snowy Mountains Australia 58 C6
Snowy River National Park Australia 58 D6
Snyder U.S.A. 62 G5
Soalala Madag. 49 E5
Soalara Madag. 49 E6
Soan r. Pak. 36 B2
Soanierana-Ivongo Madag. 49 E5
Soan-kundo is S. Korea 45 B6
Soavinandriana Madag. 49 E5
Sobat r. South Sudan 32 D8
Sobinka Rus. Fed. 12 I5
Sobradinho, Barragem de resr Brazil 69 J6
Sobral Brazil 69 J4
Sochi Rus. Fed. 13 H8
Sŏch'ŏn S. Korea 45 B5
Society Islands Fr. Polynesia 75 J7
Socorro Brazil 71 B3
Socorro Col. 68 D2
Socorro U.S.A. 62 F5
Socorro, Isla i. Mex. 66 B5
Socotra i. Yemen 33 H7
Socuéllamos Spain 25 E4
Soda Lake CA U.S.A. 65 C3
Soda Lake CA U.S.A. 65 D3
Sodankylä Fin. 14 O3
Soda Plains Aksai Chin 36 D2
Soda Springs U.S.A. 62 E3
Söderhamn Sweden 15 J6
Söderköping Sweden 15 J7
Södertälje Sweden 15 J7
Sodiri Sudan 32 C7
Sodo Eth. 48 D3
Södra Kvarken str. Fin./Sweden 15 K6
Sodus NY U.S.A. 64 C1
Soekarno, Puntjak mt. Indon. see Jaya, Puncak
Soekmekaar S. Africa 51 I2
Soerabaia Indon. see Surabaya
Sofala Australia 58 D4
Sofia Bulg. 27 J3
Sofiya Bulg. see Sofia
Sofiyevka Ukr. see Vil'nyans'k
Sofiysk Khabarovskiy Kray Rus. Fed. 44 D1
Sofiysk Khabarovskiy Kray Rus. Fed. 44 E2
Sofporog Rus. Fed. 12 F2
Sofrana i. Greece 27 L6
Softa Kalesi tourist site Turkey 39 A1
Sōfu-gan i. Japan 45 F7
Sog China 42 G6
Soğanlı Dağları mts Turkey 35 E2
Sogda Rus. Fed. 44 D2
Sogma China 36 E2
Søgne Norway 15 E7
Sognefjorden inlet Norway 15 D6
Söğüt Turkey 27 N4
Söğüt Dağı mts Turkey 27 M6
Soh Iran 35 H4
Sohâg Egypt see Sūhāj
Sohagpur India 36 D5
Soham U.K. 19 H6
Sohano P.N.G. 52 F2
Sohar Oman see Şuḥār
Sohawal India 36 E4
Sohela India 37 E5
Sŏho-ri N. Korea 45 C4
Sohŭksan-do i. S. Korea 45 B6
Soila China 37 I3
Soini Fin. 14 N5
Soissons France 24 F2
Sojat India 36 C4
Sojat Road India 36 C4
Sok r. Rus. Fed. 13 K5
Sokal' Ukr. 13 E6
Sokch'o S. Korea 45 C5
Söke Turkey 27 L6
Sokhor, Gora mt. Rus. Fed. 42 J2
Sokhumi Georgia 35 F2
Sokiryany Ukr. see Sokyryany
Sokodé Togo 46 D4
Sokol Rus. Fed. 12 I4
Sokolo Mali 46 C3
Sokoto Nigeria 46 D3
Sokoto r. Nigeria 46 D3
Sokyryany Ukr. 13 E6
Sola i. Tonga see Ata
Solan India 36 D3
Solana Beach CA U.S.A. 65 D4
Solander Island N.Z. 59 A8
Solapur India 38 B2
Soldotna U.S.A. 60 C3
Soledad CA U.S.A. 65 B2
Soledade Brazil 70 F3
Solenoye Rus. Fed. 13 I7
Solfjellsjøen Norway 14 H3
Solginskiy Rus. Fed. 12 I3
Solhan Turkey 35 F3
Soligalich Rus. Fed. 12 I4

Soligorsk Belarus see Salihorsk
Solihull U.K. 19 F6
Solikamsk Rus. Fed. 11 R4
Sol'-Iletsk Rus. Fed. 28 G4
Solimões r. S. America see Amazon
Solitaire Namibia 50 B2
Sol-Karmala Rus. Fed. see Severnoye
Şollar Azer. 35 H2
Sollefteå Sweden 14 J5
Sollum, Gulf of Egypt see Sallum, Khalīj as
Solnechnogorsk Rus. Fed. 12 H4
Solnechnyy Amurskaya Oblast' Rus. Fed. 44 A1
Solnechnyy Khabarovskiy Kray Rus. Fed. 44 E2
Solomon Islands country S. Pacific Ocean 53 G2
Solomon Sea S. Pacific Ocean 52 F2
Solor i. Indon. 54 C2
Solor, Kepulauan is Indon. 54 C2
Solothurn Switz. 24 H3
Solovetskiye Ostrova is Rus. Fed. 12 G2
Solov'yevsk Rus. Fed. 44 B1
Šolta i. Croatia 26 G3
Solṭānābād Iran 35 H4
Solṭānīyeh Iran 35 G3
Sol'tsy Rus. Fed. 12 F4
Solvay NY U.S.A. 64 C1
Sölvesborg Sweden 15 I8
Solway Firth est. U.K. 20 F6
Solwezi Zambia 49 C5
Soma Turkey 27 L5
Somalia country Africa 48 E3
Somali Basin sea feature Indian Ocean 73 L6
Somaliland reg. Somalia 48 E3
Somali Republic country Africa see Somalia
Sombo Angola 49 C4
Sombor Serbia 27 H2
Sombrio, Lago do l. Brazil 71 A5
Somero Fin. 15 M6
Somerset KY U.S.A. 63 K4
Somerset PA U.S.A. 64 B3
Somerset, Lake Australia 58 F1
Somerset East S. Africa 51 G7
Somerset Island Canada 61 I2
Somerset Reservoir VT U.S.A. 64 E1
Somerset West S. Africa 50 D8
Somersworth NH U.S.A. 64 F1
Somerton AZ U.S.A. 65 E4
Somerville NJ U.S.A. 64 D2
Someydeh Iran 35 G4
Somme r. France 19 I8
Sommen l. Sweden 15 I7
Somnath India 36 B5
Son r. India 37 F4
Sonapur India 38 D1
Sonar r. India 36 D4
Sŏnbong N. Korea 44 C4
Sŏnch'ŏn N. Korea 45 B5
Sønderborg Denmark 15 F9
Sondershausen Germany 17 M5
Søndre Strømfjord Greenland see Kangerlussuaq
Søndre Strømfjord inlet Greenland see Kangerlussuaq
Sondrio Italy 26 C1
Sonepat India see Sonipat
Sonepur India see Sonapur
Songea Tanz. 49 D5
Songhua Hu resr China 44 B4
Songhua Jiang r. Heilongjiang/Jilin China 44 D3
Songhua Jiang r. Jilin China see Di'er Songhua Jiang
Songjianghe China 44 B4
Sŏngjin N. Korea see Kimch'aek
Songkhla Thai. 31 J6
Songling China see Ta'erqi
Sŏngnam S. Korea 45 B5
Songnim N. Korea 45 B5
Songo Angola 49 B4
Songo Moz. 49 D5
Songpan China 42 I6
Songxi China 43 L7
Songyuan Fujian China see Songxi
Songyuan Jilin China 44 B3
Sonid Youqi China see Saihan Tal
Sonid Zuoqi China see Mandalt
Sonipat India 36 D3
Sonkajärvi Fin. 14 O5
Sonkovo Rus. Fed. 12 H4
Sơn La Vietnam 31 J4
Sonmiani Pak. 33 K4
Sonmiani Bay Pak. 36 A4
Sono r. Minas Gerais Brazil 71 B2
Sono r. Tocantins Brazil 69 I5
Sonoma CA U.S.A. 65 A1
Sonora r. Mex. 66 B3
Sonora CA U.S.A. 65 B2
Sonora TX U.S.A. 62 G5
Sonoran Desert AZ U.S.A. 65 F4
Sonqor Iran 35 G4
Sonsonate El Salvador 66 G6
Sonwabile S. Africa 51 I6
Soochow China see Suzhou
Soomaaliya country Africa see Somalia
Sopo watercourse South Sudan 47 F4
Sopot Bulg. 27 K3
Sopot Poland 17 Q3
Sopron Hungary 26 G1
Sopur India 36 C2
Sora Italy 26 E4
Sorab India 38 B3
Sorada India 38 E2
Söråker Sweden 14 J5
Sŏrak-san mt. S. Korea 45 C5
Sorak-san National Park S. Korea 45 C5
Sorel Canada 63 M2
Soreq r. Israel 39 B4
Sorgun Turkey 34 D3
Sorgun r. Turkey 39 B1
Soria Spain 25 E3
Sorkh, Kūh-e mts Iran 35 I4
Sorkheh Iran 35 I4
Sørli Norway 14 H4
Soro India 37 F5
Soroca Moldova 13 F6
Sorocaba Brazil 71 B3
Soroki Moldova see Soroca
Sorol atoll Micronesia 41 G7
Sorong Indon. 41 F8
Soroti Uganda 48 D3
Sørøya i. Norway 14 M1
Sorraia r. Port. 25 B4
Sørreisa Norway 14 K2
Sorrento Italy 26 F4
Sorsele Sweden 14 J4
Sorsogon Phil. 41 E6
Sortavala Rus. Fed. 12 F3
Sortland Norway 14 I2
Sortopolovskaya Rus. Fed. 12 K3
Sorūbay Afgh. 36 B2
Sorvizhi Rus. Fed. 12 K4
Sŏsan S. Korea 45 B5
Sosenskiy Rus. Fed. 13 G5
Soshanguve S. Africa 51 I3
Sosna r. Rus. Fed. 13 H5
Sosneado mt. Arg. 70 C4
Sosnogorsk Rus. Fed. 12 L3

Sosnovka Arkhangel'skaya Oblast' Rus. Fed. 12 J3
Sosnovka Kaliningradskaya Oblast' Rus. Fed. 11 K5
Sosnovka Murmanskaya Oblast' Rus. Fed. 12 I2
Sosnovka Tambovskaya Oblast' Rus. Fed. 13 I5
Sosnovo Rus. Fed. 15 Q6
Sosnovo-Ozerskoye Rus. Fed. 43 K2
Sosnovyy Rus. Fed. 14 R4
Sosnovyy Bor Rus. Fed. 12 F4
Sosnowiec Poland 17 Q5
Sosnowitz Poland see Sosnowiec
Sos'va Khanty-Mansiyskiy Avtonomnyy Okrug-Yugra Rus. Fed. 11 S3
Sos'va Sverdlovskaya Oblast' Rus. Fed. 11 S4
Sotang China 37 H3
Sotara, Volcán vol. Col. 68 C3
Sotkamo Fin. 14 P4
Sotteville-lès-Rouen France 19 I9
Souanké Congo 48 B3
Soubré Côte d'Ivoire 46 C4
Souderton PA U.S.A. 64 D2
Soufli Greece 27 L4
Soufrière St Lucia 67 L6
Soufrière vol. St Vincent 67 L6
Sougueur Alg. 25 G6
Souillac France 24 E4
Souk Ahras Alg. 26 B6
Souk el Arbaâ du Rharb Morocco 22 C5
Sŏul S. Korea see Seoul
Soulac-sur-Mer France 24 D4
Soulom France 24 D5
Souni Cyprus 39 A2
Soûr Lebanon see Tyre
Soure Brazil 69 I4
Sour el Ghozlane Alg. 25 H5
Souris Canada 62 G2
Souris r. Canada 62 H2
Souriya country Asia see Syria
Sousa Brazil 69 K5
Sousa Lara Angola see Bocoio
Sousse Tunisia 26 D7
Soustons France 24 D5
South Africa, Republic of country Africa 50 F5
Southampton Canada 63 K3
Southampton U.K. 19 F8
Southampton NY U.S.A. 64 E2
Southampton, Cape Canada 61 J3
Southampton Island Canada 61 J3
South Anna r. VA U.S.A. 64 C4
South Anston U.K. 18 F5
South Australia state Australia 52 D5
South Australian Basin sea feature Indian Ocean 73 P8
South Bank U.K. 18 F4
South Bend IN U.S.A. 63 J3
South Carolina state U.S.A. 63 K5
South China Sea N. Pacific Ocean 41 D6
South Coast Town Australia see Gold Coast
South Dakota state U.S.A. 62 G3
South Downs hills U.K. 19 G8
South Downs National Park U.K. 19 G8
South-East admin. dist. Botswana 51 G3
South East Cape Australia 57 [inset]
Southeast Cape U.S.A. 60 B3
Southeast Indian Ridge sea feature Indian Ocean 73 N8
South East Isles Australia 55 C8
Southeast Pacific Basin sea feature S. Pacific Ocean 75 M10
South East Point Australia 58 C7
Southend U.K. 20 D5
Southend-on-Sea U.K. 19 H7
Southern admin. dist. Botswana 50 G3
Southern Alps mts N.Z. 59 C6
Southern Cross Australia 55 B7
Southern Indian Lake Canada 61 I4
Southern Lau Group is Fiji 53 I3
Southern National Park South Sudan 47 F4
Southern Ocean 76 C2
Southern Rhodesia country Africa see Zimbabwe
Southern Uplands hills U.K. 20 E5
South Esk r. U.K. 20 F4
South Esk Tableland reg. Australia 54 D4
South Fiji Basin sea feature S. Pacific Ocean 74 H7
South Geomagnetic Pole (1995) Antarctica 76 F1
South Georgia terr. S. Atlantic Ocean 70 I8
South Georgia and the South Sandwich Islands terr. S. Atlantic Ocean 70 I8
South Harris pen. U.K. 20 B3
South Henik Lake Canada 61 I3
South Honshu Ridge sea feature N. Pacific Ocean 74 F3
South Island India 38 B4
South Island N.Z. 59 D7
South Korea country Asia 45 B5
South Lake Tahoe CA U.S.A. 65 B1
South Luangwa National Park Zambia 49 D5
South Magnetic Pole Antarctica 76 G2
Southminster U.K. 19 H7
South Mountains hills PA U.S.A. 64 C3
South New Berlin NY U.S.A. 64 D1
South Orkney Islands S. Atlantic Ocean 72 F10
South Platte r. U.S.A. 62 G3
South Pole Antarctica 76 C1
Southport Qld Australia 58 F1
Southport Tas. Australia 57 [inset]
Southport U.K. 18 D5
South Ronaldsay i. U.K. 20 G2
South Sand Bluff pt S. Africa 51 J6
South Sandwich Islands S. Atlantic Ocean 72 G9
South Sandwich Trench sea feature S. Atlantic Ocean 72 G9
South San Francisco CA U.S.A. 65 A2
South Saskatchewan r. Canada 62 F1
South Shetland Islands Antarctica 76 A2
South Shetland Trough sea feature S. Atlantic Ocean 76 L2
South Shields U.K. 18 F3
South Sinai governorate Egypt see Janūb Sīnā'
South Solomon Trench sea feature S. Pacific Ocean 74 G6
South Sudan country Africa 47 F4
South Taranaki Bight b. N.Z. 59 E4
South Tasman Rise sea feature Southern Ocean 74 F9
South Tons r. India 37 E4
South Tyne r. U.K. 18 E4
South Uist i. U.K. 20 B3
South Wellesley Islands Australia 56 B3
South-West Africa country Africa see Namibia
South West Cape N.Z. 59 A8
South West Entrance sea chan. P.N.G. 56 E1
Southwest Indian Ridge sea feature Indian Ocean 73 K8

South West National Park Australia 57 [inset]
Southwest Pacific Basin sea feature S. Pacific Ocean 74 I8
Southwest Peru Ridge sea feature S. Pacific Ocean see Nazca Ridge
South West Rocks Australia 58 F3
Southwold U.K. 19 I6
Southwood National Park Australia 58 E1
Soutpansberg mts S. Africa 51 I2
Souttouf, Adrar mts W. Sahara 46 B2
Soverato Italy 26 G5
Sovetsk Kaliningradskaya Oblast' Rus. Fed. 12 D5
Sovetsk Kirovskaya Oblast' Rus. Fed. 12 K4
Sovetskaya Gavan' Rus. Fed. 44 F2
Sovetskiy Khanty-Mansiyskiy Avtonomnyy Okrug-Yugra Rus. Fed. 11 S3
Sovetskiy Leningradskaya Oblast' Rus. Fed. 15 P6
Sovetskiy Respublika Mariy El Rus. Fed. 12 K4
Sovetskoye Chechenskaya Respublika Rus. Fed. see Shatoy
Sovetskoye Stavropol'skiy Kray Rus. Fed. see Zelenokumsk
Sovyets'kyy Ukr. 34 D1
Soweto S. Africa 51 H4
Sōya-kaikyō str. Japan/Rus. Fed. see La Pérouse Strait
Sōya-misaki c. Japan 44 F3
Soyana r. Rus. Fed. 12 I2
Soyma r. Rus. Fed. 12 K2
Sozh r. Europe 13 F6
Sozopol Bulg. 27 L3
Spain country Europe 25 E3
Spalato Croatia see Split
Spalatum Croatia see Split
Spalding U.K. 19 G6
Spanish Guinea country Africa see Equatorial Guinea
Spanish Netherlands country Europe see Belgium
Spanish Sahara terr. Africa see Western Sahara
Spanish Town Jamaica 67 I5
Sparks U.S.A. 62 D4
Sparta Greece see Sparti
Spartanburg U.S.A. 63 K5
Sparti Greece 27 J6
Spartivento, Capo c. Italy 26 G6
Spas-Demensk Rus. Fed. 13 G5
Spas-Klepiki Rus. Fed. 13 I5
Spassk-Dal'niy Rus. Fed. 44 D3
Spassk-Ryazanskiy Rus. Fed. 13 I5
Spata (Eleftherios Venezielos) airport Greece 27 J6
Spatha, Akrotirio pt Greece 27 J7
Spence Bay Canada see Taloyoak
Spencer IA U.S.A. 63 H3
Spencer WV U.S.A. 64 A3
Spencer Bay Namibia 50 B3
Spencer Gulf est. Australia 57 B7
Spencer Range hills Australia 54 E3
Spennymoor U.K. 18 F4
Sperrgebiet National Park Namibia 50 B4
Sperrin Mountains hills U.K. 21 E3
Sperryville VA U.S.A. 64 B3
Spessart reg. Germany 17 L6
Spétsai i. Greece see Spetses
Spetses i. Greece 27 J6
Spey r. U.K. 20 F3
Speyer Germany 17 L6
Spezand Pak. 36 A3
Spice Islands Indon. see Moluccas
Spijkenisse Neth. 16 J5
Spilimbergo Italy 26 E1
Spilsby U.K. 18 H5
Spīn Būldak Afgh. 36 A3
Spintangi Pak. 36 B3
Spirovo Rus. Fed. 12 G4
Spišská Nová Ves Slovakia 13 D6
Spiti r. India 36 D3
Spitsbergen i. Svalbard 28 C2
Spittal an der Drau Austria 17 N7
Spitzbergen i. Svalbard see Spitsbergen
Split Croatia 26 G3
Spokane U.S.A. 62 D2
Spoletium Italy see Spoleto
Spoleto Italy 26 E3
Spooner U.S.A. 63 I2
Spotsylvania VA U.S.A. 64 C3
Spratly Islands S. China Sea 41 D6
Spree r. Germany 17 N4
Springbok S. Africa 50 C5
Springdale Canada 61 M5
Springer U.S.A. 62 G4
Springfield CO U.S.A. 62 G4
Springfield IL U.S.A. 63 J4
Springfield MA U.S.A. 64 E1
Springfield MO U.S.A. 63 I4
Springfield OH U.S.A. 63 K4
Springfield WV U.S.A. 64 B3
Springfontein S. Africa 51 G6
Spring Hill U.S.A. 63 K6
Spring Mountains NV U.S.A. 65 E2
Springs Junction N.Z. 59 D6
Springsure Australia 56 E5
Spring Valley NY U.S.A. 64 D2
Springville CA U.S.A. 65 C2
Springville NY U.S.A. 64 B1
Springville PA U.S.A. 64 D2
Sprowston U.K. 19 I6
Spurn Head hd U.K. 18 H5
Squillace, Golfo di g. Italy 26 G5
Squires, Mount h. Australia 55 D6
Srbija country Europe see Serbia
Srbinje Bos.-Herz. see Foča
Srebrenica Republika Srpska 27 H2
Sredets Burgas Bulg. 27 L3
Sredets Sofiya-Grad Bulg. see Sofia
Sredinnyy Khrebet mts Rus. Fed. 29 Q4
Sredna Gora mts Bulg. 27 J3
Srednekolymsk Rus. Fed. 29 Q3
Sredne-Russkaya Vozvyshennost' hills Rus. Fed. see Central Russian Upland
Sredne-Sibirskoye Ploskogor'ye plat. Rus. Fed. see Central Siberian Plateau
Sredneye Kuyto, Ozero l. Rus. Fed. 14 Q4
Sredniy Ural mts Rus. Fed. 11 R4
Srednogorie Bulg. 27 K3
Srednyaya Akhtuba Rus. Fed. 13 J6
Sreepur Bangl. see Sripur
Sretensk Rus. Fed. 43 L2
Sri Aman Sarawak Malaysia 41 D7
Sriharikota Island India 38 D3
Sri Jayewardenepura Kotte Sri Lanka 38 C5
Srikakulam India 38 E2
Sri Kalahasti India 38 C3
Sri Lanka country Asia 38 D5
Srinagar India 36 C2
Sri Pada mt. Sri Lanka see Adam's Peak
Sripur Bangl. 37 G4
Srirangam India 38 C4
Srivardhan India 38 B2
Staaten r. Australia 56 C3
Staaten River National Park Australia 56 C3
Stabroek Guyana see Georgetown
Stade Germany 17 L4
Stadskanaal Neth. 17 K4
Staffa i. U.K. 20 C4
Stafford U.K. 19 E6
Stafford VA U.S.A. 64 C3
Stafford Springs CT U.S.A. 64 E2
Staicele Latvia 15 N8
Staines U.K. 19 G7
Stakhanov Ukr. 13 H6
Stakhanovo Rus. Fed. see Zhukovskiy
Stalbridge U.K. 19 E8
Stalham U.K. 19 I6
Stalin Bulg. see Varna
Stalinabad Tajik. see Dushanbe
Stalingrad Rus. Fed. see Volgograd
Staliniri Georgia see Ts'khinvali
Stalino Ukr. see Donets'k
Stalinogorsk Rus. Fed. see Novomoskovsk
Stalinogród Poland see Katowice
Stalinsk Rus. Fed. see Novokuznetsk
Stalowa Wola Poland 13 D6
Stamboliyski Bulg. 27 K3
Stamford Australia 56 C4
Stamford U.K. 19 G6
Stamford CT U.S.A. 64 E2
Stamford NY U.S.A. 64 D1
Stampalia i. Greece see Astypalaia
Stampriet Namibia 50 D3
Stamsund Norway 14 H2
Stanardsville VA U.S.A. 64 B3
Stancomb-Wills Glacier Antarctica 76 B1
Standerton S. Africa 51 I4
Standish U.S.A. 63 K3
Stanger S. Africa 51 J5
Stanislaus r. CA U.S.A. 65 B2
Stanislav Ukr. see Ivano-Frankivs'k
Stanke Dimitrov Bulg. see Dupnitsa
Stanley Australia 57 [inset]
Stanley Falkland Is 70 E8
Stanley U.K. 18 F4
Stanley ND U.S.A. 62 G2
Stanley VA U.S.A. 64 B3
Stanley, Mount h. N.T. Australia 54 E5
Stanley, Mount h. Tas. Australia 57 [inset]
Stanley, Mount Dem. Rep. Congo/Uganda see Margherita Peak
Stanleyville Dem. Rep. Congo see Kisangani
Stann Creek Belize see Dangriga
Stannington U.K. 18 F3
Stanovoye Rus. Fed. 13 H5
Stanovoye Nagor'ye mts Rus. Fed. 43 L1
Stanovoy Khrebet mts Rus. Fed. 29 N4
Stansmore Range hills Australia 54 E5
Stanthorpe Australia 58 E2
Stanton U.K. 19 H6
Starachowice Poland 17 R5
Stara Planina mts Bulg./Serbia see Balkan Mountains
Staraya Russa Rus. Fed. 12 F4
Stara Zagora Bulg. 27 K3
Starbuck Island Kiribati 75 J6
Starcke National Park Australia 56 D2
Stargard in Pommern Poland see Stargard Szczeciński
Stargard Szczeciński Poland 17 O4
Staritsa Rus. Fed. 12 G4
Starkville U.S.A. 63 J5
Starnberger See l. Germany 17 M7
Starobel'sk Ukr. see Starobil's'k
Starobil's'k Ukr. 13 H6
Starogard Gdański Poland 17 Q4
Starokonstantinov Ukr. see Starokostyantyniv
Starokostyantyniv Ukr. 13 E6
Starominskaya Rus. Fed. 13 H7
Staroshcherbinovskaya Rus. Fed. 13 H7
Start Point U.K. 19 D8
Starve Island Kiribati see Starbuck Island
Staryya Darohi Belarus 13 F5
Staryye Dorogi Belarus see Staryya Darohi
Staryy Kayak Rus. Fed. 29 L2
Staryy Oskol Rus. Fed. 13 H6
State College PA U.S.A. 64 C2
Staten Island Arg. see Los Estados, Isla de
Statesboro U.S.A. 63 K5
Statia i. West Indies see Sint Eustatius
Station Nord Greenland 77 I1
Staunton VA U.S.A. 64 B3
Stavanger Norway 15 D7
Staveley U.K. 18 F5
Stavropol' Rus. Fed. 13 I7
Stavropol Kray admin. div. Rus. Fed. see Stavropol'skiy Kray
Stavropol'-na-Volge Rus. Fed. see Tol'yatti
Stavropol'skaya Vozvyshennost' hills Rus. Fed. 13 I7
Stavropol'skiy Kray admin. div. Rus. Fed. 35 F1
Steadville S. Africa 51 I5
Steamboat Springs U.S.A. 62 F3
Stebbins U.S.A. 60 B3
Steele Island Antarctica 76 L2
Steenkampsberge mts S. Africa 51 J3
Steen River Canada 60 G4
Steens Mountain U.S.A. 62 D3
Steenstrup Gletscher glacier Greenland see Sermersuaq
Stefansson Island Canada 61 H2
Stegi Swaziland see Siteki
Steigerwald mts Germany 17 M6
Steinhausen Namibia 49 B6
Steinkjer Norway 14 G4
Steinkopf S. Africa 50 C5
Steinsdalen Norway 14 G4
Stella S. Africa 50 G4
Stellenbosch S. Africa 50 D7
Stello, Monte mt. Corse France 24 I5
Stelvio, Parco Nazionale dello nat. park Italy 26 D1
Stendal Germany 17 M4
Stenhousemuir U.K. 20 F4
Stenungsund Sweden 15 G7
Steornabhagh U.K. see Stornoway
Stepanakert Azer. see Xankändi
Stephens, Cape N.Z. 59 D5
Stephens City VA U.S.A. 64 B3
Stephenville U.S.A. 62 H5
Stepnoy Rus. Fed. see Elista
Stepnoye Rus. Fed. 13 J6
Sterkfontein Dam dam S. Africa 51 I5
Sterkstroom S. Africa 51 H6
Sterlibashevo Rus. Fed. 11 R5
Sterling S. Africa 50 E6
Sterling CO U.S.A. 62 G3
Sterlitamak Rus. Fed. 28 G4
Stettin Poland see Szczecin
Steubenville OH U.S.A. 64 A2
Stevenage U.K. 19 G7
Stevenson Lake Canada 63 H1
Stevens Village U.S.A. 60 D3
Stevensville PA U.S.A. 64 C2
Stewart r. Canada 60 E3
Stewart, Isla i. Chile 70 B8
Stewart Island N.Z. 59 A8
Stewart Islands Solomon Is 53 G2
Stewart Lake Canada 61 J3
Stewarton U.K. 20 E5
Stewarts Point CA U.S.A. 65 A1
Steynsburg S. Africa 51 G6
Steyr Austria 17 O6
Steytlerville S. Africa 50 G7
Stif Alg. see Sétif
Stikine r. Canada 60 E4
Stikine Plateau Canada 60 E4
Stilbaai S. Africa 50 E8
Stillwater OK U.S.A. 63 H4
Stilton U.K. 19 G6
Štip Macedonia 27 J4
Stirling Australia 54 F5
Stirling U.K. 20 F4
Stirling Creek r. Australia 54 E4
Stirling Range National Park Australia 55 B8
Stjørdalshalsen Norway 14 G5
Stockerau Austria 17 P6
Stockholm Sweden 15 K7
Stockinbingal Australia 58 C5
Stockport U.K. 18 E5
Stockton CA U.S.A. 65 B2
Stockton-on-Tees U.K. 18 F4
Stoer, Point of U.K. 20 D2
Stoke-on-Trent U.K. 19 E5
Stokesley U.K. 18 F4
Stokes Point Australia 57 [inset]
Stokes Range hills Australia 54 E4
Stokkseyri Iceland 14 [inset]
Stokkvågen Norway 14 H3
Stokmarknes Norway 14 I2
Stolac Bos.-Herz. 26 G3
Stolbovoy Rus. Fed. 77 G2
Stolbtsy Belarus see Stowbtsy
Stolin Belarus 15 O11
Stolp Poland see Słupsk
Stone U.K. 19 E6
Stoneboro PA U.S.A. 64 A2
Stonehaven U.K. 20 G4
Stonehenge Australia 56 C5
Stonehenge tourist site U.K. 19 F7
Stonewall Jackson Lake WV U.S.A. 64 A3
Stony Creek VA U.S.A. 64 C4
Stony River U.S.A. 60 C3
Stora Lulevatten l. Sweden 14 K3
Stora Sjöfallets nationalpark nat. park Sweden 14 J3
Storavan l. Sweden 14 K4
Store Bælt sea chan. Denmark see Great Belt
Støren Norway 14 G5
Storfjordbotn Norway 14 O1
Storforshei Norway 14 I3
Storjord Norway 14 I3
Storkerson Peninsula Canada 61 H2
Storm Bay Australia 57 [inset]
Stormberg S. Africa 51 H6
Storm Lake U.S.A. 63 H3
Stornosa mt. Norway 14 E6
Stornoway U.K. 20 C2
Storozhevsk Rus. Fed. 12 L3
Storozhynets' Ukr. 13 E6
Storrs CT U.S.A. 64 E2
Storseleby Sweden 14 J4
Storsjön l. Sweden 14 I5
Storskrymten mt. Norway 14 F5
Storslett Norway 14 L2
Storuman Sweden 14 J4
Storuman l. Sweden 14 J4
Storvik Sweden 15 J6
Storvorde Denmark 15 G8
Storvreta Sweden 15 J7
Stotfold U.K. 19 G6
Stour r. England U.K. 19 F6
Stour r. England U.K. 19 F8
Stour r. England U.K. 19 I7
Stour r. England U.K. 19 I7
Stourbridge U.K. 19 E6
Stourport-on-Severn U.K. 19 E6
Stout Lake Canada 63 I1
Stowbtsy Belarus 15 O10
Stowmarket U.K. 19 H6
Stoyba Rus. Fed. 44 C1
Strabane U.K. 21 E3
Stradbally Ireland 21 E4
Stradbroke U.K. 19 I6
Stradella Italy 26 C2
Strakonice Czech Rep. 17 N6
Stralsund Germany 17 N3
Strand S. Africa 50 D8
Stranda Norway 14 E5
Strangford U.K. 21 G3
Strangford Lough inlet U.K. 21 G3
Strangways r. Australia 54 F3
Stranraer U.K. 20 D6
Strasbourg France 24 H2
Strasburg VA U.S.A. 64 B3
Strassburg France see Strasbourg
Stratford Australia 58 C6
Stratford Ont. Canada 64 A1
Stratford CA U.S.A. 65 C2
Stratford TX U.S.A. 62 G4
Stratford-upon-Avon U.K. 19 F6
Strathaven U.K. 20 E5
Strathmore r. U.K. 20 E2
Strathroy Ont. Canada 64 A1
Strathspey val. U.K. 20 F3
Strathy U.K. 20 F2
Stratton U.K. 19 C8
Stratton Mountain VT U.S.A. 64 E1
Straubing Germany 17 N6
Straumnes pt Iceland 14 [inset]
Streaky Bay Australia 55 F8
Streaky Bay b. Australia 55 F8
Street U.K. 19 E7
Streetsboro OH U.S.A. 64 A2
Strehaia Romania 27 J2
Streich Mound h. Australia 55 C7
Strelka Rus. Fed. 29 Q3
Strel'na r. Rus. Fed. 12 H2
Strenči Latvia 15 N8
Streymoy i. Faroe Is 14 [inset]
Strichen U.K. 20 G3
Stroeder Arg. 70 D6
Strokestown Ireland 21 D4
Stroma, Island of U.K. 20 F2
Stromboli, Isola i. Italy 26 F5
Stromness S. Georgia 70 I8
Stromness U.K. 20 F2
Strömstad Sweden 15 G7
Strömsund Sweden 14 I5
Stronsay i. U.K. 20 G1
Stroud Australia 58 E4
Stroud U.K. 19 E7
Stroud Road Australia 58 E4
Stroudsburg PA U.S.A. 64 D2
Struer Denmark 15 F8
Struga Macedonia 27 I4
Strugi-Krasnyye Rus. Fed. 15 P7
Struis Bay S. Africa 50 E8
Struma r. Bulg. 27 J4
Strumble Head hd U.K. 19 B6
Strumica Macedonia 27 J4
Struthers OH U.S.A. 64 A2
Stryama r. Bulg. 27 K3
Strydenburg S. Africa 50 F5
Strymonas r. Greece 27 J4
Stryn Norway 14 E6
Stryy Ukr. 13 D6
Strzelecki, Mount h. Australia 54 F5
Strzelecki Regional Reserve nature res. Australia 57 B6
Stuart Lake Canada 60 F4
Stuart Range hills Australia 57 A6
Stuarts Draft VA U.S.A. 64 B3
Stuart Town Australia 58 D4
Stuchka Latvia see Aizkraukle
Stučka Latvia see Aizkraukle
Studholme Junction N.Z. 59 C7
Studsviken Sweden 14 K5
Stupino Rus. Fed. 13 H5
Sturge Island Antarctica 76 H2
Sturgis SD U.S.A. 62 G3
Sturt, Mount h. Australia 57 C6
Sturt Creek watercourse Australia 54 D4
Sturt National Park Australia 57 C6
Sturt Stony Desert Australia 57 C6
Stutterheim S. Africa 51 H7
Stuttgart Germany 17 L6
Stuttgart U.S.A. 63 I5
Stykkishólmur Iceland 14 [inset]
Styr r. Belarus/Ukr. 13 E5
Suaçuí Grande r. Brazil 71 C2
Suai East Timor 54 D2
Suakin Sudan 32 E6
Suau P.N.G. 56 E1
Subačius Lith. 15 N9
Subankhata India 37 G4
Subarnapur India see Sonapur
Sūbāshī Iran 35 H4
Şubayḩah Saudi Arabia 39 D4
Subei China 42 G5
Subotica Serbia 27 H1
Success, Lake CA U.S.A. 65 C2
Succiso, Alpi di mts Italy 26 D2
Suceava Romania 13 E7
Suchan Rus. Fed. see Partizansk
Suck r. Ireland 21 D4
Suckling, Mount P.N.G. 56 E1
Sucre Bol. 68 E7
Suczawa Romania see Suceava
Sud, Grand Récif du rf New Caledonia 53 G4
Suda Rus. Fed. 12 H4
Sudak Ukr. 34 D1
Sudan country Africa 47 F3
Suday Rus. Fed. 12 I4
Sudbury Canada 63 K2
Sudbury U.K. 19 H6
Sudd swamp South Sudan 32 C8
Sudest Island P.N.G. see Tagula Island
Sudetenland mts Czech Rep./Poland see Sudety
Sudety mts Czech Rep./Poland 17 O5
Sudislavl' Rus. Fed. 12 I4
Sudlersville MD U.S.A. 64 D3
Sudogda Rus. Fed. 12 I5
Sudr Egypt 39 A5
Suðuroy i. Faroe Is 14 [inset]
Sue watercourse South Sudan 47 F4
Sueca Spain 25 F4
Suez Egypt 39 A5
Suez, Gulf of Egypt 39 A5
Suez Bay Egypt 39 A5
Suez Canal Egypt 39 A4
Suffolk U.S.A. 63 L4
Sugarloaf Point Australia 58 F4
Sugun China 42 D5
Sūhāj Egypt 32 D4
Şuḩār Oman 33 I5
Suhaymī, Wādī as watercourse Egypt 39 A4
Sühbaatar Mongolia 42 J2
Suheli Par atoll India 38 B4
Suhl Germany 17 M5
Şuhut Turkey 27 N5
Sui Pak. 36 B3
Suibin China 44 C3
Suid-Afrika country Africa see Republic of South Africa
Suide China 43 K5
Sui Dehar Pak. 36 B3
Suidzhikurmsy Turkm. see Madaw
Suifenhe China 44 C3
Suihua China 44 B3
Suileng China 44 B3
Suining Sichuan China 42 J6
Suir r. Ireland 21 E5
Suisse country Europe see Switzerland
Suixian Hubei China see Suizhou
Suiza country Europe see Switzerland
Suizhong China 43 M4
Suizhou China 43 K6
Sujangarh India 36 C4
Sujawal Pak. 36 B4
Sukabumi Indon. 41 C8
Sukagawa Japan 45 F5
Sukarnapura Indon. see Jayapura
Sukarno, Puncak mt. Indon. see Jaya, Puncak
Sukchŏn N. Korea 45 B5
Sukhinichi Rus. Fed. 13 G5
Sukhona r. Rus. Fed. 12 J3
Sukhumi Georgia see Sokhumi
Sukhum-Kale Georgia see Sokhumi
Sukkertoppen Greenland see Maniitsoq
Sukkozero Rus. Fed. 12 G3
Sukkur Pak. 33 K4
Sukma India 38 D2
Sukpay Rus. Fed. 44 E3
Sukpay r. Rus. Fed. 44 E3
Sukri r. India 36 C4
Sukri r. India 36 C4
Suktel r. India 38 D1
Sukun i. Indon. 54 C2
Sula i. Norway 15 D6
Sula r. Rus. Fed. 12 K2
Sula, Kepulauan is Indon. 41 E8
Sulaiman Range mts Pak. 33 K3
Sulak Rus. Fed. 35 G2
Sūlār Iran 35 H5
Sula Sgeir i. U.K. 20 C1
Sulawesi i. Indon. see Celebes
Sulaymān Beg Iraq 35 G4
Sulci Sardegna Italy see Sant'Antioco
Sulcis Sardegna Italy see Sant'Antioco
Suledeh Iran 35 H3
Sule Skerry i. U.K. 20 E1
Sule Stack i. U.K. 20 E1
Sulitjelma Norway 14 J3
Sulkava Fin. 14 P6
Sullana Peru 68 B4
Sulmo Italy see Sulmona
Sulmona Italy 26 E3
Sulphur Springs U.S.A. 63 H5
Sultanabad India see Osmannagar
Sultanabad Iran see Arāk
Sultan Dağları mts Turkey 27 N5
Sultanıye Turkey see Karapınar
Sultanpur India 37 E4
Sulu Archipelago is Phil. 41 E7
Sulu Basin sea feature N. Pacific Ocean 74 E5
Sülüklü Turkey 34 D3
Sulusaray Turkey 34 E3
Sulu Sea N. Pacific Ocean 41 D7
Sulzberger Bay Antarctica 76 I1
Sumāil Oman 33 I5
Sumampa Arg. 70 D3
Sumapaz, Parque Nacional nat. park Col. 68 D3
Sūmār Iran 35 G4
Sumatera i. Indon. see Sumatra
Sumatra i. Indon. 41 B7
Šumava nat. park Czech Rep. 17 N6
Sumba i. Indon. 41 E8
Sumba, Selat sea chan. Indon. 41 D8
Sumbar r. Turkm. 35 I3
Sumbawa i. Indon. 41 D8
Sumbawabesar Indon. 54 B2
Sumbawanga Tanz. 49 D4
Sumbe Angola 49 B5
Sumbu National Park Zambia 49 D4
Sumburgh U.K. 20 [inset]
Sumburgh Head hd U.K. 20 [inset]
Sume'eh Sarā Iran 35 H3
Sumeih Sudan 32 C8
Sumgait Azer. see Sumqayıt
Sumisu-jima i. Japan 43 Q6
Summēl Iraq 35 F3
Summer Isles U.K. 20 D2
Summersville WV U.S.A. 64 A3
Summit Lake Canada 60 F4
Sumnal Aksai Chin 36 D2
Sumner N.Z. 59 D6
Sumner, Lake N.Z. 59 D6
Sumon-dake mt. Japan 45 E5
Šumperk Czech Rep. 17 P6
Sumpu Japan see Shizuoka
Sumqayıt Azer. 35 H2
Sumskiy Posad Rus. Fed. 12 G2
Sumter U.S.A. 63 K5
Sumur India 36 D2
Sumy Ukr. 13 G6
Sumzom China 37 I3
Suna Rus. Fed. 12 K4
Sunaj India 36 D4
Sunam India 36 C3
Sunamganj Bangl. 37 G4
Sunart, Loch inlet U.K. 20 D4
Sunbury Australia 58 B6
Sunbury PA U.S.A. 64 C2
Sunch'ŏn S. Korea 45 B6
Sun City S. Africa 51 H3
Sun City CA U.S.A. 65 D4
Sunda, Selat str. Indon. 41 C8
Sunda Kalapa Indon. see Jakarta
Sundance U.S.A. 62 G3
Sundarbans coastal area Bangl./India 37 G5
Sundarbans National Park Bangl./India 37 G5
Sundargarh India 37 F5
Sunda Shelf sea feature Indian Ocean 73 P5
Sunda Strait Indon. see Sunda, Selat
Sunda Trench sea feature Indian Ocean see Java Trench
Sunderland U.K. 18 F4
Sündiken Dağları mts Turkey 27 N5
Sundown National Park Australia 58 E2
Sundsvall Sweden 14 J5
Sundumbili S. Africa 51 J5
Sungari r. China see Songhua Jiang
Sungqu China see Songpan
Sungurlu Turkey 34 D2
Sun Kosi r. Nepal 37 F4
Sunndal Norway 15 E6
Sunndalsøra Norway 14 F5
Sunne Sweden 15 H7
Sunnyside U.S.A. 62 D2
Sunnyvale CA U.S.A. 65 A2
Suntar Rus. Fed. 29 M3
Suntsar Pak. 33 J4
Sunwi-do i. N. Korea 45 B5
Sunwu China 44 B2
Sunyani Ghana 46 C4
Suolijärvet l. Fin. 14 P3
Suomi country Europe see Finland
Suomussalmi Fin. 14 P4
Suō-nada b. Japan 45 C6
Suonenjoki Fin. 14 O5
Suoyarvi Rus. Fed. 12 G3
Supa India 38 B3
Supaul India 37 F4
Superior NE U.S.A. 62 H3
Superior WI U.S.A. 63 I2
Superior, Lake Canada/U.S.A. 63 J2
Süphan Dağı mt. Turkey 35 F3
Suponevo Rus. Fed. 13 G5
Support Force Glacier Antarctica 76 A1
Sūq ash Shuyūkh Iraq 35 G5
Suqian China 43 L6
Suquṭrá i. Yemen see Socotra
Şūr Oman 33 I5
Sur, Point CA U.S.A. 65 B2
Sur, Punta pt Arg. 70 E5
Sura r. Rus. Fed. 13 J4
Şuraabad Azer. 35 H2
Surabaya Indon. 41 D8
Surakarta Indon. 41 D8
Şūrān Syria 39 C2
Surat Australia 58 D1
Surat India 36 C5
Suratgarh India 36 C3
Surat Thani Thai. 31 I6
Surazh Rus. Fed. 13 G5
Surbiton Australia 56 D4
Surdulica Serbia 27 J3
Surendranagar India 36 B5
Surf CA U.S.A. 65 B3
Surgut Rus. Fed. 28 I3
Suri India see Siuri
Suriapet India 38 C2
Surigao Phil. 41 E7
Surin Thai. 31 J5
Surinam country S. America see Suriname
Suriname country S. America 69 G3
Surkhet Nepal 37 E3
Sürmene Turkey 35 F2
Surovikino Rus. Fed. 13 I6
Surpura India 36 C4
Surry VA U.S.A. 64 C4
Surskoye Rus. Fed. 13 J5
Surt Libya see Sirte
Surtsey i. Iceland 14 [inset]
Suruç Turkey 39 D1
Surud, Raas pt Somalia 48 E2
Surud Ad mt. Somalia see Shimbiris
Suruga-wan b. Japan 45 E6
Suryapet India see Suriapet
Şuşa Azer. 35 G3
Susah Tunisia see Sousse
Susaki Japan 45 D6
Susan VA U.S.A. 64 C4
Sūsangerd Iran 35 H5
Susanino Rus. Fed. 44 F1
Susanville U.S.A. 62 C3
Suşehri Turkey 34 E2
Susquehanna PA U.S.A. 64 D2
Susquehanna r. PA U.S.A. 64 C3
Susquehanna, West Branch r. PA U.S.A. 64 C2
Susques Arg. 70 C2
Sussex VA U.S.A. 64 C4
Susuman Rus. Fed. 29 P3
Susurluk Turkey 27 M5
Sutak India 36 D2
Sutherland Australia 58 E5
Sutherland S. Africa 50 E7
Sutherland Range hills Australia 55 D6
Sutjeska, Nacionalni Park nat. park Bos.-Herz. 26 H3
Sutlej r. India/Pak. 36 B3
Sütlüce Turkey 39 A1
Sutter CA U.S.A. 65 B1
Sutterton U.K. 19 G6
Sutton U.K. 19 H6
Sutton WV U.S.A. 64 A3
Sutton Coldfield U.K. 19 F6
Sutton in Ashfield U.K. 19 F5
Sutton Lake WV U.S.A. 64 A3
Suttor r. Australia 56 D4
Suttsu Japan 44 F4
Sutwik Island U.S.A. 60 C4
Sutyr' r. Rus. Fed. 44 D2
Suva Fiji 53 H3
Suvalki Poland see Suwałki
Suvorov atoll Cook Is see Suwarrow
Suvorov Rus. Fed. 13 H5
Suwa Japan 45 E5
Suwałki Poland 13 D5
Suwanose-jima i. Japan 45 C7
Suwarrow atoll Cook Is 53 J3
Suwayliḩ Jordan 39 B3
Suwayr well Saudi Arabia 35 F5
Suways, Khalīj as g. Egypt see Suez, Gulf of
Suways, Qanāt as canal Egypt see Suez Canal
Suweilih Jordan see Suwayliḩ
Suweis, Khalîg el g. Egypt see Suez, Gulf
Suweis, Qanâ el canal Egypt see Suez Canal
Suwŏn S. Korea 45 B5
Suyūl Ḩanīsh i. Yemen 32 F7
Suz, Mys pt Kazakh. 35 I2
Suzaka Japan 45 E5
Suzdal' Rus. Fed. 12 I4
Suzhou Anhui China 43 L6
Suzhou Jiangsu China 43 M6
Suzi He r. China 44 B4
Suzuka Japan 45 E6
Suzu-misaki pt Japan 45 E5
Særholthalvøya pen. Norway 14 O1
Svalbard terr. Arctic Ocean 28 C2
Svappavaara Sweden 14 L3
Svartenhuk Halvø pen. Greenland see Sigguup Nunaa
Svatove Ukr. 13 H6
Svecha Rus. Fed. 12 J4
Sveg Sweden 15 I5
Sveķi Latvia 15 O8
Svelgen Norway 14 D6
Svellingen Norway 14 F5
Švenčionėliai Lith. 15 N9
Švenčionys Lith. 15 O9
Svendborg Denmark 15 G9
Svensbu Norway 14 K2
Svenstavik Sweden 14 I5
Sverdlovsk Rus. Fed. see Yekaterinburg
Sverdlovs'k Ukr. 13 H6
Sverdrup Islands Canada 61 I2
Sverige country Europe see Sweden
Sveti Nikole Macedonia 27 I4
Svetlaya Rus. Fed. 44 E3
Svetlogorsk Belarus see Svyetlahorsk
Svetlogorsk Kaliningradskaya Oblast' Rus. Fed. 15 L9
Svetlogorsk Krasnoyarskiy Kray Rus. Fed 28 J3
Svetlograd Rus. Fed. 13 I7
Svetlovodsk Ukr. see Svitlovods'k
Svetlyy Kaliningradskaya Oblast' Rus. Fe 15 L9
Svetlyy Yar Rus. Fed. 13 J6
Svetogorsk Rus. Fed. 12 F3
Svíahnúkar vol. Iceland 14 [inset]
Svilaja mts Croatia 26 G3
Svilengrad Bulg. 27 L4
Svinecea Mare, Vârful mt. Romania 27 J
Svir Belarus 15 O9
Svir' r. Rus. Fed. 12 G3
Svishtov Bulg. 27 K3
Svitava r. Czech Rep. 17 P6
Svitavy Czech Rep. 17 P6
Svitlovods'k Ukr. 13 G6
Sviyaga r. Rus. Fed. 12 K5
Svizzer, Parc Naziunal Switz. 26 D1
Svizzera country Europe see Switzerland
Svobodnyy Rus. Fed. 44 C2
Svolvær Norway 14 I2
Svrljiške Planine mts Serbia 27 J3
Svyatoy Nos, Mys c. Rus. Fed. 12 K2
Svyetlahorsk Belarus 13 F5
Swadlincote U.K. 19 F6
Swaffham U.K. 19 H6
Swain Reefs Australia 56 F4
Swains Island atoll American Samoa 53
Swakop watercourse Namibia 50 B2
Swakopmund Namibia 50 B2
Swale r. U.K. 18 F4
Swallow Islands Solomon Is 53 G3
Swamihalli India 38 C3
Swan r. Australia 55 A7
Swanage U.K. 19 F8
Swandale WV U.S.A. 64 A3
Swan Hill Australia 58 A5
Swan Lake Man. Canada 62 G1
Swanley U.K. 19 H7
Swan Reach Australia 57 B7
Swan River Canada 62 G1
Swansea U.K. 19 D7
Swansea Bay U.K. 19 D7
Swanton CA U.S.A. 65 A2
Swartbergpas pass S. Africa 50 F7
Swart Nossob watercourse Namibia see Black Nossob
Swartruggens S. Africa 51 H3
Swatow China see Shantou
Swaziland country Africa 51 J4
Sweden country Europe 14 I5
Sweet Springs WV U.S.A. 64 A4
Sweetwater U.S.A. 62 G5
Sweetwater r. U.S.A. 62 F3
Swellendam S. Africa 50 E8
Świdnica Poland 17 P5
Świdwin Poland 17 O4
Świebodzin Poland 17 O4
Świecie Poland 17 Q4
Swift Current Canada 62 F1
Swilly r. Ireland 21 E3
Swilly, Lough inlet Ireland 21 E2
Swindon U.K. 19 F7
Swinford Ireland 21 D4
Świnoujście Poland 17 O4
Swinton U.K. 20 G5
Swiss Confederation country Europe see Switzerland
Switzerland country Europe 24 I3
Swords Ireland 21 F4
Swords Range hills Australia 56 C4
Syamozero, Ozero l. Rus. Fed. 12 G3
Syamzha Rus. Fed. 12 I3
Syang Nepal 37 E3
Syas'stroy Rus. Fed. 12 G3
Sychevka Rus. Fed. 12 G5
Sydenham atoll Kiribati see Nonouti
Sydney Australia 58 E4
Sydney Island Kiribati see Manra
Syedra tourist site Turkey 39 A1
Syeverodonets'k Ukr. 13 H6
Sykesville PA U.S.A. 64 B2
Syktyvkar Rus. Fed. 12 K3
Sylarna mt. Norway/Sweden 14 H5
Sylhet Bangl. 37 G4
Syloga Rus. Fed. 12 I3
Sylt i. Germany 17 L3

·lvester, Lake imp. l. Australia 56 A3
·mi i. Greece 27 L6
·nel'nykove Ukr. 13 G6
·ngyrli, Mys pt Kazakh. 35 I2
·nya Rus. Fed. 11 R2
·owa research stn Antarctica 76 D2
·racusae Sicilia Italy see Syracuse
·racuse Sicilia Italy 26 F6
·racuse KS U.S.A. 62 G4
·racuse NY U.S.A. 64 C1
·rdar'ya r. Asia 30 F2
·rdaryinskiy Uzbek. see Sirdaryo
·ria country Asia 34 E4
·rian Desert Asia 34 E4
·rna i. Greece 27 L6
·ros i. Greece 27 K6
·rskiy Rus. Fed. 13 H5
·smä Fin. 15 N6
·sola r. Rus. Fed. 12 K3
·umsi Rus. Fed. 12 K4
·urkum Rus. Fed. 44 F2
·urkum, Mys pt Rus. Fed. 44 F2
·zran' Rus. Fed. 13 K5
·abadka Serbia see Subotica
·czecin Poland 17 O4
·czecinek Poland 17 P4
·czytno Poland 17 R4
·echwan prov. China see Sichuan
·eged Hungary 27 I1
·ékesfehérvár Hungary 26 H1
·ekszárd Hungary 26 H1
·entes Hungary 27 I1
·entgotthárd Hungary 26 G1
·igetvár Hungary 26 G1
·olnok Hungary 27 I1
·ombathely Hungary 26 G1
·tálinváros Hungary see Dunaújváros

T

·agga Duudka reg. Somalia 48 E3
·bah Saudi Arabia 32 F4
·bajara Brazil 68 F5
·bakhmela Georgia see Kazret'i
·banan Indon. 54 A2
·bankulu S. Africa 51 I6
·baqah Ar Raqqah Syria 39 D2
·baqah Ar Raqqah Syria see Madīnat ath Thawrah
·bar Islands P.N.G. 52 F2
·barka Tunisia 26 C6
·bāsīn Iran 33 I3
·bask, Kūh-e mt. Iran 35 H5
·batinga Amazonas Brazil 68 E4
·batinga São Paulo Brazil 71 A3
·batinga, Serra da hills Brazil 69 J6
·batsquri, Tba l. Georgia 35 F2
·bayin Myanmar 37 H5
·bbita Australia 58 B5
·belbala Alg. 22 D6
·bia Tsaka salt l. China 37 F3
·biteuea atoll Kiribati 53 H2
·bivere Estonia 15 O7
·ble Cape N.Z. 59 F4
·ble Mountain Nature Reserve S. Africa 50 D8
·bligbo Togo 46 D4
·bor Czech Rep. 17 O6
·bora Tanz. 49 D4
·bou Côte d'Ivoire 46 C4
·brīz Iran 35 G3
·buaeran atoll Kiribati 75 J5
·būk Saudi Arabia 34 E5
·bulam Australia 58 F2
·bwémasana, Mount Vanuatu 53 G3
·by Sweden 15 K7
·calé Brazil 69 H3
·cheng China 42 E3
·chov Czech Rep. 17 N6
·cloban Phil. 41 E6
·cna Peru 68 D7
·coma U.S.A. 62 C2
·co Pozo Arg. 70 D3
·cuarembó Uruguay 70 E4
·dcaster U.K. 18 F5
·demaït, Plateau du Alg. 22 E6
·din New Caledonia 53 G4
·djikistan country Asia see Tajikistan
·djourah Djibouti 32 F7
·dmur Syria 39 D2
·dohae Haesang National Park S. Korea 45 B6
·doule Lake Canada 61 I4
·dpatri India 38 C3
·dwale India 38 C2
·dzhikskaya S.S.R. country Asia see Tajikistan
·aean Haean National Park S. Korea 45 B5
·ech'ŏng-do i. S. Korea 45 B5
·edasa-do N. Korea 45 B5
·edong-man b. N. Korea 45 B5
·egu S. Korea 45 C6
·ehan-min'guk country Asia see South Korea
·ehŭksan-kundo is S. Korea 45 B6
·ejŏn S. Korea 45 B5
·ejŏng S. Korea 45 B6
·aepaek S. Korea 45 C5
·'erqi China 43 M3
·f r. U.K. 19 C7
·fahi i. Tonga 53 I3
·falla Spain 25 F2
·fila Jordan see Aṭ Ṭafilah
·fi Viejo Arg. 70 C3
·fresh Iran 35 H4
·ft Iran 35 I5
·ft CA U.S.A. 65 C3
·ftān, Kūh-e mt. Iran 33 J4
·ftanāz Syria 39 C2
·ganrog Rus. Fed. 13 H7
·ganrog, Gulf of Rus. Fed./Ukr. 13 H7
·ganrogskiy Zaliv b. Rus. Fed./Ukr. see Taganrog, Gulf of
·gaung Myanmar 37 I5
·gchagpu Ri mt. China 37 E2
·gdempt Alg. see Tiaret
·ghmon Ireland 21 F5
·gtabazar Turkm. 33 J2
·gula P.N.G. 56 F1
·gula Island P.N.G. 56 F1
·gus r. Port. 25 B4
·gus r. Spain 22 C4
·ha China 44 B3
·hanroz'ka Zatoka b. Rus. Fed./Ukr. see Taganrog, Gulf of
·hat, Mont mt. Alg. 46 D2
·he China 44 B1
·heke N.Z. 59 D2
·hiti i. Fr. Polynesia 75 K7
·hlequah U.S.A. 63 I4
·hoe, Lake CA U.S.A. 65 B1
·hoe Lake Canada 61 H3
·hoe Vista CA U.S.A. 65 B1
·houa Niger 46 D3
·hrūd Iran 33 I4
·ï, Parc National de nat. park Côte d'Ivoire 46 C4
·'an China 43 L5
Taibei Taiwan see T'aipei
Taibus Qi China see Baochang
Taidong Taiwan see T'aitung
Taihang Shan mts Hebei China 43 K5
Taihang Shan mts China 43 K5
Taihape N.Z. 59 E4
Taihe Jiangxi China 43 K7
Tai Hu l. China 40 E4
Taikang China 44 B3
Tailai China 44 A3
Tailem Bend Australia 57 B7
Tain U.K. 20 F2
T'ainan Taiwan 43 M8
Tainaro, Akrotirio c. Greece 27 J6
Taiobeiras Brazil 71 C1
T'aipei Taiwan 43 M7
Taiping Malaysia 41 C7
Taipingchuan China 44 A3
Tai Poutini National Park N.Z. see Westland National Park
Tairbeart U.K. see Tarbert
Tairuq Iran 35 G4
Taishan China 43 K8
Taitao, Península de pen. Chile 70 B7
Tai Tapu N.Z. 59 D6
T'aitung Taiwan 43 M8
Taivalkoski Fin. 14 P4
Taivaskero h. Fin. 14 N2
Taiwan country Asia 43 M8
T'aiwan Haihsia str. China/Taiwan see Taiwan Strait
Taiwan Haixia str. China/Taiwan see Taiwan Strait
Taiwan Strait China/Taiwan 40 D5
Taiyuan China 43 K5
Taizhao China 37 H3
Taizhong Taiwan see Fengyüan
Taizhou Jiangsu China 43 L6
Taizhou Zhejiang China 43 M7
Taizi He r. China 44 B4
Ta'izz Yemen 32 F7
Tajamulco, Volcán de vol. Guat. 66 F5
Tajerouine Tunisia 26 C7
Tajikistan country Asia 33 L2
Tajjal Pak. 36 B4
Tajo r. Port. 25 C4 see Tagus
Tajo r. Spain see Tagus
Tajrīsh Iran 35 H4
Tak Thai. 31 I5
Takāb Iran 35 G3
Takabba Kenya 48 E3
Takahashi Japan 45 D6
Takamatsu Japan 45 D6
Takaoka Japan 45 E5
Takapuna N.Z. 59 E3
Ta karpo China 37 G4
Takatokwane Botswana 50 G3
Takatshwaane Botswana 50 E2
Takatsuki-yama mt. Japan 45 D6
Takayama Japan 45 E5
Takefu Japan 45 E6
Take-shima i. N. Pacific Ocean see Liancourt Rocks
Takestān Iran 35 H3
Takhemaret Alg. 25 G6
Takhta-Bazar Turkm. see Tagtabazar
Takhteh Iran 35 I5
Takhteh Pol Afgh. 36 A3
Takht-e Soleymān mt. Iran 35 H3
Takht-i-Bahi tourist site Pak. 36 B2
Takht-i-Sulaiman mt. Pak. 36 B3
Takijuq Lake Canada see Napaktulik Lake
Takinoue Japan 44 F3
Takla Lake Canada 60 F4
Takla Makan des. China see Taklimakan Desert
Taklimakan Desert China 36 E1
Taklimakan Shamo des. China see Taklimakan Desert
Takpa Shiri mt. China 37 H3
Takum Nigeria 46 D4
Takuu Islands P.N.G. 53 F2
Talachyn Belarus 13 F5
Talaja India 36 C5
Talakan Amurskaya Oblast' Rus. Fed. 44 C2
Talakan Khabarovskiy Kray Rus. Fed. 44 D2
Talandzha Rus. Fed. 44 C2
Talara Peru 68 B4
Talar-i-Band mts Pak. see Makran Coast Range
Talas Kyrg. 42 C4
Ṭal'at Mūsá mt. Lebanon/Syria 39 C2
Talaud, Kepulauan is Indon. 41 E7
Talavera de la Reina Spain 25 D4
Talaya Rus. Fed. 29 Q3
Talbehat India 36 D4
Talbīsah Syria 39 C2
Talbot, Mount h. Australia 55 D6
Talbragar r. Australia 58 D4
Talca Chile 70 B5
Talcahuano Chile 70 B5
Taldan Rus. Fed. 44 B1
Taldom Rus. Fed. 12 H4
Taldykorgan Kazakh. 42 D4
Taldy-Kurgan Kazakh. see Taldykorgan
Taldyqorghan Kazakh. see Taldykorgan
Tālesh Iran see Hashtpar
Talgarth U.K. 19 D7
Talguppa India 38 B3
Talia Australia 57 A7
Taliabu i. Indon. 41 E8
Talikota India 38 C2
Talin Hiag China 44 B3
Taliparamba India 38 B3
Talış Dağları mts Azer./Iran 35 H3
Talitsa Rus. Fed. 12 J4
Taliwang Indon. 54 B2
Talkeetna U.S.A. 60 C3
Talkeetna Mountains U.S.A. 60 D3
Tallacootra, Lake imp. l. Australia 55 F7
Tallahassee U.S.A. 63 K5
Tall al Aḥmar Syria 39 D1
Tall Baydar Syria 35 F3
Tall-e Ḥalāl Iran 35 I5
Tallinn Estonia 15 N7
Tall Kalakh Syria 39 C2
Tall Kayf Iraq 35 F3
Tall Kūjik Syria 35 F3
Tallow Ireland 21 D5
Tallulah U.S.A. 63 I5
Tall 'Uwaynāt Iraq 35 F3
Talmont-St-Hilaire France 24 D3
Tal'ne Ukr. 13 F6
Tal'noye Ukr. see Tal'ne
Taloda India 36 C5
Talodi Sudan 32 D7
Tālogān Afgh. 36 B1
Talos Dome Antarctica 76 H2
Talovaya Rus. Fed. 13 I6
Taloyoak Canada 61 I3
Tal Pass Pak. 36 C2
Talsi Latvia 15 M8
Taltal Chile 70 B3
Taltson r. Canada 60 G3
Talu China 37 H3
Talvik Norway 14 M1
Talwood Australia 58 D2
Talyshskiye Gory mts Azer./Iran see Talış Dağları
Talyy Rus. Fed. 12 L2
Tamala Australia 55 A6
Tamala Rus. Fed. 13 I5
Tamale Ghana 46 C4
Tamana i. Kiribati 53 H2
Tamano Japan 45 D6
Tamanrasset Alg. 46 D2
Tamanthi Myanmar 37 H4
Tamaqua PA U.S.A. 64 D2
Tamar India 37 F5
Tamar Syria see Tadmur
Tamar r. U.K. 19 C8
Tamarugal, Pampa de plain Chile 68 E7
Tamasane Botswana 51 H2
Tamatave Madag. see Toamasina
Tambacounda Senegal 46 B3
Tambaqui Brazil 68 F5
Tambar Springs Australia 58 D3
Tambelan, Kepulauan is Indon. 41 C7
Tambo r. Australia 58 C6
Tambohorano Madag. 49 E5
Tambora, Gunung vol. Indon. 54 B2
Tamboritha mt. Australia 58 C6
Tambov Rus. Fed. 13 I5
Tambovka Rus. Fed. 44 C2
Tambura South Sudan 47 F4
Tamburi Brazil 71 C1
Tâmchekkeṭ Mauritania 46 B3
Tamdybulak Uzbek. see Tomdibuloq
Tâmega r. Port. 25 B3
Tamenghest Alg. see Tamanrasset
Tamenglong India 37 H4
Tamerza Tunisia 26 B7
Tamgak, Adrar mt. Niger 46 D3
Tamgué, Massif du mt. Guinea 46 B3
Tamiahua, Laguna de lag. Mex. 66 E4
Tamil Nadu state India 38 C4
Tamitsa Rus. Fed. 12 H2
Țâmîya Egypt see Ṭāmīyah
Ṭāmīyah Egypt 34 C5
Tamkuhi India 37 F4
Tammerfors Fin. see Tampere
Tammisaari Fin. see Ekenäs
Tampa U.S.A. 63 K6
Tampa Bay U.S.A. 63 K6
Tampere Fin. 15 M6
Tampico Mex. 66 E4
Tamsagbulag Mongolia 43 L3
Tamsweg Austria 17 N7
Tamu Myanmar 37 H4
Tamworth Australia 58 E3
Tamworth U.K. 19 F6
Tana r. Fin./Norway see Tenojoki
Tana r. Kenya 48 E4
Tana Madag. see Antananarivo
Tana i. Vanuatu see Tanna
Tana, Lake Eth. 48 D2
Tanabe Japan 45 D6
Tanabi Brazil 71 A3
Tana Bru Norway 14 P1
Tanafjorden inlet Norway 14 P1
T'ana Hāyk' l. Eth. see Tana, Lake
Tanami Australia 54 E4
Tanami Desert Australia 54 E4
Tananarive Madag. see Antananarivo
Tanandava Madag. 49 E6
Tanch'ŏn N. Korea 45 C4
Tanda Côte d'Ivoire 46 C4
Tanda Uttar Prad. India 36 D3
Tanda Uttar Prad. India 37 E4
Țăndărei Romania 27 L2
Tandaué Angola 49 B5
Tandi India 36 D2
Tandil Arg. 70 E5
Tando Adam Pak. 33 K4
Tando Allahyar Pak. 36 B4
Tando Bago Pak. 36 B4
Tandou Lake imp. l. Australia 57 C7
Tandragee U.K. 21 F3
Tandur India 38 C2
Tanduri Pak. 36 A3
Tanega-shima i. Japan 45 C7
Tanezrouft reg. Alg./Mali 46 C2
Ṭanf, Jabal aṭ h. Syria 39 D3
Tanga Tanz. 49 D4
Tangail Bangl. 37 G4
Tanga Islands P.N.G. 52 F2
Tanganyika country Africa see Tanzania
Tanganyika, Lake Africa 49 C4
Tangará Brazil 71 A4
Tangasseri India 38 C4
Tangeli Iran 35 I3
Tanger Morocco see Tangier
Tanggulashan China 42 G6
Tanggula Shan mt. China 37 G2
Tanggula Shan mts China 37 G2
Tanggula Shankou pass China 37 G2
Tangguo China 37 F3
Tanghe China 43 K6
Tangier Morocco 25 D6
Tangiers Morocco see Tangier
Tang La pass China 37 G4
Tangla India 37 G4
Tangmai China 37 H3
Tangorin Australia 56 D4
Tangra Yumco salt l. China 37 F3
Tangshan Hebei China 43 L5
Tangtse India see Tanktse
Tangwanghe China 44 C2
Tangyuan China 44 C3
Tangyung Tso salt l. China 37 F3
Tanhaçu Brazil 71 C1
Tanhua Fin. 14 O3
Taniantaweng Shan mts China 42 H6
Tanimbar, Kepulauan is Indon. 41 F8
Tanjah Morocco see Tangier
Tanjay Phil. 41 E7
Tanjore India see Thanjavur
Tanjungkarang-Telukbetung Indon. see Bandar Lampung
Tanjungredeb Indon. 41 D7
Tanjungselor Indon. 41 D7
Tankse India see Tanktse
Tanktse India 36 D2
Tankwa-Karoo National Park S. Africa 50 D7
Tanna i. Vanuatu 53 G3
Tannadice U.K. 20 G4
Tännäs Sweden 14 H5
Tannu-Ola, Khrebet mts Rus. Fed. 42 G2
Tanot India 36 B4
Tanout Niger 46 D3
Tansen Nepal 37 E4
Ṭanṭā Egypt 34 C5
Tanta Egypt see Ṭanṭā
Tan-Tan Morocco 46 B2
Tantu China 44 A3
Tanuku India 38 D2
Tanumbirini Australia 54 F4
Tanumshede Sweden 15 G7
Tanzania country Africa 49 D4
Tao'an China see Taonan
Taobh Tuath U.K. see Northton
Taolanaro Madag. see Tôlañaro
Taonan China 44 A3
Taongi atoll Marshall Is 74 H5
Taos U.S.A. 62 F4
Taounate Morocco 22 D5
Taourirt Morocco 22 D5
Taoyang China see Lintao
Tapa Estonia 15 N7
Tapachula Mex. 66 F6
Tapajós r. Brazil 69 H4
Tapauá Brazil 68 F5
Tapauá r. Brazil 68 F5
Taperoá Brazil 71 D1
Tapiau Rus. Fed. see Gvardeysk
Taplejung Nepal 37 F4
Tappahannock VA U.S.A. 64 C4
Tappeh, Kūh-e h. Iran 35 H4
Taprobane country Asia see Sri Lanka
Tapti r. India 36 C5
Tapuaenuku mt. N.Z. 59 D5
Tapurucuara Brazil 68 E4
Taputeouea atoll Kiribati see Tabiteuea
Ṭaqṭaq Iraq 35 G4
Taquara Brazil 71 A5
Taquari Rio Grande do Sul Brazil 71 A5
Taquarí r. Brazil 69 G7
Taquaritinga Brazil 71 A3
Tar r. Ireland 21 E5
Tara Australia 58 E1
Ṭarābulus Lebanon see Tripoli
Ṭarābulus Libya see Tripoli
Tarahuwan India 36 E4
Tarai reg. India 37 G4
Tarakan Indon. 41 D7
Taraklı Turkey 27 N4
Taran, Mys pt Rus. Fed. 15 K9
Tarana Australia 58 D4
Taranagar India 36 C3
Taranaki, Mount vol. N.Z. 59 E4
Tarancón Spain 25 E3
Tarangambadi India 38 C4
Tarangire National Park Tanz. 48 D4
Taranto Italy 26 G4
Taranto, Golfo di g. Italy 26 G4
Taranto, Gulf of Italy see Taranto, Golfo di
Tarapoto Peru 68 C5
Tarapur India 38 B2
Tararua Range mts N.Z. 59 E5
Tarascon-sur-Ariège France 24 E5
Tarasovskiy Rus. Fed. 13 I6
Tarauacá Brazil 68 D5
Tarauacá r. Brazil 68 E5
Tarawera N.Z. 59 F4
Tarawera, Mount vol. N.Z. 59 F4
Taraz Kazakh. 42 C4
Tarazona Spain 25 F3
Tarazona de la Mancha Spain 25 F4
Tarbagatay, Khrebet mts Kazakh. 42 E3
Tarbat Ness pt U.K. 20 F3
Tarbert Ireland 21 C5
Tarbert Scotland U.K. 20 C3
Tarbert Scotland U.K. 20 D5
Tarbes France 24 E5
Tarcoola Australia 55 F7
Tarcoon Australia 58 C3
Tarcoonyinna watercourse Australia 55 F6
Tarcutta Australia 58 C5
Taree Australia 58 F3
Tarella Australia 57 C6
Tarentum Italy see Taranto
Tarfaya Morocco 46 B2
Targa well Niger 46 D3
Targan China see Talin Hiag
Targhee Pass U.S.A. 62 E3
Târgoviște Romania 27 K2
Targuist Morocco 25 D6
Târgu Jiu Romania 27 J2
Târgu Mureș Romania 27 K1
Târgu Neamț Romania 27 L1
Târgu Secuiesc Romania 27 L1
Targyailing China 37 F3
Tariat Mongolia 42 H3
Tarif U.A.E. 48 F1
Tarifa Spain 25 D5
Tarifa, Punta de pt Spain 25 D5
Tarija Bol. 68 F8
Tarikere India 38 B3
Tariku r. Indon. 41 F8
Tarīm Yemen 32 G6
Tarim Basin China 42 E5
Tarime Tanz. 48 D4
Tarim He r. China 42 F4
Tarim Pendi basin China see Tarim Basin
Tarīn Kowt Afgh. 36 A2
Taritatu r. Indon. 41 F8
Tarka r. S. Africa 51 G7
Tarkastad S. Africa 51 H7
Tarko-Sale Rus. Fed. 28 I3
Tarkwa Ghana 46 C4
Tarlac Phil. 41 E6
Tarlo River National Park Australia 58 D5
Tarma Peru 68 C6
Tarn r. France 24 E4
Tärnaby Sweden 14 I4
Tarnak Rūd r. Afgh. 36 A3
Târnăveni Romania 27 K1
Tarnobrzeg Poland 13 D6
Tarnogskiy Gorodok Rus. Fed. 12 I3
Tarnopol Ukr. see Ternopil'
Tarnów Poland 13 D6
Tarnowitz Poland see Tarnowskie Góry
Tarnowskie Góry Poland 17 Q5
Taro Co salt l. China 37 E3
Taroom Australia 57 E5
Taroudannt Morocco 22 C5
Tarpaulin Swamp Australia 56 B3
Ṭarq Iran 35 H4
Tarquinia Italy 26 D3
Tarquinii Italy see Tarquinia
Tarrabool Lake imp. l. Australia 56 A3
Tarraco Spain see Tarragona
Tarrafal Cape Verde 46 [inset]
Tarragona Spain 25 G3
Tårrajaur Sweden 14 K3
Tarran Hills h. Australia 58 C4
Tarrant Point Australia 56 B3
Tàrrega Spain 25 G3
Tarrong China see Nyêmo
Tarso Emissi mt. Chad 47 E2
Tarsus Turkey 39 B1
Tart China 37 H1
Tärtär Azer. 35 G2
Tartu Estonia 15 O7
Ṭarṭūs Syria 39 B2
Tarumovka Rus. Fed. 35 G1
Tarung Hka r. Myanmar 37 I4
Tarvisium Italy see Treviso
Tashauz Turkm. see Daşoguz
Tashi Chho Bhutan see Thimphu
Tashigang Bhutan 37 G4
Tashino Rus. Fed. see Pervomaysk
Tashir Armenia 35 G2
Tashk, Daryācheh-ye l. Iran 35 I5
Tashkent Toshkent Uzbek. see Toshkent
Tāshqurghān Afgh. see Kholm
Tashtagol Rus. Fed. 42 F2
Tashtyp Rus. Fed. 42 F2
Tasiat, Lac l. Canada 61 K4
Tasiilap Karra c. Greenland 61 O3
Tasiilaq Greenland see Ammassalik
Tasīl Syria 39 B3
Tasiujaq Canada 61 L4
Tasiusaq Greenland 61 M2
Taskala Kazakh. 11 Q5
Taşkent Turkey 39 A1
Tasker Niger 46 E3
Taskesken Kazakh. 42 E3
Taşköprü Turkey 34 D2
Tasman Abyssal Plain sea feature Tasman Sea 74 G8
Tasman Basin sea feature Tasman Sea 74 G8
Tasman Bay N.Z. 59 D5
Tasmania state Australia 57 [inset]
Tasman Islands P.N.G. see Nukumanu Islands
Tasman Mountains N.Z. 59 D5
Tasman Peninsula Australia 57 [inset]
Tasman Sea S. Pacific Ocean 52 H6
Taşova Turkey 34 E2
Tassara Niger 46 D3
Tassili n'Ajjer nat. park Alg. 46 D2
Tassili n'Ajjer plat. Alg. 46 D2
Tassili oua-n-Ahaggar plat. Alg. 46 D2
Tasty Kazakh. 42 B4
Taşucu Turkey 39 A1
Tas-Yuryakh Rus. Fed. 29 M3
Tata Morocco 22 C6
Tatabánya Hungary 26 H1
Tatamailau, Foho mt. East Timor 54 D2
Tataouine Tunisia 22 G5
Tatarbunary Ukr. 27 M2
Tatarsk Rus. Fed. 28 I4
Tatarskiy Proliv str. Rus. Fed. 44 F2
Tatar Strait Rus. Fed. see Tatarskiy Proliv
Tate r. Australia 56 C3
Tateyama Japan 45 E6
Tathlina Lake Canada 60 G3
Tathlīth Saudi Arabia 32 F6
Tathlīth, Wādī watercourse Saudi Arabia 32 F5
Tathra Australia 58 D6
Tatishchevo Rus. Fed. 13 J6
Tatkon Myanmar 37 I5
Tatra Mountains Poland/Slovakia 17 Q6
Tatry mts Poland/Slovakia see Tatra Mountains
Tatrzański Park Narodowy nat. park Poland 17 Q6
Tatsinskaya Rus. Fed. 13 I6
Tatuí Brazil 71 B3
Tatvan Turkey 35 F3
Tau Norway 15 D7
Taua Brazil 69 J5
Tauapeçaçu Brazil 68 F4
Taubaté Brazil 71 B3
Taukum, Peski des. Kazakh. 42 C4
Taumarunui N.Z. 59 E4
Taumaturgo Brazil 68 D5
Taung S. Africa 50 G4
Taungdwingyi Myanmar 37 H5
Taunggyi Myanmar 42 H8
Taung-ngu Myanmar 42 H9
Taungtha Myanmar 37 H5
Taungup Myanmar 42 G9
Taunton U.K. 19 D7
Taunton MA U.S.A. 64 F2
Taupo N.Z. 59 F4
Taupo, Lake N.Z. 59 E4
Tauragė Lith. 15 M9
Tauranga N.Z. 59 F3
Taurasia Italy see Turin
Taurianova Italy 26 G5
Tauroa Point N.Z. 59 D2
Taurus Mountains Turkey 39 A1
Taute r. France 19 F9
Tauz Azer. see Tovuz
Tavas Turkey 27 M6
Tavastehus Fin. see Hämeenlinna
Taverham U.K. 19 I6
Taveuni i. Fiji 53 I3
Tavira Port. 25 C5
Tavistock U.K. 19 C8
Tavoy Myanmar 31 I5
Tavşanlı Turkey 27 M5
Taw r. U.K. 19 C7
Tawang India 37 G4
Tawas City U.S.A. 63 K3
Tawau Sabah Malaysia 41 D7
Tawè Myanmar see Tavoy
Tawe r. U.K. 19 D7
Tawmaw Myanmar 37 I4
Taxkorgan China 42 D5
Tay r. U.K. 20 F4
Tay, Firth of est. U.K. 20 F4
Tay, Lake imp. l. Australia 55 C8
Tay, Loch l. U.K. 20 E4
Tāybād Iran 33 J3
Taybola Rus. Fed. 14 R2
Tayinloan U.K. 20 D5
Taylor AK U.S.A. 60 B3
Taylor TX U.S.A. 63 H5
Taymā' Saudi Arabia 34 E6
Taymura r. Rus. Fed. 29 K3
Taymyr, Ozero l. Rus. Fed. 29 L2
Taymyr, Poluostrov pen. Rus. Fed. see Taymyr Peninsula
Taymyr Peninsula Rus. Fed. 28 J2
Tây Ninh Vietnam 31 J5
Taypak Kazakh. 11 Q6
Taypaq Kazakh. see Taypak
Tayshet Rus. Fed. 42 H1
Taytay Phil. 41 D6
Tayuan China 44 B2
Taz r. Rus. Fed. 28 I3
Taza Morocco 22 D5
Tāza Khurmātū Iraq 35 G4
Taze Myanmar 37 H5
Tāzirbū Libya 47 F2
Tazmalt Alg. 25 I5
Tazovskaya Guba sea chan. Rus. Fed. 28 I3
Tbessa Alg. see Tébessa
T'bilisi Georgia 35 G2
Tbilisskaya Rus. Fed. 13 I7
Tchabal Mbabo mt. Cameroon 46 E4
Tchad country Africa see Chad
Tchamba Togo 46 D4
Tchibanga Gabon 48 B4
Tchigaï, Plateau du Niger 47 E2
Tchin-Tabaradene Niger 46 D3
Tcholliré Cameroon 47 E4
Tczew Poland 17 Q3
Teague, Lake imp. l. Australia 55 C6
Te Anau N.Z. 59 A7
Te Anau, Lake N.Z. 59 A7
Teapa Mex. 66 F5
Te Araroa N.Z. 59 G3
Teate Italy see Chieti
Te Awamutu N.Z. 59 E4
Tébarat Niger 46 D3
Tebay U.K. 18 E4
Tébessa Alg. 26 C7
Tébessa, Monts de mts Alg. 26 C7
Téboursouk Tunisia 26 C6
Téboursouk Tunisia 26 C6
Tebulos Mt'a Georgia/Rus. Fed. 35 G2
Tecate Baja California Mex. 65 D4
Tece Turkey 39 B1
Techiman Ghana 46 C4
Tecka Arg. 70 B6
Tecoripa Mex. 66 C3
Técpan Mex. 66 D5
Tecuala Mex. 66 C4
Tecuci Romania 27 L2
Tedzhen Turkm. see Tejen
Tees r. U.K. 18 F4
Tefé r. Brazil 68 F4
Tefenni Turkey 27 M6
Tegucigalpa Hond. 67 G6
Teguidda-n-Tessoumt Niger 46 D3
Tehachapi CA U.S.A. 65 C3
Tehachapi Mountains CA U.S.A. 65 C3
Tehachapi Pass CA U.S.A. 65 C3
Tehek Lake Canada 61 I3
Teheran Iran see Tehrān
Téhini Côte d'Ivoire 46 C4
Tehrān Iran 35 H4
Tehri India see Tikamgarh
Tehuacán Mex. 66 E5
Tehuantepec, Golfo de Mex. see Tehuantepec, Gulf of
Tehuantepec, Gulf of Mex. 66 F5
Tehuantepec, Istmo de isth. Mex. 66 F5
Teide, Pico del vol. Canary Is 46 B2
Teifi r. U.K. 19 C6
Teignmouth U.K. 19 D8
Teixeira de Sousa Angola see Luau
Teixeiras Brazil 71 C3
Teixeira Soares Brazil 71 A4
Tejakula Indon. 54 A2
Tejen Turkm. 30 F3
Tejo r. Port. 25 B4 see Tagus
Tejon Pass CA U.S.A. 65 C3
Tekapo, Lake N.Z. 59 C6
Tekax Mex. 66 G4
Tekeli Kazakh. 42 D4
Tekes China 42 E4
Tekiliktag mt. China 36 E1
Tekin Rus. Fed. 44 D2
Tekirdağ Turkey 27 L4
Tekka India 38 D2
Tekkali India 38 E2
Teknaf Bangl. 37 H5
Te Kuiti N.Z. 59 E4
Tel r. India 38 D1
Télagh Alg. 25 F6
Telanaipura Indon. see Jambi
Tel Ashqelon tourist site Israel 39 B4
Télataï Mali 46 D3
Tel Aviv-Yafo Israel 39 B3
Telč Czech Rep. 17 O6
Telchac Puerto Mex. 66 G4
Telekhany Belarus see Tsyelyakhany
Telêmaco Borba Brazil 71 A4
Teleorman r. Romania 27 K3
Telertheba, Djebel mt. Alg. 46 D2
Telescope Peak CA U.S.A. 65 D2
Teles Pires r. Brazil 69 G5
Telford U.K. 19 E6
Télimélé Guinea 46 B3
Teljo, Jebel mt. Sudan 32 C7
Tell Atlas mts Alg. see Atlas Tellien
Teller U.S.A. 60 B3
Tell es Sultan West Bank see Jericho
Tellicherry India see Thalassery
Telloh Iraq 35 G5
Tel'novskiy Rus. Fed. 44 F2
Telo Martius France see Toulon
Telpoziz, Gora mt. Rus. Fed. 11 R3
Telsen Arg. 70 C6
Telšiai Lith. 15 M9
Telukbetung Indon. see Bandar Lampung
Temagami Lake Canada 63 L2
Têmarxung China 37 G2
Temba S. Africa 51 I3
Tembagapura Indon. 52 D2
Tembenchi r. Rus. Fed. 29 K3
Tembisa S. Africa 51 I4
Tembo Aluma Angola 49 B4
Teme r. U.K. 19 E6
Temecula CA U.S.A. 65 D4
Teminabuan Indon. 52 D2
Temirtau Kazakh. 42 C2
Temmes Fin. 14 N4
Temnikov Rus. Fed. 13 I5
Temora Australia 58 C5
Tempe Downs Australia 55 F6
Temple TX U.S.A. 63 H5
Temple Bar U.K. 19 C6
Temple Dera Pak. 36 B3
Templemore Ireland 21 E5
Temple Sowerby U.K. 18 E4
Templeton watercourse Australia 56 B4
Tempué Angola 49 B5
Temryuk Rus. Fed. 13 H7
Temryukskiy Zaliv b. Rus. Fed. 13 H7
Temuco Chile 70 B5
Temuka N.Z. 59 C7
Tena Ecuador 68 C4
Tenabo Mex. 66 F4
Tenali India 38 D2
Tenbury Wells U.K. 19 E6
Tenby U.K. 19 C7
Tendaho Eth. 48 E2
Tende, Col de pass France/Italy 24 H4
Ten Degree Channel India 31 I6
Tendō Japan 45 F5
Tenedos i. Turkey see Bozcaada
Ténenkou Mali 46 C3
Ténéré reg. Niger 46 D2
Ténéré du Tafassâsset des. Niger 46 E2
Tenerife i. Canary Is 46 B2
Ténès Alg. 25 G5
Tengah, Kepulauan is Indon. 41 D8
Tengger Shamo des. China 42 I5
Tengréla Côte d'Ivoire 46 C3
Ten'gushevo Rus. Fed. 13 I5
Teni India see Theni
Teniente Jubany research stn Antarctica see Jubany
Tenke Dem. Rep. Congo 49 C5
Tenkeli Rus. Fed. 29 P2
Tenkodogo Burkina Faso 46 C3
Ten Mile Lake imp. l. Australia 55 C6
Tennant Creek Australia 54 F4
Tennessee r. U.S.A. 63 J4
Tennessee state U.S.A. 63 J4
Tennevoll Norway 14 J2
Tenojoki r. Fin./Norway 14 P1
Tenosique Mex. 66 F5
Tenterden U.K. 19 H7
Tenterfield Australia 58 F2
Tentudia mt. Spain 25 C4
Tentulia Bangl. see Tetulia
Teodoro Sampaio Brazil 70 F2
Teófilo Otoni Brazil 71 C2
Tepa Indon. 54 E1
Tepache Mex. 66 C3
Te Paki N.Z. 59 D2
Tepatitlán Mex. 66 D4
Tepehuanes Mex. 62 F6
Tepeköy Turkey see Karakoçan
Tepelenë Albania 27 I4
Tepequem, Serra mts Brazil 67 L8
Tepic Mex. 66 D4
Te Pirita N.Z. 59 C6
Teplice Czech Rep. 17 N5
Teplogorka Rus. Fed. 12 L3
Teploozersk Rus. Fed. 44 C2
Teployе Rus. Fed. 13 H5
Teploye Ozero Rus. Fed. see Teploozersk
Tequila Mex. 66 D4
Ter r. Spain 25 H2
Téra Niger 46 D3
Teramo Italy 26 E3
Terang Australia 58 A7
Tercan Turkey 35 F3
Terebovlya Ukr. 13 E6
Terekty Kazakh. 42 F3
Teresa Cristina Brazil 71 A4
Tereshka r. Rus. Fed. 13 J6
Teresina Brazil 69 J5

Teresina de Goias Brazil 71 B1
Teresita Col. 68 E3
Teresópolis Brazil 71 C3
Terezinha Brazil 69 H3
Tergeste Italy see Trieste
Teriberka Rus. Fed. 14 S2
Termez Uzbek. see Termiz
Termini Imerese Sicilia Italy 26 E6
Términos, Laguna de lag. Mex. 66 F5
Termit-Kaoboul Niger 46 E3
Termiz Uzbek. 33 K2
Termoli Italy 26 F4
Tern r. U.K. 19 E6
Ternate Indon. 41 E7
Terney Rus. Fed. 44 E3
Terni Italy 26 E3
Ternopil' Ukr. 13 E6
Ternopol' Ukr. see Ternopil'
Terpeniya, Mys c. Rus. Fed. 44 G2
Terpeniya, Zaliv g. Rus. Fed. 44 F2
Terra Alta WV U.S.A. 64 B3
Terra Bella CA U.S.A. 65 C3
Terrace Canada 60 F4
Terrace Bay Canada 63 J2
Terra Firma S. Africa 50 F3
Terråk Norway 14 H4
Terralba Sardegna Italy 26 C5
Terra Nova Bay Antarctica 76 H1
Terre Haute U.S.A. 63 J4
Terre-Neuve prov. Canada see Newfoundland and Labrador
Terre-Neuve-et-Labrador prov. Canada see Newfoundland and Labrador
Terres Australes et Antarctiques Françaises terr. Indian Ocean see French Southern and Antarctic Lands
Terskiy Bereg coastal area Rus. Fed. 12 H2
Tertenia Sardegna Italy 26 C5
Terter Azer. see Tärtär
Teruel Spain 25 F3
Tervola Fin. 14 N3
Tes Mongolia 42 H3
Tešanj Bos.-Herz. 26 G2
Teseney Eritrea 32 E6
Tesha r. Rus. Fed. 13 I5
Teshekpuk Lake U.S.A. 60 C2
Teshio Japan 44 F3
Teshio-gawa r. Japan 44 F3
Teslin Canada 60 E3
Teslin Lake Canada 60 E3
Tesouras r. Brazil 71 A1
Tessalit Mali 46 D2
Tessaoua Niger 46 D3
Tessolo Moz. 51 L1
Test r. U.K. 19 F8
Testour Tunisia 26 C6
Tetas, Punta pt Chile 70 B2
Tete Moz. 49 D5
Te Teko N.Z. 59 F4
Teteriv r. Ukr. 13 F6
Teterow Germany 17 N4
Tetiyev Ukr. see Tetiyiv
Tetiyiv Ukr. 13 F6
Tetney U.K. 18 G5
Tétouan Morocco 25 D6
Tetovo Macedonia 27 I3
Tetuán Morocco see Tétouan
Tetulia Bangl. 37 G4
Tetulia sea chan. Bangl. 37 G5
Tetyukhe Rus. Fed. see Dal'negorsk
Tetyukhe-Pristan' Rus. Fed. see Rudnaya Pristan'
Tetyushi Rus. Fed. 13 K5
Teuco r. Arg. 70 D2
Teufelsbach Namibia 50 C2
Te Urewera National Park N.Z. 59 F4
Teuva Fin. 14 L5
Tevere r. Italy see Tiber
Teverya Israel see Tiberias
Teviot r. U.K. 20 G5
Te Waewae Bay N.Z. 59 A8
Te Waiponamu i. N.Z. see South Island
Tewane Botswana 51 H2
Tewantin Australia 57 F5
Tewkesbury U.K. 19 E7
Têwo China 42 I6
Texarkana TX U.S.A. 63 I5
Texas Australia 58 E2
Texas state U.S.A. 62 H5
Texoma, Lake U.S.A. 63 H5
Teyateyaneng Lesotho 51 H5
Teykovo Rus. Fed. 12 I4
Teza r. Rus. Fed. 12 I4
Tezpur India 37 H4
Tezu India 37 I4
Tha-anne r. Canada 61 I3
Thabana-Ntlenyana mt. Lesotho 51 I5
Thaba Nchu S. Africa 51 H5
Thaba Putsoa mt. Lesotho 51 H5
Thaba-Tseka Lesotho 51 I5
Thabazimbi S. Africa 51 H3
Thabong S. Africa 51 H4
Thabyedaung Myanmar 37 I5
Tha Hin Thai. see Lop Buri
Thai Binh Vietnam 31 J4
Thailand country Asia 31 J5
Thailand, Gulf of Asia 41 C6
Thai Nguyên Vietnam 31 J4
Thakèk Laos 41 C6
Thakurgaon Bangl. 37 G4
Thakurtola India 36 E5
Thala Tunisia 26 C7
Thalassery India 38 B4
Thal Desert Pak. 33 L3
Thaliparamba India see Taliparamba
Thallon Australia 58 D2
Thamaga Botswana 51 G3
Thamar, Jabal mt. Yemen 32 G7
Thamarīt Oman 33 H6
Thame r. U.K. 19 F7
Thames r. Ont. Canada 63 K3
Thames r. Ont. Canada 64 A1
Thames N.Z. 59 E3
Thames est. U.K. 19 H7
Thames r. U.K. 19 H7
Thamesford Ont. Canada 64 A1
Thana India see Thane
Thandwè Myanmar 42 G9
Thane India 38 B2
Thanet, Isle of pen. U.K. 19 I7
Thangoo Australia 54 C4
Thangra India 36 D2
Thanh Hoa Vietnam 31 J5
Thanjavur India 38 C4
Thanlwin r. China/Myanmar see Salween
Thapsacus Syria see Dibsī
Tharad India 36 B4
Tharad India 36 B4
Thar Desert India/Pak. 33 K4
Thargomindah Australia 58 A1
Tharthār, Buḩayrat ath l. Iraq 35 F4
Thasos i. Greece 27 K4
Thaton Myanmar 31 I5
Thatta Pak. 33 K5
Thaungdut Myanmar 37 H4
Thayetmyo Myanmar 37 H6
Thazi Mandalay Myanmar 37 I5
The Aldermen Islands N.Z. 59 F3
Theba AZ U.S.A. 65 F4
The Bahamas country West Indies 67 I4
Thebes Greece see Thiva
The Broads nat. park U.K. 19 I6
The Calvados Chain is P.N.G. 56 F1
The Cheviot h. U.K. 18 E3
The Dalles U.S.A. 62 C2
Thedford U.S.A. 62 G3
The Entrance Australia 58 E4
The Faither stack U.K. 20 [inset]
The Fens reg. U.K. 19 G6
The Gambia country Africa 46 B3
The Grampians mts Australia 57 C8
The Great Oasis oasis Egypt see Khārijah, Wāḩāt al
The Grenadines is St Vincent 67 L6
The Gulf Asia 32 H4
The Hague Neth. 16 J4
The Hunters Hills N.Z. 59 C7
The Lakes National Park Australia 58 C6
Thelon r. Canada 61 I3
The Lynd Junction Australia 56 D3
Thembalihle S. Africa 51 I4
The Minch sea chan. U.K. 20 C2
The Naze c. Norway see Lindesnes
The Needles stack U.K. 19 F8
Theni India 38 C4
Thenia Alg. 25 H5
Theniet El Had Alg. 25 H6
The North Sound sea chan. U.K. 20 G1
Theodore Australia 56 E5
Theodosia Ukr. see Feodosiya
The Old Man of Coniston h. U.K. 18 D4
The Paps h. Ireland 21 C5
The Pas Canada 62 G1
The Pilot mt. Australia 58 D6
Thera i. Greece see Santorini
Thermaïkos Kolpos g. Greece 27 J4
Thermopolis U.S.A. 62 F3
The Rock Australia 58 C5
The Salt Lake salt l. Australia 57 C6
The Settlement Christmas I. 74 D4
The Skaw spit Denmark see Grenen
The Skelligs is Ireland 21 B6
The Slot sea chan. Solomon Is see New Georgia Sound
The Solent str. U.K. 19 F8
Thessalon Canada 63 K2
Thessalonica Greece see Thessaloniki
Thessaloniki Greece 27 J4
The Storr h. U.K. 20 C3
Thet r. U.K. 19 H6
The Terraces hills Australia 55 C7
Thetford U.K. 19 H6
Thetford Mines Canada 63 M2
The Triangle mts Myanmar 42 H7
The Trossachs hills U.K. 20 E4
The Twins Australia 57 A6
Theva-i-Ra rf Fiji see Ceva-i-Ra
The Valley Anguilla 67 L5
Thevenard Island Australia 54 A5
Thévenet, Lac l. Canada 61 L4
Theveste Alg. see Tébessa
The Wash b. U.K. 19 H6
The Weald reg. U.K. 19 H7
The Woodlands U.S.A. 63 H5
Thibodaux U.S.A. 63 I6
Thief River Falls U.S.A. 63 H2
Thiel Mountains Antarctica 76 K1
Thiers France 24 F4
Thiès Senegal 46 B3
Thika Kenya 48 D4
Thiladhunmathi Atoll Maldives 38 B5
Thiladhunmathi Atoll Maldives see Thiladhunmathi Atoll
Thimbu Bhutan see Thimphu
Thimphu Bhutan 37 G4
Thionville France 24 H2
Thira i. Greece see Santorini
Thirsk U.K. 18 F4
Thiruvananthapuram India 38 C4
Thiruvannamalai India see Tiruvannamalai
Thiruvarur India 38 C4
Thiruvattiyur India see Tiruvottiyur
Thisted Denmark 15 F8
Thityabin Myanmar 37 H5
Thiva Greece 27 J5
Thívai Greece see Thiva
Thoen Thai. 42 H9
Thoeng Thai. 42 I9
Thohoyandou S. Africa 51 J2
Thomas Hubbard, Cape Canada 61 I1
Thomaston CT U.S.A. 64 E2
Thomastown Ireland 21 E5
Thomasville GA U.S.A. 63 K5
Thompson Canada 61 I4
Thompson r. U.S.A. 62 I4
Thompson Falls U.S.A. 62 D2
Thompson's Falls Kenya see Nyahururu
Thompson Sound Canada 62 B1
Thoothukudi India see Tuticorin
Thorn Poland see Toruń
Thornaby-on-Tees U.K. 18 F4
Thornbury U.K. 19 E7
Thorne U.K. 18 G5
Thorne NV U.S.A. 65 C1
Thornton r. Australia 56 B3
Thorshavnfjella reg. Antarctica see Thorshavnheiane
Thorshavnheiane reg. Antarctica 76 C2
Thota-ea-Moli Lesotho 51 H5
Thouars France 24 D3
Thoubal India 37 H4
Thousand Oaks CA U.S.A. 65 C3
Thrace reg. Europe 27 L4
Thraki reg. Europe see Thrace
Thrakiko Pelagos sea Greece 27 K4
Three Gorges Reservoir resr China 43 J6
Three Hummock Island Australia 57 [inset]
Three Kings Islands N.Z. 59 D2
Three Points, Cape Ghana 46 C4
Three Springs Australia 55 A7
Thrissur India 38 C4
Throssell, Lake imp. l. Australia 55 D6
Throssel Range hills Australia 54 C5
Thrushton National Park Australia 58 C1
Thuddungra Australia 58 D5
Thul Pak. 36 B3
Thulaythawāt Gharbī, Jabal h. Syria 39 D2
Thule Greenland 61 L2
Thun Switz. 24 H3
Thunder Bay Canada 61 J5
Thurles Ireland 21 E5
Thurn, Pass Austria 17 N7
Thursday Island Australia 56 C1
Thurso U.K. 20 F2
Thurso r. U.K. 20 F2
Thurston Island Antarctica 76 K2
Thurston Peninsula i. Antarctica see Thurston Island
Thuthukudi India see Tuticorin
Thwaite U.K. 18 E4
Thwaites Glacier Tongue Antarctica 76 K1
Thyatira Turkey see Akhisar
Thyborøn Denmark 15 F8
T'ianet'i Georgia 35 G2
Tianjin China 43 L5
Tianjin mun. China 43 L5
Tianjun China 42 H5
Tianqiaoling China 44 C4
Tianshan China 43 M4
Tian Shan mts China/Kyrg. see Tien Shan
Tianshui China 42 J6
Tianshuihai Aksai Chin 36 D2
Tiantang China see Yuexi
Tianzhu Gansu China 42 I5
Tiaret Alg. 25 G6
Tiassalé Côte d'Ivoire 46 C4
Tibagi Brazil 71 A4
Tibal, Wādī watercourse Iraq 35 F4
Tibati Cameroon 46 E4
Tibba Pak. 36 B3
Tibé, Pic de mt. Guinea 46 C4
Tiber r. Italy 26 E4
Tiberias Israel 39 B3
Tiberias, Lake Israel see Galilee, Sea of
Tibesti mts Chad 47 E2
Tibet aut. reg. China see Xizang Zizhiqu
Tibet, Plateau of China 37 F2
Tibooburra Australia 57 C6
Tibrikot Nepal 37 E3
Tibro Sweden 15 I7
Tibur Italy see Tivoli
Tiburón, Isla i. Mex. 66 B3
Ticehurst U.K. 19 H7
Tîchît Mauritania 46 C3
Tichla W. Sahara 46 B2
Ticinum Italy see Pavia
Ticul Mex. 66 G4
Tidaholm Sweden 15 H7
Tiddim Myanmar 37 H5
Tidjikja Mauritania 46 B3
Tieli China 44 B3
Tieling China 44 A4
Tielongtan Aksai Chin 36 D2
Tien Shan mts China/Kyrg. 42 D4
Tientsin mun. China see Tianjin
Tierp Sweden 15 J6
Tierra del Fuego, Isla Grande de i. Arg./Chile 70 C8
Tierra del Fuego, Parque Nacional nat. park Arg. 70 C8
Tiétar r. Spain 25 D4
Tietê r. Brazil 71 A3
Tieyon Australia 55 F6
Tiflis Georgia see T'bilisi
Tifton U.S.A. 63 K5
Tiga Reservoir Nigeria 46 D3
Tigen Kazakh. 35 H1
Tigheciului, Dealurile hills Moldova 27 M2
Tighina Moldova 27 M1
Tigiria India 38 E1
Tignère Cameroon 46 E4
Tignish Canada 63 O2
Tigranocerta Turkey see Siirt
Tigre r. Venez. 68 F2
Tigris r. Asia 32 F2
Tigris r. Asia 35 G5
Tigrovaya Balka Zapovednik nature res. Tajik. 36 B1
Tiguidit, Falaise de esc. Niger 46 D3
Tīh, Gebel el plat. Egypt see Tīh, Jabal at
Tīh, Jabal at plat. Egypt 39 A5
Tijuana Mex. 66 A2
Tikamgarh India 36 D4
Tikanlik China 42 F4
Tikhoretsk Rus. Fed. 13 I7
Tikhvin Rus. Fed. 12 G4
Tikhvinskaya Gryada ridge Rus. Fed. 12 G4
Tiki Basin sea feature S. Pacific Ocean 75 L7
Tikokino N.Z. 59 F4
Tikopia i. Solomon Is 53 G3
Tikrīt Iraq 35 F4
Tiksha Rus. Fed. 14 R4
Tikse India 36 D2
Tikshozero, Ozero l. Rus. Fed. 14 R3
Tiksi Rus. Fed. 29 N2
Tiladummati Atoll Maldives see Thiladhunmathi Atoll
Tilaiya Reservoir India 37 F4
Tilbeşar Ovasi plain Turkey 39 C1
Tilbooroo Australia 58 B1
Tilburg Neth. 16 J5
Tilbury U.K. 19 H7
Tilcara Arg. 70 C2
Tilcha Creek watercourse Australia 57 C6
Tilemsès Niger 46 D3
Tilemsi, Vallée du watercourse Mali 46 D3
Tilhar India 36 D4
Tilimsen Alg. see Tlemcen
Tilin Myanmar 37 H5
Tillabéri Niger 46 D3
Tillia Niger 46 D3
Tillicoultry U.K. 20 F4
Tillsonburg Ont. Canada 64 A1
Tillyfourie U.K. 20 G3
Tilos i. Greece 27 L6
Tilothu India 37 F4
Tilpa Australia 58 B3
Tilsit Rus. Fed. see Sovetsk
Tilt r. U.K. 20 F4
Tilton NH U.S.A. 64 F1
Tim Rus. Fed. 13 H6
Ţimā Egypt 32 D4
Timakara i. India 38 B4
Timanskiy Kryazh ridge Rus. Fed. 12 K2
Timar Turkey 35 F3
Timaru N.Z. 59 C7
Timashevsk Rus. Fed. 13 H7
Timashevskaya Rus. Fed. see Timashevsk
Timbedgha Mauritania 46 C3
Timber Creek Australia 52 D3
Timber Mountain NV U.S.A. 65 D2
Timberville VA U.S.A. 64 B3
Timbuktu Mali 46 C3
Timétrine reg. Mali 46 C3
Timiaouine Alg. 46 D2
Timimoun Alg. 22 E6
Timirist, Râs pt Mauritania 46 B3
Timişoara Romania 27 I2
Timmins Canada 61 J5
Timon Brazil 69 J5
Timor i. Indon. 54 D2
Timor-Leste country Asia see East Timor
Timor Loro Sae country Asia see East Timor
Timor Sea Australia/Indon. 52 C3
Timor Timur country Asia see East Timor
Timperley Range hills Australia 55 C6
Timrå Sweden 14 J5
Tīn, Ra's at pt Libya 34 A4
Tīna, Khalīg el b. Egypt see Ţīnah, Khalīj aţ
Ţīnah Syria 39 D1
Ţīnah, Khalīj aţ b. Egypt 39 A4
Tin Can Bay Australia 57 F5
Tindivanam India 38 C3
Tindouf Alg. 22 C6
Ti-n-Essako Mali 46 D3
Tingha Australia 58 E2
Tingis Morocco see Tangier
Tingo María Peru 68 C5
Tingréla Côte d'Ivoire see Tengréla
Tingsryd Sweden 15 I8
Tingvoll Norway 14 F5
Tingwall U.K. 20 F1
Tinharé, Ilha de i. Brazil 71 D1
Tinian i. N. Mariana Is 41 G6
Tini Heke is N.Z. see Snares Islands
Tinnelvelly India see Tirunelveli
Tinogasta Arg. 70 C3
Tinos Greece 27 K6
Tinos i. Greece 27 K6
Tinrhert, Hamada de Alg. 46 D2
Tinsukia India 37 H4
Tintagel U.K. 19 C8
Tînţâne Mauritania 46 B3
Tintina Arg. 70 D3
Tintinara Australia 57 C7
Tionesta PA U.S.A. 64 B2
Tionesta Lake PA U.S.A. 64 B2
Tipasa Alg. 25 H5
Tiphsah Syria see Dibsī
Tipperary Ireland 21 D5
Tipton CA U.S.A. 65 C2
Tipton, Mount AZ U.S.A. 65 E3
Tiptree U.K. 19 H7
Tiptur India 38 C3
Tipturi India see Tiptur
Tiracambu, Serra do hills Brazil 69 I4
Tirana Albania 27 H4
Tiranë Albania see Tirana
Tirano Italy 26 D1
Tirari Desert Australia 57 B5
Tiraspol Moldova 27 M1
Tiraz Mountains Namibia 50 C4
Tire Turkey 27 L5
Tirebolu Turkey 35 E2
Tiree i. U.K. 20 C4
Tîrgovişte Romania see Târgovişte
Tîrgu Jiu Romania see Târgu Jiu
Tîrgu Mureş Romania see Târgu Mureş
Tîrgu Neamţ Romania see Târgu Neamţ
Tîrgu Secuiesc Romania see Târgu Secuiesc
Tiri Pak. 36 A3
Tirich Mir mt. Pak. 33 L2
Tirna r. India 38 C2
Tîrnăveni Romania see Târnăveni
Tírnavos Greece see Tyrnavos
Tiros Brazil 71 B2
Tirourda, Col de pass Alg. 25 I5
Tirreno, Mare sea France/Italy see Tyrrhenian Sea
Tirso r. Sardegna Italy 26 C5
Tirthahalli India 38 B3
Tiruchchendur India 38 C4
Tiruchchirappalli India 38 C4
Tiruchengodu India 38 C4
Tirunelveli India 38 C4
Tirupati India 38 C3
Tiruppattur Tamil Nadu India 38 C3
Tiruppattur Tamil Nadu India 38 C4
Tiruppur India 38 C4
Tiruttani India 38 C3
Tirutturaippundi India 38 C4
Tiruvallur India 38 C3
Tiruvannamalai India 38 C3
Tiruvottiyur India 38 D3
Tiru Well Australia 54 D5
Tisa r. Serbia 27 I2
Tisdale Canada 62 G1
Ţīsīyah Syria 39 C3
Tissemsilt Alg. 25 G6
Tisza r. Serbia see Tisa
Titalya Bangl. see Tetulia
Titan Dome Antarctica 76 H1
Titao Burkina Faso 46 C3
Tit-Ary Rus. Fed. 29 N2
Titawin Morocco see Tétouan
Titicaca, Lago Bol./Peru see Titicaca, Lake
Titicaca, Lake Bol./Peru 68 E7
Titi Islands N.Z. 59 A8
Tititea mt. N.Z. see Aspiring, Mount
Titlagarh India 38 D1
Titograd Montenegro see Podgorica
Titova Mitrovica Kosovo see Mitrovicë
Titovo Užice Serbia see Užice
Titovo Velenje Slovenia see Velenje
Titov Veles Macedonia see Veles
Titov Vrbas Serbia see Vrbas
Ti Tree Australia 54 F5
Titu Romania 27 K2
Titusville FL U.S.A. 63 K6
Titusville PA U.S.A. 64 B2
Tiumpain, Rubha an hd U.K. see Tiumpan Head
Tiumpan Head hd U.K. 20 C2
Tiva watercourse Kenya 48 D4
Tivari India 36 C4
Tiverton U.K. 19 D8
Tivoli Italy 26 E4
Ţīwī Oman 33 I5
Tizi El Arba h. Alg. 25 H5
Tizimín Mex. 66 G4
Tizi N'Kouilal pass Alg. 25 I5
Tizi Ouzou Alg. 25 I5
Tiznap He r. China 36 D1
Tiznit Morocco 46 C2
Tiztoutine Morocco 25 E6
Tjaneni Swaziland 51 J3
Tjappsåive Sweden 14 K4
Tjirebon Indon. see Cirebon
Tjolotjo Zimbabwe see Tsholotsho
Tjorhom Norway 15 E7
Tkibuli Georgia see Tqibuli
Tlahualilo Mex. 62 G6
Tlaxcala Mex. 66 E5
Tlemcen Alg. 25 F6
Tlhakalatlou S. Africa 50 F5
Tlholong S. Africa 51 I5
Tlokweng Botswana 51 G3
Tlyarata Rus. Fed. 35 G2
Toamasina Madag. 49 E5
Toano VA U.S.A. 64 C4
Toba China 37 I3
Toba and Kakar Ranges mts Pak. 33 K3
Tobago i. Trin. and Tob. 67 L6
Tobelo Indon. 41 E7
Tobercurry Ireland 21 D3
Tobermorey Australia 56 B4
Tobermory Australia 58 A1
Tobermory Canada 63 K2
Tobermory U.K. 20 C4
Tobin, Lake imp. l. Australia 54 D5
Tobin Lake Canada 62 G1
Tobi-shima i. Japan 45 E5
Tobol r. Kazakh./Rus. Fed. 30 F1
Tobol'sk Rus. Fed. 28 H4
Tobruk Libya see Tubruq
Tobseda Rus. Fed. 12 L1
Tobyl r. Kazakh./Rus. Fed. see Tobol
Tobysh r. Rus. Fed. 12 K2
Tocache Nuevo Peru 68 C5
Tocantinópolis Brazil 69 I5
Tocantins r. Brazil 71 A1
Tocantins state Brazil 71 A1
Tocantinzinha r. Brazil 71 A1
Toccoa U.S.A. 63 K5
Tochi r. Pak. 36 B2
Töcksfors Sweden 15 G7
Tocopilla Chile 70 B2
Tocumwal Australia 58 B5
Todd watercourse Australia 56 A5
Todi Italy 26 E3
Todoga-saki pt Japan 45 F5
Todos Santos Mex. 66 B4
Toe Head hd U.K. 20 B3
Tofino Canada 62 B2
Toft U.K. 20 [inset]
Tofua i. Tonga 53 I3
Togatax China 36 E2
Togian, Kepulauan is Indon. 41 E8
Togliatti Rus. Fed. see Tol'yatti
Togo country Africa 46 D4
Togtoh China 43 K4
Togton He r. China 37 H2
Togton Heyan China see Tanggulashan
Toholampi Fin. 14 N5
Toiba China 37 G3
Toijala Fin. 15 M6
Toi-misaki pt Japan 45 C7
Toivakka Fin. 14 O5
Toiyabe Range mts NV U.S.A. 65 D1
Tojikiston country Asia see Tajikistan
Tok U.S.A. 60 D3
Tokar Sudan 32 E6
Tokara-rettō is Japan 45 C7
Tokarevka Rus. Fed. 13 I6
Tokat Turkey 34 E2
Tŏkchŏk-to i. S. Korea 45 B5
Tokdo i. N. Pacific Ocean see Liancourt Rocks
Tokelau terr. S. Pacific Ocean 53 I2
Tokmak Kyrg. see Tokmok
Tokmak Ukr. 13 G7
Tokmok Kyrg. 42 D4
Tokomaru Bay N.Z. 59 G4
Tokoroa N.Z. 59 E4
Tokoza S. Africa 51 I4
Toksun China 42 F4
Tok-tō i. N. Pacific Ocean see Liancourt Rocks
Toktogul Kyrg. 42 C4
Tokto-ri i. N. Pacific Ocean see Liancourt Rocks
Tokur Rus. Fed. 44 D1
Tokushima Japan 45 D6
Tokuyama Japan 45 C6
Tōkyō Japan 45 E6
Tokzār Afgh. 36 A2
Tolaga Bay N.Z. 59 G4
Tôlañaro Madag. 49 E6
Tolbo Mongolia 42 G3
Tolbukhin Bulg. see Dobrich
Tolbuzino Rus. Fed. 44 B1
Toledo Brazil 70 F2
Toledo Spain 25 D4
Toledo OH U.S.A. 63 K3
Toledo, Montes de mts Spain 25 D4
Toletum Spain see Toledo
Toliara Madag. 49 E6
Tolitoli Indon. 41 E7
Tol'ka Rus. Fed. 28 J3
Tolmachevo Rus. Fed. 15 P7
Tolo Dem. Rep. Congo 48 B4
Tolochin Belarus see Talachyn
Tolosa France see Toulouse
Tolosa Spain 25 E2
Toluca Mex. 66 E5
Toluca de Lerdo Mex. see Toluca
To-lun Nei Mongol China see Dolonnur
Tol'yatti Rus. Fed. 13 K5
Tom' r. Rus. Fed. 44 B2
Tomah U.S.A. 63 I3
Tomakomai Japan 44 F4
Tomales CA U.S.A. 65 A1
Tomamae Japan 44 F3
Tomanivi mt. Fiji 53 H3
Tomar Brazil 68 F4
Tomar Port. 25 B4
Tomari Rus. Fed. 44 F3
Tomarza Turkey 34 D3
Tomaszów Lubelski Poland 13 D6
Tomaszów Mazowiecki Poland 17 R5
Tomatin U.K. 20 F3
Tomatlán Mex. 66 C5
Tomazina Brazil 71 A3
Tombador, Serra do hills Brazil 69 G6
Tombigbee r. U.S.A. 63 J5
Tomboco Angola 49 B4
Tombouctou Mali see Timbuktu
Tombua Angola 49 B5
Tom Burke S. Africa 51 H2
Tomdibuloq Uzbek. 33 J1
Tome Moz. 51 L2
Tomelilla Sweden 15 H9
Tomelloso Spain 25 E4
Tomi Romania see Constanţa
Tomingley Australia 58 D4
Tomini, Teluk g. Indon. 41 E8
Tominian Mali 46 C3
Tomintoul U.K. 20 F3
Tomislavgrad Bos.-Herz. 26 G3
Tomkinson Ranges mts Australia 55 E6
Tømmerneset Norway 14 I3
Tommot Rus. Fed. 29 N4
Tomo r. Col. 68 E2
Tomortei China 43 K4
Tom Price Australia 54 B5
Tomra China 37 F3
Tomsk Rus. Fed. 28 J4
Toms River NJ U.S.A. 64 D3
Tomtabacken h. Sweden 15 I8
Tomtor Rus. Fed. 29 P3
Tomur Feng mt. China/Kyrg. see Pobeda Peak
Tomuzlovka r. Rus. Fed. 13 J7
Tom White, Mount U.S.A. 60 D3
Tonalá Mex. 66 F5
Tonantins Brazil 68 E4
Tonbridge U.K. 19 H7
Tønder Denmark 15 F9
Tondi India 38 C4
Tone r. U.K. 19 E7
Toney Mountain Antarctica 76 K1
Tonga country S. Pacific Ocean 53 I4
Tongaat S. Africa 51 J5
Tongariro National Park N.Z. 59 E4
Tongatapu Group is Tonga 53 I4
Tonga Trench sea feature S. Pacific Ocean 74 I7
T'ongch'ŏn N. Korea 45 B5
Tongchuan Shaanxi China 43 J5
Tongduch'ŏn S. Korea 45 B5
Tonghae S. Korea 45 C5
Tonghe China 44 C3
Tonghua Jilin China 44 B4
Tonghua Jilin China 44 B4
Tongi Bangl. see Tungi
Tongjiang Heilong. China 44 D3
Tongking, Gulf of China/Vietnam 31 J4
Tongliao China 43 M4
Tongling China 43 L6
Tonglu China 43 L7
Tongo Australia 58 A3
Tongo Lake imp. l. Australia 58 A3
Tongren Guizhou China 43 J7
Tongren Qinghai China 42 I5
Tongsa Bhutan 37 G4
Tongtian He r. Qinghai China 37 H2
Tongtian He r. Qinghai China see Yangtze
Tongue U.K. 20 E2
Tongxin China 42 J5
T'ongyŏng S. Korea 45 C6
Tonk India 36 C4
Tonkābon Iran 35 H3
Tônlé Sab l. Cambodia see Tonle Sap
Tonle Sap l. Cambodia 31 J5
Tonopah AZ U.S.A. 65 E4
Tonopah NV U.S.A. 65 D1
Tønsberg Norway 15 G7
Tonstad Norway 15 E7
Tonzang Myanmar 37 H5
Toobeah Australia 58 D2
Toobli Liberia 46 C4
Tooele U.S.A. 62 E3
Toogoolawah Australia 58 F1
Tooma r. Australia 58 D6
Toompine Australia 58 B1
Toora Australia 58 C7
Tooraweenah Australia 58 D3
Toorberg mt. S. Africa 50 G7
Toowoomba Australia 58 E1
Tooxin Somalia 48 F2
Top Boğazı Geçidi pass Turkey 39 C1
Topeka U.S.A. 63 H4
Topia Mex. 66 C3
Topoľčany Slovakia 17 Q6
Topolovgrad Bulg. 27 L3
Topozero, Ozero l. Rus. Fed. 12 G2
Tor Eth. 47 G4
Torbalı Turkey 27 L5
Torbat-e Ḩeydarīyeh Iran 33 I2
Torbat-e Jām Iran 33 J2
Torbay Bay Australia 55 B8
Torbert, Mount U.S.A. 60 C3
Torbeyevo Rus. Fed. 13 I5
Tordesillas Spain 25 D3
Tordesilos Spain 25 F3
Töre Sweden 14 M4
Torelló Spain 25 H2
Toretam Kazakh. see Baykonyr
Torgau Germany 17 N5
Torghay Kazakh. see Turgay
Torgun r. Rus. Fed. 13 J6
Torino Italy see Turin
Tori-shima i. Japan 45 F7
Torit South Sudan 47 G4
Torkamān Iran 35 G3
Torkovichi Rus. Fed. 12 F4
Torneå Fin. see Tornio
Torneälven r. Sweden 14 N4
Torneträsk l. Sweden 14 K2
Torngat, Monts mts Canada see Torngat Mountains
Torngat Mountains Canada 61 L4
Tornio Fin. 14 N4
Toro Spain 25 D3
Torom Rus. Fed. 44 D1
Toronto Canada 61 K5
Toro Peak CA U.S.A. 65 D4
Toropets Rus. Fed. 12 F4
Tororo Uganda 48 D3
Toros Dağları mts Turkey see Taurus Mountains
Torphins U.K. 20 G3
Torquay Australia 58 B7
Torquay U.K. 19 D8
Torrance CA U.S.A. 65 C4
Torrão Port. 25 B4
Torre mt. Port. 25 C3
Torreblanca Spain 25 G3
Torrecerredo mt. Spain 25 D2
Torre del Greco Italy 26 F4
Torre de Moncorvo Port. 25 C3
Torrelavega Spain 25 D2
Torremolinos Spain 25 D5
Torrens, Lake imp. l. Australia 57 B6
Torrens Creek Australia 56 D4
Torrent Spain 25 F4
Torrente Spain see Torrent
Torreón Mex. 66 D3
Torres Brazil 71 A5
Torres del Paine, Parque Nacional nat. park Chile 70 B8
Torres Islands Vanuatu 53 G3
Torres Novas Port. 25 B4
Torres Strait Australia 52 E2
Torres Vedras Port. 25 B4
Torrevieja Spain 25 F5
Torridge r. U.K. 19 C8
Torridon, Loch b. U.K. 20 D3
Torrijos Spain 25 D4
Torrington Australia 58 E2
Torsby Sweden 15 H6
Tórshavn Faroe Is 14 [inset]
To'rtko'l Uzbek. 33 J1
Törtköl Uzbek. see To'rtko'l
Tortolì Sardegna Italy 26 C5
Tortona Italy 26 C2
Tortosa Spain 25 G3
Tortum Turkey 35 F2
Ţorūd Iran 35 I4
Torugart, Pereval pass China/Kyrg. see Turugart Pass
Torul Turkey 35 E2
Toruń Poland 17 Q4
Tory Island Ireland 21 D2
Tory Sound sea chan. Ireland 21 D2
Torzhok Rus. Fed. 12 G4
Tosa Japan 45 D6
Tosbotn Norway 14 H4
Tosca S. Africa 50 F3
Toscano, Arcipelago is Italy 26 C3
Tosham India 36 C3
Tōshima-yama mt. Japan 45 F4
Toshkent Uzbek. 33 K1
Tosno Rus. Fed. 12 F4
Toson Hu l. China 37 I1
Tostado Arg. 70 D3
Tosya Turkey 34 D2
Totapola mt. Sri Lanka 38 D5
Tôtes France 19 I9
Tot'ma Rus. Fed. 12 I4
Totness Suriname 69 G2
Tottenham Australia 58 C4
Totton U.K. 19 F8
Tottori Japan 45 D6
Touba Côte d'Ivoire 46 C4
Touba Senegal 46 B3
Toubkal, Parc National du nat. park Morocco 22 C5
Touboro Cameroon 47 E4
Tougan Burkina Faso 46 C3
Touggourt Alg. 22 F5
Tougué Guinea 46 B3
Touil Mauritania 46 B3
Toul France 24 G2
Toulon France 24 G5
Toulouse France 24 E5
Toumodi Côte d'Ivoire 46 C4
Tourane Vietnam see Đa Năng
Tourlaville France 19 F9
Tournai Belgium 16 I5
Tournon-sur-Rhône France 24 G4
Tournus France 26 A1
Touros Brazil 69 K5
Tours France 24 E3
Tousside, Pic mt. Chad 47 E2
Toussoro, Mont mt. Cent. Afr. Rep. 48 C3
Toutai China 44 B3
Touwsrivier S. Africa 50 E7
Tovarkovo Rus. Fed. 13 G5
Tovuz Azer. 35 G2
Towada Japan 44 F4
Towak Mountain h. U.S.A. 60 B3
Towanda PA U.S.A. 64 C2
Towcester U.K. 19 G6
Tower Ireland 21 D6
Townes Pass CA U.S.A. 65 D2
Townsend U.S.A. 62 E2
Townsend, Mount Australia 58 D6
Townshend Island Australia 56 E4
Townsville Australia 56 D3
Towot South Sudan 47 G4
Towr Kham Afgh. 36 B2

...vrzī Afgh. 36 A3
...vson MD U.S.A. 64 C3
...vyn U.K. see Tywyn
...ama Japan 45 E5
...ama-wan b. Japan 45 E5
...ohashi Japan 45 E6
...okawa Japan 45 E6
...onaka Japan 45 D6
...ooka Japan 45 D6
...ota Japan 45 E6
...anlı Turkey see Almus
...ê Kangri mt. China 37 E2
...eur Tunisia 22 F5
...i, Mount U.S.A. 60 C3
...ouli Georgia 35 F2
...blous Lebanon see Tripoli
...botište Macedonia 27 J4
...bzon Turkey 35 E2
...cy CA U.S.A. 65 B2
...algar, Cabo de c. Spain 25 C5
...lle, Rubha na pt U.K. 20 D5
...ll Island Greenland see Traill Ø
...ll Ø i. Greenland 61 P2
...iectum Neth. see Utrecht
...kai Lith. 15 N9
...kiya reg. Europe see Thrace
...kt Rus. Fed. 12 K3
...kya reg. Europe see Thrace
...lee Ireland 21 C5
...lee Bay Ireland 21 C5
... Lí Ireland see Tralee
...mandaí Brazil 71 A5
...mán Tepuí mt. Venez. 68 F2
... Mhór Ireland see Tramore
...more Ireland 21 E5
...nås Sweden 15 I7
...ncas Arg. 70 C3
...ncoso Brazil 71 D2
...nemo Sweden 15 H8
...nent U.K. 20 G5
...ngan i. Indon. 41 F8
...ngie Australia 58 C4
...nsantarctic Mountains Antarctica
...6 H2
...nsvase, Canal de Spain 25 E4
...nsylvanian Alps mts Romania 27 J2
...nsylvanian Basin plat. Romania 27 K1
...pani Sicilia Italy 26 E5
...pezus Turkey see Trabzon
...ralgon Australia 58 C7
...shigang Bhutan see Tashigang
...simeno, Lago l. Italy 26 E3
...tani r. Pak. 36 B3
...unsee l. Austria 17 N7
...unstein Germany 17 N7
...vellers Lake imp. l. Australia 57 C7
...vers, Mount N.Z. 59 D6
...verse City U.S.A. 63 J3
...vnik Bos.-Herz. 26 G2
...ovlje Slovenia 26 F1
...asury Islands Solomon Is 52 F2
...bević, Nacionalni Park nat. park
...os.-Herz. 26 H3
...bíč Czech Rep. 17 O6
...binje Bos.-Herz. 26 H3
...bišov Slovakia 13 D6
...bizond Turkey see Trabzon
...bnje Slovenia 26 F2
...e Island India 38 B4
...faldwyn U.K. see Montgomery
...ffynnon U.K. see Holywell
...fyclawdd U.K. see Knighton
...fynwy U.K. see Monmouth
...gosse Islets and Reefs Australia 56 E3
...inta y Tres Uruguay 70 F4
...lew Arg. 70 C6
...lleborg Sweden 15 H9
...mblant, Mont h. Canada 63 M2
...miti, Isole is Italy 26 F3
...mont PA U.S.A. 64 C2
...monton U.S.A. 62 E3
...mp Spain 25 G2
...nance U.K. 19 B8
...nčín Slovakia 17 Q6
...nque Lauquén Arg. 70 D5
...nt Italy see Trento
...nt r. U.K. 19 G5
...nto Italy 26 D1
...nton Canada 63 L3
...nton MO U.S.A. 63 I3
...nton NJ U.S.A. 64 D2
...orchy U.K. 19 D7
...passey Canada 61 M5
...s Arroyos Arg. 70 D5
...sco i. U.K. 19 A9
...s Corações Brazil 71 B3
...s Esquinas Col. 68 C3
...s Forcas, Cabo c. Morocco see
...rois Fourches, Cap des
...s Lagoas Brazil 71 A3
...s Marias, Represa resr Brazil 71 B2
...s Picos, Cerro mt. Arg. 70 D5
...s Pontas Brazil 71 B3
...s Puntas, Cabo c. Arg. 70 C7
...s Rios Brazil 71 C3
...etten Norway 15 G6
...etyy Severnyy Rus. Fed. see 3-y Severnyy
...eungen Norway 15 F7
...eves Germany see Trier
...viglio Italy 26 C2
...viso Italy 26 E2
...vose Head hd U.K. 19 B8
...anda Greece see Trianta
...angle VA U.S.A. 64 C3
...anta Greece 27 M6
...bal Areas admin. div. Pak. 36 B2
...case Italy 26 H5
...chinopoly India see Tiruchchirappalli
...chur India see Thrissur
...da Australia 58 B4
...dentum Italy see Trento
...er Germany 17 K6
...este Italy 26 E2
...este, Golfo di g. Europe see
...rieste, Gulf of
...este, Gulf of Europe 26 E2
...glav mt. Slovenia 26 E1
...glavski narodni park nat. park Slovenia
...6 E1
...kala Greece 27 I5
...kkala Greece see Trikala
...kora, Puncak mt. Indon. 41 F8
...m Ireland 21 F4
...ncomalee Sri Lanka 38 D4
...ndade Brazil 71 A2
...ndade, Ilha da i. S. Atlantic Ocean
...2 G7
...nidad Bol. 68 F6
...nidad Cuba 67 I4
...nidad i. Trin. and Tob. 67 L6
...nidad Uruguay 70 E4
...nidad U.S.A. 62 G4
...nidad country West Indies see
...rinidad and Tobago
...nidad and Tobago country West Indies
...7 L6
...nity Bay Canada 61 M5
...nity Islands U.S.A. 60 C4
...onto, Capo c. Italy 26 G5
...poli Greece 27 J6
...poli Lebanon 39 B2
Tripoli Libya 47 E1
Trípolis Greece see Tripoli
Tripolis Lebanon see Tripoli
Tripunittura India 38 C4
Tripura state India 37 G5
Tristan da Cunha i. S. Atlantic Ocean
72 H8
Trisul mt. India 36 D3
Trivandrum India see Thiruvananthapuram
Trivento Italy 26 F4
Trnava Slovakia 17 P6
Trobriand Islands P.N.G. 52 F2
Trofors Norway 14 H4
Trogir Croatia 26 G3
Troia Italy 26 F4
Troisdorf Germany 17 K5
Trois Fourches, Cap des c. Morocco 25 E6
Trois-Rivières Canada 63 M2
Troitsko-Pechorsk Rus. Fed. 11 R3
Troitskoye Altayskiy Kray Rus. Fed. 42 E2
Troitskoye Khabarovskiy Kray Rus. Fed.
44 E2
Troitskoye Respublika Kalmykiya-Khalm'g-
Tangch Rus. Fed. 13 J7
Troll research stn 76 B2
Trollhättan Sweden 15 H7
Trombetas r. Brazil 69 G4
Tromelin, Île i. Indian Ocean 73 L7
Tromen, Volcán vol. Arg. 70 B5
Tromie r. U.K. 20 E3
Trompsburg S. Africa 51 G6
Tromsø Norway 14 K2
Trona CA U.S.A. 65 D3
Tronador, Monte mt. Arg. 70 B6
Trondheim Norway 14 G5
Trondheimsfjorden sea chan. Norway
14 F5
Trongsa Bhutan see Tongsa
Troödos, Mount Cyprus 39 A2
Troödos Mountains Cyprus 39 A2
Troon U.K. 20 E5
Tropeiros, Serra dos hills Brazil 71 B1
Tropic of Capricorn 56 G4
Trosh Rus. Fed. 12 L2
Trostan h. U.K. 21 F2
Trout Lake Alta Canada 60 G4
Trout Lake N.W.T. Canada 60 F3
Trout Lake Ont. Canada 63 I1
Trout Run PA U.S.A. 64 C2
Trouville-sur-Mer France 19 H9
Trowbridge U.K. 19 E7
Troy tourist site Turkey 27 L5
Troy AL U.S.A. 63 J5
Troy NH U.S.A. 64 E1
Troy NY U.S.A. 64 E1
Troy PA U.S.A. 64 C2
Troyan Bulg. 27 K3
Troyes France 24 G2
Troy Lake CA U.S.A. 65 D3
Troy Peak NV U.S.A. 65 E1
Trstenik Serbia 27 I3
Trucial Coast country Asia see
United Arab Emirates
Trucial States country Asia see
United Arab Emirates
Trud Rus. Fed. 12 G4
Trufanova Rus. Fed. 12 J2
Trujillo Hond. 67 G5
Trujillo Peru 68 C5
Trujillo Spain 25 D4
Trujillo Venez. 68 D2
Trujillo, Monte mt. Dom. Rep. see
Duarte, Pico
Truk is Micronesia see Chuuk
Trumbull, Mount AZ U.S.A. 65 F2
Trundle Australia 58 C4
Truong Sa is S. China Sea see
Spratly Islands
Truro Canada 63 O2
Truro U.K. 19 B8
Truskmore h. Ireland 21 D3
Truth or Consequences U.S.A. 62 F5
Trutnov Czech Rep. 17 O5
Truuli Peak U.S.A. 60 C4
Truva tourist site Turkey see Troy
Trypiti, Akrotirio pt Greece 27 K7
Trysil Norway 15 H6
Trzebiatów Poland 17 O3
Tsagaan-Uul Mongolia 42 H3
Tsagan Aman Rus. Fed. 13 J7
Tsagan Nur Rus. Fed. 13 J7
Tsaidam Basin China see Qaidam Pendi
Tsaka La pass China/India 36 D2
Tsalenjikha Georgia 35 F2
Tsangbo r. China see Brahmaputra
Tsangpo r. China see Brahmaputra
Tsaratanana, Massif du mts Madag. 49 E5
Tsarevo Bulg. 27 L3
Tsaris Mountains Namibia 50 C3
Tsaritsyn Rus. Fed. see Volgograd
Tsaukaib Namibia 50 B4
Tsavo East National Park Kenya 48 D4
Tsavo West National Park Africa 48 D3
Tsefat Israel see Zefat
Tselinograd Kazakh. see Astana
Tsenhermandal Mongolia 43 J3
Tsenogora Rus. Fed. 12 J2
Tses Namibia 50 D3
Tsetseg Mongolia 42 G3
Tsetseng Botswana 50 F2
Tsetserleg Arhangay Mongolia 42 I3
Tsetserleg Mongolia 42 H3
Tshabong Botswana 50 F4
Tshad country Africa see Chad
Tshane Botswana 50 E3
Tshela Dem. Rep. Congo 49 B4
Tshibala Dem. Rep. Congo 49 C4
Tshikapa Dem. Rep. Congo 49 C4
Tshing S. Africa 51 H4
Tshipise S. Africa 51 J2
Tshitanzu Dem. Rep. Congo 49 C4
Tshofa Dem. Rep. Congo 49 C4
Tshokwane S. Africa 51 J3
Tsholotsho Zimbabwe 49 C5
Tshootsha Botswana 50 E2
Tshuapa r. Dem. Rep. Congo 47 F5
Tshwane S. Africa see Pretoria
Tsil'ma r. Rus. Fed. 12 K2
Tsimlyansk Rus. Fed. 13 I7
Tsimlyanskoye Vodokhranilishche resr
Rus. Fed. 13 I7
Tsimmermanovka Rus. Fed. 44 E2
Tsinan China see Jinan
Tsineng S. Africa 50 F4
Tsinghai prov. China see Qinghai
Tsingtao China see Qingdao
Tsining China see Jining
Tsiombe Madag. 49 E6
Tsiroanomandidy Madag. 49 E5
Tsitsihar China see Qiqihar
Tsivil'sk Rus. Fed. 12 J5
Tskhaltubo Georgia see Tsqaltubo
Ts'khinvali Georgia 35 F2
Tsna r. Rus. Fed. 13 I5
Tsnori Georgia 35 G2
Tsokar Chumo l. India 36 D2
Tsolo S. Africa 51 I6
Tsomo S. Africa 51 H7
Tsona China see Cona
Tsqaltubo Georgia 35 F2
Tsu Japan 45 E6
Tsuchiura Japan 45 F5
Tsugarū-kaikyō str. Japan 44 F4
Tsugaru Strait Japan see Tsugarū-kaikyō
Tsumeb Namibia 49 B5
Tsumis Park Namibia 50 C2
Tsumkwe Namibia 49 C5
Tsuruga Japan 45 E6
Tsurugi-san mt. Japan 45 D6
Tsurukhaytuy Rus. Fed. see Priargunsk
Tsuruoka Japan 45 E5
Tsushima is Japan 45 C6
Tsushima-kaikyō str. Japan/S. Korea see
Korea Strait
Tsuyama Japan 45 D6
Tswaane Botswana 50 E2
Tswaraganang S. Africa 51 G5
Tswelelang S. Africa 51 G4
Tsyelyakhany Belarus 15 N10
Tsypnavolok Rus. Fed. 14 R2
Tsyurupyns'k Ukr. 27 O1
Tthenaagoo Canada see Nahanni Butte
Tua Dem. Rep. Congo 48 B4
Tual Indon. 41 F8
Tuam Ireland 21 D4
Tuamotu, Archipel des is Fr. Polynesia see
Tuamotu Islands
Tuamotu Islands Fr. Polynesia 75 K6
Tuapse Rus. Fed. 13 H7
Tuath, Loch a' b. U.K. 20 C2
Tuba City U.S.A. 62 E4
Tubarão Brazil 71 A5
Tubarjal Saudi Arabia 39 D4
Tübingen Germany 17 L6
Tubmanburg Liberia 46 B4
Tubruq Libya 34 A4
Tubuai i. Fr. Polynesia 75 K7
Tubuai Islands Fr. Polynesia 75 J7
Tucano Brazil 69 K6
Tucavaca Bol. 69 G7
Tuchitua Canada 60 F3
Tuckerton NJ U.S.A. 64 D3
Tucopia i. Solomon Is see Tikopia
Tucson U.S.A. 62 E5
Tucumán Arg. see San Miguel de Tucumán
Tucumcari U.S.A. 62 G4
Tucupita Venez. 68 F2
Tucuruí Brazil 69 I4
Tucuruí, Represa de resr Brazil 69 I4
Tudela Spain 25 F2
Tuder Italy see Todi
Tüdevtey Mongolia 42 H3
Tudun Wada Nigeria 46 D3
Tuela r. Port. 25 C3
Tuensang India 37 H4
Tufts Abyssal Plain sea feature
N. Pacific Ocean 75 K2
Tugela r. S. Africa 51 J5
Tuglung China 37 H3
Tuguegarao Phil. 41 E6
Tugur Rus. Fed. 44 E1
Tujiabu China see Yongxiu
Tukangbesi, Kepulauan is Indon. 41 E8
Tukituki r. N.Z. 59 F4
Tuktoyaktuk Canada 60 E3
Tuktut Nogait National Park Canada 60 F3
Tukums Latvia 15 M8
Tukuringra, Khrebet mts Rus. Fed. 44 B1
Tukuyu Tanz. 49 D4
Tula Rus. Fed. 13 H5
Tulach Mhór Ireland see Tullamore
Tulagt Ar Gol r. China 37 H1
Tula Mountains Antarctica 76 D2
Tulancingo Mex. 66 E4
Tulare CA U.S.A. 65 C2
Tulare Lake Bed CA U.S.A. 65 C3
Tulasi mt. India 38 D2
Tulbagh S. Africa 50 D7
Tulcán Ecuador 68 C3
Tulcea Romania 27 M2
Tuléar Madag. see Toliara
Tulemalu Lake Canada 61 I3
Tulia U.S.A. 62 G5
Tulihe China 44 A2
Tulkarem West Bank see Ṭūlkarm
Ṭūlkarm West Bank 39 B3
Tulla Ireland 21 D5
Tullamore Australia 58 C4
Tullamore Ireland 21 E4
Tulle France 24 E4
Tulleråsen Sweden 14 I5
Tullibigeal Australia 58 C4
Tullow Ireland 21 F5
Tully Australia 56 D3
Tully r. Australia 56 D3
Tully U.K. 21 E3
Tulos Rus. Fed. 14 Q5
Tulqarem West Bank see Ṭūlkarm
Tulsa U.S.A. 63 H4
Tulsipur Nepal 37 E3
Tuluá Col. 68 C3
Tuluksak U.S.A. 77 B2
Tulūl al Ashāqif hills Jordan 39 C3
Tulun Rus. Fed. 42 I2
Tulu-Tuloi, Serra hills Brazil 68 F3
Tulu Welel mt. Eth. 48 D3
Tuma Rus. Fed. 13 I5
Tumaco Col. 68 C3
Tumahole S. Africa 51 H4
Tumain China 37 G2
Tumakuru India see Tumkur
Tumannyy Rus. Fed. 14 S2
Tumasik Sing. see Singapore
Tumba Dem. Rep. Congo 48 C4
Tumba Sweden 15 J7
Tumba, Lac l. Dem. Rep. Congo 48 B4
Tumbarumba Australia 58 D5
Tumbes Peru 68 B4
Tumby Bay Australia 57 B7
Tumcha r. Fin./Rus. Fed. 14 Q3
Tumen Jilin China 44 C4
Tumereng Guyana 68 F2
Tumiritinga Brazil 71 C2
Tumkur India 38 C3
Tummel r. U.K. 20 F4
Tummel, Loch l. U.K. 20 F4
Tumnin r. Rus. Fed. 44 F2
Tump Pak. 33 J4
Tumu Ghana 46 C3
Tumucumaque, Serra hills Brazil 69 G3
Tumudibandh India 38 D2
Tumut Australia 58 D5
Tuna India 36 B5
Tunbridge Wells, Royal U.K. 19 H7
Tunceli Turkey 35 E3
Tuncurry Australia 58 F4
Tunduru Tanz. 49 D5
Tunes Tunisia see Tunis
Tunga Nigeria 46 D4
Tungabhadra Reservoir India 38 C3
Tungi Bangl. 37 G5
Tungnaá r. Iceland 14 [inset]
Tungor Rus. Fed. 44 F1
Tungsten (abandoned) Canada 60 F3
Tuni India 38 D2
Tūnis country Africa see Tunisia
Tunis Tunisia 26 D6
Tunis, Golfe de g. Tunisia 26 D6
Tunisia country Africa 22 F5
Tunja Col. 68 D2
Tunkhannock PA U.S.A. 64 D2
Tunnsjøen l. Norway 14 H4
Tunstall U.K. 19 I6
Tuntsa Fin. 14 P3
Tuntsajoki r. Fin./Rus. Fed. see Tumcha
Tununak U.S.A. 60 B3
Tunxi China see Huangshan
Tuotuo He r. China see Togton He
Tuotuoheyan China see Tanggulashan
Tūp Afgh. 36 B2
Tüp Kyrg. 42 D4
Tupā Brazil 71 A3
Tupelo U.S.A. 63 J5
Tupik Rus. Fed. 43 L2
Tupinambarama, Ilha i. Brazil 69 G4
Tupiraçaba Brazil 71 A1
Tupiza Bol. 68 E8
Tüpqaraghan Tübegi pen. Kazakh. see
Mangyshlak, Poluostrov
Tupungato, Cerro mt. Arg./Chile 70 C4
Tuquan China 43 M3
Tura China 37 F1
Tura India 37 G4
Tura Rus. Fed. 29 L3
Turabah Saudi Arabia 32 F5
Turakina N.Z. 59 E5
Turan Rus. Fed. 42 G2
Turana, Khrebet mts Rus. Fed. 44 C2
Turan Lowland Asia 33 I2
Turan Oypaty lowland Asia see
Turan Lowland
Turan Pasttekisligi lowland Asia see
Turan Lowland
Turan Pesligi lowland Asia see
Turan Lowland
Turanskaya Nizmennost' lowland Asia see
Turan Lowland
Ţurāq al 'Ilab hills Syria 39 D3
Turar Ryskulov Kazakh. 33 L1
Tura-Ryskulova Kazakh. see Turar Ryskulov
Ţurayf Saudi Arabia 39 D4
Turba Estonia 15 N7
Turbat Pak. 33 J4
Turbo Col. 68 C2
Turda Romania 27 J1
Türeh Iran 35 H4
Turfan China see Turpan
Turfan Basin depr. China see Turpan Pendi
Turfan Depression China see
Turpan Pendi
Turgay Kazakh. 42 A3
Türgovishte Bulg. 27 L3
Turgutlu Turkey 27 L5
Turhal Turkey 34 E2
Türi Estonia 15 N7
Turia r. Spain 25 F4
Turin Italy 26 B2
Turiy Rog Rus. Fed. 44 C3
Turkana, Lake salt l. Eth./Kenya 48 D3
Turkestan Kazakh. 42 B4
Turkey country Asia/Europe 34 D3
Turki Rus. Fed. 13 I6
Türkistan Kazakh. see Turkestan
Türkiye country Asia/Europe see Turkey
Türkmenabat Turkm. 30 F3
Türkmen Adasy i. Turkm. see
Ogurjaly Adasy
Türkmen Aýlagy b. Turkm. 35 I3
Türkmen Aylagy b. Turkm. see
Türkmen Aýlagy
Türkmenbaşy Turkm. 35 I2
Türkmenbaşy Turkm. see Türkmenbaşy
Türkmenbaşy Aylagy b. Turkm. see
Türkmenbaşy Aýlagy
Türkmenbaşy Aýlagy b. Turkm. 35 I3
Türkmenbaşy Döwlet Gorugy nature res.
Turkm. 35 I3
Türkmen Dağı mt. Turkey 27 N5
Turkmenistan country Asia 33 I2
Turkmeniya country Asia see Turkmenistan
Türkmenostan country Asia see
Turkmenistan
Turkmenskaya S.S.R. country Asia see
Turkmenistan
Türkoğlu Turkey 34 E3
Turks and Caicos Islands terr. West Indies
67 J4
Turks Islands Turks and Caicos Is 67 J4
Turku Fin. 15 M6
Turkwel watercourse Kenya 48 D3
Turlock CA U.S.A. 65 B2
Turlock Lake CA U.S.A. 65 B2
Turmalina Brazil 71 C2
Turnagain, Cape N.Z. 59 F5
Turnberry U.K. 20 E5
Turneffe Islands atoll Belize 66 G5
Turnor Lake Canada 60 H4
Türnovo Bulg. see Veliko Türnovo
Turnu Măgurele Romania 27 K3
Turnu Severin Romania see
Drobeta-Turnu Severin
Turon r. Australia 58 D4
Turones France see Tours
Turovets Rus. Fed. 12 I4
Turpan China 42 F4
Turpan Pendi depr. China 42 F4
Turquino, Pico mt. Cuba 67 I4
Turriff U.K. 20 G3
Turris Libisonis Sardegna Italy see
Porto Torres
Tursāq Iraq 35 G4
Turtle Island Fiji see Vatoa
Turugart Pass China/Kyrg. 31 G2
Turugart Shankou pass China/Kyrg. see
Turugart Pass
Turuvanur India 38 C3
Turvo r. Brazil 71 A2
Turvo r. Brazil 71 A2
Tuscaloosa U.S.A. 63 J5
Tuscarawas r. OH U.S.A. 64 A2
Tuscarora Mountains hills PA U.S.A. 64 C2
Tuskegee U.S.A. 63 J5
Tussey Mountains hills PA U.S.A. 64 B2
Tutak Turkey 35 F3
Tutayev Rus. Fed. 12 H4
Tutera Spain see Tudela
Tuticorin India 38 C4
Tuttlingen Germany 17 L7
Tuttut Nunaat reg. Greenland 61 P2
Tutuala East Timor 54 D2
Tutubu P.N.G. 56 E1
Tutubu Tanz. 49 D4
Tutuila i. American Samoa 53 I3
Tutume Botswana 49 C6
Tuun-bong mt. N. Korea 44 B4
Tuupovaara Fin. 14 Q5
Tuusniemi Fin. 14 P5
Tuvalu country S. Pacific Ocean 53 H2
Tuwayq, Jabal hills Saudi Arabia 32 G4
Tuwayq, Jabal mts Saudi Arabia 32 G5
Ţuwayyil ash Shihāq mt. Jordan 39 C4
Tuwwal Saudi Arabia 32 E5
Tuxpan Mex. 66 E4
Tuxtla Gutiérrez Mex. 66 F5
Tuy Hoa Vietnam 31 J5
Tuz, Lake salt l. Turkey 34 D3
Tuz Gölü salt l. Turkey see Tuz, Lake
Tuzha Rus. Fed. 12 J4
Tuz Khurmātū Iraq 35 G4
Tuzla Bos.-Herz. 26 H2
Tuzla Turkey 39 B1
Tuzla Gölü lag. Turkey 27 L4
Tuzlov r. Rus. Fed. 13 I7
Tuzu r. Myanmar 37 H4
Tvedestrand Norway 15 F7
Tver' Rus. Fed. 12 G4
Twain Harte CA U.S.A. 65 B1
Tweed r. U.K. 20 G5
Tweed Heads Australia 58 F2
Tweefontein S. Africa 50 D7
Twee Rivier Namibia 50 D3
Twentynine Palms CA U.S.A. 65 D3
Twin Bridges CA U.S.A. 65 B1
Twin Falls U.S.A. 62 E3
Twin Heads h. Australia 54 D5
Twin Peak CA U.S.A. 65 B1
Twitchen Reservoir CA U.S.A. 65 B3
Twofold Bay Australia 58 D6
Two Harbors U.S.A. 63 I2
Tyan' Shan' mts China/Kyrg. see Tien Shan
Tyatya, Vulkan vol. Rus. Fed. 44 G3
Tydal Norway 14 G5
Tygart Valley val. WV U.S.A. 64 B3
Tygda Rus. Fed. 44 B1
Tygda r. Rus. Fed. 44 B1
Tyler U.S.A. 63 H5
Tym' r. Rus. Fed. 44 F2
Tymovskoye Rus. Fed. 44 F2
Tynda Rus. Fed. 43 M1
Tyndinskiy Rus. Fed. see Tynda
Tyne r. U.K. 20 G4
Tynemouth U.K. 18 F3
Tynset Norway 14 G5
Tyoploozyorsk Rus. Fed. see Teploozersk
Tyoploye Ozero Rus. Fed. see Teploozersk
Tyr Lebanon see Tyre
Tyras Ukr. see Bilhorod-Dnistrovs'kyy
Tyre Lebanon 39 B3
Tyree, Mount Antarctica 76 L1
Tyrma Rus. Fed. 44 D2
Tyrma r. Rus. Fed. 44 C2
Tyrnävä Fin. 14 N4
Tyrnavos Greece 27 J5
Tyrnyauz Rus. Fed. 35 F2
Tyrone PA U.S.A. 64 B2
Tyrrell r. Australia 58 A5
Tyrrell, Lake dry lake Australia 57 C7
Tyrrhenian Sea France/Italy 26 D4
Tyrus Lebanon see Tyre
Tysa r. Serbia see Tisa
Tyukalinsk Rus. Fed. 28 I4
Tyumen' Rus. Fed. 28 H4
Tyup Kyrg. see Tüp
Tyuratam Kazakh. see Baykonyr
Tywi r. U.K. 19 C7
Tywyn U.K. 19 C6
Tzaneen S. Africa 51 J2
Tzia i. Greece 27 K6

U

Uaco Congo Angola see Waku-Kungo
Ualan atoll Micronesia see Kosrae
Uamanda Angola 49 C5
Uarc, Ras c. Morocco see
Trois Fourches, Cap des
Uaroo Australia 55 A5
Uatumã r. Brazil 69 G4
Uauá Brazil 69 K5
Uaupés r. Brazil 68 E3
U'ayli, Wādī al watercourse Saudi Arabia
39 D4
U'aywij well Saudi Arabia 35 G5
U'aywij, Wādī al watercourse Saudi Arabia
35 F5
Ubá Brazil 71 C3
Ubaí Brazil 71 B2
Ubaitaba Brazil 71 D1
Ubangi r. Cent. Afr. Rep./
Dem. Rep. Congo 48 B4
Ubangi-Shari country Africa see
Central African Republic
Ubauro Pak. 36 B3
Ubayyid, Wādī al watercourse Iraq/
Saudi Arabia 35 F4
Ube Japan 45 C6
Úbeda Spain 25 E4
Uberaba Brazil 71 B2
Uberlândia Brazil 71 A2
Ubombo S. Africa 51 K4
Ubon Ratchathani Thai. 31 J5
Ubundu Dem. Rep. Congo 47 F5
Üçajy Turkm. 33 J2
Ucar Azer. 35 G2
Uçarı Turkey 39 A1
Ucayali r. Peru 68 D4
Uch Pak. 36 B3
Üchajy Turkm. see Üçajy
Üchān Iran 35 H3
Ucharal Kazakh. 42 E3
Uchiura-wan b. Japan 44 F4
Uchkeken Rus. Fed. 35 F2
Uchkuduk Uzbek. see Uchquduq
Uchquduq Uzbek. 33 J1
Uchur r. Rus. Fed. 29 O4
Uckfield U.K. 19 H8
Uda r. Rus. Fed. 43 J2
Uda r. Rus. Fed. 44 D1
Udachnoye Rus. Fed. 13 J7
Udachnyy Rus. Fed. 77 E2
Udagamandalam India 38 C4
Udaipur Rajasthan India 36 C4
Udaipur Tripura India 37 G5
Udanti r. India/Myanmar 37 E5
Uday r. Ukr. 13 G6
Uddevalla Sweden 15 G7
Uddingston U.K. 20 E5
Uddjaure l. Sweden 14 J4
Udgir India 38 C2
Udhagamandalam India see
Udagamandalam
Udhampur India 36 C2
Udia-Milai atoll Marshall Is see Bikini
Udimskiy Rus. Fed. 12 J3
Udine Italy 26 E1
Udmalaippettai India see Udumalaippettai
Udomlya Rus. Fed. 12 G4
Udon Thani Thai. 31 J5
Udskaya Guba b. Rus. Fed. 29 O4
Udskoye Rus. Fed. 44 D1
Udumalaippettai India 38 C4
Udupi India 38 B3
Udyl', Ozero l. Rus. Fed. 44 E1
Udzhary Azer. see Ucar
Udzungwa Mountains National Park Tanz.
49 D4
Uéa atoll New Caledonia see Ouvéa
Ueckermünde Germany 17 O4
Ueda Japan 45 E5
Uele r. Dem. Rep. Congo 48 C3
Uelen Rus. Fed. 29 U3
Uelzen Germany 17 M4
Ufa Rus. Fed. 11 R5
Ufa r. Rus. Fed. 11 R5
Uftyuga r. Rus. Fed. 12 J3
Ugab watercourse Namibia 49 B6
Ugalla r. Tanz. 49 D4
Uganda country Africa 48 D3
Ugie S. Africa 51 I6
Uglegorsk Rus. Fed. 44 F2
Uglich Rus. Fed. 12 H4
Ugljan i. Croatia 26 F2
Uglovoye Rus. Fed. 44 C2
Ugol'noye Rus. Fed. 29 P3
Ugolnyy Rus. Fed. see Beringovskiy
Ugol'nyye Kopi Rus. Fed. 29 S3
Ugra Rus. Fed. 13 G5
Uherské Hradiště Czech Rep. 17 P6
Uhrichsville OH U.S.A. 64 A2
Uibhist a' Deas i. U.K. see South Uist
Uibhist a' Tuath i. U.K. see North Uist
Uig U.K. 20 C3
Uíge Angola 49 B4
Üijŏngbu S. Korea 45 B5
Ŭiju N. Korea 45 B4
Uimaharju Fin. 14 Q5
Uinta Mountains U.S.A. 62 E3
Uis Mine Namibia 49 B6
Uitenhage S. Africa 51 G7
Ujhani India 36 D4
Uji Japan 45 D6
Uji-guntō is Japan 45 C7
Ujiyamada Japan see Ise
Ujjain India 36 C5
Ujung Pandang Indon. see Makassar
Újvidék Serbia see Novi Sad
Ukal Sagar l. India 36 C5
Ukata Nigeria 46 D3
'Ukayrishah well Saudi Arabia 35 G6
uKhahlamba-Drakensberg Park nat. park
S. Africa 51 I5
Ukholovo Rus. Fed. 13 I5
Ukhrul India 37 H4
Ukhta Respublika Kareliya Rus. Fed. see
Kalevala
Ukhta Respublika Komi Rus. Fed. 12 L3
Ukiah CA U.S.A. 65 A1
Ukkusissat Greenland 61 M2
Ukmergė Lith. 15 N9
Ukraine country Europe 13 F6
Ukrainskaya S.S.R. country Europe see
Ukraine
Ukrayina country Europe see Ukraine
Uku-jima i. Japan 45 C6
Ukwi Botswana 50 E2
Ukwi Pan salt pan Botswana 50 E2
Ulaanbaatar Mongolia see Ulan Bator
Ulaangom Mongolia 42 G3
Ulan Australia 58 D4
Ulan Bator Mongolia 42 J3
Ulanbel' Kazakh. 42 C4
Ulan Erge Rus. Fed. 13 J7
Ulanhad China see Chifeng
Ulanhot China 44 A3
Ulan Hua China 43 K4
Ulan-Khol Rus. Fed. 13 J7
Ulan-Ude Rus. Fed. 43 J2
Ulan Ul Hu l. China 37 G2
Ulaş Turkey 34 E3
Ulawa Island Solomon Is 53 G2
Ul'banskiy Zaliv b. Rus. Fed. 44 E1
Ulchin S. Korea 45 C5
Uldz r. Mongolia 43 L3
Uleåborg Fin. see Oulu
Ulefoss Norway 15 F7
Ülenurme Estonia 15 O7
Ulety Rus. Fed. 43 K2
Ulhasnagar India 38 B2
Uliastai China 43 L3
Uliastay Mongolia 42 H3
Uliatea i. Fr. Polynesia see Raiatea
Ulita r. Rus. Fed. 14 R2
Ulithi atoll Micronesia 41 F6
Ul'ken Naryn Kazakh. 42 E3
Ulladulla Australia 58 E5
Ullapool U.K. 20 D3
Ulla Ulla, Parque Nacional nat. park Bol.
68 E6
Ullava Fin. 14 M5
Ullersuaq c. Greenland 61 K2
Ullswater l. U.K. 18 E4
Ullŭng-do i. S. Korea 45 C5
Ulm Germany 17 L6
Ulmarra Australia 58 F2
Uloowaranie, Lake imp. l. Australia 57 B5
Ulricehamn Sweden 15 H8
Ulsan S. Korea 45 C6
Ulsberg Norway 14 F5
Ulster reg. Ireland/U.K. 21 E3
Ulster PA U.S.A. 64 C2
Ulster Canal Ireland/U.K. 21 E3
Ultima Australia 58 A5
Ulubat Gölü l. Turkey 27 M4
Ulubey Turkey 27 M5
Uluborlu Turkey 27 N5
Uludağ mt. Turkey 27 M4
Uludağ Milli Parkı nat. park Turkey 27 M4
Ulugqat China see Wuqia
Ulukhaktok Canada 60 G2
Ulukışla Turkey 34 D3
Ulundi S. Africa 51 J5
Ulungur Hu l. China 42 F3
Uluqsaqtuuq Canada see Ulukhaktok
Uluṟu h. Australia 55 E6
Uluṟu-Kata Tjuṯa National Park Australia
55 E6
Uluru National Park Australia see Uluṟu-
Kata Tjuṯa National Park
Ulutau Kazakh. see Ulytau
Ulutau, Gory mts Kazakh. see Ulytau, Gory
Uluyatır Turkey 39 C1
Ulva i. U.K. 20 C4
Ulverston U.K. 18 D4
Ulvsjön Sweden 15 I6
Ül'yanov Kazakh. see Ul'yanovskiy
Ul'yanovsk Rus. Fed. 13 K5
Ul'yanovskiy Kazakh. 42 C2
Ul'yanovskoye Kazakh. see Ul'yanovskiy
Ulysses KS U.S.A. 62 G4
Ulytau Kazakh. 42 B3
Ulytau, Gory mts Kazakh. 42 B3
Ulyunkhan Rus. Fed. 43 K2
Uma Rus. Fed. 44 A1
Umal'ta (abandoned) Rus. Fed. 44 D2
'Umān country Asia see Oman
Uman' Ukr. 13 F6
'Umarī, Qā' al salt pan Jordan 39 C4
Umaria India 36 E5
Umarkhed India 38 C2
Umarkot India 38 D2
Umarkot Pak. 36 B3
Umarkot Pak. 33 K4
Umaroona, Lake imp. l. Australia 57 B5
Umarpada India 36 C5
Umba Rus. Fed. 12 G2
Umbeara Australia 55 F6
Umboi i. P.N.G. 52 E2
Umeå Sweden 14 L5
Umeälven r. Sweden 14 L5
Umfolozi r. S. Africa 51 K5
Umhlanga Rocks S. Africa 51 J5
Umiiviip Kangertiva inlet Greenland
61 N3
Umingmaktok (abandoned) Canada 77 L2
Umkomaas S. Africa 51 J6
Umlaiteng India 37 H4
Umlazi S. Africa 51 J5
Umm ad Daraj, Jabal mt. Jordan 39 B3
Umm al 'Amad Syria 39 C2
Umm al Qulbān Saudi Arabia 35 F6
Umm ar Raqabah, Khabrat salt pan
Saudi Arabia 39 C5
Umm Bel Sudan 32 C7

Umm Keddada Sudan 32 C7
Umm Lajj Saudi Arabia 32 E4
Umm Nukhaylah h. Saudi Arabia 39 D5
Umm Qaşr Iraq 35 G5
Umm Quşūr i. Saudi Arabia 34 D6
Umm Ruwaba Sudan 32 D7
Umm Sa'ad Libya 34 B5
Umm Shugeira Sudan 32 C7
Umm Wa'āl h. Saudi Arabia 39 D4
Umnak Island U.S.A. 60 B4
Umpulo Angola 49 B5
Umraniye Turkey 27 N5
Umred India 38 C1
Umri India 36 D4
Umtali Zimbabwe see Mutare
Umtata S. Africa see Mthatha
Umtentweni S. Africa 51 J6
Umuahia Nigeria 46 D4
Umuarama Brazil 70 F2
Umvuma Zimbabwe see Mvuma
Umzimkulu S. Africa 51 I6
Una r. Bos.-Herz./Croatia 26 G2
Una Brazil 71 D1
Una India 36 D3
'Unāb, Jabal al h. Jordan 39 C5
'Unāb, Wādī al watercourse Jordan 39 C4
Unaí Brazil 71 B2
Unalaska Island U.S.A. 60 B4
Unapool U.K. 20 D2
'Unayzah Saudi Arabia 32 F4
'Unayzah, Jabal h. Iraq 35 E4
Uncia Bol. 68 E7
Undara National Park Australia 56 D3
Underberg S. Africa 51 I5
Underbool Australia 57 C7
Unecha Rus. Fed. 13 G5
Ungama Bay Kenya see Ungwana Bay
Ungarie Australia 58 C4
Ungava, Baie d' b. Canada see Ungava Bay
Ungava, Péninsule d' pen. Canada 61 K3
Ungava Bay Canada 61 L4
Ungava Peninsula Canada see Ungava, Péninsule d'
Ungeny Moldova see Ungheni
Ungheni Moldova 27 L1
Unguana Moz. 51 L2
Unguja i. Tanz. see Zanzibar Island
Unguz, Solonchakovyye Vpadiny salt flat Turkm. 33 I2
Üngüz Angyrsyndaky Garagum des. Turkm. 33 I1
Ungvár Ukr. see Uzhhorod
Ungwana Bay Kenya 48 E4
Uni Rus. Fed. 12 K4
União Brazil 69 J4
União da Vitória Brazil 71 A4
União dos Palmares Brazil 69 K5
Unimak Island U.S.A. 60 B4
Unini r. Brazil 68 F4
Union WV U.S.A. 64 A4
Union City PA U.S.A. 64 B2
Union City TN U.S.A. 63 J4
Uniondale S. Africa 50 F7
Uniontown PA U.S.A. 64 B3
Unionville PA U.S.A. 64 C2
United Arab Emirates country Asia 33 H5
United Arab Republic country Africa see Egypt
United Kingdom country Europe 16 G3
United Provinces state India see Uttar Pradesh
United States of America country N. America 62 F3
United States Range mts Canada 61 L1
Unity Canada 62 F1
Unjha India 36 C5
Unnao India 36 E4
Ŭnp'a N. Korea 45 B5
Unsan N. Korea 45 B4
Ŭnsan N. Korea 45 B5
Unst i. U.K. 20 [inset]
Untor, Ozero l. Rus. Fed. 11 T3
Unuli Horog China 37 G2
Unzen-dake vol. Japan 45 C6
Unzha Rus. Fed. 12 J4
Upar Ghat reg. India 37 F5
Upemba, Lac l. Dem. Rep. Congo 49 C4
Uperbada India 37 F5
Upernavik Greenland 61 M2
Upington S. Africa 50 E5
Upland CA U.S.A. 65 D3
Upleta India 36 B5
Upoloksha Rus. Fed. 14 Q3
'Upolu i. Samoa 53 I3
Upper Chindwin Myanmar see Mawlaik
Upper Hutt N.Z. 59 E5
Upper Klamath Lake U.S.A. 62 C3
Upper Lough Erne l. U.K. 21 E3
Upper Marlboro MD U.S.A. 64 C3
Upper Tunguska r. Rus. Fed. see Angara
Upper Volta country Africa see Burkina Faso
Upper Yarra Reservoir Australia 58 B6
Uppinangadi India 38 B3
Uppsala Sweden 15 J7
Upshi India 36 D2
Upton MA U.S.A. 64 F1
'Uqayqah, Wādī watercourse Jordan 39 B4
'Uqayribāt Syria 39 C2
'Uqlat al 'Udhaybah well Iraq 35 G5
Urad Houqi China see Sain Us
Urakawa Japan 44 F4
Ural h. Australia 58 C4
Ural r. Kazakh./Rus. Fed. 30 E2
Uralla Australia 58 E3
Ural Mountains Rus. Fed. 11 S2
Ural'sk Kazakh. 30 E1
Ural'skaya Oblast' admin. div. Kazakh. see Zapadnyy Kazakhstan
Ural'skiye Gory mts Rus. Fed. see Ural Mountains
Ural'skiy Khrebet mts Rus. Fed. see Ural Mountains
Urambo Tanz. 49 D4
Uran India 38 B2
Urana Australia 58 C5
Urana, Lake Australia 58 C5
Urandangi Australia 56 B4
Urandi Brazil 71 C1
Uranium City Canada 60 H4
Uranquity Australia 58 C5
Uraricoera r. Brazil 68 F3
Urartu country Asia see Armenia
Uravakonda India 38 C3
Urawa Japan 45 E6
'Urayf an Nāqah, Jabal h. Egypt 39 B4
Urbino Italy 26 E3
Urbinum Italy see Urbino
Urbs Vetus Italy see Orvieto
Urdoma Rus. Fed. 12 K3
Urdyuzhskoye, Ozero l. Rus. Fed. 12 K2
Ure r. U.K. 18 F4
Ureki Georgia 35 F2
Uren' Rus. Fed. 12 J4
Urengoy Rus. Fed. 28 I3
Uréparapara i. Vanuatu 53 G3
Urfa Turkey see Şanlıurfa
Urfa prov. Turkey see Şanlıurfa
Urga Mongolia see Ulan Bator
Urgal r. Rus. Fed. 44 D2
Urganch Uzbek. 33 J1
Urgench Uzbek. see Urganch
Ürgüp Turkey 34 D3
Urho China 42 F3
Urho Kekkosen kansallispuisto nat. park Fin. 14 O2
Urie r. U.K. 20 G3
Uril Rus. Fed. 44 C2
Urisino Australia 58 A2
Urjala Fin. 15 M6
Urkan Rus. Fed. 44 B1
Urkan r. Rus. Fed. 44 B1
Urla Turkey 27 L5
Urlingford Ireland 21 E5
Urluk Rus. Fed. 43 J2
Urmā aş Şughrá Syria 39 C1
Urmai China 37 F3
Urmia Iran 35 G3
Urmia, Lake salt l. Iran 35 G3
Uromi Nigeria 46 D4
Uroševac Kosovo see Ferizaj
Urosozero Rus. Fed. 12 G3
Urru Co salt l. China 37 F3
Urt Moron China 42 G5
Uruáchic Mex. 62 F6
Uruaçu Brazil 71 A1
Uruana Brazil 71 A1
Uruapan Michoacán Mex. 66 D5
Urubamba r. Peru 68 D6
Urucara Brazil 69 G4
Urucu r. Brazil 68 F4
Uruçuca Brazil 71 D1
Uruçuí Brazil 69 J5
Uruçuí, Serra do hills Brazil 69 I5
Urucuia Brazil 71 B2
Urucurituba Brazil 69 G4
Uruguai r. Arg./Uruguay see Uruguay
Uruguaiana Brazil 70 E3
Uruguay r. Arg./Uruguay 70 E4
Uruguay country S. America 70 E4
Uruhe China 44 B2
Urumchi China see Ürümqi
Ürümqi China 42 F4
Urundi country Africa see Burundi
Urup, Ostrov i. Rus. Fed. 43 S3
Urusha Rus. Fed. 44 A1
Urutaí Brazil 71 A2
Uryupino Rus. Fed. 43 M2
Uryupinsk Rus. Fed. 13 I6
Urzhum Rus. Fed. 12 K4
Urziceni Romania 27 L2
Usa Japan 45 C6
Usa r. Rus. Fed. 12 M2
Uşak Turkey 27 M5
Usakos Namibia 50 B1
Usarp Mountains Antarctica 76 H2
Usborne, Mount h. Falkland Is 70 E8
Ushakova, Ostrov i. Rus. Fed. 28 I1
Ushant i. France see Ouessant, Île d'
Üsharal Kazakh. see Ucharal
Ush-Bel'dir Rus. Fed. 42 H2
Ushtobe Kazakh. 42 D3
Ush-Tyube Kazakh. see Ushtobe
Ushuaia Arg. 70 C8
Ushumun Rus. Fed. 44 B1
Usinsk Rus. Fed. 11 R2
Usk U.K. 19 E7
Usk r. U.K. 19 E7
Uskhodni Belarus 15 O10
Uskoplje Bos.-Herz. see Gornji Vakuf
Üsküdar Turkey 27 M4
Usman' Rus. Fed. 13 H5
Usmanabad India see Osmanabad
Usmas ezers l. Latvia 15 M8
Usogorsk Rus. Fed. 12 K3
Usol'ye-Sibirskoye Rus. Fed. 42 I2
Ussel France 24 F4
Ussuri r. China/Rus. Fed. 44 D2
Ussuriysk Rus. Fed. 44 C4
Ust'-Abakanskoye Rus. Fed. see Abakan
Usta Muhammad Pak. 36 B3
Ust'-Balyk Rus. Fed. see Nefteyugansk
Ust'-Donetskiy Rus. Fed. 13 I7
Ust'-Dzheguta Rus. Fed. 35 F1
Ust'-Dzhegutinskaya Rus. Fed. see Ust'-Dzheguta
Ustica, Isola di i. Sicilia Italy 26 E5
Ust'-Ilimsk Rus. Fed. 29 L4
Ust'-Ilimskoye Vodokhranilishche resr Rus. Fed. 29 L4
Ust'-Ilych Rus. Fed. 11 R3
Ústí nad Labem Czech Rep. 17 O5
Ustinov Rus. Fed. see Izhevsk
Üstirt plat. Kazakh./Uzbek. see Ustyurt Plateau
Ustka Poland 17 P3
Ust'-Kamchatsk Rus. Fed. 29 R4
Ust'-Kamenogorsk Kazakh. 42 E3
Ust'-Kan Rus. Fed. 42 E2
Ust'-Koksa Rus. Fed. 42 F2
Ust'-Kulom Rus. Fed. 12 L3
Ust'-Kut Rus. Fed. 29 L4
Ust'-Kuyga Rus. Fed. 29 O2
Ust'-Labinsk Rus. Fed. 35 E1
Ust'-Labinskaya Rus. Fed. see Ust'-Labinsk
Ust'-Lyzha Rus. Fed. 12 M2
Ust'-Maya Rus. Fed. 29 O3
Ust'-Nera Rus. Fed. 29 P3
Ust'-Ocheya Rus. Fed. 12 K3
Ust'-Olenek Rus. Fed. 29 M2
Ust'-Omchug Rus. Fed. 29 P3
Ust'-Ordynskiy Rus. Fed. 42 I2
Ust'-Penzhino Rus. Fed. see Kamenskoye
Ust'-Port Rus. Fed. 28 J3
Ustrem Rus. Fed. 11 T3
Ust'-Tsil'ma Rus. Fed. 12 L2
Ust'-Uda Rus. Fed. 42 I2
Ust'-Umalta (abandoned) Rus. Fed. 44 D2
Ust'-Undurga Rus. Fed. 43 L2
Ust'-Ura Rus. Fed. 12 J3
Ust'-Urgal Rus. Fed. 44 D2
Ust'-Usa Rus. Fed. 12 M2
Ust'-Vayen'ga Rus. Fed. 12 I3
Ust'-Voya Rus. Fed. 11 R3
Ust'-Vyyskaya Rus. Fed. 12 J3
Ust'ya r. Rus. Fed. 12 I3
Ust'ye Rus. Fed. 12 H4
Ustyurt, Plato plat. Kazakh./Uzbek. see Ustyurt Plateau
Ustyurt Plateau Kazakh./Uzbek. 30 E2
Ustyurt Platosi plat. Kazakh./Uzbek. see Ustyurt Plateau
Ustyuzhna Rus. Fed. 12 H4
Usulután El Salvador 66 G6
Usumbura Burundi see Bujumbura
Usvyaty Rus. Fed. 12 F5
Utah state U.S.A. 62 E4
Utah Lake U.S.A. 62 E3
Utajärvi Fin. 14 O4
Utashinai Rus. Fed. see Yuzhno-Kuril'sk
'Utaybah, Buḩayrat al imp. l. Syria 39 C3
Utena Lith. 15 N9
Uterlai India 36 B4
Uthal Pak. 36 A4
'Uthmānīyah Syria 39 C2
Utiariti Brazil 69 G6
Utica NY U.S.A. 64 D1
Utiel Spain 25 F4
Utlwanang S. Africa 51 G4
Utrecht Neth. 16 J4
Utrecht S. Africa 51 J4
Utrera Spain 25 D5
Utsjoki Fin. 14 O2
Utsunomiya Japan 45 E5
Utta Rus. Fed. 13 J7
Uttaradit Thai. 31 J5
Uttarakhand state India see Uttaranchal
Uttaranchal state India 36 D3
Uttar Kashi India see Uttarkashi
Uttarkashi India 36 D3
Uttar Pradesh state India 36 D4
Uttoxeter U.K. 19 F6
Uttranchal state India see Uttaranchal
Utubulak China 42 F3
Utupua i. Solomon Is 53 G3
Uummannaq Greenland see Dundas
Uummannaq Fjord inlet Greenland 77 J2
Uummannarsuaq c. Greenland see Farewell, Cape
Uurainen Fin. 14 N5
Uusikaarlepyy Fin. see Nykarleby
Uusikaupunki Fin. 15 L6
Uutapi Namibia 49 B5
Uva Rus. Fed. 12 L4
Uvalde U.S.A. 62 H6
Uval Karabaur hills Kazakh./Uzbek. 35 I2
Uval Muzbel' hills Kazakh. 35 I2
Uvarovo Rus. Fed. 13 I6
Uvéa atoll New Caledonia see Ouvéa
Uvinza Tanz. 49 D4
Uvs Nuur salt l. Mongolia 42 G2
Uwajima Japan 45 D6
'Uwayrid, Ḩarrat al lava field Saudi Arabia 32 E4
Uwaysiţ well Saudi Arabia 39 D4
Uweinat, Jebel mt. Sudan 32 C5
Uxbridge U.K. 19 G7
Uxin Qi China see Dabqig
Uyar Rus. Fed. 42 G1
Uyo Nigeria 46 D4
Uyu Chaung r. Myanmar 37 H4
Uyuni Bol. 68 E8
Uyuni, Salar de salt flat Bol. 68 E8
Uza r. Rus. Fed. 13 J5
Uzbekistan country Asia 30 F2
Ŭzbekiston country Asia see Uzbekistan
Uzbekskaya S.S.R. country Asia see Uzbekistan
Uzbek S.S.R. country Asia see Uzbekistan
Uzboy Azer. 35 H3
Uzboý Turkm. 35 I3
Uzen' Kazakh. see Kyzylsay
Uzhgorod Ukr. see Uzhhorod
Uzhhorod Ukr. 13 D6
Užhorod Ukr. see Uzhhorod
Užice Serbia 27 H3
Uzlovaya Rus. Fed. 13 H5
Üzümlü Turkey 27 M6
Uzunköprü Turkey 27 L4

Vaajakoski Fin. 14 N5
Vaal r. S. Africa 51 F5
Vaala Fin. 14 O4
Vaalbos National Park S. Africa 50 G5
Vaal Dam S. Africa 51 I4
Vaalwater S. Africa 51 I3
Vaasa Fin. 14 L5
Vác Hungary 17 Q7
Vacaria Brazil 71 A5
Vacaria, Campo da plain Brazil 71 A5
Vacaville CA U.S.A. 65 B1
Vad Rus. Fed. 12 J5
Vad r. Rus. Fed. 13 I5
Vada India 38 B2
Vadakara India 38 B4
Vadla Norway 15 E7
Vadodara India 36 C5
Vadsø Norway 14 P1
Vaduz Liechtenstein 24 I3
Værøy i. Norway 14 H3
Vaga r. Rus. Fed. 12 I3
Vågåmo Norway 15 F6
Vaganski Vrh mt. Croatia 26 F2
Vágar i. Faroe Is 14 [inset]
Vägsele Sweden 14 K4
Vágur Faroe Is 14 [inset]
Váh r. Slovakia 17 Q7
Vähäkyrö Fin. 14 M5
Vaiaku Tuvalu 53 H2
Vaida Estonia 15 N7
Vail U.S.A. 62 F4
Vaitupu i. Tuvalu 53 H2
Vajrakarur India see Kanur
Vakīlābād Iran 33 I4
Valbo Sweden 15 J6
Valcheta Arg. 70 C6
Valdai Hills Rus. Fed. see Valdayskaya Vozvyshennost'
Valday Rus. Fed. 12 G4
Valdayskaya Vozvyshennost' hills Rus. Fed. 12 G4
Valdecañas, Embalse de resr Spain 25 D4
Valdemārpils Latvia 15 M8
Valdemarsvik Sweden 15 J7
Valdepeñas Spain 25 E4
Val-de-Reuil France 24 E2
Valdés, Península pen. Arg. 70 D6
Valdez U.S.A. 60 D3
Valdivia Chile 70 B5
Val-d'Or Canada 63 L2
Valdosta U.S.A. 63 K5
Valdres val. Norway 15 F6
Vale Georgia 35 F2
Valemount Canada 62 D1
Valença Brazil 71 D1
Valence France 24 G4
Valencia Spain 25 F4
València Spain see Valencia
Valencia reg. Spain 25 F4
Valencia Venez. 68 E1
Valencia, Golfo de g. Spain 25 G4
Valencia de Don Juan Spain 25 D2
Valencia Island Ireland 21 B6
Valenciennes France 24 F1
Valensole, Plateau de France 24 H5
Valentia Spain see Valencia
Valentin Rus. Fed. 44 D4
Valentine U.S.A. 62 G3
Våler Norway 15 G6
Valera Venez. 68 D2
Vale Verde Brazil 71 D2
Val Grande, Parco Nazionale della nat. park Italy 26 C1
Valjevo Serbia 27 H2
Valka Latvia 15 O8
Valkeakoski Fin. 15 N6
Valky Ukr. 13 G6
Valkyrie Dome Antarctica 76 D1
Valladolid Mex. 66 G4
Valladolid Spain 25 D3
Valle Norway 15 E7
Valle de la Pascua Venez. 68 E2
Valle del Tiétar val. Spain 25 D3
Valledupar Col. 68 D1
Vallée de l'Our val. Germany/Lux. 17 K6
Valle Fértil, Sierra de mts Arg. 70 C4
Valle Grande Bol. 68 F7
Vallejo CA U.S.A. 65 A1
Vallenar Chile 70 B3
Valletta Malta 26 F7
Valley U.K. 18 C5
Valley City U.S.A. 62 H2
Valls Spain 25 G3
Val Marie Canada 62 F2
Valmiera Latvia 15 N8
Valnera mt. Spain 25 E2
Valognes France 19 F9
Valona Albania see Vlorë
Valozhyn Belarus 15 O9
Valparai India 38 C4
Valparaíso Chile 70 B4
Valpoi India 38 B3
Valréas France 24 G4
Vals, Tanjung c. Indon. 41 F8
Valsad India 38 B1
Valspan S. Africa 50 G4
Valtimo Fin. 14 P5
Valuyevka Rus. Fed. 13 I7
Valuyki Rus. Fed. 13 H6
Vammala Fin. 15 M6
Van Turkey 35 F3
Van, Lake salt l. Turkey 35 F3
Vanadzor Armenia 35 G2
Vancouver Canada 62 C2
Vancouver U.S.A. 62 C2
Vancouver Island Canada 62 B2
Vanda Fin. see Vantaa
Vandalia IL U.S.A. 63 J4
Vanderbijlpark S. Africa 51 H4
Vandergrift PA U.S.A. 64 B2
Vanderkloof Dam dam S. Africa 50 G6
Vanderlin Island Australia 56 B2
Van Diemen, Cape N.T. Australia 54 E2
Van Diemen, Cape Qld Australia 56 B3
Van Diemen Gulf Australia 54 F2
Van Diemen's Land state Australia see Tasmania
Vändra Estonia 15 N7
Väner, Lake Sweden see Vänern
Vänern l. Sweden 15 H7
Vänersborg Sweden 15 H7
Vangaindrano Madag. 49 E6
Van Gölü salt l. Turkey see Van, Lake
Van Horn U.S.A. 62 G5
Vanikoro Islands Solomon Is 53 G3
Vanimo P.N.G. 52 E2
Vanino Rus. Fed. 44 F2
Vanivilasa Sagara resr India 38 C3
Vaniyambadi India 38 C3
Vännäs Sweden 14 K5
Vannes France 24 C3
Vannovka Kazakh. see Turar Ryskulov
Vannøya i. Norway 14 K1
Van Rees, Pegunungan mts Indon. 41 F8
Vanrhynsdorp S. Africa 50 D6
Vansbro Sweden 15 I6
Vansittart Island Canada 61 J3
Vantaa Fin. 15 N6
Van Truer Tableland reg. Australia 55 C6
Vanua Lava i. Vanuatu 53 G3
Vanua Levu i. Fiji 53 H3
Vanuatu country S. Pacific Ocean 53 G3
Vanwyksvlei S. Africa 50 E6
Vanwyksvlei l. S. Africa 50 E6
Van Zylsrus S. Africa 50 F4
Varahi India 36 B5
Varakļāni Latvia 15 O8
Varalé Côte d'Ivoire 46 C4
Varāmīn Iran 35 H4
Varanasi India 37 E4
Varandey Rus. Fed. 12 M1
Varangerfjorden sea chan. Norway 14 P1
Varangerhalvøya pen. Norway 11 L1
Varangerhalvøya pen. Norway 14 P1
Varaždin Croatia 26 G1
Varberg Sweden 15 H8
Vardar r. Macedonia 27 J4
Varde Denmark 15 F9
Vardenis Armenia 35 G2
Vardø Norway 14 Q1
Varėna Lith. 15 N9
Varese Italy 26 C2
Varfolomeyevka Rus. Fed. 44 D3
Vårgårda Sweden 15 H7
Varginha Brazil 71 B3
Varillas Chile 70 B2
Varkana Iran see Gorgān
Varkaus Fin. 14 O5
Varna Bulg. 27 L3
Värnamo Sweden 15 I8
Värnäs Sweden 15 H6
Varnavino Rus. Fed. 12 J4
Várnjárg pen. Norway see Varangerhalvøya
Varpaisjärvi Fin. 14 O5
Várpalota Hungary 26 H1
Varsh, Ozero l. Rus. Fed. 12 J2
Varto Turkey 35 F3
Várzea da Palma Brazil 71 B2
Vasa Fin. see Vaasa
Vasai India 38 B2
Vashka r. Rus. Fed. 12 J2
Vasht Iran see Khāsh
Vasilkov Ukr. see Vasyl'kiv
Vasknarva Estonia 15 O7
Vaslui Romania 27 L1
Vas-Soproni-síkság hills Hungary 26 G1
Vastan Turkey see Gevaş
Västerås Sweden 15 J7
Västerdalälven r. Sweden 15 I6
Västerfjäll Sweden 14 J3
Västerhaninge Sweden 15 K7
Västervik Sweden 15 J8
Vasto Italy 26 F3
Vasyl'kiv Ukr. 13 F6
Vatan France 24 E3
Vaté i. Vanuatu see Éfaté
Vatersay i. U.K. 20 B4
Vathar India 38 B2
Vathí Greece see Vathy
Vathy Greece 27 L6
Vatican City Europe 26 E4
Vaticano, Città del Europe see Vatican City
Vatnajökull Iceland 14 [inset]
Vatnajökull nat. park Iceland 14 [inset]
Vatoa i. Fiji 53 I3
Vatra Dornei Romania 27 K1
Vätter, Lake Sweden see Vättern
Vättern l. Sweden 15 I7
Vaughn U.S.A. 62 F5
Vaupés r. Col. 68 E3
Vauvert France 24 G5
Vavatenina Madag. 49 E5
Vava'u Group is Tonga 53 I3
Vavitao i. Fr. Polynesia see Raivavae
Vavoua Côte d'Ivoire 46 C4
Vavozh Rus. Fed. 12 K4
Vavuniya Sri Lanka 38 D4
Vawkavysk Belarus 15 N10
Växjö Sweden 15 I8
Vayenga Rus. Fed. see Severomorsk
Vazante Brazil 71 B2
Vazáš Sweden see Vittangi
Veaikevárri Sweden see Svappavaara
Vedaranniyam India 38 C4
Vedasandur India 38 C4
Veddige Sweden 15 H8
Vedea r. Romania 27 K3
Veendam Neth. 17 K4
Vega i. Norway 14 G4
Vehari Pak. 36 C3
Vehkalahti Fin. 15 O6
Vehowa Pak. 36 B3
Veinticinco de Mayo Buenos Aires Arg. see 25 de Mayo
Veinticinco de Mayo La Pampa Arg. see 25 de Mayo
Veirwaro Pak. 36 B4
Vejle Denmark 15 F9
Velbŭzhdki Prokhod pass Bulg./Macedonia 27 J3
Velddrif S. Africa 50 D7
Velebit mts Croatia 26 F2
Velenje Slovenia 26 F1
Veles Macedonia 27 I4
Vélez-Málaga Spain 25 D5
Vélez-Rubio Spain 25 E5
Velhas r. Brazil 71 B2
Velibaba Turkey see Aras
Velika Gorica Croatia 26 G2
Velika Plana Serbia 27 I2
Velikaya r. Rus. Fed. 12 K4
Velikaya r. Rus. Fed. 15 P8
Velikaya r. Rus. Fed. 29 S3
Velikaya Kema Rus. Fed. 44 E3
Veliki Preslav Bulg. 27 L3
Velikiye Luki Rus. Fed. 12 F4
Velikiy Novgorod Rus. Fed. 12 F4
Velikiy Ustyug Rus. Fed. 12 J3
Velikonda Range hills India 38 C3
Veliko Tŭrnovo Bulg. 27 K3
Velikoye Rus. Fed. 12 H4
Velikoye, Ozero l. Rus. Fed. 13 I5
Veli Lošinj Croatia 26 F2
Velizh Rus. Fed. 12 F5
Vella Lavella i. Solomon Is 53 F2
Vellar r. India 38 C4
Vellore India 38 C3
Vel'sk Rus. Fed. 12 I3
Velsuna Italy see Orvieto
Velten Germany 17 N4
Velykyy Tokmak Ukr. see Tokmak
Vel'yu r. Rus. Fed. 12 L3
Vemalwada India 38 C2
Vema Seamount sea feature S. Atlantic Ocean 72 I8
Vema Trench sea feature Indian Ocean 73 M6
Vempalle India 38 C3
Venado Tuerto Arg. 70 D4
Venafro Italy 26 F4
Venceslau Bráz Brazil 71 A3
Vendinga Rus. Fed. 12 J3
Vendôme France 24 E3
Venetia Italy see Venice
Venetie Landing U.S.A. 60 D3
Venev Rus. Fed. 13 H5
Venezia Italy see Venice
Venezia, Golfo di g. Europe see Venice, Gulf of
Venezuela country S. America 68 E2
Venezuela, Golfo de g. Venez. 68 D1
Venezuelan Basin sea feature S. Atlantic Ocean 72 D4
Vengurla India 38 B3
Veniaminof Volcano U.S.A. 60 C4
Venice Italy 26 E2
Venice U.S.A. 63 K6
Venice, Gulf of Europe 26 E2
Vénissieux France 24 G4
Venkatapalem India 38 D2
Venkatapuram India 38 D2
Vennesla Norway 15 E7
Venta r. Latvia/Lith. 15 M8
Venta Lith. 15 M8
Ventersburg S. Africa 51 H5
Ventersdorp S. Africa 51 H4
Venterstad S. Africa 51 G6
Ventnor U.K. 19 F8
Ventotene, Isola i. Italy 26 E4
Ventoux, Mont mt. France 24 G4
Ventspils Latvia 15 L8
Ventura CA U.S.A. 65 C3
Venus Bay Australia 58 B7
Vera Arg. 70 D3
Vera Spain 25 F5
Vera Cruz Brazil 71 A3
Veracruz Mex. 66 E5
Vera Cruz Mex. see Veracruz
Veraval India 36 B5
Verbania Italy 26 C2
Vercelli Italy 26 C2
Vercors reg. France 24 G4
Verdalsøra Norway 14 G5
Verde r. Goiás Brazil 71 A2
Verde r. Goiás/Minas Gerais Brazil 71 B2
Verde r. Minas Gerais Brazil 71 A2
Verde r. Minas Gerais Brazil 71 A2
Verde r. Mex. 66 C3
Verden (Aller) Germany 17 L4
Verde Pequeno r. Brazil 71 C1
Verdon r. France 24 G5
Verdun France 24 G2
Vereeniging S. Africa 51 H4
Vereshchagino Rus. Fed. 11 Q4
Véria Greece see Veroia
Verín Spain 25 C3
Veríssimo Brazil 71 A2
Verkhneimbatsk Rus. Fed. 28 J3
Verkhnekolvinsk Rus. Fed. 12 M2
Verkhnespasskoye Rus. Fed. 12 J4
Verkhnetulomskiy Rus. Fed. 14 Q2
Verkhnetulomskoye Vodokhranilishche resr Rus. Fed. 14 Q2
Verkhnevilyuysk Rus. Fed. 29 N3
Verkhneye Kuyto, Ozero l. Rus. Fed. 14 Q4
Verkhnezeysk Rus. Fed. 43 N2
Verkhniy Vyalozerskiy Rus. Fed. 12 G2
Verkhnyaya Khava Rus. Fed. 13 H6
Verkhnyaya Salda Rus. Fed. 11 S4
Verkhnyaya Tunguska r. Rus. Fed. see Angara
Verkhnyaya Tura Rus. Fed. 11 R4
Verkhoshizhem'ye Rus. Fed. 12 K4
Verkhovazh'ye Rus. Fed. 12 I3
Verkhov'ye Rus. Fed. 13 H5
Verkhoyansk Rus. Fed. 29 O3
Verkhoyanskiy Khrebet mts Rus. Fed. 29 N2
Vermelho r. Brazil 71 A1
Vermilion Canada 62 E1
Vermillion U.S.A. 63 H3
Vermont state U.S.A. 64 E1
Vernadsky research stn Antarctica 76 L2
Vernal U.S.A. 62 F3
Verneuk Pan salt pan S. Africa 50 E5
Vernon Canada 62 D1
Vernon TX U.S.A. 62 H5
Vernon Islands Australia 54 E3
Vernoye Rus. Fed. 44 C2
Vernyy Kazakh. see Almaty
Vero Beach U.S.A. 63 K6
Veroia Greece 27 J4
Verona Italy 26 D2
Verona VA U.S.A. 64 B3
Versailles France 24 F2
Versec Serbia see Vršac
Vertou France 24 D3
Verulam S. Africa 51 J5
Verulamium U.K. see St Albans
Verviers Belgium 17 J5
Vescovato Corse France 24 I5
Vesele Ukr. 13 G7
Veselyy Rus. Fed. 13 I7
Veselyy Yar Rus. Fed. 44 D4
Veshenskaya Rus. Fed. 13 I6
Veslyana r. Rus. Fed. 12 L3
Vesontio France see Besançon
Vesoul France 24 H3
Vesterålen is Norway 14 H2
Vesterålsfjorden sea chan. Norway 14 H2
Vestertana Norway 14 O1
Vestfjorddalen val. Norway 15 F7
Vestfjorden sea chan. Norway 14 H3
Véstia Brazil 71 A3
Vestmanna Faroe Is 14 [inset]
Vestmannaeyjar Iceland 14 [inset]
Vestmannaeyjar is Iceland 14 [inset]
Vestnes Norway 14 E5
Vesturhorn hd Iceland 14 [inset]
Vesuvio vol. Italy see Vesuvius
Vesuvius vol. Italy 26 F4
Ves'yegonsk Rus. Fed. 12 H4
Veszprém Hungary 26 G1
Veteli Fin. 14 M5
Vetlanda Sweden 15 I8
Vetluga Rus. Fed. 12 J4
Vetluga r. Rus. Fed. 12 J4
Vetluzhskiy Kostromskaya Oblast' Rus. Fed. 12 J4
Vetluzhskiy Nizhegorodskaya Oblast' Rus. Fed. 12 J4
Vettore, Monte mt. Italy 26 E3
Vevey Switz. 24 H3
Veyo UT U.S.A. 65 F2
Vézère r. France 24 E4
Vezirköprü Turkey 34 D2
Vialar Alg. see Tissemsilt
Viamao Brazil 71 A5
Viana Espírito Santo Brazil 71 C3
Viana Maranhão Brazil 69 J4
Viana do Castelo Port. 25 B3
Viangchan Laos see Vientiane
Viannos Greece 27 K7
Vianópolis Brazil 71 A2
Viareggio Italy 26 D3
Viborg Denmark 15 F8
Viborg Rus. Fed. see Vyborg
Vibo Valentia Italy 26 G5
Vic Spain 25 H3
Vicecomodoro Marambio research stn Antarctica see Marambio
Vicente, Point CA U.S.A. 65 C4
Vicente Guerrero Mex. 66 A2
Vicenza Italy 26 D2
Vich Spain see Vic
Vichada r. Col. 68 E3
Vichadero Uruguay 70 F4
Vichy France 24 F3
Vicksburg AZ U.S.A. 65 F4
Vicksburg MS U.S.A. 63 I5
Viçosa Brazil 71 C3
Victor, Mount Antarctica 76 D2
Victor Harbor Australia 57 B7
Victoria Arg. 70 D4
Victoria r. Australia 54 E3
Victoria state Australia 58 B6
Victoria Canada 62 C2
Victoria Chile 70 B5
Victoria Malta 26 F6
Victoria Seychelles 73 L6
Victoria TX U.S.A. 63 H6
Victoria VA U.S.A. 64 B4
Victoria prov. Zimbabwe see Masvingo
Victoria, Lake Africa 48 D4
Victoria, Lake Australia 57 C7
Victoria, Mount Fiji see Tomanivi
Victoria, Mount Myanmar 37 H5
Victoria, Mount P.N.G. 52 E2
Victoria and Albert Mountains Canada 61 K2
Victoria Falls Zambia/Zimbabwe 49 C5
Victoria Island Canada 60 H2
Victoria Land coastal area Antarctica 76 H2
Victoria Peak Belize 66 G5
Victoria Range mts N.Z. 59 D6
Victoria River Downs Australia 54 E4
Victoria West S. Africa 50 F6
Victorica Arg. 70 C5
Victorville CA U.S.A. 65 D3
Victory Downs Australia 55 F6
Vidal Junction CA U.S.A. 65 E3
Videle Romania 27 K2
Vidisha India 36 D5
Vidlin U.K. 20 [inset]
Vidlitsa Rus. Fed. 12 G3
Viedma Arg. 70 D6
Viedma, Lago l. Arg. 70 B7
Vienna Austria 17 P6
Vienne France 24 G4
Vienne r. France 24 E3
Vientiane Laos 41 C6
Vieques i. Puerto Rico 67 K5
Vieremä Fin. 14 O5
Vierzon France 24 F3
Viesīte Latvia 15 N8
Vieste Italy 26 G4
Vietas Sweden 14 K3
Việt Nam country Asia see Vietnam
Vietnam country Asia 31 J5
Việt Tri Vietnam 42 J8
Vigan Phil. 41 E6
Vigevano Italy 26 C2
Vigia Brazil 69 I4
Vignemale mt. France 22 D3
Vignola Italy 26 D2
Vigo Spain 25 B2
Vihanti Fin. 14 N4
Vihti Fin. 15 N6
Viipuri Rus. Fed. see Vyborg
Viitasaari Fin. 14 N5
Vijayadurg India 38 B2
Vijayanagaram India see Vizianagaram
Vijayapati India 38 C4
Vijayawada India 38 D2
Vík Iceland 14 [inset]
Vikajärvi Fin. 14 O3
Vikeke East Timor see Viqueque
Vikna i. Norway 14 G4
Vikøyri Norway 15 E6
Vila Vanuatu see Port Vila
Vila Alferes Chamusca Moz. see Guija
Vila Bittencourt Brazil 68 E4
Vila Bugaço Angola see Camanongue
Vila Cabral Moz. see Lichinga
Vila da Ponte Angola see Kuvango
Vila de Aljustrel Angola see Cangamba
Vila de Almoster Angola see Chiange
Vila de João Belo Moz. see Xai-Xai
Vila de María Arg. 70 D3
Vila de Trego Morais Moz. see Chókwé
Vila do Tarrafal Cape Verde see Tarrafal
Vila Fontes Moz. see Caia
Vila Franca de Xira Port. 25 B4
Vilagarcía de Arousa Spain 25 B2
Vila Gomes da Costa Moz. 51 K3
Vilalba Spain 25 C2
Vila Luísa Moz. see Marracuene
Vila Marechal Carmona Angola see Uíge
Vila Miranda Moz. see Macaloge

nandro, Tanjona c. Madag. 49 E5
nculos Moz. 51 L1
Nova de Gaia Port. 25 B3
nova i la Geltrú Spain 25 G3
Pery Moz. see Chimoio
Real Port. 25 C3
r Formoso Port. 25 C3
Salazar Angola see N'dalatando
Salazar Zimbabwe see Sango
Teixeira de Sousa Angola see Luau
Velha Brazil 71 C3
cabamba, Cordillera mts Peru 68 D6
cheka, Zemlya i. Rus. Fed. 28 H1
ed' r. Rus. Fed. 12 J3
eyka Belarus see Vilyeyka
gort Rus. Fed. 12 K3
nelmina Sweden 14 J4
nena Brazil 68 F6
ya r. Lith. see Neris
andi Estonia 15 N7
oenskroon S. Africa 51 H4
caviškis Lith. 15 M9
ija Lith. 15 M9
kitskogo, Proliv str. Rus. Fed. 29 K2
ovo Ukr. see Vylkove
Abecia Bol. 68 E8
Ahumada Mex. 66 C2
Ángela Arg. 70 D3
Bella Bol. 68 E6
Bens Morocco see Tarfaya
ablino Spain 25 C2
acañas Spain 25 E4
ach Austria 17 N7
acidro Sardegna Italy 26 C5
Cisneros W. Sahara see Ad Dakhla
Constitución Mex. see
iudad Constitución
Dolores Arg. 70 C4
agarcía de Arosa Spain see
ilagarcía de Arousa
agrán Mex. 62 H7
aguay Arg. 70 E4
ahermosa Mex. 66 F5
Insurgentes Mex. 66 B3
ajoyosa Spain see
illajoyosa-La Vila Joíosa
ajoyosa-La Vila Joíosa Spain 25 F4
María Arg. 70 D4
Montes Bol. 68 F8
Nora S. Africa 51 I2
anueva de la Serena Spain 25 D4
anueva de los Infantes Spain 25 E4
anueva-y-Geltrú Spain see
ilanova i la Geltrú
Ocampo Arg. 70 E3
Ojo de Agua Arg. 70 D3
aputzu Sardegna Italy 26 C5
Regina Arg. 70 C5
arrica Para. 70 E3
arrica, Lago l. Chile 70 B5
arrica, Parque Nacional nat. park Chile
0 B5
arrobledo Spain 25 E4
as NJ U.S.A. 64 D3
asalazar Zimbabwe see Sango
San Giovanni Italy 26 F5
Sanjurjo Morocco see Al Hoceima
San Martín Arg. 70 D3
Unión Arg. 70 C3
Unión Durango Mex. 66 D4
Unión Sinaloa Mex. 66 C4
Valeria Arg. 70 D4
avicencio Col. 68 D3
azon Bol. 68 E8
efranche-sur-Saône France 24 G4
e-Marie Canada see Montréal
ena Spain 25 F4
eneuve-sur-Lot France 24 E4
eneuve-sur-Yonne France 24 F2
ers-sur-Mer France 19 G9
eurbanne France 24 G4
ers S. Africa 51 I4
ngen Germany 17 L6
ppuram India see Villupuram
ppuram India 38 C4
a Lith. see Vilnius
ius Lith. 15 N9
yans'k Ukr. 13 G7
pula Fin. 14 N5
eyka Belarus 15 O9
uy r. Rus. Fed. 29 N3
uyskoye Vodokhranilishche resr
us. Fed. 29 M3
merby Sweden 15 I8
r. Cameroon 47 E4
del Mar Chile 70 B4
aròs Spain 25 G3
aroz Spain see Vinaròs
cennes U.S.A. 63 J4
cennes Bay Antarctica 76 F2
china Arg. 70 C3
delälven r. Sweden 14 K5
deln Sweden 14 K4
dhya Range hills India 36 C5
dobona Austria see Vienna
eland NJ U.S.A. 64 D3
Vietnam 31 J5
ta U.S.A. 63 H4
han India 36 B5
and i. Canada see Newfoundland
nitsa Ukr. see Vinnytsya
nytsya Ukr. 13 F6
ogradov Ukr. see Vynohradiv
son Massif mt. Antarctica 76 L1
stra Norway 15 F6
ukonda India 38 C2
eque East Timor 54 D2
mgam India 36 C5
ndozero Rus. Fed. 12 H3
ranşehir Turkey 35 E3
wah Pak. 36 B4
how, Mount h. Australia 54 B5
el India 36 C5
en Canada 62 G2
France 24 D2
Angola 49 B5
em da Lapa Brazil 71 C2
in r. AZ U.S.A. 65 E2
inia Ireland 21 E4
inia S. Africa 51 H5
inia state U.S.A. 64 B4
inia Beach VA U.S.A. 64 D4
inia City NV U.S.A. 65 C1
in Islands (U.K.) terr. West Indies 67 L5
in Islands (U.S.A.) terr. West Indies
L5
in Mountains AZ U.S.A. 65 E2
inópolis Brazil 71 C2
kala Fin. 15 N6
vitica Croatia 26 G2
at Fin. 14 M5
su Estonia 15 M7
dhunagar India 38 C4
dunagar India see Virudhunagar
nga, Parc National des nat. park
em. Rep. Congo 48 C4
Croatia 26 G3
ginas Lith. 15 O9
khapatnam India see Vishakhapatnam
lia CA U.S.A. 65 C2
pur India 38 B2
Visby Sweden 15 K8
Viscount Melville Sound sea chan. Canada
61 G2
Viseu Brazil 69 I4
Viseu Port. 25 C3
Vishakhapatnam India 38 D2
Vishera r. Rus. Fed. 11 R4
Vishera r. Rus. Fed. 12 L3
Viški Latvia 15 O8
Visnagar India 36 C5
Viso, Monte mt. Italy 26 B2
Visoko Bos.-Herz. 26 H3
Visp Switz. 24 H3
Vista CA U.S.A. 65 D4
Vista Lake CA U.S.A. 65 C3
Vistonida, Limni lag. Greece 27 K4
Vistula r. Poland 17 Q3
Vitebsk Belarus see Vitsyebsk
Viterbo Italy 26 E3
Vitichi Bol. 68 E8
Vitigudino Spain 25 C3
Viti Levu i. Fiji 53 H3
Vitimskoye Ploskogor'ye plat. Rus. Fed.
43 K2
Vitória Brazil 71 C3
Vitória da Conquista Brazil 71 C1
Vitoria-Gasteiz Spain 25 E2
Vitória Seamount sea feature
S. Atlantic Ocean 72 F7
Vitré France 24 D2
Vitry-le-François France 24 G2
Vitsyebsk Belarus 13 F5
Vittangi Sweden 14 L3
Vittel France 24 G2
Vittoria Sicilia Italy 26 F6
Vittorio Veneto Italy 26 E2
Viveiro Spain 25 C2
Vivero Spain see Viveiro
Vivo S. Africa 51 I2
Vizagapatam India see Vishakhapatnam
Vizcaíno, Sierra mts Mex. 66 B3
Vize Turkey 27 L4
Vize, Ostrov i. Rus. Fed. 28 I2
Vizhas r. Rus. Fed. 12 J2
Vizianagaram India 38 D2
Vizinga Rus. Fed. 12 K3
Vlaardingen Neth. 16 J5
Vlădeasa, Vârful mt. Romania 27 J1
Vladikavkaz Rus. Fed. 35 G2
Vladimir Primorskiy Kray Rus. Fed. 44 D4
Vladimir Vladimirskaya Oblast' Rus. Fed.
12 I4
Vladimiro-Aleksandrovskoye Rus. Fed.
44 D4
Vladimir-Volynskiy Ukr. see
Volodymyr-Volyns'kyy
Vladivostok Rus. Fed. 44 C4
Vlakte S. Africa 51 I3
Vlasotince Serbia 27 J3
Vlas'yevo Rus. Fed. 44 F1
Vlissingen Neth. 16 I5
Vlora Albania see Vlorë
Vlorë Albania 27 H4
Vlotslavsk Poland see Włocławek
Vltava r. Czech Rep. 17 O5
Vöcklabruck Austria 17 N6
Vodlozero, Ozero l. Rus. Fed. 12 H3
Voe U.K. 20 [inset]
Vogelkop Peninsula Indon. see
Doberai, Jazirah
Voghera Italy 26 C2
Vohémar Madag. see Iharaña
Vohibinany Madag. see Ampasimanolotra
Vohimarina Madag. see Iharaña
Vohimena, Tanjona c. Madag. 49 E6
Vohipeno Madag. 49 E6
Võhma Estonia 15 N7
Voinjama Liberia 46 C4
Vojens Denmark 15 F9
Vojvodina prov. Serbia 27 H2
Vokhma Rus. Fed. 12 J4
Voknavolok Rus. Fed. 14 Q4
Vol' r. Rus. Fed. 12 L3
Volcano Bay Japan see Uchiura-wan
Volcano Islands Japan 43 Q8
Volda Norway 14 E5
Vol'dino Rus. Fed. 12 L3
Volga Rus. Fed. 12 H4
Volga r. Rus. Fed. 13 J7
Volga Upland hills Rus. Fed. see
Privolzhskaya Vozvyshennost'
Volgodonsk Rus. Fed. 13 I7
Volgograd Rus. Fed. 13 J6
Volgogradskoye Vodokhranilishche resr
Rus. Fed. 13 J6
Völkermarkt Austria 17 O7
Volkhov Rus. Fed. 12 G4
Volkhov r. Rus. Fed. 12 G3
Volkovysk Belarus see Vawkavysk
Volksrust S. Africa 51 I4
Vol'no-Nadezhdinskoye Rus. Fed. 44 C4
Volnovakha Ukr. 13 H7
Vol'nyansk Ukr. see Vil'nyans'k
Volochanka Rus. Fed. 28 K2
Volochisk Ukr. see Volochys'k
Volochys'k Ukr. 13 E6
Volodars'ke Ukr. 13 H7
Volodarskoye Kazakh. see Saumalkol'
Volodymyr-Volyns'kyy Ukr. 13 E6
Vologda Rus. Fed. 12 H4
Volokolamsk Rus. Fed. 12 G4
Volokovaya Rus. Fed. 12 K2
Volos Greece 27 J5
Volosovo Rus. Fed. 15 P7
Volot Rus. Fed. 12 F4
Volovo Rus. Fed. 13 H5
Volozhin Belarus see Valozhyn
Volsinii Italy see Orvieto
Vol'sk Rus. Fed. 13 J5
Volta, Lake resr Ghana 46 D4
Volta Blanche r. Burkina Faso/Ghana see
White Volta
Voltaire, Cape Australia 54 D3
Volta Redonda Brazil 71 B3
Vol'tevo Rus. Fed. 12 J2
Volturno r. Italy 26 E4
Volubilis tourist site Morocco 22 C5
Volvi, Limni l. Greece 27 J4
Volzhsk Rus. Fed. 12 K5
Volzhskiy Samarskaya Oblast' Rus. Fed.
13 K5
Volzhskiy Volgogradskaya Oblast' Rus. Fed.
13 J6
Vondanka Rus. Fed. 12 J4
Vontimitta India 38 C3
Vopnafjörður Iceland 14 [inset]
Vopnafjörður b. Iceland 14 [inset]
Vörå Fin. 14 M5
Voranava Belarus 15 N9
Voreies Sporades is Greece 27 J5
Voríai Sporádhes is Greece see
Voreies Sporades
Voring Plateau sea feature
N. Atlantic Ocean 72 I1
Vorjing mt. India 37 H3
Vorkuta Rus. Fed. 28 H3
Vormsi i. Estonia 15 M7
Vorona r. Rus. Fed. 13 I6
Voronezh Rus. Fed. 13 H6
Voronezh r. Rus. Fed. 13 H6
Voronov, Mys pt Rus. Fed. 12 I2
Vorontsovo-Aleksandrovskoye Rus. Fed.
see Zelenokumsk
Voroshilov Rus. Fed. see Ussuriysk
Voroshilovgrad Ukr. see Luhans'k
Voroshilovsk Rus. Fed. see Stavropol'
Voroshilovsk Ukr. see Alchevs'k
Vorotynets Rus. Fed. 12 J4
Vorozhba Ukr. 13 G6
Vorpommersche Boddenlandschaft,
Nationalpark nat. park Germany 17 N3
Vorskla r. Rus. Fed. 13 G6
Võrtsjärv l. Estonia 15 N7
Võru Estonia 15 O8
Vosburg S. Africa 50 F6
Vose Tajik. 33 K2
Vosges mts France 24 H3
Voskresensk Rus. Fed. 13 H5
Voskresenskoye Rus. Fed. 12 H4
Voss Norway 15 E6
Vostochno-Sakhalinskiye Gory mts
Rus. Fed. 44 F2
Vostochno-Sibirskoye More sea Rus. Fed.
see East Siberian Sea
Vostochnyy Kirovskaya Oblast' Rus. Fed.
12 L4
Vostochnyy Sakhalinskaya Oblast'
Rus. Fed. 44 F2
Vostochnyy Sayan mts Rus. Fed. 42 G2
Vostok research station Antarctica 76 F1
Vostok Primorskiy Kray Rus. Fed. 44 D3
Vostok Sakhalinskaya Oblast' Rus. Fed. see
Neftegorsk
Vostok Island Kiribati 75 J6
Vostroye Rus. Fed. 12 J3
Votkinsk Rus. Fed. 11 Q4
Votkinskoye Vodokhranilishche resr
Rus. Fed. 11 R4
Votuporanga Brazil 71 A3
Voves France 24 E2
Voynitsa Rus. Fed. 12 F2
Vöyri Fin. see Vörå
Voyvozh Rus. Fed. 12 L3
Vozhayel' Rus. Fed. 12 K3
Vozhe, Ozero l. Rus. Fed. 12 H3
Vozhega Rus. Fed. 12 I3
Vozhgaly Rus. Fed. 12 K4
Voznesens'k Ukr. 13 F7
Vozonin Trough sea feature Arctic Ocean
77 F1
Vozzhayevka Rus. Fed. 44 C2
Vrangel' Rus. Fed. 44 D4
Vrangelya, Mys pt Rus. Fed. 44 E1
Vranje Serbia 27 I3
Vratnik pass Bulg. 27 L3
Vratsa Bulg. 27 J3
Vrbas Serbia 27 H2
Vrede S. Africa 51 I4
Vredefort S. Africa 51 H4
Vredenburg S. Africa 50 C7
Vredendal S. Africa 50 D6
Vriddhachalam India 38 C4
Vrigstad Sweden 15 I8
Vršac Serbia 27 I2
Vryburg S. Africa 50 G4
Vryheid S. Africa 51 J4
Vsevidof, Mount vol. U.S.A. 60 B4
Vsevolozhsk Rus. Fed. 12 F3
Vučitrn Kosovo see Vushtrri
Vukovar Croatia 27 H2
Vuktyl Rus. Fed. 11 R3
Vukuzakhe S. Africa 51 I4
Vulcan Island P.N.G. see Manam Island
Vulcano, Isola i. Italy 26 F5
Vulture Mountains AZ U.S.A. 65 F4
Vuohijärvi Fin. 15 O6
Vuolijoki Fin. 14 O4
Vuollerim Sweden 14 L3
Vuostimo Fin. 14 O3
Vurnary Rus. Fed. 12 J5
Vushtrri Kosovo 27 I3
Vvedenovka Rus. Fed. 44 C2
Vyara India 36 C5
Vyarkhowye Belarus see Ruba
Vyatka Rus. Fed. see Kirov
Vyatka r. Rus. Fed. 12 K5
Vyatskiye Polyany Rus. Fed. 12 K4
Vyazemskiy Rus. Fed. 44 D3
Vyaz'ma Rus. Fed. 13 G5
Vyazniki Rus. Fed. 12 I4
Vyazovka Rus. Fed. 13 J5
Vyborg Rus. Fed. 12 F3
Vychegda r. Rus. Fed. 12 J3
Vychegodskiy Rus. Fed. 12 J3
Vyerkhnyadzvinsk Belarus 15 O9
Vyetryna Belarus 15 P9
Vygozero, Ozero l. Rus. Fed. 12 G3
Vyksa Rus. Fed. 13 I5
Vylkove Ukr. 27 M2
Vym' r. Rus. Fed. 12 K3
Vynohradiv Ukr. 13 D6
Vypin Island India 38 C4
Vypolzovo Rus. Fed. 12 G4
Vyritsa Rus. Fed. 15 Q7
Vyrnwy, Lake U.K. 19 D6
Vyselki Rus. Fed. 13 H7
Vysha Rus. Fed. 13 I5
Vyshhorod Ukr. 13 F6
Vyshnevolotskaya Gryada ridge Rus. Fed.
12 G4
Vyshniy-Volochek Rus. Fed. 12 G4
Vyškov Czech Rep. 17 P6
Vysokaya Gora Rus. Fed. 12 K5
Vysokogorniy Rus. Fed. 44 E2
Vystupovychi Ukr. 13 F6
Vytegra Rus. Fed. 12 H3
Vyya r. Rus. Fed. 12 J3
Vyžuona r. Lith. 15 N9

W

Wa Ghana 46 C3
Waal r. Neth. 10 G5
Waat South Sudan 32 D8
Wabē Gestro r. Eth. 30 D6
Wabē Shebelē Wenz r. Eth. 48 E3
Wabowden Canada 62 H1
Wabrah well Saudi Arabia 35 G6
Waccasassa Bay U.S.A. 63 K6
Waco U.S.A. 63 H5
Wadbilliga National Park Australia 58 D6
Waddān Libya 23 H6
Waddeneilanden Neth. 16 J4
Waddenzee sea chan. Neth. 16 J4
Waddington, Mount Canada 62 B1
Wadebridge U.K. 19 C8
Wadena Canada 62 G1
Wadena U.S.A. 63 H2
Wadeye Australia 54 E3
Wadh Pak. 36 A4
Wadhwan India see Surendranagar
Wadi India 38 C2
Wādī as Sīr Jordan 39 B4
Wadi Halfa Sudan 32 D5
Wadi Howar National Park Sudan 48 C2
Wad Medani Sudan 32 D7
Wad Rawa Sudan 32 D6
Waenhuiskrans S. Africa 50 E8
Wafangdian China 43 M5
Wafra Kuwait see Al Wafrah
Wagga Wagga Australia 58 C5
Wah Pak. 36 C2
Wahai Indon. 52 C2
Wāḥāt Jālū Libya 47 F2
Wahpeton U.S.A. 63 H2
Wahran Alg. see Oran
Wah Wah Mountains UT U.S.A. 65 F1
Wai India 38 B2
Waiau N.Z. see Franz Josef Glacier
Waiau r. N.Z. 59 D6
Waidhofen an der Ybbs Austria 17 O7
Waigeo i. Indon. 41 F8
Waiheke Island N.Z. 59 E3
Waikabubak Indon. 41 D8
Waikaia r. N.Z. 59 B7
Waikari N.Z. 59 D6
Waikerie Australia 57 B7
Waikouaiti N.Z. 59 C7
Waimangaroa N.Z. 59 C5
Waimarama N.Z. 59 F4
Waimate N.Z. 59 C7
Wainganga r. India 38 C2
Waingapu Indon. 41 E8
Wainhouse Corner U.K. 19 C8
Waini Point Guyana 69 G2
Wainwright Canada 62 E1
Wainwright U.S.A. 60 C2
Waiouru N.Z. 59 E4
Waipahi N.Z. 59 B8
Waipaoa r. N.Z. 59 F4
Waipara N.Z. 59 D6
Waipawa N.Z. 59 F4
Waipukurau N.Z. 59 F4
Wairarapa, Lake N.Z. 59 E5
Wairau r. N.Z. 59 E5
Wairoa N.Z. 59 F4
Wairoa r. N.Z. 59 F4
Waitahanui N.Z. 59 F4
Waitahuna N.Z. 59 B7
Waitakaruru N.Z. 59 E3
Waitaki r. N.Z. 59 C7
Waitangi N.Z. 53 I6
Waite River Australia 54 F5
Waiuku N.Z. 59 E3
Waiwera South N.Z. 59 B8
Wajima Japan 45 E5
Wajir Kenya 48 E3
Waka Indon. 54 C2
Wakasa-wan b. Japan 45 D6
Wakatipu, Lake N.Z. 59 B7
Wakayama Japan 45 D6
Wake Atoll terr. N. Pacific Ocean see
Wake Island
WaKeeney U.S.A. 62 H4
Wakefield N.Z. 59 D5
Wakefield U.K. 18 F5
Wakefield RI U.S.A. 64 F2
Wakefield VA U.S.A. 64 C4
Wake Island terr. N. Pacific Ocean 74 H4
Wakkanai Japan 44 F3
Wakkerstroom S. Africa 51 J4
Wakool Australia 58 B5
Wakool r. Australia 58 A5
Waku-Kungo Angola 49 B5
Wałbrzych Poland 17 P5
Walcha Australia 58 E3
Wałcz Poland 17 P4
Waldburg Range mts Australia 55 B6
Walden NY U.S.A. 64 D2
Waldenburg Poland see Wałbrzych
Waldkraiburg Germany 17 N6
Waldorf MD U.S.A. 64 C3
Waldron, Cape Antarctica 76 F2
Walebing Australia 55 B7
Wales admin. div. U.K. 19 D6
Walgaon India 36 D5
Walgett Australia 58 D3
Walgreen Coast Antarctica 76 K1
Walikale Dem. Rep. Congo 47 F5
Walker r. Australia 56 A2
Walker watercourse Australia 55 F6
Walker r. NV U.S.A. 65 C1
Walker Bay S. Africa 50 D8
Walker Creek r. Australia 56 C3
Walker Lake NV U.S.A. 65 C1
Walker Pass CA U.S.A. 65 C3
Walkersville MD U.S.A. 64 C3
Wall, Mount h. Australia 54 B5
Wallaby Island Australia 56 C2
Wallal Downs Australia 54 C4
Wallangarra Australia 58 E2
Wallaroo Australia 57 B7
Wallasey U.K. 18 D5
Walla Walla Australia 58 C5
Walla Walla U.S.A. 62 D2
Wallekraal S. Africa 50 C6
Wallendbeen Australia 58 D5
Wallingford U.K. 19 F7
Wallis, Îles is Wallis and Futuna Is 53 I3
Wallis and Futuna Islands terr.
S. Pacific Ocean 53 I3
Wallis et Futuna, Îles terr. S. Pacific Ocean
see Wallis and Futuna Islands
Wallis Islands Wallis and Futuna Is see
Wallis, Îles
Wallis Lake inlet Australia 58 F4
Wallops Island VA U.S.A. 64 D4
Walls U.K. 20 [inset]
Walls of Jerusalem National Park Australia
57 [inset]
Wallumbilla Australia 57 E5
Walney, Isle of i. U.K. 18 D4
Walnut Creek CA U.S.A. 65 A2
Walnut Grove CA U.S.A. 65 B1
Walong India 37 I3
Walpole NH U.S.A. 64 E1
Walsall U.K. 19 F6
Walsenburg U.S.A. 62 G4
Waltair India 38 D2
Walterboro U.S.A. 63 K5
Walter's Range hills Australia 58 B2
Waltham MA U.S.A. 64 F1
Walton WV U.S.A. 64 A3
Walvisbaai Namibia see Walvis Bay
Walvisbaai b. Namibia see Walvis Bay
Walvis Bay Namibia 50 B2
Walvis Bay b. Namibia 50 B2
Walvis Ridge sea feature S. Atlantic Ocean
72 H8
Wāmā Afgh. 36 B2
Wamba Équateur Dem. Rep. Congo 47 F5
Wamba Orientale Dem. Rep. Congo 48 C3
Wamba Nigeria 46 D4
Wampusirpi Hond. 67 H5
Wana Pak. 33 K3
Wanaaring Australia 58 B2
Wanaka N.Z. 59 B7
Wanaka, Lake N.Z. 59 B7
Wanapitei Lake Canada 63 K2
Wanbi Australia 57 C7
Wanbrow, Cape N.Z. 59 C7
Wanda Shan mts China 44 D3
Wando S. Korea 45 B6
Wandoan Australia 57 E5
Wanganui N.Z. 59 E4
Wanganui r. N.Z. 59 E4
Wangaratta Australia 58 C6
Wangda China see Zogang
Wangdain China 37 G3
Wangdi Phodrang Bhutan 37 G4
Wanggamet, Gunung mt. Indon. 54 C2
Wang Gaxun China 37 I1
Wangkui China 44 B3
Wangmo China 42 J7
Wangqing China 44 C4
Wan Hsa-la Myanmar 42 H8
Wanie-Rukula Dem. Rep. Congo 48 C3
Wankaner India 36 B5
Wankie Zimbabwe see Hwange
Wanlaweyn Somalia 48 E3
Wanna Lakes imp. l. Australia 55 E7
Wantage U.K. 19 F7
Wanxian Chongqing China see Wanzhou
Wanyuan China 43 J6
Wanzhou Chongqing China 43 J6
Wapusk National Park Canada 61 I4
Waqf aş Şawwān, Jibāl hills Jordan 39 C4
Warab South Sudan 32 C8
Warangal India 38 C2
Waranga Reservoir Australia 58 B6
Waratah Bay Australia 58 B7
Warbreccan Australia 56 C5
Warburton Australia 55 D6
Warburton watercourse Australia 57 B5
Ward, Mount N.Z. 59 B6
Warden S. Africa 51 I4
Wardha India 38 C1
Wardha r. India 38 C2
Ward Hill h. U.K. 20 F2
Ware Canada 60 F4
Ware MA U.S.A. 64 E1
Wareham U.K. 19 E8
Waren Germany 17 N4
Warendorf Germany 17 K5
Warginburra Peninsula Australia 56 E4
Wargla Alg. see Ouargla
Warialda Australia 58 E2
Warkworth U.K. 18 F3
Warmbad Namibia 50 D5
Warmbad S. Africa 51 I3
Warmbaths S. Africa see Warmbad
Warminster U.K. 19 E7
Warminster PA U.S.A. 64 D2
Warm Springs NV U.S.A. 65 D1
Warm Springs VA U.S.A. 64 B3
Warmwaterberg mts S. Africa 50 E7
Warnes Bol. 68 F7
Warning, Mount Australia 58 F2
Waronda India 38 C2
Warora India 38 C1
Warra Australia 58 E1
Warragamba Reservoir Australia 58 E5
Warragul Australia 58 B7
Warrambool r. Australia 58 C3
Warrandirinna, Lake imp. l. Australia 57 B5
Warrandyte Australia 58 B6
Warrawagine Australia 54 C5
Warrego r. Australia 58 B3
Warrego Range hills Australia 56 D5
Warren Australia 58 C3
Warren OH U.S.A. 64 A2
Warren PA U.S.A. 64 B2
Warrenpoint U.K. 21 F3
Warrensburg MO U.S.A. 63 I4
Warrensburg NY U.S.A. 64 E1
Warrenton S. Africa 50 G5
Warrenton VA U.S.A. 64 C3
Warri Nigeria 46 D4
Warriners Creek watercourse Australia
57 B6
Warrington N.Z. 59 C7
Warrington U.K. 18 E5
Warrnambool Australia 57 C8
Warrumbungle National Park Australia
58 D3
Warsaj Afgh. 36 B1
Warsaw Poland 17 R4
Warsaw NY U.S.A. 64 B1
Warsaw VA U.S.A. 64 C4
Warshiikh Somalia 48 E3
Warszawa Poland see Warsaw
Warta r. Poland 17 O4
Warwick Australia 58 F2
Warwick U.K. 19 F6
Warwick RI U.S.A. 64 F2
Wasbank S. Africa 51 J5
Wasco CA U.S.A. 65 C3
Washburn ND U.S.A. 62 G2
Washim India 38 C1
Washington DC U.S.A. 64 C3
Washington NC U.S.A. 63 L4
Washington NJ U.S.A. 64 D2
Washington PA U.S.A. 64 A2
Washington UT U.S.A. 65 F2
Washington state U.S.A. 62 C2
Washington, Cape Antarctica 76 H2
Washington, Mount U.S.A. 63 M3
Washington Land reg. Greenland 61 L2
Washpool National Park Australia 58 F2
Wasi India 38 B2
Waskaganish Canada 63 L1
Waskaheganish Canada see Waskaganish
Waskey, Mount U.S.A. 60 C4
Wasser Namibia 50 D4
Wassuk Range mts NV U.S.A. 65 C1
Waswanipi, Lac l. Canada 63 L2
Watampone Indon. 41 E8
Watarrka National Park Australia 55 E6
Watenstadt-Salzgitter Germany see
Salzgitter
Waterbury CT U.S.A. 64 E2
Waterford Ireland 21 E5
Waterford Harbour Ireland 21 F5
Watergrasshill Ireland 21 D5
Waterloo Australia 54 E4
Waterloo Ont. Canada 64 A1
Waterloo IA U.S.A. 63 I3
Waterloo NY U.S.A. 64 C1
Waterlooville U.K. 19 F8
Watertown NY U.S.A. 63 L3
Watertown SD U.S.A. 63 H3
Watford U.K. 19 G7
Watford City U.S.A. 62 G2
Watheroo National Park Australia 55 A7
Watir, Wādī watercourse Egypt 39 B5
Watkins Glen NY U.S.A. 64 C1
Watling Island Bahamas see San Salvador
Watmuri Indon. 54 E1
Watrous Canada 62 F1
Watsi Kengo Dem. Rep. Congo 47 F5
Watson r. Australia 56 C2
Watson Lake Canada 60 F3
Watsonville CA U.S.A. 65 B2
Watten U.K. 20 F2
Watton U.K. 19 H6
Wattsburg PA U.S.A. 64 B1
Watubela, Kepulauan is Indon. 41 F8
Wau P.N.G. 52 E2
Wau South Sudan 32 C8
Wauchope N.S.W. Australia 58 F3
Wauchope N.T. Australia 54 F5
Waukaringa (abandoned) Australia 57 B7
Waukarlycarly, Lake imp. l. Australia 54 C5
Waukegan U.S.A. 63 J3
Wausau U.S.A. 63 J3
Wave Hill Australia 54 E4
Waveney r. U.K. 19 I6
Waverly NY U.S.A. 64 C1
Waverly VA U.S.A. 64 C4
Wāw al Kabīr Libya 47 E2
Waxxari China 42 F5
Way, Lake imp. l. Australia 55 C6
Waycross U.S.A. 63 K5
Waynesboro VA U.S.A. 64 B3
Waynesburg PA U.S.A. 64 A3
Waza, Parc National de nat. park
Cameroon 47 E3
Wāzah Khwāh Afgh. see Wazi Khwa
Wazi Khwa Afgh. 36 B2
Wazirabad Pak. 36 C2
W du Niger, Parc National du nat. park
Niger 46 D3
Wear r. U.K. 18 F4
Weatherford U.S.A. 62 H5
Weaverville U.S.A. 62 C3
Webb, Mount h. Australia 54 E5
Webequie Canada 63 J1
Weber Basin sea feature Laut Banda 74 E6
Webi Shabeelle r. Somalia 48 E3
Webster MA U.S.A. 64 F1
Webster SD U.S.A. 63 H2
Webster Springs WV U.S.A. 64 A3
Wedau P.N.G. 56 E1
Weddell Abyssal Plain sea feature
Southern Ocean 76 A2
Weddell Island Falkland Is 70 D8
Weddell Sea Antarctica 76 A2
Wedderburn Australia 58 A6
Weddin Mountains National Park
Australia 58 D4
Weedville PA U.S.A. 64 B2
Weenen S. Africa 51 J5
Weethalle Australia 58 C4
Wee Waa Australia 58 D3
Węgorzewo Poland 17 R3
Weichang China 43 L4
Weidongmen China see Qianjin
Weifang China 43 L5
Weihai China 43 M5
Weilmoringle Australia 58 C2
Weimar Germany 17 M5
Weinan China 43 J6
Weipa Australia 56 C2
Weir r. Australia 58 D2
Weirton WV U.S.A. 64 A2
Weißkugel mt. Austria/Italy 17 M7
Weissrand Mountains Namibia 50 D3
Weiya China 42 G4
Weiz Austria 17 O7
Wejherowo Poland 17 Q3
Wekweètì Canada 60 G3
Welbourn Hill Australia 55 F6
Weldiya Eth. 48 D2
Welford National Park Australia 56 C5
Welk'īt'ē Eth. 48 D3
Welkom S. Africa 51 H4
Welland Ont. Canada 64 B1
Welland r. U.K. 19 G6
Welland Canal Ont. Canada 64 B1
Wellesley Ont. Canada 64 A1
Wellesley Islands Australia 56 B3
Wellfleet MA U.S.A. 64 F2
Wellingborough U.K. 19 G6
Wellington Australia 58 D4
Wellington N.Z. 59 E5
Wellington S. Africa 50 D7
Wellington England U.K. 19 D8
Wellington England U.K. 19 E6
Wellington NV U.S.A. 65 C1
Wellington, Isla i. Chile 70 B7
Wellington Range hills N.T. Australia
54 F3
Wellington Range hills W.A. Australia
55 C6
Wells U.K. 19 E7
Wells, Lake imp. l. Australia 55 C6
Wellsboro PA U.S.A. 64 C2
Wellsburg WV U.S.A. 64 A2
Wellsford N.Z. 59 E3
Wells-next-the-Sea U.K. 19 H6
Wellsville NY U.S.A. 64 C1
Wellton AZ U.S.A. 65 E4
Wels Austria 17 O6
Welshpool U.K. 19 D6
Welwitschia Namibia see Khorixas
Welwyn Garden City U.K. 19 G7
Wem U.K. 19 E6
Wembesi S. Africa 51 I5
Wemindji Canada 63 L1
Wenatchee U.S.A. 62 C2
Wenbu China see Nyima
Wenchang Hainan China 43 K9
Wenchow China see Wenzhou
Wenden Latvia see Cēsis
Wenden AZ U.S.A. 65 F4
Wendover U.S.A. 62 E3
Wenlock r. Australia 56 C2
Wenquan Qinghai China 37 G2
Wenquan Xinjiang China 42 E4
Wenshan China 42 I8
Wensum r. U.K. 19 I6
Wentworth Australia 57 C7
Wenzhou China 43 M7
Wepener S. Africa 51 H5
Wer India 36 D4
Werda Botswana 50 F3
Werdēr Eth. 48 E3
Werder Germany 17 N4
Werra r. Germany 17 L5
Werris Creek Australia 58 E3
Wesel Germany 17 K5
Weser r. Germany 17 L4
Wessel, Cape Australia 56 B1
Wessel Islands Australia 56 B1
Wesselsbron S. Africa 51 H4
Wesselton S. Africa 51 I4
Westall, Point Australia 55 F8
West Antarctica reg. Antarctica 76 J1
West Australian Basin sea feature
Indian Ocean 73 O7
West Bank terr. Asia 39 B3
West Bend U.S.A. 63 J3
West Bengal state India 37 F5
West Bromwich U.K. 19 F6
West Burra i. U.K. see Burra
Westbury U.K. 19 E7
West Cape Howe Australia 55 B8
West Caroline Basin sea feature
N. Pacific Ocean 74 F5
West Chester PA U.S.A. 64 D3
West Coast National Park S. Africa 50 D7
Westerland Germany 17 L3
Westerly RI U.S.A. 64 F2
Western Australia state Australia 55 C6
Western Cape prov. S. Africa 50 E7
Western Desert Egypt 34 C6
Western Dvina r. Europe see
Zapadnaya Dvina
Western Ghats mts India 38 B3
Western Port b. Australia 58 B7
Western Sahara terr. Africa 46 B2
Western Samoa country S. Pacific Ocean
see Samoa
Western Sayan Mountains reg. Rus. Fed.
see Zapadnyy Sayan
West Falkland i. Falkland Is 70 D8
Westfield MA U.S.A. 64 E1
Westfield NY U.S.A. 64 B1
Westfield PA U.S.A. 64 C2
West Frisian Islands Neth. see
Waddeneilanden
Westgate Australia 58 C1
West Hartford CT U.S.A. 64 E2

West Haven CT U.S.A. 64 E2
Westhill U.K. 20 G3
Westhope U.S.A. 62 G2
West Ice Shelf Antarctica 76 E2
West Indies is Caribbean Sea 67 J4
West Kazakhstan Oblast admin. div. Kazakh. see Zapadnyy Kazakhstan
West Kingston RI U.S.A. 64 F2
Westland Australia 56 C4
Westland National Park N.Z. 59 C6
Westleigh S. Africa 51 H4
Westleton U.K. 19 I6
West Linton U.K. 20 F5
West Loch Roag b. U.K. 20 C2
West Lorne Ont. Canada 64 A1
West Lunga National Park Zambia 49 C5
West MacDonnell National Park Australia 55 F5
West Malaysia pen. Malaysia see Peninsular Malaysia
Westmar Australia 58 D1
West Mariana Basin sea feature N. Pacific Ocean 74 F4
Westminster MD U.S.A. 64 C3
Westmoreland Australia 56 B3
Westmorland CA U.S.A. 65 E4
Weston WV U.S.A. 64 A3
Weston-super-Mare U.K. 19 E7
West Palm Beach U.S.A. 63 K6
West Plains U.S.A. 63 I4
West Point pt Australia 57 [inset]
West Point CA U.S.A. 65 B1
West Point VA U.S.A. 64 C4
Westport Ireland 21 C4
Westport N.Z. 59 C5
Westray Canada 62 G1
Westray i. U.K. 20 F1
Westray Firth sea chan. U.K. 20 F1
West Rutland VT U.S.A. 64 E1
West Siberian Plain Rus. Fed. 28 J3
West Union WV U.S.A. 64 A3
West Virginia state U.S.A. 64 A3
West Wyalong Australia 58 C4
West York PA U.S.A. 64 C3
Wetar i. Indon. 41 E8
Wetar, Selat sea chan. Indon. 54 D2
Wetaskiwin Canada 62 E1
Wete Tanz. 49 D4
Wetzlar Germany 17 L5
Wewak P.N.G. 52 E2
Wexford Ireland 21 F5
Wexford Harbour b. Ireland 21 F5
Weyakwin Canada 62 F1
Weybridge U.K. 19 G7
Weyburn Canada 62 G2
Weymouth U.K. 19 E8
Weymouth MA U.S.A. 64 F1
Whakaari i. N.Z. 59 F3
Whakatane N.Z. 59 F3
Whalan Creek r. Australia 58 D2
Whalsay i. U.K. 20 [inset]
Whangamata N.Z. 59 E3
Whanganui National Park N.Z. 59 E4
Whangarei N.Z. 59 E2
Wharfe r. U.K. 18 F5
Wharfedale val. U.K. 18 F4
Wharton U.S.A. 63 H6
Wharton Lake Canada 61 I3
Whatî Canada 60 G3
Wheatland WY U.S.A. 62 F3
Wheaton-Glenmont MD U.S.A. 64 C3
Wheeler Peak NM U.S.A. 62 F4
Wheeler Peak NV U.S.A. 65 E1
Wheeling WV U.S.A. 64 A2
Whernside h. U.K. 18 E4
Whinham, Mount Australia 55 E6
Whitburn U.K. 20 F5
Whitby U.K. 18 G4
Whitchurch U.K. 19 E6
White r. Canada/U.S.A. 60 E3
White r. AR U.S.A. 63 I5
White r. IN U.S.A. 64 B2
White, Lake imp. l. Australia 54 E5
White Bay Canada 61 M5
White Butte mt. U.S.A. 62 G2
Whitecourt Canada 60 G4
Whitehall Ireland 21 E5
Whitehall U.K. 20 G1
Whitehall NY U.S.A. 64 E1
Whitehaven U.K. 18 D4
Whitehead U.K. 21 G3
Whitehill U.K. 19 G7
Whitehorse Canada 60 E3
White Horse, Vale of val. U.K. 19 F7
White Island Antarctica 76 D2
White Island N.Z. see Whakaari
White Lake LA U.S.A. 63 I6
Whitemark Australia 57 [inset]
White Mountain Peak CA U.S.A. 65 C2
White Mountains National Park Australia 56 D4
White Nile r. South Sudan/Uganda 47 G3
White Nossob watercourse Namibia 50 D2
White Pine Range mts NV U.S.A. 65 E1
White Plains NY U.S.A. 64 E2
White River Valley val. NV U.S.A. 65 E1
White Rock Peak NV U.S.A. 65 E1
White Russia country Europe see Belarus
White Sea Rus. Fed. 12 H2
White Stone VA U.S.A. 64 C4
White Sulphur Springs WV U.S.A. 64 A4
Whiteville U.S.A. 63 L5
White Volta r. Burkina Faso/Ghana 46 C4
Whitewater Baldy mt. U.S.A. 62 F5
Whitewood Australia 56 C4
Whitewood Canada 62 G1
Whitfield U.K. 19 I7
Whithorn U.K. 20 E6
Whitianga N.Z. 59 E3
Whitland U.K. 19 C7
Whitley Bay U.K. 18 F3
Whitmore Mountains Antarctica 76 K1
Whitney, Mount CA U.S.A. 65 C2
Whitstable U.K. 19 I7
Whitsunday Group is Australia 56 E4
Whitsunday Island National Park Australia 56 E4
Whitsun Island Vanuatu see Pentecost Island
Whittlesea Australia 58 B6
Whittlesey U.K. 19 G6
Whitton Australia 58 C5
Wholdaia Lake Canada 60 H3
Whyalla Australia 57 B7
Wichita U.S.A. 63 H4
Wichita Falls U.S.A. 62 H5
Wick U.K. 20 F2
Wick r. U.K. 20 F2
Wickenburg AZ U.S.A. 65 F4
Wickford U.K. 19 H7
Wickham r. Australia 54 E4
Wickham, Cape Australia 57 [inset]
Wickham, Mount h. Australia 54 E4
Wicklow Ireland 21 F5
Wicklow Head hd Ireland 21 G5
Wicklow Mountains Ireland 21 F5
Wicklow Mountains National Park Ireland 21 F4
Widerøe, Mount Antarctica 76 C2
Widerøefjellet mt. Antarctica see Widerøe, Mount
Widgeegoara watercourse Australia 58 B1
Widgiemooltha Australia 55 C7
Widnes U.K. 18 E5
Wi-do i. S. Korea 45 B6
Wielkopolskie, Pojezierze reg. Poland 17 O4
Wielkopolski Park Narodowy nat. park Poland 17 P4
Wieluń Poland 17 Q5
Wien Austria see Vienna
Wiener Neustadt Austria 17 P7
Wiesbaden Germany 17 L5
Wieżyca h. Poland 17 Q3
Wigan U.K. 18 E5
Wight, Isle of i. England U.K. 19 F8
Wigierski Park Narodowy nat. park Poland 15 M9
Wigston U.K. 19 F6
Wigton U.K. 18 D4
Wigtown U.K. 20 E6
Wigtown Bay U.K. 20 E6
Wilberforce, Cape Australia 56 B1
Wilcannia Australia 58 A3
Wilcox PA U.S.A. 64 B2
Wilczek Land i. Rus. Fed. see Vil'cheka, Zemlya
Wildcat Peak NV U.S.A. 65 D1
Wild Coast S. Africa 51 I6
Wildspitze mt. Austria 17 M7
Wildwood NJ U.S.A. 64 D3
Wilge r. S. Africa 51 I4
Wilge r. S. Africa 51 I3
Wilgena Australia 55 F7
Wilhelm, Mount P.N.G. 52 E2
Wilhelm II Land reg. Antarctica see Kaiser Wilhelm II Land
Wilhelmina Gebergte mts Suriname 69 G3
Wilhelmshaven Germany 17 L4
Wilhelmstal Namibia 50 C1
Wilkes-Barre PA U.S.A. 64 D2
Wilkes Coast Antarctica 76 G2
Wilkes Land reg. Antarctica 76 G2
Wilkie Canada 62 F1
Wilkins Coast Antarctica 76 L2
Wilkins Ice Shelf Antarctica 76 L2
Wilkinson Lakes imp. l. Australia 55 F7
Willand U.K. 19 D8
Willandra Billabong watercourse Australia 58 B4
Willandra National Park Australia 58 B4
Willcox U.S.A. 62 F5
Willemstad Curaçao 67 K6
Willeroo Australia 54 E3
William, Mount Australia 57 C8
William Creek Australia 57 B6
Williams AZ U.S.A. 62 E4
Williams CA U.S.A. 65 A1
Williamsburg VA U.S.A. 64 C4
Williams Lake Canada 62 C1
Williamson NY U.S.A. 64 C1
Williamson WV U.S.A. 63 K4
Williamsport PA U.S.A. 64 C2
Williamstown NJ U.S.A. 64 D3
Williamstown NY U.S.A. 64 D1
Willimantic CT U.S.A. 64 E2
Willis Group atolls Australia 56 E3
Williston S. Africa 50 E6
Williston ND U.S.A. 62 G2
Williston Lake Canada 60 F4
Williton U.K. 19 D7
Willmar U.S.A. 63 H2
Willow Beach AZ U.S.A. 65 E3
Willow Hill PA U.S.A. 64 C2
Willowmore S. Africa 50 F7
Willowra Australia 54 F5
Willowvale S. Africa 51 I7
Wills, Lake imp. l. Australia 54 E5
Wilmington DE U.S.A. 64 D3
Wilmington NC U.S.A. 63 L5
Wilmslow U.K. 18 E5
Wilno Lith. see Vilnius
Wilpattu National Park Sri Lanka 38 D4
Wilson watercourse Australia 57 C5
Wilson NC U.S.A. 63 L4
Wilson NY U.S.A. 64 B1
Wilson, Mount NV U.S.A. 65 E1
Wilsonia CA U.S.A. 65 C2
Wilson's Promontory pen. Australia 58 C7
Wilson's Promontory National Park Australia 58 C7
Wilton r. Australia 54 F3
Wiluna Australia 55 C6
Wimereux France 19 I8
Wina r. Cameroon see Vina
Winbin watercourse Australia 57 D5
Winburg S. Africa 51 H5
Wincanton U.K. 19 E7
Winchester U.K. 19 F7
Winchester KY U.S.A. 63 K4
Winchester NH U.S.A. 64 E1
Winchester VA U.S.A. 64 B3
Windau Latvia see Ventspils
Windber PA U.S.A. 64 B2
Windermere U.K. 18 E4
Windermere l. U.K. 18 E4
Windhoek Namibia 50 C2
Windlestraw Law h. U.K. 20 G5
Windom U.S.A. 63 H3
Windorah Australia 56 C5
Wind River Range mts U.S.A. 62 F3
Windrush r. U.K. 19 F7
Windsor Australia 58 E4
Windsor Ont. Canada 63 K3
Windsor U.K. 19 G7
Windsor NY U.S.A. 64 D1
Windsor VT U.S.A. 64 E1
Windsor Locks CT U.S.A. 64 E2
Windward Islands Caribbean Sea 67 L5
Windward Passage Cuba/Haiti 67 J5
Windy U.S.A. 60 D3
Winfield KS U.S.A. 63 H4
Wingate U.K. 18 F4
Wingen Australia 58 E3
Wingham Australia 58 F3
Winisk r. Canada 61 J4
Winisk (abandoned) Canada 61 J4
Winisk Lake Canada 63 J1
Winneba Ghana 46 C4
Winnecke Creek watercourse Australia 54 E4
Winnemucca U.S.A. 62 D3
Winner U.S.A. 62 G3
Winnfield U.S.A. 63 I5
Winning Australia 55 A5
Winnipeg Canada 61 I5
Winnipeg, Lake Canada 61 I4
Winnipegosis, Lake Canada 61 H4
Winnipesaukee, Lake NH U.S.A. 64 F1
Winona MN U.S.A. 63 I3
Winona MS U.S.A. 63 J5
Winsford U.K. 18 E5
Winslow AZ U.S.A. 62 E4
Winsted CT U.S.A. 64 E2
Winston-Salem U.S.A. 63 K4
Winters CA U.S.A. 65 B1
Winterthur Switz. 24 I3
Winterton S. Africa 51 I5
Winton Australia 56 C4
Winton N.Z. 59 B8
Winwick U.K. 19 G6
Wirral pen. U.K. 18 D5
Wirrulla Australia 57 A7
Wisbech U.K. 19 H6
Wisconsin state U.S.A. 63 J3
Wisconsin Rapids U.S.A. 63 J3
Wiseman U.S.A. 60 C3
Wishaw U.K. 20 F5
Wisil Dabarow Somalia 48 E3
Wisła r. Poland see Vistula
Wismar Germany 17 M4
Witbank S. Africa 51 I3
Witbooisvlei Namibia 50 D3
Witham U.K. 19 H7
Witham r. U.K. 19 H6
Withernsea U.K. 18 H5
Witjira National Park Australia 57 A5
Witney U.K. 19 F7
Witrivier S. Africa 51 J3
Witteberg mts S. Africa 51 H6
Wittenberg Germany see Lutherstadt Wittenberg
Wittenberge Germany 17 M4
Wittenburg Germany 17 M4
Wittlich Germany 17 K6
Wittstock Germany 17 N4
Witu Islands P.N.G. 52 E2
Witvlei Namibia 50 D2
Wivenhoe, Lake Australia 58 F1
Władysławowo Poland 17 Q3
Włocławek Poland 17 Q4
Wodonga Australia 58 C6
Wohlthat Mountains Antarctica 76 C2
Wōjjā atoll Marshall Is see Wotje
Wokam i. Indon. 41 F8
Woken He r. China 44 C3
Wokha India 37 H4
Woking U.K. 19 G7
Wokingham watercourse Australia 56 C4
Wokingham U.K. 19 G7
Woko National Park Australia 58 E3
Wolcott NY U.S.A. 64 C1
Wolfenbüttel Germany 17 M4
Wolf Point U.S.A. 62 F2
Wolfsberg Austria 17 O7
Wolfsburg Germany 17 M4
Wolfville Canada 63 O2
Wolgast Germany 17 N3
Wolin Poland 17 O4
Wollaston Lake Canada 60 H4
Wollaston Lake Canada 60 H4
Wollaston Peninsula Canada 60 G3
Wollemi National Park Australia 58 E4
Wollongong Australia 58 E5
Wolmaransstad S. Africa 51 G4
Wolseley Australia 57 C8
Wolseley S. Africa 50 D7
Wolsingham U.K. 18 F4
Wolverhampton U.K. 19 E6
Wonarah Australia 56 B3
Wonay, Kowtal-e Afgh. 36 B2
Wondai Australia 57 E5
Wongalarroo Lake salt l. Australia 58 B3
Wongarbon Australia 58 D4
Wŏnju S. Korea 45 B5
Wonowon Canada 60 F4
Wŏnsan N. Korea 45 B5
Wonthaggi Australia 58 B7
Wonyulgunna, Mount h. Australia 55 B6
Woocalla Australia 57 B6
Woodbine NJ U.S.A. 64 D3
Woodbridge U.K. 19 I6
Woodbridge VA U.S.A. 64 C3
Wood Buffalo National Park Canada 60 G4
Woodbury NJ U.S.A. 64 D3
Wooded Bluff hd Australia 58 F2
Woodlake CA U.S.A. 65 C2
Woodland CA U.S.A. 65 B1
Woodland PA U.S.A. 64 B2
Woodlark Island P.N.G. 52 F2
Woodroffe watercourse Australia 56 B4
Woodroffe, Mount Australia 55 E6
Woods, Lake imp. l. Australia 54 F4
Woods, Lake of the Canada/U.S.A. 63 I2
Woodsfield OH U.S.A. 64 A3
Woodside Australia 58 C7
Woodstock Ont. Canada 64 A1
Woodstock VA U.S.A. 64 B3
Woodstock VT U.S.A. 64 E1
Woodward U.S.A. 62 H4
Woody CA U.S.A. 65 C3
Wooler U.K. 18 E3
Woolgoolga Australia 58 F3
Wooli Australia 58 F2
Woollard, Mount Antarctica 76 K1
Woolyeenyer Hill h. Australia 55 C8
Woomera Australia 57 B6
Woomera Prohibited Area Australia 55 F7
Woonsocket RI U.S.A. 64 F1
Woorabinda Australia 56 E5
Wooramel r. Australia 55 A6
Wooster U.S.A. 63 K3
Worbody Point Australia 56 C2
Worcester S. Africa 50 D7
Worcester U.K. 19 E6
Worcester MA U.S.A. 64 F1
Worcester NY U.S.A. 64 D1
Wörgl Austria 17 N7
Workington U.K. 18 D4
Worksop U.K. 18 F5
Worland U.S.A. 62 F3
Worms Head hd U.K. 19 C7
Wortel Namibia 50 C2
Worthing U.K. 19 G8
Wotje atoll Marshall Is 74 H5
Wotu Indon. 52 C2
Wowoni i. Indon. 41 E8
Wrangel Island Rus. Fed. 29 T2
Wrangell Mountains U.S.A. 77 B3
Wrangell-St Elias National Park and Preserve U.S.A. 60 D3
Wrath, Cape U.K. 20 D2
Wray U.S.A. 62 G3
Wreake r. U.K. 19 F6
Wreck Point S. Africa 50 C5
Wreck Reef Australia 56 F4
Wrecsam U.K. see Wrexham
Wrexham U.K. 19 E5
Wrightwood CA U.S.A. 65 D3
Wrigley Canada 60 F3
Wrigley Gulf Antarctica 76 J2
Wrocław Poland 17 P5
Września Poland 17 P4
Wubin Australia 55 B7
Wuchang Heilong. China 44 B3
Wuchow China see Wuzhou
Wudalianchi China 44 B2
Wudinna Australia 55 F8
Wuhai China 42 J5
Wuhan China 43 K6
Wuhu China 43 L6
Wüjang China 36 D2
Wujin Jiangsu China see Changzhou
Wukari Nigeria 46 D4
Wuli China 42 G6
Wuliang Shan mts China 42 I8
Wuliaru i. Indon. 54 E1
Wulur Indon. 54 E1
Wundwin Myanmar 37 I5
Wunnummin Lake Canada 61 J4
Wuppertal Germany 17 K5
Wuppertal S. Africa 50 D7
Wuqi China 43 J5
Wuqia China 42 C5
Wuranga Australia 55 B7
Wurno Nigeria 46 D3
Würzburg Germany 17 L6
Wusuli Jiang r. China/Rus. Fed. see Ussuri
Wuvulu Island P.N.G. 52 E2
Wuwei China 42 I5
Wuxi Jiangsu China 43 M6
Wuxian China see Suzhou
Wuxing China see Huzhou
Wuyang Guizhou China see Zhenyuan
Wuyiling China 44 C2
Wuyi Shan mts China 43 L7
Wuyuan Nei Mongol China 43 J4
Wuzhong China 42 J5
Wuzhou China 43 K8
Wyalkatchem Australia 55 B7
Wyalong Australia 58 C4
Wyandra Australia 58 B1
Wyangala Reservoir Australia 58 D4
Wyara, Lake imp. l. Australia 58 B2
Wycheproof Australia 58 A6
Wylliesburg VA U.S.A. 64 B4
Wyloo Australia 54 B5
Wylye r. U.K. 19 F7
Wymondham U.K. 19 I6
Wynbring Australia 55 F7
Wyndham Australia 54 E3
Wyndham-Werribee Australia 58 B6
Wynyard Canada 62 G1
Wyola Lake imp. l. Australia 55 E7
Wyoming state U.S.A. 62 F3
Wyong Australia 58 E4
Wyperfeld National Park Australia 57 C7
Wysox PA U.S.A. 64 C2
Wyszków Poland 17 R4
Wythall U.K. 19 F6
Wytheville VA U.S.A. 64 A4

X

Xaafuun Somalia 48 F2
Xaafuun, Raas pt Somalia 32 H7
Xabyaisamba China 37 I3
Xaçmaz Azer. 35 H2
Xago China 37 G3
Xagquka China 37 H3
Xaidulla China 36 D1
Xaignabouli Laos 42 I9
Xaignabouri Laos see Xaignabouli
Xainza China 37 G3
Xai-Xai Moz. 51 K3
Xalapa Mex. 66 E5
Xambioá Brazil 69 I5
Xam Nua Laos 40 C5
Xá-Muteba Angola 49 B4
Xanagas Botswana 50 E2
Xangda China see Nangqên
Xangdin Hural China 43 K4
Xangdoring China 37 E2
Xangongo Angola 49 B5
Xankändi Azer. 35 G3
Xanthi Greece 27 K4
Xarag China 37 I1
Xarardheere Somalia 48 E3
Xàtiva Spain 25 F4
Xavantes, Serra dos hills Brazil 69 I6
Xaxa China 37 E2
Xayar China 42 E4
Xela Guat. see Quetzaltenango
Xelva Spain see Chelva
Xero Potamos r. Cyprus see Xeros
Xeros r. Cyprus 39 A2
Xhora S. Africa see Elliotdale
Xiabole Shan mt. China 44 B2
Xiaguan China see Dali
Xiamen China 43 L8
Xi'an China 43 J6
Xianfeng China 43 J7
Xiangfan China 43 K6
Xianggang H.K. China see Hong Kong
Xianggang Tebie Xingzhengqu aut. reg. China see Hong Kong
Xianggelila China 42 H7
Xiangning China 43 K5
Xiangquan He r. China see Langqên Zangbo
Xiangride China 37 I2
Xiangtan China 43 K7
Xiangyang China see Xiangfan
Xiangyang Hu l. China 37 G2
Xianning China 43 K7
Xiantao China 43 K6
Xianxia Ling mts China 43 L7
Xianyang China 43 J6
Xiaocaohu China 42 F4
Xiao'ergou China 44 A2
Xiaogan China 43 K6
Xiao Hinggan Ling mts China 44 B2
Xiaonanchuan China 37 H2
Xiaoshi China see Benxi
Xiao Surmang China 37 I2
Xiashan China see Zhanjiang
Xiayingpan Guizhou China see Liupanshui
Xichang China 42 I7
Xieng Khouang Laos see Phônsavan
Xifeng Liaoning China 44 B4
Xigazê China 37 G3
Xi Jiang r. China 43 K8
Xijir China 37 G2
Xijir Ulan Hu salt l. China 37 G2
Xiliao He r. China 44 A4
Xilinhot China 43 L4
Ximiao China 42 I4
Xinavane Moz. 51 K3
Xin Barag Zuoqi China see Amgalang
Xincun China see Dongchuan
Xindian China 44 B3
Xindu Sichuan China see Luhuo
Xing'an Shaanxi China see Ankang
Xingba China see Lhünzê
Xingguo Jiangxi China 43 L7
Xinghai China 42 H5
Xingkai China 44 D3
Xingkai Hu l. China/Rus. Fed. see Khanka, Lake
Xinglong China 44 B2
Xinglong Heilong. China 44 B3
Xingning Guangdong China 43 L8
Xingtai China 43 K5
Xingu r. Brazil 69 H4
Xingu, Parque Indígena do res. Brazil 69 H6
Xinguara Brazil 69 H5
Xingyi China 42 I7
Xinhua Yunnan China see Funing
Xinhuang China 43 J7
Xining China 42 I5
Xinjiang aut. reg. China see Xinjiang Uygur Zizhiqu
Xinjiang Uygur Zizhiqu aut. reg. China 36 E1
Xinkai He r. China 44 A4
Xinkou Shanxi China 43 K5
Xinlitun China 44 B2
Xinmin China 44 B2
Xinning Gansu China see Ningxian
Xinqing China 44 C2
Xintai China 43 L5
Xinxian Shanxi China see Xinkou
Xinxiang China 43 K5
Xinyang Henan China 43 K6
Xinyi Jiangsu China 43 L6
Xinyu China 43 K7
Xinyuan Qinghai China see Tianjun
Xinyuan Xinjiang China 42 E4
Xinzhangfang China 44 A2
Xinzhu Taiwan see Hsinchu
Xinzo de Limia Spain 25 C2
Xiongzhou China see Nanxiong
Xiqing Shan mts China 42 I6
Xique Xique Brazil 69 J6
Xisha Qundao is S. China Sea see Paracel Islands
Xiugu China see Jinxi
Xi Ujimqin Qi China see Bayan Ul Hot
Xiwu China 37 I2
Xixabangma Feng mt. China 37 F3
Xixiang China 43 J6
Xixiu China see Anshun
Xixón Spain see Gijón-Xixón
Xizang aut. reg. China see Xizang Zizhiqu
Xizang Gaoyuan plat. China see Tibet, Plateau of
Xizang Zizhiqu aut. reg. China 37 G3
Xo'jayli Uzbek. 33 I1
Xorkol China 42 G5
Xuanhua China 43 L4
Xuanwei China 42 I7
Xucheng China see Xuwen
Xuddur Somalia 48 E3
Xugui China 42 H5
Xuguit Qi China see Yakeshi
Xümatang China 37 I2
Xungba China see Xangdoring
Xungmai China 37 G3
Xunhe China 44 B2
Xun He r. China 44 C2
Xun Jiang r. China 40 D5
Xúquer, Riu r. Spain 25 F4
Xuru Co l. China 37 F3
Xuwen China 31 K4
Xuyong China 42 J7
Xuzhou China 43 L6

Ya'an China 42 I6
Yabanabat Turkey see Kızılcahamam
Yabēlo Eth. 48 D3
Yablonovyy Khrebet mts Rus. Fed. 43 J2
Yabuli China 44 C3
Yacuma r. Bol. 68 E6
Yadgir India 38 C2
Yadrin Rus. Fed. 12 J5
Yaeyama-rettō is Japan 43 M8
Yafa Israel see Tel Aviv-Yafo
Yagaba Ghana 46 C3
Yagan China 42 I4
Yağda Turkey see Erdemli
Yaghan Basin sea feature S. Atlantic Ocean 72 D9
Yagman Turkm. 35 I3
Yagodnoye Rus. Fed. 29 P3
Yagodnyy Rus. Fed. 44 E2
Yagoua Cameroon 47 E3
Yagra China 37 E3
Yagradagzê Feng mt. China 37 H2
Yahualica Mex. 66 D4
Yahyalı Turkey 23 L4
Yaizu Japan 45 E6
Yakacık Turkey 39 C1
Yakeshi China 43 M3
Yakima U.S.A. 62 C2
Yako Burkina Faso 46 C3
Yakovleva Rus. Fed. 44 D3
Yaku-shima i. Japan 45 C7
Yakutat U.S.A. 60 E4
Yakutat Bay U.S.A. 60 D4
Yakutsk Rus. Fed. 29 N3
Yakymivka Ukr. 13 G7
Yala Thai. 31 J6
Yalai China 37 F3
Yala National Park Sri Lanka see Ruhuna National Park
Yalan Dünya Mağarası tourist site Turkey 39 A1
Yalgoo Australia 55 B7
Yalleroi Australia 56 D5
Yaloké Cent. Afr. Rep. 48 B3
Yalova Turkey 27 M4
Yalta Ukr. 34 D1
Yalu Jiang r. China/N. Korea 44 B4
Yalujiang Kou r. mouth China/N. Korea 45 B5
Yalvaç Turkey 27 N5
Yamagata Japan 45 F5
Yamaguchi Japan 45 C6
Yamal, Poluostrov pen. Rus. Fed. see Yamal Peninsula
Yam-Alin', Khrebet mts Rus. Fed. 44 D1
Yamal Peninsula Rus. Fed. 28 H2
Yamanie Falls National Park Australia 56 D3
Yamba Australia 58 F2
Yambarran Range hills Australia 54 E3
Yambi, Mesa de hills Col. 68 D3
Yambio South Sudan 47 F4
Yambol Bulg. 27 L3
Yamdena i. Indon. 54 E1
Yamethin Myanmar 37 I5
Yamkanmardi India 38 B2
Yamkhad Syria see Aleppo
Yamm Rus. Fed. 15 P7
Yamma Yamma, Lake imp. l. Australia 57 C5
Yamoussoukro Côte d'Ivoire 46 C4
Yampil' Ukr. 13 F6
Yampol' Ukr. see Yampil'
Yamuna r. India 36 E4
Yamunanagar India 36 D3
Yamzho Yumco l. China 37 G3
Yana r. Rus. Fed. 29 O2
Yanam India 38 D2
Yan'an China 43 J5
Yanaoca Peru 68 D6
Yanaon India see Yanam
Yanaul Rus. Fed. 11 Q4
Yanbu' al Baḩr Saudi Arabia 32 E5
Yancheng Jiangsu China 43 M6
Yanchep Australia 55 A7
Yanco Australia 58 C5
Yanco Creek r. Australia 58 B5
Yanco Glen Australia 57 C6
Yanda watercourse Australia 58 B3
Yandama Creek watercourse Australia 57 C6
Yandun China 42 G4
Yanfolila Mali 46 C3
Ya'ngamdo China 37 H3
Yangdok N. Korea 45 B5
Yang Hu l. China 37 F2
Yangī Qal'eh Afgh. 36 B1
Yangiqishloq Uzbek. 33 K1
Yangirabot Uzbek. 33 K1
Yangiyo'l Uzbek. 42 B4
Yangjiang China 43 K8
Yangôn Myanmar see Rangoon
Yangquan China 43 K5
Yangtze r. Qinghai China 42 H6
Yangtze r. China 42 I7
Yangtze r. China 43 M6
Yangtze Kiang r. Qinghai China see Yangtze
Yangtze Kiang r. China see Yangtze
Yangudi Rassa National Park Eth. 48 E
Yangyang S. Korea 45 C5
Yangzhou Jiangsu China 43 L6
Yanhuqu China 37 E2
Yanishpole Rus. Fed. 12 G3
Yanis"yarvi, Ozero l. Rus. Fed. 14 Q5
Yanji China 44 C4
Yanjiang China see Ziyang
Yanjing Sichuan China see Yanyuan
Yankari National Park Nigeria 46 E4
Yankton U.S.A. 63 H3
Yannina Greece see Ioannina
Yano-Indigirskaya Nizmennost' lowlar Rus. Fed. 29 P2
Yanrey r. Australia 55 A5
Yanshan Yunnan China 42 I8
Yanshiping China 37 H2
Yanskiy Zaliv g. Rus. Fed. 29 O2
Yantabulla Australia 58 B2
Yantai China 43 M5
Yantongshan China 44 B4
Yany-Kurgan Kazakh. see Zhanakorgan
Yanyuan China 42 I7
Yao Chad 47 E3
Yaoundé Cameroon 46 E4
Yaoxiaoling China 44 B2
Yap i. Micronesia 41 F7
Yapen i. Indon. 41 F8
Yappar r. Australia 56 C3
Yap Trench sea feature N. Pacific Ocean 74 F5
Yaqui r. Mex. 66 B3
Yar Rus. Fed. 12 L4
Yaraka Australia 56 D5
Yarangüme Turkey see Tavas
Yaransk Rus. Fed. 12 J4
Yardea Australia 57 A7
Yardımcı Burnu pt Turkey 27 N6
Yardımlı Azer. 35 H3
Yardymly Azer. see Yardımlı
Yare r. U.K. 19 I6
Yarega Rus. Fed. 12 L3
Yaren Nauru 53 G2
Yarensk Rus. Fed. 12 K3
Yariga-take mt. Japan 45 E5
Yarīm Yemen 32 F7
Yarımca Turkey see Körfez
Yarkand China see Shache
Yarkant China see Shache
Yarkant He r. China 33 M2
Yarkhun r. Pak. 36 C1
Yarlung Zangbo r. China see Brahmapu
Yarmouth Canada 63 N3
Yarmouth England U.K. 19 F8
Yarmouth England U.K. see Great Yarmouth
Yarmuk r. Asia 39 B3
Yarnell AZ U.S.A. 65 F3
Yaroslavl' Rus. Fed. 12 H4
Yaroslavskiy Rus. Fed. 44 D3
Yarra r. Australia 58 B6
Yarra Junction Australia 58 B6
Yarram Australia 58 C7
Yarraman Australia 58 E1
Yarrawonga Australia 58 B6
Yarra Yarra Lakes imp. l. Australia 55 A
Yarronvale Australia 58 B1
Yarrowmere Australia 56 D4
Yartö Tra La pass China 37 H3
Yartsevo Krasnoyarskiy Kray Rus. Fed. 28 J3
Yartsevo Smolenskaya Oblast' Rus. Fed. 13 G5
Yarumal Col. 68 C2
Yaş Romania see Iaşi
Yasawa Group is Fiji 53 H3
Yashkul' Rus. Fed. 13 J7
Yasin Pak. 36 C1
Yasnogorsk Rus. Fed. 13 H5
Yasnyy Rus. Fed. 44 C1
Yass Australia 58 D5
Yass r. Australia 58 D5
Yassı Burnu c. Cyprus see Plakoti, Cape
Yāsūj Iran 35 H5
Yasuní, Parque Nacional nat. park Ecuador 68 C4
Yatağan Turkey 27 M6
Yaté New Caledonia 53 G4
Yathkyed Lake Canada 61 I3
Yatsushiro Japan 45 C6
Yatta West Bank 39 B4
Yatton U.K. 19 E7
Yauca Peru 68 D7
Yavari r. Brazil/Peru 68 E4
Yavatmal India 38 C1
Yavi Turkey 35 F3
Yaví, Cerro mt. Venez. 68 E2
Yavoriv Ukr. 13 D6
Yavuzlu Turkey 39 C1
Yawatongguzlangar China 37 E1
Yaw Chaung r. Myanmar 37 H5
Yaxian China see Sanya
Yayladağı Turkey 39 C2
Yazd Iran 35 I5
Yazd-e Khvāst Iran 35 I5
Yazıhan Turkey 34 E3
Yazoo City U.S.A. 63 I5
Y Bala U.K. see Bala
Yding Skovhøj h. Denmark 17 L3
Ydra i. Greece 27 J6
Y Drenewydd U.K. see Newtown
Yea Australia 58 B6
Yealmpton U.K. 19 D8
Yebawmi Myanmar 37 H4
Yebbi-Bou Chad 47 E2
Yecheng China 42 D5
Yécora Mex. 62 F6
Yedi Burun Başı pt Turkey 27 M6
Yeeda River Australia 54 C4
Yefremov Rus. Fed. 13 H5
Yeghegnadzor Armenia 35 G3
Yegindykol' Kazakh. 42 B2
Yegorlykskaya Rus. Fed. 13 I7
Yegorova, Mys pt Rus. Fed. 44 E3
Yegor'yevsk Rus. Fed. 13 H5
Yei South Sudan 47 G4
Yei r. South Sudan 47 G4
Yeji China 43 L6
Yejiaji China see Yeji
Yekaterinburg Rus. Fed. 28 H4
Yekaterinodar Rus. Fed. see Krasnodar
Yekaterinoslav Ukr. see Dnipropetrovs
Yekaterinoslavka Rus. Fed. 44 C2
Yekhegnadzor Armenia see Yeghegnad
Yekibastuz Kazakh. 42 D2
Yelabuga Khabarovskiy Kray Rus. Fed. 44 D2
Yelabuga Respublika Tatarstan Rus. Fed. 12 K5
Yelan' Rus. Fed. 13 I6
Yelan' r. Rus. Fed. 13 I6

landur India 38 C3
lantsy Rus. Fed. 42 J2
larbon Australia 58 E2
lenovskiye Kar'yery Ukr. see Dokuchayevs'k
lets Rus. Fed. 13 H5
limané Mali 46 B3
lizavetgrad Ukr. see Kirovohrad
lkhovka Rus. Fed. 13 K5
ll i. U.K. 20 [inset]
llabina Regional Reserve nature res. Australia 55 F7
llandu India 38 D2
llapur India 38 B3
llow r. China 43 L5
llowknife Canada 60 G3
llow Mountain h. Australia 58 C4
llow Sea N. Pacific Ocean 43 N5
llowstone r. U.S.A. 62 G2
llowstone Lake U.S.A. 62 E3
ll Sound str. U.K. 20 [inset]
lovo Rus. Fed. 11 Q4
l'sk Belarus 13 F6
lva r. Rus. Fed. 12 K3
matan China 37 I2
men country Asia 32 G6
metsk Rus. Fed. 12 I3
mişenbükü Turkey see Taşova
mmiganur India see Emmiganuru
mtsa Rus. Fed. 12 I3
mva Rus. Fed. 12 K3
na Rus. Fed. 14 Q3
nagoa Nigeria 46 D4
nakiyeve Ukr. 13 H6
nakiyevo Ukr. see Yenakiyeve
nangyat Myanmar 37 H5
nangyaung Myanmar 37 H5
nanma Myanmar 37 H6
nda Australia 58 C5
ngisar China 33 M2
ngo National Park Australia 58 E4
nice Turkey 27 L5
nidamlar Turkey see Demirtaş
nıhan Turkey see Yıldızeli
nije-i-Vardar Greece see Giannitsa
nişehir Greece see Larisa
nişehir Turkey 27 M4
nisey r. Rus. Fed. 28 J2
nisey-Angara-Selenga r. Rus. Fed. 28 J2
niseysk Rus. Fed. 28 K4
niseyskaya Ravnina ridge Rus. Fed. 28 K4
niseyskiy Zaliv inlet Rus. Fed. 77 F2
niyol Turkey see Borçka
notayevka Rus. Fed. 13 J7
ola India 38 B1
o Lake imp. l. Australia 55 D6
otmal India see Yavatmal
oval Australia 58 D4
ovil U.K. 19 E8
o Yeo r. Australia see Bland
ppoon Australia 56 E4
raliyev Kazakh. see Kuryk
rbogachen Rus. Fed. 29 L3
rcaud India 38 C4
revan Armenia 35 G2
reymentau Kazakh. 42 C2
rgara India 38 C2
rgeni hills Rus. Fed. 13 J7
rgoğu Romania see Giurgiu
riho West Bank see Jericho
rilla Australia 55 C7
rington NV U.S.A. 65 C1
rköy Turkey 34 D3
rla r. India 38 B2
rmak Kazakh. see Aksu
rmakovo Rus. Fed. 44 B1
rmak Plateau sea feature Arctic Ocean 77 H1
rmentau Kazakh. see Yereymentau
rmo CA U.S.A. 65 D3
rofey Pavlovich Rus. Fed. 44 A1
rsha r. Rus. Fed. 12 L2
rshov Rus. Fed. 13 K6
rtsevo Rus. Fed. 12 I3
ruḥam Israel 39 B4
rupaja mt. Peru 68 C6
rushalayim Israel/West Bank see Jerusalem
ruslan r. Rus. Fed. 13 J6
sagyo Myanmar 37 H5
san S. Korea 45 B5
sil' Kazakh. 30 F1
şilhisar Turkey 34 D3
şilırmak r. Turkey 34 E2
şilova Burdur Turkey 27 M6
şilova Yozgat Turkey see Sorgun
ssentuki Rus. Fed. 35 F1
ssey Rus. Fed. 29 L3
s Tor h. U.K. 19 C8
tatang China see Baqên
tman Australia 58 E2
-U Myanmar 37 H5
u, Île d' i. France 24 C3
vdokimovskoye Rus. Fed. see Krasnogvardeyskoye
vlakh Azer. see Yevlax
vlax Azer. 35 G2
vpatoriya Ukr. 34 D1
vreyskaya Avtonomnaya Oblast' admin. div. Rus. Fed. 44 D2
xian China see Laizhou
yik China 37 E1
ysk Rus. Fed. 13 H7
yungou China 42 F4
zhuga r. Rus. Fed. 12 J2
zo i. Japan see Hokkaidō
zyaryshcha Belarus 12 F5
Fenni U.K. see Abergavenny
Fflint U.K. see Flint
Gelli Gandryll U.K. see Hay-on-Wye
alí i. Greece see Gyali
allaq, Gebel mt. Egypt see Yu'alliq, Jabal
alousa Cyprus see Aigialousa
'an China 44 B3
anisádha i. Greece see Gianisada
annitsá Greece see Giannitsa
bin Sichuan China 42 I7
oug Caka salt l. China 37 F2
chang Hubei China 43 K6
cheng Henan China see Zhumadian
chun Heilong. China 44 C3
chun Jiangxi China 43 K7
laha China 44 B2
lan China 44 C3
lan Taiwan see Ilan
ldız Dağları mts Turkey 27 L4
ldızeli Turkey 34 E3
Yilehuli Shan mts China 44 A2
Yilong Heilong. China 44 B3
Yimianpo China 44 C3
Yinchuan China 42 J5
Yindarlgooda, Lake imp. l. Australia 55 C7
Yinggehai China 43 J9
Yingkiong India 37 H3
Yingkou China 43 M4
Yingtan China 43 L7
Yining Xinjiang China 42 E4
Yinmabin Myanmar 37 H5
Yin Shan mts China 43 J4
Yinxian China see Ningbo
Yirga Alem Eth. 48 D3
Yirol South Sudan 47 G4
Yishan Guangxi China see Yizhou
Yishui China 43 L5
Yíthion Greece see Gytheio
Yitiaoshan China see Jingtai
Yitong He r. China 44 B3
Yixing China 43 L6
Yiyang China 43 K7
Yizhou China 43 J8
Yizra'el country Asia see Israel
Yläne Fin. 15 M6
Ylihärmä Fin. 14 M5
Yli-Ii Fin. 14 N4
Yli-Kärppä Fin. 14 N4
Ylikiiminki Fin. 14 O4
Yli-Kitka l. Fin. 14 P3
Ylistaro Fin. 14 M5
Ylitornio Fin. 14 M3
Ylivieska Fin. 14 N4
Ylöjärvi Fin. 15 M6
Ymer Ø i. Greenland 61 P2
Ynys Enlli i. U.K. see Bardsey Island
Ynys Môn i. U.K. see Anglesey
Yogan, Cerro mt. Chile 70 B8
Yogyakarta Indon. 41 D8
Yokadouma Cameroon 47 E4
Yokkaichi Japan 45 E6
Yoko Cameroon 46 E4
Yokohama Japan 45 E6
Yokosuka Japan 45 E6
Yokote Japan 45 F5
Yola Nigeria 46 E4
Yolo CA U.S.A. 65 B1
Yolombo Dem. Rep. Congo 48 C4
Yolöten Turkm. 33 J2
Yomou Guinea 46 C4
Yŏnan N. Korea 45 B5
Yonezawa Japan 45 F5
Yong'an Fujian China 43 L7
Yŏnghŭng N. Korea 45 B5
Yŏnghŭng-man b. N. Korea 45 B5
Yŏngil-man b. S. Korea 45 C6
Yongjing Liaoning China see Xifeng
Yŏngju S. Korea 45 C5
Yongkang Yunnan China 42 H8
Yongning Sichuan China see Xuyong
Yongxiu China 43 L7
Yongzhou China 43 K7
Yonkers NY U.S.A. 64 E2
Yopal Col. 68 D2
Yopurga China 42 D5
Yordu India 36 C2
York Australia 55 B7
York Ont. Canada 64 B1
York U.K. 18 F5
York PA U.S.A. 64 C3
York, Cape Australia 56 C1
York, Kap c. Greenland see Innaanganeq
York, Vale of val. U.K. 18 F4
Yorke Peninsula Australia 57 B7
Yorkshire Dales National Park U.K. 18 E4
Yorkshire Wolds hills U.K. 18 G5
Yorkton Canada 62 G1
Yorktown VA U.S.A. 64 C4
Yorosso Mali 46 C3
Yosemite National Park CA U.S.A. 65 C2
Yoshkar-Ola Rus. Fed. 12 J4
Yos Sudarso i. Indon. see Dolok, Pulau
Yŏsu S. Korea 45 B6
Yotvata Israel 39 B5
Youghal Ireland 21 E6
Young Australia 58 D5
Younghusband, Lake imp. l. Australia 57 B6
Younghusband Peninsula Australia 57 B7
Young Island Antarctica 76 H2
Youngstown OH U.S.A. 64 A2
Youssoufia Morocco 22 C5
Youvarou Mali 46 C3
Youyang China 43 J7
Youyi China 44 C3
Youyi Feng mt. China/Rus. Fed. 42 F3
Yowah watercourse Australia 58 B2
Yozgat Turkey 34 D3
Yreka U.S.A. 62 C3
Yrghyz Kazakh. see Irgiz
Yrgyz r. Kazakh. 42 A3
Yr Wyddfa mt. U.K. see Snowdon
Ystad Sweden 15 H9
Ystwyth r. U.K. 19 C6
Ysyk-Köl Kyrg. see Balykchy
Ysyk-Köl salt l. Kyrg. 42 D4
Ythan r. U.K. 20 G3
Y Trallwng U.K. see Welshpool
Ytyk-Kyuyel' Rus. Fed. 29 O3
Yu'alliq, Jabal mt. Egypt 39 A4
Yuanlin China 44 A2
Yuanquan China see Anxi
Yub'ā i. Saudi Arabia 34 D6
Yuba City CA U.S.A. 65 B1
Yucatán pen. Mex. 66 F5
Yucatan Channel Cuba/Mex. 67 G4
Yucca AZ U.S.A. 65 E3
Yucca Lake NV U.S.A. 65 D2
Yucca Valley CA U.S.A. 65 D3
Yucheng Sichuan China see Ya'an
Yuci China see Jinzhong
Yudi Shan mt. China 44 A1
Yuelai China see Huachuan
Yueliang Pao l. China 44 A3
Yuendumu Australia 54 E5
Yuexi China 43 L6
Yueyang Hunan China 43 K7
Yug r. Rus. Fed. 12 J3
Yugorsk Rus. Fed. 11 S3
Yugoslavia
Yuhu China see Eryuan
Yuin Australia 55 B6
Yukagirskoye Ploskogor'ye plat. Rus. Fed. 29 Q3
Yukamenskoye Rus. Fed. 12 L4
Yukarı Sakarya Ovaları plain Turkey 27 N5
Yukarısarıkaya Turkey 34 D3
Yukon admin. div. Canada 60 E3
Yukon r. Canada/U.S.A. 60 B3
Yüksekova Turkey 35 G3
Yulara Australia 55 E6
Yule r. Australia 54 B5
Yuleba Australia 58 D1
Yulin Guangxi China 43 K8
Yulin Shaanxi China 43 J5
Yulong Xueshan mt. China 42 I7
Yuma AZ U.S.A. 65 E4
Yumen China see Laojunmiao
Yumenguan China 42 G4
Yumurtalık Turkey 39 B1
Yuna Australia 55 A7
Yunak Turkey 34 C3
Yunaska Island U.S.A. 60 A4
Yuncheng China 43 K6
Yundamindera Australia 55 C7
Yungas reg. Bol. 68 E7
Yungui Gaoyuan plat. China 42 I7
Yunjinghong China see Jinghong
Yunkai Dashan mts China 43 K8
Yunling China see Yunxiao
Yunnan prov. China 42 I8
Yunta Australia 57 B7
Yunt Dağı mt. Turkey 39 A1
Yunxiao China 43 L8
Yuraygir National Park Australia 58 F2
Yurba Co l. China 37 F2
Yürekli Turkey 39 B1
Yurga Rus. Fed. 28 J4
Yuriria Mex. 66 D4
Yurungkax He r. China 36 E1
Yur'ya Rus. Fed. 12 K4
Yur'yakha r. Rus. Fed. 12 L2
Yuryev Estonia see Tartu
Yur'yevets Rus. Fed. 12 I4
Yur'yev-Pol'skiy Rus. Fed. 12 H4
Yü Shan mt. Taiwan 43 M8
Yushino (abandoned) Rus. Fed. 12 L1
Yushkozero Rus. Fed. 12 G2
Yushu Jilin China 44 B3
Yushu Qinghai China 42 H6
Yushuwan China see Huaihua
Yusufeli Turkey 35 F2
Yus'va Rus. Fed. 11 Q4
Yuta West Bank see Yatta
Yuxi Yunnan China 42 I8
Yuzawa Japan 45 F5
Yuzha Rus. Fed. 12 I4
Yuzhno-Kamyshovyy Khrebet ridge Rus. Fed. 44 F3
Yuzhno-Kuril'sk Rus. Fed. 44 G3
Yuzhno-Muyskiy Khrebet mts Rus. Fed. 43 K1
Yuzhno-Sakhalinsk Rus. Fed. 44 F3
Yuzhno-Sukhokumsk Rus. Fed. 13 J7
Yuzhnoukrayinsk Ukr. 13 F7
Yuzhnyy Rus. Fed. see Adyk
Yuzhou Chongqing China see Chongqing
Yuzhou Henan China 43 K6
Yuzovka Ukr. see Donets'k
Yverdon Switz. 24 H3
Yvetot France 24 E2
Ywamun Myanmar 37 H5
Z
Zaandam Neth. 16 J4
Zab, Monts du mts Alg. 25 I6
Zabaykal'sk Rus. Fed. 43 L3
Zabīd Yemen 32 F7
Zābol Iran 33 J3
Zacapa Guat. 66 G5
Zacatecas Mex. 66 D4
Zacharo Greece 27 I6
Zacoalco Mex. 66 D4
Zacynthus i. Greece see Zakynthos
Zadar Croatia 26 F2
Zadoi China 42 H6
Zadonsk Rus. Fed. 13 H5
Za'farâna Egypt see Za'farānah
Za'farānah Egypt 34 D5
Zafer Adaları is Cyprus see Kleides Islands
Zafer Burnu c. Cyprus see Apostolos Andreas, Cape
Zafra Spain 25 C4
Zagazig Egypt see Az Zaqāzīq
Zaghouan Tunisia 26 D6
Zagorsk Rus. Fed. see Sergiyev Posad
Zagreb Croatia 26 F2
Zagros Mountains Iran 35 G4
Za'gya Zangbo r. China 37 G3
Zāhedān Iran 33 J4
Zahir Pir Pak. 36 B3
Zaḥlah Lebanon see Zahlé
Zahlé Lebanon 39 B3
Ẓahrān Saudi Arabia 32 F6
Zahrez Chergui salt pan Alg. 25 H6
Zahrez Rharbi salt pan Alg. 25 H6
Zainsk Rus. Fed. see Novyy Zay
Zaire country Africa see Congo, Democratic Republic of the
Zaïre r. Congo/Dem. Rep. Congo see Congo
Zaječar Serbia 27 J3
Zaka Zimbabwe 49 D6
Zakamensk Rus. Fed. 42 I2
Zakataly Azer. see Zaqatala
Zakháro Greece see Zacharo
Zākhō Iraq 35 F3
Zakhodnyaya Dzvina r. Europe see Zapadnaya Dvina
Zákinthos i. Greece see Zakynthos
Zakopane Poland 17 Q6
Zakouma, Parc National de nat. park Chad 47 E3
Zakynthos Greece 27 I6
Zakynthos i. Greece 27 I6
Zalaegerszeg Hungary 26 G1
Zalai-domsag hills Hungary 26 G1
Zalamea de la Serena Spain 25 D4
Zalantun China 44 A3
Zalari Rus. Fed. 42 I2
Zalău Romania 27 J1
Ẓalim Saudi Arabia 32 F5
Zalingei Sudan 47 F3
Ẓalmā, Jabal aẓ mt. Saudi Arabia 32 E4
Zambeze r. Africa 49 C5 see Zambezi
Zambezi r. Africa 49 C5
Zambezi Zambia 49 C5
Zambia country Africa 49 C5
Zamboanga Phil. 41 E7
Zamfara watercourse Nigeria 46 D3
Zamkog China see Zamtang
Zamora Ecuador 68 C4
Zamora Spain 25 D3
Zamora de Hidalgo Mex. 66 D5
Zamość Poland 13 D6
Zamost'ye Poland see Zamość
Zamtang China 42 I6
Zamuro, Sierra del mts Venez. 68 F3
Zanaga Congo 48 B4
Zancle Sicilia Italy see Messina
Zandamela Moz. 51 L3
Zanesville U.S.A. 63 K4
Zangguy China 36 D1
Zangsêr Kangri mt. China 37 F2
Zangskar reg. India see Zanskar
Zangskar Mountains India see Zanskar Mountains
Zanjān Iran 35 H3
Zanjān Rūd r. Iran 35 G3
Zanskar reg. India 36 D2
Zanskar Mountains India 36 D2
Zante i. Greece see Zakynthos
Zanthus Australia 55 C7
Zanzibar Tanz. 49 D4
Zanzibar Island Tanz. 49 D4
Zaouatallaz Alg. 46 D2
Zaouet el Kahla Alg. see Bordj Omer Driss
Zaozernyy Rus. Fed. 29 K4
Zaozhuang China 43 L6
Zapadnaya Dvina r. Europe 12 F5
Zapadnaya Dvina Rus. Fed. 12 G4
Zapadni Rodopi mts Bulg. 27 J4
Zapadno-Kazakhstanskaya Oblast' admin. div. Kazakh. see Zapadnyy Kazakhstan
Zapadno-Sakhalinskiy Khrebet mts Rus. Fed. 44 F2
Zapadno-Sibirskaya Nizmennost' plain Rus. Fed. see West Siberian Plain
Zapadno-Sibirskaya Ravnina plain Rus. Fed. see West Siberian Plain
Zapadnyy Chink Ustyurta esc. Kazakh. 35 I2
Zapadnyy Chink Ustyurta esc. Kazakh. 35 I2
Zapadnyy Kazakhstan admin. div. Kazakh. 11 Q6
Zapadnyy Kil'din Rus. Fed. 14 S2
Zapadnyy Sayan reg. Rus. Fed. 42 F2
Zapata U.S.A. 62 H6
Zapiga Chile 68 E7
Zapolyarnyy Rus. Fed. 12 F1
Zapol'ye Rus. Fed. 12 H4
Zaporizhzhya Ukr. 13 G7
Zaporozh'ye Ukr. see Zaporizhzhya
Zapug China 36 E2
Zaqatala Azer. 35 G2
Zaqên China 37 H2
Za Qu r. China 42 H6
Zaqungngomar mt. China 37 G2
Zara China see Moinda
Zara Croatia see Zadar
Zara Turkey 34 E3
Zarafshan Uzbek. see Zarafshon
Zarafshon Uzbek. 33 J1
Zaragoza Spain 25 F3
Zarand Iran 33 I3
Zarang China 36 D3
Zaranikh Reserve nature res. Egypt 39 B4
Zaranj Afgh. 33 J3
Zarasai Lith. 15 O9
Zárate Arg. 70 E4
Zaraysk Rus. Fed. 13 H5
Zaraza Venez. 68 E2
Zärdab Azer. 13 J8
Zarechensk Rus. Fed. 14 Q3
Zāreh Iran 35 H4
Zargun mt. Pak. 36 A3
Zaria Nigeria 46 D3
Zarichne Ukr. 13 E6
Zarifète, Col des pass Alg. 25 F6
Zaring China see Liangdaohe
Zarinsk Rus. Fed. 42 E2
Zarneh Iran 35 G4
Zărneşti Romania 27 K2
Zarqā' Jordan see Az Zarqā'
Zarqā', Nahr az r. Jordan 39 B3
Zarubino Rus. Fed. 44 C4
Żary Poland 17 O5
Zarya Rus. Fed. 13 H5
Zarzis Tunisia 22 G5
Zasheyek Rus. Fed. 14 Q3
Zaskar reg. India see Zanskar
Zaskar Range mts India see Zanskar Mountains
Zaslawye Belarus 15 O9
Zastron S. Africa 51 H6
Za'tarī, Wādī az watercourse Jordan 39 C3
Zaterechnyy Rus. Fed. 13 J7
Zavetnoye Rus. Fed. 13 I7
Zavety Il'icha Rus. Fed. 44 F2
Zavidovići Bos.-Herz. 26 H2
Zavitaya Rus. Fed. see Zavitinsk
Zavitinsk Rus. Fed. 44 C2
Zavolzhsk Rus. Fed. 12 I4
Zavolzh'ye Rus. Fed. see Zavolzhsk
Závora, Ponta pt Moz. 51 L3
Zawiercie Poland 17 Q5
Zawīlah Libya 47 E2
Zāwiyah, Jabal az hills Syria 39 C2
Zaydī, Wādī az watercourse Syria 39 C3
Zaysan Kazakh. 42 E3
Zaysan, Lake Kazakh. 42 E3
Zaysan, Ozero l. Kazakh. see Zaysan, Lake
Zayü China see Gyigang
Zayü Qu r. China/India 37 I3
Žďár nad Sázavou Czech Rep. 17 O6
Zdolbuniv Ukr. 13 E6
Zdolbunov Ukr. see Zdolbuniv
Zealand i. Denmark 15 G9
Zeerust S. Africa 51 H3
Ẓefat Israel 39 B3
Zeil, Mount Australia 55 F5
Zela Turkey see Zile
Zelennik Rus. Fed. 12 J3
Zelenoborsk Rus. Fed. 11 S3
Zelenoborskiy Rus. Fed. 14 R3
Zelenodol'sk Rus. Fed. 12 K5
Zelenogorsk Rus. Fed. 12 F3
Zelenograd Rus. Fed. 12 H4
Zelenogradsk Rus. Fed. 15 L9
Zelenokumsk Rus. Fed. 35 F1
Zelentsovo Rus. Fed. 12 J4
Zelenyy, Ostrov i. Rus. Fed. 44 G4
Zell am See Austria 17 N7
Žemaitijos nacionalinis parkas nat. park Lith. 15 L8
Zemetchino Rus. Fed. 13 I5
Zémio Cent. Afr. Rep. 48 C3
Zemmora Alg. 25 G6
Zempoaltépetl, Nudo de mt. Mex. 66 E5
Zenica Bos.-Herz. 26 G2
Zenifim watercourse Israel 39 B4
Zennor U.K. 19 B8
Zenta Serbia see Senta
Zenzach Alg. 25 H6
Zerenike Reserve nature res. Egypt see Zaranikh Reserve
Zernograd Rus. Fed. 13 I7
Zernovoy Rus. Fed. see Zernograd
Zêtang China 37 G3
Zeven Germany 17 L4
Zevgari, Cape Cyprus 39 A2
Zeya Rus. Fed. 44 B1
Zeya r. Rus. Fed. 44 B2
Zeyskiy Zapovednik nature res. Rus. Fed. 44 B1
Zeysko-Bureinskaya Vpadina depr. Rus. Fed. 44 C2
Zeyskoye Vodokhranilishche resr Rus. Fed. 44 B1
Zeytin Burnu c. Cyprus see Elaia, Cape
Zêzere r. Port. 25 B4
Zgharta Lebanon 39 B2
Zghorta Lebanon see Zgharta
Zgierz Poland 17 Q5
Zhabdün China see Zhongba
Zhabinka Belarus 15 N10
Zhaggo China see Luhuo
Zhaksy Sarysu watercourse Kazakh. see Sarysu
Zhalpaktal Kazakh. 11 P6
Zhalpaqtal Kazakh. see Zhalpaktal
Zhaltyr Kazakh. 28 H4
Zhambyl Karagandinskaya Oblast' Kazakh. 42 C3
Zhambyl Zhambylskaya Oblast' Kazakh. see Taraz
Zhamo China see Bomi
Zhanakorgan Kazakh. 42 B4
Zhanaozen Kazakh. 30 E2
Zhanatas Kazakh. 42 B4
Zhanbei China 44 B2
Zhangaözen Kazakh. see Zhanaozen
Zhanga Qazan Kazakh. see Novaya Kazanka
Zhangaqorghan Kazakh. see Zhanakorgan
Zhangatas Kazakh. see Zhanatas
Zhangbei China 43 K4
Zhangde China see Anyang
Zhangdian China see Zibo
Zhangguangcai Ling mts China 44 C3
Zhanghua Taiwan see Changhua
Zhangjiajie China 43 K7
Zhangjiakou China 43 K4
Zhangjiapan China see Jingbian
Zhangling China 44 A1
Zhangqiangzhen China 44 A4
Zhangshu China 43 L7
Zhangye China 42 I5
Zhangzhou China 43 L8
Zhanhe China see Zhanbei
Zhanibek Kazakh. 11 P6
Zhanjiang China 43 K8
Zhaodong China 44 B3
Zhaoqing China 43 K8
Zhaotong China 42 I7
Zhaoyuan China 44 B3
Zhaozhou China 44 B3
Zhari Namco salt l. China 37 F3
Zharkent Kazakh. 42 E4
Zharkovskiy Rus. Fed. 12 G5
Zharma Kazakh. 42 E3
Zhashkiv Ukr. 13 F6
Zhashkov Ukr. see Zhashkiv
Zhaslyk Uzbek. see Jasliq
Zhaxi Co salt l. China 37 F2
Zhaxigang China 36 D2
Zhaxizê China 37 I3
Zhaxizong China 37 F3
Zhayyq r. Kazakh./Rus. Fed. see Ural
Zhdanov Ukr. see Mariupol'
Zhdanovsk Azer. see Beyläqan
Zhejiang prov. China 43 M7
Zhelaniya, Mys c. Rus. Fed. 28 H2
Zheleznodorozhnyy Rus. Fed. see Yemva
Zheleznodorozhnyy Uzbek. see Qo'ng'irot
Zheleznogorsk Rus. Fed. 13 G5
Zheltyye Vody Ukr. see Zhovti Vody
Zhem Kazakh. see Emba
Zhengjiatun China see Shuangliao
Zhengzhou China 43 K6
Zhenlai China 44 A3
Zhenyuan China 43 J7
Zherdevka Rus. Fed. 13 I6
Zheshart Rus. Fed. 12 K3
Zhetikara Kazakh. see Zhitikara
Zhetysuskiy Alatau mts China/Kazakh. 42 D4
Zhêxam China 37 F3
Zhezkazgan Kazakh. 42 B3
Zhezqazghan Kazakh. see Zhezkazgan
Zhidoi China 42 H6
Zhigalovo Rus. Fed. 42 J2
Zhigansk Rus. Fed. 29 N3
Zhigung China 37 G3
Zhi Qu r. China see Yangtze
Zhitikara Kazakh. 30 F1
Zhitkovichi Belarus see Zhytkavichy
Zhitkur Rus. Fed. 13 J6
Zhitomir Ukr. see Zhytomyr
Zhīvār Iran 35 G4
Zhlobin Belarus 13 F5
Zhmerinka Ukr. see Zhmerynka
Zhmerynka Ukr. 13 F6
Zhob Pak. 33 K3
Zhob r. Pak. 36 B2
Zhongba Xizang China 37 F3
Zhongduo China see Youyang
Zhongguo country Asia see China
Zhongguo Renmin Gongheguo country Asia see China
Zhongping China see Huize
Zhongshan research stn Antarctica 76 E2
Zhongshan Guizhou China see Liupanshui
Zhongwei China 42 J5
Zhongxin Yunnan China see Xianggelila
Zhongyaozhan China 44 B2
Zhosaly Kazakh. see Dzhusaly
Zhoujiajing China 42 I5
Zhoukou Henan China 43 K6
Zhoushan Dao i. China 43 M6
Zhovti Vody Ukr. 13 G6
Zhuanghe China 45 A5
Zhubgyügoin China 37 I2
Zhuchengzi China 44 B3
Zhukovka Rus. Fed. 13 G5
Zhukovskiy Rus. Fed. 13 H5
Zhumadian China 43 K6
Zhuxi China 43 J6
Zhuzhou Hunan China 43 K7
Zhydachiv Ukr. 13 E6
Zhympity Kazakh. 11 Q5
Zhytkavichy Belarus 15 O10
Zhytomyr Ukr. 13 F6
Zīā'ābād Iran 35 H4
Žiar nad Hronom Slovakia 17 Q6
Zibā salt pan Saudi Arabia 39 D4
Zībāk Afgh. 36 B1
Zibo China 43 L5
Zicheng China see Zijin
Zidi Pak. 36 A4
Ziel, Mount Australia see Zeil, Mount
Zielona Góra Poland 17 O5
Ziemelkursas augstiene hills Latvia 15 M8
Ziftá Egypt 34 C5
Zīghan Libya 47 F2
Zigong China 42 I7
Ziguey Chad 47 E3
Ziguinchor Senegal 46 B3
Žīguri Latvia 15 O8
Zihuatanejo Mex. 66 D5
Zijin China 43 L8
Ziketan China see Xinghai
Zile Turkey 34 D2
Žilina Slovakia 17 Q6
Zillah Libya 47 E2
Zima Rus. Fed. 42 I2
Zimba Zambia 49 C5
Zimbabwe country Africa 49 C5
Zimi Sierra Leone see Zimmi
Zimmerbude Rus. Fed. see Svetlyy
Zimmi Sierra Leone 46 B4
Zimnicea Romania 27 K3
Zimniy Bereg coastal area Rus. Fed. 12 H2
Zimovniki Rus. Fed. 13 I7
Zimrīn Syria 39 B2
Ẓin watercourse Israel 39 B4
Zinave, Parque Nacional do nat. park Moz. 49 D6
Zinder Niger 46 D3
Ziniaré Burkina Faso 46 C3
Zinjibār Yemen 32 G7
Zinoyevsk Ukr. see Kirovohrad
Zion National Park UT U.S.A. 65 F2
Zippori Israel 39 B3
Ziqudukou China 37 H2
Zirc Hungary 26 G1
Ziro India 37 H4
Zīr Rūd Iran 32 H4
Zi Shui r. China 43 K7
Zistersdorf Austria 17 P6
Zitácuaro Mex. 66 D5
Zito China see Lhorong
Zittau Germany 17 O5
Ziyang Sichuan China 42 I6
Ziyaret Dağı h. Turkey 39 B1
Ziz, Oued watercourse Morocco 22 D5
Zlatoustovsk Rus. Fed. 44 D1
Zlín Czech Rep. 17 P6
Zmeinogorsk Rus. Fed. 42 E2
Zmiyevka Rus. Fed. 13 H5
Znamenka Rus. Fed. 13 I5
Znamenka Ukr. see Znam"yanka
Znam"yanka Ukr. 13 G6
Znojmo Czech Rep. 17 P6
Zoar S. Africa 50 E7
Zogang China 42 H7
Zoigê China 42 I6
Zoji La pass India 36 C2
Zola S. Africa 51 H7
Zolochev Kharkivs'ka Oblast' Ukr. see Zolochiv
Zolochev L'vivs'ka Oblast' Ukr. see Zolochiv
Zolochiv Kharkivs'ka Oblast' Ukr. 13 G6
Zolochiv L'vivs'ka Oblast' Ukr. 13 E6
Zolotonosha Ukr. 13 G6
Zolotoye Rus. Fed. 13 J6
Zolotukhino Rus. Fed. 13 H5
Zomba Malawi 49 D5
Zombor Serbia see Sombor
Zongga China see Gyirong
Zonguldak Turkey 27 N4
Zongxoi China 37 G3
Zorgho Burkina Faso 46 C3
Zorgo Burkina Faso see Zorgho
Żory Poland 17 Q5
Zouar Chad 47 E2
Zouérat Mauritania 46 B2
Zousfana, Oued watercourse Alg. 22 D5
Zrenjanin Serbia 27 I2
Zubālah, Birkat waterhole Saudi Arabia 35 F5
Zubillaga Arg. 70 D5
Zubova Polyana Rus. Fed. 13 I5
Zubtsov Rus. Fed. 12 G4
Zuénoula Côte d'Ivoire 46 C4
Zug Switz. 24 I3
Zugdidi Georgia 35 F2
Zugspitze mt. Austria/Germany 17 M7
Zugu Nigeria 46 D3
Zuider Zee l. Neth. see IJsselmeer
Zújar r. Spain 25 D4
Zumba Ecuador 68 C4
Zunheboto India 37 H4
Zuni Mountains U.S.A. 62 F4
Zunyi Guizhou China 42 J7
Županja Croatia 26 H2
Zūrābād Āzarbāyjān-e Gharbī Iran 35 G3
Zürich Switz. 24 I3
Zuru Nigeria 46 D3
Zurzuna Turkey see Çıldır
Zuwārah Libya 46 E1
Zuyevka Rus. Fed. 12 K4
Zvishavane Zimbabwe 49 D6
Zvolen Slovakia 17 Q6
Zvornik Bos.-Herz. 27 H2
Zwedru Liberia 46 C4
Zweletemba S. Africa 50 D7
Zwelitsha S. Africa 51 H7
Zwettl Austria 17 O6
Zwickau Germany 17 N5
Zwolle Neth. 17 K4
Zyablovo Rus. Fed. 12 L4
Zygi Cyprus 39 A2
Zyryan Kazakh. see Zyryanovsk
Zyryanka Rus. Fed. 29 Q3
Zyryanovsk Kazakh. 42 E3
Zyyi Cyprus see Zygi